Global
Biogeochemical
Cycles in the
Climate System

Global Biogeochemical Cycles in the Climate System

Edited by

**Ernst-Detlef Schulze, Martin Heimann,
Sandy Harrison, Elisabeth Holland,
Jonathan Lloyd, Iain Colin Prentice,
and David Schimel**

Max-Planck-Institute for Biogeochemistry
Jena, Germany

ACADEMIC PRESS
A Harcourt Science and Technology Company

San Diego   San Francisco   New York   Boston   London   Sydney   Tokyo

Academic Press
A Harcourt Science and Technology Company
525 B Street, Suite 1900, San Diego, California 92101-4495, USA
http://www.academicpress.com

Academic Press
Harcourt Place, 32 Jamestown Road, London NW1 7BY, UK
http://www.academicpress.com

Library of Congress Catalog Card Number: 00-109575

International Standard Book Number: 0-12-631260-5

Printed and bound by CPI Group (UK) Ltd, Croydon, CR0 4YY

Transferred to Digital Printing, 2013

Contents

Introduction

Contributors

Allison, C. E.
CSIRO Atmospheric Research,
PMB1 Aspendale, Victoria 3195
Australia

Archer, S.
Texas A&M University
Dept. of Rangeland Ecology and Management
College Station, Texas 77843-2126

Baldocchi, D. D.
University of California
Department of Environmental Science, Policy and
Management, Berkeley, California 94720

Beese, F.
Universität Göttingen
Institut für Bodenkunde und Waldernährung
Büsgenweg 2 37077, Göttingen, Germany

Benedick, R. E.
Wissenschaftszentrum Berlin für Sozialforschung
Reichpietschufer 50, 10785 Berlin, Germany

Bengtsson, L.
Max-Planck-Institut für Meteorologie
Bundesstr.55, 20146 Hamburg, Germany

Bird, M. I.
Australian National University Research School
of Earth Sciences Environmental Processes
Group, Canberra 0200, Australia

Boutton T. W.
Texas A&M University
Dept.of Rangeland Ecology and Management
College Station, Texas 77843-2126

Brasseur, G. P.
Max-Planck-Institut für Meteorologie
Bundesstr.55, 20146 Hamburg, Germany

Buchmann, N.
Max-Planck-Institut für Biogeochemie
Postfach 100164, 07701 Jena, Germany

Busch, G.
Institut für Bodenkunde und Waldernährung
Büsgenweg 2, 37077 Göttingen, Germany

Cerling, T. E.
Department of Geology and Geophysics
University of Utah, Salt Lake City, Utah 84112

Chapin III, F. S.
University of Alaska Institute of Arctic Biology
Fairbanks, Alaska 99775

Claussen, M.
Potsdam-Institut für Klimafolgenforschung
P.O.Box 601203, 14412 Potsdam, Germany

Coe, M. T.
Center for Climatic Research Climate,
People, and Environment Program,
University of Wisconsin, Madison,
Wisconsin 53706

Crutzen, P.Z.
Max-Planck-Institut für Chemie
Postfach 3060, 55020 Mainz, Germany

Czimczik, C.
Max-Planck-Institut für Biogeochemie
Postfach 100164, 07701 Jena, Germany

Davis, M. B.
University of Minnesota Department of Ecology,
Evolution and Bahavior, St. Paul,
Minnesota 55108

Ehleringer, J. R.
University of Utah Department of Biology
Salt Lake City, Utah 84112

Field, C. B.
Carnegie Institution of Washington Depatment of
Plant Biology, Stanford, California 94305

Francey, R. J.
CSIRO Atmospheric Research, PMB 1 Aspendale,
Victoria 3195, Australia

Gleixner, G.
Max-Planck-Institut für Biogeochemie
Postfach 100164, 07701 Jena, Germany

Harrison, S. P.
Max-Planck-Institut für Biogeochemie
Postfach 100164, 07701 Jena, Germany

Hasselmann, K.
Max-Planck-Institut für Meteorologie
Bundesstr.55, 20146 Hamburg, Germany

Hibbard, K. A.
University of New Hampshire Climate Change
Research Center, GAIM Task Force Institute for
the Study of Earth, Oceans, and Space (EOS)
Morse Hall, Durham,
New Hampshire 03824-3525

Högberg, P.
SLU Department of Forest Ecology, Section of Soil
Science, S-901 83 Umeå, Sweden

Holland, E. A. H.
Max-Planck-Institut für Biogeochemie
Postfach 100164, 07701 Jena, Germany

Jonasson, S.
University of Copenhagen Department of Plant
Ecology, Botanical Institute, Øster Farimagsgade
2D, DK-1353 Copenhagen, Denmark

Kaplan, J. O.
Max-Planck-Institut für Biogeochemie
Postfach 100164, 07701 Jena, Germany

Kelliher, F. M.
Manaaki Whenua-Landcare Research
P.O.Box 69, Lincoln, New Zealand

Klinke, A.
Center of Technology Assessment in
Baden-Württemberg, 70565
Stuttgart, Germany

Kramer C.
Max-Planck-Institut für Biogeochemie
Postfach 100164, 07701 Jena, Germany

Kruijt, B.
Alterra Green World Research,
6700 AA Wageningen, The Netherlands

Kutzbach, J. E.
Center for Climatic Research, Madison,
Wisconsin 53706

Lammel, G.
Max-Planck-Institut für Meteorologie
Bundesstr.55, 20146 Hamburg, Germany

Lloyd, J.
Max-Planck-Institut für Biogeochemie
Postfach 100164, 07701 Jena, Germany

Lühker, B. M.
Max-Planck-Institut für Biogeochemie
Postfach 100164, 07701 Jena, Germany

Mooney, H. A.
Stanford University, Department of Biological
Sciences, Stanford, California 94305

Prentice, I. C.
Max-Planck-Institut für Biogeochemie
Postfach 100164, 07701 Jena, Germany

Raupach, M. R.
CSIRO Land and Water, Canberra,
ACT 2601, Australia

Raynaud, D.
Laboratoire de Glaciologie et de Geophysique de
l' Environnement, St. Martin d' Hères, F-38402
France

Rayner, P. J.
Cooperative Research Centre for Southern
Hemisphere Meteorology Clayton, Victoria 3168,
Australia

Rebmann, C.
Max-Planck-Institut für Biogeochemie
Postfach 100164, 07701 Jena, Germany

Renn, O.
*Center of Technology Assessment in
Baden-Wüttemberg Industriestrasse 5,
70565 Stuttgart, Germany*

Santrùcková, H.
*University of South Bohemia Institute of Soil
Biology AS CR and Faculty of Biological Sciences
Na Sàdkách 7, CZ-370 05 Ceské Budìjovice,
Czech Republic*

Schimel, D. S. S.
*Max-Planck-Institut für Biogeochemie
Postfach 100164, 07701 Jena, Germany*

Schimel, J.
*University of California
Department of Ecology, Evolution and Marine*

Biology, Santa Barbara, California 93106

Schmidt, M. W. I.
*Universität zu Köln Geographisches Institut
Zuelpicher Str. 49a, 50674 Koeln, Germany*

Schulze, E.-D.
*Max-Planck-Institut für Biogeochemie,
Postfach 100164, 07701 Jena, Germany*

Shaver, G. R.
*The Ecosystems Center Marine Biological
Laboratory, Woods Hole,
Massachusetts 02543*

Veenendaal, E. M.
*Harry Oppenheimer Okavango Research Centre,
P. Bag 285 Maun, Botswana*

Vitousek, P.
*Stanford University
Department of Biological Sciences, Stanford,
California 94305*

Wirth, C.
*Max-Planck-Institut für Biogeochemie,
Postfach 100164, 07701 Jena, Germany*

Wolfrum, R.
*Max-Planck-Institut für Ausländisches
Öffentliches Recht und Völkerrecht Im
Neuenheimer Feld 535, 69120 Heidelberg,
Germany*

Especially during the past century, land use changes and agricultural and industrial activities have been growing so rapidly that their effects on the environment, including the chemical composition of the global atmosphere have become clearly noticeable on all scales. The first realization of the possibility of global effects was connected with the growth of the "greenhouse" gas carbon dioxide measured by C. D. Keeling and R. Revelle, on the basis of these measurements they stated that humanity had embarked on a global geophysical experiment potentially leading to climate warming. Other human-caused global disturbances in the atmosphere were discovered thereafter. In 1971 attention was called to the possible loss of stratospheric ozone, caused by NO_x catalysts in the exhaust of supersonic aviation. The projected large fleets of aircraft were never built. However, in 1974 an already existing, but late recognized threat to the ozone layer by ClO_x radicals produced in the stratosphere by the photochemical destruction of entirely man-made chlorofluorocarbon (CFC) gases was hypothesized and later confirmed by atmospheric observations. In fact, in 1985, scientists were caught totally by surprise when researchers of the British Antarctic survey reported much larger springtime ozone depletions, than originally estimated, on the order of 30%. It was found that the ozone loss was largest at altitudes between about 12 and 22 km, exactly the height region in which, under undisturbed conditions, maximum ozone concentrations had always been measured. At this location, it had always been thought that ozone was chemically inert. Since then, the "ozone hole" has grown in area and depth, so that by this year's spring total ozone had declined by more than 50% over a region three times the size of the United States. A couple of years of intensive research efforts showed that a chemical instability had developed, involving formation of ClO_x catalysts on ice particles under sunlit conditions, followed by rapid ozone destruction. The combination of special natural factors in early spring, cold temperatures, and availability of sunlight, together with about six times larger than natural loadings of chlorine gases, had led to this chemical instability over the Antarctic. Since 1996 the production of CFC gases on the industrial world has been forbidden. I have dwelled in this issue in some detail for two reasons. First, international political action would not have been taken without convincing scientific evidence that the CFC emissions were the cause of the heavy ozone loss. Second, it will be particularly important to determine where the world's complex environmental system may be most vulnerable to human perturbation. For this purpose, modeling alone will be far from sufficient. Surprises are not excluded, as the ozone hole story so drastically has demonstrated.

In the 1970s the substantial impact of the bioshpere on atmospheric chemistry was also realized. First, the main natural loss of stratospheric ozone occurs through reactions involving NO_x radi-cals that derived photochemically from the oxidation of N_2O, a by-product of the biological nitrogen cycle in soils and waters. Second, it was discovered that tropospheric ozone and its photochemical by-product, hydroxyl, are much influenced by chemical chain reactions involving CH_4 and other hydrocarbons, carbon monoxide, and NO_x. All these gases have both natural and anthropogenic sources. This is of the greatest importance, as the hydroxyl radicals, also called the "detergent of the atmosphere," to a large degree determine the chemical composition of the atmosphere by reacting with almost all gases that are emitted by natural processes and human activities.

In addition to being chemically active in the stratosphere and troposphere, several of the afore-mentioned and other gases serve as "greenhouse gases," thereby significantly adding to the climate warming caused by CO_2. On the other hand, aerosol particles, in particular, sulfates derived by the oxidation of largely anthropogenic SO_2 from oil and coal burning, have a cooling effect on climate.

The estimation of the impact of various kinds of human activities on atmospheric chemistry and climate clearly requires a good understanding of the natural and anthropogenic sources of large number of trace gases, as well as particulate matter, and the biological processes creating them. This research not only deals with the present and future, but also profits much from the vast amount of information regarding climate parameters and chemical composition of the atmosphere cores that is deposited in sediments and in ice. The latter data clearly show that the biosphere does not counteract climate change in some Gaian fashion. On the contrary, during earlier glacial periods all greenhouse gases were less abundant in the atmosphere than during the interglacials. This research has received special international, political attention in connection with the proposed Kyoto protocol to reduce the emissions of CO_2 caused by fossil fuel burning and deforestation. As was the case with the CFC regulations, effective CO_2 emission control measures will rely also on a strong scientific base. It was the realization of these strong needs, requiring improved knowledge especially about the biogeochemical cycles of C, N, S, P, and trace compounds such as iron, that led to the creation of a Max Planck Institute for Biogeochemistry. During the initial discussions, involving the cream of the international, biogeochemical, and climate community, the proposal received enthusiastic endorsement, emphasizing the uniqueness of the institute on the global scence. The proposal was also well received by the scientific members and the senate of the Max Planck Society. The search for directors and key scientific personel of the institute proved highly successful, with several key recruitments coming from overseas, clearly showing the enthusiasm accompanying the creation of the institute.

The MPI for Biogeochemistry in Jena is one of several Max Planck Institutes involved in global change research. This book, based on the presentations given to celebrate the first anniversary of the institute shows many important examples of the breadth and excitement of Global Change research around the world, including legal/political aspects. I hope that the so successful creation sets an example and promotes initiatives elsewhere to enhance biogeochemical research efforts, and its connections to ecology, climate and atmospheric chemistry. Many Happy Returns.

Dr. Paul Crutzen

Preface

Biogeochemistry: The Jena Perspective

In the late 20th century, biogeochemistry emerged as a new discipline in which the biological, physical, and human sciences collaborate (CGCR, 1999; Schlesinger, 1997). Biological, because the chemical cycles of the planet are mediated by life (Table 1). Physical, because of the strong coupling between climate and atmospheric composition so evident in the glacial–interglacial record of the ice cores (Fig. 1). And, human, because of the massive human disruption of the planet's carbon and nitrogen cycles by fossil fuel burning (which produces CO_2 and a range of volatile nitrogen compounds) (Fig. 2).

From the three figures, one gets an overview of the way in which the field of biogeochemistry has emerged. The evidence for the importance of biology in the composition of the atmosphere (Fig. 2) was deduced from geochemical measurements of air enabled by advances in analytical technology. The chemistry of the atmosphere and the discipline of atmospheric chemistry provided a view of the biosphere not accessible from "within" the discipline. The atmosphere reflects biotic processes operating over "deep" time as well as processes operating on rapid time scales (especially with respect to the oxidized N species). Some compounds, especially the hydrocarbons, may reflect plant–insect coevolution, and so to understand the atmosphere requires a deep understanding of biology. When insights into atmospheric chemistry were combined with emerging ecosystem studies of nitrogen and other elements (e.g., Vitousek and Reiners, 1977), a paradigm emerged that enriched both ecology and geophysics (Andreae and Schimel, 1989).

The realization that ecosystem biogeochemistry and climate were dynamically coupled was nascent for most of the 20th century. The ice-core records showing the coordinated rhythm of temperature, CO_2, and methane provided conclusive evidence of interactions (Fig. 2). The ice cores show coupled changes in trace gases and climate. They preserve a tantalizing body of information about leads, lags, and amplification that is not yet fully unravelled. While variations in CO_2 are strongly governed by changes in ocean circulation, mass balance considerations and isotopes suggest land-ecosystem changes as well (Indermuhle *et al.*, 1999). Climate effects on terrestrial biogeochemistry are demonstrated by the patterns in methane (produced in terrestrial wetlands and ungulate mammals) and nitrous oxide. High-resolution records showing high-frequency changes in ice cores, and detailed records of the Holocene provide information on timescales tractable, or nearly so, in analysis using today's biogeochemical models. Again, the perspective from geophysical records provides a view of ecosystem processes different from, and most strongly complementary to, the paradigms emerging from within the discipline.

The scientific community was galvanized by the Mauna Loa curve of increasing carbon dioxide and the political ramifications of this scientific result will echo for the foreseeable future (Benedick, Chapter 26 of this volume). Geophysical measurements provide a trans-disciplinary view of human processes. Since biogeochemistry has a "basic science" character and remains concentrated in academia, the carbon and nitrogen cycles would be of far less interest without the challenges of carbon and climate change, acid rain, and tropospheric ozone increase. The Mauna Loa curve challenges both the policy-relevant and intellectual sides of biogeochemistry. The policy side is obvious—the rate of increase in atmospheric CO_2 is the index of humanity's export of carbon to the atmosphere.

Scientifically, the fraction of CO_2 released to the atmosphere that remains as CO_2 in the air (about half) is not yet explained on the basis of incontrovertible measurements. While the holy grail of explaining the "missing sink" grows asymptotically closer, the political stakes and hence the standard of proof required are growing. The interannual variability of the growth rate of CO_2 gives evidence of climate–carbon interactions. Subtle year-to-year variations in the increase in CO_2 reflect changes in land and ocean uptake. The measurement and modeling tools to understand these changes are emerging and provide a direct means of understanding how climate affects the carbon system at large scales. Changes in the carbon system are reflected also in changes in the pole-to-pole gradient of CO_2. That gradient reflects the balance of sources and sinks on large scales. Because the equilibration time (the interhemispheric transport time) is about a year, changes in the gradient are another source of information about interannual variability. The seasonal cycle of CO_2 provides information about the seasonal activity of the biosphere. Because the phase and amplitude of the seasonal cycle vary spatially (Fig. 3), they provide rich information about land ecosystems. To date, we cannot fully separate changes in carbon uptake (photosynthesis) and release (respiration) to provide unique explanations for the seasonal cycle and its variation. This remains a research challenge.

1. Research Challenges

The discipline of biogeochemistry confronts a wide array of scientific and methodological challenges, as is evident in the balance of this book. These are not limited to the cycles of carbon and nitrogen, but include the role of phosphorus, iron, calcium, aluminum, and acidity, to name just a few. In this section, I will identify four cross-cutting challenges that illustrate aspects of the science.

1.1 Large-Scale Carbon Sinks: Detection and Attribution

The problem of the terrestrial missing sink remains. Where and why is there net uptake in terrestrial systems? The two questions,

TABLE 1   Chemical Composition of the Atmosphere

Constituent	Chemical formula	Volume mixing ratio in dry air	Major sources and remarks
Nitrogen	N_2	78.084%	Biological
Oxygen	O_2	20.948%	Biological
Argon	Ar	0.934%	Inert
Carbon dioxide	CO_2	360 ppmv	Combustion, ocean, biosphere
Neon	Ne	18.18 ppmv	Inert
Helium	He	5.24 ppmv	Inert
Methane	CH_4	1.7 ppmv	Biogenic and anthropogenic
Hydrogen	H_2	0.55 ppmv	Biogenic, anthropogenic, and photochemical
Nitrous oxide	N_2O	0.31 ppmv	Biogenic and anthropogenic
Carbon monoxide	CO	50–200 ppbv	Photochemical and anthropogenic
Ozone (troposphere)	O_3	10–500 ppbv	Photochemical
Ozone (stratosphere)	O_3	0.5–10 ppm	Photochemical
Nonmethane hydrocarbons		5–20 ppbv	Biogenic and anthropogenic
Halocarbons (as chlorine)		3.8 ppbv	85% anthropogenic
Nitrogen species	NO_y	10 ppt–1 ppm	Soils, lightning, anthropogenic
Ammonia	NH_3	10 ppt–1 ppb	Biogenic
Particulate nitrate	NO_3^-	1 ppt–10 ppb	Photochemical, anthropogenic
Particulate ammonium	NH_4^+	10 ppt–10 ppb	Photochemical, anthropogenic
Hydroxyl	OH	0.1–10 ppt	Photochemical
Peroxyl	HO_2	0.1–10 ppt	Photochemical
Hydrogen peroxide	H_2O_2	0.1–10 ppb	Photochemical
Formaldehyde	CH_2O	0.1–1 ppb	Photochemical
Sulfur dioxide	SO_2	10 ppt–1 ppb	Photochemical, volcanic, anthropogenic
Dimethyl sulfide	CH_3SCH_3	10–100 ppt	Biogenic
Carbon disulfide	CS_2	1–300 ppt	Biogenic, anthropogenic
Carbonyl sulfide	OCS	500 pptv	Biogenic, volcanic, anthropogenic
Hydrogen sulfide	H_2S	5–500 ppt	Biogenic, volcanic
Particulate sulfate	SO_4^{2-}	10 ppt–10 ppb	Photochemical, anthropogenic

where and why, cannot be separated. Different parts of the world and differing ecosystem types are influenced by differing nitrogen additions, disturbance, and pollution. Answering the question "why is there a sink" requires explaining the differences between climate zones, management, and disturbance regimes and chemical climate. This is a practical problem because, in the future, there will be increasing pressure to manage carbon sinks. How can sinks best be induced and sustained? How can the effects of intentional measures be quantified and verified? What impacts does managing ecosystems for carbon storage have on other ecosystem goods and services, including diversity? Without scientific understanding, no intelligent design of management systems can emerge. Equally important is the fact that without scientific consensus there can be no political will to implement expensive management systems. Carbon science must integrate a basic understanding of process with powerful measurement techniques. Models are also required that have the credibility to be used in what-if exercises to aid in designing new management systems. Local models are crucial because sinks must be long lasting and management systems to store carbon must aim at a decadal to centennial timescale. The agronomic paradigm of "test plots" is needed but limited in utility because of timescale. Large-scale models are needed to test the global effect of an international regime, including the stability of induced ecosystem sinks to potential changes in the chemical, physical, and human environment.

1.2 New Methods for Measurement

Measurement capability has been a continual challenge to the carbon research community in accomplishing the ambitious goals. The foundation of carbon cycle research lies in stable absolute calibration, an initial priority of the Keeling Mauna Loa effort and a persistent feature of the community. New measurements, such as of stable and radio-isotopes, the O_2/N_2 ratio, and remote sensing have been developed and adopted by the community. Techniques for dealing with spatial heterogeneity are also in rapid evolution: these include ecosystems studies, local eddy covariance flux measures, mesoscale aircraft and tall-tower techniques, and continental to global inverse modeling techniques (Valentini *et al.*, 2000). As the scope of large-scale biogeochemical research expands beyond a carbon cycle and greenhouse gas focus, techniques will need to be developed for the spatial-temporal integration of a range of processes. It is likely that new techniques will be needed for the study of airborne and waterborne nutrient transport as in gas, suspended and dissolved water-borne and aerosol phases. Techniques for spatial integration of belowground processes are crucial—there still exist only rudimentary measures for root growth and soil C and N turnover at or above the plot scale (Valentini *et al.*, 2000). Continuing adoption and endogenous development of measurement and data analytical techniques is a priority for biogeochemistry.

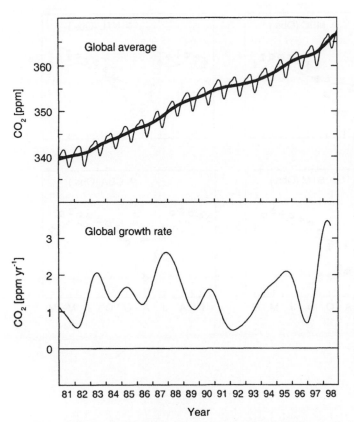

FIGURE 1 Global average atmospheric carbon dioxide mixing ratios and long-term trend determined using measurements from the NOAA CMDL cooperative air sampling network. Also shown is the global average growth rate for carbon dioxide; the variability in this is diagnostic of changes in biospheric and oceanic exchange. Data from National Oceanic and Atmospheric Administration's (NOAA) Climate Monitoring and Diagnostics Laboratory (CMDL), Carbon Cycle-Greenhouse Gas Group.

1.3 Biological Diversity and Evolution

The roots of biogeochemistry are in geochemistry and ecosystem ecology. Most work in biogeochemistry has followed chemical fluxes and treated ecosystems as series of linked compartments rather than as associations of species. In a sense, this always represented an operational convenience more than a hypothesis that species characteristics were irrelevant. The global loss of species diversity raises the concerns that critical thresholds of diversity may exist below which the functioning of ecosystems or their reliable delivery of ecosystem goods and services will be impaired. All but the most aggregated ecosystem models recognize the role of different functional types of plants, and some recognize at least implicitly the distinction between major microbial functional types (bacteria and fungi). Several questions remain open and contentious. The hypothesis that organisms in particular ecosystems are optimally adapted to local conditions (or statistics of those conditions) is often used as an operating rule in biogeochemistry. This is a defensible assumption in steady-state conditions: can it be assumed during changing times? Second, what role does the diversity of organisms in a given ecosystem play in system function? Can all variation be explained based on losses or gains of particular functional types, or does diversity itself play a role? These topics remain controversial to say the least (Hector *et al.*, 1999; Huston, 1997), and it is not yet clear whether the community has even formulated the right questions to ask about diversity and ecosystem function. It is clear that this area, the role of genetic, phenotypic, and taxonomic diversity in biogeochemistry, requires investigation. It poses an immense scale and measurement challenge because, while diversity varies on the smallest scales, ecosystem function (productivity, hydrology, nutrient cycling) intrinsically occurs in the aggregate. Measuring the rela-

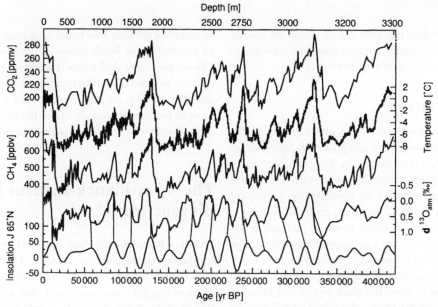

FIGURE 2 Climate and atmospheric composition over the past 420,000 years from the Vostok core (Petit *et al.*, 1999).

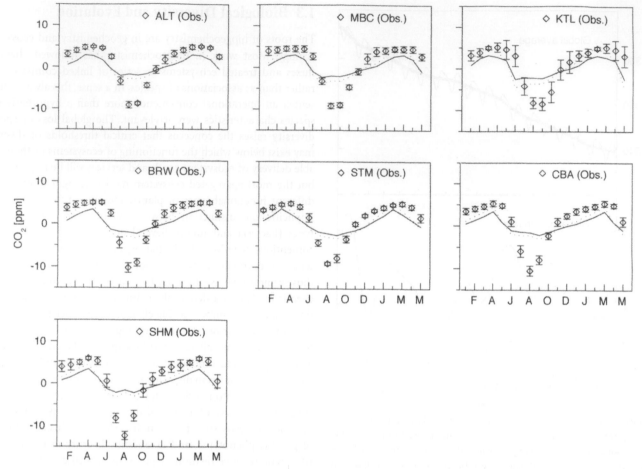

FIGURE 3   Comparison between the observed seasonal cycle of CO_2 and the simulated seasonal cycle produced by coupling the monthly estimates of net ecosystem production estimated by the Century model and fossil fuel emissions with the Hamburg ocean and atmospheric transport models for each of the seven high-latitude monitoring stations. The first six months of each cycle are displayed twice to reveal the annual variation more clearly. Mean and standard deviation are shown for the observed data (McGuire *et al.*, 2000).

tionships for testing diversity–function theory based on models and model ecosystems will require another quantum advance in the field's ability to make integrative measurements.

1.4 Belowground Processes

Our understanding of processes occurring above-ground is far more complete than our knowledge of soil processes. Soils contain 2–3 times more carbon than vegetation does. They are the primary reservoir of long-lived organic matter in the terrestrial biosphere and provide critical resources needed for photosynthesis. They remain persistently difficult to study, being impenetrable to remote sensing, locally variable such that meter-to-meter changes in organic matter or microbial activity can approximate the mean changes across continents, and of secondary interest to many academic researchers. New techniques involving isotopes, especially ^{14}C, ^{13}C, and ^{15}N are beginning to open the black box of soil turnover times, as are new techniques for chemical analysis of soil organic matter (Schulze *et al.*, 2000). No theory of ecosystem behavior will

be complete without a far better understanding of soil biology than we currently have. While current models of soil processes have significant predictive skill, many crucial processes are represented empirically with no real understanding of the underlying biology and chemistry. This is an important step in integrating the largest reserve of biodiversity—soils—and the longest-lived-ecosystem organic-matter reserves (also in soils) into ecosystem theory.

The "Max-Planck-Institut für Biogeochemie"

The Max-Planck-Gesellschaft (Society) has a number of institutes conducting biogeochemical research. The Institutes for Chemistry and Meteorology have long had programs addressing atmosphere–biosphere exchange and the carbon cycle. The Max-Planck-Institut for Biogeochemistry was set up to provide an intellectual focus for research on the global biogeochemical cycles.

Biogeochemistry had emerged in the 1990s as a center of both intellectual and policy ferment and yet the discipline lacked a center like the great institutions in meteorology and oceanography. The goal of the MPI for Biogeochemistry was to combine the biological and geophysical components of the field in a balanced way, and to bring together empirical, theoretical, and modeling groups. While biogeochemistry has flourished internationally, few biogeochemists, especially the more biologically oriented, find themselves close to the center of their institution's interests.

The Institute combines measurements and experimental studies with theory, simulation, and diagnostic research. The initial staff in the Institute brings together young scientists trained in both the geophysical and ecological disciplines, together in the same Institute, Departments and even offices. The combination of perspectives will lead to all sorts of turmoil (hopefully) with modelers suggesting new measurements, experimentalists identifying new modeling approaches, and diagnosticians identifying inconsistency between theory and observation (the latter requiring new data and models). This type of ferment has long been a highlight of the field, and the MPI will accelerate the pace of discovery in an already dynamic area.

References

Andreae, M. O., and Schimel, D. S. (1989). "Exchange of Trace Gases between Terrestrial Ecosystems and the Atmosphere." John Wiley.

Brasseur, G. P., Orlando, J. J., and Tyndall, G. S. (Eds.) (1999). "Atmospheric Chemistry and Global Change., Oxford University Press, Oxford, United Kingdom.

Committee on Global Change Research. (1999). Changes to the biology and biochemistry of ecosystems. In: "Global Environmental Change: Research Pathways for the Next Decade" National Research Council Press, Washington DC.

Hector, A., Schmid, B., Beierkuhnlein, C., Caldeira, M. C., Diemer, M., Dimitrakopoulos, P. G., Finn, J. A., Freitas, H., Giller, P. S., Good, J., Harris, R., Hogberg, P., Huss-Danell, K., Joshi, J., Jumpponen, A., Korner, C., Leadley, P. W., Loreau, M., Minns, A., Mulder, C. P. H., O'-Donovan, G., Otway, S. J., Pereira, J. S., Prinz, A., Read, D. J. *et al.* (1999). Plant diversity and productivity experiments in European grasslands. Science 286, 1123–1127.

Huston, M. A. (1997). Hidden treatments in ecological experiments—Re-evaluating the ecosystem function of biodiversity. *Oecologia* **110**(4), 449–460.

Indermhle, A., Joos, F., Fischer, H., Smith, H. J., Wahlen, M., Deck, B., Mastroianni, D., Tschumi, J., Blunier, T., Meyer, R., Stauffer, B., Stocker, T. F., (1999). Holocene carbon-cycle dynamics based on CO_2 trapped in ice at Taylor Dome, Antarctica. *Nature* **398** (6723), 121–126.

McGuire, A. D., Melillo, J. M., Randerson, J. T., Parton, W. J., Heimann, M., Meier, R. A., Clein, J. S., Kicklighter, D. W., and Sauf., W. (2000). Modeling the effect of snowpack on heterotrophic respiration across northern temperate and high latitude regions: comparison with measurments of atmospheric carbon dioxide in high latitudes. Biogeo*chemistry* **4**, 91–114.

Petit, J. R. *et al.* (1999). Climate and atmospheric history of the past 420,000 years from the Vostok Ice Core, Antarctica.

Schlesinger, W. H. (1997). "Biogeochemistry: An Analysis of Global Change." Academic Press, London.

Schulze, E-D. *et al.* (the CANIF community). (2000). Evaluation and significance of mean residence times for budgeting forest carbon sinks. Nature (submitted).

Valentini, R. *et al.* (the Euroflux community). (2000). Respiration as the main determinant of carbon balance in European forests. Nature (accepted).

Vitousek, P. M., and Reiners, W. (1977). Ecosystem succession and nutrient retention: A hypothesis. *BioScience* **25**, 376–381.

Biogeochemistry had emerged in the 1980s as a center of both intellectual and policy ferment and yet the discipline lacked a center like the great institutions in meteorology and oceanography. The goal of the MPI for biogeochemistry was to combine the biological and geophysical components of the field in a balanced way, and to bring together empirical, theoretical, and modeling groups. While biogeochemistry has flourished internationally, few biogeochemists, especially the more biologically oriented, find themselves close to the center of their institution's interests.

The Institute combines measurements and experimental and... with theory, simulation, and diagnostic research. The initial staff in the Institute brings together young scientists trained in both the geophysical and ecological disciplines, together in one institute. The measurements and even offices. The combination of approaches allows the sort of mutual (insecurity) with modern... theory suggests some measurements, experimental... identifying new measures... theories and observation that are... requiring new data and methods. This type of ferment has long been a highlight of the field, and the MPI will ascertain they are at discovery in an already enriched area.

References

Anders, E.T. and Schield, P. ...

Benson, T.F., Farmer, E., ...

Chen, L. ...

Introduction

I

1.1

Uncertainties of Global Biogeochemical Predictions

E.-D. Schulze,
D. S. S. Schimel
Max-Planck-Institute for
Biogeochemistry
Jena

1. Introduction

In the past few years, application of improved measurements and models suggests a robust partitioning of CO_2 emissions from fossil fuel consumption and land use: about one-third remains in the atmosphere, one-third is reassimilated by land surfaces, and one-third is absorbed by the oceans (Keeling et al., 1996). The terrestrial component of the sink has special political interest, because it is that part of the global carbon which can most directly be managed. If we were able to change the large fluxes of assimilation and respiration, as they were summarized by Schimel (1996), a tiny bit towards assimilation, we would be able, in principle, to compensate for fossil fuel emissions. The Kyoto Protocol (1997) is based on this assumption, and mirrors the attempt of mankind to actually manage a major global biogeochemical cycle (Schellnhuber, 1999). The commercial idea to market carbon sinks has initiated a major discussion about where on earth the largest sink capacity exists. Ciais et al. (1995) had proposed that the sink exists in the Northern hemisphere with its center in the Eurasian region. This was countered by Fan et al. (1998) who propose, based on analysis of gradients of CO_2 in the atmosphere, that continental USA was the major carbon sink in the Northern hemisphere. Schimel et al. (2000) argue against Fan's result based on models and *in situ* measurements. Lloyd (1999), on the other hand, predicts that the main terrestrial carbon sink is in the tropics. It is surprising that despite the intensive research going on in the field of production biology at global scale since the International Biological Programme in the sixties, there is so little consensus about the ability of terrestrial surfaces to absorb CO_2.

In the following sections, we will discuss the problems in locating a sink, and we will emphasize the spatial, temporal, and biological variability of processes on local as well as continental scale.

2. The IGBP Transect Approach

Although new satellite images allow a more and more detailed observation of the earth surface (e.g., Defries et al., 2000), the ground truth remains essential. This was recognized by IGBP as a significant problem because it remains impossible to study processes with global coverage. Continental transects were suggested that represent the major climatic regions of the globe and allow repeated observations at the same time (Walker et al., 1999). Continental transects were suggested for the boreal region (Alaska, West and East Siberia), the temperate arid regions (USA, China, Europe), and the subtropical climate (West Africa, Patagonia, South America, Australia). It was recognized that land-use change and not climate is the main driver in the humid tropics, and thus the "transect" consists of different land use types in that region (Steffen et al., 1999).

The transect approach was just one of many solutions to cope with the problem of spatial variability of processes and their integration to larger scales. Figure 1 summarizes a whole suit of approaches that were used to understand and to integrate highly variable processes at the landscape scale (Schulze et al., 1999).

In the following we will present ecological process data, to demonstrate the variability or constancy of processes which may or may not correlate with vegetation, plant functional types or species.

2.1 The Patagonian Transect

A range of plant parameters were studied in Patagonia including vegetation types ranging from tall forest to desert (Fig. 2, Schulze et al., 1996). Only root biomass and density in the top soil decreased linearly with rain fall. All other parameters, either changed in a threshold manner (such as LAI) or remained constant. $\delta^{13}C$

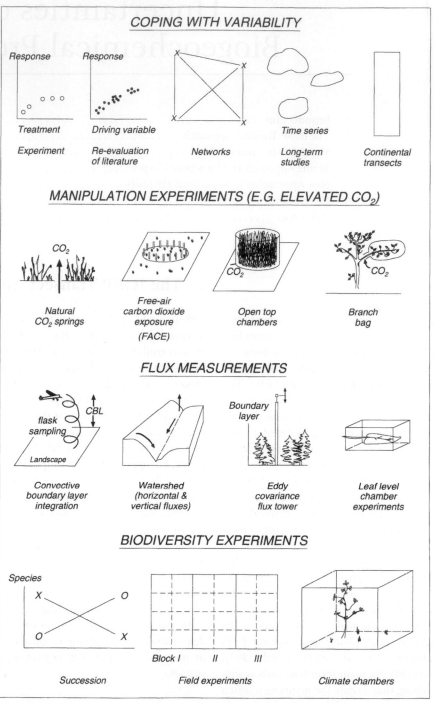

FIGURE 1 A summary of approaches to integrate highly variable processes of ecosystems at the plot and at the landscape scale in the context of global change (Schulze *et al.*, 1998).

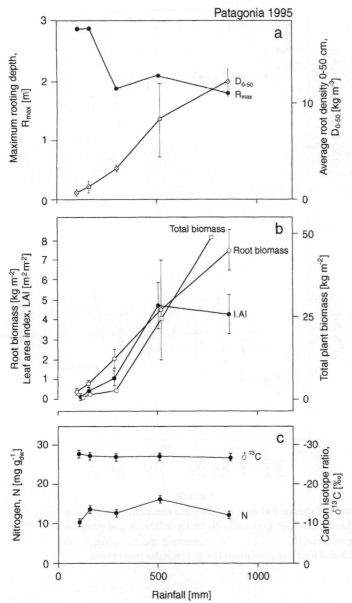

FIGURE 2 A range of plant parameters studied in Patagonia along a transect of decreasing rainfall. Vegetation types range from tall forest to desert (Schulze *et al.*, 1996).

was −27‰ independent of rainfall, and leaf nitrogen remained at 13 mg g^{-1}. There was no distinct relation to vegetation type or functional types of the vegetation. Assuming that a correlation exists between biomass and carbon storage in soils (Schulze et al., 2000), it would seem to be impossible to infer from $\delta^{13}C$ on the sink capacity of the underlying vegetation (Ciais et al., 1995). The constancy in $\delta^{13}C$ and leaf nitrogen concentration is caused by a change in species composition which maintains intrinsic water

use constant at decreasing rainfall due to changes in leaf morphology, independent of productivity.

2.2 The Australian Transect

The North Australian Transect extends about 300 km from Darwin to the interior of Australia. A study was undertaken (Schulze et al., 1998) in which this transect was extended to the higher (1800 mm) and lower (216 mm) rainfall regions along a transect extending to about 1000 km. The most striking observation was that the $\delta^{13}C$ value remained essentially constant at −28.1‰ (Fig. 3). Only when rainfall decreased below 400 mm, an effect was seen on the carbon isotope ratio. Again, this conservative response of the carbon isotope ratio was associated with a decrease in specific leaf area, and this leaf property was associated with a fivefold change in leaf nitrogen concentrations, depending on plant functional types. Plant species that were potentially fixing atmospheric N_2 had higher N concentrations and a higher specific leaf area than spinescent species. However, also the classification of plant functional types did not describe the functional process involved. The constant $\delta^{13}C$ value was associated with, and is most likely the result of, a change in species within each functional type; only when the diversity of species decreased to a single species in the dry region, the $\delta^{13}C$ values changed. At any one site, the local variability between species was as large as the continental variability along the transect.

The Australian transect extended from tall monsoon forest (21 m canopy height) to scattered dwarf trees (6 m height) in a subtropical grassland. We expect that the atmospheric isotope signal (not measured) would increase due to the increasing proportion of C4 grass photosynthesis, despite the fact that the total sink capacity of the region would decrease. This would make it very hard to infer from the isotopic signal on the sink capacity at the land surface.

2.3 The European Transect

Forest sites were selected across Europe ranging from North Sweden to Central Italy in order to study the effect of nitrogen deposition on ecosystem processes in coniferous and deciduous plantations. This study was designed such that the edaphic conditions were maintained as constant as possible, i.e., acid soils were chosen when available in order to detect effects of nitrogen. Figure 4 shows that the nitrogen concentration in needles and leaves were remarkably constant for conifers and deciduous trees, although *Fagus sylvatica* had a higher N concentration than *Picea abies*. Also, the $\delta^{13}C$ and $\delta^{15}N$ concentrations were remarkably constant despite the large variation in climate and plant species (deciduous vs conifer). There was no obvious relation to NPP or leaf area index. Besides the fact that the carbon isotope ratios were increasing from north to south despite increasing NPP, the most remarkable observation was,

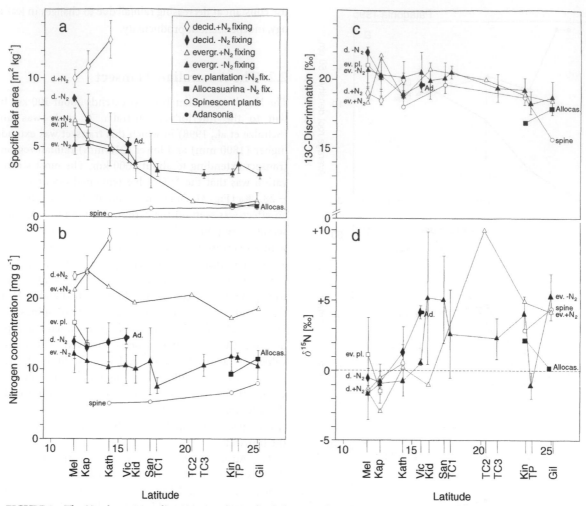

FIGURE 3 The Northern Australian Transect: Latitudinal changes of specific leaf area, leaf nitrogen concentration, δ^{13}C-isotope discrimination and of the δ^{15}N-isotope ratio in the plant functional types: potentially N_2 fixing deciduous and evergreen trees (d. + N_2, ev. + N_2) and non-N_2 fixing deciduous and evergreen trees (d. − N_2, ev. − N_2), spinescent species (spine), bottle tree *Adansonia* (Ad.), *Allocasuarina* (Allocas.), and evergreen cultivated fruit tree plantations (ev. pl.) (Schulze *et al.*, 1998).

that by selection of specific sites, the N concentration in the foliage was responding neither to N deposition nor to climate. However, at each site, the local variation in N concentration showed a range larger than concentrations along the whole transect. For instance, at the German "Waldstein" site, N concentrations in needles vary between habitats from 0.54 to 2.12 mmol g^{-1}, while the whole continental transect varied between 0.5 and 1.1 mmol g^{-1}. Again, inferring from isotope ratios or from N concentrations or LAI or C-fixation would be difficult, and it would not recover the local mosaic type variation. In addition, there was no significant difference between conifers and deciduous trees with respect to NPP despite the difference in foliage N concentration. The complex basis for the

observed homeostatic response is explained by Schulze *et al.* (2000).

3. Variability in Processes

The Kyoto protocol allows the compensation of fossil fuel emission by biological sinks without defining its components. In contrast to plant physiologists who are mainly concerned with photosynthesis, land managers are mainly interested in growth of products (timber, grain), but the atmosphere integrates carbon assimilation and respiration which includes the soil. In addition, carbon is released from ecosystems not only by respira-

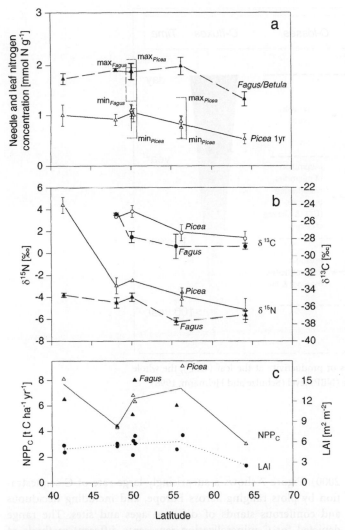

FIGURE 4 The European Transect: Latitudinal changes of needle and leaf nitrogen concentration, δ^{13}C-isotope ratio and δ^{15}N-isotope ratio, net primary productivity (NPP) and leaf area index (LAI) for conifers (*Picea abies*) and deciduous trees (*Fagus sylvatica*). In Fig. 4a minimum (min) and maximum (max) values show the absolute range of data.

tion but also by harvest and fire. In the case of harvest, respiration may take place elsewhere on the globe due to the globalization of trade. Schulze and Heimann (1998) illustrated the different views of productivity and quantified the consequences thereof in a flux scheme (Fig. 5), where the input by photosynthesis represents 100% and this quantity is generally termed gross primary productivity (GPP). Since the plant needs energy for its own growth and maintenance metabolism, 50% of GPP is used by the plant itself. The resultant quantity is termed net primary production (NPP) which includes growth of all components, especially leaves, stems, fine roots, and fruits. The harvestable fraction may only be a small quantity of NPP and the

resultant biomass. The so-called harvest index (crop/biomass) is 50% in high yielding crop varieties, but generally averages 30%. Also, in trees, timber production is generally less than 20% of NPP. The largest quantity of NPP is not retained by the plant but shed as litter. This may take place in a seasonal rhythm or continuously, but it is a rejuvenescence process and compensates for aging of organelles and organs. The litter of roots and foliage reaches the ground and is decomposed by heterotrophic organisms which use the litter as the sole carbon source. Thus, the largest fraction of the litter returns to the atmosphere as CO_2 and some undigestable carbon remains as humus. The balance between assimilation and ecosystem respiration is termed net ecosystem productivity (NEE). However, also this fraction may be remobilized and converted to CO_2 by disturbance, or by fire. The remaining carbon, mainly in the form of recalcitrant humus and charcoal, contributes to the net biome productivity (NBP Schulze *et al.*, 2000). The definitions are based on the assumption, that the observations are made on an increasing area, namely, it moves from the leaf level (GPP) to the plant cover (NPP) and to the stand level (NEP), and finally reaches the landscape level (NBP).

The terrestrial surface looks quite different, depending on which quantity we chose (Schulze and Heimann, 1998). Photosynthesis is related to leaf structure and available nitrogen, and reaches highest rates in the temperate climate and in regions with intensive agriculture (Eastern US, Europe, India, East Asia). In contrast, NPP, which depends on the length of the growing season and on leaf biomass, reaches highest rates in the humid tropics and monsoon climates. Predictions on NEP are problematic, because according to ecological theory respiration should balance assimilation in the long term. However, a disequilibrium exists between assimilation and decomposition due to a continuous increase in atmospheric CO_2. Based on this effect, NEP would reach a maximum in subtropical and temperate regions, not in the boreal climate.

Schulze *et al.* (1999) compared a European Picea forest with a Siberian Pine forest. The European forest has a high NPP (15 m³ stem growth per year) but also high respiration, while the Siberian forest has a low NPP (1 m³ stem growth per year) but also low respiration. Integrated NEP over the growing season of both sites was surprisingly similar. Both sites assimilated about 15 mol m⁻² during the summer. However, it would not be appropriate to generalize from this observation, because a high variability exists on a landscape basis in Siberia (Fig. 6), ranging from plots that are carbon neutral or carbon sources after logging to very effective carbon sinks, such as old growth-unmanaged forests. In fact, sphagnum bogs, representing a totally different plant cover than forest, reach rates of net carbon sequestration similar to those of a forest.

The main natural factor that disturbs the Siberian forest are fires which either occur as repeated ground fires (fire frequency about 50 years) or burn the whole forest (crown fires, every 200–300 years). The study of ground fires shows, that the forest ecosystem

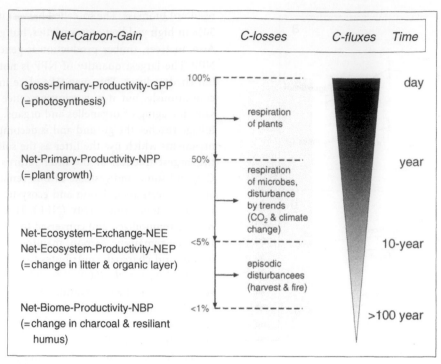

FIGURE 5 Schematic explanation and estimates of productivity at the leaf (GPP), the whole plant (NPP), the ecosystem (NEP) and the biome (NBP) level (Schulze and Heimann, 1998).

accumulates organic material in the organic layer. This material is being decomposed to a level, that the local decomposers cannot digest this material any further, and it thus builds up an organic layer with increasing thickness. Groundfires will burn this layer and return it as CO_2 to the atmosphere, but groundfires will also produce charcoal with higher longevity. On a landscape level, the organic layer will probably be carbon neutral, although a specific plot shows a distinct rate of accumulation, depending on the time since the last fire. The only component, which accumulates at the landscape level is most likely charcoal, but also this component appears to undergo a decomposition process (Czimczik et al., 2000). Nevertheless, some of this carbon is stabilized in soils, and based on this fraction Schulze *et al.* (1999) calculated a rate of NBP in the order of 13–130 mmol m^{-2} y^{-1}.

The accumulation of carbon in the organic layer after fire points at a basic problem of the simplified flux scheme of Schulze and Heimann (1998), namely, that intermediate pools exist at each level with different mean residence time, and depending on the level of spatial integration these pools may or may not average out. The problem is illustrated in Figure 7, where an inventory-process type approach (in contrast to eddy covariance flux measurements) was chosen to calculate the carbon sink capacity of European forest ecosystems. If NPP was plotted against C-mineralization, then NPP minus C-mineralization would represent NEP at the plot scale (Schulze *et al.*,

2000). Figure 7 shows a surprisingly large rate of C-sequestration by plots ranging across Europe, and including deciduous and coniferous stands of different ages and sites. The range depicted for C-mineralization represents different methods of assessing C-mineralization. This large rate of NEP includes wood growth at each plot. However, on a landscape scale, wood would be harvested in managed European forests, and this compartment would thus be carbon neutral. If this fraction is removed, the rate of carbon sequestration, here termed NBP, would only be a small fraction of NEP measured at the plot scale. We are not sure, if this rate of NBP is permanent. Harrison et al. (2000) determined the mean residence time of the organic layer in these habitats, and found that this layer may be very short lived (L + F layer = 5 to 6 years), but also the following 0–5 cm layer (A + Oh horizon) was not very long lived and showed a mean residence time of 35 years in France and 340 years in North Sweden, which is the same order of magnitude as that of the tree cover.

The implications of this observation are manifold. The "Kyoto forest" is planned to contain 0.5 ha plots, which will sequester carbon at the rate as it is shown in Figure 7a until harvest. However, if the "Kyoto forest" would extend across a landscape, the forest would most likely be carbon neutral, and only a very small fraction will be stabilized as "recalcitrant" soil organic matter. The other implication is that the flux measurement at the plot scale

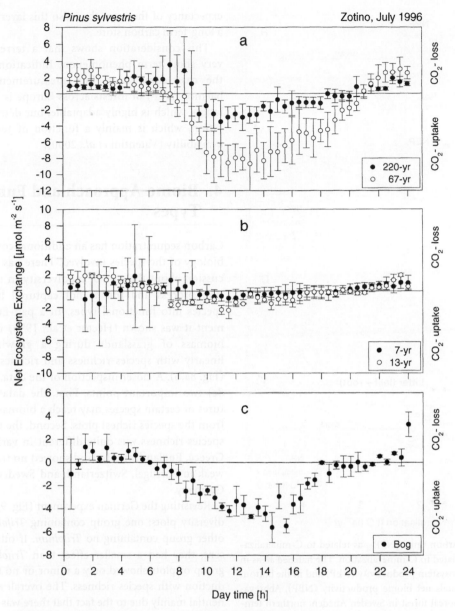

FIGURE 6 Average daily courses of ecosystem CO_2 exchange of (a) Siberian pristine pine forest and natural regeneration after fire, (b) regrowth after logging, and (c) a bog, all located near Zotino, Central Yenisey River, July 1996 (Schulze et al., 1999).

does not measure the long-term carbon sequestration, but mainly tree growth at the plot scale. Also, the change in C-pools at the plot scale does not represent the C-sequestration at the landscape scale, because some of the compartments remain at an average level as long as management is constant. However, changes in management may cause short-term variations in carbon pools, which do not reflect the long-term C-balance

$$\Delta C = \Delta C_{biomass} + \Delta C_{organic\ layer} + \Delta C_{soil}.$$

The change in biomass takes into account carbon that has not yet been stabilized to a degree for which it could be counted as long-term C-sequestration. After tree death, this carbon would still decompose, and at a landscape level remain carbon neutral.

The change in carbon stocks of the biomass has a mean residence time related to the change in harvest by management, i.e., if harvest is delayed by 20 years (commitment period of the Kyoto protocol from 1990 to 2010) and timber demand is

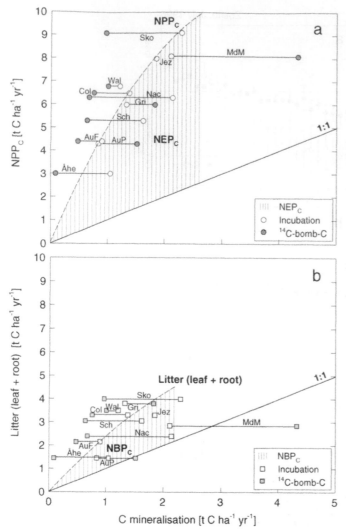

FIGURE 7 (a) NPP in carbon units (NPP$_C$) as related to C-mineralization and (b) litter fall as related to C-mineralization. The hatched areas in the top panel equals net ecosystem productivity (NEP) and in the bottom panel the hatched area equals net biome productivity (NBP). Abbreviations refer to study sites: boreal forest in Sweden Åheden, northern temperate forest of Denmark *Sko*gaby (*Picea*) and *Grib*skov (*Fagus*), temperate forest in Germany *Wal*dstein (*Picea*) and *Sch*acht (*Fagus*), in Czech Republik *Nac*etin (*Picea*) and *Jez*ery (*Fagus*), in France *Au*bure *Picea* and *Fagus*, and montan mediterranean forest in Italy *Col*lelongo (*Fagus*) and *Monte di Mezzo* (*Picea*) (Schulze et al., 2000).

expectancy of the trees, then also this layer cannot be counted as a long-term carbon store.

This consideration shows that a terrestrial sink may be a very short term phenomenon. Indications that this is indeed the case come from flux measurements. The net carbon balance (NEP) of forests across Europe is not related to assimilation, which is highly adaptable, but determined by soil respiration, which is mainly a function of temperature and water availability (Vatentini *et al.*, 2000).

4. Biome Approach and Functional Types

Carbon sequestration has an additional component, namely, the biology of the species involved. There has been an ongoing discussion about the role of biodiversity in the carbon sequestration process and this is not captured by a classification of species into functional types. In a pan-European field experiment it was shown (Hector et al., 1999) that the aboveground biomass of grasslands during a growing season increased linearly with species richness and richness of functional types (Fig. 8a,b). A closer inspection of the data, however, shows (Fig. 8c) two important points. First, the data show that monocultures of certain species may reach a biomass that is not different from the species richest plots. Second, the trend of biomass with species richness was quite different in various countries. While Greece, England, and Ireland showed no trend, the response was weak in Portugal, Switzerland, and Sweden, and exponential in Germany.

Revisiting the German experiment (Fig. 9), we find two types of diversity plots: one group containing *Trifolium pratense* and the other group containing no *Trifolium*. If other legumes were present, they had a smaller effect than *Trifolium*. Each individual group of plots showed only a minor or no trend of biomass production with species richness. The overall average trend is exponential mainly due to the fact that there was no high-diversity plot without *Trifolium*.

This result indicates, that functional groups (Cramer et al., 1998) and species richness (Hector et al., 1999) may not be driving biomass production, but the presence of certain genera or species, we may call them keystone species (Bond, 1994), determines the NPP of a system.

5. New Approaches to Functional Diversity

While traditional functional groups and species concepts have identified problems, other approaches show promise, at least for identifying sensitivities. Global applications of the Century ecosystem model (Parton et al. 1994; Schimel et al., 1996) were carried out. The model was set up with a relatively large number

equilibrated by import, then the change in average biomass would be equivalent to the average growth rate of this forest (after 10 years) in that period. However, it cannot be expected that this forest will last for the next 100 years, because that would lead to exploitation of the timber reservoirs of other countries while this forest would reach the life expectancy of some of the species in European climate. If the mean residence time in the organic layer of soils is indeed of the same order of magnitude as the life

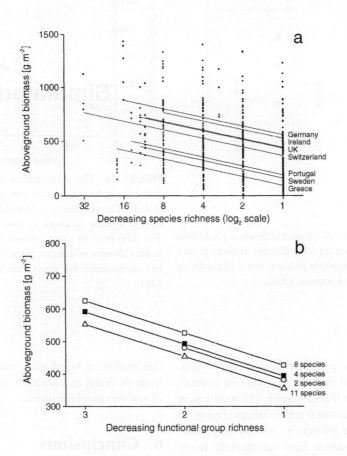

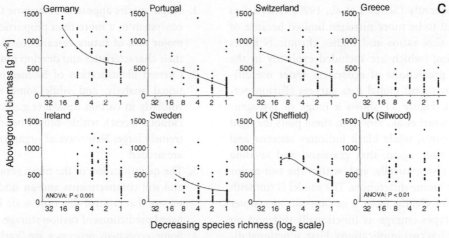

FIGURE 8 A pan-European field experiment: (a) Overall log-linear reduction of above ground biomass with the simulated loss of plant species richness. (b) Linear reduction with the loss of functional group richness within species richness levels. (c) Biomass patterns at each site (displayed with species richness on a log$_2$ scale for comparison with panel (a)) (Hector et al., 1999).

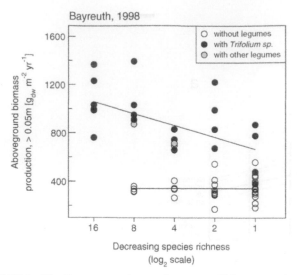

FIGURE 9   The German experiment: above ground biomass production as related to decreasing species richness on three different diversity plots: a plot without legumes, a plot with *Trifolium pratense*, and a plot with no *Trifolium* but other legumes (Scherer-Lorenzen, 1999).

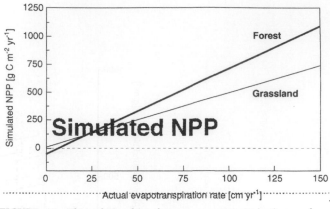

FIGURE 10   The relationships between evapotranspiration and net primary production as they emerge from the Century model, applied globally. The near-linear realtionship between evaporation and NPP, very similar to those observed in semi-arid lands, is an emergent property. The differences in slope between biomes are largely due to differences in the C:N ratio of different plant functional types, indicating that ecosystem composition has direct effects on biogeochemistry (Scholes et al., 1999).

of biomes, where biome-specific parameterizations were based, as much as possible, on site-level data. Biomes differed in functional plant physiology, and disturbance regime. The results were analyzed in terms of major functional relationships. Figure 10 shows ecosystem-level water use efficiency (NPP vs evapotranspiration) identifying that grasslands have substantially lower ecosystem water use efficiency, despite higher leaf-level water use efficiency deserved frequently (Scholes et al., 1999). This arises because grasslands tend to be more nitrogen limited because of narrower whole-plant C:N ratios and higher chronic N losses due to fires and grazing (which are included explicitly in the model). These model predictions of ecosystem water use efficiency are borne out by recent global data surveys (Parton personal communication). Figure 11 shows a range of nitrogen-related efficiencies (Schimel et al., 1999). In these plots the light gray indicates forest points, while black indicates savanna and grassland ecosystems. Note again, that grassland and savanna systems tend to differ systematically, and, within the two major types, there are further biome differences. The model is run with 28 different biomes, but clearly from a carbon-nitrogen-water point of view, 5 or 6 types emerge as functionally different on large scales. This result has two implications. First, functional diversity matters even on the global scale: the biosphere cannot be treated as a uniform black box. Second, the dimensionality of functional diversity in carbon–nitrogen–water processes is much less than that of the number of species or even biomes. As a caution, for other processes, the dimensionality could be higher than that even of species, e.g., for pathogen resistance where both species and population variability are high. While

this analysis is based on a model, sufficient data are emerging from the IGBP and scientific community to perform these types of analyses on observations.

6. Conclusions

1. Single species appear to determine the NPP processes in the ecosystem in a more than proportional manner. It will be a major task of future research to identify these species and their characteristics and develop modeling approaches.

2. Current identification of biomes is not based on a functional analysis, and while some major biomes function similarly in carbon uptake (e.g., European conifer and deciduous forest), within-biome or-species effects can be extremely large. New ways of organizing ecological variability are needed.

3. The quantification of the mean residence time of vegetation and soil compartments and an understanding of the parameters that control this time-scale is necessary for process-based predictions of carbon storage.

4. Some ecosystem processes are "carbon neutral" on a landscape scale because of disturbance-related variability but show distinct trends (positive and negative) at plot within landscapes. Relatively few types of direct measurements are possible at the landscape scale and so the differences in plot and landscape behavior will be a major problem in predicting the effects of the implementation of the Kyoto protocol with respect to mitigation of the increase in atmospheric CO_2.

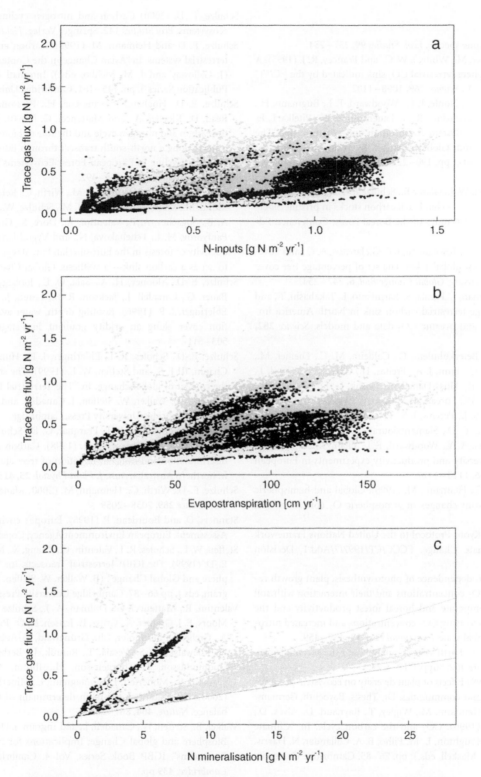

FIGURE 11 (a) Sum of annual N gas fluxes (N_2 + NO + N_2O) vs N inputs from a global Century model simulation. N inputs result from wet and dry deposition and biological nitrogen fixation. Yellow points are grassland ecosystems, green points are forests, and black points are "mixed" ecosystems such as savannas. (b) N trace gases vs evapotranspiration. (c) N trace gases vs annual N-mineralization. Lines indicate regressions computed from Matson and Vitousek (1990) based on data from the Amazon Basin (Schimel and Panikov, 1999). See also color insert.

References

Bond, W. J. (1994). Keystone species. *Ecol. Studies* **99**, 237–254.

Ciais, P., Tans, P. P., Trolier, M., White, J. W. C., and Francey, R. J. (1995). A large northern hemisphere terrestrial CO_2 sink indicated by the $^{13}C/^{12}C$ ratio of atmospheric CO_2. *Science* **269**, 1098–1102.

Cramer, W., Shugart, H. H., Nonle, I. R., Woodward, F. I., Bugmann, H., Bondeau, A., Foley, J. A., Gardner, R. H., Lauenroth, W. K., Pitelka, L. F., and Sutherst. (1998). Ecosystem composition and structure. In: "The Terrestrial Biosphere and Global Change." B. Walker, W. Steffen, J. Canadell, J. Ingram, eds., pp. 190–228. Cambridge University Press, Cambridge.

Czimczik, C., Schmidt, M. W. I., Glaser B., Schulze E.-D. (2000). The inert carbon pool in boreal soils—char black carbon stocks in pristine Siberian Scots pine forest. Proceedings of the Boreal Forest Conference. Edmonton, May 2000.

Defries, R. S., Hansen, M. C., Townshend, J. R. G., Janetos, A. C., and Loveland, T. R. (2000). A new global 1-km data set of percentage tree cover derived from remote sensing. *Global Change Biol.* **6**, 247–254.

Fan, S., Gloor, M., Mahlman, J., Pacala, S., Sarmiento, J., Takahashi, T., and Tans, P. (1998). A large terrestrial carbon sink in North America implied by atmospheric and oceanic CO_2 data and models. *Science* **282**, 442–446.

Hector, A., Schmid, B., Beierkuhnlein, C., Caldeira, M. C., Diemer, M., Dimitrakopouluos, P. G., Finn, J. A., Freitas, H., Giller, P. S., Good, J., Harris, R., Högberg, P., Huss-Danell, K., Joshi, J., Jumpponen, A., Körner, C., Leadley, P. W., Loreau, M., Minns, A., Mulder, C. P. H., O'-Donovan, G., Otway, S. J., Pereira, J. S., Prinz, A., Read, D. J., Scherer-Lorenzen, M., Schulze, E. D., Siamantziouras, A. S. D., Spehn, E. M., Terry, A. C., Troumbis, A. Y., Woodward, F. I., Yachi, S., and Lawton, J. H. (1999). Plant diversity and productivity experiments in European Grasslands. *Science* **286**, 1123–1127.

Keeling, R. F., Piper, S. C., Heimann, M. (1996). Global and hemispheric CO_2 sinks deduced from changes in atmospheric O_2 concentrations. *Nature* **381**, 218–221

Kyoto Protocol. (1997). Kyoto Protocol to the United Nations Framework Convention on Climate Change. FCCC/CP/1997/7/Add.1, Decision 1/CP.3, Annex, 7.

Lloyd, J. (1999). The CO_2 dependence of photosynthesis, plant growth responses to elevated CO_2 concentrations and their interaction with soil nutrient status. II. Temperate and boreal forest productivity and the combined effects of increasing CO_2 concentrations and increased nitrogen deposition at a global scale. *Functional Ecol.* **13**, 439–459.

Schellnhuber, H. J. (1999). "Earth system" analysis and the second Copernican revolution. *Nature* **402**: Supp. C19–C23.

Scherer-Lorenzen, M.(1999) Effects of plant diversity on ecosystem processes in experimental grassland communities. Dr. Thesis. Bayoeuth, Germany.

Schimel, D., Enting, I., Heimann, M., Wigley, T., Raynaud, D., Alves, D., and Siegenthaler, U. (1996). CO_2 and the carbon cycle. In: "Climate Change 1995." (J. T. Houghton, L. M. Filho, B. A. Callandar, N. Harris, A. Kattenberg, and K. Maskell, eds.), pp. 76–85. Cambridge University Press, Cambridge.

Schulze, E. D. (2000) Carbon and nitrogen cycling in European forest ecosystems. *Ecol Studies* **142**. Springer Verlag, *Heidelberg,* **498**pp.

Schulze, E. D. and Heimann, M. (1998). Carbon and water exchange of terrestrial systems. In "Asian Change in the Context of Global Change". (J. Galloway and J. M. Melillo, eds.), Internatl Geosphere-Biosphere Publication Series 3, pp. 145–161. Cambridge University Press.

Schulze, E.-D., Högberg, P., vanOene, H., Persson, T., Harrison, A. F., Read, D., Kjoeller, A., and Matteucci, G. (2000). Interactions between the carbon and nitrogen cycle and the role of biodiversity: A synopsis of a study along a north-south transect through Europe. In: "Carbon and Nitrogen Cycling in European Forest Ecosystems". (E. D. Schulze, ed.), *Ecological Studies* **142**, 1468–492.

Schulze, E. D., Lloyd, J., Kelliher, F. M., Wirth, C., Rebmann, C., Lühker, B., Mund, M., Knohl, A., Milukova, I. M., Schulze, W., Ziegler, W., Varlagin, A. B., Sogachev, A. F., Valentini, R., Dore, S., Grigoriev, S., Kolle, O., Panfyorov, M. I., Tchebakova, N., and Vygodskaya, N. N. (1999). Productivity of forests in the Eurosiberian boreal region and their potential to act as a carbon sink—a systhesis. *Global Change Biol.* **5**, 703–722.

Schulze, E.-D., Mooney, H. A., Sala, O. E., Jobbagy, E., Buchmann, N., Bauer, G., Canadell, J., Jackson, R. B., Loreti, J., Oesterheld, M., and Ehleringer, J. R. (1996). Rooting depth, water availability, and vegetation cover along an aridity gradient in Patagonia. *Oecologia* **108**, 503–511.

Schulze, E.-D., Scholes, R. J., Ehleringer, J. R., Hunt, L. A., Canadell, J., Chapin, III F. S., and Steffen, W. L. (1999). The study of ecosystems in the context of global change. In "The Terrestrial Biosphere and Global Change". (B. Walker, W. Steffen, J. Canadell, and J. Ingram, eds.), pp. 19–44. Cambridge University Press, Cambridge.

Schulze, E.-D., Williams, R. J., Frarqhar, G. D., Schulze, W., Langridge, J., Miller, I. M., and Walker, B. H. (1998). Carbon and nitrogen isotope discrimination and nutrient nutrition of trees along a rainfall gradient in northern Australia. *Aust. J. Plant Physiol.* **25**, 413–425.

Schulze, E.-D., Wirth, C., Heimann, M. (2000) Managing forests after Kyoto. *Science* **289**, 2058–2059

Stanners, D. and Bourdeau, P. (1995). Europe's environment. The Dobris Assessment. European Environment Agency, Copenhagen.

Steffen, W. L., Scholes, R. J., Valentin, C., Zhang, X., Menaut, J. C., Schulze, E.-D. (1999). The IGBP Terrestrial Transects. In: "The Terrestrial Biosphere and Global Change". (B. Walker, W. Steffen, J. Canadell, and J. Ingram, eds.), pp 66–87. Cambridge University Press, Cambridge.

Valentini, R., Matteucci, G., Dolman, A. J., Schulze, E. D., Rebmann, C., Moors, E. J, Granier, A., Gross, P., Jensen, N. O., Pilegaard, K., Lindroth, A., Grelle, A., Bernhofer, Ch., Grünewald, T., Aubinet, M., Ceulemans, R., Koewalski, A. S., Vesala, T., Rannik, Ü. Berbigier, P., Loustau, D., Gudmundsson, J., Thorgeirsson, H., Ibrom, A., Morgenstern, K., Clement, R., Moncrieff, J., Montagnani, L., Minerbi, S., and Jarvis, P. G. (2000). Respiration as the main determinant of the European carbon balance. *Nature,* **404**, 861–865.

Walker, B., Steffen, W., Canadell, J., and Ingram, J. (1999). "The Terrestrial Biosphere and global Change; Implications for Natural and Managed Ecosystems". IGBP Book Series, Vol 4. Cambridge University Press, *Cambridge,* **439** pp.

1.2

Uncertainties of Global Climate Predictions

Lennart Bengtsson
*Max Planck Institut for
 Meteorology*
Hamburg, Germany

1. Introduction

Our subject of discussion in this chapter is global climate prediction and the uncertainties of such predictions. What do we mean by a prediction of climate? E. Lorenz, the father of the chaos theory (Gleick, 1988), once clarified the important difference between forecasts of climate anomalies, such as the one caused by the El Niño phenomenon, and forecasts of the state of climate caused by changes in the solar forcing or by changes in the composition of the atmosphere. The first of these phenomena can in principle be predicted per se with useful skill, while in the second case only changes in the statistical structure of climate can be predicted. We will not be able to say whether a particular summer or winter will be warmer or colder than normal but only say, for example, that the number of summers with a temperature above a certain value will be more common than what it was previously. In this chapter I will use the expression climate prediction only in the context of the ability to simulate or predict the overall statistics of climate.

Even if the second kind of prediction is less precise, it is nevertheless very important, since knowledge of the average condition of climate, including its statistical structure in space and time, is of importance to the society and to the environment. Only a modest change in the average temperature or precipitation may imply changes in the statistical distribution of extremes. Precautions for flooding in most cities and municipalities, for example, have been designed to withstand extreme events, which on average occur only once a century. A warming on the order of one degree may lead to changes in the moisture content of the atmosphere by some 6%, with the consequences that extreme rainfall may be more common.

A fundamental question that first needs to be addressed is whether climate is at all unique or *transitive*, that is, for a given set of external forcing it follows that there exists a unique set of climate statistics, or whether several possible sets of climate statistics are possible for a given set of forcings? Such an *intransitive* climate would then be a priori unpredictable, since infinitesimal changes in the initial data or in the forcing may change the climate in ways similar to the chaotic processes that limit the length of useful weather forecasts (Lorenz, 1968). We believe, based on numerical modeling studies, that this is not the case with the atmosphere when forced from prescribed boundary conditions, but there is no indication that it will not be the case when we incorporate the full feedback with the oceans and the land surfaces. In fact, we have several indications of the nonuniqueness of the earth's climate, one of them related to the thermohaline circulation of the ocean. Such a mechanism, indicated by Stommel (1961), arises from the influence of the ocean salinity on the vertical heat exchange with the deep ocean, whereas salinity does not influence the interaction with the atmosphere (Bryan, 1986; Maier-Reimer and Mikolajewicz, 1989). There are other similar examples from the interaction between the atmosphere and the land vegetation which can change the regional climate at least as significantly (Claussen, 1998; Brovkin *et al.*, 1998).

These considerations and the additional fact that climate is the integral of weather over long periods of time, the weather itself being unpredictable, mean that the predictability of climate is a fundamental issue. The uncertainties of global climate prediction are a broad subject and I will here restrict my presentation to a time scale of a few hundred years and thereby concentrate on the time from early industrialization to the middle of the next century or so.

In Section 2, I will discuss the observational evidence of climate change. As a suitable point of reference in my review, I will use the recent attempts by Mann *et al.*, (1998, 1999) to reconstruct the surface temperature of the Northern Hemisphere for the last millennium. In Section 3, I will address the physical rationale underpinning climate change modeling, and finally in Sections 4 and 5,

I will be concerned with the modeling aspects and present and analyze some general results of numerical experiments.

2. Observational Evidence

While the variations of climate over time scales of ten thousand to a hundred thousand years in all likelihood are caused by variations in solar irradiation over the year due to orbital effects, the so-called Milankovitch effect (Milankovitch, 1920, 1941; Berger, 1988), climate variations on shorter time scales are still rather mysterious. The most spectacular of these variations, at least as interpreted from ice-core measurements, appear to have amplitudes of several degrees Kelvin and were particularly common during the last glaciation (e.g., Alley *et al.*, 1993). Occurrence of such extreme events during the Holocene, at least for the last 8000 years or so, has so far not been reported. Also, the less extreme climate fluctuations are of considerable importance to society. There have been numerous reports of climate variations over the last several hundred years, including a period of relatively warm climate, at

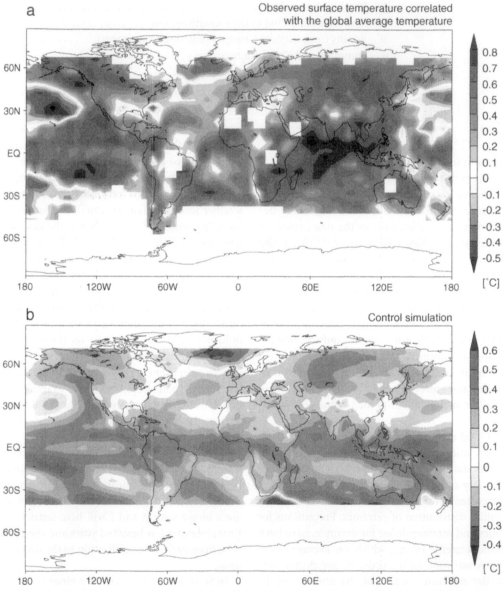

FIGURE 1 (a) Observed pointwise correlation of the annual surface temperature with the global averaged temperature based on observations from the period 1950–1995. (b) The same for a 300-year control simulation with ECHAM4/OPYC3 coupled model. Note the area of slight negative correlation in the North-Atlantic Greenland area both in the observations and in the model results. Similar patterns are found in the model also when averaged over longer periods, for at least until 50 years means. See also color insert.

least in Europe, during the 11–13th centuries, and a relatively long period of cold climate, the so-called *little ice age* from the 14th to the end of the 19th century.

Available observational records, instrumental as well as indirect information on past climate, are spatially rather restricted. Before the end of the 18th century they are available mainly from Europe and central China, together covering only some 3% of the earth surface. Furthermore, available data as well as model simulation studies show that the patterns of surface temperature anomalies have rather distinct signatures, with some areas of the earth in fact being negatively correlated with the global average temperature (Fig. 1). A notable region is the Atlantic–Arctic sector, including parts of Northern Europe, which in fact is slightly *negatively correlated* with the global average temperature. This has the surprising effect that Iceland, Greenland, and Northern Scandinavia generally are colder than normal when the average temperature of the earth is higher than normal. The reverse can be seen over the tropical part of the Pacific and the Indian ocean and is strongly correlated with the global averaged temperature. It is interesting to note that climate models are capable of reproducing this particular pattern rather well (Fig. 1b). Model experiments also suggest (e.g., Fig. 8 in Bengtsson, 1997) that climate anomalies over large geographical regions can continue over several decades due to internal low-frequency variations in the climate system. This means that it may be quite misleading to rely too heavily on observational information which is restricted by geography and time when we wish to draw general conclusions on climate events in the past and relate such events to specific external forcing mechanisms such as variations in solar irradiation or atmospheric changes due to volcanic eruptions.

Mann *et al.* (1998, 1999) have addressed this problem in a commendable systematic and comprehensive way. By combining available instrumental and palaeo-data at annual resolution, they have produced a continuous record of the annually averaged surface temperature of the Northern Hemisphere for the period 1000 until present. The method is based on the determination of the characteristic empirical orthogonal functions (EOFs) for the present climate and then the projection of the available palaeo-data onto these modes (Fig. 2). For information before the middle of the 18th century, one must rely on palaeo-data such as those from ice cores, tree-rings, and corals. Before 1450 even such data at an annual resolution are sparse, so the reconstructed temperature evolution has large error bars. An important aspect of the methodology used in the study of Mann *et al.* is that such error bars follow a priori. The reduction in the size of the error bars with time reflects the steadily improved data set and its geographical coverage, making it possible to determine more EOFs.

Three important aspects in Figure 2 need to be highlighted. First, there is an indication of a general ongoing cooling on the order of 0.1 K until ca. 1900. (This cooling trend is more clearly seen in Mann *et al.* (1999), where the record is extended over the whole period 1000–1998.) It is concluded that this cooling trend is in broad agreement with the Milankovitch forcing. Second, there are characteristic temperature fluctuations from year to year but with

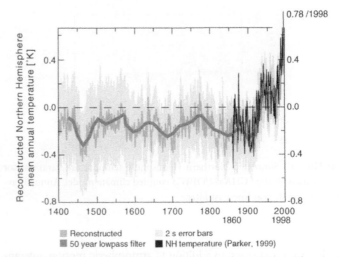

FIGURE 2   Reconstructed surface temperature from 1400 until present (after Mann *et al.*, 1998). Observed surface temperature data from Parker (1999, personal communication) have beeen inserted.

typical low-frequency variations of several decades. These fluctuations extend over the whole record. Third, there is a pronounced warming from the early part of the 20th century, reaching large values in the past few years. The warmest decade of the last 1000 years is the 1990s, with 1995, 1997, and 1998 being the warmest years in the whole record, with more than 3 standard errors than any year back to 1400. There are two circumstances in Figure 2 which require a more substantial analysis. These are the low-frequency temperature variations and the steep temperature warming taking place during the last century. Mann *et al.* (1998) have offered a set of explanations based on a simple correlation with the assumed solar variations, a volcanic index, and a simplified expression for the greenhouse gas forcing.

I will discuss this in more detail using some recent climate simulation experiments (Roeckner *et al.*, 1999; Bengtsson *et al.*, 1999) as tools in such an evaluation. However, first I will discuss the possible mechanisms responsible for the variation of the Northern Hemisphere temperature.

3. Physical Rationale

3.1 Stochastic Forcing

Stochastic forcing as originally suggested by Hasselmann (1976), is a mechanism that can generate low-frequency variations in the climate system. How does it work? The atmosphere is constantly in motion and, while we do not think of atmospheric motion as being decadal in nature, atmospheric motion can readily induce decadal and longer motions in the more slowly varying systems (such as the ocean) that are coupled to the atmosphere. Sarachik *et al.* (1996) have suggested the analogy of the tossing of a coin, which generates arbitrary long fluctuations depending on the number of tossings. In a coupled system, damping mechanisms prevent arbitrary long time

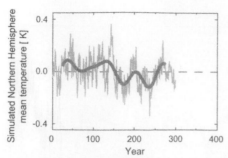

FIGURE 3 Simulated Northern Hemisphere temperature varations for 300 years with the ECHAM4/OPYC3 coupled climate model. Annually averaged and 50-year low-pass filter.

scales from occurring. In addition to atmospheric motion, interannual forcing of irregularly occurring El Niño events can similarly generate suitable ultra-low-frequency fluctuations in the coupled system. I fully share the view of Wunsch (1992), who proposed that stochastic forcing could preferably be considered a null hypothesis for decadal to centennial variability unless proven otherwise.

Is it possible to reproduce this type of variability with a climate model? I will here show results from the Hamburg coupled ocean atmosphere GCM (Roeckner *et al.*, 1999) using the present concentration of greenhouse gases. Figure 3 shows the results of a 300-year-long integration. It shows the variation of the Northern Hemisphere surface temperature as well as the 50-year low-frequency variability. As can be seen by directly comparing this result with Figure 2 the internal variability follows very closely the observational estimate by Mann *et al.* (1998). It therefore appears likely that the internal variability of the coupled climate system can give rise to the kind of variations that have occurred in the climate system from 1000 to 1900. However, the model cannot reproduce the accelerated warming trend of almost 1 K over the past 100 years. Another modeling experiment using a mixed layer model did not generate the large low-frequency variability as the fully coupled model. This suggests that the stronger El Niño events than normal (ENSO)-type phenomena which are realistically reproduced by the Hamburg GCM (Roeckner *et al.*, 1999; Oberhuber *et al.*, 1998) apparently are required to generate realistic low-frequency variability.

We may therefore conclude that stochastic forcing is the most likely explanation for the natural variability in the period prior to 1900. It is also concluded that stochastic forcing for this model cannot explain the large, sustained warming during the 20th century. Similar results have been obtained in other model studies (Manabe and Stouffer, 1997).

3.2 Solar Irradiation Changes

The forcing of the climate processes of the earth through radiation processes is, as far as we know, remarkably stable even if seen in a very long perspective. The variability of the solar irradiation cannot accurately be determined from earth-based observations

since clouds, aerosols, ozone, and other radiatively active gases interfere with the solar beam in the atmosphere. Observations from satellites have only been available for some 20 years. Satellite observations reveal that solar irradiation varies on very short time scales as well as with the 11-year solar cycle. The magnitude of this decadal variation is $1-2$ W m^{-2} compared to the solar constant of 1367 W m^{-2}, a variation of 0.1%. The radiation which reaches the earth must be spread over the whole area of the earth (which is 4 times larger than the interception area) and the planetary albedo is 0.3 (mostly due to clouds), which means that the solar variations translate to a variability of about 0.2 W m^{-2}. Numerical experiments undertaken by Cubasch *et al.* (1997) suggest that such a small forcing may not be detectable in the troposphere. The reason is presumably that the damping influence of the oceans cancels the positive and negative parts of the signal.

The question of longer periods of solar irradiance has been hotly debated in recent years. Such possible variations are inferred from historical records of variations in sunspots, the so-called Maunder Minimum in the late 1600 (Eddy, 1976), analogues with other sunlike stars, and paleo measurements of radioactive isotopes supposedly coupled to solar variations. Cubasch *et al.* (1997) forced a coupled climate model with data provided by Lean *et al.* (1995) as well as by Hoyt and Schatten (1993) for the period 1700 until the present. As would be expected, when the fluctuations are on a time scale of centuries or so, the model response broadly follows the forcing. The linear warming trend for the 100 years 1893–1992 is 0.19 and 0.17 K, respectively.

We may conclude that if the estimated variations in the solar forcing are correct, they can explain global temperature changes at a level of a few tenths of a degree, although the actual patterns differ between the two data sets and are different from the pattern provided by Mann *et al.* (1998). However, the main concern is that we currently have no observational evidence of any low-frequency variations in solar irradiation, since reliable data exist only for some 20 years. It is essential to stress that the available long-term datasets of solar forcing are based essentially on the *hypothesis* that the sun is an analogue to certain stars, which may show such characteristic variations in their irradiation. Therefore, our ability to say more about solar effects will crucially depend on obtaining reliable observations of the solar irradiation over longer periods.

If the low-frequency variations in the solar irradiation were correct, they could explain the climate variability in the period before 1900 or so, but because of the small amplitudes they cannot explain the rapid warming during this century. The solar forcing must therefore with high probability be excluded as the major cause of climate warming during the 20th century. Neither is the solar variability *required* to explain the variability of climate as documented over the last millennium, since this can be explained by internal variability of the climate system.

3.3 Volcanic Effects

Volcanic aerosols (mostly sulfate) have been suggested to have global effects on the climate when ejected in sufficient amounts into the stratosphere. If the aerosols do not enter the stratosphere, they will be rapidly removed by precipitation, and hence the effect on climate can probably be ignored. The major eruption from Mount Pinatubo on the Philippine Island Luzon on June 15–16, 1991, provided an opportunity to quantify the effect fairly accurately. The eruption was one of the largest in the 20th century. It is estimated (Krueger *et al.*, 1995) that 14–21 million tons of SO_2 were ejected into the stratosphere. The volcanic cloud moved eastward by some 20 m s^{-1}, thus encircling the earth in 3 weeks, whereby SO_2 was converted into sulfate aerosols (Bluth *et al.*, 1995).

In the first month most of the aerosol mass was located in a band between 20°S and 30°N, and then the cloud gradually spread to finally encircle the whole global stratosphere. Radiosonde observations as well as measurements from the microwave sounding unit (MSU) indicated a global stratospheric warming of about 2 K. The observations also suggested a cooling of the lower global

troposphere and the surface of the earth by about 0.5 K (Dutton and Christy, 1992).

There have been several attempts to calculate the climate effect of Mount Pinatubo, for example, those by Hansen *et al.* (1992). Bengtsson *et al.* (1999) recently carried out an experiment with the MPI high-resolution coupled ocean–atmosphere model. In this experiment the aerosol clouds were introduced into the stratosphere month by month for over 2 years and the corresponding change in radiation was calculated by the model. As observed, a rapid warming occurred in the stratosphere and a corresponding cooling in the troposphere. In Figure 4 we compare the results with observed microwave radiation data from the polar orbiting satellite microwave sounding unit. The model results have been expressed in the same units as the MSU data. To ensure that the model-calculated results were representative, an ensemble with six different integrations was carried out. The figure shows a

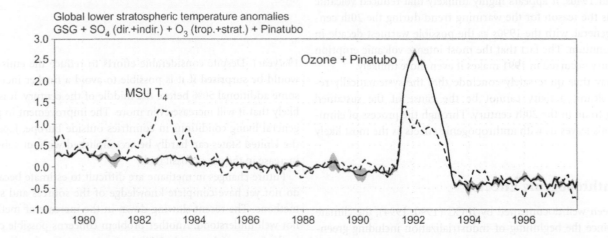

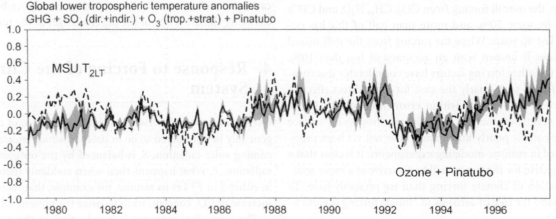

FIGURE 4 Observed MSU temperature, shown as dashed line, for channel 4 (top) for the period 1979–1997 and the equivalent for the simulations with Mt. Pinatubo and stratospheric ozone. The mean value obtained from the six realizations is denoted by the solid line, whereas the shaded area represents this value plus and minus one standard deviation of the individual simulations, respectively. The same for channel 2LT (bottom). Note the steady response in the stratosphere and the large variability in the lower troposphere. (from Bengtsson *et al.*, 1999).

model integration from 1979, an integration where the observed stratospheric ozone data were also considered. The effect of the eruption of El Chichon in 1983 was not incorporated.

As can be seen, the predicted tropospheric cooling is rather close to the observed temperature reduction. It is also quite robust since very similar results were obtained from all integrations. The stratospheric warming is somewhat overpredicted. The effect of the eruption lasted 5 years and was apparently prolonged due to delayed effects of the oceans. In conclusion it seems that major volcanic eruptions will affect global climate, but the cooling effect disappears comparatively fast. Only series of major eruptions are therefore likely to cool the global temperature on decadal and longer time scales and thus probably could explain at least part of the variations in the climate as occurred over the Northern Hemisphere from 1000 until 1900 (Lindzen and Giannitsis, 1998).

However, the rapid warming during the last century can hardly be attributed to quiescent volcanic activity. Although a systematic decrease in volcanic activity from the late 19th and early 20th centuries may have contributed to the relatively fast warming in the 1930s and 1940s, it appears highly unlikely that reduced volcanic activity is the reason for the warming trend during the 20th century in general, with the 1990s as the possible warmest decade in this millennium. The fact that the most intense volcanic eruption this century occurred in 1991 makes it even more unlikely.

We may thus quite safely conclude that the systematically reduced volcanic activity cannot be the cause of the sustained warming trend in the 20th century. Through the process of elimination, this leaves us with anthropogenic effects as the most likely cause.

3.4 Anthropogenic Effects

As has been well documented by IPCC (1990, 1994), the climate forcing since the beginning of industrialization including greenhouse gases (GHGs), aerosols, and land use has changed and continues to do so at an accelerating pace. Since the beginning of industrialization, the overall forcing from CO_2, CH_4, N_2O, and CFCs has increased by some 50%, and more than half of this has occurred in the last 40 years. While the forcing from the well-mixed greenhouse gases is known with an accuracy of less than 10%, practically all the other forcing factors have considerable inaccuracies (Fig. 5). This is particularly the case for the indirect effect of aerosols, which is only known within error limits of 50–100%. The effects of vegetation changes and other anthropogenic surface alterations are equally poorly known and have not yet been properly investigated in realistic modeling experiments. It is clear that a important objective for the future will be to arrive at a more accurate determination of climate forcing than we presently have. To better understand the role of aerosols in climate forcing is particularly important.

IPCC has tried to estimate the future change in the atmospheric concentrations of the well-mixed greenhouse gases. Present projections used by modelers are essentially based on an extrapolation of the increase during the last decades (for CO_2 it is about

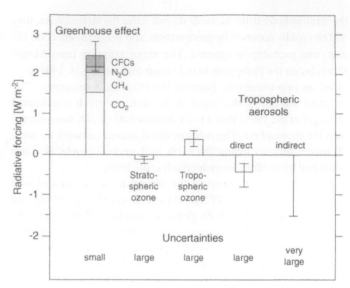

FIGURE 5 The direct greenhouse effect in W m^{-2} at the top of the tropopause from the beginning of industrialization until present. After IPCC.

1%/year). Despite considerable efforts to reduce the emissions, I would be surprised if it is possible to avoid a further increase by some additional 50% before the middle of the century. It is rather likely that it will increase even more. The improvement in present general living conditions in countries outside Europe, Japan, and the United States can hardly be accomplished without substantial increase in the use of fossil fuels.

Future changes in methane are difficult to estimate because we do not yet have complete knowledge of the sources and sinks of methane. The recent slowing down in the increase of methane is not well understood. Another problem concerns possible changes in the future carbon cycle. Will the future uptake of carbon in the terrestrial biosphere increase faster or slower than the emission of carbon dioxide or not? To better understand the carbon cycle in a changing climate is one of the big challenges of the future.

4. Response to Forcing of the Climate System

Over a time span of a few years the heat balance of the earth can generally be considered to be in balance, which means that the incoming solar radiation, S, is balanced by the outgoing long wave radiation, F. What happens then when suddenly there is a change in either S or F? Let us assume, for example, that there is a sudden increase in CO_2 concentration to twice the present value.

The immediate response is a reduction in the outgoing longwave radiation at the tropopause of about 3.1 W m^{-2} and an increase in the downward emission from the stratosphere by about 1.3 W m^{-2}. The sum of the two, 4.4 W m^{-2}, is the net instantaneous forcing at the tropopause.

Following this immediate shock the stratosphere cools. The increased CO_2 in the stratosphere enhances the thermal emission. Because the stratospheric temperature increases with altitude, this has the effect that the cooling into space is larger than the absorption from layers below. This is in fact the fundamental reason for the CO_2-induced cooling in the stratosphere. After stratospheric cooling a new radiative equilibrium develops with the new doubled CO_2 concentration. This reduces the increased downward emission at the tropopause by about 0.2 W m^{-2} and the tropopause forcing is adjusted accordingly.

The surface–troposphere system will continue to warm until the entire system reaches a new equilibrium. This may take a considerable time due to the very high heat capacity of the ocean and it will certainly last several decades before an equilibrium is reached, if at all.

Why does the surface–troposphere system warm at all, since in the end the radiation emission from the earth must balance the incoming solar radiation which stays the same? The reason is that the negative vertical temperature gradient in the troposphere has the effect that the equivalent level of outgoing radiation is successively lifted and the levels below are warmed due to hydrostatic influences (Fig. 6). If there were no vertical temperature gradients in the atmosphere, the surface emission would be equal to the outgoing emission at the top of the atmosphere and the greenhouse effect would consequently disappear.

However, this cannot happen in the present atmosphere so the direct warming effect at the surface, assuming no feedback, would amount to about 1.3 K (Ramanathan, 1981). Now it appears that the atmosphere is close to conserving relative humidity, so a warming would increase the water vapor in the atmosphere and hence further increase the warming, thus creating a *positive feedback* effect. It is interesting to note that even Arrhenius (1896) included the feedback from water vapor.

Empirical studies (Hense *et al.*, 1988; Flohn *et al.*, 1989; Raval and Ramanathan, 1989; Gaffen *et al.*, 1991; Inamdar and Ramanathan, 1998) show that temperature and water vapor changes are positively correlated and so are results from model studies (Manabe and Wetherald, 1967; Mitchell, 1989). In summary, it has been shown that both studies by simple models and GCMs and observations from independent sources (Inamdar and Ramanathan, 1998) all converge in the range of a positive feedback factor of 1.3–1.7 from water vapor. The only deviating results are those from Lindzen (1990, 1994), which suggest a negative feedback with water vapor due to a drying out effect of the upper troposphere caused by enhanced deep convection.

Inamdar and Ramanathan (1998) have shown that there are considerable geographical variations in water vapor feedback, with the dominating effect in the equatorial ocean region. In this area the greenhouse feedback exceeds the blackbody emission, reproducing the so-called super-greenhouse effect (Ramanathan and Collins, 1991). The overall results demonstrate the importance of realistically reproducing the three-dimensional atmospheric circulation and the associated water distribution for a credible water vapor feedback.

While models generally agree in reproducing the water vapor feedback, the cloud feedback is much more complex. The overall effect of clouds is to cool the surface and the troposphere since the albedo effect (reflection of solar radiation) is larger than the enhanced absorption of long-wave radiation by clouds. The difference is substantial and amounts to some 20 W m^{-2}. The change in cloud forcing due to enhanced greenhouse forcing is strongly model-dependent, with some models giving positive feedback and others negative (Cess *et al.*, 1997).

The ECHAM4/OPYC model discussed below has a negative cloud feedback, with the transient integration having a stronger negative cloud feedback than the equilibrium model (Bengtsson, 1997). The cloud feedback depends, though, to a considerable degree on changes of the lower boundary. Clouds over open water (more common in a warmer climate) have a strong negative forcing, while clouds over sea ice and snow (more common in a cold climate) generate practically no feedback because of similar albedo.

Surface processes such as the melting of snow and ice at higher temperatures will decrease the surface albedo, leading to a positive feedback, while changes in cloud cover and cloud distribution can give rise to either a negative or a positive feedback. Other feedback processes depend on changes in the general circulation, such as those in the dominating storm tracks and in the vertical stability of the atmosphere, affecting the surface temperature. For this reason, as will be demonstrated below, it is not possible to infer from a certain forcing pattern what the climate response would be. This is one of the reasons realistic climate models must be used in such an evaluation. This can be illustrated by comparing the geographical distribution of forcing here taken from the Hamburg climate model (Roeckner *et al.*, 1999) and the corresponding temperature change (Fig. 7 and Table 1). The actual forcing was taken from an equilibrium climate change experiment including the anthropogenic effect

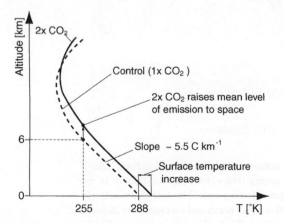

FIGURE 6 Illustration of the greenhouse effect. The height of the equivalent outgoing radiation is around 6 km with a temperature of ca. 255 K (global average). A doubling of the CO_2 will raise the height of the outgoing radiation by a few hundred meters and thus warm the surface accordingly (extrapolated via an averaged lapse rate of 5.5°Ckm^{-1}).

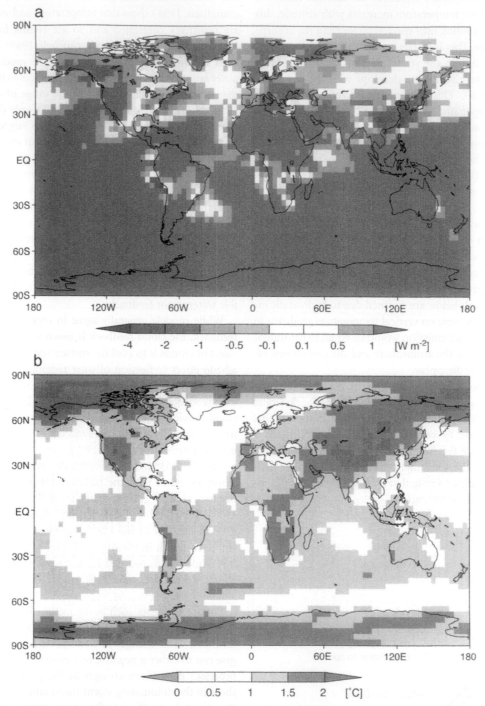

FIGURE 7 (a) Radiative forcing from greenhouse gases, sulfate aerosols (direct and indirect effect), and tropospheric ozone from the anthropogenic emission during 1860–1990. See also Table 1. In the Northern Hemisphere there are widespread areas with negative forcing caused by sulfate aerosols. (b) Equilibrium response calculated from the ECHAM4 coupled to a slab ocean and averaged over 20 years. Note the differences between the forcing and the response pattern. For further information see Roeckner *et al.* (1999). See also color insert.

TABLE 1 Global Annual Mean Radiative Forcing at the Top of the Tropopause and Equilibrium Response in Global Annual Mean Surface Air Temperature*

Experiment No.	Historical Forcing Experiments (1860–1990)	Radiative Forcing (Wm^{-2})	Temperature Response (°C)	Climate Sensitivity (°C/W m^{-2})
1	Well-mixed greenhouse gases (CO_2, CH_4, N_2O, CFCs)	2.12 (2.45)[a]	1.82	0.86
2	Tropospheric ozone	0.37 (0.2 to 0.6)	0.34	0.91
3	Direct sulfate aerosol	−0.34 (−0.2 to −0.8)	−0.24	0.71
4	Indirect sulfate aerosol	−0.89 (0 to −1.5)	−0.78	0.87
	Sum (1 to 4)	1.26	1.15	0.91
5	Effects (1 to 4) included	1.26	1.13	0.90

* Forcing data in brackets indicate range of forcing provided by IPCC.
[a] IPCC value from 1750 to 1994

of the well-mixed greenhouse gases, sulfate aerosols, and tropospheric ozone from the beginning of the industrialization until present. As can be seen, there is practically no correlation between the pattern of forcing and the pattern of temperature response. The areas of net negative forcing over large parts of Eurasia, for example, are becoming significantly warmer. The reason is that warming from other regions, such as from the tropical oceans, transports heat toward the higher latitudes and thus gives rise to a warmer climate.

It follows from this discussion that climate response to external forcing is rather complex and hence, as can be seen from a recent study by Le Treut and McAvaney (1999), strongly model-dependent. Figure 8 shows the equilibrium response in global surface temperature and precipitation to a doubling of CO_2 for 11 different "state-of-the art" climate models. As can be seen, the temperature increase varies between 2.1 and 4.8 K and the pre-

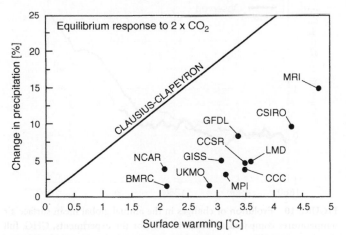

FIGURE 8 Equilibrium response to $2 \times CO_2$ for 11 GCM coupled to a mixed layer ocean. For further information see text (after Le Treut and McAvaney, 1999).

cipitation between 1 and 15%. It can further be seen that the increase in precipitation as a function of temperature is significantly less than that from the Clausius–Clapeyrons equation. The reason is that global precipitation must balance global evaporation. Global evaporation in turn is controlled by the net radiative forcing at the ground, which apparently increases more slowly than the availability of moisture in the free atmosphere.

In conclusion, we must still count on considerable inaccuracy even in such general quantities as the change in global average temperature and precipitation—and this is when the forcing of climate is known exactly!

5. Results from Climate Change Prediction Experiments

As we have seen in the previous section, there is still a considerable spread between different climate models in the equilibrium response to a given forcing. Similar differences can be found in transient experiments. The main reason is that the degree of climate change strongly depends on the dynamical response of the coupled system. The marked surface warming of the Northern Hemisphere during the past 20 years, for example, is strongly influenced by a positive phase of both ENSO (stronger El Niño events than normal) and NAO (stronger westerlies over the North Atlantic), both of which have contributed to milder winters over the land areas of the Northern Hemisphere (Hurrell, 1995; Wallace *et al.*, 1995).

If, for example, both ENSO and NAO are chaotic events and hence unpredictable, this could cause long-term differences between models since they could then statistically correctly simulate these features out of phase with each other. Alternatively, it could also happen that both ENSO and NAO respond to the increased forcing of the greenhouse gases so that there is a systematic change in their probability distribution and then the positive phase we have seen in recent decades is a physically correct

TABLE 2 List of Experiments

Name	Forcing Due to Changing Atmospheric Concentrations of . . .	Years
GHG	CO$_2$ and other well-mixed greenhouse gases	1860–2100
GSD	GHG plus sulfate aerosols (direct effect only)	1860–2050
GSDIO	GHG plus sulfate aerosols (direct and indirect effect) plus tropospheric ozone	1860–2050
CTL	Unforced control experiment	300

response. However, at present we cannot answer this important question. Some models indicate a successive increase in the positive phase of NAO; others like the MPI model do not show any distinct response at all. At the same time the MPI model (Timmermann *et al.*, 1998) suggests a slow increase in the amplitude of ENSO events, which is not so clearly seen in other models.

It also follows from this general discourse that regional climate is even more strongly model-dependent, since small geographical changes in predominant weather patterns such as the stormtrack between different models may create huge differences. This is confirmed by Räisänen (1998), who has compared results from 12 transient coupled GCMs for Northern Europe and the eastern North Atlantic.

With these general reservations, we will now describe results from a recent series of transient experiments by Roeckner *et al.* (1999)(Table 2). The experiments start in the year 1860. Observed concentrations of greenhouse gases and sulfate aerosols were used until 1990 and thereafter changed according to the IPCC scenario IS92a. Tropospheric ozone changes have been calculated from precursor gases.

In the first experiment, GHG, the concentrations of the following greenhouse gases were prescribed as a function of time: CO$_2$, CH$_4$, and N$_2$O, as well as a series of industrial gases, including CFCs and HCFCs. The absorptive properties of each gas constituent were calculated separately. Furthermore, the radiative forcing was practically identical to the narrow band calculations. This meant an increase in the radiative forcing by some 10% compared to the actual broad band calculation in the radiation code of the model.

In the second experiment, GSD, Table 2 the greenhouse gases were treated as in GHG but with the additional incorporation of the tropospheric sulfur cycle as due to anthropogenic sources only. Natural biogenic and volcanic sulfur emissions were neglected, and the aerosol radiative forcing was generated through the anthropogenic part of the sulfur cycle only. The space/time evolution in the sulfur emissions was derived from actual emission records. The full anthropogenic sulfur cycle was integrated into the atmospheric model, including the actual geographical emission of SO$_2$, chemical transformation to sulfate, semi-Lagrangian transport of the sulfate aerosols, and finally the dry and wet disposition of sulfate particles from the atmosphere.

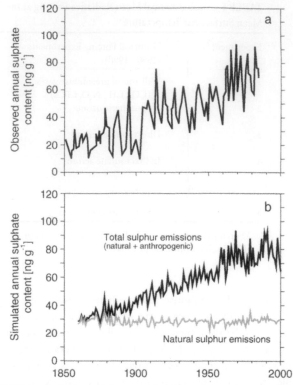

FIGURE 9 Evolution of the annual sulfate content in snow/ice at the Dye 3 site in southern Greenland (65°N, 43°W). (a) Observed (Legrand, 1955). (b) GSDIO simulations for the nearest grid point with prescribed natural sulfur emissions only (gray line) and total emission (natural plus anthropogenic (dash line)). After Roeckner *et al.* (1999).

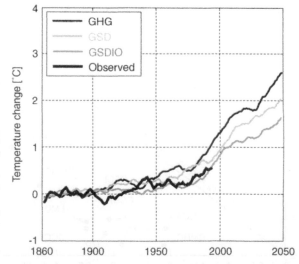

FIGURE 10 Evolution of changes in the annual global mean surface air temperatures compared to observations (for the experiments, GHG, full thin line, GSD light gray thin line, and GSDIO, gray thin line). Observational data from 1860 until present is shown by a heavy dark line. A 5-year running mean is applied (after Roeckner *et al.*, 1999).

In the third simulation, GSDIO, table 2 the indirect aerosol effect on cloud albedo was added. The tropospheric ozone distribution was also changed as a result of the prescribed anthropogenic emission of precursor gases. Figure 9 shows an attempt to validate the deposition of sulfate in the wet and dry deposition in ice core measurements at the Dye 3 on Greenland. Figure 9a shows the measured concentration of sulfate (in ng g^{-1}) according to Legrand (1995), and Figure 9b shows the results from the corresponding control integration and from the GSDIO experiment. The agreement between the calculated depositions is in broad agreement with the measurements.

The global annual mean temperature change from the three experiments, GHG, GSD, and GSDIO, is shown in Figure 10. As can

be expected, the long-term warming is largest in experiment GHG and smallest in experiment GSDIO. Until 1980 or so the simulated temperatures are more or less within the range of natural variability of the control integration (not shown). However, the simulated temperature patterns undergo large low-frequency variations on a multidecadal time scale, in broad agreement with the estimated observed temperature pattern. In the model simulations, there are pronounced ultra-low fluctuations at higher latitudes of the Southern Hemisphere, but it is not possible to say whether these fluctuations are realistic or simply an artifact of the coupled model. However, when we compare the long-term trends in observations and simulation, a reasonably good agreement is found (Fig. 11).

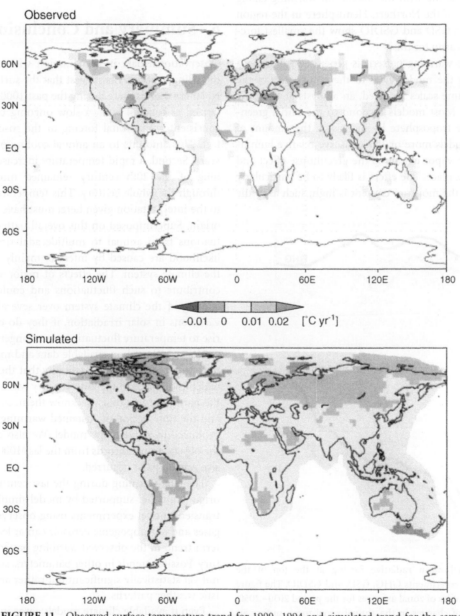

FIGURE 11   Observed surface temperature trend for 1900–1994 and simulated trend for the same period with the ECHAM4/OPYC3 coupled climate model.

We will next investigate the geographical relation between the forcing and the response to forcing in the three transient experiments. This is done by comparing the meridional profiles of zonally averaged forcing for the period 2030–2050 to the corresponding meridional profile of the surface temperature (Figs. 12a and 12b, respectively). The result is very much of the same type as in the equilibrium experiment with a slab ocean, suggesting that atmospheric processes are important for the response pattern. The experiment with greenhouse gases only, GHG, has a maximum forcing at around 20°, decreasing both toward the equator and toward higher latitudes. The other experiments have a reduced forcing increasing toward middle latitudes at the Northern Hemisphere due to the emission of SO_2 in these regions. The meridional profile of response to forcing looks very much different from that of the forcing with the maximum warming taking place at high latitudes of the Northern Hemisphere in the region where the experiments GSD and GSDIO show the smallest forcing! What can be the reason for this?

We believe that the warming pattern is generated following a series of feedbacks in the model. Due to the complexity of the model and the long time scales involved, an analysis at this time can only be tentative. Most models respond to the initial greenhouse warming in the troposphere by increasing the amount of water vapor (most models more or less conserve relative humidity). The altered water vapor enhances the greenhouse effect and positive feedback takes place. The effect is likely to be particularly strong in areas where the moisture content is high, such as in the intertropical convergence zones. Warming over land areas is larger than that over oceans due to the large ocean heat capacity, which delays the warming considerably.

The delayed warming is particularly strong in the southernmost oceans, with their strong oceanic vertical heat exchange. Finally, in the climate warming experiments the storm tracks are moved slightly poleward, particularly over the Northern Hemisphere. The feedback at high-latitude land areas is also enhanced through albedo feedback due to reduced snow cover on the ground in the climate change experiments. We thus anticipate that complex feedbacks such as those suggested here are the probable reasons for the distribution of the warming. Many of the feedback processes are model-dependent and the main cause of the large model variability as that shown in Figure 8.

6. Summary and Conclusions

Observational data in combination with theoretical studies and modeling experiments suggest that the surface temperature of the Northern Hemisphere during the past 1000 years could be characterized as follows: First, a slow ongoing cooling is by and large consistent with orbital forcing or the so-called Milankovitch effect. This amounts to an annual cooling of some 0.2 K over 900 years. Second, a rapid temperature increase starting at the beginning of the 20th century remained more or less unchanged through the whole century. This temperature increase, according to the interpretation given here, must have been of anthropogenic origin. Superimposed on this overall pattern are substantial fluctuations from annual to multidecadal time scales, which in all likelihood are caused by internal, mainly stochastic processes in the climate system. The effects of major volcanic eruptions also contribute to such fluctuations and could have caused notable cooling of the climate system over several years. Low-frequency variations in solar irradiation, if they do exist, can similarly give rise to temperature fluctuations over longer time scales.

However, based on available data and model studies, it appears that we can rule out the possibility that the unparalleled warming which took place in the 20th century was a consequence of any of the natural processes as we know them, since both the amplitude and the time period of sustained warming are too large to be reproduced by any climate model. We also can find no support in the observational records from the last 1000 years that such a massive warming has occurred.

That the warming during the last century is of anthropogenic origin is further supported by model simulation studies. Coupled transient model experiments using observed data for greenhouse gases and anthropogenic aerosols can at least reproduce the long-term trend in the observed warming pattern during the 20th century. Possible trends in other parameters, such as precipitation, are not yet statistically significant but rather are changes in characteristic weather patterns.

Even if available climate models agree in simulating a warming, there are considerable model differences, particularly in the pat-

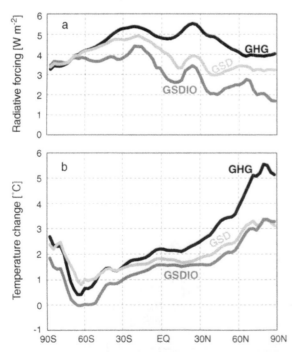

FIGURE 12 (a) Annual mean radiative forcing at the top of the tropopause in the three experiments GHG, GSD, and GSDIO. The figure shows the meridional profiles of zonal averages for the period 2030–2050. (b) Meridional profiles of changes in the annual zonal mean surface air temperatures for the same period (after Roeckner *et al.*, 1999).

tern and speed of warming. Results obtained so far are strongly model-dependent, suggesting the importance of dynamical and physical feedbacks in determining regional changes in surface temperature, precipitation, and weather pattern. The result of a climate model calculation is currently determined more by the model than by the details of the particular forcing being used. This strongly suggests that climate models have to be realistic and rather detailed, since any systematic model deficiency could create an erroneous response pattern. At least a necessary condition must be that climate models be able to realistically reproduce the present climate and its characteristic variations. Simple models could in this context be quite misleading.

Present models still have major deficiencies due to insufficient horizontal and vertical resolution, which leads to difficulties in representing orography and coastlines as well as limitations in reproducing realistic weather patterns. This affects not only the ability to simulate regional climate, but also to some extent to correctly maintain the large-scale atmospheric and ocean circulation, since the large-scale circulation in turn is partly driven dynamically by smaller weather systems.

Another major problem concerns the representation of physical processes in large-scale models. Radiation and clouds, deep convection, and near surface- and free atmospheric turbulence are examples of atmospheric processes that are extremely difficult to handle, partly due to the lack of suitable observational data as well as to lack of proper understanding of complex atmospheric processes. Similar difficulties occur in ocean and land surface modeling, where the processes regulating the exchange of heat, water, and momentum on the scale of climate models are not completely known.

The coupling between the atmosphere and the ocean is a particular problem. Minor changes in cloud cover and sea ice distribution drastically influence the exchange of heat and water between the atmosphere and the ocean. Small systematic errors in the atmospheric and ocean model components can then generate an erroneous climate drift in a long integration over several centuries. Many models handle this by introducing a small systematic correction of the surface fluxes to ensure that no systematic errors will occur in the equilibrium case when no climate change occurs. Climate change model studies are therefore in essence perturbation studies and are likely to become misleading when the perturbations become too large. For the global change studies assuming a doubling or a trebling of the greenhouse gas concentration, this does not appear to be a serious restriction.

The so-called flux adjustment (Sausen *et al.*, 1988) has been criticized and used as an example that coupled models making use of this assumption are less credible. Recently, there have been a few integrations where the flux adjustment has been significantly relaxed (e.g., only using annual means; Roeckner *et al.*, 1999) or even fully eliminated (Mitchell, personal communication). However, it appears that whether or not a flux adjustment was used has no apparent effect on the overall result. It is nevertheless to be expected that the new generation of coupled models are able to reduce systematic error to such a low level that flux adjustment or any other empirical correction is no longer required.

An outstanding issue, which finally must be stressed, is the inherent stochastic variability in models and as I believe also in nature itself. This means that just by chance we can have climate anomalies lasting for several decades influencing regional climate in a significant way. Such anomalies are often mistakenly taken for a genuine climate change as happened after the warm period in the 1930s and again after the cold periods in the 1960s and 1970s when even some climate scientists seriously suggested the possibility of a persistent change toward a much colder climate.

However, we still do not know whether characteristic climate anomalies like ENSO and NAO will systematically change in a warmer climate. This cannot be ruled out at present, and in fact some models indicate that the present probability distribution of both ENSO and NAO may be different (Timmermann *et al.*, 1998). Such a change may have more serious consequences for the regional climate than an overall superimposed warming.

Finally, let me return to the question of the overall dynamical stability of the earth's climate, that is, whether climate is transient or intransient. The most likely event creating a switch to another regime of climate, although essentially only of regional influence, could be caused by a reduction or even a halt in the thermohaline circulation of the North Atlantic, leading to a situation with reduced sea surface temperatures in that region (Marotzke and Willebrand, 1991; Rahmstorf, 1996). This is currently an issue of considerable interest and concern, since models have indicated that such an instability could be initiated (e.g., Manabe and Stouffer, 1988) by increased precipitation in the North Atlantic storm track or by increased melting of glaciers in southern Greenland. Whether such event may take place in reality or not is still an open question and more advanced modeling studies are urgently required here.

Acknowledgments

Let me first use this opportunity to congratulate Detlef Schulze, Colin Prentice, David Schimel, and their co-workers on the determination and speed they have demonstrated in putting the Institute of Biogeochemistry in Jena on the scientific map. As chairman of the MPG planning committee for this institute, I put my heart into this venture and I am very grateful to the MPG, which managed to find ways to support its establishment. The author acknowledges Dr. Erich Roeckner and Mr. Bernard Reichert, who were most helpful in providing part of the material used in this study. Ms. Kornelia Müller, Ms. Karin Niedl, and Mr. Norbert Noreiks kindly assisted with text and graphics.

References

Alley, R. B., Meese, D. A., Shuman, C. A., Gow, A. J., Taylor, K. C., Grootes, P. M., White, J. W. C., and Ram, M. (1993). Abrupt increase in Greenland snow accumulation at the end of the Younger Dryas event. *Nature* **362,** 527–529.

Arrhenius, S. (1896). On the influence of carbonic acid in the air upon the temperature of the ground. *Philos. Mag.* **41**, 237–276.

Bengtsson, L. (1997). A numerical simulation of anthropogenic climate change. *Ambio.* **26**(1), 58–65.

Bengtsson, L., Roeckner, E., and Stendel, M. (1999). Why is the global warming proceeding much slower than expected? *J. Geophys. Res.* **104**, 3865–3876.

Berger, A. (1988). Milankovitch theory and climate. *Rev. Geophys.* 26, **4**, 624–657.

Bluth, G. J. S., Doiron, S. D., Schnetzler, C. C., Krueger, A. J., and Walter, L. S. (1992). Global tracking of the SO_2 clouds from the June 1991 Mounth Pinatubo eruptions. *Geophys. Res. Lett.* **19**, 151–154.

Bryan, F. (1986). High latitude salinity effects and interhemispheric thermohaline circulations. *Nature* **305**, 301–304.

Brovkin, V., Claussen, M., Petoukhov, V., and Ganopolski, A. (1998). On the stability of the atmosphere-vegetation system in the Sahara/Sahel region. *J. Geophys. Res.* **103**, D24, 31613–31624.

Cess, R. D, Zhang, M. H., Potter, G. L., Alekseev, V., Barker, H. W., Bony, S., Colman, R. A., Dazlich, D. A., Del Genio, A. D., Deque, M., Dix, M. R., Dymnikov, V., Esch, M., Fowler, L. D., Fraser, J. R., Galin, V., Gates, W. L., Hack, J. J., Ingram, W. J., Kiehl, J. T., Kim, Y., Le Treut, H., Liang,X.-Z., McAvaney, B. J., Meleshko, Morcrette, J.-J., Randall, D. A., Roeckner, E., Schlesinger, M. E., Sporyshev, P. V., Taylor, K. E., Timbal, B., Volodin, E. M., Wang, W., Wang, W. C., and Wetherald, R. T., (1997). Comparison of the seasonal change in cloud-radiative forcing from atmospheric general circulation models and satellite observations. *J. Geophys. Res.* **102**, 16593–16603.

Claussen, M. (1998). On multiple solutions of the atmosphere-vegetation system in present-day climate. *Global Change Biol.* **4**, 549–559.

Cubasch, U., Voss, R., Hegerl, G. C., Waszkewitz, J., and Crowley, T. J. (1997). Simulation of the influence of solar radiation variations on the global climate with an ocean-atmosphere general circulation model. *Climate Dynamics* **13**, 757–767.

Dutton, E. G. and Christy, J. R. (1992). Solar radiative forcing at selected locations and evidence for global lower tropospheric cooling following the eruptions of El Chichònn and Pinatubo. *Geophys. Res. Lett.* **19**, 2313–2316.

Eddy, J. A., (1976). The Maunder minimum. *Science* **192**, 1189–1202.

Flohn, H. and Kapala, A. (1989). Changes in tropical sea-air interaction processes over a 30-year period. *Nature* **338**, 244–245.

Gaffen, D. J., Barnett, T. P., and Elliott, W. P. (1991). Spaces and timescales of global tropospheric moisture. *J. Climate* **4**, 989–1008.

Gleick, J. (1988). Chaos—Making a new science. Willam Heinemann Ltd., ISBN: 043429554x, 353 p.

Hansen, J., Fung, I., Ruedy, R., and Sato, M. (1992). Potential climate impact of Mount Pinatubo eruption. *Geophys. Res. Lett.* **19**, 215–218.

Hasselmann, K. (1976). Stochastic climate models I, Theory. *Tellus* **28**, 473–485.

Hense, A., Krahe, P., and Flohn, H. (1988). Recent fluctuations of tropospheric temperature and water vapor content in the Tropics. *Meteorol. Atmos. Phys.* **38**, 215–227.

Hoyt, D. V. and Schatten, K. H. (1993). A discussion of plausible solar irradiance variations, 1700–1992. *J. Geophys. Res.* **98**, 18895–18906.

Hurrell, J. (1995). Decadal trends in the north Atlantic oscillation. Regional temperatures and precipitations. *Science* **269**, 676–679.

Inamdar, A. K. and Ramanathan, V. (1998). Tropical and global scale interactions among water vapor, atmospheric greenhouse effect, and surface temperature. *J Geophys. Res.* **103**(D24), 32177–32194.

IPCC. (1990). "Climate change, The IPCC Scientific Assessments."

(J. Houghton, G. J. Jenkins, and J. J. Ephraums, Eds.). Cambridge University Press.

IPCC. (1994). "Climate change." (J. Houghton, L. K. Meira Filho, J. Bruce, H. Lee, B. A. Callender, E. Haites, N. Harris, and K. Maskell, Eds.). Cambridge University Press.

Krueger, A. J., Walter, S. L., Bhartia, P. K., Schnetzler, C. C., Krotkov, N. A., Sprod, I., and Bluth, G. J. S. (1995). Volcanic sulfur dioxide measurements from the total ozone mapping spectrometer instruments. *J. Geophys. Res.* **100**, 14057–14076.

Lean, J., Beer, J., and Bradley, R. (1995). Reconstruction of solar irradiance since 1610: implications for climate change. *Geophys. Res. Lett.* **22**, 3195–3198.

Legrand, M. (1955). Atmospheric chemistry changes versus post climate inferred from polar ice cores. In "Aerosol Forcing of Climate." (R. J. Charlson and J. Heintzenberg, Eds.), pp. 123–151. John Wiley, Chichester.

LeTreut, H. and McAvaney, B. J. (1999). Model intercomparison: Slab Ocean $2 \times CO_2$ Equilibrium Experiments. Submitted.

Lindzen, R. S. (1990). Some coolness concerning global warming. *Bull. Am. Meteor. Soc.* **71**, 288–299.

Lindzen, R. S. (1994). On the scientific basis for global warming scenarios. *Environ. Pollut.* **83**, 125–134.

Lindzen, R. S. and Giannitsis, C. (1998). On the climatic implications of volcanic cooling. *J. Geophys. Res.* **103**, 5929–5941.

Lorenz, N. (1968). Climate determinism. *Meteor. Monogr.* **8**, 30, 1–3.

Maier-Reimer, E. and Mikolajevicz, U. (1989). Experiments with an OGCM on the cause of the Younger Dryas. In "Oceanography 1988." (A. Ayala-Castanares, W. Wooester, and A. Yane-Arancibia, Eds.), pp. 87–100. UNAM Press, Mexico.

Manabe, S. and Stouffer, R. J. (1998). Two stable equilibria of a coupled ocean-atmosphere model. *J. Climate.* **1**(9), 841–866.

Manabe, S. and Stouffer, R.J. (1997). Climate variability of a coupled ocean-atmosphere-land surface model: implication for the detection of global warming. *Bull. Am. Met. Soc.* **78**(6), 1177–1185.

Marotzke, J. and Willebrand, J. (1991). Multiple equilibria of the global thermohaline circulation. *Journal of Physical Oceanography* **21**(9), 1372–1385.

Manabe, S. and Wetherald, R. T. (1967). Thermal equilibrium of the atmosphere with a given distribution of relative humidity. *J. Atmos. Sci.* **24**, 241–259.

Mann, M., Bradley, R., and Hughes, M. (1998). Global-scale temperature patterns and climate forcing over the past six centuries. *Nature* **392**, 779–787.

Mann, M., Bradley, R., and Hughes, M. (1999). Northern hemisphere temperatures during the post millenium: inferences, uncertainties, and limitations. *Geophys. Res. Lett.* **26**, 759–762.

Milankovitch, M. (1920). Théorie mathématique des phénomènes thermiques produits par la radiation solaire, Académie Yugoslave des Sciences et des Art de Zagreb, Gauthier-Villars.

Milankovitch, M. (1941). Kanon der Erdbestrahlung und seine Anwendung auf das Eiszeitenproblem. Royal Serbian Sciences, Spec. pub. 132, Section of Mathematical and Natural Sciences, vol. 33, Belgrade, 633 p. ("Canon of Insolation and the Ice Age Problem", English Translation by Israel Program for Scientific Translation and published for the U. S. Department of Commerce and National Science Foundation, Washington D. C., 1969).

Mitchell, J. F. B., (1989). The "greenhouse effect" and climate change. *Rev. Geophys.* **27**, 115–139.

Oberhuber, J. M., Roeckner, E., Christoph, M., Esch, M., and Latif, M.

(1998). Predicting the '97 El Niño event with a global climate model. Max-Planck-Institut für Meteorologie Report No. 254, Hamburg. Shortened version in Geophys. *Res. Lett.* **25(13)**, 2273–2276.

Räisänen, J. (1998). "CMIP2 Subproject Climate Change in Northern Europe: Plans and First Results." Proceedings, Coupled Model Intercomparison Project Workshop, Melbourne, Australia, 14–15 October 1998.

Rahmstorf, S. (1996). On the freshwater forcing and transport of the Atlantic thermohaline circulation. *Climate Dynamics* **12**(12), 799–811.

Rahmstorf, S. (1997). Risk of sea-change in the Atlantic. *Nature* **388**, 825–826.

Ramanathan, V. (1981). The role of ocean-atmosphere interactions in the CO$_2$ climate problem. *J. Atmos. Sci.* **38**, 918–930.

Ramanathan, V. and Collins, W. (1991). Thermodynamic regulation of ocean warming by cirrus clouds deduced from observations of the 1987 EL Niño. *Nature* **351**, 27–32.

Raval, A. and Ramanathan, V. (1989). Observational determination of the greenhouse effect. *Nature* **342**, 758–761.

Roeckner, E., Bengtsson, L., Feichter, J., Lelieveld, J., and Rodhe, H. (1999). Transient climate change simulations with a coupled atmosphere-ocean GCM including the tropospheric sulfur cycle. Max-Planck-Institut für Meteorologie Report No. 266, Hamburg. (Accepted for publication in J. Climate).

Sarachik, E. S., Winton, M., and Yin, F. L. (1996). Mechanisms for decadal-to-centennial climate variability. In "Decadal Climate Variability—Dynamics and Predictabilities." (D. Anderson and J. Willebrand, Eds.). NATO ASI Series I: *Global Environmental Change,* **Vol. 44,** 157–210.

Sausen, R., Barthels, R. K., and Hasselmann, K. (1988). Coupled ocean- and atmospheric models with flux corrections. *Climate Dynamics* **2**, 154–163.

Stommel, H. (1961). Thermohaline convection with two stable regimes of flow. *Tellus* **13(2)**, 224–230.

Timmermann, A., Oberhuber, J., Bacher, A., Esch, M., Latif, M., and Roeckner, E. (1998). ENSO response to greenhouse warming. Max-Planck-Institut für Meteorologie Report 251, Hamburg.

Wallace, J. M., Zhang, Y., and Renwick, J. A. (1995). Dynamical contribution to hemispheric mean temperature trends. *Science* **270**, 780–783.

Wunsch, C. (1992). Decade-to-century changes in the ocean circulation. *Oceanography* **5**, 99–106.

Uncertainties in the Atmospheric Chemical System

Guy. P. Brasseur
Max Planck Institute for
Meteorology
Hamburg, Germany

Elisabeth A. Holland
Max Planck Institute for
Biogeochemistry
Jena, Germany

1. Introduction

Since the preindustrial era, the chemical composition of the atmosphere has changed dramatically. For example, the concentration of carbon dioxide (CO_2) has increased from approximately 280 ppmv in 1850 to 367 ppmv in 1999, that of methane (CH_4) from 700 to 1745 parts per billion (ppbv), and that of nitrous oxide (N_2O) from 270 to 314 ppbv (IPCC, 1996; WMO, 1999) . In addition, since 1950, large quantities of industrially manufactured chlorofluorocarbons (CFCs) have leaked to the atmosphere. These CFCs have been the major cause of the formation of the observed depletion of stratospheric ozone (O_3). Because of their long lifetimes (several decades), the effects of these gases will be felt for many years to come, despite the implementation of the Montreal Protocol and other international agreements (WMO, 1999). Fossil fuel combustion, a primary anthropogenic perturbation of the 20th century, has produced not only large amounts of carbon dioxide, but also substantial quantities of shorter lived trace gases, including nitrogen oxides (NO_x), carbon monoxide, and volatile organic carbon compounds. The release of these short-lived compounds has contributed to substantial (but hard to quantify) changes in tropospheric ozone at the global scale (see, e.g., WMO, 1999).

Radiatively active gases, including CO_2, CH_4, N_2O, CFCs, and O_3, contribute to the so-called "greenhouse effect" of the atmosphere, and the observed perturbations in their atmospheric concentrations have led to significant "climate forcing." For the period 1850–2000, this forcing is estimated to be around 2.5 W m^{-2} (IPCC, 1996).

At the same time, large amounts of sulfur dioxide have been released to the atmosphere, primarily as a result of coal burning. These emissions are most intense in the urbanized and industrialized regions of Asia, Europe, and North America. Sulfur dioxide is rapidly converted into tiny sulfate aerosol particles (0.1–1 μm in size) which scatter a relatively large fraction of the incoming radiation back to space, resulting in a substantial cooling in the regions where the particles are particularly abundant (Erisman and Draaijeers, 1995; Roeckner *et al.*, 1999). The inclusion of sulfate aerosols into general circulation models has resulted in substantial improvement in these models' ability to correctly capture the spatial pattern of global increases in temperature (Kiehl and Briegleb, 1993). The presence of sulfate aerosols also tends to modify the optical properties and lifetime of clouds, providing an additional regional cooling mechanism (Santer *et al.*, 1995). The magnitude of this indirect effect, however, is, poorly quantified.

Changes in the chemical composition of the atmosphere both affect the physical climate system and disrupt biogeochemical cycles, which are central to the "health" of the biosphere. For example, the sulfate aerosols, mentioned previously, together with the enhanced concentrations of nitrates, constitute the major sources of acid rain. Acid precipitation can destroy aquatic ecosystems and has contributed to the well-known phenomenon of *waldsterben* in Europe and North America (Schulze, 1989; Aber, 1989?). Clean air acts implemented in Europe and North America have been remarkably successful at reducing the sulfate content of rainwater but the regulation of nitrogen oxides has proven to be less tractable. Enhanced concentrations of nitrogen oxides constitute a second major source of acid rain. They also lead to fertilization of the biosphere, which has a direct impact on the global carbon cycle. In addition, the enhanced concentrations of nitrogen oxides in the atmosphere has led to enhanced ozone concentrations in the boundary layer and probably in the free troposphere. Surface ozone concentrations greater than 40 ppbv damage plant leaves and decrease plant productivity (Reich, 1987). Clearly, changes in the chemical composition of the atmosphere have multifaceted counteracting effects.

In spite of the measures taken to reduce anthropogenic emissions of chemical compounds, the impact of regional and global perturbations on atmospheric composition remains large and is expected to intensify in the next decades. Economic development, expansion of urbanization, and the accompanying rise in the emissions of greenhouse gases and of ozone precursors in Asia and South America are expected to be rapid over the next decades.

In this chapter, we focus on processes that affect the budget of ozone in the troposphere at the global scale. We use the global chemical transport model of the troposphere called IMAGES to assess the importance of various factors that influence the global ozone budget. In Section 2, we provide a synthetic overview of the chemical processes that affect O_3 and several of its precursors in the atmosphere. In Section 3, we provide a brief description of the IMAGES model, and in Section 4, we discuss some results obtained by using this model. Conclusions are provided in Section 5.

2. Synthetic View of Chemical Processes in the Troposphere

An important property of the atmosphere is its ability to oxidize chemical compounds, and to cleanse itself of natural and anthropogenic substances. The most efficient oxidizing agent is the hydroxyl radical (OH). This radical is produced in the atmosphere by the oxidation of water vapor (and to a lesser extent of methane and molecular hydrogen). OH reacts with a large number of chemical compounds, and its reactivity determines in large part the atmospheric lifetimes of these compounds. The reaction with methane, for example, is considerably slower than that with isoprene, so that the global lifetime of CH_4 is on the order of 8 years, while that of isoprene is less than 1 day (see, e.g., Brasseur et al., 1999). Other powerful oxidants include ozone, the nitrate radical (most abundant during nighttime), and hydrogen peroxide.

Without human-induced perturbations, the chemical composition of the atmosphere would be strongly determined by biological processes at the earth's surface. Photosynthesis, respiration, matter decomposition, microbial activity produce intensive exchanges of chemical elements at the earth's surface with a deep influence on the atmospheric composition. Volcanic eruptions, which have played a major role in the evolution of the earth's atmosphere at geological time scales, have limited effects in the contemporary atmosphere, except episodically after major events such as the eruption of Mount Pinatubo. Eruptions may enhance dramatically the aerosol load of the stratosphere, leading to a visible signature in the climate system and in the chemical composition of the lower stratosphere. Many compounds released at the surface are oxidized, leading to the formation of chemical intermediates and eventually longer-lived chemical reservoirs. These are progressively eliminated from the atmosphere either by dry deposition on the surface or by wet scavenging in precipitation.

An interesting example of the biological influence on the atmospheric composition is provided by the processes affecting the formation and destruction of atmospheric ozone in the natural atmosphere. Although the formation of stratospheric ozone is a photochemical process acting through the dissociation of molecular oxygen (O_2) by solar ultraviolet radiation, the concentration of O_2 in the atmosphere is determined by a balance between photosynthesis and respiration processes on land and in the ocean. The major stratospheric ozone loss mechanism is provided by a catalytic cycle involving the presence of the reactive nitrogen oxides ($NO_x = NO + NO_2$). These are produced by oxidation of nitrous oxide, a long-lived compound released at the surface as a result of nitrification and denitrification in soils.

In the troposphere, the production of ozone results from the day-time oxidation of methane, nonmethane hydrocarbons, and carbon monoxide in the presence of nitrogen oxides. Under natural conditions, methane, produced in oxygen-deficient environments, is released primarily by wetlands, lakes, and rivers. Nonmethane hydrocarbons, such as isoprene and terpenes, are emitted by various types of trees. Nitric oxide is released by soils as a result of microbial activity and is produced in the atmosphere by lightning in thunderstorm systems.

Another interesting example of the biological influence on atmospheric chemistry is provided by sulfur. Under natural conditions, sulfur compounds in the atmosphere are provided by the oceanic emission of dimethyl disulfide (DMS). This biogenic emission results from the breakdown of sulfoniopropionate (DMSP), which is thought to be used by marine phytoplankton to control their osmotic pressure. The oxidation of DMS leads to the formation of sulfur dioxide, which is further converted to sulfate particles. As indicated above, these particles, by scattering back to space some of the incoming solar radiation, tend to cool the earth's surface. Their presence also affects the optical properties of the clouds, which introduces an indirect climatic effect.

During the 20th century, the chemical composition of the atmosphere has been altered, sometimes in a major way, by human activities. For example, in the troposphere, the atmospheric concentrations of ozone precursors have increased substantially primarily as a result of industrialization and land-use changes. Intensification of biomass burning, mostly in the tropics, and increase in fossil fuel consumption have profoundly modified the source strengths of volatile organic carbon, carbon monoxide, and reactive nitrogen oxides. As a result, the level of ozone and the concentration of OH, and hence the oxidizing potential of the atmosphere, have been modified and are expected to continue to change in the future. An important scientific issue is to quantify the magnitude of these changes and to assess their impact on climate and on the biosphere.

Today, the anthropogenic emissions of SO_2, primarily from fossil fuel combustion, largely dominate the sulfur flux into in the atmosphere on the global scale. Climate models have determined the corresponding direct and indirect impacts on radiative forcing, but large uncertainties remain in these estimates. In fact, predictions of future climate need to account not only for the effects of sulfate aerosols, but also for the contributions of mineral dust, black carbon, organic carbon, and sea salt. The current view is that atmospheric particles should be treated as multicomponent, mul-

TABLE 1   Contemporary Surface Emissions Used in IMAGES (Müller, 1992)

	Technological Source (%)	Biomass Burning (%)	Biogenic[a] (%)	Oceans	Total (Tg)
NO_x–N	59	15	18	$0.1*E^{-5}$	37[b]
SO_2–S	100				91[c]
DMS				100%	21[c]
CO	27	51	11	11%	1440
CH_4	26	11	61	2%	506
VOCs	13	7	67	4–40%	750

[a]Includes animal, microbial and foliage emissions.
[b]Tg N.
[c]Tg S.
[d]Tg CO, CH_4, or VOCs, respectively.

tisize aerosols. Much remains to be done to properly treat these mixtures of particles with varying chemical compositions and physical properties in climate and earth system models.

3. The IMAGES Model

One of the exciting challenges for the scientific community in the coming decade will be the development of coupled earth system models that account for the interactions between the biogeochemical cycles and the physical climate system. At present, most simulations of the atmospheric composition are performed using chemical transport models in which the atmospheric dynamics are prescribed (based on meteorological analyses) or calculated by an atmospheric general circulation model. Like most chemical transport models, IMAGES, which is used in the present study, incorporates several basic elements: surface emissions, atmospheric transport, chemical transformations, and surface deposition. The horizontal resolution is 5° in longitude and latitude with finely resolved vertical layering (25 layers) and an atmosphere that extends 22.5 km high. The transport time step for IMAGES is typically 6 h. A global climatology of wind from the European Center for Medium-Range Weather Forecasts was used to drive transport. The model represents advection (Smolarkiewicz and Rasch, 1991) and accounts for subgrid transport through diffusive mixing in the boundary layer (Müller, 1993), deep convection (Costen, 1988) in the free troposphere, and eddy diffusive mixing to account for unresolved wind variability. IMAGES derives the concentration of 41 species, including seven different hydrocarbons, and several oxygenated organics, including PAN and MPAN (Müller and Brasseur, 1995). It includes a relatively detailed chemical scheme with 125 reactions (including 26 photolytic reactions and a few heterogeneous reactions),

The spatial distributions of the deposited species, NO, NO_2, HNO_3, and O_3, depend on interactions of the transport and chemical schemes with both wet and dry deposition. IMAGES parameterizes wet deposition or wash out as a first-order loss rate calculated as a function of the precipitation rate (Müller and

Brasseur, 1995). The precipitation rate was taken from the climatology of Shea (1986). IMAGES expresses dry deposition as a function of a prescribed deposition velocity (which is specific to vegetation type). Wet deposition depends on the rate of precipitation, but the formulation accounts for the different types of precipitations and the species-dependent solubility coefficients. Dry deposition velocities vary considerably from species to species. HNO_3 has by far the largest deposition velocity.

Emission estimates for a number of relevant chemical species considered in modeling the contemporary atmosphere are provided in Table 1 (Muller, 1992; Muller and Brasseur, 1995). The relative contributions of the anthropogenic and biogenic sources vary a great deal depending on the chemical compounds considered. Sulfur dioxide emissions are almost entirely produced by fossil fuel burning, while technological sources account for only 13% of volatile organic carbon emissions. For carbon monoxide and methane, technological sources provide only 25% of the total budget. For the remainder of the CO budget, biomass burning constitutes the bulk of the respective budgets. For NO_x emissions, fossil fuel combustion is the largest term in the budget but biogenic production, biomass burning, and lightning all contribute a substantial portion of the total budget. For modeling the future atmosphere, we use the IS92a scenario developed for IPCC 1995.

4. Changes in the Chemical Composition of the Global Troposphere

The IMAGES model has been used to assess the impact of human activities on the chemical composition of the global troposphere. To quantify past, current, and future changes in tropospheric composition, IMAGES was used to simulate the preindustrial atmosphere (year 1850), a contemporary atmosphere (year 1990), and a future atmosphere (year 2050), respectively. As expected, the largest increase in ozone occurs in the lower troposphere in the Northern Hemisphere with changes of more than 70% at mid-

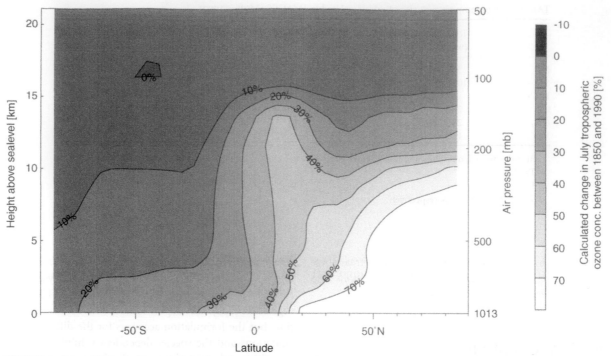

FIGURE 1   The calculated change (%) in the July zonally averaged concentration of tropospheric ozone between 1850 and 1990 simulated by the IMAGES model using the IS92a emission scenario.

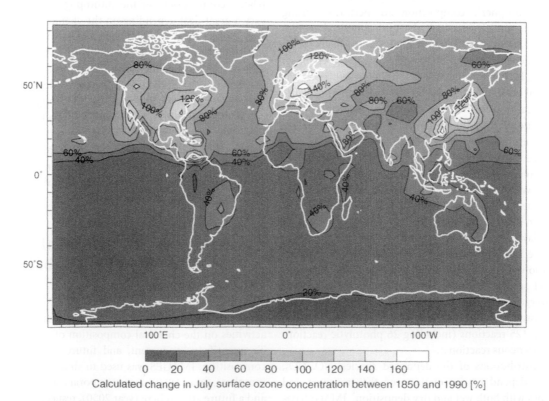

FIGURE 2   The calculated change (%) in July surface ozone concentrations between 1850 and 1990 simulated by IMAGES using the IS92a emission scenario.

and high latitudes (Fig. 1). In the Southern Hemisphere, the estimated ozone increase is typically 10–20%, and in the tropics 30–50%. The change is most intense near the pollution sources and becomes more uniform with height in the atmosphere: the distribution of the change is relatively uniform with longitude in the upper troposphere. Examination of the surface ozone concentrations shows increases of more than 120% over the east and west coasts of the United States, over Europe, and over China and Japan (Fig. 2). These estimated changes are relatively consistent with the limited information available on the evolution of ozone in Europe during the 20th century (Fishman and Brackett, 1997; Hudson and Thompson, 1998; Logan, 1994).

Future ozone changes are difficult to predict because they depend directly on the future evolution of emissions and hence of population growth and economic development. Such predictions must therefore be based on a series of scenarios. In the present study, we simply adopt the IS92a scenario developed by IPCC (1996) as well as the NASA estimates of future growth in aviation. The projected increase in the zonally averaged ozone concentration (July conditions) is highest in the tropics at all altitudes and low in the vicinity of the tropopause in the Northern Hemisphere (data not shown). The remarkable increase in surface ozone predicted for tropical and subtropical regions, ranging from 10 to 75% (Fig. 3), is associated with rapid economic development in the region. In the upper troposphere, the projected increase in aircraft traffic over the next 50 years has a large impact on upper tropospheric ozone (Brasseur *et al.*, 1998; data not shown). When expressed in absolute O_3 concentrations, rather than in percentages, the projected increase in ozone is highest near the mid- and high latitude tropopause (data not shown). It is interesting to note that during January (Southern Hemisphere summer; see Fig. 3), large increases in surface ozone are predicted not only in the tropics but also in Brazil, South Africa, and Southern Asia. These predictions suggest that ozone pollution events are likely to become more frequent in the populated areas of the Southern Hemisphere during the next decades.

Over the past hundred years, changes in sulfate concentrations have been greatest in Central and Eastern Europe and in China (Fig. 4). The percentages of change in sulfate concentrations over the United States and Canada have been relatively smaller. Changes in sulfate concentrations are also expected to increase in the future, as suggested by Figure 5, especially as a result of massive coal burning. The effect is predicted to be most intense in China and northern India where coal is widely used. Thus, anthropogenic pollution will not only enhance the ozone concentration in these regions, but also increase the aerosol load. Note that, based on the IS92a scenario, the change in sulfate concentration is greater than 400% near the surface over the Asian continent. Little change is expected in the industrialized Northern Hemisphere, where the use of coal has dramatically decreased, and new tech-

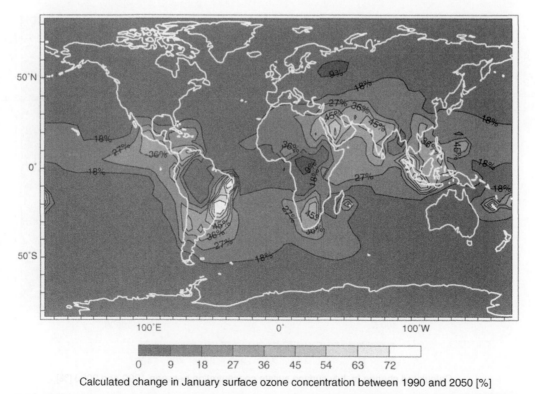

Calculated change in January surface ozone concentration between 1990 and 2050 [%]

FIGURE 3 The calculated change (%) in the July surface ozone concentrations between 1990 and 2050 simulated by IMAGES. January was chosen to better represent the dramatic changes in the Southern Hemisphere.

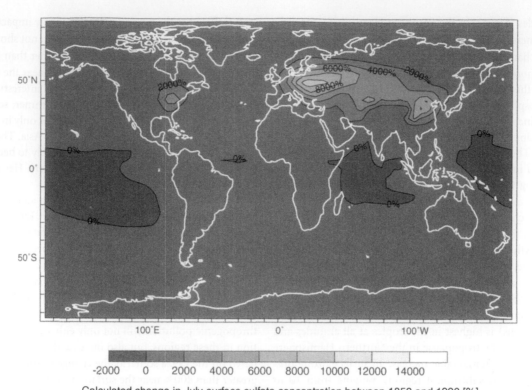

Calculated change in July surface sulfate concentration between 1850 and 1990 [%]

FIGURE 4   The calculated change (%) in July surface sulfate concentrations between 1850 and 1990 atmospheres simulated by IMAGES using the IS92a emission scenario.

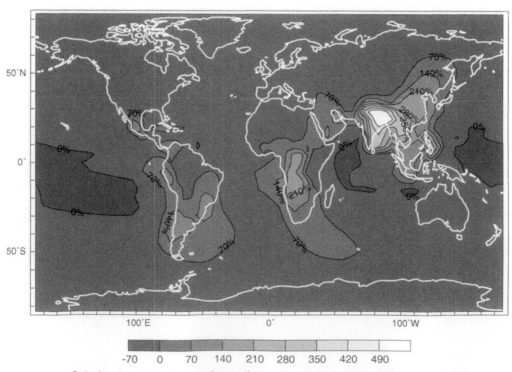

Calculated change in July surface sulfate concentration between 1990 and 2050 [%]

FIGURE 5   The calculated change (%) in the July surface sulfate concentrations between 1990 and 2050 simulated by IMAGES using the IS92a emission scenario.

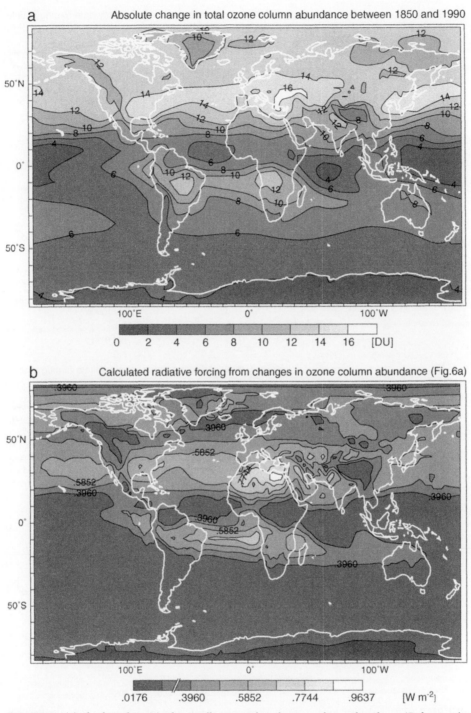

FIGURE 6 (a) Absolute change in the zonally averaged total ozone column abundance (Dobson units, DU) from 1850 to 1990 simulated by IMAGES. (b) Calculated radiative forcing resulting from the changes in ozone column abundance as shown in (a).

nologies have been developed to reduce the emissions of sulfur compounds.

Several attempts have been made to assess the climatic impact of chemical compounds. In the case of long-lived greenhouse gases, IPCC (1996) has estimated the radiative forcing to be approximately 2.5 W m^{-2}. The climate impact of ozone and sulfate is more difficult to quantify due to the nonuniform nature of the perturbation and its seasonal variability. Figures 6a and 6b show the estimated change in the tropospheric ozone column (estimated by IMAGES in October) and the corresponding change

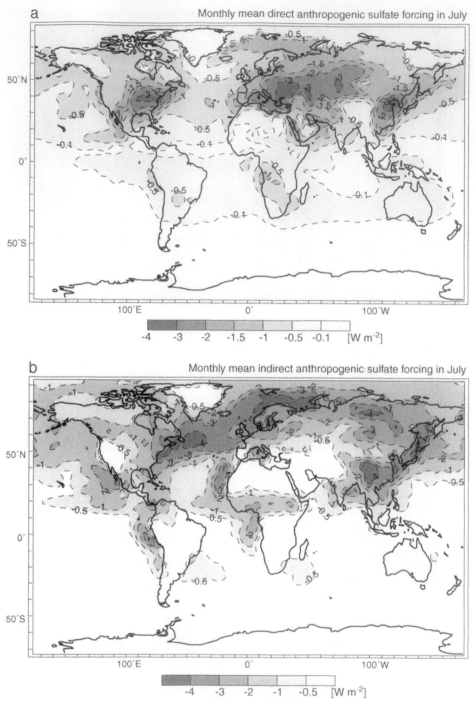

FIGURE 7 The radiative forcing for contemporary production of sulfate aerosols simulated by ECHAM (Roeckner *et al.*, 1999).

in radiative forcing. These graphs show an increase in ozone of typically 10 Dobson units in the Northern Hemisphere (resulting primarily from fossil fuel combustion) and of typically 8–13 Dobson units in the tropics (resulting primarily from biomass burning effects). The corresponding radiative forcing varies from about 0.4 to 0.8 W m^{-2} in the Northern Hemisphere. Changes are small in the Southern Hemisphere. The globally averaged radiative forcing is estimated to be 0.37 W m^{-2}.

In the case of sulfate aerosol particles, the cooling between the preindustrial era and the present period is estimated to be on the order of 1–3 W m^{-2} (IPCC, 1995; Santner *et al.*, 1995). The values are highest over the eastern portion of North America and the southeastern part of Europe and Asia. The global cooling due to the direct sulfate effect is estimated by Roeckner et al. (1999) to be 0.35 W m^{-2} and thus on the same order (but with opposite sign) of the mean warming by ozone. Note, however, that Roeckner *et al.*, (1999) estimate the indirect cooling effect to be close to 1 W m^{-2} (Fig. 7). The uncertainty of these numbers is close to a factor of 2–5.

5. Concluding Remarks

Current chemical transport models of the atmosphere, which typically include 50–100 chemical compounds and 150–250 chemical reactions, reproduce with reasonable success the global behavior of the chemical system in the atmosphere. Differences between the results provided by these models remain substantial and will have to be addressed in the future. These models are used to explain the dramatic changes that have occurred in the chemical composition of the atmosphere over the last century and to predict changes in the future on the basis of plausible emission scenarios.

- Increases in the atmospheric concentration of long-lived greenhouse gases since the preindustrial era have led to a climate forcing of about 2.5 W m^{-2} (IPCC, 1996). The interannual variability in this trend is not well understood and involves complex interactive processes between the atmosphere, the ocean, and the continental biosphere. Coupled earth system models with a detailed representation of global biogeochemical cycles will help address these issues.
- The aerosol load of the atmosphere has also increased as a result of human activities, specifically biomass burning and fossil fuel combustion. Sulfate aerosols have produced a direct cooling effect that can reach locally more than 2 W m^{-2} over industrialized areas, but is much smaller on the global scale. The indirect radiative effects of aerosols (through changes in the optical properties and lifetimes of the clouds) remain rather uncertain, but could be larger than the direct effects (IPCC, 1996). Changes in upper level clouds (i.e., cirrus) could lead to a warming of the earth's surface. Progress in this area requires a better understanding of aerosol microphysics and chemistry. The role of nonsulfate aerosols (and specifically multicomponent aerosols) will also have to be included in comprehensive model calculations.
- The oxidizing power of the atmosphere has likely decreased significantly, especially in the Northern Hemisphere, as a result of human activities. As a result, the lifetime of methane may have increased by 10–15% since the preindustrial era. At the same time, the abundance of tropospheric ozone has increased perhaps by as much as a factor of 2–3 in the Northern Hemisphere. Enhanced biomass burning fluxes of NO$_x$, CO, and hydrocarbons from tropical ecosystems are likely to be important. Future changes in tropospheric ozone are predicted to be largest in the tropics (India, China). These projected increases in tropical emissions are likely to have a disproportionate impact on global atmospheric chemistry because of the vigorous upward transport that characterizes the region. The global budget of ozone, however, remains, rather uncertain due to the lack of systematic observations, especially in the tropics and in the Southern Hemisphere. The impact of future commercial aircraft operations on upper tropospheric ozone in the Northern Hemisphere will probably become significant during the 21st century.

- In the stratosphere, the link between long-term buildup of anthropogenic chlorine and ozone decline is now firmly established (WMO, 1999). Recently, the decline in mid-latitude ozone has slowed, but late winter/early spring ozone values in the Arctic were often unusually low during the 1990s. The Antarctic ozone hole, which is observed in September/October, continues unabated. Increasing concentrations of carbon dioxide together with the observed stratospheric ozone losses have caused a cooling of the lower stratosphere and a negative radiative forcing of the climate system. All of these highlight the existing link between ozone and climate issues that society has been facing.

Biogeochemistry is inherently a broad subject and clearly requires interdisciplinary approaches. Today, as the community regards the earth as a complex nonlinear system, studies of atmospheric chemistry and biogeochemistry cannot be dissociated from studies of the physical climate system. Interactions between the ocean, the continental biosphere, and the atmosphere are therefore central themes for the science of the 21st century. The challenges for the new Max Planck Institute for Biogeochemistry in Jena are particularly exciting.

References

Aber, J. D., Nadelhoffer, K. J., Steudler, P, and Melillo, J. M. (1989). Nitrogen saturation in northern forest ecosystems. *Bioscience* **39**, 378–386.

Brasseur, G. P., Kiehl, J. T., Müller, J. F., Schneider, T., Granier, C., Tie, X. X., and Hauglustaine, D. (1998). Past and future changes in global tropospheric ozone: Impact on radiative forcing. *Geophys. Res. Lett.* **25**, 3807–3810.

Brasseur, G. P., Orlando, J. J., and Tyndall, G. S. (1999). "Atmospheric Chemistry and Global Change." 654 p. Oxford University Press.

Costen, R. C., Tennille, G. M., Levine, J. S. (1988). Cloud pumping in a one-dimensional model. *J. Geophys. Res.* **93**, 941–954.

Fishman, J. and Brackett, V. G. (1997). The climatological distribution of tropospheric ozone derived from a satellite measurements using version 7 Total Ozone Mapping Spectrometer and Stratospheric Aerosol and Gas Experiment data sets. *J. Geophys. Res.* **102**, 19275–19278.

Hudson, R. D. and Thompson, A. M. (1998). Tropical tropospheric ozone (TTO) form Toms by a modified residual method. *J. Geophys. Res.* **103**, 22129–22145.

IPCC. (1996). "Climate Change 1995." Cambridge University Press, Cambridge.

Logan, J. A. (1994). Trnads in the vertical distribution of ozone: an analysis of ozonesonde data. *J. Geophys. Res.* **99**, 25553–25585.

Müller, J. F. (1992) Geographical distribution and seasonal variation of surface emissions and deposition velocities of atmospheric trace gases. *J. Geophys. Res.* **97**, 3787–3804.

Müller, J. F. (1993). Modélisation tri-dimensionelle globale de la chimie et du transport des gaz en trace dans la troposphere. PhD thesis, Belgian Institute for Space Aeronautics, Brussels.

Müller, J. F., and Brasseur, G. (1995). IMAGES: a three-dimensional chemical transport model of the global troposphere. *J. Geophys. Res.* **100,** 445–490.

Roeckner, E., Bengtsson, L., Feichter, J., Lelieveld, J., and Rohde, H. (1999). Transient climate change simulations with a coupled atmosphere-ocean GCM including the tropospheric sulfur cycle. *J. Climate* **12,** 3004–3032.

Santer, B. D., Taylor, K. E., Wigley, T. M. L., Penner, J. E., Jones, P. D., and Cubasch, U. (1995). Towards the detection and attribution of an an-thropogenic effect on climate. *Climate Dynamics* **12,** 77–100.

Schulze, E. D. (1989). Air pollution and forest decline in a spruce (*Picea abies*) forest. *Science* **244,** 776–783.

Shea, R. C. (1986). " Climatological Atlas: 1950–1979." NCAR Technical Note, NCAR/TN-269+STR.

Smolarkiewicz, P. K. and Rash, P. J. (1991). Monotone advection on the sphere: an Eulerian versus semi-Lagrangian approach. *J. Atmos. Sci.* **48,** 793–810.

WMO. (1999). "Scientific Assessment of Ozone Depletion: 1998. Global Ozone Research and Monitoring Project—Report No. 44." World Meteorological Organization, Geneva.

1.4

Inferring Biogeochemical Sources and Sinks from Atmospheric Concentrations: General Considerations and Applications in Vegetation Canopies

M. R. Raupach
CSIRO Land and Water,
Canberra,
Australia

This chapter is a review of the principles and application of atmospheric inverse methods in vegetation canopies. These methods enable the source–sink distributions of biogeochemically active entities (such as water, carbon dioxide, non-CO_2 greenhouse gases, and aerosols) to be inferred from measurements of their atmospheric concentrations. The chapter covers four topics. First, canopy-scale inverse methods are placed in the context of atmospheric inverse methods in general, at scales from small chambers to the globe. Next, because these methods depend on mass conservation, the balance equations for scalar mole fraction (c), iso-concentration ($c\delta$) and isotopic composition (δ) are analyzed in single-point Eulerian form and in Lagrangian form. This leads to the third topic, inverse Lagrangian methods for inferring source–sink profiles from concentration measurements in vegetation canopies. The theory of the approach is reviewed and results from several field experiments are summarised, showing that this approach is a practically useful tool for inferring the canopy source–sink distributions of scalars such as water vapor, heat, CO_2 and ammonia. However, there is a continuing need for improvement in the knowledge of the turbulence field in the canopy. The fourth topic is the extension of inverse Lagrangian analysis to describe the relationship between profiles of isotopic composition in canopy air and the profiles of isotopic sources and sinks. Lagrangian analysis is shown to provide a general basis for the Keeling plot and the Yakir–Wang expression for distinguishing assimilation and respiration, in the case where the isotopic composition of the exchanged scalar (δ_P) is constant through the canopy. Using the Inverse Lagrangian approach, this analysis is extended to explain air isotopic composition profiles in circumstances where δ_P is strongly nonuniform through the canopy.

1. Introduction

Exchange between the earth's surface and the atmosphere is a crucial part of the cycles of almost all biogeochemically active entities, including water, carbon dioxide, methane, oxides of nitrogen, volatile organic compounds, and others. As surface–atmosphere exchanges occur, the atmospheric concentrations of these entities are altered in both space and time, providing a transient imprint of both the magnitudes and the distributions of their sources and sinks at the surface. The information in this imprint can be used to infer the source–sink distribution of an entity from measurements of its atmospheric concentration field, by "inversion" of a "forward model" which specifies the concentration field in terms of the source–sink distribution. The forward model is based (in different ways according to the application) on mass-balance principles augmented by knowledge of atmospheric advection and dispersion processes, which together determine the concentration field produced by a specified source–sink distribution. Such methods can be generically called atmospheric inverse methods. This chapter reviews some general principles underlying atmospheric inverse methods, and then applies these principles to the specific problem of inferring biogeochemical sources and sinks in vegetation canopies, from measurements of concentration and isotopic composition profiles in the air.

The plan of the chapter is as follows. This introductory section surveys atmospheric inverse methods from small-chamber to global scales, indicating the commonalities and differences among methods at various scales. Because of the fundamental reliance of atmospheric inverse methods on mass conservation, Section 2 discusses the generic balance equations for the mole fraction and isotopic composition of a scalar entity, considering both Eulerian (fixed) and Lagrangian (fluid-following) frameworks. The chapter then moves to its specific focus on vegetation canopies: Section 3 describes the methodology and application of methods for inferring scalar source–sink distributions from measured concentration profiles, and Section 4 extends this treatment to isotopic composition. Section 5 provides a summary and conclusions.

We begin with a general survey of atmospheric inverse methods. In all cases, the broad goal is to use concentration measurements in the air, together with information about atmospheric flow, to infer sources and sinks of entities at the earth's surface. Since the key concentration observations are remote from the surface sources and sinks, this entire class of methods relies explicitly or implicitly on an atmospheric mass or molar balance for the entity being measured, within a specified control volume. Such a balance can be either in an Eulerian framework, in which the control volume is fixed in space, or in a Lagrangian framework, in which the control volume moves with the flow. Considering the Eulerian framework first, the molar balance for a scalar entity can be written informally as

$$\frac{\partial c}{\partial t} = -u\,\frac{c(x_1) - c(x_0)}{x_1 - x_0} - \frac{F(z_1)}{\rho z_1} + \phi \qquad (1)$$

$$\underset{\text{I}}{} \qquad \underset{\text{II}}{} \qquad \underset{\text{III}}{} \quad \underset{\text{IV}}{}$$

where c is the average mole fraction (specific concentration) of the entity in a control region extending from x_0 to x_1 in the direction of the mean flow and from the surface to height z_1; u is the mean flow velocity; ρ is air density; F is the vertical flux density of the scalar; and ϕ is the source (or sink) of the scalar in the control region, including contributions from fluxes at the surface. All quantities are suitably averaged in space (see next section and appendixes for details). The four terms respectively represent (I) storage change, (II) advection, (III) flux out of the top of the control region, and (IV) sources within the region. The aim is to infer the source term ϕ by choosing the control region so that most other terms in the balance are small, leaving only one other major term (sometimes two) which can be measured to infer ϕ. Such an

Eulerian approach is used at the scales of small chambers, small plots, vegetation canopies, the atmospheric surface layer, the atmospheric boundary layer (both convective and stable), large regions, and the globe. Reviews of micrometeorological methods are offered by Denmead and Raupach (1993), Denmead (1994, 1995), and Denmead *et al.* (1999), while aspects of methods at global scales are described by Enting *et al.* (1995), Ciais and Meijer (1998), Bousquet *et al.* (1999a, b), Enting (1999), and Rayner *et al.* (1999).

Commonalities and differences between these methods are highlighted by grouping the methods not by scale, but according to the term(s) in Eq. (1) used to infer ϕ. Table 1 shows such a grouping, in which four classes of Eulerian method are identified:

1. *Methods based on spatial gradients in the direction of diffusion*: These include the standard gradient and Bowen-ratio methods of surface-layer flux measurement in micrometeorology. The key assumption is that terms I and II are small, so that the source term ϕ (in this case identifiable with the flux at the surface) equals the flux F at a measurement height z_1.

2. *Methods based on spatial gradients in the direction of advection*: These include the small-plot techniques reviewed by Denmead (1994, 1995) and also chambers operated in a continuously ventilated, quasi-steady-state mode. The control region is designed so that all terms except II and IV are small.

3. *Methods based on temporal gradients*: This approach is used at all scales, including small chambers operated in fully enclosed mode for short periods, boundary-layer budget methods in both the daytime convective boundary layer (CBL) and the nocturnal stable boundary layer (SBL), and global atmospheric budget methods. The dominant terms are I and IV.

4. *Flux-resolving, fluctuation-based methods*: In these methods the vertical turbulent eddy flux F is measured as $\rho\overline{w'c'}$, where overbars and primes denote time means and fluctuations, respectively, and w is the vertical velocity component. They include eddy-covariance and eddy-accumulation methods for measuring fluxes from towers and aircraft. They are not normally regarded as atmospheric inverse methods, since that term usually refers to methods which relate a mean concentration field to a source density or surface flux distribution.

Lagrangian inverse methods also rely on a balance equation. In this case, the principle is to consider individual "marked fluid particles" which (with sufficient knowledge of the velocity field) can be followed as they move. The balance equation in a Lagrangian framework is simply

$$Dc/Dt = \phi, \qquad (2)$$

where D/Dt is the material derivative, or time derivative following the motion of the fluid particle. This says that the concentration

TABLE 1 Atmospheric Inverse Methods, Grouped According to the Term(s) in the Scalar Conservation Equation Used as Surrogates for the Source–Sink Term at the Surface.

Method	Quantity Measured	Key Assumptions and Constraints	Spatial Footprint (X) and Time Resolution (T)
Eulerian methods based on spatial gradients in the direction of diffusion			
Gradient, aerodynamic, Bowen ratio	Surface flux	Steady, horizontally uniform	$X = 1$ km, $T = 30$ min
Eulerian methods based on spatial gradients in the direction of advection			
Ventilated chamber	Surface flux	Chamber does not alter flux	$X = 1$ m, $T = 30$ min
Small plot	Surface flux	Control region is deep enough to neglect flux through top	$X = 30$ m, $T = 30$ min
Eulerian methods based on temporal gradients			
Enclosed chamber	Surface flux	Chamber does not alter flux	$X = 1$ m, $T = 30$ min
CBL or SBL budget	Surface flux (regional average)	Semi-Lagrangian and Eulerian averages are equivalent	$T = 0.3$ day $X = 100$ km (CBL) $X = 10$ km (SBL)
Global budget	Surface flux (global average)	Well-mixed global atmosphere	$X = $ globe, $T > 1$ year
Eulerian flux-resolving, fluctuation-based methods			
Eddy covariance, eddy accumulation, relaxed eddy accumulation	Surface flux (sensor above canopy) or flux profile (sensor within canopy)	Steady, horizontally uniform	$X = 1$ km, $T = 30$ min
Lagrangian methods			
Inverse Lagrangian methods in canopies	Source–sink profile $\phi(z)$	Steady, horizontally uniform	$X = 100$ m, $T = 30$ min
Global or regional synthesis inversion	Surface flux (regional distribution)		$X = 3000$ km, $T = 1$ month

in the fluid particle changes in response to the sources or sinks it encounters along its path. Two problems can then be defined. The "forward" problem is to find the concentration field from the source–sink distribution and is solved by tracking the scalar added to (or removed from) each fluid particle by the sources or sinks and then calculating statistically the points to which the flow moves the scalar. To solve the opposite "inverse" problem, finding sources and sinks from concentrations, it is necessary to find the concentrations produced by a large number of point sources and then find the mix of point sources or sinks which best matches the observed concentration field.

Lagrangian inverse methods are listed as a fifth group in Table 1. There are several methods in this class, including the "synthesis inversion" methods for inferring the global distribution of atmospheric sources and sinks for entities such as CO_2 and its isotopes (Enting *et al.*, 1995; Bousquet *et al.*, 1999a, b; Enting, 1999a, b; Rayner *et al.*, 1999) and the "inverse Lagrangian" methods for inferring scalar source–sink distributions in vegetation canopies which are the primary specific focus in this paper.

In this brief survey, atmospheric inverse methods are defined broadly to include both Lagrangian and Eulerian approaches because their common foundation is the inference of sources and sinks from atmospheric concentration measurements. The two broad streams offer different strengths: Eulerian approaches, by using tightly defined control volumes, can give quite precise estimates of average sources or sink strengths (ϕ) within the control volume, subject to the requirement that the neglected terms in Eq. (1) are indeed small. Their demands for information about the

velocity field are usually modest, but they inherently produce results for ϕ which are averaged through the control volume. By contrast, Lagrangian approaches offer much more resolution of the space–time distribution of ϕ at the expense of far greater requirements for information about the velocity field.

2. Scalar and Isotopic Molar Balances

2.1 General Principles

This section considers the scalar and isotopic balance equations in general, with attention to the source terms for scalars and isotopes which arise in vegetation canopies. After general principles are set out in Section 2.1, the Eulerian framework is described in Section 2.2 and the ensuing Lagrangian framework in Section 2.4. Section 2.3 discusses the source terms for isotopes.

We are concerned with the molar balance of a scalar entity with mole fraction c, with a minor isotopic constituent such as ^{13}C or ^{18}O in CO_2. The isotopic ratio R is the molar ratio of the minor (heavier) to the major (lighter) isotope, and the isotopic composition δ is the normalized departure of R from its value R_* in a standard reference material, so that $\delta = R/R_* - 1$ and $R = R_*(1 - \delta)$. Henceforth c will denote the mole fraction of the major isotope unless otherwise stated, so the mole fraction of the minor isotope is cR. The molar balance applies in general to a region with volume $V(t)$ enclosed by a bounding surface $S(t)$ which may be moving, so the region may deform, expand, or contract. Usually, a part (say

S_O) of the surface S occurs in the open air, and the remainder (say S_P) coincides with plant, soil, or water surfaces bounding V. Thus, $S = S_O + S_P$. It is useful to consider a general moving control volume in order to support Lagrangian or semi-Lagrangian applications in which V or some of its boundaries move with the flow, and also because major applications occur in the atmospheric convective boundary layer (CBL) which grows through the day.

For the major isotope, the molar balance equation is

$$\frac{\partial}{\partial t} \iiint_{V(t)} \rho c\, dx = \iint_{S_o(t)} [\mathbf{F} + \rho c(\mathbf{u} - \mathbf{v})] \cdot \mathbf{n}\, dS + \iint_{S_p(t)} \mathbf{F} \cdot \mathbf{n}\, dS, \quad (3)$$

where $\mathbf{F}$ is the flux density of the major isotope (mol m^{-2} s^{-1}), ρ is the molar air density, $\mathbf{u}$ and $\mathbf{v}$ are respectively the vector velocities of the fluid and the moving surface S, and $\mathbf{n}$ is the *inward* unit normal vector on S (pointing into V). The term $\rho c(\mathbf{u} - \mathbf{v})$ is the advective scalar flux. The first integral covers the open-air part S_O of S and the second the surface part S_P. The equation for the overall scalar entity is the same as Eq. (3), with appropriate redefinition of c and $\mathbf{F}$.

Equation (3) takes a similar form for both time-averaged and instantaneous quantities. In the time-averaged case, $\mathbf{F}$ is a turbulent eddy flux $\overline{\rho \mathbf{u}'c'}$, where overbars and primes denote time means and fluctuations, respectively. In the instantaneous case, $\mathbf{F}$ is a molecular diffusive flux which in practice can be neglected in the open air (on S_O) relative to fluxes arising from fluid motion (the high Péclet number approximation). In this case the fluid-motion fluxes appear in the term $\rho c(\mathbf{u} - \mathbf{v})$, which is unaveraged and includes transport by turbulent fluctuations as well as by the mean flow. In contrast with the situation in the open air (on S_O), molecular fluxes can never be neglected at solid boundaries (S_P), where they are responsible for all the scalar transport.

Turning to the minor isotope, the ratio of minor to major isotope fluxes in the open air (on S_O) is the same as the isotopic concentration ratio R, because there is negligible discrimination by fluid motion and mixing. Hence, on S_O, the advective flux for the minor isotope is $\rho c R(\mathbf{u} - \mathbf{v})$ and the molecular or turbulent flux is $R\mathbf{F}$. However, transport across plant or other solid surfaces (S_P) does discriminate between minor and major isotopes. The flux of minor isotope across these surfaces is $R_P\mathbf{F}$, where R_P is the isotopic ratio of the scalar exchanged (transported across S_P) by the flux $\mathbf{F}$. Hence the molar balance for the minor isotope can be written in either of the forms

$$\frac{\partial}{\partial t} \iiint_{V(t)} \rho c R\, dx = \iint_{S_o(t)} R[\mathbf{F} + \rho c(\mathbf{u} - \mathbf{v})] \cdot \mathbf{n}\, dS + \iint_{S_p(t)} R_P \mathbf{F} \cdot \mathbf{n}\, dS$$

$$(4)$$

$$\frac{\partial}{\partial t} \iiint_{V(t)} \rho c \delta\, dx = \iint_{S_o(t)} \delta[\mathbf{F} + \rho c(\mathbf{u} - \mathbf{v})] \cdot \mathbf{n}\, dS + \iint_{S_p(t)} \delta_P \mathbf{F} \cdot \mathbf{n}\, dS,$$

where $\delta_P = R_P/R_* - 1$ is the isotopic composition of the scalar exchanged across plant surfaces. The first form reverts to Eq. (3), when $R = R_P = 1$.

Balance equations are needed mainly for three quantities: c, $c\delta$, and δ. The quantity $c\delta = (cR - cR_*)/R_*$ is the "iso-concentration," a linear combination of the mole fractions for the minor and major isotopes which measures the departure of the minor isotope mole fraction cR from its value cR_* when the isotopic composition is that of the reference material.

2.2 Single-Point Eulerian Equations

In the limit $V \to 0$ at a single fixed point $\mathbf{x}$, Eqs. (3) and (4) reduce to conventional Eulerian differential conservation equations. Anticipating application in vegetation canopies, we consider that some of the region V contains plant leaves and stems. Provided that the canopy is sufficiently finely textured to be regarded as a porous continuum from the standpoint of the airflow, a scalar source density in Eq. (3) can be defined by

$$\Phi(\mathbf{x}, t) = \rho \phi(\mathbf{x}, t) = \lim_{V \to 0} \frac{1}{V} \iint_{S_p(t)} \mathbf{F} \cdot \mathbf{n}\, dS, \quad (5)$$

where $\Phi(\mathbf{x}, t)$ is the source density for absolute molar concentration and $\phi(\mathbf{x}, t)$ the source density for mole fraction. If there is no porous canopy at $\mathbf{x}$, then $\Phi = \phi = 0$. The equivalent "iso-source" term for the iso-concentration, in Eq. (4), is

$$\delta_P \Phi(\mathbf{x}, t) = \rho \delta_P \phi(\mathbf{x}, t) = \lim_{V \to 0} \frac{1}{V} \iint_{S_p(t)} \delta_P \mathbf{F} \cdot \mathbf{n}\, dS. \quad (6)$$

The "continuum porous canopy" assumption underlying Eqs. (5) and (6) is a simple approximation which gives the same result in the present case as the more formally rigorous procedure of taking finite volume averages over thin horizontal slabs with horizontal length scales large enough to average over the local heterogeneity of the canopy (Wilson and Shaw, 1977; Raupach and Shaw, 1982; Finnigan, 1985).

It is shown in Appendix A that in the limit $V \to 0$, Eqs. (2) to (4) yield the following conservation equations for scalar mole fraction (c), iso-concentration ($c\delta$), and isotopic composition (δ) at a fixed point $\mathbf{x}$:

$$\left(\frac{\partial c}{\partial t} + \mathbf{u} \cdot \nabla c \right) + \frac{\nabla \cdot \mathbf{F}}{\rho} = \phi \quad (7)$$

$$\left(\frac{\partial (c\delta)}{\partial t} + \mathbf{u} \cdot \nabla (c\delta) \right) + \frac{\nabla \cdot (\delta \mathbf{F})}{\rho} = \delta_P \phi \quad (8)$$

$$\left(\frac{\partial \delta}{\partial t} + \mathbf{u} \cdot \nabla \delta \right) + \frac{\mathbf{F} \cdot \nabla \delta}{\rho c} = \frac{(\delta_P - \delta)\phi}{c}. \quad (9)$$

Equations (7) and (8) (for c and $c\delta$) include the usual time derivative and advection terms (the bracketed terms on the left-hand side), flux divergence terms, and source terms. Under steady, horizontally homogeneous conditions, both equations reduce to balances between vertical flux divergence and source density, so that (in the case of Eq. (7)) $dF/dz = \rho\phi$, where z is height and $\mathbf{F} = (0, 0, F)$. Equation (9) (for δ) also includes familiar time derivative, advection, and source terms, but now the source term is nonzero only if $\delta_P - \delta \neq 0$, that is, if the scalar transferred across plant surfaces has an isotopic composition different from the air. The other significant term in Equation (9) is $(\mathbf{F} \cdot \nabla\delta)/(\rho c)$, the counterpart of the flux divergence term in Eqs. (7) and (8). It represents the contribution to local changes in δ from the interaction of a scalar flux $\mathbf{F}$ with a gradient in δ and is not a divergence. It balances the source term under steady, horizontally homogeneous conditions, so $F(d\delta/dz) = (\delta_P - \delta)\rho\phi$.

2.3 Source Terms for CO$_2$

At this point it is useful to identify the isotopic source terms more precisely. If the scalar is CO_2, the source density for CO_2 mole fraction (ϕ) has an assimilation component (ϕ_A) and a respiration component (ϕ_R), so $\phi = \phi_A + \phi_R$. The sign convention that ϕ is positive into V (that is, into the air) is retained, so $\phi_A < 0$ and $\phi_R > 0$. The source terms in Eqs. (8) and (9), for $c\delta$ and δ, then become

$$\delta_P\phi = \delta_A\phi_A + \delta_R\phi_R$$

$$\phi(\delta_P - \delta) = \phi_A(\delta_P - \delta) + \phi_R(\delta_R - \delta), \tag{10}$$

where δ_A is the isotopic composition of the CO_2 assimilated into plants by current photosynthesis, and δ_R is the isotopic composition of respired CO_2. The iso-source $\delta_P\phi$ (the source density for the iso-concentration $c\delta$) is therefore the sum of contributions from assimilation and respiration, each the product of a CO_2 flux and the isotopic composition of the CO_2 exchanged (transported across the bounding surface S_P) by the flux.

For the assimilation component, δ_A can be quantified in terms of the discrimination Δ or Δ^*, defined by

$$\Delta = \frac{R_{reactant}}{R_{product}} - 1 = \frac{\delta_{reactant} - \delta_{product}}{1 + \delta_{product}};$$

$$\Delta^* = 1 - \frac{R_{product}}{R_{reactant}} = \frac{\delta_{reactant} - \delta_{product}}{1 + \delta_{reactant}}, \tag{11}$$

where Δ is the definition of Farquhar and Richards (1984), convenient for calculating discrimination by a series of sequential processes, and Δ^* is an alternative definition, convenient for processes acting in parallel (Farquhar et al., 1989a, Appendix Part III). The two are related by $\Delta^* = \Delta/(1 + \Delta)$. For the present case, the product is the carbon fixed by current photosynthesis (composition δ_A) and the reactant is the CO_2 in the air at the point $\mathbf{x}$ (composition δ), so that

$$\Delta = \frac{\delta - \delta_A}{1 + \delta_A}; \quad \Delta^* = \frac{\delta - \delta_A}{1 + \delta}; \quad \delta_A = \delta - \Delta^* - \delta\Delta^*. \tag{12}$$

Neglecting second-order ($\Delta\delta$, Δ^2) and higher-order terms, the last of Eq. (12) simplifies to $\delta_A \approx \delta - \Delta^*$. This approximate expression, or the exact Eq. (12), specifies δ_A in terms of Δ^* and the air isotopic composition δ. The source terms in Eq. (10) now become

$$\delta_P\phi = \phi_A(\delta - \Delta^* - \delta\Delta^*) + \phi_R\delta_R$$
$$\approx \phi_A(\delta - \Delta^*) + \phi_R\delta_R$$

$$\tag{13}$$

$$\phi_A(\delta_P - \delta) = -\phi_A(\Delta^* + \delta\Delta^*) + \phi_R(\delta_R - \delta)$$
$$\approx -\phi_A\Delta^* + \phi_R(\delta_R - \delta)$$

The discrimination Δ^* (rather than Δ) is appropriate because surface elements or sources act in parallel rather than in series to influence ambient atmospheric isotopic composition.

It remains to specify δ_R and Δ^*. For the respiration component, δ_R can be taken as the isotopic composition of the plant material, provided that discrimination during respiration can be ignored (Lloyd et al., 1996; Lin and Ehleringer, 1997), and discrimination on translocation of carbon within the plant (such as export of assimilated carbon from leaves) can also be ignored. For the assimilation component, net discrimination during photosynthesis depends on whether the photosynthetic pathway is C$_3$ or C$_4$ (Lloyd and Farquhar, 1994) and involves different mechanisms for discrimination against ^{13}C in CO_2 (Farquhar et al., 1989a, b) and against ^{18}O in CO_2 (Farquhar et al., 1993). For C$_3$ photosynthetic discrimination against ^{13}C, Δ is around about 20‰, decreasing with increasing water use efficiency (Farquhar and Richards, 1984; Farquhar et al., 1989b), and Δ^* is about 0.4‰ smaller. For C$_4$ photosynthesis, discrimination against ^{13}C is much lower with typical Δ values around 4–5‰.

2.4 Single-Point Lagrangian Equations

To identify the Lagrangian or fluid-following forms of the balance equations, we consider an arbitrary scalar entity with concentration $a(\mathbf{x}, t)$ and source density $\varphi(\mathbf{x}, t, a)$ which may depend on the scalar concentration a. Here a can stand for any of c, cR, or $c\delta$ (but not R or δ, for reasons given shortly). Table 2 gives the source density φ for each choice of a. Imagine an ensemble of realizations of the turbulent flow so that the instantaneous concentration a^ω, the instantaneous turbulent velocity $\mathbf{u}^\omega$, and the instantaneous source density $\varphi^\omega = \varphi(\mathbf{x}, t, a^\omega)$ are variables which differ randomly among realizations (distinguished by a superscript ω). The relationship between a^ω, $\mathbf{u}^\omega$, and φ^ω is given by the instantaneous form of Eq (7) in which there is no Reynolds decomposition to produce eddy fluxes, so that the flux $\mathbf{F}^\omega$ accounts for scalar transfer by molecular diffusion only. In the body of the fluid, molecular fluxes are negligible relative to fluxes due to fluid motion and can safely be ignored (the high Péclet number approximation), while at source surfaces (S_P), molecular fluxes are described by the source density. Under these conditions, the instantaneous equation (7) is

TABLE 2 Choices of the Arbitrary Concentration a and Source Density φ for the Scalars c, cR, $c\delta$, R, and δ

Single-Point Eulerian Equation	Balance of	a	φ	Lagrangian Methods Applicable?
(7)	Scalar mole fraction	c	ϕ	Yes
(8) with $\delta \rightarrow R$	Mole fraction of minor isotope	cR	$R_P\phi$	Yes
(8)	Iso-concentration	$c\delta$	$\delta_P\phi$	Yes
(9) with $\delta \rightarrow R$	Isotopic ratio	R	$\phi(R_P-R)/c$	No
(9)	Isotopic composition	δ	$\phi(\delta_P-\delta)/c$	No

The last column indicates the applicability of the Lagrangian equation (16) and its canopy version, Eq. (17).

$$\frac{Da^\omega}{Dt} = \frac{\partial a^\omega}{\partial t} + \mathbf{u}^\omega \cdot \nabla a^\omega = \varphi(\mathbf{x}, t, a^\omega), \qquad (14)$$

where D/Dt denotes the material derivative, or derivative following the motion.

The average scalar concentration field is found in the Lagrangian approach by tracking the motion of "marked fluid particles", connected parcels of fluid containing many molecules but small enough to be regarded as single points from the standpoint of the continuum turbulent flow. Equation (14) can be integrated along the wandering path $\mathbf{X}^\omega(t)$ of a single fluid particle to give

$$a^\omega(t_1) - a^\omega(t_0) = \int_{t_0}^{t_1} \varphi(\mathbf{X}^\omega, t)\, dt, \qquad (15)$$

which says that the concentration in the fluid particle changes in response to the source densities it encounters along its path. A fluid passing through the space–time point $(\mathbf{x}_0, t_0)$ is therefore "marked" with a in proportion to the source density $\varphi(\mathbf{x}_0, t_0)$, and then contributes to the concentration $a(\mathbf{x}, t)$ at other points $(\mathbf{x}, t)$ according to the transition probability $P(\mathbf{x}, t \,|\, \mathbf{x}_0, t_0)$, the probability that fluid motion transports the particle from $(\mathbf{x}_0, t_0)$ to $(\mathbf{x}, t)$. By considering a large number of fluid particles and taking an ensemble average, it can be shown (see Appendix B) that the mean concentration $a(\mathbf{x}, t)$ is given by

$$a(\mathbf{x}, t) = \int_{(t_0 < t)} \int_{(V)} P(\mathbf{x}, t \,|\, \mathbf{x}_0, t_0)\, \varphi(\mathbf{x}_0, t_0, a_0)\, d\mathbf{x}_0\, dt_0. \qquad (16)$$

The quantities a, φ, and P are all ensemble-averaged, an operation which is the same as the more familiar time average in a stationary (statistically steady) flow, but not in a nonstationary flow.

The transition probability $P(\mathbf{x}, t \,|\, \mathbf{x}_0, t_0)$ carries all information about the velocity field that is needed to deduce $a(\mathbf{x}, t)$. In practice there are three main ways of determining P: first, one may obtain an exact or approximate analytic solution of the stochastic differential equations for the velocities of an ensemble of marked fluid particles. This approach was the basis of the Taylor (1921) solution for scalar dispersion in homogeneous turbulence and has been applied in vegetation canopies (Raupach, 1989a). A second approach is "random-flight" simulation, in which the stochastic differential equations for the marked particle velocities are solved by numerically constructing an ensemble of particle trajectories (Thomson, 1984, 1987; Sawford and Guest, 1987; Baldocchi, 1992; Katul et al., 1997). Third, one may obtain P from an Eulerian model for the velocity field, using higher-order closure or other methods. Examples of higher-order closure models in vegetation canopies include Wilson and Shaw (1977), Wilson (1988), Katul and Albertson (1998), Ayotte et al. (1999), and Massman and Weil (1999).

The arbitrary scalar entity a can be any conserved entity satisfying the superposition principle (that if source densities φ_1 and φ_2 produce concentration fields a_1 and a_2, then the source density $\varphi_1 + \varphi_2$ produces the concentration field $a_1 + a_2$). This is true for the scalars c, cR, and $c\delta$, but not for R (isotopic ratio) or δ (isotopic composition), because these scalars are ratios of the concentrations of the minor and major isotopes and so have a nonlinear dependence on sources. Hence, Eq. (16) cannot be applied to R or δ (see Table 2).

The source density φ can depend on the concentration a at the point $(\mathbf{x}, t)$, provided that φ is statistically independent of the wind field (see Appendix B). Such a dependence, so that $\varphi = \varphi(\mathbf{x}, t, a)$, is indicated in Eq. (16). In this case Eq. (16) becomes an integral equation in a which must be solved by recursive or other means, rather than an explicit solution for a. This issue does not arise when the source density is specified independently, for instance, by the locations and strengths of point sources of air pollutants. However, it is crucial in most biogeochemical applications, because the source or sink densities for entities such as heat, water vapor, CO_2, and aerosols depend on the ambient concentrations of those entities. The inclusion of this dependence in Eq. (16) is considered in Appendix B, and the practical implications are further discussed in Section 3.5.

3. Inverse Methods for Inferring Scalar Sources and Sinks in Canopies

3.1 General Principles

Under steady conditions in a uniform, horizontally homogeneous vegetation canopy, the scalar source density $\phi(z)$ and concentration $c(z)$ are functions only of height z. Equation (16) (with $a = c$ and $\varphi = \phi$, from Table 2) can then be written in the discrete form (Raupach, 1989b),

$$c_i - c_r = \sum_{j=1}^{m} D_{ij} \phi_j \Delta z_j, \tag{17}$$

where ϕ_j is the source density in canopy layer j, Δz_j is the thickness of layer j, m is the total number of source layers in the canopy, c_i is the concentration at height z_i, c_r is the concentration at a reference height z_r above the canopy, and D_{ij} is the dispersion matrix, with n rows ($i = 1$ to n) corresponding to concentration measurement heights, and m columns ($j = 1$ to m) corresponding to source layers. The lowest source layer includes the ground, so that $\phi_1 \Delta z_1$ is the sum of the scalar fluxes from the ground and the lowest canopy layer. Once D_{ij} is known, inversion of Eq. (17) provides a solution to the inverse problem of calculating sources ϕ_j from measured concentrations c_i. The linearity of Eq. (17) means that this solution is unique.

The dispersion matrix D_{ij} is a discrete form of the transition probability P in Eq. (16) and thus carries all the required information about the velocity field. Its elements have the dimension of aerodynamic resistance (s m^{-1}). The elements of column j of D_{ij} are found by considering $\phi_j \Delta z_j$ to be a steady unit source, with sources in all other canopy layers set to zero. A theory of turbulent dispersion is used to calculate the concentration field $c(z)$ resulting from this source distribution. The elements of column j of D_{ij} are then given by

$$D_{ij} = \frac{c_i - c_r}{\phi_j \Delta z_j} \quad (\text{all other } \phi \Delta z = 0). \tag{18}$$

The heights of the sources (Z_j) and the concentrations (z_i) need not be the same in general.

Canopy-scale atmospheric inverse methods are all based explicitly or implicitly on Eq. (17). There are three main methods in this class, distinguished by the turbulent dispersion theory used to calculate D_{ij} and aligning with the three means outlined in Section 2.4 for obtaining the transition probability P. Because all three rely on Eq. (17) and its Lagrangian foundation, the label "inverse Lagrangian" is appropriate in all three cases. The methods are (1) approximate analytic solution of Lagrangian equations for the velocities of an ensemble of marked fluid particles, leading to "localized near field" (LNF) theory (see below) and thence to an analytic specification for D_{ij}; (2) random-flight numerical solution of stochastic differential equations for marked-

particle velocities, leading to numerical specification of D_{ij} (Baldocchi, 1992; Katul *et al.*, 1997); and (3) definition of D_{ij} by the solution of an Eulerian model for the velocity field and scalar transfer (Katul and Albertson, 1998; Massman and Weil, 1999; Katul *et al.*, 2000). Specific calculations in the following use method (1) to find D_{ij}.

3.2 Localized Near Field Theory

LNF theory (Raupach, 1989a, b) is a semi-Lagrangian theory which provides an approximate means of calculating the concentration profile $c(z)$ from a given source density profile $\phi(z)$, recognizing the large-scale, coherent nature of turbulent eddies in vegetation canopies. Because of dominant role of these eddies, which have length scales on the same order as the strong shear layer at the top of the canopy (Raupach *et al.*, 1996), the vertical turbulent transfer of scalars in the canopy cannot be described by gradient-diffusion theory. A clear indication of the failure of gradient-diffusion in canopies is provided by observations of counter-gradient or zero-gradient vertical fluxes of heat, water vapor, and CO_2 in forest canopies (Denmead and Bradley, 1987).

The theory centers on the evaluation of the transition probability $P(\mathbf{x}, t | \mathbf{x}_0, t_0)$ in Eq. (16). The basic idea is that a cloud of marked fluid particles dispersing from an instantaneous point source at $(\mathbf{x}_0, t_0)$ can be regarded as undergoing random motions (in the sense that particle position is a Markov process) in the "far field" when the travel time $(t - t_0)$ is large compared with the Lagrangian time scale of the turbulence (T_L). By contrast, in the "near field" when $t - t_0$ is comparable with or smaller than T_L, the dispersion of the cloud is governed by the persistence of the Lagrangian velocities of marked fluid particles close to the source (Taylor, 1921). Therefore, the dispersion of the cloud of marked particles (or of the scalar with which the particles are marked) is described by gradient-diffusion theory in the far field but not in the near field. Scalar sources in a vegetation canopy are spread throughout the canopy volume, so concentrations at any point $\mathbf{x}$ are made up of superposed contributions from sources at all travel times, including both far-field and near-field. Contributions from the latter are responsible for the observed failure of gradient-diffusion theory, including counter-gradient fluxes. A formal theory can be developed by splitting the transition probability $P(\mathbf{x}, t | \mathbf{x}_0, t_0)$ into a diffusive "far-field" part P_F and a nondiffusive "near-field" part P_N, so that $P = P_F + P_N$. The diffusive part P_F is described by a diffusion equation at all travel times. Two postulates are then made: first, that $P \rightarrow P_F$ at large travel times $t - t_0$, and second, that $P_N = P - P_F$ can be described by its value in locally homogeneous turbulence. The motivation for the second postulate is that for the small travel times for which P_N is significant ($t - t_0$ comparable with or less than T_L) the particles are still quite close to the source and are in a turbulence field approximately the same as that at the source.

Corresponding to the partition $P = P_F + P_N$ of the transition probability, the concentration c can be broken into two parts: $c = c_F + c_N$. Since the far-field part c_F obeys a gradient-diffusion relationship between flux and concentration, it satisfies that

$$F(z) = -\rho K_F(z)\frac{dc_F}{dz} \qquad K_F(z) = \sigma_w^2(z)T_L(z), \qquad (19)$$

where $\sigma_w(z)$ is the standard deviation, $T_L(z)$ is the Lagrangian time scale of the vertical velocity, K_F is the far-field eddy diffusivity, F is the vertical eddy flux of the scalar, and z is height above the ground surface ($z = 0$). The expression for K_F is the result of Lagrangian and Eulerian analyses in homogeneous turbulence (Taylor, 1921; Batchelor, 1949). The scalar flux $F(z)$ is related to the source density by Eq. (7), which simplifies under steady, horizontally homogeneous conditions to

$$\frac{dF}{dz} = \rho\phi, \qquad F(z) = F_0 + \int_0^z \phi(z)dz, \qquad (20)$$

where F_0 is the flux at the ground surface. The flux at the top of the canopy is $F_h = F(h)$, and a concentration scale can be defined as $c_* = F_h/u_*$.

The near-field part c_N of the concentration accounts for the nondiffusive character of near-field dispersion. It is given by

$$c_N(z) = \int_0^\infty \frac{\phi(z')}{\sigma_w(z')}\left[k_N\left(\frac{z - z'}{\sigma_w(z')T_L(z')}\right)\right.$$
$$\left. + k_N\left(\frac{z + z'}{\sigma_w(z')T_L(z')}\right)\right]dz', \qquad (21)$$

where k_N is the "near-field kernel," a weighted version of P_N which can be calculated from the theory of dispersion in homogeneous turbulence. A good approximation (Raupach, 1989a) is

$$k_N(\zeta) = c_1\ln(1 - e^{-|\zeta|}) + c_2 e^{-|\zeta|}$$

$$\left[c_1 = -1/\sqrt{2\pi} = -0.39894,\right.$$

$$\left.c_2 = \frac{1}{2} - \frac{\pi^2}{6\sqrt{2\pi}} = -0.15623\right]. \qquad (22)$$

These equations provide a complete solution to the forward problem of calculating $c(z)$ from a given $\phi(z)$ in a uniform canopy. The required turbulence properties are canopy profiles of the standard deviation $\sigma_w(z)$ and the Lagrangian time scale $T_L(z)$ for vertical velocity.

3.3　The Dispersion Matrix

The specification of D_{ij} is given here in more detail than in previous descriptions. From Eq. (18), D_{ij} may be split into far-field term and near-field terms in the same way as c,

$$D_{ij} = D_{ij}^{(F)} + D_{ij}^{(N)}, \qquad (23)$$

where all elements are functions only of $\sigma_w(z)$ and $T_L(z)$. These elements can be determined as follows: Let concentration measurement heights be z_i, so $c_i = c(z_i)$. For source layers, let z_j be the top of layer j and $Z_j = (z_{j-1} + z_j)/2$ the center of layer j, with $j = 0$ at the ground so $z_0 = Z_0 = 0$. The far-field part of D_{ij} is given by Eqs. (18) to (20):

$$D_{ij}^{(F)} = \int_{\max(z_i, Z_j)}^{z_r} \frac{dz'}{\sigma_w^2(z')T_L(z')} \qquad (24)$$

The near-field part needs to be calculated carefully because of the logarithmic singularity of k_N at $\zeta = 0$. If z_i is not between z_{j-1} and z_j (that is, not within source layer j), then the singularity is not a problem. By replacing $c_i - c_r$ with $c_N(z_i) - c_N(z_r)$ in Eq. (18), using Eq. (21) to find c_N, and evaluating the integrals with rectangular approximations, $D_{ij}^{(N)}$ is found to be

$$D_{ij}^{(N)} = \frac{1}{\sigma_{wj}}\left[k_N\left(\frac{z_i - Z_j}{\sigma_{wj}T_{Lj}}\right) + k_N\left(\frac{z_i + Z_j}{\sigma_{wj}T_{Lj}}\right) - k_N\left(\frac{z_r - Z_j}{\sigma_{wj}T_{Lj}}\right)\right.$$
$$\left. - k_N\left(\frac{z_r + Z_j}{\sigma_{wj}T_{Lj}}\right)\right], \qquad (25)$$

for z_i not in (z_{j-1}, z_j)

where $\sigma_{wj} = \sigma_w(Z_j)$ and $T_{Lj} = T_L(Z_j)$. However, if z_i is between z_{j-1} and z_j (within source layer j), then the first term in this expression must be replaced. We define $I_N(\zeta)$ as the integral of $k_N(\zeta)$ from 0 to ζ, and evaluate it using a small-ζ expansion (Raupach, 1989a, Eq. (A10)). This gives

$$I_N(\zeta) = \int_0^\zeta k_N(\zeta')d\zeta' \approx c_3\zeta((\ln\zeta) - 1) + c_4\zeta + \frac{c_5\zeta^2}{2} + \cdots$$

$$[c_3 = c_1 = -0.39894, c_4 = -0.14127, c_5 = 0.33333]. \qquad (26)$$

Then, $D_{ij}^{(N)}$ can be evaluated as

$$D_{ij}^{(N)} = \frac{T_{Lj}}{\Delta z_j}\left[I_N\left(\frac{z_i - z_{j-1}}{\sigma_{wj}T_{Lj}}\right) + I_N\left(\frac{z_j - z_i}{\sigma_{wj}T_{Lj}}\right)\right] +$$
$$+ \frac{1}{\sigma_{wj}}\left[k_N\left(\frac{z_i + Z_j}{\sigma_{wj}T_{Lj}}\right) - k_N\left(\frac{z_r - Z_j}{\sigma_{wj}T_{Lj}}\right)\right.$$
$$\left. - k_N\left(\frac{z_r + Z_j}{\sigma_{wj}T_{Lj}}\right)\right] \qquad (27)$$

for z_i in (z_{j-1}, z_j).

Figure 1 illustrates the dispersion matrix by plotting the elements D_{ij}, normalized as $D_{ij}u_*$ where u_* is the friction velocity.

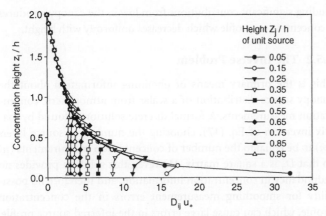

FIGURE 1 The normalised dispersion matrix D_{ij} u_*. Each profile is a column of the matrix, equal to the concentrations $c_i - c_r$ at heights z_i ($i = 1$ to n) produced by a unit source in the layer centered at Z_j ($j = 1$ to m). The number of concentration heights (n) is 20, and the number of source layers (m) is 10. The reference height z_r is $2h$. Turbulence profiles $\sigma_w(z)/u_*$ and $T_L(z)u_*/h$ are from Eqs. (28) and (29), with $c_{sw} = 1.6$, $a_{3(h)} = 1.1$, $a_{3(i)} = 1.25$, $c_{TL} = 0.3$, $z_1/h = 0.2$, $d/h = 0.75$.

Each profile is the set of concentrations $c_i - c_r$ ($i = 1$ to n) produced by a unit source in the layer centered at Z_j ($j = 1$ to m). The near-field term $D_{ij}^{(N)}$ is responsible for the peak in each profile at the level of the unit source.

3.4 Turbulent Velocity Field

The turbulence properties $\sigma_w(z)$ and $T_L(z)$ can be specified with the aid of the observation that, for thermally neutral flow within and just above a vegetation canopy, the profiles of $\sigma_w(z)$ and $T_L(z)$ are largely governed by a single velocity scale and a single length scale (Raupach, 1988; Kaimal and Finnigan, 1994; Raupach *et al.*, 1996). Such a scaling is suggested by the mixing-layer analogy for flow within and just above the canopy (Raupach *et al.*, 1996), which proposes that the turbulence structure in the strong shear layer near the top of the canopy is patterned on a plane mixing layer rather than on a boundary layer. This hypothesis implies that the velocity scale for the canopy turbulence is the friction velocity u_*, a measure of the turbulent momentum flux to the canopy, and that the length scale is the thickness of the mixing layer. This can be characterized by the length $L_s = u(h)/u'(h)$ (where u is the mean flow velocity, h is the canopy height, and $u'(z) = (du/dz)$), which is quite well determined by $h - d$ (where d is the zero-plane displacement of the canopy). Data from a wide range of canopies show that $L_s/(h - d)$ is close to 2. Also, d/h is well constrained (Raupach, 1994),[1] being about 0.75 for typical canopies but depending slowly on the leaf area index and its profile with height in the canopy. Therefore L_s can also be related to h, and in typical

canopies, L_s/h is about 0.5. However, the ratio L_s/h shows more variation with canopy density than does $L_s/(h - d)$.

Several empirical forms have been proposed for $\sigma_w(z)$. A typical choice is

$$\frac{\sigma_w(z)}{u_*} = \begin{cases} a_{3(h)} \exp(c_{sw}(z/h - 1)) & \text{for } z \leq h \\ a_{3(h)} + (a_{3(i)} - a_{3(h)})\dfrac{(z - h)}{z_{ruff} - h} & \text{for } h < z \leq z_{ruff}, \\ a_{3(i)} & \text{for } z_{ruff} < z \end{cases}$$

(28)

where c_{sw} is an empirical constant (typically around 1.5), $a_{3(h)}$ (typically about 1.1) is the ratio σ_w/u_* at the top of the canopy, $a_{3(i)}$ (typically about 1.25) is the ratio σ_w/u_* in the inertial sublayer above the roughness sublayer, and z_{ruff} is the height of the roughness sublayer (the layer just above the canopy within which the mixing-layer analogy applies). This profile assumes that σ_w/u_* takes a slightly lower value at the top of the canopy than in the inertial sublayer well above the canopy, as implied by the mixing-layer analogy and observed in practice (Raupach *et al.*, 1996). Within the canopy, σ_w/u_* is assumed to have an exponential form.

For $T_L(z)$, the principle of the parameterization is that u_*T_L/L_s is constant with height within the canopy (except very close to the ground) and in the roughness sublayer. This is a consequence of the mixing-layer analogy. Constancy of u_*T_L/L_s leads to two alternative parameterizations

$$\frac{u_*T_L}{h} = c_{TL}, \qquad \frac{u_*T_L}{h - d} = c'_{TL},$$

(29)

depending on whether h or $h - d$ is used as a surrogate for the shear length scale L_s. Appropriate values for the constants are $c_{TL} = 0.3$ and $c'_{TL} = 1.2$. As argued above, the mixing-layer analogy implies that the second form (involving $(h - d)$) is physically preferable, but it carries the penalty that d as well as h must be determined or estimated. Equation (29) is applicable in the range $z_1 < z < z_{ruff}$, where z_1 is a low level in the canopy (typically about $0.25h$) below which T_L decreases with proximity to the ground. Above z_{ruff}, T_L is given by standard inertial-sublayer expressions and increases linearly with $z - d$ under thermally neutral conditions.

The height of the roughness sublayer, z_{ruff}, can be estimated by noting that the far-field eddy diffusivity K_F is given everywhere by $\sigma_w^2 T_L$ (Eq. (19)). In particular, within the roughness sublayer ($z \leq z_{ruff}$), Eq. (29) implies that $K_F = c_{TL}h\sigma_w^2/u_*$. Also, K_F in the neutral inertial sublayer ($z \geq z_{ruff}$) is given by $K_F = \kappa u_*(z - d)$, where κ is the von Karman constant (assumed to be 0.4). Equating these two estimates for K_F at z_{ruff}, it is seen that $z_{ruff} = d + (a_{3(i)}^2/\kappa)c_{TL}h$, where $a_{3(i)} = \sigma_w/u_*$ in the inertial sublayer. Typically, z_{ruff} is about $2h$.

Recent works by Leuning *et al.* (2000) and Leuning (2000) have led to improvements in these parameterizations in two respects. First, smoothing of the slope discontinuities in $\sigma_w(z)$ and $T_L(z)$ (implied by Eqs. (28) and (29)) produces more stable behavior in source distributions inferred from concentration profiles by

---

[1] In Fig. 1 of Raupach (1994), the horizontal axis is incorrectly labeled as Λ (Leaf Area Index). It should be λ (Frontal Area Index). The relationship assumed in that paper between the two is $\Lambda = 2\lambda$.

inversion of Eq. (17). Second, Leuning (2000) introduced thermal stability into the parameterizations of $\sigma_w(z)$ and $T_L(z)$ by assuming that the stability parameter throughout the roughness sublayer is h/L_{MO} (where L_{MO} is the Monin-Obhukov length) and that the effect of this on T_L below z_{ruff} can be described by the factor $[K_H(h/L_{MO})\sigma_w{}^2(0)]/[K_H(0)\sigma_w{}^2(h/L_{MO})]$, where K_H is a scalar eddy diffusivity. Standard, inertial-sublayer forms were used to describe the stability dependencies of K_H and $\sigma_w{}^2$.

3.5 Solutions for Forward, Inverse, and Implicit Problems

Having specified the turbulence properties σ_w and T_L and thence the dispersion matrix D_{ij}, the apparatus is now in place to use Eq. (17) to solve three generic kinds of problem: the forward problem of determining the scalar concentration profile $c(z)$ from a specified source density profile $\phi(z)$, the inverse problem of determining $\phi(z)$ from specified or measured information about $c(z)$, and the implicit or coupled problem of determining both $c(z)$ and $\phi(z)$ together when ϕ is a given function of c.

3.5.1 The Forward Problem

Of the three kinds of problem, this is the easiest theoretically but its main practical use is as a stepping stone in the solution of the other two problems. The solution in discrete form is given directly by Eq. (17). Sample results are shown in Figure 2, which shows the concentration profiles calculated from Eq. (17) (with D_{ij} specified using Eqs. (24) to (27)) for three assumed source profiles. A sharp, strongly peaked elevated source profile produces a peak in the concentration profile and hence a countergradient flux in the layer just below the peak where the flux is upward but dc/dz is positive, because of the strongly localized near-field contribution to the concentration field. By contrast, a more uniform source profile, in-

cluding significant contributions from low in the canopy, produces a concentration profile which decreases uniformly with height.

3.5.2 The Inverse Problem

This is the primary means of obtaining information about the canopy source distribution of a scalar from atmospheric concentration measurements. A formal discrete solution is found by matrix inversion of Eq. (17), choosing the number of source layers (m) to be equal to the number of concentration measurements (n) so that D_{ij} is a square matrix. However, this solution provides no redundancy in concentration information, and therefore no possibility for smoothing measurement errors in the concentration profile, which can cause large errors in the inferred source profile. A simple means of overcoming this problem is to include redundant concentration information, and then find the sources ϕ_j which produce the best fit to the measured concentrations c_i by maximum-likelihood estimation. By minimizing the squared error between measured values and concentrations predicted by Eq. (17), ϕ_j is found (Raupach, 1989b) to be the solution of m linear equations

$$\sum_{k=1}^{m} A_{jk}\phi_k = B_j \quad \text{with} \quad \begin{cases} A_{jk} = \sum_{i=1}^{n} D_{ij}\Delta z_j D_{ik}\Delta z_k \\ B_j = \sum_{i=1}^{n} (c_i - c_r)_{(means)}D_{ij}\Delta z_j. \end{cases} \quad (30)$$

This solution uses a nonsquare dispersion matrix in which $n > m$, to obtain the necessary redundancy in concentration information. The solution is valid whether or not ϕ is dependent on c, because it determines the values of ϕ consistent with the current c field. The result is the source density profile ϕ_j in a (small) number of canopy layers, with the lowest layer including the ground source (F_0). The total flux from the canopy is also obtained as the sum of $\phi_j\Delta z_j$ over all layers.

Figure 3 shows a test of the sensitivity of this method to two key factors, the size of D_{ij} (determined by the number of concentration measurements, n, and source layers, m) and the presence or absence of the near-field term in $D_{ij}{}^{(N)}$ in the dispersion matrix D_{ij}. Figure 3a shows the assumed concentration field, based on forward calculation of the concentration profile c_i ($i = 1$ to n) from a specified source profile using Eq. (17), with D_{ij} as in Figure 1, and with $n = 20$, $m = 10$. This concentration field is then used to reconstruct the source profile $\phi(z)$ with Eq. (30), and the flux $F(z)$ from Eq. (20), under three scenarios: (1) $(n,m) = (20,10)$, with $D_{ij}{}^{(N)}$ included; (2) $(n,m) = (10,5)$, with $D_{ij}{}^{(N)}$ included; and (3) $(n,m) = (10,5)$, with $D_{ij}{}^{(N)}$ omitted. Figures 3b and 3c respectively show the inferred profiles ϕ_j and F_j at layers centered on heights Z_j. Scenario (1) represents optimum information, with (n, m) identical with that used in the forward calculation of c_i. Not surprisingly, the inverse method exactly recovers the profile of ϕ_j initially assumed for the forward calculation, shown as a heavy line in Figures 3b and 3c. Scenarios (2) and (3) attempt to recover the initially assumed source profile with progressively more degraded

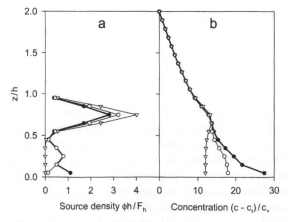

FIGURE 2 (a) Assumed normalized source profiles $\phi(z)h/F_h$ and (b) predicted normalized concentration profiles $(c - c_r)/c_*$ (where $c_* = F_h/u_*$), illustrating a typical solution to the forward problem. Turbulence profiles $\sigma_w(z)/u_*$ and $T_L(z)u_*/h$ are from Eqs. (28) and (29), with $c_{sw} = 1.6$, $a_{3(h)} = 1.1$, $a_{3(i)} = 1.25$, $c_{TL} = 0.3$, $z_1/h = 0.2$, $d/h = 0.75$. The dispersion matrix is as in Figure 1.

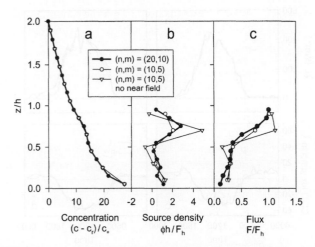

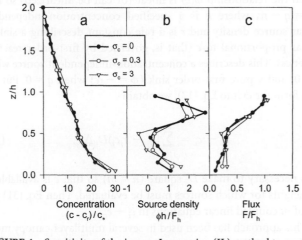

FIGURE 3 Sensitivity of the inverse Lagrangian (IL) method to the number of concentration measurements (n) and source points (m), and to the presence or absence of the near-field term in the dispersion matrix. The dispersion matrix is as in Figure 1. (a) Assumed normalized concentration field $(c - c_r)/c_*$; (b) normalized source profile $\phi(z)h/F_h$; (c) normalized flux profile $F(z)/F_h$. Scenarios are: (1) $(n, m) = (20, 10)$, with $D_{ij}^{(N)}$ included; (2) $(n, m) = (10, 5)$, with $D_{ij}^{(N)}$ included; (3) $(n, m) = (10, 5)$, with $D_{ij}^{(N)}$ omitted.

FIGURE 4 Sensitivity of the inverse Lagrangian (IL) method to errors in concentration measurements. To each c_i is added a random error, uniformly distributed over the range $[-\sigma_c c_*, \sigma_c c_*]$. Scenarios are: (1) $(n, m) = (20, 10)$ and $\sigma_c = 0$; (2) $(n, m) = (10, 5)$ and $\sigma_c = 0.3$; (3) $(n, m) = (10, 5)$ and $\sigma_c = 3$. Other details are as for Figure 3.

information about the concentration field and the dispersion matrix. In scenario (2), alternate concentration points are omitted and the number of source layers is also reduced to achieve redundancy, so that $(n, m) = (10, 5)$. Scenario (3) uses $(n, m) = (10, 5)$ as in scenario (2), but also uses a degraded dispersion matrix in which the near-field component $D_{ij}^{(N)}$ is omitted, so that the inversion is done by solving Eq. (30) with $D_{ij} = D_{ij}^{(F)}$. Figures 3b and 3c show the effects of these losses in information. Reduction of (n, m) from (20, 10) to (10, 5) in scenario (2) still produces a reasonable approximation to the initially assumed ("correct") source and flux profiles, but simultaneously omitting the near-field term in the dispersion matrix, in scenario (3), produces clearly unphysical source and flux profiles with large overshoots on both profiles.

Figure 4 shows a further test of the effect of imperfect information on the inferred source profiles. This time, random noise is added to the concentration profile before carrying out the inversion with Eq. (30), to test the ability of the method to cope with random errors in concentration measurement. Again, three scenarios are used: the first is identical with scenario (1) above. The other two use $(n, m) = (10, 5)$ and the full D_{ij} as in scenario (2), but also add to each c_i a random error, uniformly distributed over the interval $[-\sigma_c c_*, \sigma_c c_*]$, where σ_c is either 0.3 or 3 (and $c_* = F_h/u_*$). The smaller error produces little effect on the inferred profiles ϕ_j and F_j, but the larger one leads to significant degradation in both profiles.

These illustrative calculations have been carried out with a very simple inversion procedure based on Eq. (30). Rapid recent developments in the application inverse theory offer several possibilities for improvement. First, the use of singular-value decomposition as a formal inversion framework (e. g., Press *et al.*, 1992)

provides means for assessing the effect of measurement errors and guarding against degeneracies in the solution caused by the inability of the measurements to distinguish between alternative different solutions to the inverse problem. Second, Bayesian synthesis approaches (Tarantola, 1987) offer means for making the inversion process more robust through the application of prior constraints. Third, there are possibilities for further improving the robustness of the method by fitting for parameters in an empirical or process-based model for the source profile, rather than for the layer-discretized source–sink profile directly.

3.5.3 The Implicit or Coupled Problem

This problem is important in modeling the canopy concentration (c) and source–sink (ϕ) profiles of entities for which ϕ depends on c, including most biogeochemically significant entities exchanged between vegetation and the atmosphere: heat, water vapor, aerosols, CO_2, other trace gases which react with stomatal or other canopy surfaces, and isotopes (to be discussed in more detail in the next section). In these cases both $\phi(z)$ and $c(z)$ are unknowns, and the prescribed constraints in the problem are the concentration c_r at a reference height above the canopy and forcing or surface parameters determining the relationship $\phi(c)$. The biogeochemical importance of such situations is the reason for careful consideration of concentration-dependent sources in the Lagrangian analysis (Section 2.4 and Appendix B). An Eulerian version of an analogous problem for diffusive flow in the atmospheric surface layer is defined by the scalar conservation equation (7) and mixed or radiation boundary conditions (Philip, 1959). A simple solution for the implicit problem in a canopy can be obtained when the relationship $\phi(c)$ is linear (or can be linearized) so that $\phi = q - rc$, where q is a specified concentration-independent scalar source density and r is a rate constant describing a sink of scalar proportional to c (that is, governed by first-order reaction kinetics). This describes a concentration-independent source when $r = 0$, and a pure first-order sink ($\phi = -rc$) when $q = 0$. Putting this form for ϕ into Eq. (17), we obtain

$$c_i - c_r = \sum_{j=1}^{m} (q_j - r_j c_j) D_{ij} \Delta z_j. \tag{31}$$

It is necessary to make D_{ij} square ($n = m$) so that c is available at all heights for which sources must be evaluated. Then Eq. (31) is a set of m coupled linear equations in $c_i - c_r$.

This approach has been used in several multilayer canopy models for coupled heat and water vapor transfer (Dolman and Wallace, 1991; van den Hurk and McNaughton, 1995; McNaughton and van den Hurk, 1995). The solution in this case can be obtained with a linear recombination of the primary scalar variables (temperature and humidity) into alternative variables (available energy and saturation deficit) which have separated or decoupled boundary conditions (McNaughton, 1976; McNaughton and van den Hurk, 1995; Raupach, 2000). Coupling of this generic kind also occurs for canopy–atmosphere exchanges of aerosols (Davidson and Wu, 1990) and of many trace gases (Hicks et al., 1985).

3.6 Field Tests

A number of experiments have now been carried out to apply or test canopy-scale inverse methods in the field (Raupach et al., 1992; Denmead and Raupach, 1993; Denmead et al., 1997; Katul et al., 1997, 2000; Leuning et al., 2000; Leuning, 2000; Denmead et al., 2000; Harper et al., 2000) with generally good results. Many, but not all, of these have used LNF theory to obtain D_{ij}.

Raupach et al. (1992) and Denmead and Raupach (1993) reported inferences of the source–sink profiles for water vapor and CO_2 in a wheat canopy, comparing the inverse Lagrangian (IL) method (with D_{ij} from LNF theory) with simultaneous measurements by both eddy covariance (EC) and Bowen ratio methods. Good agreement was found (over half-hour intervals through a day) between the total water vapor and CO_2 fluxes from the canopy obtained from all three methods. More importantly, there was also good agreement between the water vapor source from the lowest layer (up to $0.25h$) and measurements of soil evaporation by mini-lysimeters, available as morning and afternoon averages. In a later experiment on a similar wheat canopy over a much longer measurement period of seven weeks, Denmead et al. (1997) reported similarly good agreement between total water vapor fluxes above the canopy measured by IL, EC, and a variance-ratio method (a relative of the Bowen ratio approach using variances rather than gradients of temperature and water vapor). Good agreement was also obtained between IL estimates of the water vapor source below $0.25h$ and measurements from mini-lysimeters.

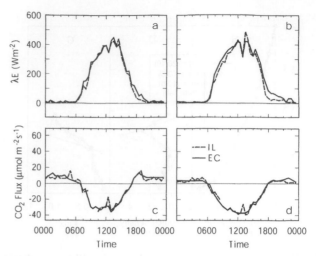

FIGURE 5 Comparison of time series for the upward fluxes of latent heat (λE) and CO_2 above a rice canopy on 8 August 1996 (a and c) and 11 August 1996 (b and d), from the EC and IL methods (after Leuning, 2000).

Recently, Leuning et al. (2000) and Leuning (2000) measured the source–sink profiles for heat, water vapor, CO_2, and methane in a rice canopy in Japan, comparing the IL method with EC measurements above the canopy. As mentioned earlier, two new features were introduced into their application of the IL method through revised assumptions about the profiles of $\sigma_w(z)$ and $T_L(z)$: smoothing of slope discontinuities in both profiles, and introduction of thermal stability into the parameterizations. Excellent agreement was obtained between the IL and EC methods for water vapor and CO_2 fluxes above the canopy under both daytime and nocturnal conditions, as shown in Figure 5 for 2 days. The main effect of introducing the effect of stability was to improve the agreement at night. The source–sink profiles for heat, water vapor, and

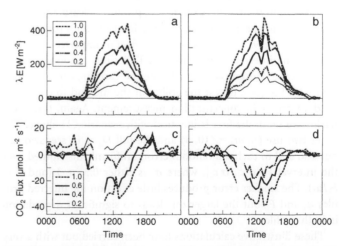

FIGURE 6 Time series for the upward fluxes of latent heat (λE) and CO_2 from the IL method, at levels $z/h = 0.2$, 0.4, 0.6, 0.8, and 1.0 in a rice canopy on 8 August 1996 (a and c) and 11 August 1996 (b and d) (after Leuning, 2000).

CO_2 inferred by the IL method are shown for the same 2 days in Figure 6, by plotting the time course of the inferred flux $F(z)$ at five heights from $0.2h$ to h. These source–sink distributions are in accord with expectations, for instance, in showing CO_2 efflux from the ground surface (greater when the rice paddy was drained on 8 August than when it was flooded on 11 August) and a water vapor source below $0.2h$ of about 20% of the total source (slightly higher when the paddy was flooded than when it was drained). Leuning (2000) compared these inferences for the water vapor and CO_2 source–sink distributions with predictions from a multilayer canopy model incorporating radiation absorption, energy partition, and CO_2 exchanges by photosynthesis and respiration. Broadly good agreement was found, though with differences of detail in that the model predicted higher water vapor fluxes than the IL inferences in the lower canopy, and a negative daytime CO_2 flux (indicating assimilation) in the lowest canopy layer on 11 August.

In another recent experiment, Denmead *et al.* (2000) and Harper *et al.* (2000) measured land–air exchanges of heat and ammonia within and above a corn canopy after application of effluent fertilizer, using the IL and EC methods and an aerodynamic method based on above-canopy gradients. For heat, they found that the bulk of the heat source was in the upper half of the canopy with the lower canopy being a weak heat sink (implying an evaporation rate greater than the available energy in that layer). On average, the total IL heat flux was slightly greater than the EC measurements but exhibited similar trends in time, unlike the aerodynamic estimate which was highly scattered. For ammonia, IL inference of the source–sink distribution showed that the soil was efficient in retaining nitrogen applied by sprinkler irrigation of effluent, but large losses occurred from the foliage. This experiment demonstrates the benefit of incorporating within-canopy as well as above-canopy concentration data in estimates of total fluxes, both because of the extra information and because gradients are larger within than above the canopy. In the case of ammonia, the IL approach provided the only useful estimates.

In summary, these experiments indicate that the IL method can yield practically useful information about canopy source distributions from atmospheric concentration measurements. However, the method depends on several factors, any of which may be limiting: (1) measured concentration profiles of sufficient accuracy and density in the vertical dimension; (2) adequate and sufficiently accurate measurements of the turbulence field; (3) an adequate theory of turbulent transport and dispersion in canopies, to calculate the dispersion matrix; (4) an adequate procedure for doing the inversion itself. Evidence presented in the next subsection indicates that all of these aspects can be combined successfully, though there is undoubtedly room for improvement especially in the latter two factors.

4. Inverse Methods and Isotopes in Canopies

This section considers the application of inverse methods to measurements of the distribution of isotopic composition in the air,

particularly profiles of c and δ in the air within and above vegetation canopies. Examples of such measurements include profiles of $\delta^{13}C$ within and above Amazonian rain forest (Lloyd *et al.*, 1996), of $\delta^{13}C$ and $\delta^{18}O$ above a wheat field (Yakir and Wang, 1996), and of $\delta^{13}C$ and $\delta^{18}O$ above C_3 (alfalfa) and C_4 (corn) canopies (Buchmann and Ehleringer, 1998). The aim is to assess what biogeochemically useful information can be gleaned from such measurements by inverse analysis, both in principle and in practice.

4.1 Path Integrals and Keeling Plots

The starting point is a simple Lagrangian analysis, considering a marked fluid particle as it wanders through a vegetation canopy (or in general, any region V containing sources and sinks of a scalar with major and minor isotopes). The changes in the concentration c and iso-concentration $c\delta$ of the moving fluid particle are given by Eqs. (14) and (15), which show that for a particular realization ω of the turbulent flow, c and $c\delta$ in the fluid particle are given by

$$c_1^\omega - c_0^\omega = \int_{t_0}^{t_1} \phi^\omega dt; \quad \delta_1^\omega c_1^\omega - \delta_0^\omega c_0^\omega = \int_{t_0}^{t_1} \delta_P^\omega \phi^\omega dt, \quad (32)$$

where the subscripts 0 and 1 denote values at two times (t_1 later than t_0). Equation (32) is a path integral, taken over the trajectory of the fluid particle. As before, δ_P is the isotopic composition of the scalar exchanged (released or fixed) by the source ϕ.

Consideration is temporarily restricted to the case where δ_P is constant through the canopy. Taking the ensemble average of Eq. (32) and assuming (reasonably) that the covariance of fluctuations in c and δ can be neglected relative to the mean products $c_1\delta_1$ and $c_0\delta_0$, Eq. (32) yields

$$\delta_1 c_1 - \delta_0 c_0 = \delta_P \int_{t_0}^{t_1} \phi dt = \delta_P(c_1 - c_0) \quad (33)$$

so that

$$\delta_1 = \delta_P + \frac{c_0}{c_1}(\delta_0 - \delta_P). \quad (34)$$

Hence, in the moving fluid particle, δ_1 is linearly dependent on $1/c_1$. This is the well-known Keeling relationship for a mixture of two pools with different isotopic compositions (Keeling, 1961), usually derived for the (very special) case of complete mixing from pools with compositions δ_0 and δ_P (for example, Lloyd *et al.*, 1996; Yakir and Wang, 1996). The present derivation shows that Eq. (34) is a general consequence of the Lagrangian scalar and isotopic balances for fluid particles moving through a region (such as a vegetation canopy or a landscape) in which the exchanged scalar is *uniformly* labeled with isotopic composition δ_P. During daytime, the initial properties c_0 and δ_0 can be regarded as those of the well-mixed convective boundary layer above the surface, and the trajectories of the fluid particles are any which bring air down from this layer to contact the surface. The final points are arbitrary and can be (for example) points on a vertical profile within and/or above the canopy.

The analysis can be extended to consider the separate contributions of assimilation (subscript A) and respiration (subscript R). The source term for c ($= [^{12}CO_2]$) is now $\phi = \phi_A + \phi_R$ and that for $c\delta$ is $\phi_P = \phi_A\delta_A + \phi_R\delta_R$. With δ_A and δ_R constant, Eq. (33) becomes

$$\delta_1 c_1 - \delta_0 c_0 = \delta_A \int_{t_0}^{t_1} \phi_A dt + \delta_R \int_{t_0}^{t_1} \phi_R dt. \qquad (35)$$

Let f_A and f_R be the fractions of the total CO_2 exchange due to assimilation and respiration, respectively, so that for a moving fluid particle, $f_A = (\int \phi_A dt)/(\int \phi dt)$ and similarly for f_R. The definitions imply $f_A + f_R = 1$, with f_A usually larger than 1 and f_R negative (when respiration is smaller in magnitude than assimilation). Using Eq. (32), it follows that

$$\frac{\delta_1 c_1 - \delta_0 c_0}{c_1 - c_0} = \delta_K = \delta_A f_A + \delta_R f_R, \qquad (36)$$

where δ_K is an atmospheric property defined by the slope of the Keeling plot of δ_1 against $1/c_1$, as in Eq. (34). It is conceptually different from δ_P (the isotopic composition of the scalar exchanged by the source) which is a local biogeochemical property of a particular canopy element or set of elements. Only when δ_P is uniform does δ_P equal δ_K. Equation (36) implies

$$f_A = \frac{\delta_K - \delta_R}{\delta_A - \delta_R}, \qquad f_R = \frac{\delta_A - \delta_K}{\delta_A - \delta_R}. \qquad (37)$$

Equation (37) generalizes a relationship by Yakir and Wang (1996), who considered gradients above the canopy. It implies that if δ_A and δ_R are uniform through the canopy (including both plant and soil surfaces), then measurements of δ_A, δ_R, and δ_K can be used to partition the overall CO_2 exchange into assimilation to plants and respiration from soil and plants. Of the three isotopic compositions, δ_K is obtained from profiles of δ and c in the air with Eq. (36), and δ_A and δ_R can be obtained from measurements on plant and soil material in the case of ^{13}C. The situation for ^{18}O is more complicated (Farquhar et al., 1993; Ciais and Meijer, 1998).

4.2 Inverse Lagrangian Analysis of Isotopic Composition

The above analysis has shown that two expressions for interpreting gradients in air isotopic composition, the Keeling plot and the Yakir–Wang expression for distinguishing assimilation and respi-

ration, are consequences of a general Lagrangian argument in the case where δ_A and δ_R (and thence δ_P) are constant through the canopy. This suggests the application of inverse Lagrangian analysis in the situation where δ_A and δ_R are not constant. To approach this question, Eq. (17) is written for the iso-concentration $c\delta$,

$$c_i\delta_i - c_r\delta_r = \sum_{j=1}^{m} D_{ij}\,\delta_{Pj}\,\phi_j\,\Delta z_j, \qquad (38)$$

where $\delta_{Pj} = (\delta_{Aj}\phi_{Aj} + \delta_{Rj}\phi_{Rj})/(\phi_{Aj} + \phi_{Rj})$ is the isotopic composition of the CO_2 exchanged by the source $\phi_j = \phi_{Aj} + \phi_{Rj}$, with the index j representing a layer in the canopy. In principle, Eqs. (38) and (17) can be used together in both forward and inverse modes. The forward calculation is to find the profiles of c_i and δ_i (in the air) implied by specified profiles of ϕ_j and δ_{Pj}. In the inverse mode, ϕ_j and δ_{Pj} can be found from measurements of c_i and δ_i.

When $\delta_{Pj} = \delta_P$, a constant throughout the canopy, Eq. (38) reduces to the linear Keeling relationship, Eq. (34) (with δ_i, c_i, and c_r replacing δ_1, c_1, and c_0, respectively). Hence, when δ_P is uniform, its value can be found from measurements of δ_i (in the canopy air) by a Keeling plot of δ_i against $1/c_i$, from which δ_P is determined by both the intercept and the slope. Conversely, if the uniform value of δ_P is known, then δ_i is exactly predicted by Eq. (34). Isotopic inverse methods in canopies are therefore of value when δ_P is not uniform through the canopy, and the information they contain is related to the departure from a straight-line Keeling plot of δ_i against $1/c_i$.

Exploratory calculations have been carried out for ^{13}C in CO_2, using Eqs. (38) and (17) both in forward mode to find the profiles of c_i and δ_i from specified ϕ_j and δ_{Pj} profiles and in inverse mode to find ϕ_j and δ_{Pj} from specified or measured c_i and δ_i. The assumed scenario, chosen to provide a significant spatial variation of δ_{Pj}, is based on measurements by Buchmann and Ehleringer (1998). They measured $\delta^{13}C$ in a C_3 (alfalfa) canopy, under a farming system which used a C_4 (corn) crop in previous rotations so that a significant proportion of the soil carbon was derived from C_4 material with a much lower $\delta^{13}C$ than the C_3 material in the canopy. The difference arises because of the much lower discrimination against ^{13}C by photosynthesis in C_4 plants than in C_3 plants. Buchmann and Ehleringer measured $\delta^{13}C$ values close to $-28‰$ in the canopy and $-22‰$ in the soil.

To establish the initial specified profile of δ_{Pj} for the present scenario, it was assumed that for plant elements, $\delta_P = \delta_A = \delta_R = -28‰$. This means that departures of δ_A in current photosynthesis from the long-term mean are ignored, and discrimination on respiration is assumed to be zero. For the soil, $\delta_P = \delta_R$ was taken as $-22‰$. A profile for CO_2 source distribution was constructed by assuming a net canopy assimilation of 40 μmol m^{-2} s^{-1} and a soil respiration of 10 μmol m^{-2} s^{-1}, with the canopy CO_2 source density ϕ having a truncated Gaussian distribution in height. The dispersion matrix was based on the same assumptions as used in Figure 1, with $u_* = 0.3$ m s^{-1}. Reference concentrations (at twice the canopy height) were taken as $c_r = 360$ μmol mol^{-1} and

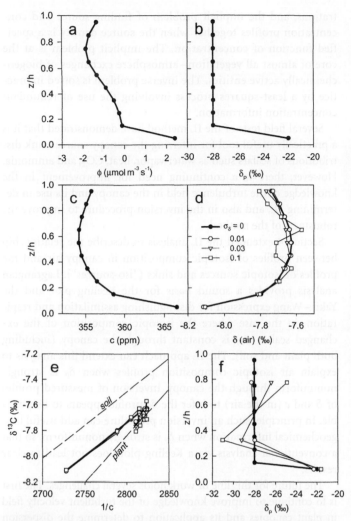

FIGURE 7 Application of IL method to calculate profiles of δ (air) and δ_P (plant) in a canopy, for $^{13}CO_2$ and $^{12}CO_2$. The dispersion matrix is as in Figure 1. (a) Assumed profile of CO_2 source density ϕ (μmol m^{-3} s^{-1}); (b) assumed profile of δ_P; (c) inferred CO_2 concentration profile c, from Eq. (17) in forward mode; (d) inferred profile of δ(air), from Eq. (38) in forward mode (solid symbols), and profiles of δ with added random noise to simulate measurement error (open symbols); (e) Keeling plot of δ against $1/c$, using values in (c) and (d); (f) inverse inference of δ_P (plant), using profiles of δ and c in air from (d) and (c).

$\delta_r = -8‰$. Figures 7a and 7b show the initial assumed profiles of ϕ_j and δ_{Pj}, which were calculated for 10 equispaced layers through the canopy ($m = 10$). The large positive soil respiration source was combined with the small negative canopy source in the lowest layer, accounting for the departure of ϕ_j from a Gaussian shape in that layer.

Figures 7c and 7d show the result of the forward calculation, in which the canopy profiles of c_i (the CO_2 concentration) and δ_i (for ^{13}C) were inferred from ϕ_j and δ_{Pj} at 10 heights coincident with the centeres of the source layers so that $(n, m) = (10, 10)$. As anticipated, c_i is lower through most of the canopy than the above-canopy reference value, because of uptake by assimilation, and δ_i

is less negative than the reference value (more enriched in ^{13}C) because of discrimination against ^{13}C during assimilation. These trends are reversed near the ground because of the soil source of ^{13}C-depleted CO_2. The profiles in Figures 7c and 7d qualitatively reproduce the main features of the profiles of CO_2 concentration and $\delta^{13}C$ measured by Buchmann and Ehleringer (1998).

To test the inverse predictions of ϕ_j and δ_{Pj} from c_i and δ_i, random noise was added to the δ_i profile predicted by the forward calculation. This simulates a random error in the measurement of δ_i. The random noise was uniformly distributed over the interval $[-\sigma_\delta, \sigma_\delta]$, where σ_δ is 0, 0.01, 0.03, or 0.1%. The perturbed, noisy profiles of δ_i used for the inverse analysis are shown in Figure 7d. (No noise was added to c_i, because the effect of such noise has already been tested in Figure 4.) When a direct inversion of Eq. (38) was used to calculate ϕ_j and δ_{Pj}, with $(n, m) = (10, 10)$, the initial assumed CO_2 source profile δ_j was recovered exactly (because no noise was added to c_i) but the profile of δ_{Pj} inferred from the noisy profile of δ_i showed extremely large random noise even at very small assumed random measurement errors (not shown). To obtain an acceptably smooth profile of δ_{Pj}, the least-squares error minimization procedure of Eq. (30) was used with $(n, m) = (10, 3)$, so that δ_i (concentration) data from 10 heights were combined to infer δ_{Pj} in just three levels. The result, shown in Figure 7f, was a good recovery of the initial profile of δ_{Pj} when the assumed noise in δ_i was small (0.01%) and a reasonable recovery (though with some overshoot in the upper canopy) even when the assumed noise was larger. Notably, the lowest point was well recovered in all cases. Other tests also confirmed this property.

Figure 7e shows the Keeling plot of δ_i against $1/c_i$ from the profiles in Figures 7c and 7d (including the random noise in the latter). The Keeling plot is far from a straight line. It falls roughly into two segments corresponding to the influence of soil-respired CO_2 on the lower canopy and the influence of canopy assimilation on boundary-layer air parcels, associated with the multiple marking of air parcels with CO_2 with different isotopic signatures.

These calculations illustrate the way that a Lagrangian analysis can explain isotopic composition profiles in canopy air (δ_i) in terms of the isotopic composition of sources or sinks of the exchanged scalar (δ_{Pj}), especially when δ_{Pj} is strongly nonuniform through the canopy. Inversion of measured profiles of δ_i and c_i to infer δ_{Pj} is possible, certainly in principle and probably in practice, though practical challenges include not only those associated with the IL method in general (in particular the determination of the turbulent velocity field and thence the dispersion matrix) but also the measurement of small gradients in air isotopic composition.

5. Summary and Conclusions

This chapter has reviewed canopy-scale inverse methods for inferring distributions of sources and sinks from concentration profiles in the air and has further developed these methods for the interpretation of measurements of profiles of isotopic composition. Canopy-scale inverse methods have been placed in context with an initial discussion (in Section 1) of atmospheric inverse methods in general, at scales from small chambers to the globe. Methods can be classified as either Eulerian (with a fixed control volume) or Lagrangian (using a control volume which moves with the flow, thereby tracking the scalar carried by fluid particles). Eulerian methods are further distinguished by the dominant term(s) in the scalar balance equation used as surrogates for the scalar source–sink density or surface flux (see Table 1). The two broad streams have different strengths: Eulerian approaches yield good estimates of averaged source or sink densities within the control volumes and have modest requirements for information about the velocity field, whereas Lagrangian approaches offer much more resolution of source or sink densities but require far more velocity information.

Section 2 considered the balance equations for scalar mole fraction (c), iso-concentration ($c\delta$), and isotopic composition (δ), in both Eulerian and Lagrangian forms. This systematized many conventional and some less well known results. In an Eulerian framework, an asymmetry emerges between single-point conservation equations for c and δ, with the flux divergence term in the c equation being replaced in the δ equation by a term of the form (in one dimension) $F\partial\delta/\partial z$, where F is the flux of the major isotope. In a Lagrangian framework, the single-point balance equation has an integral solution for the mean concentration $a(\mathbf{x}, t)$ of a conserved scalar entity satisfying the superposition principle, which admits $a = c$, cR, and $c\delta$, but not $a = R$ or δ. The velocity field in this solution is described by the transition probability P, identifiable with the ensemble average of the Green's function for a single realization of the flow. In the Lagrangian integral solution, the source density φ can depend on the concentration a provided that φ is statistically independent of the wind field. When φ depends on a in this way, the solution becomes an integral equation in a rather than an explicit solution.

Canopy-scale inverse methods for scalar entities were discussed in Section 3. The transition probability here reduces in discrete form to a dispersion matrix D_{ij}, dependent only on the turbulent velocity field in the canopy, which relates the profiles of scalar concentration and source density. This defines the inverse Lagrangian (IL) approach. The dispersion matrix can be found in several ways, including (but not only) with the analytic, semi-Lagrangian LNF theory, which permits D_{ij} to be calculated from profiles of turbulence statistics in the canopy (the standard deviation, σ_w, and Lagrangian time scale, T_L, of the vertical velocity). These profiles can be quite well constrained by current knowledge of turbulent flow in canopies, in particular by the mixing-layer analogy which proposes that the turbulence structure in the strong shear layer near the top of the canopy is patterned on a plane mixing layer rather than on a boundary layer (Raupach *et al.*, 1996). Thermal stability effects have also been incorporated into this picture (Leuning, 2000). The IL approach, irrespective of the means of finding D_{ij}, permits solutions to three related problems: the forward problem of finding concentration profiles from specified source profiles, the inverse problem of finding sources from measured concen-

trations, and the implicit problem of finding source and concentration profiles together when the source density is a specified function of concentration. The implicit problem is at the core of almost all vegetation–atmosphere exchanges of biogeochemically active entities. The inverse problem is solved in practice by a least-squares process involving the use of redundant concentration information.

Several field tests of the IL method have demonstrated that it is a practically useful tool for inferring the canopy source–sink distributions of scalars such as water vapor, heat, CO_2, and ammonia. However, there is a continuing need for improvement in the knowledge of the turbulence field in the canopy and its use in determining D_{ij}, and also in the inversion procedure to improve the robustness of the method.

Section 4 extended the IL analysis to describe the relationship between profiles of isotopic composition in canopy air and the profiles of isotopic sources and sinks ("iso-sources"). Lagrangian analysis provides a sound basis for the Keeling plot and the Yakir–Wang expression for distinguishing assimilation and respiration, in the case where the isotopic composition of the exchanged scalar (δ_P) is constant through the canopy (including both plant and soil). The IL approach can extend this analysis to explain air isotopic composition profiles when δ_P is strongly nonuniform through the canopy. Inversion of measured profiles of δ and c (in the air) to infer the δ_P profile appears to be possible. In principle, such an inversion procedure can add useful biogeochemical information when δ_P is strongly nonuniform, so that a conventional analysis via a Keeling plot does not lead to clear results.

The future for this line of work holds several challenges. The first is to continue to improve knowledge of the turbulent velocity field in plant canopies and its application to determine the dispersion matrix D_{ij}. Particular physical problems which continue to need attention in this respect are the nature of the turbulent exchange of scalars between the ground surface and the lowest canopy layer, and the effect of stability within and just above the canopy. It is likely that LNF theory (despite its convenience) will be augmented or replaced as a means of determining D_{ij} by other approaches. Second, the inversion process can be improved, particularly by applying suitable constraints. Both of these potential developments are likely to contribute to the progress of the canopy IL method from research tool to robust technique. Finally, the application of IL methods to air isotopic measurements in canopies remains an entirely open issue.

Acknowledgments

I am grateful to Nina Buchmann, Graham Farquhar, Jon Lloyd, and Julie Styles for interaction on isotopic issues, to Brian Sawford for very helpful interactions on Lagrangian analysis when sources are concentration-dependent, to Ray Leuning and Tom Denmead for many discussions and for access to field data, and to Damian

Barrett and Ray Leuning for comments on a draft of this chapter. I thank the Max-Planck Institut für Biogeochimie for making possible the trip which prompted this chapter.

Appendix A: Single-Point Eulerian Molar Balance Equations

This Appendix supplements Section 2.2. It suffices to work with the first of Eq. (4) for the molar balance of the minor isotope, since the molar balance of the major isotope is obtained when $R = R_P = 1$. The derivation of Eqs. (7) to (9) proceeds by the following steps:

(a) The open-air surface S_O is converted to a surface fully enclosing the region V by continuation through plant elements where necessary, with all integrated quantities being set to zero on these continuations so that the integrals are unaffected.

(b) Equations (5) or (6) are used to convert the surface integral over plant surfaces (S_P) to a volume integral over a source density ($\Phi = \rho\phi$ for a scalar, $R_P\Phi = R_P\rho\phi$ for isotopic ratio).

(c) We use the vector Leibnitz rule for time differentiation of an integral over a moving region,

$$\frac{\partial}{\partial t} \iiint_{V(t)} a(\mathbf{x}, t)d\mathbf{x} = \iiint_{V(t)} \frac{\partial a}{\partial t}d\mathbf{x} - \iiint_{S(t)} a\mathbf{v} \cdot \mathbf{n}dS, \quad (A1)$$

where $a(\mathbf{x}, t)$ is a scalar, $S(t)$ is a surface enclosing a region $V(t)$, $\mathbf{n}$ is the inward unit normal vector, and $\mathbf{v}$ is the velocity vector of the surface element dS. Steps (a) to (c) convert the first of Eq. (4) to

$$\iiint_{V(t)} \frac{\partial(\rho cR)}{\partial t}d\mathbf{x} = \iiint_{S_o(t)} [R\mathbf{F} + \rho cR\mathbf{u}] \cdot \mathbf{n}dS + \iiint_{V(t)} R_P \Phi d\mathbf{x}. \quad (A2)$$

(d) The divergence theorem is used to write Eq. (A2) as a volume integral

$$\iiint_{V(t)} \left\{ \frac{\partial(\rho cR)}{\partial t} + \nabla \cdot (R\mathbf{F} + \rho cR\mathbf{u}) - R_P\Phi \right\} d\mathbf{x} = 0. \quad (A3)$$

(e) Since this is true independent of the choice of $V(t)$, the integrand in braces must be zero. This yields the following sequence of balance equations for the minor isotope, the scalar, and air,

$$\frac{\partial(\rho cR)}{\partial t} + \nabla \cdot (\rho cR\mathbf{u}) + \nabla \cdot (R\mathbf{F}) = R_P\Phi \quad (A4)$$

$$\frac{\partial(\rho c)}{\partial t} + \nabla \cdot (\rho c\mathbf{u}) + \nabla \cdot \mathbf{F} = \Phi \quad (A5)$$

$$\frac{\partial \rho}{\partial t} + \nabla \cdot (\rho\mathbf{u}) = 0, \quad (A6)$$

where Eq. (A5) follows by putting $R = R_P = 1$ in Eq. (A4), and Eq. (A6) (the continuity equation) follows by writing Eq. (A5) for air ($c = 1, \mathbf{F} = 0, \Phi = 0$).

(f) Writing the left-hand side of Eq. (A5) as $\partial(\rho c)/\partial t = \rho\partial c/\partial t + c\partial\rho/\partial t$ and substituting for $\partial\rho/\partial t$ with Eq. (A6), we obtain the balance equation for scalar mole fraction, Eq. (7).

(g) Expanding the left-hand side of Eq. (A4) as $\partial(\rho cR)/\partial t = \rho\partial(cR)/\partial t + (cR)\partial\rho/\partial t$ and using Eq. (A6) to eliminate $\partial\rho/\partial t$, we obtain the balance equation for cR and thence the corresponding equation for the iso-concentration ($c\delta$). The latter is Eq. (8) in the main text, and the former is identical in under the substitutions of δ for R and δ_P for R_P.

(h) Similarly, by expanding $\partial(\rho cR)/\partial t$ as $R\partial(\rho c)/\partial t + (\rho c)\partial R/\partial t$ and using Equation (A5), Equation (A4) yields the balance equations for isotopic ratio (R) and thence isotopic composition (δ). The latter is Equation (9) in the main text.

Appendix B: Lagrangian Molar Balance Equations and Green's Functions

This Appendix supplements Section 2.4. The problem is to find the mean concentration field $a(\mathbf{x}, t)$ for an arbitrary scalar entity, given a turbulent velocity field $\mathbf{u}(\mathbf{x}, t)$, a specified source density $\varphi(\mathbf{x}, t, a)$ which may depend on the scalar concentration a, and homogeneous initial and boundary conditions $a = 0$ on the outer boundary S_O of a region V. Hence, all scalar is introduced into the flow by the sources φ within V. We consider an ensemble of realizations of the turbulent flow, denoted by a superscript ω, so that a^ω, $\mathbf{u}^\omega$ and $\varphi^\omega = \varphi(\mathbf{x}, t, a^\omega)$ are variables which differ randomly among realizations. In the high Péclet number limit, the relationship between a^ω, $\mathbf{u}^\omega$ and φ is given by Eq. (14), here rewritten as

$$L^\omega(a^\omega) = \frac{\partial a^\omega}{\partial t} + \mathbf{u}^\omega \nabla a^\omega = \varphi(\mathbf{x}, t, a^\omega), \quad (B1)$$

where the linear operator $L^\omega = \partial/\partial t + \mathbf{u}^\omega\nabla$ is prescribed (though complicated) once the turbulent velocity field $\mathbf{u}^\omega$ is specified for the realization ω. Given $\mathbf{u}^\omega$, the function φ, and the initial and boundary conditions on a^ω, this inhomogeneous linear equation has a solution $a^\omega (\mathbf{x}, t)$ which can be written in terms of Green's function $G^\omega (\mathbf{x}, t|\mathbf{x}_0, t_0)$ for the realization ω:

$$a^\omega(\mathbf{x}, t) = \int_{(t_0 < t)} \int_{(V)} G^\omega(\mathbf{x}, t|\mathbf{x}_0, t_0)\varphi(\mathbf{x}_0, t_0, a_0^\omega)d\mathbf{x}_0 dt_0. \quad (B2)$$

The Green's function is the solution of Eq. (B1) when the source density φ is a unit point source at $(\mathbf{x}_0, t_0)$, with initial and boundary conditions on G^ω identical with those on a^ω. That is,

$$L^\omega(G^\omega) = \frac{\partial G^\omega}{\partial t} + u^\omega\nabla G^\omega = \hat{\delta}(\mathbf{x} - \mathbf{x}_0, t - t_0), \quad (B3)$$

where $\hat{\delta}$ is the Dirac delta function (a spike at $(\mathbf{x}_0, t_0)$, zero at all other $\mathbf{x}$ and t, integrating to 1 over $\mathbf{x}$ and t). Equation (B2) is easily shown to be the required solution, since by applying the linear operator L^ω to both sides,

$$
\begin{aligned}
L^\omega(a^\omega) &= L^\omega\left(\iint G^\omega(\mathbf{x}, t \,|\, \mathbf{x}_0, t_0)\, \varphi(\mathbf{x}_0, t_0, a_0^\omega)d\mathbf{x}_0 dt_0\right) \\
&= \iint L^\omega(G^\omega)\varphi(\mathbf{x}_0, t_0, a_0^\omega)d\mathbf{x}_0 dt_0 \qquad \text{(B4)} \\
&= \iint \hat{\delta}(\mathbf{x}, t \,|\, \mathbf{x}_0, t_0)\varphi(x_0, t_0, a_0^\omega)\, d\mathbf{x}_0 dt_0 \\
&= \varphi(\mathbf{x}, t, a^\omega),
\end{aligned}
$$

where the linearity of L^ω is used to reach the second line, and Eq. (B3) is used to reach the third. Hence, Eq. (B2) satisfies Eq. (B1). The information in the velocity field $\mathbf{u}^\omega$ is now contained in G^ω.

It remains to take the ensemble average over all realizations ω. We can write

$$
a = \langle a^\omega \rangle, \quad \mathbf{u} = \langle \mathbf{u}^\omega \rangle, \quad \varphi(\mathbf{x}, t, a) = \langle \varphi(\mathbf{x}, t, a^\omega) \rangle,
$$

$$
P(\mathbf{x}, t \,|\, \mathbf{x}_0, t_0) = \langle G^\omega(\mathbf{x}, t \,|\, \mathbf{x}_0, t_0) \rangle, \qquad \text{(B5)}
$$

where P is the transition probability defined in Section 2.4 and angle brackets denote the ensemble average. The ensemble average of Eq. (B2) is

$$
a(\mathbf{x}, t) = \int_{(t_0 < t)} \int_{(V)} P(\mathbf{x}, t \,|\, \mathbf{x}_0, t_0)\varphi(\mathbf{x}_0, t_0, a_0)d\mathbf{x}_0\, dt_0, \qquad \text{(B6)}
$$

provided that G^ω and $\varphi^\omega = (\mathbf{x}, t, a^\omega)$ are statistically independent, that is, provided that the source density is statistically independent of the wind field. Equation (B6) is the same as Eq. (16).

The derivation highlights two results: First, the transition probability can be identified with the ensemble average of Green's function G^ω for a single realization. Second, when $\varphi(\mathbf{x}, t, a)$ depends on a, Eq. (B6) becomes an integral equation in a which must be solved by recursive or other means, rather than an explicit solution for a. This issue does not arise when the source density is specified independently, for instance, by the locations and strengths of point sources of air pollutants. However, it is the norm rather than the exception in biogeochemical applications, as discussed in Sections 2.4 and 3.5.

References

Ayotte, K. W., Finnigan, J. J., and Raupach, M. R. (1999). A second-order closure for neutrally stratified vegetative canopy flows. *Boundary-Layer Meteorol.* **90**, 189–216.

Baldocchi, D. D. (1992). A Lagrangian random-walk model for simulating water vapour, CO_2 and sensible heat densities and scalar profiles over and within a soybean canopy. *Boundary-Layer Meteorol.* **61**, 113–144.

Batchelor, G. K. (1949). Diffusion in a field of homogeneous turbulence. I. Eulerian analysis. Aust. *J. Sci. Res.* **2**, 437–450.

Bousquet, P., Ciais, P., Peylin, P., Ramonet, M., and Monfray, P. (1999a). Inverse modelling of annual atmospheric CO_2 sources and sinks. 1. Method and control inversion. *J. Geophys. Res.* **104**, 26161–26178.

Bousquet, P., Peylin, P., Ciais, P., Ramonet, M., and Monfray, P. (1999b). Inverse modelling of annual atmospheric CO_2 sources and sinks, 2, Sensitivity study. *J. Geophys. Res.* **104**, 26179–26193.

Buchmann, N. and Ehleringer, J. R. (1998). CO_2 concentration profiles, and carbon and oxygen isotopes in C_3 and C_4 crop canopies. *Agric. Forest Meteorol.* **89**, 45–58.

Ciais, P. and Meijer, H. A. J. (1998). The $^{18}O/^{16}O$ isotope ratio of atmospheric CO_2 and its role in global carbon cycle research. In "Stable Isotopes: Integration of Biological, Ecological and Geochemical Processes." (H. Griffiths, Ed.), pp. 409–431. Bios Scientific Publishers Ltd., Oxford.

Davidson C. I., and Y. L. Wu. (1990). Dry deposition of particles and vapors. p. 103–216. In *"Acidic Precipitation,* **Vol. 3**: Sources, Deposition, and Canopy Interactions." (S. E. Lindberg, A. L. Page, and S. A. Norton, Eds.), Springer-Verlag, New York.

Denmead, O. T. and Bradley, E. F. (1987). On scalar transport processes in plant canopies. *Irrig. Sci.* **8**, 131–149.

Denmead, O. T., Dunin, F. X., Leuning, R., and Raupach, M. R. (1997). Measuring and modelling soil evaporation in wheat crops. *Phys. Chem. Earth* **21**, 97–100.

Denmead, O. T. and Raupach, M. R. (1993). Methods for measuring atmospheric gas transport in agricultural and forest systems. In "Agricultural Ecosystem Effects on Trace Gases and Global Climate Change." (D. E. Rolston, L. A. Harper, A. R. Mosier, and J. M. Duxbury, Eds.), *ASA Special Publication no.* **55**, pp. 19–43. American Society of Agronomy, Madison.

Denmead, O. T. (1994). Measuring fluxes of CH_4 and N_2O between agricultural systems and the atmosphere. In "CH_4 and N_2O: Global Emissions and Controls from Rice Fields and Other Agricultural and Industrial Sources." (K. Minami, A. Mosier, and R. Sass, Eds.), pp. 209–234. Yokendo Publishers, Tokyo.

Denmead, O. T. (1995). Novel micrometeorological methods for measuring trace gas fluxes. *Phil. Trans. Roy. Soc. Lond. A.* **351**, 383–396.

Denmead, O. T., Leuning, R., Griffith, D. W. T., and Meyer, C. P. (1999). Some recent developments in trace gas flux measurement techniques. In "Approaches to Scaling of Trace Gas Fluxes in Ecosystems." (A. F. Bouwman, Ed.), pp. 69–84. Elsevier, Amsterdam.

Denmead, O. T., Harper, L. A., and Sharpe, R. R. (2000). Identifying sources and sinks of scalars in a corn canopy with inverse Lagrangian dispersion analysis. *I. Heat. Agric. Forest Meteorol.* **104**, 67-73.

Dolman, A. J. and Wallace, J. S. (1991). Lagrangian and K-theory approaches in modelling evaporation from sparse canopies. *Quart. J. Roy. Meteorol. Soc.* **117**, 1325–1340.

Enting, I. G. (1999). Characterising the temporal variability of the global carbon cycle. Tech. Paper 40, CSIRO Atmospheric Research, Melbourne, Australia.

Enting, I. G., Trudinger, C. M., and Francey, R. J. (1995). A synthesis inversion of the concentration and $\delta^{13}C$ of atmospheric CO_2. *Tellus* **47B**, 35–52.

Farquhar, G. D. and Richards, R. A. (1984). Isotopic composition of plant carbon correlates with water use efficiency of wheat genotypes. *Aust. J. Plant Physiol.* **11**, 539–552.

Farquhar, G. D., Ehleringer, J. D., and Hubick, K. T. (1989a). Carbon isotope discrimination and photosynthesis. *Annu. Rev. Plant Physiol.* **40**, 503–537.

Farquhar, G. D., Hubick, K. T., Condon, A. G., and Richards, R. A. (1989b). Carbon isotope fractionation and plant water-use efficiency. In "Stable Isotopes in Ecological Research." (P. W. Rundel, J. R. Ehleringer, and K. A. Nagy, Eds.) pp. 21–40. Springer-Verlag, Berlin.

Farquhar, G. D., Lloyd, J., Taylor, J. A., Flanagan, L. B., Syvertsen, J. P., Hubick, K. T., Wong, S. C., and Ehleringer, J. R. (1993). Vegetation effects on the isotope composition of oxygen in atmospheric CO_2. *Nature* **363,** 439–443.

Finnigan, J. J. (1985). Turbulent transport in flexible plant canopies. In "The Forest-Atmosphere Interaction." (B. A. Hutchison, and B. B. Hicks, Eds.), pp. 443–480. D. Reidel Publishing Co., Dordrecht.

Harper, L. A., Denmead, O. T., and Sharpe, R. R. (2000). Identifying sources and sinks of scalars in a corn canopy with inverse Lagrangian dispersion analysis. *I. Ammonia. Agric. Forest Meteorol.* **104,** 75–83.

Hicks, B. B., Baldocchi, D. D., Hosker, R. P. Jr., Hutchison, B. A., Matt, D. R., McMillen, R. T., and Satterfield, L. C. (1985). On the use of monitored air concentrations to infer dry deposition. *NOAA Tech. Memorandum ERL ARL.* **141,** Silver Spring, MD.

Kaimal, J. C. and Finnigan, J. J. (1994). Atmospheric Boundary Layer Flows. Oxford University Press, New York, Oxford. 289 p.

Katul, G. G. and Albertson, J. D. (1998). An investigation of higher-order closure models for a forested canopy. *Boundary-Layer Meteorol.* **89,** 47–74.

Katul, G. G., Oren, R., Ellsworth, D., Hseih, C. I., Phillips, N., and Lewin, K. (1997). A Lagrangian dispersion model for predicting CO_2 sources, sinks and fluxes in a uniform loblolly pine (*Pinus taeda* L.) stand. J. Geophys. Res. **102,** 9309–9321.

Katul, G. G., Leuning, R., Kim, J., Denmead, O. T., Miyata, A., and Harazono, Y. (2001). Estimating CO_2 source/sink distributions within a rice canopy using higher-order closure models. *Boundary-Layer Meteorol.* **98,** 103–125.

Keeling, C. D. (1961). The concentrations and isotopic abundances of atmospheric carbon dioxide in rural and marine air. *Geochimica et Cosmochimica Acta* **24,** 277–298.

Leuning, R. (2000). Estimation of scalar source/sink distributions in plant canopies using Lagrangian dispersion analysis: corrections for atmospheric stability and comparison with a multilayer canopy model *Boundary-Layer Meteorol.* **96,** 293–314.

Leuning, R., Denmead, O. T., Miyata, A., and Kim, J. (2000). Source-sink distributions of heat, water vapour, carbon dioxide and methane in rice canopies estimated using Lagrangian dispersion analysis. Agric. *Forest Meteorol.* **104,** 233–249.

Lin, G. and Ehleringer, J. R. (1997). Carbon isotopic fractionation does not occur during dark respiration in C_3 and C_4 plants. *Plant Physiol.* **114,** 391–394.

Lloyd, J. and Farquhar, G. D. (1994). ^{13}C discrimination during CO_2 assimilation by the terrestrial biosphere. *Oecologica* 99, 201–215.

Lloyd, J., Kruijt, B., Hollinger, D. Y., Grace, J., Francey, R. J., Wong, S. C., Kelliher, F. M., Miranda, A. C., Farquhar, G. D., Gash, J. H. C., Vygodskaya, N. N., Wright, I. R., Miranda, H. S., and Schulze, E. D. (1996). Vegetation effects on the isotopic composition of atmospheric CO_2 at local and regional scales: theoretical aspects and a comparison between a rainforest in Amazonia and a boreal forest in Siberia. *Aust. J. Plant Physiol.* **23,** 371–399.

Massman, W. J. and Weil, J. C. (1999). An analytical one-dimensional second-order closure model of turbulence statistics and the Lagrangian time scale within and above plant canopies of arbitrary structure. *Boundary-Layer Meteorol.* **91,** 81–107.

McNaughton, K. G. (1976). Evaporation and advection I: evaporation from extensive homogeneous surfaces. *Q. J. Roy. Meteorol. Soc.* **102,** 181–191.

McNaughton, K. G. and van den Hurk, B. J. J. M. (1995). A 'Lagrangian' revision for the resistors in a two-layer model for calculating the energy budget of a plant canopy. *Boundary-Layer Meteorol.* **74,** 261–288.

Philip, J. R. (1959). The theory of local advection: *I. J. Meteorol.* **16,** 535–547.

Press, W. H., Teukolsky, S. A., Vetterling, W. T. and Flannery, B. P. (1992). Numerical Recipes in Fortran 77: the Art of Scientific Computing. 2 ed. Cambridge University Press, Cambridge.

Raupach, M. R. (1988). Canopy transport processes. In "Flow and Transport in the Natural Environment: Advances and Applications." (W. L. Steffen and O. T. Denmead, Eds.), pp. 95–127. Springer-Verlag, Berlin.

Raupach, M. R. (1989a). A practical Lagrangian method for relating scalar concentrations to source distributions in vegetation canopies. *Quart. J. Roy. Meteorol. Soc.* **115,** 609–632.

Raupach, M. R. (1989b). Applying Lagrangian fluid mechanics to infer scalar source distributions from concentration profiles in plant canopies. Agric. *Forest Meteorol.* **47,** 85–108.

Raupach, M. R. (1994). Simplified expressions for vegetation roughness length and zero-plane displacement as functions of canopy height and area index. *Boundary-Layer Meteorol.* **71,** 211–216. [Corrigendum: *Boundary-Layer Meteorol.* **76,** 303–304 (1995)].

Raupach, M. R. (2000). Equilibrium evaporation and the convective boundary layer. *Boundary-Layer Meteorol.* **96,** 107–141.

Raupach, M. R., Denmead, O. T., and Dunin, F. X. (1992). Challenges in linking atmospheric CO_2 concentrations to fluxes at local and regional scales. *Aust. J. Bot.* **40,** 697–716.

Raupach, M. R., Finnigan, J.J., and Brunet, Y. (1996). Coherent eddies and turbulence in vegetation canopies: the mixing layer analogy. *Boundary-Layer Meteorol.* **78,** 351–382.

Raupach, M. R. and Shaw, R. H. (1982). Averaging procedures for flow within vegetation canopies. *Boundary-Layer Meteorol.* **22,** 79–90.

Rayner, P. J., Enting, I. G., Francey, R. J., and Langenfelds, R. (1999). Reconstructing the recent carbon cycle from atmospheric CO_2, $\delta^{13}C$ and O_2/N_2 observations. *Tellus* **51B,** 213–228.

Sawford, B. L. and Guest, F. M. (1987). Lagrangian stochastic analysis of flux-gradient relationships in the convective boundary layer. *J. Atmos. Sci.* **44,** 1152–1165.

Tarantola, A. (1987). Inverse Problem Theory: Methods for Data Fitting and Model Parameter Estimation. Elsevier, Amsterdam.

Taylor, G. I. (1921). Diffusion by continuous movements. *Proc. London Math. Soc.* **A20,** 196–211.

Thomson, D. J. (1984). Random walk modelling of diffusion in inhomogeneous turbulence. *Quart. J. Roy. Meteorol. Soc.* **110,** 1108–1120.

Thomson, D. J. (1987). Criteria for selection of stochastic models of particle trajectories in turbulent flows. *J. Fluid Mech.* **180,** 529–556.

van den Hurk, B. J. J. M. and McNaughton, K. G. (1995). Implementation of near-field dispersion in a simple two-layer surface resistance model. *J. Hydrol.* **166,** 293–311.

Wilson, J. D. (1988). A second-order closure model for flow through vegetation. *Boundary-Layer Meteorol.* **42,** 371–392.

Wilson, N. R. and Shaw, R. H. (1977). A higher order closure model for canopy flow. *J. Appl. Meteorol.* **16,** 1198–1205.

Yakir, D. and Wang, X. F. (1996). Fluxes of CO_2 and water between terrestrial vegetation and the atmosphere estimated from isotope measurements. *Nature* **380,** 515–517.

Biogeophysical Feedbacks and the Dynamics of Climate

M. Claussen
*Potsdam-Institute for
    Klimafolgenforschung*
Potsdam, Germany

1. Introduction

Traditionally, vegetation has been considered a more or less passive component of climate. For example, Alexander von Humboldt (1849) imagined the desertification of North Africa to be caused by an oceanic impact. He argues that somewhere in the "dark past," the subtropical Atlantic gyre was much stronger and flooded the Sahara, thereby washing away vegetation and fertile soil. When examining different theories of ice ages, DeMarchi (1885) concluded that the occurrence of glacial epochs does not depend on changes in the "covering of the earth's surface (vegetation)." Köppen (1936) described vegetation as "crystallized, visible climate" and referred to it as an indicator of climate much more accurate than our instruments. I interpret Köppen's statement in the sense that he considered vegetation as being completely determined by climate. If Köppen would have taken into account the possibility that vegetation could affect atmospheric and oceanic circulation, then he certainly would have sought a more "objective" parameter. In the same line of thinking, coupled atmosphere–ocean models were regarded as state-of-the-art climate models (see, for example, Cubasch et al., 1995). Global vegetation patterns in these models are kept constant in time. Only short-term plant physiology and, to some extent, fractional vegetation and leaf area are allowed to change with meteorological conditions.

Today, a more general definition of climate in terms of state and ensemble statistics of the climate system is generally accepted (see Peixoto and Oort, 1992). The climate system encompasses not only the abiotic world (atmosphere, hydrosphere, cryosphere, pedosphere) but also the living world, the biosphere. Interestingly, the IPCC (Houghton et al., 1997) defines a climate model as a model which "include(s) enough of the components of the climate system to be useful for simulating the climate." This defini-

tion is misleading. One can successfully simulate the observed state of a system with a reduced model, e.g., the present-day climate using atmosphere–ocean models. However, the nonlinearity of the climate system could lead to multiple states under the same external forcing owing to feedbacks between all components of the system. Hence, when operating with a subset of the complete model, one could miss important aspects of the dynamics of the entire system, which I discuss for the case of vegetation–climate interaction.

A number of studies reveal that predictions of global atmospheric models are highly sensitive to prescribed large-scale changes in vegetation cover, such as removal of tropical (e.g., Henderson-Sellers et al., 1993; Polcher and Laval, 1994; Zheng and Elthair, 1997) and boreal (e.g., Bonan et al., 1992) forests. Although these studies illustrate the potential effects of massive vegetation changes on the climate system, they can hardly be validated. Therefore, Foley et al. (1994) suggest investigation of past environments such as the climate of the early to middle Holocene, some 6000–9000 years ago, for which strong differences in global vegetation pattern are amply documented (see below). I follow their reasoning and discuss mainly palaeo climate.

Generally, I review the state of the art of our knowledge of vegetation–climate interaction, where I will restrict myself to biogeophysical aspects. First, I discuss synergisms of feedbacks between various components of the climate system, with emphasis on the inclusion of vegetation. Second, I explore the nonlinear character of vegetation–climate interaction: the possibility of multiple solutions to the vegetation–climate system and, third, its consequences for the transient vegetation–climate dynamics. I do not try to seek a complete-as-possible summary; instead, I focus on gaps and perspectives in biogeophysical modeling.

2. Synergisms

2.1 High Northern Latitudes

Palaeobotanic evidence indicates that during the early to middle Holocene, boreal forests extended north of the modern treeline (Frenzel *et al.*, 1992; TEMPO, 1996; Cheddadi *et al.*, 1997). It is suggested that this migration was triggered by changes in the earth's orbit. Moreover, the migration of boreal trees is assumed to amplify the initial warming owing to the so-called taiga–tundra feedback, first discussed by Otterman *et al.* (1984) and Harvey (1988, 1989a,b). The albedo of snow-covered vegetation is much lower for forests than for low vegetation such as tundra, which can readily be seen from a bird's-eye view. Hence the darker, snow-covered taiga receives more solar energy than the snow-covered tundra, which, in turn, favors the growth of taiga. Later, Foley *et al.* (1994) analyzed the vegetation–snow–albedo feedback in more detail. By imposing an increase in forest area of some 20% as a surface condition, they find that changes in land surface conditions give rise to an additional warming of some 4 °C in spring and about 1 °C in the other seasons. Orbital forcing would produce only some 2 °C. The additional warming is caused mainly by a reduction of snow and sea-ice volume by nearly 40% and subsequent reduction in surface albedo. Further simulations using similar experimental setups but different models (TEMPO, 1996) corroborate the earlier results. These studies clearly point at the importance of vegetation–climate interaction at high northern latitudes in amplifying climate change triggered by some external forcing. Unfortunately, no attempt has been made to isolate the effect of a decrease in vegetation–snow–albedo and the sea-ice–albedo feedback. So it was not clear how much the biospheric process actually contributes to the mid-Holocene warming at high northern latitudes or whether the warming was mainly caused by a synergism between vegetation change and oceanic feedback.

Only coupled atmosphere–vegetation models can analyze the dynamics of the feedback, i.e., the interaction between changes in vegetation structure and climate. For example, Gutman *et al.* (1984) and Gutman (1984, 1985) explored the idea of relating the surface parameters of an atmospheric model (e.g., albedo and water availability) to climatic variables. They used the Budyko (1974) radiative index of dryness, D, to characterize the geobotanic type of a climate zone and proposed a simple relation between albedo, water availability, and D. Later, Henderson-Sellers (1993) and Claussen (1994) coupled comprehensive atmospheric circulation models with (diagnostic) biome models, i.e., with models of macro ecosystems assuming an equilibrium with climate. Hence these asynchronously coupled atmosphere–biome models can be used to assess equilibrium solutions of the system, but not system dynamics. Nevertheless, the idea of developing such models turned out to be a valuable extension of sensitivity studies based on one-way coupled models and the more simple models of Gutman *et al.* (1984).

Returning to the problem of the vegetation–snow–albedo feedback at high northern latitudes, i.e., the taiga–tundra feedback, one would expect this feedback to be a positive one: a reduction in surface albedo increases near-surface temperatures, which, in turn, favors growth of taller vegetation, reducing surface albedo further (see Otterman *et al.*, 1984). The feedback is limited by topographical constraints, e.g., coast lines, or by the insolation. The studies of Claussen and Gayler (1997) and Texier *et al.* (1997), using different atmospheric models but the same biome model of Prentice *et al.* (1992), confirm the earlier assertion that the vegetation–snow–albedo feedback is positive. However, both models show a rather small northward expansion of boreal forests. This is not surprising, as the annual cycle of sea-surface temperatures (SSTs) and Arctic sea-ice volume are kept constant. Obviously, the synergism between terrestrial and marine feedbacks is missing. This has clearly been demonstrated in a study by Ganopolski *et al.* (1998) using a coupled atmosphere–ocean–vegetation model. They find a summer warming over the Northern Hemisphere continents of some 1.7 °C (in comparison with present-day climate) owing to orbital forcing on the atmosphere alone. Inclusion of ocean–atmosphere feedbacks (but keeping vegetation structure constant in time) reduces this signal to some 1.2 °C, whereas the taiga–tundra feedback (but now without any oceanic feedback) enhances summer warming to 2.2 °C. In the full system (including all feedbacks) this additional warming is not reduced, as one would expect from linear reasoning, but it is increased to 2.5 °C as a result of a synergism between the taiga–tundra feedback and the Arctic sea-ice–albedo feedback. Likewise, orbital forcing alone induces a wintertime cooling of some −0.8 °C. The biogeophyscial feedbacks alone reduce this cooling to −0.7 °C, and the atmosphere–ocean interaction, to −0.5 °C. The synergism between the two feedbacks, however, causes a winter warming of some 0.4 °C. The warming of Northern Hemisphere winters, which is supported by reconstructions (e.g., Cheddadi *et al.*, 1997), is often referred to as the "biome paradox." From the results of Ganopolski *et al.* (1998) one can conclude that the biome paradox is not a pure biospheric feedback, but it is caused mainly by the synergism between this feedback and the oceanic feedback.

During the mid-Holocene, orbital forcing triggered a warming of the Northern Hemisphere in summer, whereas the opposite was valid for the end of the Eemian warm period some 115 ka B. P., as pointed out by Harvey (1989b) and subsequently by Gallée *et al.* (1992), Berger *et al.* (1992, 1993), and Gallimore and Kutzbach (1996). These studies show that the taiga–tundra feedback contributes significantly to the temperature response to orbital forcing. Gallimore and Kutzbach (1996) state that even a prescribed increase in surface albedo which is deduced from a biome model estimate of tundra expansion at 115 ka B. P. is sufficient to induce glaciation over northeastern Canada. (Actually, Gallimore and Kutzbach (1996) did not simulate glacial inception, just the occurrence of permanent snow cover.) deNoblet *et al.* (1996) support this hypothesis by using a coupled atmosphere–biome model, although they obtain just a substantial increase in snow depth, but no large-scale perennial snow cover over North Canada was obtained. Moreover, they restrict themselves to the biospheric feedbacks ignoring any synergism between land surface and sea ice (which presumably could help to get perennial snow cover).

2.2 Subtropics

While most researchers in the field agree on the relative importance of biospheric feedbacks operating at high northern latitudes, the discussion becomes more interesting and diverse as the subtropics are concerned. Climate reconstructions and data on fossil pollen compiled by Jolly *et al.* (1998), Hoelzmann *et al.* (1998), Petit-Maire (1996), and Anhuf *et al.* (1999) indicate that North Africa was much greener in the mid-Holocene than today. The Saharan desert was, presumably to a large extent, covered by annual grasses and low shrubs. The Sahel reached at least as far north as 23 °N, more so in the western than in the eastern part.

In their model, Texier *et al.* (1997) yield a positive feedback between vegetation and precipitation in this region, which is, however, much too weak to get any substantial greening (Fig. 1A). They suggest an additional (synergistic) feedback between sea-surface temperature (SST) and land-surface changes. By modifying surface conditions in North Africa (increased vegetation cover, increased areas of wetlands and lakes) Kutzbach *et al.* (1996) obtain some change in their model that leads to an increase in precipitation in the southeastern part of the Sahara, but almost none in the western part (Fig. 1B). An upgraded version of the model used by Kutzbach *et al.* (1996) reveals a northward spread of vegetation also in the western part of north Africa according to Broström *et al.* (1998). Claussen and Gayler (1997) find a strong feedback between vegetation and precipitation and an almost complete greening in the western Sahara and some in the eastern part (Fig. 1C). By and large the latter model results, although far from perfect, seem to agree best with the data. Claussen and Gayler (1997) and Claussen *et al.* (1998) explain the positive feedback by an interaction between high albedo of Saharan sand deserts and atmospheric circulation as hypothesized by Charney (1975). They extend Charney's theory by accounting for atmospheric hydrology, i.e., moisture convergence and associated convective precipitation. [For present-day climate this feedback, or "Charney's loop," was discussed in detail by Lofgren (1995) and, independently by Claussen (1997).]

Now the question of which model is "correct" arises. To tackle this problem, deNoblet *et al.* (2000) compare the "extreme"— concerning the magnitude of Saharan greening—models of Claussen and Gayler (1997) and Texier *et al.* (1997). Both groups use the same biome model, but different atmospheric models. Moreover, the atmospheric model and the biome model are asynchronously coupled in different manners: Claussen and Gayler (1997) use the output of the climate model directly to drive the biome model, while Texier *et al.* (1997) take the difference between model results and a reference climate as input to the biome model. The latter, the so-called anomaly approach, prevents the coupled model from drifting to an unrealistic climate which could be induced by some positive feedbacks between biases in either model. Hence this method is similar to the "flux correction"

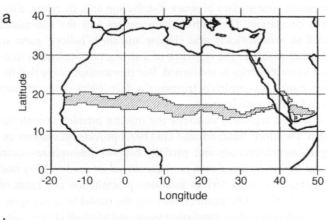

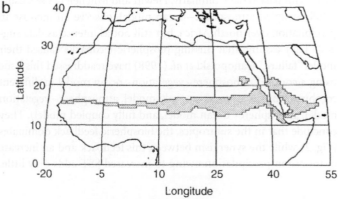

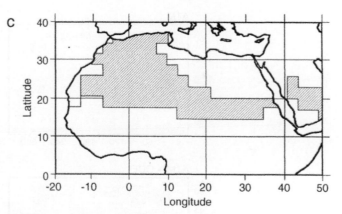

FIGURE 1 Reduction of desert from present-day climate to mid-Holocene climate simulated by (a) the models of Texier *et al.* (1997), (b) Kutzbach *et al.* (1996), and (c) Claussen and Gayler (1997). (a, c) are taken with modifications from deNoblet *et al.* (2000) and (b) with modifications from Kutzbach *et al.* (1996).

in coupled atmosphere–ocean models. It turns out that the difference between coupling procedures affects the results of the coupled atmosphere–biome model only marginally. Hence deNoblet *et al.* (2000) conclude that the differences in north Africa greening cannot be attributed to the coupling procedure; it can be traced back to different representations of the atmospheric circulation in the tropics. The atmospheric model of Claussen and Gayler (1997) somewhat overestimates the duration of the north African monsoon, while the other model of Texier *et al.* (1997) yields an unrealistic near-surface pressure distribution and, therefore, a too zonal circulation. The authors demonstrate why the one model yields an unrealistically arid climate and they "believe" more in the other model as the existence of a strong biogeophysical feedback in north Africa is concerned. But they cannot prove that the latter model is completely trustworthy. Hence this issue certainly needs further consideration.

A second argument concerns the missing interaction with the ocean. Therefore, Kutzbach and Liu (1997) provide simulations using an asynchronously and partially coupled atmosphere–ocean model (no freshwater fluxes, no dynamic sea-ice model). They find an increase in north African monsoon precipitation as a result of increased SST in late summer bringing the model in closer agreement with palaeo data. Similarly, Hewitt and Mitchell (1998), using a fully coupled atmosphere–ocean model, observe an increase in precipitation over north Africa, but still not as intense as data suggest. They assume that missing biospheric feedbacks caused their model "failure." Ganopolski *et al.* (1998) have readdressed this issue using a coupled atmosphere–vegetation–ocean model in different combinations (as atmosphere-only model, atmosphere–vegetation model, atmosphere–ocean model, and fully coupled model). They conclude that in the subtropics, the biospheric feedback dominates (Fig. 2) while the synergism between this feedback and an increase in monsoon precipitation owing to increased SST adds only little.

The model of Ganopolski *et al.* (1998) is the only "true" climate model according the IPCC definition as it includes all components of the climate system relevant to describe mid-Holocene climate. However, it has a rather coarse horizontal resolution. Hence to be certain of their results, one must confirm that these results are independent of the model resolution.

3. Multiple Equilibria

As the interaction between components of the climate system is nonlinear, one might expect multiple equilibrium solutions. Gutman *et al.* (1984) and Gutman (1984, 1985) found only unique, steady-state solutions in their zonally averaged model. (Actually, they regarded their results as "tentative and merely as an illustration of the suggested approach," because of the simplicity of their model.) The possibility of multiple equilibria in the 3-dimensional atmosphere–vegetation system was discovered later by Claussen (1994) and subsequently analyzed in detail by Claussen (1997, 1998) for present-day climate, i.e., present-day insolation and SST. Two solutions to the atmosphere–vegetation system appear: the arid, present-day climate and a humid solution resembling more the mid-Holocene climate, i.e., with a Sahara greener than today, albeit less green than in the mid-Holocene (Fig. 3).

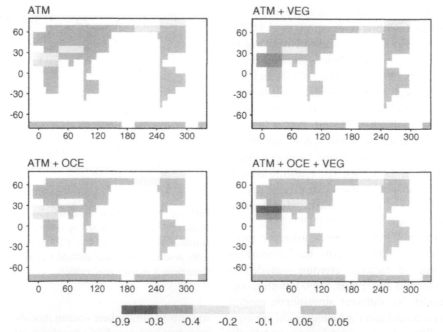

FIGURE 2 Reduction of desert from present-day climate to mid-Holocene climate simulated by Ganopolksi *et al.* (1998). The color labels refer to differences in (nondimensional) fractional coverage of desert between today and 6000 years before present. Desert fractions are diagnosed from annual mean precipitation and temperature obtained by the atmosphere-only model (ATM) and the atmosphere–ocean model (ATM + OCE) using present-day land-surface conditions. Desert fractions are predicted from vegetation dynamics by using the atmosphere–vegetation model (ATM + VEG) and the fully coupled model (ATM + OCE + VEG). See also color insert.

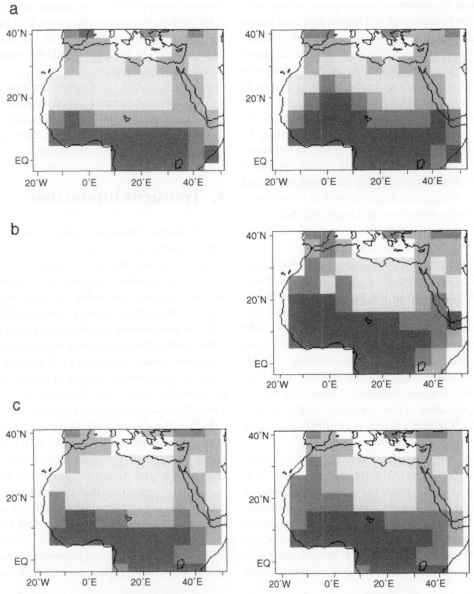

FIGURE 3 Multiple equilibria computed for present-day climate (a) and for the climate of the last glacial maximum (c). For mid-Holocene conditions, only one solution is obtained (b). A summary of the results of Claussen (1997) (a), Claussen and Gayler (1997) (b), and Kubatzki and Claussen (1998) (c).

The two solutions differ mainly in the subtropical areas of north Africa and, but only slightly, in central east Asia. The possibility of multiple equilibria in the atmosphere–vegetation system of North-west Africa has recently been corroborated by Wang and Eltahir (2000) and Zeng and Neelin (2000) by using completely different models of the tropical atmosphere and dynamic vegetation.

Interestingly, the stability of the atmosphere–vegetation system seems to change with time: experiments with mid-Holocene vegetation yield only one solution, the green Sahara (Claussen and Gayler, 1997), while two solutions exist for the Last Glacial Maximum (LGM) (Kubatzki and Claussen, 1998).

So far, no other regions on earth in which multiple equilibria could evolve on a large scale have been identified Levis *et al.* (1999) seek multiple solutions to the atmosphere–vegetation–sea-ice system at high northern latitudes. Their model converges to one solution in this region corroborating the earlier assertion (Claussen, 1998) that multiple solutions manifest themselves in the subtropics, mainly in north Africa.

Why do we find multiple solutions in the subtropics, but none at high latitudes—and why for the present-day and LGM climates, but not for mid-Holocene climate? Claussen *et al.* (1998) analyze large-scale atmospheric pattern in present-day, mid-Holocene, and LGM climates. They find that velocity potential patterns, which indicate divergence and convergence of large-scale atmospheric flow, differ between arid and humid solutions mainly in the tropical and subtropical regions. It appears that the Hadley-Walker circulation slightly shifts to the west. This is consistent with Charney's (1975) theory of albedo-induced desertification in the subtropics. Moreover, changes in surface conditions directly influence vertical motion, and thereby large-scale horizontal flow, in the tropics (Eltahir, 1996), but hardly at middle and high latitudes (e.g., Lofgren, 1995a,b). For the mid-Holocene climate, the large-scale atmospheric flow is already close to the humid mode, even if one prescribes present-day land surface conditions. This is caused by differences in insolation: in the mid-Holocene boreal summer, the Northern Hemisphere received up to 40 W m^{-2} more energy than today, thereby strengthening African and Asian summer monsoon (Kutzbach and Guetter, 1986). During the LGM, insolation was quite close to present-day conditions.

A more ecological interpretation of multiple equilibria is given by Brovkin *et al.* (1998). They develop a conceptual model of vegetation–precipitation interaction in the western Sahara which is applied to interpret the results of comprehensive models. The conceptual model finds three solutions for present-day and LGM climate; one of these, however, is unstable to infinitesimally small perturbations. The humid solution is shown to be less probable than the arid solution, and this explains the existence of the Sahara desert as it is today. For mid-Holocene climate, only one solution is obtained. Application of the conceptual model to biospheric feedbacks at high latitudes (Levis *et al.*, 1999) yields only one solution for the present-day conditions.

Are multiple equilibria just a matter of the atmosphere–vegetation system, or do they occur also in the atmosphere–ocean–vegetation system? So far, we have not yet found multiple solutions in the model of Ganopolski *et al.* (1998). (The model attains multiple solutions associated with multiple states of the thermohaline convection.) I blame this deficit on the coarse resolution of this model, because north Africa is represented by just three grid boxes, Sahara, Sudan, and tropical north Africa. Subsequently,

Saharan precipitation in the coarse model of Ganopolski *et al.* (1998) is less sensitive to changes in land-surface conditions than the west Saharan precitpitation in the model used by Claussen (1997, 1998). On the other hand, the study of Ganopolski *et al.* (1998) shows that the biogeophysical feedback in north Africa is mainly a vegetation–atmosphere feedback. Therefore, I assume that our conclusion from coupled vegetation–atmosphere models should generally be valid, i.e., also vegetation–atmosphere–ocean models (with finer horizontal resolution) should exhibit multiple equilibria in the north African region.

4. Transient Interaction

The discussion of multiple equilibria seems to be somewhat academic. However, the existence of these could explain abrupt transitions in vegetation structure (Claussen *et al.*, 1998; Brovkin *et al.*, 1998). If global stability changes in the sense that one equilibrium solution becomes less stable to finite amplitude perturbations than the others, then an abrupt change of the system from the less stable to a more stable equilibrium is to be expected. Brovkin *et al.* (1998) find in their conceptual model that the green solution becomes less stable around 3.6 ka B.P. Keeping in mind that the variability of precipitation is larger in humid regions than in arid regions of north Africa (e.g., Eischeid *et al.*, 1991), one would expect a transition roughly between 6 and 4 ka BP.

In fact, there is evidence that the mid-Holocene wet phase in north Africa ended around 5.0–4.5 ka B.P. even in the high continental position of the east Sahara (Pachur and Wünnemann, 1996; Pachur and Altmann, 1997). Petit-Maire and Guo (1996) present data suggesting that the transition to present-day's arid climate did not occur gradually, but in two steps with two arid periods, at 6.7–5.5 and 4–3.6 ka B.P. Other reconstructions indicate that freshwater lakes in the eastern Sahara began to disappear from 5.7 to 4 ka B. P., when recharge of aquifers ceased at the end of the wet phase (Pachur and Hoelzmann, 1991). Pachur and

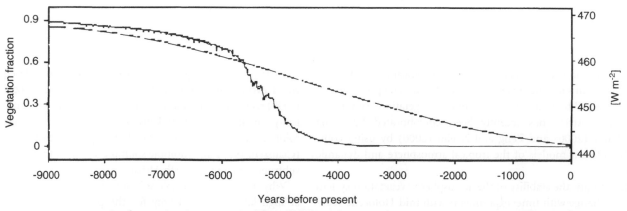

FIGURE 4 Development of vegetation fraction in the Sahara (full line, left ordinate) as response to changes in insolation of the Northern Hemisphere during boreal summer (dashed line, right ordinate). The abscissa indicates the number of years before present. Figure 4 is taken with modifications from Claussen *et al.* (1999).

Hoelzmann (personal communication) suggest that climate change at the end of the mid-Holocene was faster in the western than in the eastern Sahara. Indeed, deMenocal *et al.* (2000) report of an abrupt decline in aeolian dust transport off the Northwest African Atlantic coast 5500 years ago. This reconstruction is consistent with the hypothesis of multiple equilibria in the western, not in the eastern Sahara.

The arguments above are based on studies of the system at or in the vicinity of an equilibrium state. Only with fully coupled, dynamic vegetation models can one explore the time evolution of biogeophysical feedbacks. Claussen *et al.* (1999) analyze the transient structures in global vegetation pattern and climate using the coupled atmosphere–ocean–vegetation model of Ganopolski *et al.* (1998), but with a dynamic vegetation module. Their simulations clearly show (not just suggest) that subtle changes in orbital forcing triggered changes in north African climate which were then strongly amplified by biogeophysical feedbacks in this region. The timing of the transition, which started at around 5.5 ka B.P. in the model (Fig. 4), was governed by a global interplay between atmosphere, ocean, sea ice, and vegetation. The interplay is affected by a change in tropical SST and by the synergisms between biospheric and oceanic feedbacks, mentioned in Section 2.1, which influence the large-scale meridional temperature gradient. Hence the abrupt desertification—abrupt in comparison with the subtle change in orbital forcing—is a regional effect. The timing of it depends, however, on global processes. Whether tropical SST or biospheric feedbacks at high northern latitudes dominate the latter has still to be evaluated.

5. Perspectives

The investigation of biospheric feedbacks using coupled vegetation–climate models has just started. Therefore, it is too early to arrive at a conclusion, which, in its true sense, always implies some "closure." Instead, I try to "open" this issue further.

So far it has been recognized that there are biogeophysical feedbacks which affect the (global) climate system. However, as outlined above, theoretical analyses of biogeophysical *feedbacks* often focus on *synergisms* instead of feedbacks. The influence of several biogeophysical feedbacks, having included their synergism with other, for example, oceanic feedbacks, on the climate system is simulated without paying attention to the role of individual feedbacks. To illustrate the problem, I briefly recall the classical feedback analysis presented by, for example, Schlesinger (1988) and Peixoto and Oort (1992), and I extend their analysis to include synergisms.

Let us assume that the state of the climate system depends on external forcing, E, such as insolation and anthropogenic land cover change, and internal processes H_i. Any external forcing E will change the state of the climate system defined in terms of extensive variables S. Hence $S = G E$, where G is a sensitivity factor or sometimes referred to as a gain. Without any feedback, the response of the system would be $S_0 = G_0 E$. With feedbacks,

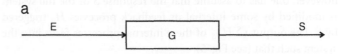

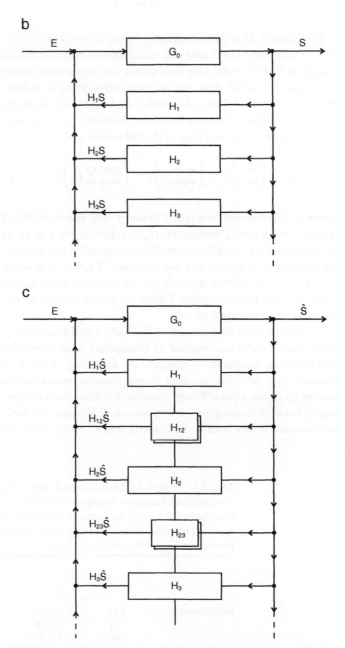

FIGURE 5 A schematic view on the linear feedback analysis (a,b) and its extension to synergisms (c). G and G_0 represent the gain of a system with and without any feedback, respectively. S is the response of the system to an external forcing E. H_i ($i = 1, 2, 3$) are internal or feedback processes. H_{12} and H_{23} are synergistic processes between H_1 and H_2 and H_2 and H_3, respectively, which modify the output $H_i S$ ($i = 1,2,3,$). $\hat{S}$ is the response of the nonlinear system. Synergisms between more than two internal processes are omitted in this sketch. (a) and (b) are taken with modifications from Peixoto and Oort (1992).

however, one has to assume that the response S of the full system is modified by some internal or feedback processes H_i, triggered by S. The output $(\Sigma H_i)_S$ of these internal processes feeds into the system such that (see Fig. 5)

$$S = G_0 E + G_0\left(\sum_i H_i\right) S.$$

The factor $G_0 H_i$ is called feedback f_i. Hence $G = G_0 / (1 - \Sigma f_i)$. Peixoto and Oort (1992) note that this analysis is based on the assumption that there exist no synergisms, i.e., interaction among feedbacks. To extend their analysis one could define a multidimensional transfer function $\hat{G}$ which includes not only feedbacks but also synergisms between feedbacks. Formally, one may write $\hat{S} = \hat{G}E$, where $\hat{S}$ is the response of the full system, and

$$\hat{S} = G_0 E + \left(\sum_i f_i\right)\hat{S} + \left(\sum_i \sum_j f_{ij}\right)\hat{S} + \left(\sum_i \sum_j \sum_k f_{ijk}\right)\hat{S} + \cdots,$$

where f_{ij} indicate the synergism between two processes H_i and H_j (with $i \neq j$), and f_{ijk} between H_i, H_j, and H_k (with $i \neq j \neq k$). As for feedbacks, we can differentiate between positive, i.e., amplifying synergisms, if $f_{ij} > 0$, and negative ones, if $f_{ij} < 0$. It is worth noting that this analysis depends on the reference state chosen. Generally, gain G and response $\hat{S}$ differ no matter whether we apply the external forcing E or $-E$.

For illustration, I have calculated feedbacks f_i and synergisms f_{ij} from model results summarized in Ganopolski *et al.* (1998) for mid-Holocene temperature changes. By inspecting Table 1, it becomes clear why the so-called biome paradox, mentioned in Section 2.1, is not a pure "biome" paradox, but arises from the synergism between biogeophysical and oceanic feedbacks. The feedback analysis shows positive feedbacks f_1 and f_2. Hence both, the

atmosphere–ocean feedback f_1 and the atmosphere–vegetation feedback f_2, tend to "oppose" wintertime cooling by enhancing radiative forcing; however, they are not strong enough to produce a warming. The response of the system without synergisms would produce a cooling with respect to today's climate, i.e., $\Delta S = S(6k) - \hat{S}(0k) < 0$, where $S(6k) = S_0(6k)/(1-f_1-f_2)$. It is the synergism f_{12} between these feedbacks that produces a wintertime warming, indicated in Table 1 by $\Delta \hat{S} = \hat{S}(6k) - \hat{S}(0k) > 0$ and $f_{12} >> f_1, f_2$.

Berger (1999) uses the factor-separation technique proposed by Stein and Alpert (1993) to exlore feedbacks and synergisms. Work in progress suggests that both methods, my extension of the classical feedback analysis and the factor-separation technique, are similar and that they yield the same results if properly normalized.

In this context, it should be emphasised that the *problem of synergisms* has been overlooked in the investigation of anthropogenic land cover change. Generally these experiments are undertaken as sensitivity experiments, i.e., the response of the atmosphere to (prescribed) changes in land cover is analyzed. Hence these experiments do not really belong to the category of feedback experiments. However, if longer time scales are considered, the (prescribed) changes in land cover could trigger changes in natural vegetation in regions not directly affected by anthropogenic land use (e.g., Brovkin *et al.*, 1999) and, perhaps even more importantly, they could trigger synergisms with other feedbacks. For example, work in progress (Ganopolski *et al.*, 2000) suggests that, using the model of Ganopolski *et al.* (1998), the effect of tropical deforestation differs, if we allow for oceanic feedbacks. In temperate regions we find a summer warming in the case of fixed SST, but a summer cooling in the case of an interactive ocean.

Biogeophysical feedbacks can lead to *multiple equilibria of the climate system* and they influence the (transient) dynamics of the climate system. This has been shown—meanwhile by three com-

TABLE 1 Oceanic Feedback Factors, f_1, Biogeophysical Feedback Factors, f_2, and their Synergism, f_{12}, for Temperature Changes on Average over the Northern Hemisphere (NH), the Northern Hemisphere Continents (NH$_L$), and the Southern Hemisphere (SH) during Boreal Summer (June, July, August), Boreal Winter (December, January, February) and on Annual Average in Response to a Change in Orbital Parameters from 6000 Years Ago to Today

		Temperature				
		f_1	f_2	f_{12}	ΔS	$\Delta \hat{S}$
Boreal summer	NH$_L$	− 1.61	+ 1.65	+ 2.75	+ 1.72	+ 2.54
	NH	− 1.38	+ 0.95	+ 2.42	+ 0.85	+ 1.57
	SH	− 1.41	+ 0.21	+ 3.13	− 0.11	+ 0.78
Boreal winter	NH$_L$	+ 0.84	+ 0.47	+ 2.93	− 0.41	+ 0.39
	NH	+ 0.73	+ 0.20	+ 2.80	− 0.21	+ 0.58
	SH	+ 0.50	+ 0.02	+ 2.39	− 0.16	+ 0.53
Annual average	NH$_L$	− 0.30	+ 1.06	+ 2.82	+ 0.39	+ 1.19
	NH	− 0.19	+ 0.56	+ 2.63	+ 0.20	+ 0.96
	SH	− 0.55	+ 0.11	+ 2.77	− 0.11	+ 0.69

Values of f_1, f_2, f_{12} are scaled by a factor of 10^3. ΔS indicates the difference between the response of the linear system without any synergism and the present-day signal. $\Delta \hat{S}$ is the difference between the mid-Holocene and present-day response of the full system.

pletely different models—for the atmosphere–vegetation system, but not yet for the complete climate system. Hence, we must consider the existence of multiple equilibria as a hypothesis awaiting further analysis and palaeo climate simulations.

Validation, of course, is a major problem in this field. So far, I have discussed mainly palaeo climate simulations. For a good reason: in many papers on biosphere–climate interaction, validation is not really considered. Instead, models being calibrated to present-day climate are applied to scenario experiments. These experiments are interesting from the academic point of view. However, their value for an assessment of future climate is limited.

Often, validation is done separately. On the one side, modules which simulate near-surface energy, moisture, and momentum fluxes in an atmospheric model are evaluated against data (e.g., in the frame of PILPS, the Project for the Intercomparsion of Land surface Parameterizations Schemes; Henderson-Sellers *et al.*, 1995). On the other side, vegetation models are tested by intercomparison with other models (Cramer *et al.*, 2000). This can be only a first step, which is quite appropriate as long as atmospheric models and vegetation models are not directly coupled; for example, if the two models do not share the same module of soil hydrology. If validation of fully coupled models is considered, in particular validation of continental-scale vegetation dynamics, then comparison of model results with palaeo climate reconstructions is the only way. As a side effect, this approach has the advantage that climate modelers do not need to rely on "soft" data, i.e., proxy-data from which the state of the atmosphere is derived indirectly. Instead, biospheric variables appear as (prognostic) state variables of the climate system model and can be used for direct validation. The PMIP, the Palaeoclimate Modelling Intercomparison Project (Joussaume and Taylor, 1995), provides a proper framework for this effort.

Finally, biogeophysical feedbacks and *biogeochemical feedbacks* are closely related (Schimel, 1998). Ignoring biogeochemical feedbacks seems to be reasonable for periods of nearly constant atmospheric composition, which presumably do not exist. Even throughout the last 6000–8000 years, atmospheric CO_2 concentrations have increased by some 20 ppm (Indermühle *et al.*, 1999). The assumption that this increase is caused by the decline in boreal forests and subtropical (mainly north African) grassland and savanna is not at variance with reconstructions of $\delta^{13}C$ values by Indermühle *et al*. Hence one may suspect that the decline in vegetation during the last 6000–8000 years which has amplified the long-term cooling via biogeophysical feedbacks and synergisms has also weakend the cooling trend via biogeochemical feedbacks.

The interaction between biogeophysical and biogeochemical feedbacks is quite subtle: while the latter tends to be negative through its interaction with greenhouse gases, the former can be either positive or negative, depending on whether changes in vegetation structure affect evaporation or albedo more strongly. It would be an interesting task to explore the spatial and temporal dynamics of the biogeophysical–biogeochemical interplay. Presumably, there are regions on earth in which, depending on external forcing and earth's history, the one or the other dominates. I

bet that, by solving this riddle, will we will find the answer to the question of climate-system stability which is a prerequisite for assessing the resilience of the present-day climate to large-scale perturbation such as the continuing release of fossil fuel combustion products into the atmosphere or the fragmentation of terrestrial vegetation cover.

Acknowledgments

This chapter could not have been written without the fruitful discussion within the CLIMBER group, in particular with Victor Brovkin, Andrey Ganopolski, Claudia Kubatzki, Stefan Rahmstorf, and Vladimir Petoukhov. Furthermore, I thank André Berger, Université Catholique Louvain la Neuve, for constructive comments. This work is partially funded by the European Union, Contract ENV4-CT97-0696.

References

Anhuf, D., Frankenber, P., and Lauer, W. (1999). Die postglaziale Warmphase vor 8000 Jahren. *Geologische Rundschau*. **51**, 454–461.

Berger, A. (1999). The role of CO_2, sea-level and vegetation during the Milankovitch-forced glacial-interglacial cycles. in: Bengtsson, L.O., Hammer, C. U. (eds.): Geosphere-Biosphere Interactions and Climate. Cambridge Univ. Press, New York, in press.

Berger, A., Fichefet, T., Gallée H., Tricot, Ch., and van Ypersele, J. P. (1992). Entering the glaciation with a 2-D coupled climate model. *Quaternary Sci. Rev.* **11(4)**, 481–493.

Berger, A., Gallée, H., and Tricot, Ch. (1993). Glaciation and deglaciation mechnisms in a coupled two-dimensional climate–ice-sheet model. *J. Glaciol.* **39(131)**, 45–49.

Bonan, G. B., Pollard, D., and Thompson, S. L. (1992). Effects of boreal forest vegetation on global climate. *Nature* **359**, 716–718.

Broström, A., Coe, M., Harrison, S., Gallimore., R., Kutzbach, J.E., Foley, Prentice, I.C., Beh-ling, P., 1998: Land surface feedbacks and palaeomonsoons in northern Africa. *Geophys. Res. Letters*, **25(No 19)**, 3615–3618

Brovkin, V., Claussen, M., Petoukhov, V., and Ganopolski, A. (1998). On the stability of the atmosphere-vegetation system in the Sahara/Sahel region. *J.Geophys. Res.* **103 (D24)**, 31613–31624.

Brovkin, V., Ganopolski, A., Claussen, M., Kubatzki, C., and Petoukhov, V. (1999). Modelling climate response to historical land cover change. *Global Ecol. Biogeography*, **8(6)**, 509–517.

Budyko, M. I. (1974). "Climate and Life." International Geophysical Series, Vol. 18. Academic Press, New York. 508 p.

Charney, J. G. (1975). Dynamics of deserts and drought in the Sahel. *Quart. J. R. Met. Soc.* **101**, 193–202.

Cheddadi, R., Yu, G., Guiot, J., Harrison, S. P., and Prentice, I. C. (1997). The climate of Europe 6000 years ago. *Climate Dynamics* **13**, 1–9.

Claussen, M. (1994). On coupling global biome models with climate models. *Climate. Res.* **4**, 203–221.

Claussen, M. (1997). Modeling biogeophysical feedback in the African and Indian Monsoon region. *Climate Dynamics*. **13**, 247–257.

Claussen, M. (1998). On multiple solutions of the atmosphere-vegetation system in present-day climate. *Global Change Biol.* **6**, 369–377.

Claussen, M. and Gayler, V. (1997). The greening of Sahara during the mid-Holocene: results of an interactive atmosphere-biome model. *Global Ecol. Biogeography Lett.* **6**, 369–377.

Claussen, M., Brovkin, V., Ganopolski, A., Kubatzki, C., and Petoukhov, V. (1998). Modeling global terrestrial vegetation-climate interaction. *Phil. Trans. R. Soc. Lond.* **B 353**, 53–63.

Claussen, M., Kubatzki, C., Brovkin, V., Ganopolski, A., Hoelzmann, P., and Pachur, H. J. (1999). Simulation of an abrupt change in Saharan vegetation at the end of the mid-Holocene. *Geophys. Res. Lett.* **24 (14)**, 2037–2040.

Cubasch, U., Santer, B. D., and Hegerl, G. C. (1995). Klimamodelle—wo stehen wir? *Phys. Bl.* **51**, 269–276.

Cramer, W., Bondeau, A., Woodward, F. I., Prentice, I. C., Betts, R. A., Brovkin, V., Cox, P. M., Fisher, V., Foley, J., Friend, A. D., Kucharik, C., Lomas, M. R., Ramankutty, N., Sitch, S., Smith, B., White, A., and Young-Molling, C. (2000). Global response of terrestrial ecosystem structure and function to CO_2 and climate change: results from six dynamic global vegetation models. *Global Change Biol.* in press.

Eischeid, J. D., Diaz, H. F., Bradley, R. S., and Jones, P. D. (1991). A comprehensive precipitation data set for global land areas. DOE/ER-6901T-H1, TR051. United States Dep. of Energy, Carbon Dioxide Research Program, Washington, DC.

Eltahir, E. A. B. (1996) Role of vegetation in sustaining large-scale atmospheric circulation in the tropics *J. Geophys. Res.* **101 (D2)**, 4255–4268

Foley, J., Kutzbach, J. E., Coe, M. T., and Levis, S. (1994). Feedbacks between climate and boreal forests during the Holocene epoch. *Nature* **371**, 52–54.

Frenzel, B., Pesci, M., and Velichko, A. A. (1992). Atlas of paleoclimates and paleoenviroments of the Northern Hemisphere: Late Pleistocene-Holocene. Geographical Research Institute, Budapest.

Gallée, H., van Ypersele, J.P., Fichefet, T., Marsiat, I., Tricot, C., Berger, A. 1992: Simulation of the last glacial cycle by a coupled, sectorial averaged climate-ice sheet model. 2. response ot insolation and CO2 variations *J. Geosphys. Res.* **97**, No. D14, 15713–15740

Ganopolski, A., Kubatzki, C., Claussen, M., Brovkin, V., and Petoukhov, V. (1998). The influence of vegetation-atmosphere-ocean interaction on climate during the mid-Holocene. *Science* **280**, 1916–1919.

Ganopolski, A., Petoukhov, V., Rahmstorf, S., Brovkin, V., Claussen, M., Eliseev, A, and Kubatzki, C. (2000). CLIMBER-2: a climate system model of intermediate complexity. Part II: Validation and sensitivity tests. Climate Dynamics, (submitted).

Gallimore, R. G. and Kutzbach, J. E. (1996). Role of orbitally-induced vegetative changes on incipient glaciation. *Nature* **381**, 503–505.

Gutman, G. (1984). Numerical experiments on land surface alterations with a zonal model allowing for interaction between the geobotanic state and climate. *J. Atmos. Sci.* **41**, 2679–2685.

Gutman, G. (1985). On modeling dynamics of geobotanic state-climate interaction. *J. Atmos. Sci.* **43**, 305–306.

Gutman, G., Ohring, G., and Joseph, J. H. (1984). Interaction between the geobotanic state and climate: a suggested approach and a test with a zonal model. *J. Atmos. Sci.* **41**, 2663–2678.

Harvey, L. D. D. (1988). A semianalytic energy balance climate model with explicit sea ice and snow physics. *J Climate* **1**, 1065–1085.

Harvey, L. D. D. (1989a). An energy balance climate model study of radiative forcing and temperature response at 18 ka. *J. Geophys. Res.* **94(D10)**, 12873–12884.

Harvey, L. D. D. (1989b). Milankovitch forcing, vegetation feedback, and North Atlantic deep-water formation. *J. Climate* **2**, 800–815

Henderson-Sellers, A. (1993). Continental vegetation as a dynamic component of global climate model: a preliminary assessment. *Climatic Change* **23**, 337–378.

Henderson-Sellers, A., Dickinson, R. E., Durbridge, T. B. Kennedy, P. J., McGuffie, K., and Pitman, A. J. (1993). Tropical deforestation: Modeling local- to regional-scale climate. *J.Geophys. Res.* **98(D4)**, 7289–7315.

Henderson-Sellers, A., Pitman, A. J., Love, P. K., Irannejad, P., and Chen, T. (1995). The project for the intercomparison of land surface parameterisation schemes (PILPS) phases 2 and 3. *Bull. Am. Meteorol. Soc.* **76**, 489–503.

Hewitt C. D. and Mitchell, J. F. B. (1998). A fully coupled GCM simulation of the climate of the mid-Holocene. *Geophys. Res. Lett.* **25(3)**, 361–364.

Hoelzmann, P., Jolly, D., Harrison, S. P., Laarif, F., Bonnefille, R., and Pachur, H.-J. (1998). Mid-Holocene land-surface conditions in northern Africa and the arabian peninsula: a data set for the analysis of biogeophysical feedbacks in the climate system. *Global Biogeochem. Cycles* **12**, 35–51.

Houghton, J. T., Filko, L. G. M., Griggs, D. J., and Maskell, K. (1997). An introduction to simple climate models used in the IPCC second assessment report. *IPCC Technical Paper II.* von Humboldt, A. (1849) Ansichten der Natur. 3rd ed. J. G. Cotta, Stuttgart and Tübingen, Reprint 1969. P. Reclam, Stuttgart.

Indermuehle, A., Stocker, T. F., Joos, F., Fischer, H, Smith, H. J., Wahlen, M, Deck B., Mastroianni, D., Tschumi, J., Blunier, T., Meyer, and R., Stauffer, B., 1999, Holocence carbon-cycle dynamics based on CO2 trapped in ice at Taylor Dome, Antarctica. *Nature* **398**, 121–126.

Jolly, D., Harrison, S. P., Damnati, B., and Bonnefille, R. (1998). Simulated climate and biomes of Africa during the late quarternary: comparison with pollen and lake status data. *Quaternary Sci. Rev.* **17(6–7)** 629–657.

Jousaume, S. and Taylor, K. (1995). Status of the paleoclimate modeling intercomparison project. In "Proceedings of the first international AMIP scientific conference." pp. 425–430. WCRP Report.

Köppen, W. (1936). Das geographische System der Klimate. In "Handbuch der Klimatologie." (W. Köppen and R. Geiger, Eds.), Band 5, Teil C. Gebrüder Bornträger, Berlin.

Kubatzki, C. and Claussen, M. (1998). Simulation of the global biogeophysical interactions during the last glacial maximum. *Climate Dynamics.* **14**, 461–471.

Kutzbach, J. E., Bonan, G., Foley, J., and Harrison, S. P. (1996). Vegetation and soil feedbacks on the response of the African monsoon to orbital forcing in the early to middle Holocene. *Nature* **384**, 623–626.

Kutzbach, J. E. and Guetter, P. J. (1986). The influence of changing orbital parameters and surface boundary conditions on climate simulations for the past 18,000 years. *J. Atmos. Sci.* **43**, 1726–1759.

Kutzbach, J. E. and Liu, Z. (1997). Response of the African monsoon to orbital forcing and ocean feedbacks in the middle Holocene. *Science* **278**, 440–443.

Levis, S., Foley, J. A., Brovkin, V., and Pollard, D. (1999). On the stability of the high-latitude climate-vegetation system in a coupled atmosphere-biosphere model. *Global Ecol Biogeography*, **8(6)**, 489–500

Lofgren, B. M. (1995a). Sensitivity of land-ocean circulations, precipitation, and soil moisture to perturbed land surface albedo. *J. Climate* **8(10)**, 2521–2542.

Lofgren, B. M. (1995). Surface albedo-climate feedback simulated using two-way coupling. *J. Climate* **8(10)**, 2543–2562.

deMarchi, L. (1885). in "Arrhenius, S. (1896)." On the influence of cabonic acid in the air upon the temperature of the ground. *The London, Edin-*

burgh, and Dublin Philosophical Magazine and Journal of Science **41**, 237–276.

deMenocal, P. B, Ortiz, J., Guilderson, T., Adkins, J., Sarnthein, M., Baker, L., and Yarusinski, M. (2000). Abrupt onset and termination of the African Humid Period: rapid climate response to gradual insolation forcing. *Quat. Sci. Rev*, **19**, 347–361.

deNoblet, N., Prentice, I. C., Jousaume, S., Texier, D., Botta, A., and Haxeltine, A. (1996). Possible role of atmosphere-biosphere interactions in triggering the last glaciation. *Geophys. Rev. Lett.* **23(22)**, 3191–3194.

deNoblet, N., Claussen, M., and Prentice, I. C. (2000). Mid-Holocene greening of the Sahara: first results of the GAIM 6000 year BP Experiment with two asynchronously coupled atmosphere/biome models. *Climate Dyn.* **16(9)**, 643–659

Otterman, J., Chou, M.-D., and Arking, A. (1984). Effects of nontropical forest cover on climate. *J. Climate Appl. Met.* **23**, 762–767.

Pachur, H.-J. and Altmann, N. (1997). The Quaternary (Holocene, ca. 8000a BP). In "Palaeo-geographic-Palaeotectonic atlas of North-Eastern Africa, Arabia, and adjacent areas Late Neoproterozoic to Holocene." (H. Schandelmeier and P.-O. Reynolds, Eds.), pp. 111–125.

Pachur, H.-J. and Hoelzmann, P. (1991). Paleoclimatic implications of Late Quaternary Lacustrine Sediments in Western Nubia, Sudan. *Quat. Res.* **36**, 257–276.

Pachur, H.-J. and Wünnemann, B. (1996). Reconstruction of the palaeoclimate along 30 °E in the eastern Sahara during the Pleistocene/Holocene transition. In "Palaeoecology of Africa and the Surrounding Islands." (K. Heine, Ed.), pp. 1–32.

Peixoto, J. P., and Oort, A. H. (1992). "Physics of Climate." *American Institute of Physics, New York.*

Petit-Maire, N. and Guo, Z. (1996). Mise en evidence de variations climatiques holocenes rapides, en phase dans les deserts actuels de Chine et du Nord de l'Afrique. *Sciences de la Terre et des Planetes* **322**, 847–851.

Polcher, J. and Laval, K. (1994). The impact of African and Amazonian deforestation on tropical climate. *J. Hydrol.* **155**, 389–405.

Prentice, I. C., Cramer, W., Harrison, S. P., Leemans, R., Monserud, R. A., and Solomon, A. M. (1992). A global biome model based on plant physiology and dominance, soil properties and climate. *J. Biogeography* **19**, 117–134.

Schimel, D. (1998). The carbon equation. *Nature* **393**, 208–209.

Schlesinger, M. E. (1988). Quantitative analysis of feedbacks in climate model simulations of CO_2-induced warming. In "Physically-based modelling and simulation of climate and climate change, part 2." (M. E. Schlesinger, Ed.), pp. 653–735. NATO ASI Series.

Stein, U. and Alpert, P. (1993). Factor separation in numerical simulations. *J. Atmos. Sci.*, **50**, 2107–2115.

TEMPO. (1996). Potential role of vegetation feedback in the climate sensitiviy of high-latitude regions: A case study at 6000 years B. P. *Global Biogeochem. Cycles* **10(4)**, 727–736.

Texier, D. de Noblet, N., Harrison, S. P., Haxeltine, A., Jolly, D., Joussaume, S., Laarif, F., Prentice, I. C., and Tarasov, P. (1997). Quantifying the role of biosphere-atmosphere feed-backs in climate change: coupled model simulations for 6000 years BP and comparison with palaeodata for northern Eurasia and northern Africa. *Climate Dynamics.* **13**, 865–882.

Wang, G. and Eltahir, E. A. B. (2000). Biosphere-atmosphere interactgions over west Africa. 2. Multiple Equilibira. *Q. J. R. Meteorol. Soc.* **126**, 1261–1280.

Zeng, N. and Neelin, J. D. (2000). The role of vegetation-climate interaction and interannual variability in shaping the African savanna. *J. Climate.* in press.

Zheng X. and Elthair, E. A. B. (1997) The response to deforestation and desertification in a model of West African monsoons. *Geophys. Res. Lett* **24(2)**, 155–158.

1.6

Land–Ocean–Atmosphere Interactions and Monsoon Climate Change: A Paleo-Perspective

John E. Kutzbach
*Center for Climatic Research,
    University of
    Wisconsin–Madison
Madison, Wisconsin*

Sandy P. Harrison
*Max Planck Institute for
    Biogeochemistry,
Jena, Germany*

Michael T. Coe
*Center for Sustainability and the
    Global Environment, University
    of Wisconsin-Madison,
Madison, Wisconsin*

1. Introduction

The climate system involves multiple interactions between the atmosphere, the land surface, and the oceans. Understanding both the physical and the biogeochemical linkages between these components is a fundamental challenge for earth system science. In addition to the complexity of the linkages and the existence of synergistic relationships between them (see, e.g., Berger, in press), there are very real difficulties in studying interactions which operate on timescales ranging from seconds to many millennia and in which, as a consequence, the relationship between cause and effect can be reversed (Schumm and Lichty, 1965). The seasonal cycle of atmospheric CO_2, for example, is controlled by changes in the terrestrial biosphere as a function of plant phenology (Knorr and Heimann, 1995). On longer (i.e., multimillennial to glacial–interglacial) timescales, changes in atmospheric $[CO_2]$ affect the competitive balance between C_3 and C_4 plants, which decreases the productivity of the terrestrial biosphere and causes massive redistributions of major vegetation types (Crowley and Baum, 1997; Levis *et al.*, 1999a). It is clear that a complete understanding of the physical and biogeochemical linkages in the earth system must include how they have operated on longer timescales and the consequences of changes in their operation.

Furthermore, the recent geological past offers significant opportunities for using climate and earth system models to study geosphere–biosphere interactions. First, the observed changes in climate and paleoenvironmental conditions were large (COHMAP Members, 1988; Wright *et al.*, 1993). We can therefore expect to be able to resolve these changes even with the present somewhat limited earth system modeling capability. Second, the fundamental cause of these changes lies in changes in earth's orbital geometry (Hays *et al.*, 1976; Imbrie, 1985; Berger, 1988), and the consequent changes in the seasonal and latitudinal distribution of incoming solar radiation (insolation) can be precisely specified (Berger, 1978; Berger and Loutre, 1991). Finally, at least for the most recent 10,000–30,000 years of the earth's history, the continental-scale to global-scale databases that have been assembled provide spatially- explicit reconstructions of climate and environmental parameters that can be used to benchmark model simulations (see, e.g., Kohfeld and Harrison, 2000). Thus, we can anticipate being able to study the changing role of physical and biogeochemical linkages between the atmosphere, the land, and the ocean within the climate system in an increasingly detailed fashion by combining pale o-observations and carefully designed model experiments.

Nevertheless, these studies (and more specifically the models required to make them) are still in their infancy. Until recently,

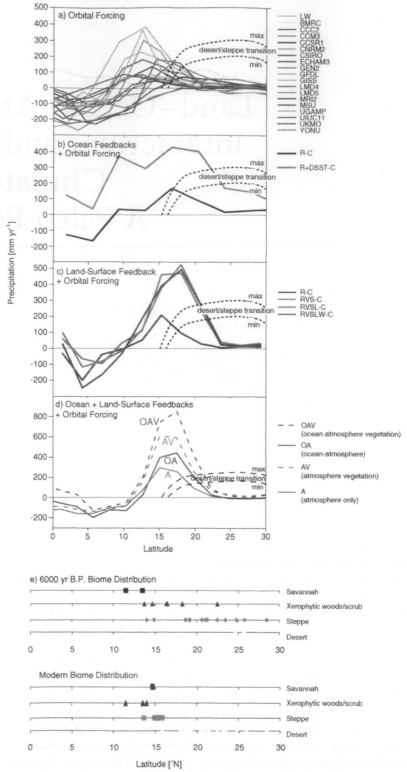

FIGURE 1 Zonally averaged simulated annual precipitation anomalies (6000 year B.P.—Control) versus latitude for northern Africa (land grid cells between 20°W and 30°E). Precipitation anomalies include the effects of: (a) radiative forcing (R) alone for the 18 climate models participating in the Paleoclimate Modeling Intercomparison Project (Joussaume *et al.,* 1999); (b) radiative forcing plus ocean feedbacks (ΔSST) for an asynchronous coupling of GENESIS2 and MOM1 (Kutzbach and Liu, 1997); (c) radiative forcing plus land-surface feedbacks (soil, S; vegetation, V; lakes, L; and wetlands, W) simulated using CCM3 (Broström *et al.,* 1998); and (d) radiative forcing (A) plus ocean feedbacks (OA) from a fully coupled simulation with the IPSL AOGCM, radiative forcing plus vegetation feedbacks (AV) from an AGCM simulation forced with 6000 yr B.P. vegetation derived by forcing BIOME1 with the output from the OAGCM simulation, and radiative forcing plus ocean- and land-surface feedbacks from an asynchronous coupling of the IPSL AOGCM and BIOME1 (Braconnot *et al.,* 1999). The hatched lines in (a–d) represent upper and lower estimates of the additional precipitation (excess over modern) required to support the grassland vegetation observed in northern Africa at 6000 yr B.P. (see Joussaume *et al.,* 1998). (e) Latitudinal distribution of biome types (desert, steppe, xerophytic, and savannah) for 6000 yr B.P. and 0 yr B.P. over the longitudes 20W–30E (Joussaume *et al.,* 1999). See also color insert.

past global changes have been primarily studied with atmospheric general circulation models (AGCMs). In simulations made with these models, other components of the earth system that were thought to have been important at a particular time (e.g., changes in the extent of land, ocean, and ice cover at the last glacial maximum, ca. 21,000 yr B.P.) are specified from observations. This approach is limited, both because of our inability to completely specify many paleoenvironmental boundary conditions (see, e.g., Broccoli and Marciniak, 1996) and, perhaps more importantly, because it ignores potential feedbacks between, e.g., land-surface, biospheric, or ocean changes on the atmosphere. The advent of fully coupled ocean–atmosphere models (e.g., Meehl, 1995; Murphy, 1995; Johns et al., 1997; Braconnot et al., 1997; Gent et al., 1998) and atmosphere–vegetation models (e.g., Foley et al., 1998, in press; Levis et al., 1999a,b) allows these feedbacks to be included. However, full coupling between the ocean, atmosphere, and vegetation has only been achieved (to date) in very much simplified models (the so-called EMICs or models of intermediate complexity: see, e.g., Gallée et al., 1992; Ganopolski et al., 1998). EMICs do not incorporate even all of those land-surface processes that are thought to impact on climate, and there are many other aspects of the interaction between the land, ocean, and atmosphere that have not been addressed in any coupled modeling scheme.

Although we are still a long way from having fully functional earth system models, some significant progress toward understanding the physical and biogeochemical linkages between the atmosphere, the land surface, and the oceans on geological timescales has been made during the past years. This chapter is not meant to provide a complete review of the state of knowledge, but rather (a) to illustrate the gains that have been made in understanding the linkages and feedbacks within the earth system by focusing particularly on the tropical monsoons and (b) to suggest some areas for future interdisciplinary work in this area.

2. Response of the Monsoon to Orbital Forcing

The expansion of the area influenced by the Afro-Asian summer monsoons during the early to mid-Holocene is one of the most striking features shown by palaeoenvironmental data (see, e.g., Street and Grove, 1976; Street-Perrott and Harrison, 1985; Street-Perrott et al., 1989; Roberts and Wright, 1993; Street-Perrott and Perrott, 1993; Winkler and Wang, 1993; Gasse and van Campo, 1994; Jolly et al., 1998a,b; Prentice and Webb, 1998; Yu et al., 1998, 2000; Kohfeld and Harrison, 2000; Prentice et al., 2000). The fundamental mechanism underlying these changes is well known (see, e.g., Kutzbach, 1981; Kutzbach and Otto-Bleisner, 1982; Kutzbach and Street-Perrott, 1985; Kutzbach and Guetter, 1986; Kutzbach and Gallimore, 1998; COHMAP Members, 1988; Kutzbach et al., 1993): the orbitally induced enhancement of

Northern Hemisphere summer insolation during the early to mid-Holocene (Berger, 1978) resulted in increased heating over the Northern Hemisphere continents and thus intensified the thermal contrast between the land and the ocean. The increased heating over the continents resulted in the northward displacement of the intertropical convergence zone (ITCZ) and hence of the monsoon front, while the enhanced land–sea contrast increased the flux of moisture from the ocean to the continent.

The response of the climate system to orbital forcing during the mid-Holocene (ca. 6000 Yr B.P.) has been investigated by a range of atmospheric general circulation models (AGCMs) within the Palaeoclimate Modelling Intercomparison Project (PMIP: Joussaume and Taylor, 1995; 2000). In these simulations, the atmospheric [CO_2] was reduced (from 345 to 280 ppmv), but land-surface conditions and sea-surface temperatures (SSTs) were prescribed to be the same as today. The effect of the reduction in atmospheric [CO_2] is negligible given that the simulations were run with fixed (modern) SSTs; thus, the experiment can be viewed primarily as an examination of role of orbitally induced insolation changes on climate. The PMIP simulations confirm that orbital changes produce a significant enhancement of the Afro-Asian monsoons but show that the magnitude of the enhancement varies from model to model (Joussaume et al., 1999; see also individual simulations: Dong et al., 1996; Hewitt and Mitchell, 1996; Lorenz et al., 1996; Hall and Valdes, 1997; Masson and Joussaume, 1997; Vettoretti et al., 1998). The sensitivity of the monsoonal response to orbital forcing is a function of the climatological characteristics of the model: models whose African summer monsoon limit is farther north in the control simulation tend to demonstrate a larger northward extension of the monsoon limit in response to 6000 yr B.P. orbital forcing (Joussaume et al., 1999). One of the controls on the variation in the magnitude of the response between models appears to be the dynamical structures of regional subsidence and the subtropical anticyclone over northern Africa, which in turn are influenced by global-scale dynamics (de Noblet-Ducoudré et al., 2000) and are ultimately tied to the global-scale response to orographic and diabatic forcing. These differences in AGCM base-state dynamics apparently play a dominant role in determining the model response to changes in forcing (see, e.g., Masson et al., 1998) even when other components of the climate system, such as the ocean, are included in the simulation (Harrison et al., unpublished analyses).

Comparison of the simulated enhancement of the African monsoon with a variety of pale o-observations shows that the PMIP simulations (in common with earlier simulations of the response to orbital forcing) consistently underestimate both the northward shift in the monsoon belt shown by paleoenvironmental data and the magnitude of the precipitation required to produce the observed lake and vegetation changes in northern Africa. Comparisons of the spatial patterns in the simulated P − E fields with lake data from northern Africa (Yu and Harrison, 1996), for example, indicate that the PMIP simulations consistently underestimate the northward shift in the monsoon front. Similarly, when the changes in precipitation simulated in the PMIP experiments are used to drive an equilibrium vegetation model (BIOME3: Haxeltine and Prentice, 1996) in order to evaluate the likely response of vegetation to the simulated change in climate, the simulations consistently fail to reproduce the observed northward shift in the Sahara/Sahel boundary (Harrison *et al.,* 1998). The precipitation required to generate the observed latitudinal distribution of grassland (steppe) vegetation in northern Africa at 6000 yr B.P. has been estimated using a combination of forward-modeling and inverse techniques. Joussaume *et al.* (1999) showed that the PMIP simulations underestimate the required precipitation at ca. 23°N by at least 100 mm (Fig. 1a), i.e., by ca. 50% of the minimum amount required to support grassland. When output from the PMIP experiments is used to simulate the extent of lakes across northern Africa using the HYDRA model (Coe, 1998; 2000), the observed area of Lake Chad (350,000 km²: Schneider, 1967; Pias, 1970) during the mid-Holocene is significantly underestimated by all of the models (Coe and Harrison, 2000). The failure of the PMIP simulations (in common with earlier AGCM simulations of the response to orbital forcing) to reproduce the observed changes in the African monsoon during the mid-Holocene provides strong support for the argument that the response to orbital forcing is mediated by feedbacks associated with changes in either the ocean or the land-surface.

3. Ocean Feedbacks on the Monsoon

Several studies indicate that ocean processes can produce feedbacks that enhance the monsoon response to 6000 yr B.P. orbital forcing (e.g., Kutzbach and Liu, 1997; Hewitt and Mitchell, 1998; Liu *et al.,* 1999a,b; Otto-Bliesner, 1999; Texier *et al.,* 2000; Braconnot *et al.,* 2000). In the asynchronously coupled atmosphere–ocean experiments performed by Kutzbach and Liu (1997), for example, precipitation over northern Africa increases by 25% compared to simulations made with prescribed modern SSTs (Fig. 1b). Monsoon enhancement is expressed by a northward shift of the monsoon front, increased precipitation and, at least in some cases, by an extension in the length of the monsoon season. A number of different processes appear to be involved. Radiative forcing alone, operating in a static column energy budget, cools the tropical Atlantic both north and south of the equator by as much as 0.5 °C in the spring (February through May) and raises the temperature by a comparable amount in autumn (August through November). The cooler ocean in the spring and early summer can enhance land–sea temperature contrast and thereby strengthen the African monsoon at onset (Hewitt and Mitchell, 1998). The fundamental changes in surface windflow associated with the orbitally forced enhancement of the southwesterly atmospheric inflow to West Africa can also act to decrease the normal north-easterly trades of the tropical North Atlantic, thereby reducing the total wind speed

over the eastern Atlantic and consequently reducing evaporative cooling of the ocean surface. This change in the column energy budget preferentially increases SSTs north of the equator during the summer/autumn (Kutzbach and Liu, 1997). Most models show enhanced warming to the north of the equator, and some experiments even show a slight cooling south of the equator, during the summer/autumn months (Kutzbach and Liu, 1997; Hewitt and Mitchell, 1998; Otto-Bleisner, 1999; Braconnot *et al.,* in press; Liu *et al.,* 1999a). As shown by Hastenrath (1985) and others, a changed cross-equatorial SST gradient (warmer to the north) is of importance in producing a northward shift of the ITCZ in the North Atlantic and thereby increased advection of moisture into northern Africa from the west. Braconnot *et al.* (2000) have shown that increased south to north advection of heat within the upper ocean can also contribute to this dipole structure. Although most attention has been paid to the effects of orbitally forced ocean changes in the Atlantic on the African monsoon, it is possible that orbitally forced changes in the mean climate of the Pacific or Indian Oceans could, via teleconnection, influence the climate of northern Africa (Otto-Bleisner, 1999). In summary, models agree that there is a positive SST feedback effect on African monsoon precipitation in the mid-Holocene, although the relative importance of the various mechanisms that might contribute to this feedback requires further analysis. Furthermore, although the SST-driven enhancement is significant, the precipitation increase induced by the combined effect of orbital forcing and ocean feedbacks is not enough to support the observed grassland vegetation as far north as 23°N (Fig. 1b).

All of the studies that have been conducted to date, whether the simulations are simple sensitivity tests made by prescribing stylized changes in ocean temperature in an AGCM (Texier *et al.,* 2000) or with fully coupled ocean-atmosphere general circulation models (OAGCMs) (Hewitt and Mitchell, 1998; Otto-Bliesner, 1999; Braconnot *et al.,* in press), show that ocean feedbacks enhance the northern African monsoon, although the relative importance of the mechanisms by which this enhancement is produced may vary from model to model. However, there is far less agreement about the role of ocean feedbacks on other monsoon systems. In coupled OAGCM simulations for both the early Holocene (11,000 yr B.P.: Liu *et al.,* 1999a) and the mid-Holocene (6000 yr B.P.: Liu *et al.,* 2000), the enhancement of the Indian monsoon is less than that produced by orbital changes alone (Fig. 2). In the 11,000 yr B.P. simulation, the reduction in precipitation due to ocean feedbacks is ca. 30% of the simulated increase due to the direct radiation effect. At 6000 yr B.P. the reduction is ca. 12% of the simulated increase due to the direct radiation effect. The negative ocean feedback on Indian monsoon rainfall appears to be caused by warming of the tropical Indian Ocean, which causes anomalous convergent flow over the Indian Ocean and hence increases precipitation over the ocean while decreasing precipitation over the Indian subcontinent. The simulated warming of the tropical Indian Ocean is partly a direct response to increased summer insolation, but is also partly due

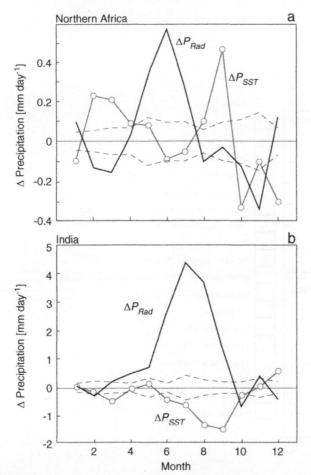

FIGURE 2 Changes in the annual cycle of monthly mean precipitation (mm day^{-1}) over (a) northern Africa (land only, 5°W–35°E, 5–30°N) and (b) India (land only, 75–90°E, 10–25°N). The solid line (no circles) is the precipitation change forced by 11,000 year B.P. insolation (ΔP_{rad}), obtained by holding SSTs at the modern (control) value. The solid line with circles is the precipitation change forced by the SST changes associated with 11,000 yr B.P. insolation (ΔP_{SST}). This SST feedback was isolated by differencing two 11,000 yr B.P. simulations: (1) an 11,000 yr B.P. simulation with SSTs specified from the results of an 11,000 yr B.P. coupled atmosphere–ocean simulation, and (2) an 11,000 yr B.P. simulation with SSTs specified from the coupled modern control simulation. All simulations for 11 ka used solar radiation values 11,000 yr B.P. based 11,000 yr B.P. orbital parameters, but other boundary conditions (atmospheric [CO_2], ice sheets) were set at modern values. The dashed lines represent the standard deviation based on the internal variability of a long control simulation. The coupled atmosphere–ocean simulations used equilibrium asynchronous coupling (Liu *et al.*, 1999a). The 11,000 year B.P. orbital forcing alone acts to enhance summer monsoon precipitation in both regions. The SST feedback is generally positive in the case of the African summer monsoon, with the main effects concentrated in spring and autumn, thereby increasing the length of the rainy season and the total precipitation. The SST feedback is generally negative in the case of the Indian summer monsoon.

to reduced evaporative cooling consequent on the weakening of the surface monsoon winds caused by the direct insolation response.

According to these AOGCM experiments, then, ocean feedbacks appear to somewhat damp the monsoon response to orbital forcing in India and increase the response in Africa. This difference in behavior may go some way to explaining why the observational evidence of monsoon changes in Africa is stronger than the response in India and over Asia more generally (though, admittedly, the amount of data from India is limited). A similar response (i.e., amplification of African monsoon precipitation and suppression of Indian monsoon precipitation) to ocean feedbacks is also shown in the experiments with prescribed SST changes by Texier *et al.* (2000). However, other OAGCM simulations (e.g., Hewitt and Mitchell, 1998; Braconnot *et al.*, 2000) apparently do not demonstrate a comparable reduction in the strength of orbitally induced enhancement of the Indian monsoon. Rather, in these simulations, ocean feedbacks further enhance the orbitally induced increase in the Indian monsoon in a fashion comparable to the enhancement of the African monsoon. The reasons for these differences between the response of the Indian monsoon to ocean feedbacks still need to be examined.

4. Land–Surface Feedbacks on the Monsoon

Changes in vegetation (and hence some soil characteristics, including organic matter content, and hence water-holding capacity and albedo) or the extent of surface water (lakes and wetlands) affect land-surface conditions through changing albedo (which determines the surface energy balance and hence surface heating), surface roughness (which affects both the water and energy fluxes between the land and the atmosphere), and moisture availability for recycling. The role of vegetation changes (and vegetation-induced soil changes) in enhancing orbitally induced changes in the monsoon circulation over northern Africa was originally studied by sensitivity experiments with stylized or quasi-realistic changes in vegetation and soil characteristics (e.g., Street-Perrott *et al.*, 1990; Kutzbach *et al.*, 1996; Broström *et al.*, 1998; Texier *et al.*, 2000). These experiments produce a significant enhancement of the African monsoon compared to the effects of orbital-forcing alone. The vegetation-induced lowering of albedo increases surface heating (and hence amplifies the land–sea contrast, promoting increased advection of moisture into the continent). At the same time, the presence of vegetation and changes in the water-holding capacity of the soils leads to increased moisture recycling. In the Broström *et al.* (1998) experiments, the presence of vegetation (experiment RVS) leads to a substantial warming over northern Africa during the spring and early summer. As a result (Fig. 1c), the onset of the monsoon occurs 2 months earlier than it does in response to orbital forcing alone (experiment R).

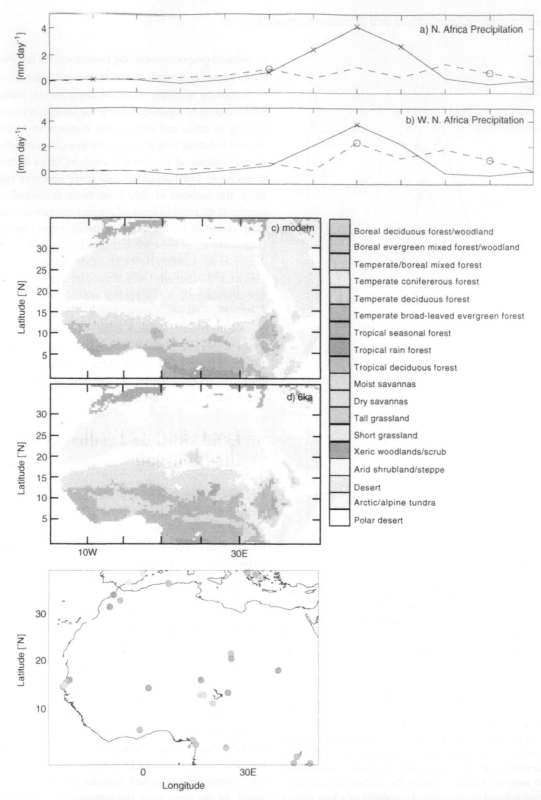

FIGURE 3 Simulated changes in precipitation and changes in biome distributions at 6000 year B.P. relative to 0 year B.P., obtained with and without vegetation feedbacks: (a) annual cycle of monthly mean precipitation change (mm day^{-1}) between simulations for 6000 yr B.P. and 0 year B.P. over northern Africa (land area, 11°W–34°E and 11–20°N) and (b) western portion of northern Africa (land area, 11°W–11°E, 11–20°N). The solid line is the precipitation change forced by 6000 yr B.P. insolation with vegetation set at the modern control values. The dashed line is the precipitation change at 6000 yr B.P. forced by vegetation feedback. The vegetation feedback was isolated by differencing two 6000 yr B.P. simulations: (1) a 6000 yr B.P. simulation with vegetation specified from the results of a 6000 yr B.P. coupled atmosphere–vegetation simulation, and (2) a 6000 yr B.P. simulation with vegetation specified from the modern coupled control simulation (Doherty *et al.*, 2000). The 6000 yr B.P. orbital forcing enhances the summer monsoon precipitation. Vegetation feedback enhances the precipitation in summer and autumn, with the largest effects in the western region. The overall effect is to lengthen the rainy season and increase the total precipitation. (c) Modern simulated biomes using BIOME 3 driven by present-day climatological values of the monthly mean annual cycle of temperature, precipitation, and solar radiation, and (d) 6000 yr B.P. simulated biomes using BIOME 3 driven by 6000 yr B.P. minus 0 yr B.P. differences in these climatic variables, taken from the coupled atmosphere–terrestrial vegetation model, and then combined with the present-day climatology (Doherty *et al.*, 2000). (e) Observed distribution of vegetation during the mid-Holocene (from Jolly *et al.*, 1998a,b). See also color insert.

The role of vegetation feedbacks has also been examined in a number of experiments with asynchronous coupling between an equilibrium vegetation model and an AGCM (e.g., Texier *et al.*, 1997; Claussen and Gaylor, 1997; Pollard *et al.*, 1998; de Noblet-Ducoudré *et al.*, 2000) and, most recently, using the dynamically coupled GENESIS-IBIS atmosphere–vegetation model (Doherty *et al.*, 2000; Fig. 3). A number of robust conclusions emerge from analyses of these simulations. Vegetation feedbacks increase precipitation during the peak of the monsoon season (July–August), by an amount comparable to the increase produced by orbital forcing alone but the absolute magnitude of this increase in precipitation is rather small. However, vegetation feedbacks have a significant impact on total precipitation by causing an extension of the monsoon season. Specifically, changes in albedo caused by the presence of vegetation lead to warming of the continent, thus enhancing land–sea contrast and increasing onshore advection, in spring and early summer and the onset of the monsoon therefore occurs 1–2 months (depending on the simulation) earlier than with orbital forcing alone (see, e.g., Broström *et al.*, 1998; Texier *et al.*, 2000; Doherty *et al.*, 2000). Vegetation feedbacks also tend to prolong the monsoon into the autumn. Advection is relatively weak during the autumn, so that the extension of monsoon rains into the autumn appears to reflect enhanced moisture recycling in these simulations. In general, vegetation feedbacks appear to increase the importance of moisture recycling in maintaining the monsoon. The relative importance of the contributions of advection and recycling can be estimated from $\Delta P = \Delta(P - E) + \Delta E$) (Braconnot *et al.*, 1999), where $\Delta(P - E)$ represents the advection term and ΔE the recycling term. In the Doherty *et al.* (2000) experiments, the advection and recycling terms are of comparable magnitude in response to orbital changes (0.39 mm day^{-1} vs 0.42 mm day^{-1}) but the increase in recycling (0.25 mm day^{-1}) in response to vegetation feedbacks is larger than the increase in advection (0.18 mm day^{-1}). The relative importance of advection and recycling in monsoon regions appears to vary between models, and the simulation of this partitioning has not been adequately evaluated. However, as models increasingly incorporate improved representations of vegetation and soil processes, there appears to be a significant increase in the importance of recycling relative to advection (see, e.g., Masson and Joussaume, 1997; Kleidon *et al.*, 2000). Finally, vegetation feedbacks have a more significant role in enhancing monsoon precipitation in West Africa than in the eastern Sahara/Sahel. In the coupled GENESIS-IBIS simulations (Doherty *et al.*, 2000), for example, vegetation feedbacks produce an increase in annual average precipitation of 0.63 mm day^{-1} comparable to the increase due to orbital forcing (0.66 mm day^{-1}) over West Africa, whereas the comparable estimates over the eastern Sahara/Sahel region are 0.17 and 1.01 mm day^{-1}. The regional differences in the impact of vegetation on the northern African monsoon are not so pronounced in e.g.,

the Texier *et al.* (1997) simulations but are much larger in, e.g., the Claussen and Gaylor (1997) experiments.

Although most attention has been directed toward studying the possible feedbacks associated with vegetation, these were not the only landscape changes that would have affected land-surface characteristics. The mid-Holocene landscape of northern Africa was likely a mosaic of lakes, wetlands, and grasslands (Hoelzmann *et al.*, 1998). Coe and Bonan (1997) used a model sensitivity experiment to illustrate that expanded lakes, specifically Palaeolake Chad and the lakes north of the Niger Bend, cause localized changes in circulation and some small enhancement of precipitation above and beyond that caused by orbital forcing alone. Prescribed additions of wetlands (with and without wetland vegetation) to a landscape of expanded grasslands also produced only small adjustments in large-scale precipitation (Carrington *et al.*, in press). Broström *et al.* (1998) analyses of the relative contribution of sequential changes in the extent of lakes (RVSL) and wetlands (RVSLW) compared to vegetation and vegetation-induced soil changes (RVS) across northern Africa confirm that the largest precipitation enhancement came from orbital forcing combined with changes in vegetation and soils; there was little or no additional enhancement of precipitation from wetlands or lakes (Fig. 1c). It is perhaps not surprising that the relatively small areas covered by wetlands and lakes seem to produce only small-scale (local) climate perturbations. In some situations, $P - E$ actually decreases over the expanded water surfaces because of increased evaporation, but the increase in precipitation over the surrounding catchment may be sufficient to maintain the lake and wetlands through local catchment-scale recycling. Nonetheless, these studies underscore the likelihood that only the relatively large-scale vegetation changes (and associated soil changes) interact to enhance regional precipitation, while the more localized areas of enhanced lakes and wetlands play only a minor role by comparison.

The amplification of the monsoon response to orbital forcing by land-surface feedbacks is apparently insufficient to explain the full observed expansion of the African monsoon. Thus, comparisons of the Broström *et al.* (1998) simulations with benchmark data show that even when all possible land-surface feedbacks are taken into account (RVSLW), they are insufficient to produce the full observed northward expansion of grasslands into regions occupied today by desert (Fig. 1c). There remains one further change in environmental conditions that might impact this question, however, namely the possibility that northward transport of excess runoff from the zone under the direct influence of the monsoon front could play a role. Simulations with the HYDRA model, driven by output from the GENESIS-IBIS coupled atmosphere–vegetation model (Figs. 4a, 4b) simulations for 6000 yr B.P., show that runoff is transported north from the zone of the monsoon front by a pale o-river network that is more extensive than today. Palaeodata (e.g., Hoelzmann *et al.*, 1998) confirm the existence of a more extensive and active

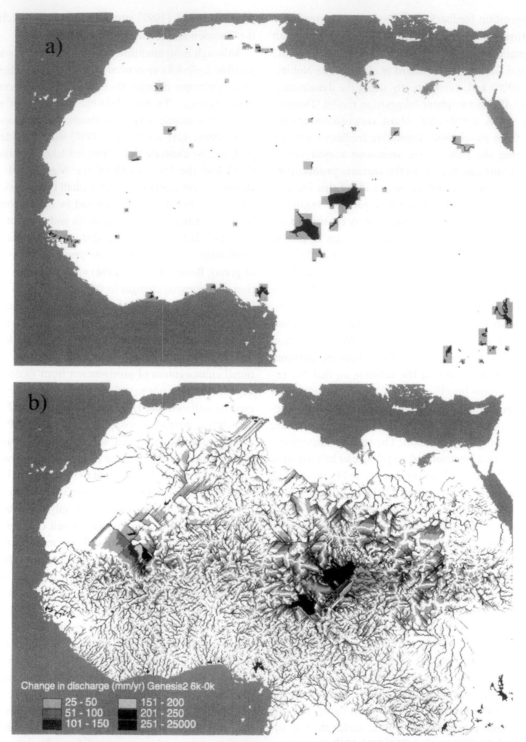

FIGURE 4 Surface hydrology of northern Africa, simulated by HYDRA forced by runoff generated by the GENESIS2 AGCM coupled to the IBIS ecosystem model. HYDRA operates on a 5′ × 5′ (ca. 10 km) global grid to simulate the flow of water from land surfaces through a complex of rivers, lakes, and wetlands to the ocean or to inland drainage basins (such as closed lakes and interdunal wetlands). (a) Surface water area for 6000 year B.P. simulated by HYDRA at the 5′ × 5′ horizontal (in black) showing pale o-lake Chad and other expanded pale o-lakes; and smoothened to 0.5° resolution (in pink) showing all regions with surface water area in excess of 10% of the 0.5° grid cell. The sum of the water areas at both resolutions is identical. (b) Change in annual mean discharge (in mm yr⁻¹) between simulations for 6000 and 0 yr B.P. over northern Africa. Only positive differences are shown. The colors represent those stream channels for which the discharge is increased in the 6000 yr B.P. experiment compared to modern. The results show the relatively large increase in runoff and stream flow in northern Africa (from 25–3000 mm yr⁻¹ increase). Greatest increases in discharge occur between about 15°–25°N and in Algeria. Paleostream channels occur throughout northern Africa where none exist today. Sheet-flow discharge across very flat terrain is also present in central Mali and in the northern basin of paleo-lake Chad. Simulated water areas of 6000 yr B.P. are shown in black. See also color insert.

river system in northern Africa during the mid-Holocene. It has been questioned whether the apparent expansion of steppe vegetation in northern Africa during the mid-Holocene reflected the presence of vegetation along water courses or in other well-watered locations, rather than the more general expansion implied by, e.g., the maps in Hoelzmann *et al.* (1998). If this were true, then the northward transport of excess runoff from the zone of the monsoon front, as shown in our simulations (Fig. 4b), would be a significant factor explaining the apparent mismatch between model simulations and observations. Although this suggestion cannot be entirely dismissed (given the limited number of sites documenting the mid-Holocene vegetation in northern Africa), it is unlikley that the northward expansion of steppe vegetation was confined to water courses or other preferred locations. Vegetation records south of ca. 23°N, for example, do not contain pollen from any obligate desert species (I. C. Prentice and D. Jolly, unpublished analyses cited in Joussaume *et al.*, 1999). Some desert indicators would be expected to be present if the sites south of 23°N were representative of a landscape in which islands of steppe (in more well-watered sites) were set in a matrix of desert. On the other hand, the northward transport of excess runoff from further south could be responsible for the maintenance of wetlands and even lakes well beyond the limits of the monsoon front (Fig. 4a). Insofar as wetlands and lakes have an impact on local moisture recycling, as shown by Coe and Bonan (1997), Broström *et al.* (1998), and Carrington *et al.* (2000), any mechanism which increases their extent north of the monsoon front could potentially lead to feedbacks which might affect the spatial expression of monsoon enhancement. Additional experiments which take into account changes in the surface hydrological network, through either asynchronous or explicit coupling of a terrestrial hydrology model like HYDRA with an atmosphere–vegetation model (AVCGM), are required to test whether this mechanism might have an impact on the simulated monsoon climate.

5. Synergies between the Land, Ocean, and Atmosphere

Since neither land-surface nor ocean-surface feedbacks alone are sufficient to explain the observed expansion of the African monsoon during the mid-Holocene, synergistic feedbacks involving land–atmosphere–ocean interactions are likely to be involved (Ganopolski *et al.*, 1998; Braconnot *et al.*, 1999; Berger, in press). There have been only two attempts to examine this question. In simulations with an intermediate-complexity model, Ganopolski *et al.* (1998) showed that vegetation feedbacks were more important than ocean feedbacks in the amplification of the African monsoon. This simulation may not be realistic, however, because the ocean model does not re-

solve the full dynamics of the equatorial ocean and the coarse spatial resolution of the atmospheric part of the model prevents the simulation of detailed regional monsoon changes. Braconnot *et al.* (1999) used asynchronous coupling between the IPSL OAGCM and an equilibrium biome model (BIOME1: Prentice *et al.*, 1992) to examine the synergistic relationships between land and ocean feedbacks. This simulation makes it clear that incorporation of both kinds of feedbacks amplifies the orbitally induced enhancement of the African monsoon (Fig. 1d). However, comparison of the amplification due either to the ocean alone (AO) or to vegetation alone (AV) in this model with the comparable feedback effects simulated by other models (e.g., respectively Fig. 1b: Kutzbach and Liu, 1997; and Fig. 1c: Broström *et al.*, 1998) suggests that there is a strong model dependence in the magnitude of the simulated response. Thus, these experiments need to be revisited using a number of other coupled ocean-atmosphere-vegetation (OAVGCM) models to assess their robustness.

The omission of land–surface and ocean feedbacks and their possible interactions have been invoked to explain other mismatches between observations and climate simulations of the effects of orbital changes at 6000 yr B.P., including the degree of warming in the northern high latitudes (e.g., Foley *et al.*, 1994; TEMPO, 1996; Texier *et al.*, 1997) and the anomalous (i.e., opposite to the orbital forcing) winter warming in Europe (Prentice *et al.*, 1998; Masson, 1998). Further improvements in the simulation of mid-Holocene climate changes will likely require the use of fully coupled OAVGCM models, which are now under development by several modeling groups.

6. The Role of Climate Variability

Changes in the mean climate state may be accompanied by changes in short-term (i.e., interannual to interdecadal) variability. The relationship between mean climate state and climate variability has not been extensively investigated, despite the fact that the impacts of climate change on earth systems may derive more from changes in variability than in the mean state. In large part this reflects the history of paleoclimate modeling. Simulations of Holocene paleoclimates using prescribed SSTs (as in PMIP) require only a short time (ca. 10–20 years) to reach equilibrium, and therefore attention naturally focused on describing only the mean climate since the simulation length was too short to permit studies of variability. As model simulations are necessarily extended with the advent of coupled OAGCMs and coupled atmosphere-vegetation general circulation models (AVGCMs), it has become both natural and important to focus on climate variability and the estimation of changes in climate variability.

Palaeoclimatic evidence suggests that short-term climate variability may have been substantially different from today during

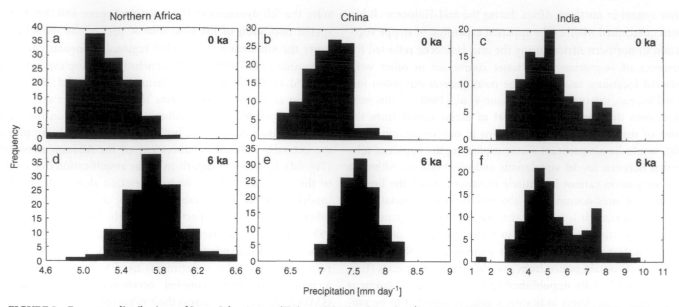

FIGURE 5 Frequency distributions of June–July–August (JJA) precipitation (mm day⁻¹) at 0 year B.P. for (a) northern Africa (10°W–20°E and 10–20°N), (b) China (100–120°E and 35–45°N), and (c) India (75–85°E and 25–30°N), for (d) northern Africa, (e) China and (f) India at 6000 year B.P. The simulated precipitation values were taken from the last 120 years of 150-year simulations with FOAM. The mean values (M) and standard deviation (SD) for each frequency distribution are shown. Especially in the case of the summer monsoon rains in northern Africa and China, the changes in the overall frequency distributions are very large. In northern Africa and China, the increases in mean JJA precipitation at 6000 year B.P. compared to 0 year B.P. pass two-tailed *t*-tests at the 95% level and the difference in the variances between 6000 yr B.P. and 0 yr B.P. pass an *F*- test at the 90% level. The largest variance (and the largest standard deviation) in northern Africa occurs at 6000 yr B.P., and the largest variance in China occurs at 0 yr B.P. The changes in mean and variance in the Indian region are not statistically significant.

the early to mid-Holocene. Time-series of archaeological deposits in northern Peru (Sandweiss *et al.*, 1996) and clastic deposits in an Andean lake in Ecuador (Rodbell *et al.*, 1999) indicate less severe flooding events along the west coast of tropical South America during the early Holocene. The δ¹⁸O records from the Sajama ice core in the tropical Andes also show less variability during the early Holocene than in the later Holocene (Thompson *et al.*, 1998; Thompson, 2000). Records of fires in Australia (McGlone *et al.*, 1992) and isotopic records from fossil corals in the western tropical Pacific (Gagan *et al.*, 1998) have been interpreted as showing that monsoon rainfall was less variable during the first half of the Holocene than today.

Liu *et al.* (1999b), in simulations examining the response of the Fast Ocean Atmosphere Model (FOAM: Jacob, 1998) to 11,000 yr B.P. orbital forcing, have shown that El Niño variability is reduced by ca. 20% and the spectral bandwidth of El Niño changes from the broad (3–10 year) peak characteristic of the modern simulations to a narrower peak (2–3 year) at 11,000 yr B.P. This reduction in the variability appears to be associated with the simulated increase in the Indian summer monsoon. There are, however, a number of mechanisms through which changes in the Indian summer monsoon circulation appear to impact the El Niño signal. First, the enhanced Indian monsoon

strengthens the deep convection in the eastern Indian Ocean and western Pacific warm pool, increasing the strength of the easterly trades (by ca. 1 m s⁻¹), and hence increasing the upwelling (and cooling) in the central and eastern Pacific. Ocean feedbacks further enhance this wind-driven cooling. In these simulations, the combination of forcing by changes in the Indian monsoon and positive ocean feedbacks leads to an SST cooling in the eastern Pacific of ca. 0.5°C in May–June–July. This cooling tends to suppress the growth of warm El Niño events during the Southern Hemisphere spring and therefore reduces their final amplitude later in the year. In the Liu *et al.* (1999b) simulations, the annual mean trades are stronger (by ca. 2 s⁻¹) than in the control simulation. This results in enhanced upwelling and therefore enhanced SST cooling throughout the year and provides a further mechanism for reducing El Niño variability. More recent simulations with the FOAM model (Liu *et al.*, 2000) show that reduced El Niño variability is also produced in response to 6000 yr B.P. orbital forcing.

Otto-Bliesner (1999) found that the teleconnections relating the patterns of Pacific ENSO to Sahelian rainfall in the 6000 yr B.P. experiment are different from those in the control (modern) simulation. We have therefore reexamined the FOAM simulations specifically to determine whether there are changes in precipitation variability associated with the mid-Holocene enhancement of the Afro-Asian monsoons. In northern Africa, the increased summer precipitation at 6000 yr B.P. is associated with a significant increase in interannual precipitation variability (Fig. 5). In India, the increase in mean precipitation during the monsoon season is also accompanied by increased interannual variability. However, the reverse is true in China. In our simulations, the enhancement of the Pacific monsoon leads to increased summer precipitation and reduced interannual variability (Fig. 5). The causes of these regional differences in the relationship between mean climate and climate variability required further analysis. However, it is clear that the increased/decreased variability at 6000 yr B.P. has the potential to significantly impact the regional paleoenvironmental response to the change in mean climate. For example, in relatively arid environments with comparable mean annual rainfall, sparse shrub or open woodland vegetation tends to be favored in regions with high interannual variability whereas steppe grasslands occur where the variability is less.

7. Final Remarks

The ability to correctly simulate past climates bears directly on whether we can confidently predict future climates (Joussaume, 1999). Comparisons of climate experiments with paleoenvironmental data have clearly demonstrated that the observed large changes in climate during the mid-Holocene (or at the LGM: see discussion in Kohfeld and Harrison, 2000) cannot be simulated without explicitly considering the feedbacks associated with the ocean, vegetation, and other components of the land-surface. There are a number of other feedbacks (e.g., radiative forcing by mineral aerosols at the LGM: Harrison *et al.*, in press; Claquin *et al.*, submitted) that are potentially important. COHMAP results (Kutzbach *et al.*, 1998) indicating that the apparent mismatches between observed and simulated climate changes during the transition from glacial to interglacial conditions are greater than at either the LGM or the mid-Holocene suggest that the incorporation of these feedbacks may be even more important in attempts to simulate times of rapid climate change when there is a strong disequilibrium between insolation and other conditions. Thus, simulations of potential future climate changes need to be made using fully coupled ocean–atmosphere–biosphere models, taking into account the potential additional impact of changes in surface conditions on atmospheric aerosols.

References

Berger, A. (1978). Long-term variation of daily insolation and Quaternary climatic changes. *J. Atmos. Sci.* **35**, 2362–2367.

Berger, A. (1988). Milankovitch theory and climate. *Rev. Geophys.* **26**, 624–657.

Berger, A. (in press). The role of CO$_2$ and of the geosphere-biosphere interactions during the Milankovitch-forced glacial-interglacial cycles. In *"Workshop on Geosphere-Biosphere Interactions and Climate."* Pontifical Academy of Sciences, Vatican City.

Berger, A. and Loutre, M.-F. (1991). Insolation values for the climate of the last million years. *Quaternary Sci. Rev.* **10**, 297–317.

Braconnot, P., Marti, O., and Joussaume, S. (1997). Adjustments and feedbacks in a global coupled ocean-atmosphere model. *Climate Dynamics.* **13**, 507–519.

Braconnot, P., Joussaume, S., Marti, O., and de Noblet, N. (1999). Synergistic feedbacks from ocean and vegetation on the African monsoon response to mid-Holocene insolation. *Geophys. Res. Lett.* **16**, 2481–2484.

Braconnot, P., Marti, O, Joussaume, S., and Leclainche, Y. (2000). Ocean feedback in response to 6 kyear BP insolation. *J. Climate.* **13**, 1537-1553.

Broccoli, A. J. and Marciniak, E. P. (1996). Comparing simulated glacial climate and paleodata: a reexamination. *Paleoceanography* **11**, 3–14.

Broström, A., Coe, M., Harrison, S. P., Gallimore, R., Kutzbach, J. E., Foley, J., Prentice, I. C., and Behling, P. (1998). Land surface feedbacks and paleomonsoons in northern Africa. *Geophys. Res. Lett.* **25**, 3615– 3618.

Carrington, D., Gallimore, R. G., and Kutzbach, J. E. (2000). Climate sensitivity to wetlands and wetland vegetation in mid–Holocene north Africa. *Climate Dynamics.* **17**, 151–157.

Claquin, T., Roelandt, C., Kohfeld, K. E., Harrison, S. P., Prentice, I. C., Balkanski, Y., Bergametti, G., Hansson, M., Mahowald, N., Rodhe, N., and Schulz, M. (submitted). Radiative forcing of climate by ice-age dust. *Nature.*

Claussen, M. and Gaylor, V. (1997). The greening of the Sahara during the mid-Holocene: results of an interactive atmosphere-biome model. *Global Ecol. Biogeography Lett.* **6**, 369–377.

Coe, M. T. (1998). A linked global model of terrestrial hydrologic processes: simulation of modern rivers, lakes, and wetlands. *J. Geophys. Res.* **103**, 8885–8899.

Coe, M. T. (2000). Modeling terrestrial hydrological systems at the continental scale: testing the accuracy of an atmospheric GCM. *J. Climate* **13**, 686–704.

Coe, M. T. and Bonan, G. B. (1997). Feedbacks between climate and surface water in Northern Africa during the Middle Holocene. *J. Geophys. Res.* **102**, 11087.

Coe, M. T. and Harrison, S. P. (2000). A comparison of the simulated surface water area in norhtern Africa for the 6000 year B.P. PMIP experiments. In *"Paleoclimate Modeling Intercomparison Project, Proceedings of the Third Conference."* (P. Braconnot, Ed.), WCRP.

COHMAP Members (1988). Climatic changes of the last 18,000 years: observations and model simulations. *Science* **241**, 1043–1052.

Crowley, T. J. and Baum, S. K. (1997). Effect of vegetation on an ice-age climate model simulation. *J. Geophys. Res.* **102**, 16463–16480.

Doherty, R., Kutzbach, J. E., Foley, J., and Pollard, D. (2000). Fully-coupled climate/dynamical vegetation model simulations over northern Africa during the mid-Holocene. *Climate Dynamics.* **16**, 561–573.

Dong, B., Valdes, P. J., and Hall, N. M. J. (1996). The changes in monsoonal climates due to Earth's orbital perturbations and ice age boundary conditions. *Palaeoclimates: Data and Modelling* **1**, 203–240.

Foley, J. A., Kutzbach, J. E., Coe, M. T., and Levis, S. (1994). Climate and vegetation feedbacks during the mid-Holocene. *Nature* **371**, 52–54.

Foley, J. A., Levis, S., Prentice, I. C., Pollard, D., and Thompson, S .L. (1998). Coupling dynamic models of climate and vegetation. *Global Change Biol.* **4**, 561–579.

Foley, J. A., Levis, S., Costa, M. H., Doherty, R., Kutzbach, J. E., and Pollard, D. (in press). Vegetation as an interactive part of global climate models. *Ecol. Applications.*

Gagan, M. K., Ayliffe, L., Hopley, D., Cali, J. A., Mortimer, G. E., Chappell, J., McCulloch, M. T., and Head, M .J. (1998). Temperature and surface-ocean water balance of the Mid-Holocene tropical Western Pacific. *Science* **279**, 1014–1018.

Gallée, J.-F., van Ypersele, J. P., Fichefet, T., Marsiat, I., Tricot, C., and Berger, A. (1992). Simulation of the last glacial cycle by a coupled, sectorially averaged climate-ice sheet model. Part 2. Response to insolation and CO$_2$ variations. *J. Geophys. Res.* **97**, 15713–15740.

Ganopolski, A., Kubatzki, C., Claussen, M., Brovkin, V., and Petoukhov, V. (1998). The influence of vegetation-atmosphere-ocean interaction on climate during the mid-Holocene. *Science* **280**, 1916–1919.

Gasse, F. and van Campo, E. (1994). Abrupt post–glacial climate events in West Asia and North Africa monsoon domains. *Earth and Planetary Sci. Lett.* **126**, 435–456.

Gent, P. R., Bryan, F. O., Danabasoglu, G., Doney, S.C., Holland, W. R., Large, W. G., and McWilliams, J. C. (1998). The NCAR Climate System Model global ocean component. *J. Climate* **11**, 1287–1306.

Hall, N. M. and Valdes, P. J. (1997). A GCM simulation of the climate 6000 years ago. *J. Climate* **10**, 3–17.

Harrison, S. P., Jolly, D., Laarif, F., Abe-Ouchi, A., Dong, B., Herterich, K., Hewitt, C., Joussaume, S., Kutzbach, J. E., Mitchell, J., de Noblet, N., and Valdes, P. (1998). Intercomparison of simulated global vegetation distribution in response to 6 kyr B.P. orbital forcing. *J. Climate* **11**, 2721–2742.

Harrison, S. P., Kohfeld, K. E., Roelandt, C., and Claquin, T. (in press). The role of dust in climate today, at the last glacial maximum and in the future. *Earth Sci. Rev.*

Hastenrath, S. (1985). *"Climate and Circulation in the Tropics."* Reidel, Norwell, MA., 455 p.

Haxeltine, A. and Prentice, I. C. (1996). BIOME3: An equilibrium terrestrial biosphere model based on ecophysiological constraints, resource availability, and competition among plant functional types. *Global Biogeochem. Cycles* **10**, 693–710.

Hays, J. D., Imbrie, J., and Shackleton, N. J. (1976). Variations in the earth's orbit: pacemakers of the ice ages. *Science* **194**, 1121–1132.

Hewitt, C. D. and Mitchell, J. F. B. (1996). GCM simulations of the climate of 6 k year BP: mean changes and interdecadal variability. *J. Climate* **9**, 3505–3529.

Hewitt, C. D. and Mitchell, J. F. B. (1998). A fully coupled GCM simulation of the climate of the mid-Holocene. *Geophys. Res. Lett.* **25**, 361–364.

Hoelzmann, P., Jolly, D., Harrison, S. P., Laarif, F., Bonnefille, R., and Pachur, H.-J. (1998). Mid-Holocene land-surface conditions in northern Africa and the Arabian peninsula: a data set for the analysis of biogeophysical feedbacks in the climate system. *Global Biogeochem. Cycles* **12**, 35–51.

Imbrie, J. (1985). A theoretical framework for the Pleistocene Ice Ages. *J. Geol. Soc. (London)* **142**, 417–432.

Jacob, R. L. (1998). *Low frequency variability in a simulated atmosphere ocean system.* PhD Thesis, University of Wisconsin-Madison, 155 p.

Johns, T. C., Carnell, R. E., Crossley, J. F., Gregory, J. M., Mitchell, J. F. B., Senior, C. A., Tett, S. F. B., and Wood, R. A. (1997). The second Hadley Centre coupled ocean-atmosphere GCM: model description, spinup and validation. *Climate Dynamics* **13**, 103–134.

Jolly, D., Harrison, S. P., Damnati, D. and Bonnefille, R. (1998a). Simulated climate and biomes of Africa during the Late Quaternary: comparison with pollen and lake status data. *Quaternary Sci. Rev.* **17**, 629–657.

Jolly, D., Prentice, I. C., Bonnefille, R., Ballouche, A., Bengo, M., Brénac, P.,

Buchet, G., Burney, D., Cazet, J.-P., Cheddadi, R., Edorh, T., Elenga, H., Elmoutaki, S., Guiot, J., Laarif, F., Lamb, H., Lézine, A. M., Maley, J., Mbenza, M., Peyearon, O., Reille, M., Reynaud-Ferrara, I., Riollet, G. Ritchie, J. C., Roche, E., Scott, L., Ssemmanda, I., Staka, H., Umer, M., Van Campo, E., Vilimumbala, S., Vincens, A., and Waller, M. (1998b). Biome reconstruction from pollen and plant macrofossil data from Africa and the Arabian peninsula at 0 and 6 ka. *J. Biogeography* **25**, 1007–1027.

Joussaume, S. (1999). Modeling extreme climates of the past 20,000 years with general circulation models. In *"Modeling the Earth's Climate and its Variability."* (W. R. Holland, S. Joussaume, and F. David, Eds.), pp. 527–565, Elsevier, Amsterdam, Netherlands.

Joussaume, S. and Taylor, K. E. (1995). Status of the Paleoclimate Modeling Intercomparison Project (PMIP). In *"Proceedings of the First International AMIP Scientific Conference, 15–19 May 1995."* (W. L. Gates, Ed.), p. 532. Monterey, CA.

Joussaume, S. and Taylor, K. E. (2000). The Paleoclimate Modeling Intercomparison Project. In *"Paleoclimate Modeling Intercomparison Project, Proceedings of the Third Conference."* (P. Braconnot, Ed.), WCRP.

Joussaume, S., Taylor, K. E., Braconnot, P., Mitchell, J. F. B., Kutzbach, J. E., Harrison, S. P., Prentice, I. C., Broccoli, A. J., Abe-Ouchi, A., Bartlein, P. J., Bonfils, C., Dong, B., Guiot, J., Herterich, K., Hewitt, C. D., Jolly, D., Kim, J. W., Kislov, A., Kitoh, A., Loutre, M. F., Masson, V., McAvaney, B., McFarlane, N., deNoblet, N., Peltier, W. R., Peterschmitt, J. Y., Pollard, D., Rind, D., Royer, J. F., Schlesinger, M. E., Syktus, J., Thompson, S., Valdes, P., Vettoretti, G., Webb, R. S., and Wyputta, U. (1999). Monsoon changes for 6000 years ago: results of 18 simulations from the Paleoclimate Modeling Intercomparison Project (PMIP). *Geophys. Res. Lett.* **26**, 859–862.

Kleidon, A., Fraedrich, K., and Heimann, M. (2000). A green planet versus a desert world: estimating the maximum effect of vegetation on the land surface climate. *Climatic Change* **44**, 471–493.

Knorr, W. and Heimann, M. (1995). Impact of drought stress and other factors on seasonal land biosphere CO$_2$ exchange studied through an atmospheric tracer transport model. *Tellus Series B–Chem. Phys. Meteorol.* **47**, 471–489.

Kohfeld, K. E. and Harrison, S. P. (2000). How well can we simulate past climates? Evaluating earth system models using global paleoenvironmental datasets. *Quaternary Sci. Rev.* **19**, 321–346.

Kutzbach, J. E. (1981). Monsoon climate of the eartly Holocene: climate experiment with the earth's orbital parameters for 9000 years ago. *Science* **214**, 59–61.

Kutzbach, J. E. and Gallimore, R. G. (1998). Sensitivity of a coupled atmosphere/mixed layer ocean model to changes in orbital forcing at 9000 years B.P. *J. Geophys. Res.* **93**, 803–821.

Kutzbach, J. E., and Guetter, P. J. (1986). The influence of changing orbital parameters and surface boundary conditions on climate simulations for the past 18000 years. *J. Atmos. Sci.* **4**, 1726–1759.

Kutzbach, J. E. and Liu, Z. (1997). Response of the African monsoon to orbital forcing and ocean feedbacks in the Middle Holocene. *Science* **278**, 440–443.

Kutzbach, J. E. and Otto-Bleisner, B. L. (1982). The sensitivity of the African-Asian monsoonal climate to orbital parameter changes for 9000 yr B.P. in a low-resolution general circulation model. *J. Atmos. Sci.* **39**, 1177–1188.

Kutzbach, J. E. and Street-Perrott, F. A. (1985). Milankovitch forcing of fluctuations in the level of tropical lakes from 18 to 0 kyear B.P. *Nature* **317**, 130–134.

Kutzbach, J. E., Guetter, P. J., Behling, P. J., and Selin, R. (1993). Simulated climatic changes: Results of the COHMAP climate-model experiments. In *"Global Climates since the Last Glacial Maximum."* (H. E. Wright Jr., J. E. Kutzbach, T. Webb III, W. F. Ruddiman, F. A. Street-Perrott, and P. J. Bartlein, Eds.), pp. 24–93. University of Minnesota Press, Minneapolis.

Kutzbach, J. E., Bonan, G., Foley, J., and Harrison, S. P. (1996). Vegetation and soil feedbacks on the response of the African monsoon to forcing in the early to middle Holocene. *Nature* **384,** 623–626.

Kutzbach, J. E., Gallimore, R., Harrison, S. P., Behling, P., Selin, R., and Laarif, F. (1998). Climate and biome simulations for the past 21,000 years. *Quaternary Sci. Rev.* **17,** 473–506.

Levis, S., Foley, J. A., and Pollard, D. (1999a). CO$_2$, climate, and vegetation feedbacks at the Last Glacial Maximum. *J. Geophys. Res.* **104,** 31191–31198.

Levis, S., Foley, J. A., and Pollard, D. (1999b). Potential high-latitude vegetation feedbacks on CO$_2$-induced climate change. *Geophys. Res. Lett.* **26,** 747–750.

Liu, Z., Gallimore, R., Kutzbach, J. E., Xu, W., Golubev, Y., Behling, P., and Selin, R. (1999a). Modeling long-term climate changes with equilibrium asynchronous coupling. *Climate Dynamics* **15,** 325–340.

Liu, Z., Jacob, R., Kutzbach, J. E., Harrison, S. P., and Anderson, J. (1999b). Monsoon impact on El Niño in the early Holocene. *PAGES Newslett.* **7,** 16–17.

Liu, Z., Kutzbach, J. E., and Wu, L. (2000). Modeling climatic shift of El Niño variability in the Holocene. *Geophys. Res. Lett.* **27,** 2265–2268.

Lorenz, S., Grieger, B., Helbig, P., and Herterich, K. (1996). Investigating the sensitivity of the Atmospheric General Circulation Model ECHAM 3 to paleoclimatic boundary conditions. *Geol. Rundschau* **85,** 513–524.

Masson, V., Joussaume, S., Pinot, S., and Ramstein, G. (1998). Impact of parameterizations on simulated winter mid-Holocene and Last Glacial Maximum climatic changes in the northern hemisphere. *J. Geophys. Res.* **103,** 8935–8946.

Masson, V. and Joussaume, S. (1997). Energetics of 6000 B.P. atmospheric circulation in boreal summer, from large scale to monsoon areas. *J. Climate* **10,** 2888–2903.

McGlone, M. S., Kershaw, A. P., and Markgraf, V. (1992). El Niño/Southern oscillation climatic variability in Australian and South American paleocnvironmental records. In *"El Niño—historical and paleoclimatic aspects of the Southern Oscillation.* (H. F. Diaz and V. Markgraf, Eds.). Univ. Press cambridge.

Meehl, G. A. (1995). Global coupled general circulation models. *Bull. Meteorol. Soc. Am.* **76,** 951–957.

Murphy, J. M. (1995). Transient response of the Hadley Centre coupled ocean-atmosphere model to increasing carbon dioxide. Part 1. Control climate and flux adjustment. *J. Climate* **8,** 36–56.

de Noblet, N., Claussen, M., and Prentice, I. C. (2000). Mid-Holocene greening of the Sahara: first results of the GAIM 6000 year BP experiment with two asynchronously coupled atmosphere/biome models. *Climate Dynamics.* **16,** 643–659

Otto-Bleisner, B. (1999). El Niño/La Niña and Sahel precipitation during the middle Holocene. *Geophys. Res. Lett.* **26,** 87–90.

Pias, J. (1970). Les formation sédimentaires tertiares et quaternaires de la cuvette tchadienne et les sols qui en derivent. *Mémoires, ORSTOM* **43,** 1–408.

Pollard, D., Bergengren, J. C., Stillwell-Stoller, L. M., Felzer, B., and Thompson, S. L. (1998). Climate simulations for 10000 and 6000 years BP using the GENESIS global climate model. *Palaeoclimates: Data and Modelling* **2,** 183–218.

Prentice, I. C., and Webb III, T. (1998). BIOME 6000: reconstructing global mid-Holocene vegetation patterns from paleoecological records. *J. Biogeography* **25,** 997–1005.

Prentice, I. C., Cramer, W., Harrison, S. P., Leemans, R., Monserud, R. A., and Solomon, A. M. (1992). A global biome model based on plant physiology and dominance, soil properties, and climate. *J. Biogeography* **19,** 117–134.

Prentice, I. C., Harrison, S. P., Jolly, D., and Guiot, J. (1998). The climate and biomes of Europe at 6000 year BP: comparison of model simulations and pollen-based reconstructions. *Quaternary Sci. Rev.* **17,** 659–668.

Prentice, I. C., Jolly, D., and BIOME 6000 Members (2000). Mid-Holocene and Glacial Maximum vegetation geography of the northern continents and Africa. *J. Biogeography.* **27,** 507–519

Roberts, N. and Wright Jr, H. E. (1993). Vegetational, lake-level, and climatic history of the Near East and Southwest Asia. In *"Global Climates since the Last Glacial Maximum."* (H. E. Wright Jr., J. E. Kutzbach, T. Webb III, W. F. Ruddiman, F. A. Street-Perrott, and P. J. Bartlein, Eds.), pp. 221–264. University of Minnesota Press, Minneapolis.

Rodbell, D. T., Seltzer, G. O., Anderson, D. M., Abbott, M. B., Enfield, D. B., and Newman, J. H. (1999). An 15,000 year record of El Niño-driven alluviation in southwestern Ecuador. *Science* **283,** 516–520.

Sandweiss, D. H., Richardson III, J. B., Reitz, E. J., Rolins, H. B., and Maasch, K. A. (1996). Geoarchaeological evidence from Peru for a 5000 years B.P. onset of El Niño. *Science* **273,** 1531–1533.

Schneider, J. L. (1967). Evolution du dernier lacustre et peuplements préhistoriques aux Bas-Pays du Tchad. *Bull. ASEQUA,* Dakar, 18–23.

Schumm, S.A. and Lichty, R.W. (1965). Time, Space and causality in geomorphology Am. *J. Science* **263,** 110–119.

Street, F. A. and Grove, A. T. (1976). Environmental and climatic implications of late Quaternary lake-level fluctuations in Africa. *Nature* **261,** 385–390.

Street-Perrott, F. A. and Harrison, S. P. (1985). Lake levels and climate reconstruction. In *"Paleoclimate Analysis and Modeling."* (A. D. Hecht, Ed.), pp. 291–340. John Wiley, New York.

Street-Perrott, F. A. and Perrott, R. A. (1993). Holocene vegetation, lake levels, and climate of Africa.. In *"Global Climates since the Last Glacial Maximum."* (H. E. Wright Jr., J. E. Kutzbach, T. Webb III, W. F. Ruddiman, F. A. Street-Perrott, and P. J. Bartlein, Eds.), pp. 221–264. University of Minnesota Press, Minneapolis.

Street-Perrott, F. A., Marchand, D. S., Roberts, N., and Harrison, S. P. (1989). *Global lake-level variations from 18,000 to 0 years ago: a paleoclimatic analysis.* U.S. Department of Energy, Washington, DC.

Street-Perrott, F. A., Mitchell, J. F. B., Marchand, D. S., and Brunner, J. S. (1990). Milankovitch and albedo forcing of the tropical monsoons: a comparison of geologic evidence and numerical simulations for 9000 y BP. *Trans. R. Soc. Edinburgh* **81,** 407–427.

TEMPO (1996). Potential role of vegetation feedback in the climate sensitivity of high-latitude regions: a case study at 6000 years B.P. *Global Biogeochem. Cycles* **10,** 727–736.

Texier, D., de Noblet, N., Harrison, S. P., Haxeltine, A., Jolly, D., Joussaume, S., Laarif, F., Prentice, I.C., and Tarasov, P. E. (1997). Quantifying the role of biosphere-atmosphere feedbacks in climate change: coupled model simulation for 6000 years BP and comparison with paleodata for northern Eurasia and northern Africa. *Climate Dynamics* **13,** 865–882.

Texier, D., de Noblet, N., and Braconnot, P. (2000). Sensitivity of the African and Asian monsoons to mid-Holocene insolation and data inferred surface changes. *J. Climate* **13,** 164–181.

Thompson, L. G. (2000). Ice core evidence for climate change in the tropics: implications for our future. *Quaternary Sci. Rev.* **19,** 19–35.

Thompson, L. G., Davis, M. E., Mosley-Thompson, E., Sowers, T. A., Henderson, K. A., Zagorodnov, V. S., Lin, P.-N., Mikhalenko, V. N., Campen, R. K., Bolzan, J. F., and Cole-Dai, J. A. (1998). 25,000 year tropical climate history from Bolivian ice cores. *Science* **282**, 1858–1864.

Vettoretti, G., Peltier, W. R., and McFarlane, N. A. (1998). Simulations of mid-Holocene climate using an atmospheric general circulation model. *J Climate* **11**, 2607–2627.

Winkler, M. G. and Wang, P. K. (1993). The late-quaternary vegetation and climate of China. In *"Global Climates since the Last Glacial Maximum."* (H. E. Wright Jr., J. E. Kutzbach, T. Webb III, W. F. Ruddiman, F. A. Street-Perrott, and P. J. Bartlein, Eds.), pp. 221–264. University of Minnesota Press, Minneapolis.

Wright, Jr., H. E., Kutzbach, J. E., Webb, III, T., Ruddiman, W. F., Street-Perrott, F. A., and Bartlein, P. J. (1993). *"Global Climates since the Last Glacial Maximum."* University of Minnesota Press, Minneapolis, MN.

Yu, G. and Harrison, S. P. (1996). An evaluation of the simulated water balance of Eurasia and northern Africa at 6000 year BP using lake status data. *Climate Dynamics* **12**, 723–735.

Yu, G., Prentice, I. C., Harrison, S. P., and Sun, X. J. (1998). Pollen-based biome reconstructions for China at 0 and 6000 years. *J. Biogeography* **25**, 1055–1069.

Yu, G., Chen, X., Ni, J., Cheddadi, R., Guiot, J., Han, H., Harrison, S.P., Huang, C., Ke, M., Kong, Z., Li, S., Li, W., Liew, P., Liu, G., Liu, J., Liu, Q., Liu, K.-B., Prentice, I.C., Qui, W., Ren, G., Song, C., Sugita, S., Sun, X., Tang, L., Van Campo, E., Xia, Y., Xu,, Q., Yan, S., Yang, X., Zhao J., and Zheng, Z. (2000). Palaeovegetation of China: a pollen data-based synthesis for the mid-Holocene and the last glacial maximum. *J. Biogeography.* **27**, 635–664.

1.7

Paleobiogeochemistry

I. C. Prentice
Max Planck Institut for Biogeochemistry Jena, Germany

D. Raynaud
Laboratoire de Glaciologie et de Geophysique de l' Environnement St. Martin d' Hères, France

1. Introduction

Paleobiogeochemistry—this word, as far as we know, did not previously exist. It defines a newly emerging research field that, we believe, will within the next decade come to play a central role in our understanding of the earth system and of how human activities are modifying that system.

Paleobiogeochemistry draws its inspiration and challenge from the polar ice-core records, which have provided a window on the natural dynamics of atmospheric composition and the relationship of atmospheric composition to an ever-changing climate. Technological developments in ice-core drilling systems, together with the associated systems that had to be developed to allow the secure transport and laboratory sampling of ice, have made the ice-core records possible. The increasing refinement of analytical methods has allowed us to determine concentrations of the impurities and gases incorporated in ice. These include not only the relatively abundant atmospheric components such as CO_2 and mineral dust, but also many other atmospheric constituents present in far lower amounts such as CH_4, N_2O, light carboxylic acids, and isotopes such as $^{13}CO_2$, $^{18}O^{16}O$, and even $^{17}O^{16}O$ and $^{13}CH_4$.

High-precision measurements in ice cores now provide a rich source of information about most of the significant, radiatively active constituents of the atmosphere—greenhouse trace gases and aerosols—and, along with this information, data on numerous tracers that help to elucidate mechanisms associated with natural changes in the abundances of these constituents. The overarching challenges posed by the ice-core records of the changing atmospheric composition can be summarized as follows.

- The greenhouse gases CO_2, CH_4, and N_2O and aerosols containing SO_4^{2-}, volatile organic compounds, and mineral dust have varied in abundance during the past half-million years in a systematic manner, showing periodicities characteristic of the earth's orbital variations (the Milankovitch

frequencies ≈ 20, ≈ 40, ≈ 100 ka; 1 ka = 1000 years) and a clear association with glacial–interglacial cycles on the 100-ka time scale (Fig. 1) (Petit *et al.*, 1999).

- The changes in greenhouse gases and aerosols could in principle represent either *effects* or *drivers* of climate change. Calculations suggest that the total contribution of natural variations in some atmospheric constituents, such as CH_4, to global radiative forcing of climate cannot be large. The observed natural variations in CH_4 concentration prior to the present human perturbation of the CH_4 cycle, represent a response to climate change, and not a significant driver of climate change (Lorius and Oeschger, 1994). Glacial–interglacial CO_2 changes are large enough to be significant drivers of climate (Raynaud *et al.*, 2000, and references therein). On the other hand, temporal patterns of change in atmospheric CO_2 concentration over the past four glacial–interglacial cycles suggest that CO_2 (as well as CH_4) concentrations respond to climate change (Fischer *et al.*, 1999). Atmospheric dust loading clearly responds to climate change, but inclusion of dust as a radiatively active atmospheric constituent may be needed to produce an accurate simulation of glacial climates (Kohfeld and Harrison, 2000; Claquin *et al.*, submitted). In other words, these atmospheric constituents (CO_2 and dust) represent *interactive components* of the earth system that both influence and are influenced by the changing climate (Prentice, in press). We cannot properly understand the dynamics of climate unless we understand not only how these atmospheric constituents influence climate, but also how climate change influences their atmospheric concentrations by altering the strengths of their natural sources and sinks.

- Despite the richness of the records, ice-core data alone can give only clues and not definitive answers concerning the mechanisms of the observed changes. This is because climate and atmospheric composition are multidimensional phenomena that cannot be adequately indexed by measurements from one or two regions. Changes in climate have distinctive spatial

GLOBAL BIOGEOCHEMICAL CYCLES IN THE CLIMATE SYSTEM

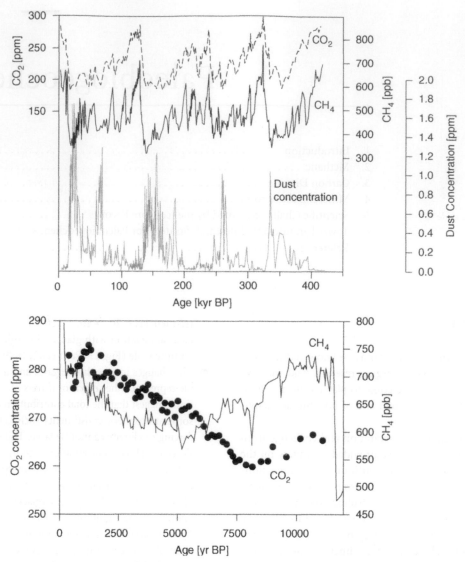

FIGURE 1 Top: Records of CO_2, CH_4, and dust concentrations from the Vostok Antarctic ice core (Petit *et al.*, (1999) covering four glacial–interglacial cycles. Bottom: Holocene records of CO_2 and CH_4 concentrations. CO_2 from Taylor Dome, Antarctica (Indermuehle *et al.*, 1999); CH_4 from the Greenland summit (Chapellaz *et al.*, 1997a).

signatures that cannnot be revealed by records confined to the polar regions (or extremely high elevations) where ice can persist through interglacial periods. The spatial heterogeneity of climate change is an important determinant of the effects of climate change on ecosystem activity and trace-gas production; different ecosystems in different climatic zones can react in quite distinct ways to a given change in climate (Melillo *et al.*, 1993; Cramer *et al.*, in press). Similarly, important atmospheric constituents with short lifetimes in the atmosphere, including reactive trace gases and aerosols, have heterogeneous spatial distributions and their spatial interactions are crucially important in determining the nature of their effect on climate (Dentener *et al.*, 1996).

Thus, *explanation* of the ice-core records demands a creative interaction with other fields of earth system science. We identify two such fields as crucial. The first is global modeling, which is a standard approach to climate change problems (both present/future and past) and is being actively extended to include interactions of the biogeochemical cycles with climate. The second is Quaternary science in the broad sense, especially insofar as Quaternary paleoenvironmental records have been standardized and compiled on a global scale. Quaternary scientists have amassed data from widely distributed natural archives such as marine and lacustrine sediments which, when suitably compiled, yield spatially extensive information about many aspects of climate, ecosystem composition, and even some indicators of atmospheric composition such as the flux density of mineral dust, over time scales (and, in some cases, temporal resolution) similar to those of the ice-core records (Kohfeld and Harrison, 2000).

This chapter is not by any means intended to be a review of ice-core research. Our purpose is rather to select a few illustrations of the specific scientific challenges raised by the ice-core records—especially the most recent, high-resolution records from Antarctica and the Greenland Summit—and to outline an interdisciplinary research strategy for tackling these challenges.

2. Methane

Despite its low natural concentration (<1 ppm), CH_4 can be measured in the air bubbles trapped in fossil ice with high precision and reliability. Once isolated from the "scavenging" effect of oxygen-containing free radicals in the atmosphere, CH_4 ceases to be oxidized and its concentration becomes constant. The main features of the ice-core CH_4 records are as follows:

- A clear pattern of variation in concentration between glacial (≈ 300 ppb) and interglacial periods (600–700 ppb), with additional variations at the Milankovitch frequencies of (≈ 20 and ≈ 40 ka (Petit *et al.*, 1999).
- Higher-frequency ("sub-Milankovitch") variability (Chappellaz *et al.*, 1993a) associated with climatic fluctuations in the form of Dansgaard–Oeschger events, including the Younger Dryas cold interval that interrupted the last deglaciation.
- An extremely close tie to climate variations as indexed by isotopic signals in the ice cores, so that any possible lead or lag between CH_4 and climate is within the ≈ 50-year detectability horizon caused by the finite time taken for occlusion of air bubbles in the firn.
- A distinctive pattern (Blunier *et al.*, 1995; Chappellaz *et al.*, 1997a) of relatively smooth changes in concentration during the Holocene (the present interglacial, starting at ≈ 11.6 ka), with amplitude ≈ 100 ppb. CH_4 concentration was relatively high near the beginning of the Holocene, low during the middle Holocene (minimum around 6 ka B.P., before present), and rising again after 6 ka B.P.
- Changes in the interhemispheric gradient, as shown by the difference in concentration between Greenland and Antarctic ice cores, suggesting that both tropical and northern high-latitude sources are involved in the glacial–interglacial, sub-Milankovitch glacial, and Holocene changes in the concentration of CH_4 (Chappellaz *et al.*, 1997a; Dällenbach *et al.*, 2000).

This last point could be established because of the high measurement precision that can be achieved for CH_4 in ice cores. Its fast atmospheric mixing time and its fast response to climate change allow the qualitative features of the CH_4 records to be used as a way of synchronizing records from the two hemispheres. These features have been useful in various other contexts. For ex-

ample, using CH_4 and $\delta^{18}O$ of O_2 measurements from Antarctic ice, it was possible to show that the lowest layers of the GRIP and GISP2 cores from the Greenland summit, representing the last or Eemian interglacial, were not in a correct temporal sequence (Chappellaz *et al.*, 1997b).

3. Carbon Dioxide

CO_2 concentrations in the atmosphere have generally been 3 orders of magnitude higher than CH_4 concentrations, yet paradoxically it has proved far more difficult to obtain a reliable, high-resolution record for CO_2. The problem is the presence of mineral and/or organic impurities in the ice. The much higher content of impurities in Greenland ice than in Antarctic ice can lead to *in situ* CO_2 production via acid–carbonate interactions or oxidation of organic material and significantly alter the original concentration of CO_2 in the ancient air (Delmas, 1993; Anklin *et al.*, 1997; Haan and Raynaud, 1998).

The Vostok core, from the central part of East Antarctica, provided conclusive evidence for glacial–interglacial changes in atmospheric CO_2 concentration (Barnola *et al.*, 1987; Petit *et al.*, 1999), but the low precipitation rate limits the achievable temporal resolution at Vostok. The Byrd ice-core record from coastal Antarctica (Neftel *et al.*, 1988), with higher precipitation rates, yielded an apparent Holocene signal with a sharp CO_2 concentration peak at the beginning of the Holocene followed by a drawdown and subsequent recovery (by about 6 ka B.P.) to approximately the "preindustrial" concentration of ≈ 280 ppm. This record was long regarded as problematic, because the main changes occurred in a part of the core where the ice was exceptionally brittle. More recent measurements from Taylor Dome (Indermuehle *et al.*, 1999), another high-resolution site lacking the problems of the Byrd core, have shown that the Byrd measurements overestimated the amplitude of Holocene variability in CO_2 concentration. Based on present knowledge, the major features of the ice-core CO_2 records are as follows:

- On glacial–interglacial time scales, there is systematic variation between <200 ppm during glacial periods (185 ppm at the last glacial maximum around 21 ka B.P.) and ≈ 280 ppm during interglacial periods, with additional variability particularly in the ≈ 40-ka band (Petit *et al.*, 1999).
- There is some higher-frequency variability (about ± 10 ppm) associated with Heinrich iceberg-discharge events during the last glacial period. However, the CO_2 record essentially lacks the imprint of the faster Dansgaard–Oeschger events (Stauffer *et al.*, 1998). During the Younger Dryas, deglacial warming of the Northern Hemisphere was interrupted while CO_2 concentration continued to rise (Blunier *et al.*, 1997).
- There has been relatively small variability (range 260–280 ppm) during the Holocene, but with a distinctive pattern: a slight fall in concentration to a minimum of ≈ 260 ppm at

≈ 8 ka B.P., followed by a steady rise toward 280 ppm in preindustrial time (Indermuehle *et al.*, 1999). Note that although this pattern shares some features of the Holocene changes in CH_4, its amplitude (relative to the average concentration of the two gases) is smaller, the pattern for CO_2 is less temporally symmetric than that for CH_4, and the timings of the Holocene minima for the two gases are 2000 years apart.

- A higher-resolution record of the past 1000 years provides evidence for a slight (amplitude < 10 ppm), temporary lowering of CO_2 concentration during the Little Ice Age (Etheridge *et al.*, 1996).

We have no information on past changes in the interhemispheric gradient of atmospheric CO_2 because accurate determination of past CO_2 concentrations from Northern-Hemisphere ice is still impossibile.

It is possible to measure the $\delta^{13}C$ of CO_2 in ice. This measurement is potentially important because the ^{13}C fractionation associated with CO_2 exchanges between the atmosphere and ocean is very different from the fractionation due to the predominant C_3 pathway of photosynthesis (Ehleringer, this volume; Kaplan and Buchmann, this volume). Recent measurements suggest that whereas the increase of CO_2 concentration following the last glacial maximum has the isotopic signature characteristic of a predominantly oceanic source of atmospheric CO_2 (Smith *et al.*, 1999), the changes after 8 ka B.P. bear a distinct signature consistent with progressive loss of carbon (amounting to ≈ 200 Pg C altogether) from the terrestrial biosphere (Indermuehle *et al.*, 1999). However, the measurement errors in the published $\delta^{13}C$ measurements are uncomfortably large compared to the signal, indicating the importance of improving the repeatability of these measurements as well as increasing the sampling density for the Holocene. The $\delta^{13}C$ record of the Little Ice Age is also consistent with the small CO_2 anomaly being due to a temporary increase in carbon storage on land (Francey *et al.*, 1999).

The distinctive spatial pattern of ^{18}O in precipitation, caused by fractionation in evaporation and condensation, applies to atmospheric CO_2 when CO_2 dissolves in leaf water; this fact has been exploited to provide "top-down" estimates of global gross primary productivity (Farquhar *et al.*, 1993; Ciais *et al.*, 1999). The oxygen isotope composition of CO_2 is not preseved in ice cores because CO_2 exchanges oxygen atoms with the ice. However, O_2 propagates the ^{18}O signature of CO_2 because photosynthesis and respiration (including photorespiration) uniquely transfer oxygen atoms from CO_2 to O_2 (Bender *et al.*, 1996). The difference between the $\delta^{18}O$ of seawater (determined mainly by continental ice volume) and the $\delta^{18}O$ of O_2 in air bubbles trapped in ice is known as the "Dole effect" and show promise as a palaeotracer for primary productivity. Additional, independent information on productivity is provided by the ^{17}O content of O_2 (Barkan *et al.*, 1999). Due to differences in the mechanisms of fractionation of the 3 stable oxygen isotopes in reactions involving stratospheric O_3, these isotopes together can be used to infer O_2 turnover by the biosphere (terrestrial and marine).

4. Mineral Dust Aerosol

Aerosols include both soluble and insoluble components. The insoluble component, which consists of mineral dust of terrestrial origin, can be measured directly in ice cores. The soluble component, which includes components from sulfate aerosol and sea salt, can be assessed by measuring the concentration of base cations such as Na^+ and Mg^{2+} and anions such as SO_4^{2-} and Cl^-.

Unlike CH_4 and CO_2, terrigenous dust is not a well-mixed atmospheric constituent; atmospheric concentrations of dust vary spatially and temporally by several orders of magnitude. Similarly, ice-core records show enormous variation compared to that showed by trace gases (Thompson and Mosley-Thompson, 1981; De Angelis *et al.*, 1997; Petit *et al.*, 1999). Also, what is measured in the ice cores is not directly the atmospheric concentration, but rather the concentration in the ice, which (taking into account the variable precipitation rate) can be used to infer the flux density of mineral dust reaching the ice surface (Mahowald *et al.*, 1999). The main features of the ice-core dust records may be summarized as follows.

- There is a very strong glacial–interglacial pattern, with glacial periods typically showing an order of magnitude higher dust deposition in both Antarctica and Greenland. The dust records from Antarctica show spectral power in all of the Milankovitch bands (Petit *et al.*, 1999).
- A trend during glacial periods, such that higher dust deposition occurs during the latter (colder) part of each glacial period. During the last glacial period, the highest dust deposition rates did not occur until ≈ 70 ka B.P. (marine isotope stage 4).
- High variability within glacial periods, associated with the recorded rapid climate changes and also showing considerable amplitude on interannual time scales.
- Very low deposition rates during the Holocene.

Heavy-element abundances and isotopic composition of dust in polar ice cores were compared with signatures of aeolian dust from different regions to assign likely source areas for the glacial dust. This exercise pointed to Central Asia as a likely major source region for glacial-age dust in Greenland (Biscaye *et al.*, 1997), and to Patagonia for glacial-age dust in Antarctica (Basile *et al.*, 1997). Particle size distribution is a useful ancillary statistic. Particle size spectra from Greenland dust support the hypothesis that average wind speeds during dust transport at the last glacial maximum were not much greater than the present, because the mean particle size hardly changed, although the data do indicate a slight increase in the amount of dust in the largest size classes (Steffensen, 1997).

5. Scientific Challenges Posed by the Ice-Core Records

Although the ice-core records illustrate the pervasive Milankovitch periodicities and has yielded a great deal of information about natural changes in climate and biogeochemical cycles occurring at the Milankovitch and higher frequencies, the underlying causal sequences are hardly known. Herein lie the challenges of paleobiogeochemistry. We present some illustrative examples here, focusing as above on CH_4, CO_2, and mineral dust aerosol.

5.1 Methane

Natural variations in atmospheric CH_4 concentration, as observed to date in ice cores, show no evidence for catastrophic CH_4 hydrate release (Raynaud *et al.*, 1998), and consequently do not support the speculation (Nisbet, 1992) that CH_4 releases from marine clathrates were implicated in triggering the last deglaciation. Normally, the main natural sources of CH_4 are on land, and the largest source component is due to methanogenesis under anaerobic conditions in seasonal or permanent wetlands (Melillo *et al.*, 1996). The glacial–interglacial changes in CH_4 concentrations are too large to be fully accounted for by plausible variations in the atmospheric chemical sink, and must therefore be explained, at least in part, in terms of the changing areas and activities of natural CH_4 sources (Pinto and Khalil, 1991, Chappellaz *et al.*, 1993b, Crutzen and Bruhl, 1993, Thompson *et al.*, 1993, Martinerie *et al.*, 1995). It is plausible that the wetland extent was less during glacial times and also that net primary productivity of terrestrial wetlands was less, allowing the production of less substrate for methanogenic microorganisms. But it is not clear whether the rapid climate response of atmospheric CH_4 concentration is due to effects of temperature on substrate formation and methanogenesis or to rapid changes in the areas of wetlands. Holocene changes in CH_4 concentration were at first attributed to a balance of declining tropical monsoons (implying a reduction in the area of tropical wetlands) and later to an increasing build-up of boreal wetlands (e.g., on the Hudson Bay lowlands, which became exposed due to isostatic uplift only during the latter part of the Holocene) (Blunier *et al.*, 1995). Analysis of the changing interhemispheric gradient of CH_4 adds important information to these, showing that this picture is probably too simplistic (Chappellaz *et al.*, 1997a). Other quantitative constraints may in future be brought to bear on this issue by measuring the isotopic composition of CH_4 in ice, while better quantification and modeling is needed for the impacts of climate change and atmospheric CO_2 concentration on wetland extent, CH_4 release from wetlands, and sources and sinks of other biogenic trace gases (NO_x, CO, and volatile organic compounds) that also affect the strength of the atmospheric sink for CH_4.

5.2 Carbon Dioxide

Ice-core records have confirmed Svante Arrhenius' prescient hypothesis that variations in atmospheric CO_2 concentration were associated with glacial–interglacial cycles, but we are still uncertain about the primary cause of these variations. The first-order explanation must come from the ocean, where more than 90% of the total inventory of carbon in the ocean–atmosphere–terresterial biosphere system. (According to several lines of evidence, carbon storage on land was substantially less during glacial periods, e.g., Shackleton, 1977; Crowley, 1995; Friedlingstein *et al.*, 1995; Bird *et al.*, 1994; Peng *et al.*, 1998.) One family of hypotheses to explain glacial–interglacial CO_2 variations relies on changes in the dissolution of CO_2 in the ocean. The effect of increased solubility of CO_2 in the ocean at low temperatures is insufficient and is counteracted by the effect of higher salinity during glacial periods. Stephens and Keeling (2000) proposed that the extended winter sea ice cover around Antarctica prevented the outgassing of upwelled, CO_2-rich water in glacial times. This is an attractive hypothesis in that it explains the synchroneity of increasing CO_2 concentration and Antarctic warming during deglaciations, as shown in Antarctic ice-cores. On the other hand, it postulates far less upwelling in low latitudes than most of the current ocean models allow. A second family of hypotheses relies on changes in nutrient supply or the efficiency of its utilization to increase marine biological productivity, thereby increasing the sinking flux of organic carbon and maintaining a stronger gradient of dissolved inorganic concentration away from the sea surface. The currently popular explanations along these lines invoke increased mineral dust acrosol input as an external source of Fe, which has been shown to limit production in the equatorial Pacific and Southern Oceans and which in addition may be generally limiting for nitrogen fixation in the open ocean (Martin, 1990; Falkowski, 1997; Broecker and Henderson, 1998; Pedersen and Bertrand, 2000). These explanations provide a putative link between CO_2 and mineral dust, over and above the fact that both have significant radiative forcing effects. A third family of hypotheses relies on various mechanisms that alter the alkalinity of the ocean. Both the nutrient and alkalinity hypotheses have problems to explain the full magnitude of the change without violating the constraints revealed by other information about marine sediments: proxy data for nutrients do not support great increases in productivity while calcium carbonate dissolution patterns do not support a large alkalinity change. It seems likely that more than one mechanism may be involved, but all of them need to be quantified better than they are at present. Useful additional information on the ice-core records is likely to come from geochemical and isotopic proxies for marine productivity, while a better quantification of the processes will require improved modeling of physical and biological processes in the ocean.

Holocene variations in CO_2 are also poorly understood. Even if it is true that terrestrial ecosystems lost carbon progressively since 8 ka B.P., the required magnitude of loss cannot be accounted for by the disappearance of vegetation in the Sahara (Indermuehle *et al.*, 1999)—although this is by far the most extensive vegetation change that has occurred during the past 6000 years, according to pollen and plant macrofossil data assembled by the BIOME 6000 project (Prentice *et al.*, 2000). It is not clear *a priori* whether terrestrial carbon storage would be expected to increase or decline in response to the changing orbital configuration during the Holocene; this too requires quantification. More (and if possible, more precise) measurements of $\delta^{13}C$ in CO_2 from ice cores are clearly required, and the possible contribution of changes in marine chemistry through the Holocene needs to be more exactly calculated by ocean models.

5.3 Mineral Dust Aerosol

Analyses of dust from the polar ice cores have yielded information about candidate source areas. Although the changes observed in Greenland and Antarctica are large in a relative sense, in an absolute sense even the highest deposition rates to these remote areas are tiny compared to the contemporary rates in regions close to major dust sources, such as the Sahara or the Chinese loess plateau. Modeling is essential to establish links between the ice-core records and dust distribution outside the polar regions (Krogh-Andersen *et al.*, 1998). Dust records have now been obtained from tropical ice cores and show mutually contradictory results, apparently reflecting predominance of large nearby sources with different histories and pointing to a high degree of spatial heterogeneity in the change of atmospheric dust content between glacial and interglacial periods. Such heterogeneity is evident in spatially distributed record of mineral dust deposition in the ocean, as compiled by the DIRTMAP project (Kohfeld and Harrison, 2000). Spatial heterogeneity is particularly crucial to determining the climatic effect of dust, because the sign of the radiative forcing due to dust is a function of the underlying surface albedo and is usually opposite over oceans and land (Tegen and Fung, 1995; Claquin *et al.*, 1998). The lofting of dust from the land surface is itself dependent on climate and atmospheric CO_2 concentration, insofar as these variables control soil moisture and vegetation structure (Mahowald *et al.*, 1999); therefore, a reciprocal relationship between mineral dust and atmospheric CO_2 may exist. Advances in our understanding of the controls on mineral dust aerosol may occur in part through systematic "sourcing" of the dust in ice cores and marine sediments and will also require improved models for changing dust source areas and emission strengths.

6. Toward an Integrated Research Strategy for Paleobiogeochemistry

We propose the following general strategy to test hypotheses about the natural dynamics of the earth system.

1. The first step is to define clear *data targets* in the ice-core records. For example, the glacial–interglacial difference in CH_4 concentration, the increase in CO_2 concentration since the early Holocene, and the various time scales of response and time sequences of changes in atmospheric CO_2, CH_4, and mineral dust.

2. The second step is to define *model experiments* that could be performed by (preferably several) earth system models. The models should include whatever components and linkages are hypothesized to be crucial to explaining the target data. (The models should also be applied without the specified components or linkages, to test their importance in the modeled world.) The target data will provide an immediate assessment of the extent to which the model experiments are successful.

3. Given that complex models can easily be right for the wrong reasons, especially when a global three-dimensional model is called upon to predict a single number, the third and essential step is to define the *ancillary tests* of model performance against (a) spatially distributed data (such as vegetation types, dust deposition fields, marine biogeochemical tracers) and (b) additional data from ice cores (such as isotopic measurements) that are relevant to the modeled processes. For example, a model to explain CH_4 changes should be shown to generate realistic spatial distributions of wetlands; a model to account for CO_2 changes should also be able to hindcast ice-core measurements of $\delta^{13}C$ and the Dole effect; a model to account for temporal changes in atmospheric dust loading at the poles should be called upon to reproduce global patterns of dust fluxes to the oceans and loess accumulation on land. We note that this strategy implies running considerably more highly coupled models of the earth system than is possible today. Harrison *et al.* (this volume) and Claussen (this volume) show that the incorporation of both ocean–atmosphere and vegetation–atmosphere feedbacks into atmospheric general circulation models is a prerequisite for the correct simulation of Holocene palaeoclimates. Without physical coupling of atmopshere, oceans, and land, climate models cannot simulate past climates with sufficient accuracy to make our strategy viable. Furthermore, biogeochemical interactions through the carbon cycle and atmospheric chemistry have to be coupled to climate models, and specifically modules describing the sources and sinks of all key reactive chemical species and aerosol precursors have to be coupled to models of terrestrial and marine ecosystems. No existing model is sufficiently comprehensive to do all of these things; yet rapid progress is being made toward the development of true earth system models through separate activities such as coupled climate–carbon modeling, trace-gas source modeling in ecosystem models, and coupled climate–atmospheric-chemistry–transport models. Our strategy therefore relies on the continuation of a trend that already exists in global modeling and is being strongly promoted by the International Geosphere-Biosphere Programme (IGBP).

Earth system models in the limited form that they exist today fall into two major categories, namely, full three-dimensional

(3-D) models (based on atmospheric and ocean general circulation models) and reduced-form models or "models of intermediate complexity" in which the spatial resolution is generally lower (Kutzbach *et al.*, this volume) and atmospheric and ocean dynamics are represented in a parameterized, computationally efficient form (Schellnhuber, 1999). (Hybrids between these two types of model are beginning to appear, but this does not affect our argument.) The strategy we envisage allows an immediate role for both types of model, because for computational reasons experiments with 3-D models are likely to focus mainly on quasi-equilibrium conditions centered on canonical "time slices" while reduced-form models can far more readily perform multiple, transient simulations of long periods (e.g., the whole Holocene: Clanssen *et al.*, 1999). For certain time slices, data already exist as convenient global summaries for comparison with model output (Kohfeld and Harrison, 2000). Transient analyses pose additional challenges to the Quaternary data community, to process the data into a suitable, synthetic form.

In conclusion, by insisting that the model results are routinely tested against the full spectrum of available palaeodata (from ice cores and other natural archives), we suggest that the study of past biogeochemical cycles can provide both a unique means to test complex earth system models and a powerful stimulus to their further development.

Acknowledgments

We thank the many scientists with whom we have discussed these matters, including Jérôme Chapellaz, Torben Christensen, Frank Dentener, Sandy Harrison, Ivar Isaksen, Sylvie Joussaume, Karen Kohfeld, Corinne Le Quéré, Particia Martinerie, Nathalie de Noblet, Henning Rodhe, Doug Wallace and many more. Karen Kohfeld and Jean-Robert Petit commented on an earlier draft.

References

Anklin, M., Schwander, J., Stauffer, B., Tschumi, J., Fuchs, A., Barnola J. M., and Raynaud, D. (1997). CO_2 record between 40 and 8 kyr BP from the Greenland Ice Core Project ice core. *J. Geophys. Res.–Oceans* **102**, 26539–26545.

Barnola, J. M., Raynaud, D., Korotkevich, Y. S., and Lorius, C. (1987). Vostok ice core provides 160,000-year record of atmospheric CO_2. *Nature* **329**, 408–414.

Barkan, E., Bender, M. L., Thiemens, M. H., Boering, K. A., and Luz, B. (1999). Triple-isotope composition of atmospheric oxygen as a tracer of biosphere productivity. *Nature* **400**, 547–550.

Basile, I., Grousset, F. E., Revel, M., Petit, J. R., Biscaye, P. E., and Barkov, N. I. (1997). Patagonian origin of glacial dust deposited in East Antarctica (Vostok and Dome C) during glacial stages 2, 4 and 6. *Earth Planet. Sci. Lett.* **146**, 573–589.

Bender, M., Ellis, T., Tans, P., Francey, R., and Lowe, D. (1996). Variability in the O_2/N_2 ratio of southern hemisphere air, 1991–1994—implications for the carbon cycle. *Global Biogeochem. Cycles* **10**, 9–21.

Biscaye, P. E., Grousset, F. E., Revel, M., van der Gaast, S., Zielinski, G. A., Vaars, A., and Kukla, G. (1997). Asian provenance of glacial dust (stage 2) in the Greenland Ice Sheet Project 2 Ice Core, Summit, Greenland. *J. Geophys. Res.* **102**, 26765–26781.

Bird , M. I., Lioyd. J., and Farquhas, F. d. (1994). Tereshial carbon storage at the LGM. *Nature* **37**, 556.

Blunier, T., Chappellaz, J., Schwander, J., Stauffer, B., and Raynaud, D. (1995). Variations in the atmospheric methane concentration during the Holocene epoch. *Nature* **374**, 46–49.

Blunier, T., Schwander, J., Stauffer, B., Stocker, T., Dällenbach, A., Indermühle, A., Tschumi, J., Chappellaz, J., Raynaud, D., and Barnola, J.-M. (1997). Timing of the Antarctic Cold Reversal and the atmospheric CO_2 increase with respect to the Younger Dryas event. *Geophys. Res. Lett.* **24**, 2683–2686.

Broecker, W. S., and Henderson, G. M. (1998). The sequence of events surrounding Termination II and their implications for the cause of glacial-interglacial CO_2 changes. *Paleoceanography* **13**, 352–364.

Chappellaz, J., Blunier, T., Kints, S., Dällenbach, A., Barnola, J.-M., Schwander, J., Raynaud, D., and Stauffer, B. (1997a). Changes in the atmospheric CH_4 gradient between Greenland and Antarctica during the Holocene. *J. Geophys. Res.* **102**, 15987–15997.

Chappellaz, J., Brook, E., Blunier, T., and Malaize, B. (1997b). CH_4 and $\delta^{18}O$ of O_2 records from Antarctic and Greenland ice: A clue for stratigraphic disturbance in the bottom part of the Greenland Ice Core project and the Greenland Ice Sheet Project 2 ice cores. *J. Geophys. Res.* **102**, 26547–26557.

Chappellaz, J., Blunier, T., Raynaud, D., Barnola, J.-M., Schwander, J., and Stauffer, B. (1993a). Synchronous changes in atmospheric CH_4 and Greenland climate between 40 and 8 kyr BP. *Nature* **366**, 443–445.

Chappellaz, J. A., Fung, I. Y., and Thompson, A. M. (1993b). The atmospheric CH_4 increase since the Last Glacial Maximum—(1) Source estimates. *Tellus* **45B**, 228–241.

Ciais, P., Friedlingstein, P., Schimel, D. S., and Tans, P. P. (1999). A global calculation of the delta C-13 of soil respired carbon: Implications for the biospheric uptake of anthropogenic CO_2. *Global Biogeochem. Cycles* **13**, 519–530.

Claquin, T., Roelandt, C., Kohfeld, K. E., Harrison, S. P., Prentice, I. C., Balkanski, Y., Bergametti, G., Hansson, M., Mahowald, N., Rodhe, N., and Schulz, M., (submitted). Radiative forcing of climate by ice-age dust.

Claquin, T., Schulz, M., Balkanski, Y., and Boucher, O. (1998). Uncertainties in assessing radiative forcing by mineral dust. *Tellus Series B—Chem. Phys. Meteorol.* **50**, 491–505.

Claussen, M., Kubatzki, C., Brovkin, V., Ganopolski, A., Hoelzmann, P., and Pachur, H. J. (1999). Simulation of an abrupt change in Saharan vegetation in the mid-Holocene. *Geophys. Res. Lett.* **26**, 2037–2040.

Cramer, W., Bondeau, A., Woodward, F. I., Prentice, I. C., Betts, R. A., Brovkin, V., Cox, P. M., Fisher, V., Foley, J. A., Friend, A. D., Kucharik, C., Lomas, M. R., Ramankutty, N., Sitch, S., Smith, B., White, A., and Young-Molling, C. (in press). Global response of terrestrial ecosystem structure and function to CO_2 and climate change: resutls from six dynamic global vegetation models. *Global Change Biol.*

Crowley, T. J., (1995). Ice-age terrestrial carbon changes revisited. *Global Biogeochem. Cycles* **9**, 377–389.

Crutzen, P. J. and Bruhl, C. (1993). A model study of atmospheric temperatures and concentrations of ozone, hydroxyl, and some other photochemically active gases during the glacial, the preindustrial Holocene, and the present. *Geophys. Res. Lett.* **20**, 1047–1050.

Dällenbach, A., Blunier, T., Flückiger, J., Stauffer, B., Chappellaz, J., and Raynaud, D. (2000). Changes in the atmospheric CH_4 gradient between Greenland and Antarctica during the Last Glacial and the transition to the Holocene, *Geophys. Res. Lett.* **27**, 1005–1008.

De Angelis, M., Steffensen, J. P., Legrand, M. R., Clausen, H. B., and Hammer, C. U. (1997). Primary aerosol (sea salt and soil dust) deposited in Greenland ice during the last climatic cycle: Comparison with east Antarctic records. *J. Geophys. Res.* **102**, 26681–26698.

Delmas, R. J., (1993). A natural artefact in Greenland ice-core CO_2 measurements. *Tellus* **45B**, 391–396.

Dentener, F. J., Carmichael, G. R., Zhang, Y., Lelieveld, J., and Crutzen, P. J., (1996). Role of mineral aerosol as a reactive surface in the global troposphere. *J. Geophys. Res.* **101**, 22869–22889.

Etheridge, D. M., Steele, L. P., Langenfelds, R. L., Francey, R. J., Barnola, J. M., and Morgan, V. I. (1996). Natural and anthropogenic changes in atmospheric CO_2 over the last 1000 years from air in Antarctic ice and firn. *J. Geophys. Res–Atmos.* **101**, 4115–4128.

Falkowski, P. G. (1997). Evolution of the nitrogen cycle and its influence on the biological sequestration of CO_2 in the ocean. *Nature* **387**, 272–275.

Farquhar, G. D., Lloyd, J., Taylor, J. A., Flanagan, L. B., Syvertsen, J. P., Hubick, K. T., Wong, S. C., and Ehleringer, J. R. (1993). Vegetation effects on the isotope composition of oxygen in atmospheric CO_2. *Nature* **363**, 439–443.

Fischer, H., Wahlin, M., Smith, J., Mastroianni, D., and Deck, B. (1999). Ice core records of atmospheric CO_2 around the last three glacial terminations. *Science* **283**, 1712–1714.

Francey, R. J., Allison, C. E., Etheridge, D. M., Trudinger, C. M., Enting, I. G., Leuenberger, M., Langenfelds, R. L., Michel, E., and Steele, L. P. (1999). A 1000-year high precision record of delta C-13 in atmospheric CO_2. *Tellus Series B—Chem. Phys. Meteorol.* **51**, 170–193.

Friedlingstein, P., Prentice, K. C., Fung, I. Y., John, J. G., and Brasseur, G. P. (1995). Carbon-biosphere-climate interactions in the last glacial maximum climate. *J. Geophys. Res–Atmos.* **100**, 7203–7221.

Haan, D., and Raynaud, D. (1998). Ice core record of CO_2 variations during the two last millennia: Atmospheric implications and chemical interactions within the Greenland ice. *Tellus* **50B**, 253–262.

Indermuehle, A., Stocker, T. F., Joss, F., Fischer, H., Smith, H. J., Wahlen, M., Deck, B., Mastroianni, D., Tschumi, J., Blunier, T., Meyer, R., and Stauffer, B. (1999). Holocene carbon-cycle dynamics based on CO_2 trapped in ice at Taylor Dome, Antarctica. *Nature* **398**, 121–126.

Kohfeld, K. E., and Harrison, S. P., (2000). How well can we simulate past climates? Evaluating the models using global palaeoenvironmental datasets. *Quat. Sci. Rev.* **19**, 321–346.

Krogh-Andersen K., Armengaud, A., and Genthon, C. (1998). Atmospheric dust under glacial and interglacial conditions. *Geophys. Res. Lett.* **25**, 2281–2284.

Lorius, C., and Oeschger, H. (1994). Paleo-perspectives–Reducing uncertainties in global change. *Ambio* **23**, 30–36.

Mahowald, N., Kohfeld, K. E., Hansson, M., Balkanski, Y., Harrison, S. P., Prentice, I. C., Rodhe, H., and Schulz, M. (1999). Dust sources and deposition during the Last Glacial Maximum and current climate: a comparison of model results with paleodata from ice cores and marine sediments. *J. Geophys. Res.* **104**, 15895–16436.

Martin, J. (1990). Glacial-interglacial CO_2 change: the iron hypothesis. *Paleoceanography* **5**, 1–13.

Martinerie, P., Brasseur, G. P., and Granier, C. (1995). The chemical composition of ancient atmospheres: a model study constrained by ice core data. *J. Geophys. Res.* **100**, 14291–12304.

Melillo, J. M., McGuire, A. D., Kicklighter, D. W., C.J. Vörösmarty III, B. M., and Schloss, A. L. (1993). Global climate change and terrestrial net primary production. *Nature* **363**, 234–240.

Melillo, J. M., Prentice, I. C., Farquhar, G. D., Schulze, E.-D., and Sala, O. E.

(1996). Terrestrial biotic response to environmental change and feedbacks to climate, In "Climate Change 1995. The Science of Climate Change." (Houghton, J. T., Filho, L. G. M., Callander, B. A., Harris, N., Kattenberg, A. and Maskell, K. Eds.), pp. 449–481. University Press, Cambridge.

Neftel, A., Oeschger, H., Staffelbach, T., and Stauffer, B.; (1988). CO_2 record in the Byrd ice core 50,000–5,000 years BP. *Nature* **331**, 609–611.

Nisbet, E. G. (1992). Sources of atmospheric CH_4 in early postglacial time. *J. Geophys. Res.* **97**, 12859–12867.

Pedersen, T. F. and Bertrand, P. (2000). Influences of oceanic rheostats and amplifiers on atmospheric CO_2 content during the Late Quaternary. *Quat. Sci. Rev.* **19**, 273–283.

Peng, C. H., Guiot, J., and Van Campo, E. (1998). Estimating changes in terrestrial vegetation and carbon storage: using palaeoecological data and models. *Quat. Sci. Rev.* **17**, 719–735.

Petit, J. R., Jouzel, J., Raynaud, D., Barkov, N. I., Barnola, J. M., Basile, I., Bender, M., Chappellaz, J., Davis, M., Delaygue, G., Delmotte, M., Kotlyakov, V. M., Legrand, M., Lipenkov, V. Y., Lorius, C., Pepin, L., Ritz, C., Saltzman, E., and Stievenard, M. (1999). Climate and atmospheric history of the past 420,000 years from the Vostok ice core, Antarctica. *Nature* **399**, 439–436.

Pinto, J. P. and Khalil, M. A. K. (1991). The stability of tropospheric Oh during Ice Ages, inter-glacial epochs and modern times. *Tellus Series B–Chem. Phys. Meteorol.* **43**, 347–352.

Prentice, I. C., Jolly, D., and BIOME 6000 participants. (2000). Mid-Holocene and glacial-maximum vegetation geography of the northern continents and Africa. *J. Biogeography.* **27**, 507–519.

Raynaud, D., Barnola, J-M., Chappellaz, J., Blunier, T., Indermühle, A., and Stauffer, B. (2000). The ice record of greenhouse gases: a view in the context of future changes. *Quat. Sci. Rev.* **19**, 9–17.

Raynaud, D., Chappellaz, J., and Blunier, T. (1998). Ice core record of atmospheric methane changes: relevance to climatic changes and possible gas hydrate sources. *Geol. Soc. Special Publ.* **137**, 327–331.

Schellnhuber, H. J. (1999). 'Earth system' analysis and the second Copernican revolution. *Nature* **402**, C19–C23.

Shackleton, N. J., (1977). Carbon-13 in Uvigerina: tropical rainforest history and the equatorial Pacific carbonate dissolution cycles. In "The Fate of Fossil Fuel CO_2 in the Oceans." Anderson, N. R. and Malahoff, A. (Eds.). *Marine Sci.* 401–427.

Smith, H. J., Fischer, H., Wahlen, M., Mastroianni, D., and Deck, B. (1999). Dual modes of the carbon cycle since the Last Glacial Maximum. *Nature* **400**, 248–250.

Stauffer, B., Blunier, T., and Dallenbach, A. (1998). Atmospheric CO_2 concentration and millennial-scale climate change during the last glacial period. *Nature* **392**, 59–62.

Steffensen, J. P. (1997). The size distribution of microparticles from selected segments of the Greenland Ice Core Project ice core representing different climatic periods. *J. Geophys. Res.* **102**, 26755–26763.

Stephens, B. B., Keeling, R. F. (2000). The influence of of Antarctic sea ice on glacial-interglacial CO2 variations, *Nature* **404**, 171–174.

Tegen, I., and Fung, I. (1995). Contribution to the atmospheric mineral aerosol load from land surface modification. *J. Geophys. Res.* **100**, 18707–18726.

Thompson, A. M., Chapellaz, J. A., Fung, I. Y., and Kuscera, T. L. (1993). The atmospheric CH_4 increase since the last glacial maximum: 2. Interaction with oxidents. *Tellus* **45B**, 242–257.

Thompson, L. G. and Mosley-Thompson, E. (1981). Microparticle concentration variations linked with climatic change: evidence from polar ice. *Science* **212**, 812–815.

Should Phosphorus Availability Be Constraining Moist Tropical Forest Responses to Increasing CO$_2$ Concentrations?

J. Lloyd
Max Planck Institute for
* Biogeochemistry*
Jena, Germany

M.I. Bird
Australian National University
Canberra, Australia

E.M. Veenendaal
Harry Oppenheimer Okavango
* Research Centre*
Maun, Botswana

B. Kruijt
Alterra Green World Research
* Foundation*
Wageningen, The Netherlands

Moist tropical forests account for a substantial amount of global plant productivity. And several lines of evidence suggest that they may be sequestering significant amounts of anthropogenically released carbon at the present time. But there are also indications that the productivity of many of these forests is limited by low phosphorus availability. This has led to suggestions that moist tropical forests may be constrained in their ability to increase their growth rates in response to increases in atmospheric carbon dioxide concentrations. This notion is examined in this chapter.

Several factors should prevent low levels of available phosphorus significantly constraining moist tropical forest [CO$_2$]/growth responses. One of the main reasons for low soil-solution P concentrations in many tropical soils is the adsorption of most of the phosphate ions onto iron and aluminum oxides and clay minerals. This adsorption is, to a large extent, reversible. This means that, in response to increased rates of removal of P from the soil solution, such as would be required to sustain faster plant growth with increasing [CO$_2$], phosphate ions should be desorbed from their fixation sites and released into the soil solution, thus maintaining the concentration of P in the soil solution at a more-or-less constant level. This contrasts with the situation for nitrogen in temperate and boreal forests, where the rate of entry of nitrogen into the soil solution is closely linked to the rate of carbon mineralization.

The nature of P mineralization in soils is a second factor mediating towards phosphorus availability not constraining tropical forest [CO$_2$] responses. This is because, unlike nitrogen, phosphorus is mineralized independent of carbon in most soils. Thus, it has less potential to be "locked up" in the larger soil carbon pool that should occur as a result of increased plant productivity at higher [CO$_2$].

Third, most of the available evidence suggests that, at a given soil P concentration, plants growing at elevated [CO$_2$] are capable of maintaining their tissue phosphorus concentrations. This is in contrast to nitrogen and occurs because of the positive effects of larger root systems on the extent of root mycorrhizal colonization, root organic acid efflux per plant, and root acid phosphatase activity. All three processes play important roles in phosphorus acquisition.

An additional phenomenon may also be important in the tropical forest C/P interaction. Humic molecules and organic acids actively compete with phosphorus for soil fixation sites. This means that increases in soil carbon density at higher [CO$_2$] may serve to displace phosphate ions from sorption sites and into the soil solution, where they can then be utilized by plants. It is not inconceivable that this effect could give rise to a "runaway" positive feedback: CO$_2$-induced increases in tropical forest plant growth giving rise to increases in soil carbon content, which in turn liberates previously adsorbed phosphorus—this in turn giving rise to even more substantial increases in plant growth.

1. Introduction

By virtue of their large area and year-round favorable growing conditions, moist tropical forests may account for as much as 50% of the global net primary productivity (Grace *et al.*, 2000). This high productivity combined with reasonably long carbon residence times means that tropical forests are likely to be a substantial component of the terrestrial sink for anthropogenic CO_2 (Lloyd and Farquhar, 1996; Lloyd, 1999a; Malhi *et al.*, 1999).

Although this notion contrasts with some global $[CO_2]/\delta^{13}C$ inversion studies (e.g., Ciais *et al.*, 1995), it has received experimental support from eddy covariance studies above undisturbed forests in Brazil (Grace *et al.*, 1995, 1996; Malhi *et al.*, 1998) and from biomass inventory data (Phillips *et al.*, 1998). It is also in accordance with some synthesis atmospheric inversions (Enting *et al.*, 1995; Rayner *et al.*, 1999) and current interpretations of the rate of increase and latitudinal gradients in atmospheric O_2/N_2 ratios (Keeling *et al.*, 1996). In both Keeling *et al.* (1996) and Rayner *et al.* (1999), the carbon balance of tropical regions appears to be more or less netural, despite the fact that many tropical regions are significant sources of anthropogenic CO_2 because of deforestation associated with land-use change or shifting cultivation. The magnitude of this source is substantial, currently an estimated 0.1–0.2 Pmol C year^{-1} (Houghton, 1996). By simple mass balance, then, undisturbed tropical regions must be substantial sinks for anthropogenically released CO_2 of about the same amount.

But the dogma also has it that tropical forests are usually found on heavily weathered soils that are low in nutrients, particularly phosphorus. In a manner similar to possible interactions between the nitrogen and carbon cycles in temperate and boreal forests (McGuire *et al.*, 1995) it has thus been suggested that low P availability may limit the extent to which tropical forests are able to increase their productivity in response to increases in atmospheric CO_2 concentrations (Friedlingstein *et al.*, 1995; McKane *et al.*, 1995).

We therefore examine here in some detail the relationships between the biogeochemical cycling of carbon and phosphorus in moist tropical forests. Our main purpose is to examine whether low P availability should actually be constraining the ability of moist tropical forests to increase their productivity and carbon stocks as a consequence of increasing atmospheric $[CO_2]$.

2. Phosphorus in the Soils of the Moist Tropics

That tropical soils are highly weathered and infertile is a generalization at best. Indeed, only about 50% of the total tropical soil area can be considered to consist of highly weathered leached soils such as oxisols, ultisols, and alfisols (Sanchez, 1976; Richter and ie. ultisol Babbar, 1991). But when one considers the moist tropical regions with an annual precipitation greater than 1500 mm only, these soils account for about 75% of the total area (Sanchez, 1976).

That such soils should be phosphorus deficient is a prediction from pedogenic theory (Walker and Syers, 1976). This is because, in contrast to carbon, nitrogen, and sulfur, P is cycled mainly on geological time scales. That is to say, the only substantial primary source of P for plants is from the weathering of parent material at the base of the soil. As soil development proceeds, there is a loss of this weathered P as a consequence of leaching. The rate of leaching is quite small on an annual basis, even in the tropics (typically 0.1–1 mmol P m^{-2} year^{-1}: Bruijnzeel, 1991) but it occurs over several thousand years. Moreover, as soils become older, not only does the total amount of phosphorus decline, but there is also a transfer of phosphate from labile pools to nonlabile pools (Walker and Syers, 1976).

The chemistry of soil phosphorus transformations giving rise to this situation is complex (Sanyal and DeDatta, 1991). But even for the simplest understanding of soil phosphorus, it is necessary to consider labile and nonlabile pools of phosphorus in both the organic and inorganic forms as well as the significant fluxes through the microbial pool (Brookes *et al.*, 1984; Singh *et al.*, 1989; Lodge *et al.*, 1994; Gijsman *et al.*, 1996).

From the outset, we need to define the terms used to define the states and fluxes of phosphorus in the soil. Following Barrow (1999), *sorption* is taken to mean the transfer of a material from a liquid phase (such as the soil solution) to the solid phase, the soil itself. Sorption includes *ad*sorption, which means that the sorbed material is on the outside of the soil particle.

When a *labile* pool is being discussed it is considered that the material in question is liable to displacement or change. Likewise, the *nonlabile* pool is considered to be in a stable state for the time scales of interest here (years to centuries).

It should also be emphasized from the outset that, in both a physical and a chemical sense, soils are strongly heterogeneous media and elements such as phosphorus do not really partition into such simple compartmented states (Barrow, 1999). Likewise when attempts are made to fractionate P into pools of varying stability, the exact nature of the different P pools within the soils that these chemically isolated fractions represent is also not entirely clear (Gijsman *et al.*, 1996).

2.1 Soil Organic Phosphorus

For soils underneath moist tropical forests, organically bound phosphorus generally accounts for 20–80% of the total P (Westin and de Brito, 1969; Sanchez, 1976; Tiessen *et al.*, 1994a; Newberry *et al.*, 1997). This organic P represents a wide spectrum of compounds, reflecting the diverse biological origins of soil organic matter (Magid *et al.*, 1995). Labile forms include nucleic acids and phospholipids (of primarily bacterial origin). Inositol phosphates often constitute the bulk of the nonlabile organic-P pool, forming sparingly soluble salts with ions such as iron, aluminum, and calcium. They can also form strong complexes with proteins and can be strongly adsorbed by clay minerals, typically constituting about 50% of organic P (McLaren and Cameron, 1996).

Organic phosphorus is considered to play a key role as a source of P for plants in tropical soils (Sanchez *et al.*, 1976; Sec. 2.3). In this context it is important to note that, in contrast to nitrogen, phosphorus is to a large degree mineralized independent of carbon (McGill and Cole, 1981). This is a result of the production of phosphatases by plant roots, mycorrhizae, and microbes. These specifically hydrolyze phosphate ester linkages on soil organic compounds, releasing phosphorus and making it available for plant uptake (Sec. 3.3.3).

According to Gijsman *et al.* (1996), data of Ognalaga *et al.* (1994) also suggest that organic P can be stabilized into nonlabile forms independently of organic carbon. A similar conclusion was also reached by McGill and Cole (1981). This means that there is much more chance for variation in C/P ratios of the labile soil organic pool than is the case for C/N ratios. This has important implications for the response of P-limited systems to increases in atmospheric carbon dioxide concentrations (Sec. 4.3).

2.2 Soil Inorganic Phosphorus

The labile component of the inorganic phosphorus pool is generally taken to comprise calcium-bonded phosphates, aluminum-bonded phosphates, and iron-bonded phosphates. For highly acid and highly weathered tropical soils such as oxisols and ultisols, iron and aluminum phosphates tend to dominate and thus adsorption capacity for P is usually quite high (Sanchez, 1976). Crystalline clay minerals are also able to specifically adsorb P through a ligand-exchange reaction with the (OH)H groups coordinated with the Al ion on the edge of the crystal (Muljadi *et al.*, 1966).

The high content of aluminum and iron oxides in the oxisols and ultisols typically found underneath moist tropical forests is the reason for the ability of these soils to "fix" significant amounts of phosphorus when applied as a fertilizer after conversion of these systems to agriculture. Most of the added phosphorus is adsorbed within the first few days of application, although subsequent continued long-term sorption also occurs (Sample *et al.*, 1980; Barrow, 1999). It is important to recognize that this adsorption is a more or less reversible reaction, with the amount of sorbed phosphorus being dependent on the soil solution P concentration (Barrow, 1983). This accounts for the long-term beneficial effects of massive initial applications of phosphorus fertilizers applied to some tropical soils (e.g., Younge and Plucknett, 1966). Phosphorus "fixed" by these soils is subsequently released to the soil solution and utilized for plant growth over many years. This is because desorption occurs in response to the diffusion gradient that typically occurs around any plant root or microbe actively acquiring P (Mattingly, 1975).

Barrow (1983) has pointed out the complexities of the sorption process onto and within soil particles. He suggests that the relatively rapid adsorption of P onto the soil surface is followed by a slow diffusive penetration. Support for this idea comes from the observation that the relative rates of penetration of different adsorbed ions into reacting particles are correlated with the affinity of the surface (Barrow and Whelan, 1989). More recently, Strauss *et al.* (1997) have shown that the extent of the slow reaction between goethite and phosphate depends on the crystallinity of the geothite. Strong evidence was provided that the mechanism for slow phosphate sorption was a slow penetration of the spaces between the crystal domains. Importantly, the longer a sorption reaction takes to occur, the slower the subsequent desorption reaction and the smaller the amount desorbed after a given period of time (Barrow, 1999).

For many tropical soils, the amount of labile inorganic phosphorus in the sorbed form is typically more than a thousand times greater than the amount of P in the soil solution (Sanchez, 1976). As the net movement of phosphate ions between these pools will always be toward a new equilibrium, this much greater amount of sorbed P means that soil solution P concentrations are strongly buffered against any changes in the rates of entry of P into, or removal of P from, this pool, such as changes that might occur due to changes in P mineralization rates or variations in plant P uptake rates. Thus, the inorganic labile phosphorus pool in many tropical soils can almost be looked upon as a slow-release fertilizer pool whose rate of release is determined by the rate of plant phosphorus utilization. As is shown in Sec. 4.3, this has important implications for the ability of tropical forests to maintain increasing growth rates in response to increases in atmospheric [CO₂].

The nonlabile fraction of inorganic phosphorus not available to plants is sometimes divided into the occluded and reductant soluble forms. Occluded phosphorus consists of aluminum- and/or iron-bonded phosphates surrounded by an inert coat of another material such as oxides or hydrous oxides of iron or aluminum. Reductant soluble forms are covered by a coat that may be partially or totally dissolved under anaerobic conditions (Uehara and Gillman, 1981). The opportunities for occlusions to occur increase dramatically with soil age (Walker and Syers, 1976). This is because substantial amounts of Fe and Al oxides tend to be present only in heavily weathered soils in which the secondary silicate minerals have already dissolved (Fox *et al.*, 1991). Data from tropical forest chronosequence studies in Hawaii are more or less in accordance with this view: the fraction of P present in the "occluded" form increases with soil age (Crews *et al.*, 1995). Nevertheless, that study also showed high amounts of nonoccluded (i.e., labile and accessible) inorganic phosphorus to be present, even in forests growing on the oldest soils.

2.3 Soil Carbon/Phosphorus Interactions

Tropical agronomists have long realized the importance of organic phosphorus as the main source of phosphorus in nonfertilizer agriculture, such as that occurs in traditional systems (Nye and Bertheux, 1957; Sanchez, 1976). In addition to being a source of phosphorus for plant uptake after mineralization, the importance of organic matter in tropical crop productivity is associated with the critical relationship between organic matter content and soil fertility in highly weathered tropical soils (Tiessen *et al.*, 1994b). These soils typically have a very low cation-exchange capacity (CEC) or even a dominant anion-exchange capacity (Sanchez, 1976; Sollins *et al.*, 1988). Soil organic matter performs a vital function in these soils by reacting with Fe and Al oxides, coating the surfaces of oxide particles. This gives rise to a net negative charge and hence a dominant cation-exchange capacity (Uehara and Gilman, 1981; Sollins *et al.*, 1988). This strong association between organic matter content and soil fertility has led to the suggestion that the rapid decline in soil carbon stocks after conversion of forest to agriculture is the prime cause for the subsequent leaching of essential elements out of the active rooting zone (Tiessen *et al.*, 1994b).

The coating of Al - and Fe-oxides by soil organic matter in many tropical soils probably increases phosphorus availability as well. Adherence of large humic molecules to the surfaces of clays and metal hydrous oxide particles (Hughes, 1982; Bonde *et al.*, 1992) should mask the phosphorus fixation sites and prevent oxide particles from interacting with phosphorus ions in solution. In addition, the organic acids in the soil that are produced during microbial degradation of organic matter and directly by plants themselves (Sec. 3.3.2) actively compete with phosphorus ions for soil fixation sites (Dalton *et al.*, 1952; Lopez-Hernandez *et al.*, 1986; Sibanda and Young, 1989; Fox *et al.*, 1990; Bhatti *et al.*, 1998; Jones, 1998). To date, the relationship between tropical soil organic matter content and plant phosphorus availability has concentrated mostly on the significant declines in soil C and P that usually occur after forest clearance (Mueller-Harvey *et al.*, 1985; Tiessen *et al.*, 1992, 1994b). This decline in soil organic matter has also been associated with an increase in the proportion of phosphorus in less labile forms (Tiessen *et al.*, 1992).

If tropical forests are indeed responding to increases in $[CO_2]$ by increasing their growth (Grace *et al.*, 1995; Phillips *et al.*, 1998), then much of this extra carbon fixed will eventually end up in the soil (Lloyd and Farquhar, 1996). Thus, a crucial question is whether the positive relationship between soil organic matter and soil phosphorus fertility will hold when soil carbon stocks are increasing? If this were the case, then irrespective of the mechanisms tropical trees may employ to acquire the extra phosphorus needed for increased growth in response to increases in $[CO_2]$ (Sec. 3.3), improved phosphorus fertility would currently be occurring, merely by virtue of increases in soil carbon density.

Indeed, there is some evidence that the relationship between soil organic matter and phosphorus fertility holds for natural rain forest as well as for degrading systems. For example, there are strong correlations between plant available phosphorus and soil organic matter concentration where natural spatial variability is the primary source of variation (Burghouts *et al.*, 1998; Silver *et al.*, 1999). However, this relationship might also arise from a stimulating effect of soil phosphorus availability on above-ground carbon acquisition being reflected in the soil carbon pool. Likewise, correlations between soil organic matter content and maxi-

mum degree of phosphorus adsorption for tropical soils (Sanyal and De Datta, 1991) may reflect effects of phosphorus availability on plant productivity and hence soil carbon content rather than vice versa. A more specifically targeted experiment is therefore required to test for this phenomenon. The possible magnitude of the effect is modeled in Sec. 4.3.

3. States and Fluxes of Phosphorus in Moist Tropical Forests

From Sec. 2 it can be concluded that, due to the highly weathered state and high phosphorus sorption capacity of many moist tropical forests soils, the level of readily plant available phosphorus is low. Discussion on whether this means that phosphorus availability actually *limits* productivity of moist tropical forests is reserved until Sec. 4.1. Here we limit our concerns to a discussion of the phosphorus cycle in moist tropical forests and methods by which plant phosphorus acquisition can occur in environments characterized by low levels of available P. The main aim of this section is to quantify the amounts and annual input/output fluxes of P for leaves, branches, boles, and roots of moist tropical vegetation. The inputs of phosphorus into moist tropical forests from rock weathering and wet and dry deposition, as well as from leaching losses, are also considered. This information is then used for model simulations in Sec. 4.3.

3.1 Inputs and Losses of Phosphorus through Rainfall, Dry Deposition, and Weathering: Losses via Leaching

3.1.1 Atmospheric Deposition

Atmospheric inputs of mineral elements into tropical rain forests may constitute an important input of plant nutrients, especially for soils of low inherent fertility (Proctor, 1987; Bruijnzeel, 1991). Such atmospheric inputs are traditionally divided into wet deposition (input of mineral elements dissolved in rainwater) and dry deposition (inputs from deposited aerosol particles or as dust). For large particles such a distinction may be obvious, for example, in examining effects of Saharan dust on overall forest nutrient balances in West African rain forests (Stoorvogel *et al.*, 1997) or in examining long-range advection of particles such as the deposition of Saharan dust into the vegetation of the Amazon Basin (Swap *et al.*, 1992). But for marine, anthropogenic, and biogenic aerosols, entrainment into atmospheric water vapor may occur during the convective mixing of the lower troposphere, with the elements of such particles then being deposited during rainfall events as well as by dry deposition. The separation of dry versus wet deposition is fraught with technical difficulties (Lindberg *et al.*, 1986), but for many tropical forest studies a simple combined measure of the two has been obtained by sampling in a forest clearing or sometimes above the canopy. In this way the bulk nutrient content of the precipitation has been obtained, at least for the collector itself (Bruijnzeel, 1991). As is also discussed by Bruijnzeel (1989; 1991), due to several complications, this method does not necessarily give the amounts as the amounts of nutrients deposited on the proximal forest canopy. In attempts to deduce external nutrient inputs into a forest, a further complication may be that tropical forests themselves produce aerosols (Crozat 1979; Artaxo *et al.*, 1988, 1990).

Given the above uncertainties, and even after unreasonably high values have been excluded, the high variability in reported rates of P deposition onto tropical forests, 0.3–7 mmol P m^{-2} year^{-1} (Bruijnzeel, 1991; Lesak and Melack, 1996; Stoorvogel *et al.*, 1997; Williams *et al.*, 1997), is not all that surprising. What is surprising is the magnitude of this input relative to the annual litter fall flux, which, from the summary of Proctor (1987), typically ranges from 19 to 44 mmol P m^{-2} year^{-1} (see also Sec. 3.2). Indeed, comparisons of lowland forest sites where both bulk precipitation inputs and litter fall measurements have been made suggest that the input of P into tropical ecosystems from the atmosphere above is 0.27 ± 0.17 ($n = 6$) of the annual litter fall P (Nye, 1961; Bernhard-Reverset, 1975; Golley *et al.*, 1975; Brinkmann, 1985). These relatively high rates of P deposition onto tropical forests contrast with the standard view that atmospheric inputs of P into these ecosystems are not significant (Vitousek *et al.*, 1988; Kennedy *et al.*, 1998).

This atmospheric P deposition cannot be supported by long-term transport of P from tropical oceans, as these typically have very low P concentrations in their surface waters (Graham and Duce, 1979). One possibility is the intrusion of dust from arid regions (Swap *et al.*, 1992). The importance of dust as a nutrient source is likely for West African rain forests (Stoorvogel *et al.*, 1997) but the significance of occasional long-term transport of Saharan dust into Amazonia has been questioned (Lesak and Melack, 1996). For Amazonia, it appears that biogenic emissions from the tropical forests themselves are the main source of atmospheric P in the region (Artaxo *et al.*, 1998; Echalar *et al.*, 1998).

3.1.2 Retention of Atmospherically Derived P

Irrespective of the source(s), hydrological studies have shown that a significant proportion of the atmospherically derived phosphorus appears to be retained by moist tropical forests, rather than being leached out of the system (Bruijnzeel, 1991). This rate of retention seems to be between 0.05 and 0.95 of the rate of input. This probably reflects, as much as anything else, the many sources of error in making such measurements. Bruijnzeel (1991) suggests that this general pattern of phosphorus accumulation in the forest/soil system is real and that it may arise as a consequence of P "fixation" onto iron and aluminum oxides (Sec. 2.3). But although the sorption mechanism is undoubtedly complex (Barrow, 1999) and perhaps less rapidly reversible in highly leached tropical soils than elsewhere (Gijsman *et al.*, 1996), the rate at which P is actually transformed into nonlabile forms is likely to be substantially less than this rate of atmospheric input. Indeed, it is not at all clear whether this external phosphorus arriving at the forest floor

would even reach the soil sorption sites. This is because of the extensive root mat near and above the soil surface in many rain forests that can effectively trap dissolved and fine-litter nutrient inputs (Sec. 3.3.1).

3.1.3 Throughfall and Stemflow

In addition to substantial inputs of phosphorus occurring as a consequence of wet and dry deposition, substantial enrichment of rainwater phosphorus concentrations occurs during the passage of rainwater through tropical forest canopies (Vitousek and Sanford, 1986; Proctor 1987; Veneklass, 1990; Forti and Moreira-Nordemann, 1991; McDowell, 1998). Again, exact values for the enrichment in this throughfall are subject to considerable uncertainties as a consequence of methodological problems. For example, it is not always clear whether this enrichment estimate includes accumulation of elements deposited during dry deposition. But both Marschner (1995) and Richards (1996) consider "canopy leaching" to provide the main source of nutrient additions to rainfall as it passes through the canopy. From data summarized by Vitousek and Sanford (1986) and Proctor (1987), canopy leaching can contribute as much as 20 mmol P m^{-2} year^{-1}, with average values around 8 mmol P m^{-2} year^{-1}. As for the P input in rainfall itself, this amount is significant compared with an average litterfall value around 25 mmol P m^{-2} year^{-1} (Sec. 3.2). Such rates of P leaching are higher than those that typically occur in temperate regions (Parker, 1983; Marschner, 1995), as would be anticipated on the basis of the much higher rainfall amounts and intensities in tropical regions. According to Marschner (1995) canopy leaching for elements such as P can arise as a consequence of the passage of water through the apoplast of intact leaf tissue as well as through damaged leaf areas, with rates of leaching greater at high temperatures. Proctor (1987) has also pointed out the possible importance of insect frass. Generally speaking, nutrient enrichments during stemflow are much less significant source of nutrients to the soil than canopy throughfall (Parker, 1983; Vitousek and Sanford, 1986; Proctor 1987; Richards, 1996).

3.1.4 Weathering as a Source of Biologically Available Phosphorus

From basin-wide studies in South America, phosphorus weathering rates of 0.3–1.0 mmol P m^{-2} year^{-1} have been reported (Lewis *et al.*, 1987; Gardner, 1990). The degree to which such weathering of parent material may supply nutrients for plant growth in moist tropical forests has been considered by Burnham (1989) and Bruijnzeel (1989). They point out that for already highly weathered soils, the active zone of rock weathering occurs a considerable distance below the zone where active root uptake of any nutrients released by the weathering process is likely. Nevertheless, there are some cases where moist tropical forest roots can penetrate the underlying weathered rock (Ballie and Mamit, 1983), and this would certainly be expected to be the case for montane forests. Clearly more experimental work is required, but available evidence indicates that because of the great depth at which weathering generally occurs in moist lowland tropical forests, it is unlikely to be a significant source of biologically available phosphorus in most cases.

3.2 Internal Phosphorus Flows in Moist Tropical Forests

The subject of the cycling of mineral nutrients in tropical forests, particularly the degree to which systems are closed with little leakage of nutrients out of them, is a long-standing area of interest and controversy for tropical ecologists (Hardy, 1935; Walter, 1936, 1971; Jordon and Herrera, 1981; Vitousek and Sanford, 1986; Proctor, 1989; Whitmore, 1989; Silver, 1994; Richards, 1996). In general, the earlier paradigm of closed nutrient cycles with little or no leakage out of them (Hardy, 1935; Walter, 1936) has given way to an appreciation of the diversity of nutrient cycles in different tropical forests, with effects of natural variations in soil fertility now being a central emphasis (Vitousek and Sanford, 1986; Whitmore, 1989).

3.2.1 Above-Ground Phosphorus Stocks and Soil Fertility

Vitousek and Sanford (1986) grouped lowland forests according to the underlying soil fertility and showed that forests growing on moderately fertile soils (about 15% of the total moist tropical forest area) tend to have foliar N, P, K, Ca, and Mg concentrations higher than do those growing on the more common oxisol or ultisol soil types of moderate to low fertility (63% of the total moist tropical forest area). Forests on the latter tend to have foliar nutrient concentrations not very different from forests growing on the very low-fertility spodosol or psamment soil types (7% of the total moist tropical forest area: Sanchez, 1976; Vitousek and Sanford, 1986).

The relationship between above-ground carbon density and above-ground phosphorus density (taken from Table 2 of Vitousek and Sanford (1986) with additional data from Hughes *et al.* 1999) is shown in Figure 1. This shows a remarkably strong relationship between the two parameters, but with a different relationship for moderately fertile soils versus the infertile oxisols/ultisols. For both forests the relationship between above-ground carbon density and above-ground phosphorus density is stronger than that for other nutrients such as nitrogen (not shown). Importantly, forests growing on soils with a low level of phosphorus availability are still capable of achieving substantial above-ground carbon densities, despite having much lower phosphorus stocks than forests growing on more fertile soils. As has been pointed out by Vitousek and Sanford (1986), at least part of this difference in phosphorus is due to much higher foliar P concentrations for trees growing on more fertile soils (1.1 ± 0.2 mmol P mol^{-1} C) than for those growing on the less fertile soils (0.5 ± 0.1 mmol P mol^{-1} C).

Foliar phosphorus concentrations typically decline with canopy depth in tropical rain forests (Lloyd *et al.*, 1995) and so it is not straightforward to relate bulked canopy values to physiological measurements made on individual leaves. But similar

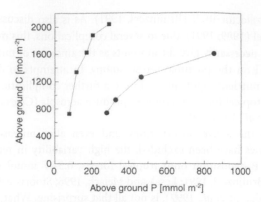

FIGURE 1 The relationship between above ground carbon density and above ground phosphorus density for moist tropical forests growing on moderately fertile (●) and infertile(■) soils.

magnitude differences in foliar nutrient concentrations often occur between primary and secondary successional rain forest species and this is reflected in differences in plant photosynthetic rates (Raaimakers *et al.*, 1995; Reich *et al.*, 1995). Given the laboratory gas-exchange data of the phosphorus dependency of photosynthesis for leaves of warmer-climate trees (Kirschbaum and Tompkins, 1990; Cromer *et al.*, 1993; Sec. 4.1) it seems likely that rain forests growing on more fertile soils have higher gross primary productivities.

Most likely there are also differences in the general growth strategies employed by trees on the different soil types. For example, Veenendaal *et al.* (1996) examined growth responses of tropical tree seedlings from low- and high-fertility soils in Ghana. Although there were some exceptions, they found that seedlings whose natural distribution was limited to low-fertility soils were not capable of faster growth rates when grown on the higher nutrient soil. Likewise, species restricted to high-fertility soils grew poorly on the lower fertility soil.

Along with faster growth rates and higher phosphorus and nitrogen requirements for the species from the higher nutrient soil (Veenendaal *et al.*, 1996), a picture emerges of species adapted to higher nutrient soils being successful by virtue of high potential growth rates and an ability to rapidly acquire nutrients. Likewise, the moist tropical forest species usually found growing on poorer soils are probably successful on these soils as a consequence of low nutritional requirements, particularly with respect to phosphorus. Also associated with these plants should be specific physiological adaptions allowing high phosphorus uptake rates despite low levels of readily available P (Sec. 3.3). This is similar to the relationships between plant growth strategy and soil fertility proposed for temperate, arctic, and boreal ecosystems (Chapin, 1980).

Tropical forest foliage typically accounts for less than 15% of the above-ground P pool (Fölster *et al.*, 1976; Klinge, 1976; Hase and Fölster, 1982; Uhl and Jordan 1984). Therefore, most of the differences between the C/P relationships in Figure 1 are attributable to differences in the phosphorus concentrations in twigs, branches, and boles. For example, the average concentration of phosphorus in the boles of the forests growing on the more fertile soils is 0.30 ± 0.06 mmol P^{-1} mol^{-1}c ($n = 4$: Greenland and Kowal, 1960; Golley *et al.*, 1975; Hase and Fölster, 1982; Hughes *et al.*, 1999), whereas for above-ground woody tissues on the less fertile oxisols/ultisols this figure is 0.16 ± 0.05 mmol P mol^{-1} C ($n = 6$: Bernhard-Reversat, 1975; Fölster *et al.*, 1976; Klinge, 1976; Uhl and Jordan, 1984).

Explaining this twofold difference between the two forest types is difficult. Despite the fact that woody components constitute the dominant above-ground pool for P in moist tropical forests, the role of P in woody tissue is not well defined. Most likely its functions relate to its being a structural constituent of the growing sapwood, as well as inorganic phosphorus being associated with general energy transfer reactions in sapwood and phloem-associated cells. In both cases, a general positive relationship between high plant growth rates and woody tissue P concentrations would be expected.

As is the case for nitrogen (Lloyd and Farquhar, 1996), one might expect a decrease in bole P content with increasing plant size. This is because most of the P would be expected to be in the physiologically active sapwood tissue. This constitutes a progressively smaller portion of the total stemwood as trees become bigger. Nevertheless, when compared across sites, there seems to be no general pattern of lower bole P concentrations in forests with increasing carbon density (data not shown). However, for individual tropical forest species, such a trend of deceasing P concentrations with increasing bole size has been observed (Grubb and Edwards, 1982).

Along with the likely higher photosynthetic rates discussed above, the greater phosphorus content of woody tissue from forests growing on more fertile soils suggests higher potential gross and net primary productivities than those of less fertile forests. This then begs the question of how the above-ground carbon density of nutrient-poor forests can generally be higher than that of forests growing on more nutrient-rich soils (Figure 1).

In considering the observed lack of correlation between forest biomass and soil nutrient status for moist tropical forests, Vitousek and Sanford (1986) proposed that previous natural and anthropogenic stand-level disturbances may have been responsible. Differences in site water balance might also be important. Three of the four sites in Figure 1 growing on moderately fertile soils are moist semideciduous forests and are characterized by the presence of some drought deciduous species (Greenland and Kowal, 1960; Golley *et al.*, 1975; Hase and Fölster, 1982). This reflects a greater than average seasonality in water supply. Even for moist evergreen forests, marked effects of soil water deficit on photosynthetic productivity during the dry season can occur (Malhi *et al.*, 1998). This observation, combined with the observation that biomass of dry tropical forests is positively related with annual rainfall up to at least 1500 mm per annum (Martínez-Yrízar, 1995), suggests that the lower biomass of "moist" forests on the more fertile soils could in some cases be a consequence of more prolonged soil water deficits during the dry season than is the case for the forests growing on the more highly leached oxisol

and ultisol soil types. Indeed a negative association between soil fertility and soil water balance (Veenendaal *et al.*, 1996) is likely. Soils in areas exposed to lower rainfalls are likely to be less leached and therefore higher in nutrient status (Burnham, 1989). Confounding this rainfall/fertility correlation at the stand productivity level is the observation that even on the same soil, tropical drought-deciduous species typically have higher N and P concentrations than do proximal evergreen species (Medina, 1984).

It is also possible that plant growth traits associated with potentially faster growing trees on higher nutrient soil predispose such forests to lower carbon densities. For example, Phillips *et al.* (1994) showed a positive relationship between soil fertility and tree turnover rates for tropical forests. Likewise, leaves of inherently slower growing species tend to be longer lived (Chabot and Hicks, 1992). Studies with different successional species have confirmed this pattern for moist tropical forests (Reich *et al.*, 1995). Thus, despite their lower productivities, slower tree turnover rates might contribute to the attainment of high above-ground carbon densities for forests growing on nutrient-poor soils.

3.2.2 Phosphorus Content of Coarse and Fine Root Tissue

Not surprisingly, the available information on root P content is less than that on the above-ground biomass. Nevertheless, the available data suggest that the effects of soil fertility on root P concentrations are similar to those discussed above for leaves and above-ground woody tissue. For the two high-fertility sites where data are available (Greenland and Kowal, 1960; Golley, 1975), the average value is 0.37 ± 0.04 mmol P mol^{-1} C, whereas for the low-fertility oxisol sites for which data are available (Klinge, 1976; Uhl and Jordan, 1984) the average value is 0.15 ± 0.05 mmol P mol^{-1} C. These values are remarkably similar to the average values for above-ground woody tissue given above: 0.30 ± 0.05 and 0.16 ± 0.05 mmol P mol^{-1} C, respectively.

The values cited above represent a pooled average for coarse and fine roots. Greenland and Kowal (1960) separated out roots of varying diameter from a forest in Ghana. They showed that P concentration increased with decreasing root size with the finest size category (< 6 mm) having a concentration of 0.59 mmol P mol^{-1} C, much higher than their coarsest size category (> 25 mm), which contained 0.10 mmol P mol^{-1} C. For fine roots in poorer soils, fine root P concentrations seem to be similar. For a forest growing on an oxisol in Venezuela, Medina and Cuevas (1989) give a fine root concentration of 0.85 mmol P mol^{-1} C. Vitousek and Sanford (1986) cite a fine root concentration of 0.55 mmol P mol^{-1} C, also for a Venezuelan forest. Thus, unlike foliar tissue or structural woody biomass, it seems that there is little systematic effect of soil fertility on fine root P concentrations. This, along with the tendency of lower fertility sites to have a greater proportion of their total biomass below ground (Vitousek and Sanford, 1986), indicates a need for plants on low-nutrient soils to allocate a greater proportion of their carbon and nutrient resources to the acquisition of limiting elements (Chapin, 1980).

3.3 Mechanisms for Enhanced Phosphorus Uptake in Low P Soils

Even in very fertile soils, phosphorus concentrations in the soil solution are low, rarely exceeding 10 μM. This is several orders of magnitude lower than the concentration of phosphorus in plant tissues, typically 5–20 mM (Marschner, 1995; Raghothama, 1999). It is therefore not surprising that plants have developed several specialized physiological and biochemical mechanisms for acquiring and utilizing phosphorus. Our purpose here is to consider these mechanisms, especially as they relate to tropical forests. An emphasis is also placed on interactions between plant carbon supply and phosphorus acquisition. For more detailed recent reviews on plant P uptake, the reader is referred to Schachtman *et al.* (1998) and Raghothama (1999).

3.3.1 Distribution of Fine Roots and Mycorrhizal Associations

Generally speaking, most fine roots in tropical forest soils are found in the upper 0.5 m (Kerfoot, 1963), with a marked concentration of roots into a "root mat" close to the soil surface and within the litter layer being especially common on low-fertility soils (Stark and Jordan, 1978; Medina and Cuevas, 1989). It is generally considered that these root mats serve to ensure the maximum retention of nutrients by the vegetation and to minimize any leaching losses. Surveys of tropical forests have indicated almost ubiquitous mycorrhizal associations for such roots (Alexander, 1989; Janos, 1989).

As for temperate plants, it is widely assumed that mycorrhizal associations in tropical forests serve to improve the uptake of mineral nutrients, particularly phosphorus (Bolan, 1991; Koide, 1991; Smith and Read, 1997). Growth stimulations and enhanced P uptake in response to mycorrhizal infection have been reported for tropical tree seedlings (Janos, 1989; Lovelock *et al.*, 1996, 1997).

Several mechanisms may be involved in enhanced P uptake by mycorrhizal symbioses. First, the extensive network of fungal hyphae enables plants to explore a greater volume of soil, thereby overcoming limitations associated with the relatively slow diffusion of P in the soil solution (Marschner, 1995; Smith and Read, 1997). Second, although mycorrhizae often access phosphorus from the same labile pool as nonmycorrhizal roots, there is also some evidence that they are capable of accessing forms of phosphorus not generally available to the host plant (Marschner, 1995). Whether the mycorrhizae actually serve to increase the affinity of a root system for phosphorus or to allow plants to compete more effectively for phosphorus with soil microbes is unclear. For example, Thompson *et al.* (1990) reported that mycorrhizal roots and isolated hyphae have P uptake kinetics similar to those of nonmycorrhizal roots and other fungi.

This improved P uptake occurs in exchange for the provision of C from the host plant, and the carbon requirements of the mycorrhizal association can be substantial. For example, Baas *et al.* (1989) showed "root" respiration rates of mycorrhizal plant to be 20–30% higher than those of nonmycorrhizal plants. Similarly,

Jakobsen and Rosendahl (1990) observed 20% of plant carbon to be allocated below ground for nonmycorrhizal cucumber plants and 44% for those with mycorrhizal associations. In both cases, about half of this was respired. Working with subtropical *Citrus* species, Peng *et al.* (1993) suggested that root respiration rates were about 35% higher for mycorrhizal than for nonmycorrhizal roots.

The high carbon requirements of the mycorrhizal symbiosis have led to the suggestion that such symbioses may be enhanced when plant carbon supply is improved (Díaz, 1996). Nevertheless, as has been pointed out by Staddon and Fitter (1998), although increases in atmospheric $[CO_2]$ no doubt enhance (vesicular–arbuscular) mycorrhizal infection on a per plant per unit time basis, this may be a simple consequence of bigger plants at higher $[CO_2]$. That is, there may be no direct effect of carbohydrate supply on mycorrhizal colonization rates per unit root length once faster plant growth rates at elevated $[CO_2]$ are taken into account (Staddon and Fitter, 1998).

Lovelock *et al.* (1996; 1997) investigated the interaction between myccorrhizal infection and ambient $[CO_2]$ in the shade-tolerant tropical tree *Beilschmiedia pendula*. They found mycorrhizal infection to stimulate growth and phosphorus uptake at both ambient and elevated $[CO_2]$. Mycorrhizal plants had similar, if not higher, tissue P concentrations at the higher $[CO_2]$. This indicates an ability to maintain or perhaps even increase the degree of mycorrhizal infection per unit root length. This increased root system P uptake capacity seems to occur to nearly the same degree as the overall increase in plant growth. This is different from the situation for nitrogen/CO_2 interactions, where tissue N concentrations nearly always decline with increasing $[CO_2]$ (Drake *et al.*, 1997). Phosphorus uptake rates are therefore generally able to keep pace with metabolic requirements when growth is stimulated by increased $[CO_2]$.

3.3.2 Organic Acid Exudation

It is now well documented that plant roots, bacteria, and fungi (including those involved in mycorrhizal associations) can all excrete organic acids into the soil solution (Marschner, 1995; Jones, 1998). As discussed in Sec. 2.3, some of these organic acids are capable of mobilizing sorbed P mainly by ligand exchange and occupation of P sorption sites (Lopez-Hernandez *et al.*, 1986; Fox *et al.*, 1990; Jones and Darrah, 1994; Bhatti *et al.*, 1998). Consistent with this role for organic acids is the frequent observation that rates of organic acid exudation tend to increase in response to low levels of phosphorus availability (Jones, 1998). Although we know of no reports of organic acid exudation by plants native to moist tropical forests, there is no reason to suspect that this does not occur. In that context, the extent to which this organic efflux is modified by plant carbon supply is of relevance to the current analysis.

Changes in organic acid efflux at elevated $[CO_2]$ have been reported by Whipps (1985), Gifford *et al.* (1996), DeLucia *et al.* (1997), Barrett and Gifford (1999), and Watt and Evans (1999). On balance, these observations suggest that, when expressed per unit root length, there is little or no change in organic acid exudation rates (Watt and Evans, 1999). This situation is similar to the probable [CO₂]-independent plant-mycorrhizal infection rate when expressed per unit root length as discussed in Sec. 3.3.1. Similarly, this maintenance of the exudation rate per unit root length should allow plant phosphorus concentrations to be maintained at elevated [CO₂] (DeLucia *et al.*, 1997; Barrett and Gifford, 1999).

3.3.3 Acid Phosphatase Exudation

As discussed in Secs. 2.1 and 2.3, soil phosphorus mineralization is governed by plant and microbial extracellular phosphatases which hydrolyze the ester bonds of organic P compounds. As for organic acid exudation, the extent to which plant extracellular phosphatases are active in improving the phosphorus nutrition of tropical forests is unknown. Nevertheless, we also note that, as for organic acids, the rate of root phosphatase activity increases with decreasing soil P availability (Barrett *et al.*, 1998; Almeida *et al.*, 1999), and rates of activity per unit root length are maintained under CO₂ enrichment (Gifford *et al.*, 1996; Barrett *et al.*, 1998) or nearly so (Almeida *et al.*, 1999). Thus, as for mycorrhizally mediated P uptake as well as P release mediated by organic acids, there is no reason to suspect that tropical forest phosphatase exudation rates per unit root length (or per unit root mass) should be reduced as CO₂ concentrations increase. To date there has been only one report of acid phosphatase activity for tropical forest soils (Olander and Vitousek, 2000). Working in Hawaii, they observed unusually high activities for this enzyme.

4. Linking the Phosphorus and Carbon Cycles

4.1 To What Extent Does Phosphorus Availability Really Limit Moist Tropical Forest Productivity?

The idea that it is phosphorus that specifically limits the production of many tropical rainforests was first discussed at length by Vitousek (1984). He provided two lines of evidence to support the hypothesis. First, he showed that, after fitting a simple statistical model for the relationship between annual litterfall rate and climate, the residuals of this regression were positively correlated with phosphorus concentrations, but not the nitrogen concentrations of the litterfall. Second, he demonstrated that for moist tropical forests, the annual litterfall rate (and hence by implication net primary production) was significantly correlated with litterfall phosphorus concentrations. The general idea that phosphorus, rather than nitrogen, constrains the productivity of lowland tropical forests is consistent with high amounts and cycling rates of nitrogen in tropical ecosystems (Vitousek and Sanford, 1986; Bruijnzeel, 1991; Neill *et al.*, 1995) and relatively high rates of emission of N-containing trace gases (Keller, *et al.* 1986; Matson and Vitousek, 1987). Most recently, Martinelli *et al.* (1999)

have used foliar δ^{15}N abundances to show that, compared to temperate forests at least, nitrogen is relatively abundant in many tropical forest ecosystems.

It is, however, clear that this generalization does not apply to all moist forests. For example, Tanner *et al.* (1998) argue that, due to slow rates of nitrogen mineralization at high altitudes, nitrogen, rather than phosphorus, is likely to constrain production of many montane tropical forests. A further refinement has come from chronosequence studies in Hawaii (Crews *et al.*, 1995; Herbert and Fownes, 1995; Raich *et al.*, 1996; Vitousek and Farrington, 1997). From studies of soil phosphorus biogeochemistry, tree nutrient status, and fertilization experiments, this work suggests that forests on younger soils are limited by both nitrogen and phosphorus. But consistent with the theory of Walker and Syers (1976), discussed also in Sec. 2.3, forests on older soils seem to be limited only by low phosphorus availability (Herbert and Fownes, 1995; Vitousek and Farrington, 1997).

This conclusion is based, at least in part, on the observation that forests on older soils show increased growth in response to phosphorus but not to nitrogen fertilization (Herbert and Fownes, 1995). But at the individual plant level, it is also often observed that climax-tree species native to both moist and dry tropical forests may show little if any growth response to increased soil phosphorus availability (Rincón and Huante, 1994; Huante *et al.*, 1995; Raaimakers and Lambers, 1996; Veenendaal *et al.*, 1996). Thus, as was discussed in Sec. 3.2, it may actually be that most plants adapted to low-phosphorus tropical forest soils, while having adaptions to such soils such as lower inherent growth rates and higher root–shoot ratios may not be able to substantially increase that growth in response to higher phosphorus levels (Veenendaal *et al.*, 1996). In this context, one can still regard the low productivity of some tropical forests as being a consequence of low nutrient availability, but analyzing the extent of nutrient constraints on ecosystem productivity by means of fertilizer experiments might be misleading. Perhaps the exceptionally low species diversity in tropical forests in Hawaii is the reason for the large growth responses in response to P fertilization observed by Herbert and Fownes (1995) for trees growing on the older soil there.

Despite probably being well adapted to nutrient-poor soils, there is no doubt that trees growing on oxisols or ultisols have very low foliar P concentrations (Sec. 3.2). For example, their average value of 0.5 mmol P mol⁻¹ C is substantially less than the P requirement for adequate growth of most plants, which is considered to range from 2 to 4 mmol P mol⁻¹ C (Marschner, 1995). Indeed, even when compared to other tropical tree species, these values are in the range generally considered to be "deficient" (Drechsel and Zech, 1991).

It is likely that these low concentrations of leaf phosphorus are limiting for photosynthesis. This is shown in Figure 2, where the photosynthesis/phosphorus relationship is shown for the laboratory studies of Cromer *et al.* (1993) and Lovelock *et al.* (1997) and the field studies of Raich *et al.* (1995) and Raaimakers *et al.* (1995). The curve for Cromer *et al.* (1993) comes from their Figure 6, viz.

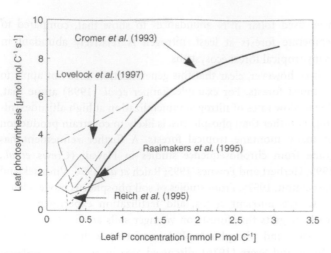

FIGURE 2 The relationship between leaf photosynthetic rate and leaf phosphorus concentration of tropical trees (from Cromer *et al.*, 1993; Raaimakers *et al.*, 1995; Raich *et al.*, 1995, Lovelock *et al.*, 1997).

$$A = 9.67[1 - e^{-0.8379(|P| - 0.404)}], \qquad (1)$$

where A is the photosynthetic rate in μmol mol^{-1} C s^{-1} and $[P]$ is the leaf P concentration in mmol P mol^{-1} C.

Both of the field studies were carried out in moist tropical forests growing on oxisols and these can thus be considered representative of the sort of relationships that can be expected for plants growing on highly weathered soils in the moist tropics. By contrast, the relationship of Cromer *et al.* (1993) comes from a nutrition experiment using *Gmelina arborea*, a fast-growing seasonally deciduous tree often characterized by exceptionally high levels of foliar P (Drechsel and Zech, 1991). Of an intermediate nature is *Beilschmiedia pendula*, a common species of the humid forests of Panama (Lovelock *et al.*, 1997). Other things being equal, the methodology of Lovelock *et al.* (1997) would have yielded higher rates, as photosynthesis was measured in a leaf disc electrode at saturating [CO$_2$]. Despite different methodologies and genetically different plant material, the photosynthesis versus leaf phosphorus relationships are surprisingly consistent, especially when it is considered that the methodology of Lovelock *et al.* (1997) should have given rates higher than the other studies.

Given that a typical foliar P concentration for a moist tropical forest growing on an oxisol soil is only 0.5 mmol P mol^{-1} C, the low phosphorus concentrations typically encountered in moist tropical forests on infertile soils are thus almost certainly limiting their rates of carbon acquisition. Nevertheless, moist tropical forests growing on infertile soils also have a high leaf area index (LAI) and rates of carbon acquisition are remarkably similar to those in temperate zone broadleaf forests (Malhi *et al.*, 1999). But the relationship of Figure 2 also suggests that for many tropical forests which have foliar nutrient concentrations of about 0.5 mmol P mol^{-1} C, these nutrient concentrations are close to being critically low. Any reduction in phosphorus availability would therefore be expected to be reflected in dramatic reductions in

canopy photosynthetic capacity, as a consequence of either a reduction in the total leaf area or a decrease in foliar P concentrations. From studies of *bana* forests that grow on very low fertility spodosol or psamment soil types in Venezuela, it would seem that the former explanation is the case. Total canopy phosphorus is much lower for forests growing on oxisols or utisols, with leaves of these forests typically having phosphorus concentrations similar to those of forests growing on the relatively more fertile oxisol/ultisol soil types (Vitousek and Sanford, 1986). Thus, rather than reducing their foliar P concentrations below about 0.5 mmol P mol^{-1} C, these forests adjust to very low levels of phosphorus availability by having a relatively low LAI (Medina and Cuevas, 1989).

4.2 Tropical Plant Responses to Increases in Atmospheric CO$_2$ Concentrations

From Sec. 4.1 it can be concluded that plants typically found on nutrient-poor tropical soils may be relatively slow-growing, at least by the standards of the moist tropics. And also given that many species associated with these forests do not appear to be able to respond significantly to phosphorus fertilization (Sec. 4.1) it might also be argued that these species may not be able to significantly increase their growth in response to increases in ambient [CO$_2$] (Poorter, 1993, 1998). On the other hand, Lloyd and Farquhar (1996) have argued that tropical plants typically have high respiratory costs. Any increase in photosynthesis in response to increased [CO$_2$] should therefore result in a greater than average increase in growth (Lloyd and Farquhar, 2000).

Unfortunately, there are only scant experimental data with which to make a judgment on this. The [CO$_2$] growth responses of potted tropical trees have also been investigated by Oberbauer *et al.* (1985), Reekie and Bazzaz (1989), Ziska *et al.* (1991), Lovelock *et al.* (1996), Winter and Lovelock (1999), and Carswell *et al.* (2000). With the exception of Reekie and Bazzaz (1989) and Carswell *et al.* (2000), substantial increases in the rate of plant biomass accumulation have been observed, with Ziska *et al.* (1991) observing a massive dry-weight increase of 164% at harvest for *Tabebuia rosea*, a canopy tree species native to Panama. Substantial growth enhancement stimulations in response to CO$_2$ enrichment have also been observed for another Panamanian tree species, *Beilschmiedia pendula* (Lovelock *et al.*, 1996). Lovelock *et al.* (1997) also showed that the presence of mycorrhizae serves to significantly increase rates of P uptake per plant and maintain high photosynthetic capacities of *B. pendula* when grown under CO$_2$ enrichment. Hogan *et al.* (1991) have suggested that the lack of positive growth enhancement observed by Reekie and Bazzaz (1989) was a consequence of the use of very small pots in that experiment. Working with *Cedrel odorata* from Costa Rica, Carswell *et al.* (2000) observed substantial stimulations of leaf photosynthetic rates for high [CO$_2$]-grown plants at both high and low nutrient supply rates. Although they observed a trend toward increased growth at higher [CO$_2$] at both nutrient supply rates, this effect was not statistically significant. This was considered to be mostly a consequence of an unusually high within-treatment variation in plant growth rates.

The idea that a strong growth stimulation to elevated CO_2 concentrations should be seen in understory plants in tropical forests which typically grow close to their light compensation points was tested by Würth *et al.* (1998). Investigating a range of species, growth enhancements ranging from 25 to 76% in response to a doubling of $[CO_2]$ were observed.

A second approach has been the use of "model" tropical rain forest communities (Körner and Arnone, 1992; Arnone and Körner, 1995). In their first experiment plants were exposed to reasonably well-fertilized soil but without any attempt to ensure adequate mycorrhizal infection, and only a modest difference of 11% was observed in final harvest biomass between ambient $[CO_2]$ and $2 \times [CO_2]$ treatments (Körner and Arnone, 1992). After accounting for the significant biomass of both communities before the instigation of treatments, this does, however, suggest an overall growth stimulation of about 20% in response to CO_2, not greatly different from the enhancements typically associated with woody C3 species (Poorter, 1993). In a follow-up experiment, where nutrients where purposely kept low, this growth stimulation was much reduced, though still significant (Arnone and Körner, 1995). But the "soil" used in this circumstance was a C-free quartz sand. It is thus hard to relate the results of such an experiment to rain forests *in situ*, where significant amounts of inorganic and organic phosphorus are available in various forms. Lovelock *et al.* (1998) did, however, observe that for communities of tropical forest tree seedlings grown at ambient and elevated $[CO_2]$ in open-top chambers at the edge of a forest in Panama, no enhancements in plant biomass occurred under elevated $[CO_2]$, either for the whole communities or in the growth of individual species. But, particularly as several different successional types were present, the extent to which the intense interplant competition in that experiment was actually representative of a typical tropical forest regeneration pattern is unclear.

Such methodological concerns with "model communities" aside, the tendency seems to be for potted tropical plants to show significant growth responses to elevated $[CO_2]$, but for model communities these responses are much reduced or absent. Clearly, much more work is required to elucidate the basis of these contradictory results. Nevertheless in what follows, we assume that moist tropical forest trees increase their productivity in a manner typical of C3 plants, investigating the extent to which changes in phosphorus availability might modify that response.

4.3 Using a Simple Model to Examine CO_2/Phosphorus Interactions in Tropical Forests

In this section we use a simple model to examine the possible magnitude of phosphorus constraints on the CO_2 fertilization response of tropical rain forests. The model presented is based on that originally developed by Lloyd and Farquhar (1996) and modified by Lloyd (1999a). In brief, the model consists of a simulation of ecosystem plant growth in response to changing CO_2 concen-

trations coupled with the Rothamstead model of soil carbon dynamics (Jenkinson and Rayner, 1977). In Lloyd (1999a), simulations of nitrogen/carbon interactions were undertaken for temperate and boreal forests and the same principles are applied to phosphorus and moist tropical forests here. We consider the effects of the release of adsorbed phosphorus into the soil solution in response to higher rates of P removal, as a consequence of the CO_2-induced growth stimulation. The possible release of additional phosphorus into the soil solution as a consequence of increased plant growth eventually leading to increases in soil carbon density (Sec. 2.3) is also considered.

There is good evidence that soil microbes can actively compete with plants for soil P (Singh *et all*, 1989). Thus, even though there is also considerable evidence for wide-ranging C/P ratios in microbial biomass (McGill and Cole, 1981; Gijsman *et al.*, 1996) we also consider the extent to which plant phosphorus availability might be reduced by the phosphorus demands of soil microbial biomass. The soil microbial pool increases its size and activity in these simulations as increased plant growth in response to elevated CO_2 is rapidly translated into increased litter inputs into the soil.

Based on recent data, some modifications of model parameters used in Lloyd and Farquhar (1996) and Lloyd (1999a) have been made. In particular, based on recently reported foliar and woody respiration rates (Meir, 1996; Malhi *et al.*, 1999) we change our estimates for the proportion of the total plant maintenance respiration for leaves, branches, and boles from 0.36, 0.10, and 0.18 to 0.30, 0.10, and 0.10, respectively. Malhi *et al.* (1999) also suggested a much lower proportion of Gross Primary Productivity, G_P, being lost as plant respiration (0.49) than has been reported for other studies in tropical forests (0.67–0.87: for a summary, see Medina and Klinge, 1983; Lloyd *et al.*, 1995; Lloyd and Farquhar, 1996). The low estimate of Malhi *et al.* (1999) is, however, made on the unverified assumption that root respiration and root detritus production are approximately in balance. But this means that root respiration would represent only 41% of the total soil respiration and only 22% of G_P. Given the respiratory costs associated with the extensive mycorrhizal symbioses in tropical soils (Sec. 3.3), this seems too low. But we must also take into account that a previous estimate of autotrophic respiratory losses accounting for 75% of G_P (Lloyd and Farquhar, 1996) could be too high. This is because some estimates of N_P giving rise to that number probably ignored fine root production (Lloyd, 1999a). We therefore reduce our estimate of total plant respiration in tropical forests to 65% of G_P, but we allocate 50% of the total maintenance respiration to the roots. Overall, this gives a respiration rate for coarse and fine roots which is about 30% of G_P.

Also, the tropical forest molar leaf area ratio used in Lloyd and Farquhar (1996) and Lloyd (1999a) of 0.34 m^{-2} mol^{-1} C is probably most applicable to the deciduous leaves of drier tropical forests or to the pioneer species of moist tropical forests (Medina and Klinge, 1983: Reich *et al.*, 1995; Raaimakers *et al.*, 1995). For the moist forest climax species of interest here we therefore use an amended value of 0.20 m^{-2} mol^{-1} C.

To account for the importance of sorbed phosphorus, we first characterize the relationship between the sorbed phosphorus concentration (ground area basis) and the concentration of P in the soil solution. Rather than using the Langmuir model (Sanyal and DeDatta, 1991), we use.

$$[P_{sorb}] = \frac{S_{max}[P_{sol}]}{K_S + [P_{sol}]}, \quad (2)$$

where $[P_{sorb}]$ is the amount of P sorbed on or in the soil particles (ground area basis) S_{max} is the maximum absorption, $[P_{sol}]$ is the concentration of P in the soil solution and k_S is a constant relating to the P binding energy. Values of K_S and S_{max} for the simulation here are taken from a fit to the data for a Zimbabwe oxisol presented by Sibanda and Young (1989) for which, assuming an active soil rooting depth of 0.5 m and a bulk density of 1.3 g cm^{-3}, we estimate $S_{max} = 5$ mol m^{-2} and $k_S = 1$ mmol m^{-2}. Based on soil P data for an adystrophic rainforest in Venezuela (Tiessen *et al.*, 1994a), we take an initial estimate for $[P_{sorb}]$ of 350 mmol m^{-2} for which the equivalent $[P_{sol}] = 0.075$ mmol m^{-2}, about 1.5 μM.

There is good evidence that high-affinity phosphorus uptake by plants can be well represented by Michaelis–Menton kinetics (McPharlin and Bieleski, 1989; Jungk *et al.*, 1990). And, as discussed in Secs. 3.3 and 4.2, there are good reasons to suppose that, irrespective of the mechanism of phosphorus mineralization and uptake, phosphorus uptake rates per unit fine root density are maintained as ambient CO_2 concentrations increase. We therefore allow the maximum phosphorus uptake rate to increase linearly with increases in fine root density and thus write the rate of plant phosphorus uptake as

$$U_P = \gamma M_{fr} \frac{[P_{sol}]}{K_u + [P_{sol}]}, \quad (3)$$

where U_P is the rate of uptake of P by the vegetation (mol m^{-2} year^{-1}), γ is the maximum (P saturated) rate per unit fine root density, M_{fr} is the fine root carbon density, and K_u is a constant. There is very little information on the kinetics of P uptake by the roots of tropical plants and so it is hard to determine a priori a reasonable value for K_u. Working mostly with crop species, researchers have reported values between 1 and 10 μM (Marschner, 1995; Schachtman *et al.*, 1998; Raghothama, 1999). Given that we expect the trees adapted to these nutrient-poor tropical soils to have developed high-affinity P uptake systems, we assume a value of 2.0 μM.

Here, we are mostly interested in calculating "generic" values from which phosphorus fluxes and their dependence on soil availability and internal plant physiological status can be quantified. Because of the effects of soil fertility on phosphorus concentrations discussed above, we consider only moist tropical forests growing on the more abundant but lower nutrient oxisol/ultisol soils. This is because the $[CO_2]$/phosphorus interaction we are seeking to model is likely to be most marked on these low-fertility soils. Moreover, compared to other soil types, they tend to dominate the moist tropics (Sanchez, 1976).

Based on the discussion in Sec. 3.2 we take the following phosphorus concentrations for use in the model simulation:

1. Foliage: 0.50 mmol P mol^{-1} C,
2. Branches: 0.16 mmol P mol^{-1} C,
3. Boles: 0.16 mmol P mol^{-1} C,
4. Coarse roots: 0.10 mmol P mol^{-1} C,
5. Fine roots: 0.50 mmol P mol^{-1} C.

In order to estimate the phosphorus fluxes in the various plant tissues, we assume that 68% of leaf and fine root P is retranslocated prior to abscission (Vitousek and Sanford, 1986). The distribution of P between the various tissues is achieved using a scheme similar to that used for nitrogen in temperate and boreal forests (Lloyd, 1999a). Phosphorus is first distributed according to assumed C/P ratios in branches, boles, and roots. The variable remainder is then allocated to the leaf tissue. From Figure 2, the P concentration of leaves of moist tropical rainforest species is already very low and probably strongly limiting for photosynthesis (Sec. 4.1). Moreover, on a canopy basis it seems reasonably constant, despite variations in total leaf carbon density (Vitousek and Sanford, 1986). The model therefore assumes that foliar P concentrations are maintained at 0.5 mmol mol^{-1}, with variations in the total foliar P content being reflected in changes in the total leaf carbon density (and hence leaf area), rather than by differences in the photosynthetic rate per unit leaf area. As was discussed in Sec. 4.1, it seems likely that some plants in moist tropical forests may not be capable of significantly enhanced growth responses to increased phosphorus availability. Variations in leaf area but not leaf photosynthetic capacity per unit leaf area are therefore also assumed when total canopy phosphorus content is either increasing or decreasing. The latter turns out to be the case in some scenarios examined below.

The rate of change in the concentration of P in the soil solution is written as

$$\frac{\partial[P_{sol}]}{\partial t} = I - k_L[P_{sol}] + L_P - \frac{\partial[P_{org}]}{\partial t} - U_P - \frac{\partial[P_{sorb}]}{\partial t}, (4)$$

where I is the atmospheric input, k_L is a constant relating the rate of leaching to the soil solution phosphorus concentration, L_P is the rate of P input into the soil through litterfall, $[P_{org}]$ is the concentration of organic phosphorus in litter and soil, and U_P is the rate of plant P uptake as given in Eq. (3).

In the simulations here, there are two components to I: atmospheric deposition, taken here as 1.5 mmol m^{-2} year^{-1}, and canopy leaching, which is calculated on the basis of the canopy P content assuming a rate of 6 mmol $^{-2}$ year^{-1} in 1730 (see Sec. 3.1.3). To estimate k_L in our standard cases, we assume that the input of phosphorus through atmospheric deposition was exactly balanced by leaching losses in 1730. This is somewhat at odds with several observations suggesting that much of the atmospherically derived P deposited onto tropical forests is retained rather than being leached out of the system (Sec. 3.1.2). Nevertheless, as was discussed in Sec. 3.1.1 almost all of this atmospherically derived P probably comes from the forests themselves and thus does not represent a net positive P input.

The term $\partial[P_{org}]/\partial t$ is calculated assuming that the concentration of phosphorus in all decomposing litter is 0.16 mmol mol^{-1}. This is based on the 68% retranslocation of P from leaves and fine roots and the average branch, bole, and coarse root P concentrations (Sec. 3.2). Where the sensitivity of the model to P accumulation in the microbial carbon pool is tested, based on data summarized by Gijsman *et al.* (1996) we use a tissue P concentration for microbes of 6.4 mmol P mol^{-1} C. In all simulations, it is assumed that soil phosphorus mineralization proceeds with a rate constant of 0.5 year^{-1}, with phosphorus mineralization proceeding independently of carbon mineralization. This is on the basis of the evidence discussed in Sec. 2.1. Indeed, inflexible soil carbon pool C/P ratios which effectively link phosphorus mineralization rate to the carbon mineralization rate in models such as CENTURY (Parton *et al.*, 1988) have been strongly criticized by some tropical soil chemists (Gijsman *et al.*, 1996).

In order to estimate the last term of Eq. (4), we differentiate Eq. (2) and then write

$$\frac{\partial[P_{sorb}]}{\partial t} = \frac{\partial[P_{sorb}]}{\partial[P_{sol}]} \cdot \frac{\partial[P_{sol}]}{\partial t} = \frac{S_{max}K_s}{(K_s + [P_{sol}])^2} \cdot \frac{\partial[P_{sol}]}{\partial t}. \quad (5)$$

Combining Eqs. (4) and (5) then gives

$$\frac{\partial[P_{sol}]}{\partial t} = \frac{I - k_L[P_{sol}] + L_P - \dfrac{\partial[P_{org}]}{\partial t} - U_P}{1 + \dfrac{S_{max}K_S}{(K_S + [P_{sol}])^2}} \quad (6)$$

The second term in the denominator is typically around 5000 and represents the "buffering" effect of the sorbed P. That is, in the presence of an appreciable sorbed P pool, the soil solution P concentration is extremely insensitive to the rate of removal of phosphorus into or from it. This is because increased rates of removal of P are almost totally balanced by desorption. Likewise, increased rates of P input result in large increases in $[P_{sorb}]$, but with very little change in $[P_{sol}]$. That latter case represents, of course, the tropical soil phosphorus fertilizer "fixation" problem discussed in Sec. 2.2.

As was discussed in Sec. 2.3 there are some indications that increases in soil carbon density could release adsorbed phosphorus into the soil solution, and available evidence suggests that S_{max} declines and k_S increases with increasing soil carbon density (Sibanda and Young, 1989). A precise understanding of the mechanisms involved is still lacking, as is any general quantitative description of the nature of the relationship. So based on the data of Sibanda and Young (1989) we simply assume for a simple sensitivity study $S_{max} = \alpha/[C_{hum}]$ and $k_S = \beta[C_{hum}]$, where α and β are fitted constants and $[C_{hum}]$ is the modeled soil humus carbon density. We then write

$$\frac{\partial[P_{sorb}]}{\partial t} = \frac{\partial[P_{sorb}]}{\partial[P_{sol}]} \cdot \frac{\partial[P_{sol}]}{\partial t} + \frac{\partial[P_{sorb}]}{\partial[C_{hum}]} \cdot \frac{\partial[C_{hum}]}{\partial t}. \quad (7)$$

This gives an alternative to Eq. (6) but now with a dependence on $[C_{hum}]$,

$$\frac{\partial[P_{sol}]}{\partial t} = \frac{\left\{\begin{array}{l} I - k_L[P_{sol}] + L_P - \dfrac{\partial[P_{org}]}{dt} - U_P \\[2mm] + \dfrac{\alpha[P_{sol}](2\beta[C_{hum}] + [P_{sol}])}{(\beta[C_{hum}]^2 + [P_{sol}][C_{hum}])^2} \cdot \dfrac{\partial[C_{hum}]}{\partial t} \end{array}\right\}}{1 + \dfrac{S_{max}k_S}{(K_S + [P_{sol}])^2}}. \quad (8)$$

The value of $[C_{hum}]$ in 1730 is used in conjunction with $S_{max} = 5$ mol m^{-2} and $K_S = 1$ mmol^{-2} to determine α and β.

The simulations that have been undertaken are as follows:

(A) No consideration is given to plant phosphorus requirements (i.e., the growth rate of tropical forests is affected only by the atmospheric CO_2 concentration).

(B) Plant phosphorus uptake is dependent on $[P_{sol}]$ but the sorption and desorption of P are ignored (Eq. (4), but with $\partial[P_{sorb}]/\partial t = 0$).

(C) Plant phosphorus uptake is dependent on $[P_{sol}]$. Sorption and desorption of P in response to changes in $[P_{sol}]$ are considered (Eq. (6).

(D) Plant phosphorus uptake is dependent on $[P_{sol}]$. Sorption and desorption of P in response to changes in $[P_{sol}]$ and $[C_{hum}]$ are considered (Eq. (8)).

(E) Plant phosphorus uptake is dependent on $[P_{sol}]$. Sorption and desorption of P occur in response to changes in $[P_{sol}]$ and $[C_{hum}]$ (Eq. (8)). Increases in microbial biomass remove phosphorus from the soil solution.

Results of these simulations are shown in Table 1. For Scenario A (essentially as in Lloyd, 1999a, but with minor changes as discussed above) the simulated increases in G_P and N_P are substantial, being 35 and 44% higher for the period 1981–1990 (average value), respectively. This large modeled stimulation of productivity arises mostly because of the high sensitivity of G_P to CO_2 concentrations at warmer temperatures that occur in moist tropical forests (Lloyd and Farquhar, 1996). As must happen with finite turnover times for plant and soil carbon, increases in rates of litterfall always lag behind the CO_2-induced increase in G_P and N_P and the soil respiration rate always lags behind litterfall (Lloyd and Farquhar, 1996; Lloyd 1999a). Thus a substantial sink of carbon is modeled to be occurring in this situation: 8.3 mol m^{-2} year^{-1}. Spread across the moist tropics (ca. 12×10^{12} m^{-2}), such uptake would be significant: 0.1 Pmol year^{-1} or about 50% of most current estimates of the terrestrial carbon sink (Lloyd, 1999b).

But this modeled stimulation of enhanced productivity in response to increasing $[CO_2]$ can be greatly modified when variations in phosphorus availability are considered. This is shown in Table 1 for Scenario B. In the absence of a resupply of phosphorus

TABLE 1 Effect of Phosphorus Availability Model Assumptions on Simulated Gross Primary Productivity, Net Primary Productivity, Litterfall, Soil Respiration Rate, Rate of Net Ecosystem Carbon Accumulation, Plant Carbon Density, and Soil Carbon Density for 1730 and 1981–1990 (Average Value)

Submodel	Gross Primary Production (mol C m^{-2} year^{-1})	Net Primary Production (mol C m^{-2} year^{-1})	Litterfall (mol C m^{-2} year^{-1})	Soil Respiration (mol C m^{-2} year^{-1})	Rate of Ecosystem C Accumulation (mol m^{-2} year^{-1})	Plant Carbon Density (mol C m^{-2})	Soil Carbon Density (mol C m^{-2})
1730: No P constraints	164.0	57.4	57.4	57.4	0.0	696	1218
A: 1981–1990: No P constraints	220.7	82.7	77.3	74.4	8.3	978	1402
B: 1981–1990: No P sorption or desorption	197.0	69.4	66.8	65.8	3.6	890	1368
C: 1981–1990: With P sorption and desorption (Eq. 6)	214.9	78.6	74.0	72.0	6.5	963	1394
D: 1981–1990: With P sorption and desorption + soil C effect (Eq. 8)	221.2	82.9	77.5	74.5	8.3	980	1403
E: 1981–1990: With P sorption and desorption + soil C effect + P sequestration and desorption in microbes	221.2	82.9	77.5	74.5	8.3	980	1403

Note. See text for a full description of model structure and Scenarios-A to -E.

from the labile pool, the increased P uptake required to sustain the extra growth in response to increasing [CO_2] results in soil solution phosphorus concentrations being rapidly depleted, being reduced by about 25% for 1981–1990 compared to 1730 in this scenario. This offsets to some degree the increased intrinsic phosphorus uptake ability of the larger trees (due to more fine roots) and thus the CO_2-induced increases in G_P and N_P are only about half the magnitude of the no-P-constraint case (Scenario A). Accordingly, the rate of net carbon accumulation by the ecosystem is only 3.6 mol^{-2} year^{-1}, less than half the value for Scenario A.

This picture of a substantial phosphorus constraint is drastically altered when the presence of the inorganic labile (i.e., sorbed) phosphorus pool is taken into account (Scenario C). Desorption of phosphate occurs in response to increased rates of removal from the soil solution. Consequently, the reduction in soil solution phosphorus concentration over 1730 levels is only 9% for 1981–1990. This contrasts with the 25% reduction in Scenario B. Consequently, the enhancements of G_P and N_P are more similar to the no-P-constraint case, though a full expression of the CO_2-induced growth response is still not possible. Accordingly, the rate of net carbon accumulation by the ecosystem is 6.5 mol m^{-2} year^{-1}, substantially more than Scenario B, but about 20% less than what is modeled to be the case if no phosphorus limitations to plant production occurred.

When the potential positive feedback between increased soil carbon densities and the desorption of phosphorus is considered (Scenario D), all phosphorus constraints on the CO_2-induced growth response or the rate of net ecosystem carbon accumulation disappear. Indeed, the rates of G_P and N_P are actually slightly higher than those in Scenario A. Indeed, according to Scenario D, the average soil solution phosphorus concentration for 1981–1990 is actually higher than the 1730 value. Moreover, including some sequestration of P into the increasing soil microbe pool (Scenario E) has absolutely no effect. This is because the rate of desorption is substantially greater than the rate of sequestration into this pool. But in the absence of the positive feedback between increased soil carbon densities and the desorption of phosphorus, sequestration of P into the microbe pool does have a small effect, reducing fluxes by about 5% for 1981–1990.

Thus, the simulations here suggest that for tropical soils such as oxisols and ultisols which contain an appreciable pool of labile sorbed phosphorus, the transfer of phosphorus from this pool to the soil solution in response to increased rates of P uptake by plants serves to more or less maintain soil phosphorus concentrations. This allows the increased rates of phosphorus uptake by faster growing vegetation to continue.

Kirschbaum *et al.* (1998) used a somewhat different modeling approach to simulate the effects of phosphorus availability on temperate forest CO_2-induced growth responses. But similar to the results here, they concluded that the presence of the "secondary" (labile) pool means that, in the short term, phosphorus availability should not constrain the ability of these forests to respond to [CO_2]. They also concluded, however, that marked phosphorus constraints should become apparent on a time scale of centuries. This is of course also possible here, as the size of the sorbed phosphorus pool is not infinite. A second uncertainty is related to the degree to which sorption is indeed a reversible process on the time scales of interest here. A traditional view had been that sorption of phosphorus is a more or less irreversible process. But as discussed by Barrow (1983, 1999) and Sanyal and De Datta (1991), this apparent irreversibility more likely reflects the relatively slow time frame over which desorption occurs. The degree to which the sorption of phosphorus onto tropical soils occurs is a truly reversible process remains an important research issue (Gijsman *et al.*, 1996).

As discussed in Sec. 2.3, there are some indications that increased soil carbon density should act to release previously adsorbed phosphorus, making it available for plant growth. The simulation here suggests that this effect is potentially quite important (Scenario D in Table 1). Indeed, as our parameterization of the dependence of canopy photosynthetic rate on canopy phosphorus content is conservative (being modulated solely by changes in leaf area without any changes in the rate per unit leaf area) it is quite possible that this effect may be even more potent than modeled here. Indeed, it is not inconceivable that a "runaway" positive feedback could occur. This would involve CO_2-induced increases in tropical forest plant growth giving rise to increases in soil carbon content, which in turn liberates previously sorbed phosphorus, which then gives rise to yet more increased plant growth. In that context, the very high rate of ecosystem carbon sequestration observed by Malhi *et al.* (1998) for a mature moist tropical forest near Manaus in Brazil, 49 mol m^{-2} year^{-1}, may not be as unexplicably high as it first seems.

Acknowledgments

We thank Jim Barrow, John Grace, and Michelle Watt for useful comments.

References

Alexander, I. (1989). Mycorrhizas in tropical forests. In *"Mineral Nutrients in Tropical Forest and Savanna Ecosystems"* (J. Proctor, Ed.), pp. 169–188. Blackwell Scientific Publications, Oxford.

Almeida, J. P. F., Lüscher, A., Frehner, M., Oberson, A., and Nösberger, J. (1999). Partitioning of P and the activity of root acid phosphatase in white clover (*Trifolium repens* L.) are modified by increased atmospheric CO_2 and P fertilisation. *Plant and Soil* **210,** 159–166.

Arnone, J. A., III and Körner (1995). Soil and biomass carbon pools in model communities of tropical plants under elevated CO_2. *Oecologia* **104,** 61–71.

Artaxo, P., Fernandes, E. T., Martins, J. V., Yamasoe, M. A., Hobbs, P. V., Maenhaut, W., Longo, K. M., and Castanho, A. (1998). Large scale aerosol source apportionment in Amazonia. *J. Geophys. Res.* **103,** 31837–31847.

Artaxo, P., Maenhaut, W., Storms, H., and Van Grieken, R. (1990). Aerosol characteristics and sources for the Amazon Basin during the wet season. *J. Geophys. Res.* **95,** 16971–16985.

Artaxo, P., Storms, H., Bruynseels, F., Van Grieken, R., and Maenhaut, W. (1988). Composition and sources of aerosols from the Amazon Basin. *J. Geophys. Res.* **93,** 1605–1615.

Baas, R., van der Werf, A., and Lambers, H. (1989). Root respiration and growth in *Plantago major* as affected by vesicular-arbuscular mycorrhizal infection. *Plant Physiol.* **91,** 227–232.

Ballie, I. C., and Mamit, J. D. (1983). Observations on rooting depth in mixed dipterocarp forest. *Malayan Forester* **46,** 369–374.

Barrett, D. J. and Gifford, R. M. (1999). Increased C-gain by an endemic Australian pasture grass at elevated atmospheric CO_2 concentration when supplied with non-labile inorganic phosphorus. *Aust. J. Plant Physiol.* **26,** 443–451.

Barrett, D. J., Richardson, A. E., and Gifford, R. M. (1998). Elevated atmospheric CO_2 concentrations increase wheat root phosphatase activity when growth is limited by phosphorus. *Aust. J. Plant Physiol.* 25, 87–93.

Barrow, N. J. (1983). On the reversibility of phosphate sorption by soils. *J. Soil Sci.* **34,** 751–758.

Barrow, N. J. (1999). The four laws of soil chemistry: the Leeper lecture 1998. *Aust. J. Soil Res.* **37,** 787–829.

Barrow, N. J., and Whelan, B. W. (1989). Testing a mechanistic model. VIII. The effects of time and temperature of incubation on the sorption and subsequent desorption of selenite and selenate by a soil. *J. Soil Sci.* **40,** 29–37.

Bernhard-Reversat, F. (1975). Recherches sur l'écosystème del la fôret subequatoriale de base Côte-d'Ivorie. Les cycles des macroelements. *La Terre et la Vie* **29,** 229–254.

Bhatti, J. S., Comerford, N. B., and Johnston, C. T. (1998). Influence of oxalate and soil organic matter on sorption and desorption of phosphate onto a spodic horizon. *Soil Sci. Soc. Am. J.* **62,** 1089–1095.

Bolan, N. S. (1991). A critical review on the role of mycorrhizal fungi in the uptake of phosphorus by plants. *Plant and Soil* **134,** 189–207.

Bonde, T. A., Christensen, B. T., and Cerri, C. C. (1992). Dynamic of soil organic matter as reflected by natural ^{13}C abundance in particle size fractions of forested and cultivated oxisols. *Soil Biol. and Biochem.* **24,** 275–277.

Brinkmann, W. L. F. (1985). Studies on hydrobiogeochemistry of a tropical lowland forest system. *Geojournal* **11,** 89–101.

Brookes, P., Powlson, D. S., and Jenkinson, D. S. (1984). Phosphorus in the soil microbial biomass. *Soil Biol. Biochem.* **16,** 169–175.

Bruijnzeel, L. A. (1989). Nutrient cycling in moist tropical forests: the hydrological framework. In *"Mineral Nutrients in Tropical Forest and Savanna Ecosystems."* (J. Proctor, Ed.), pp. 383–415. Blackwell Scientific Publications, Oxford.

Bruijnzeel, L. A. (1991). Nutrient input-output budgets of tropical forest ecosystems: a review. *J. Tropical Ecol.* **7,** 1–24.

Burghouts, T. B. A., Van Straalen, N. M., and Bruijnzeel, L. A. (1998). Spatial heterogeneity of element and litter turnover in a Bornean rain forest. *J. Tropical Ecol.* **14,** 477–506.

Burnham, C. P. (1989). Pedological processes and nutrient supply from parent material in tropical soils. In *"Mineral Nutrients in Tropical Forest and Savanna Ecosystems"* (J. Proctor, Ed.), pp. 27–41. Blackwell Scientific Publications, Oxford.

Carswell, F. E., Grace, J., Lucas, M. E., and Jarvis, P. G. (2000). The interaction of nutrient limitation and elevated CO_2 on carbon assimilation of a tropical tree seedling (*Cedrela odorata* L.) *Tree Physiol.* **20,** 000–000.

Chabot, B. F. and Hicks, D. J. (1992). The ecology of leaf life spans. *Annu. Rev. Ecol. Systematics* **13,** 229–259.

Chapin, F. S. III. (1980). The mineral nutrition of wild plants. *Annu. Rev. Ecol. Systematics* **11,** 233–260.

Ciais, P., Tans, P. P., Trolier, M., White, J. W. C., and Francey, R. J. (1995). A large northern hemisphere CO_2 sink indicated by the $^{13}C/^{12}C$ ratio of atmospheric CO_2. *Science* **269,** 1098–1102.

Crews, T. E., Kitayama, K., Fownes, J. H., Riley, R. H., Herbert, D. A., Mueller-Dombois, D., and Vitousek, P. M. (1995). Changes in soil phosphorus fractions and ecosystem dynamics across a long chronosequence in Hawaii. *Ecology* **76,** 1407–1424.

Cromer, R. N., Kriedemann, P. E., Sands, P. J., and Stewart, L. G. (1993). Leaf growth and photosynthetic response to nitrogen and phosphorus in seedling trees of *Gmelina arborea*. *Aust. J. Plant Physiol.* 20, 83–98.

Crozat, G. (1979). Sur l'émission d'un aérosol riche en potassium par lâ foret tropicale. *Tellus* **31B,** 52–57.

Dalton, J. D., Russell, G. C., and Sieling, D. H. (1952). Effect of organic matter on phosphate availability. *Soil Sci.* **73,** 173–177.

DeLucia, E. H., Callaway, R. M., Thomas, E. M., and Schlesinger, W. H. (1997). Mechanisms of phosphorus acquisition for Ponderosa pine seedlings under high CO_2 and temperature. *Ann. Bot.* **79,** 111–120.

Díaz, S. (1996). Effects of elevated $[CO_2]$ at the community level mediated by root symbiosis. *Plant and Soil* **187,** 309–320.

Drake, B. G., González-Meler, M. A., and Long, S. P. (1997). More efficient plants: A consequence of rising atmospheric CO_2? *Annu. Rev. Plant Physiol. Plant Mol. Biol.* **48,** 609–639.

Drechsel, P., and Zech, W. (1991). Foliar nutrient levels of broad-leaved tropical trees: A tabular review. *Plant and Soil* **131,** 29–46.

Echalar, F., Artaxo, P., Martins, J. V., Yamasoe, M., and Gerab, F. (1998). Long-term monitoring of atmospheric aerosols in the Amazon Basin: Source identification and apportionment. *J. Geophys. Res.* **103,** 31849–31864.

Enting, I. G., Trudinger, C. M., and Francey, R. J. (1995). A synthesis inversion on the concentration and $\delta^{13}C$ of atmospheric CO_2. *Tellus* **47B,** 35–52.

Fölster, H., De Las Salas, G., and Khanna, P. (1976). A tropical evergreen forest site with perched water table, Magdalena valley, Columbia. Biomass and bioelement inventory of primary and secondary vegetation. *CEcologia Plantarum* **11,** 297–320.

Forti, C., and Moreira-Nordemann, L. M. (1991). Rainwater and throughfall chemistry in a 'terra-firme' rain forest: Central Amazonia. *J. Geophys. Res.* **96,** 7415–7421.

Fox, R. L., De La Pena, R. S., Gavenda, R. T., Habte, M, Hue, N. V., Ikawa, H., Jones, R. C., Plucknett, D. L., Silva, J. A., and Soltanpour, P. (1991). Amelioration, revegetation and subsequent soil formation in denuded bauxite materials. *Allertonia* **6,** 128–184.

Fox, T. R., Comerford, N. B., and McFee, W. W. (1990). Phosphorus and aluminium release from a spodic horizon mediated by organic acids. *Soil Sci. Soc. Am. J.* **54,** 1763–1767.

Friedlingstein, P., Fung, I., Holland, E., John, J., Brasseur, G., Erickson, D., and Schimel, D. (1995). On the contribution of CO_2 fertilization to the missing biospheric sink. *Global Biogeochem. Cycles* **9,** 541–556.

Gardner, L. R. (1990). The role of rock weathering in the phosphorus budget of terrestrial watersheds. *Biogeochemistry* **11,** 97–110.

Gifford, R., Lutze, J. L., and Barrett, D. (1996). Global atmospheric change effects on terrestrial carbon sequestration: Exploration with a global C- and N-cycle model (CQUESTN). *Plant and Soil* **187,** 369–387.

Gijsman, A. J., Oberson, A., Tiessen, H., and Friesen, D. K. (1996). Limited applicability of the CENTURY model to highly weathered tropical soils. *Agronomy J.* **88,** 894–903.

Golley, F. B., McGinnis, J. T., Clements, R. G., Child, G. I., and Deuver, M. J. (1975). "*Mineral Cycling in a Tropical Moist Forest System.*" University of Georgia Press, Athens.

Grace, J., Lloyd, J., McIntyre, J., Miranda, A. C., Meir, P., Miranda, H. S., Nobre, C., Moncrieff, J. M., Massheder, J., Wright, I. R., and Gash, J. (1995). Carbon-dioxide uptake by an undisturbed tropical rain forest in southwest Amazonia, 1992–1993. *Science* **270**, 778–780.

Grace, J., Malhi, Y., Higuchi, N., and Meir, P. (2000). Productivity and carbon fluxes of tropical rain forests. In "*Global Terrestrial Productivity: Past, Present and Future*" (H. A. Mooney, J. Roy, and B. Saugier, Ed.). Springer-Verlag, Heidelberg.

Grace, J., Malhi, Y., Lloyd, J., McIntyre, J., Miranda, A. C., Meir, P., and Miranda, H. S. (1996). The use of eddy-covariance to infer the net carbon-dioxide uptake of Brazilian rain forest. *Global Change Biol.* **2**, 209–218.

Graham, M. T., and Duce, R. A. (1979). Atmospheric pathways of the phosphorus cycle. *Geochem. Cosmochem. Acta* **43**, 1195–1208.

Greenland, D. J., and Kowal, J. M. L. (1960). Nutrient content of the moist tropical forest of Ghana. *Plant and Soil* **12**, 154–174.

Grubb, P. J., and Edwards, P. J. (1982). Studies of mineral cycling in a montane rain forest in New Guinea. III. The distribution of mineral elements in the above-ground material. *J. Ecol.* **70**, 623–648.

Hardy, F. (1935). Some aspects of tropical soils. *Trans. 3rd Int. Congr. Soil Sci. (Oxford)* **2**, 150–163.

Hase, H., and Fölster, H. (1982). Bioelement inventory of a tropical (semi-) evergreen seasonal forest on eutrophic alluvial soils, West Llanos, Venezuela. *Acta CEcolog./CEcolog. Plantarum* **3**, 331–346.

Herbert, D. A., and Fownes, J. H. (1995). Phosphorus limitation of forest leaf area and net primary production on a highly weathered soil. *Biogeochemistry* **29**, 223–235.

Hogan K. P., Smith A. P., and Ziska L. H. (1991). Potential effects of elevated CO₂ and changes in temperature on tropical plants. *Plant Cell Environ.* **14**, 763–778.

Houghton, R. A. (1996). Land-use change and terrestrial carbon: The temporal record. In "*Forest Ecosystems, Forest Management and the Global Carbon Cycle*" (M. J. Apps and D. T. Price, Ed.), pp. 117–134. NATO ASI Series, Springer-Verlag, Berlin, Heidelberg.

Huante, P., Rincón, E., and Chapin, F. S. III. (1995). Responses to phosphorus of contrasting successional tree-seedling species from the tropical deciduous forest of Mexico. *Functional Ecol.* **9**, 760–766.

Hughes, J. C. (1982). High gradient separation of some soil clays from Nigeria, Brazil and Columbia. I. The interrelationships between iron and aluminium extracted by acid ammonium oxalate and carbon. *J. Soil Sci.* **33**, 509–519.

Hughes, R. F., Kauffman, J. B., and Jaramillo, V. (1999). Biomass, carbon and nutrient dynamics of secondary forests in a humid tropical region of Mexico. *Ecology* **80**, 1892–1907.

Jakobsen, I., and Rosendahl, L. (1990). Carbon flow into soil and external hyphae from roots of mycorrhizal cucumber plants. *New Phytol.* **120**, 77–83.

Janos, D. P. (1989). Tropical mycorrhizas, nutrient cycles and plant growth. In "*Mineral Nutrients in Tropical Forest and Savanna Ecosystems*" (J. Proctor, Ed.), pp. 327–345. Blackwell Scientific Publications, Oxford.

Jenkinson, D. S., and Rayner, J. H. (1977). The turnover of soil organic matter in some of the Rothamsted classical experiments. *Soil Sci.* **123**, 298–305.

Jones, D. L. (1998). Organic acids in the rhizosphere—A critical review. *Plant and Soil* **205**, 25–44.

Jones, D. L., and Darrah, P. R. (1994). Role of root derived organic acids in the mobilization of nutrients from the rhizosphere. *Plant and Soil* **166**, 247–257.

Jordon, C. F., and Herrera, R. (1981). Tropical rain forests: Are nutrients really critical? *Am. Nat.* **117**, 167–180.

Jungk, A., Asher, C. J., Edwards, D. G., and Meyer, D. (1990). Influence of phosphate status on phosphate kinetics of maize (*Zea mays*) and soybean (*Glycine max*) roots. *Plant and Soil* **124**, 175–182.

Keeling, R. F., Piper, S. C., and Heimann, M. (1996). Global and hemispheric CO₂ sinks deduced from changes in atmospheric O₂ concentration. *Nature* **381**, 218–221.

Keller, M., Kaplan, W. A., and Wofsey, S. C. (1986). Emissions of N₂O, CH₄ and CO₂ from tropical soils. *J. Geophys. Res.* **91**, 11791–11802.

Kennedy, M. J., Chadwick, O. A., Vitousek, P. M., Derry, L. A., and Hendricks, D. M. (1998). Changing sources of base cations during ecosystem development, Hawaiian islands. *Geology* **26**, 1015–1018.

Kerfoot, O. (1963). The root systems of tropical forest trees. *Empire Forestry Rev.* **42**, 19–26.

Kirschbaum, M. U. F., Medlyn, B. E., King, D. A., Pongracic, S., Murty, D., Keith, H., Khanna, P. K., Snowdon, P., and Raison, R. J. (1998). Modelling forest growth responses to increasing CO₂ concentration in relation to various factors affecting nutrient supply. *Global Change Biol.* **4**, 23–41.

Kirschbaum, M. U. F., and Tompkins, D. (1990). Photosynthetic responses to phosphorus nutrition in *Eucalyptus grandis* seedlings. *Aust. J. Plant Physiol.* **17**, 527–535.

Klinge, H. (1976). Bilanzierung von hauptnährstoffen im ökosystem tropischer regenwald (Manaus). vorläufige daten. *Biogeographica* **9**, 59–77.

Koide, R. T. (1991). Nutrient supply, nutrient demand and plant responses to mycorrhizal infection. *New Phytol.* **117**, 365–386.

Körner, C., and Arnone, J. A., III (1992). Responses to elevated carbon-dioxide in artificial tropical ecosystems. *Science* **257**, 1672–1675.

Lesack, L. F. W., and Melack, J. M. (1996). Mass balance of major solutes in a rainforest catchment in the Central Amazon. Implications for nutrient budgets in tropical rainforests. *Biogeochemistry* **32**, 115–142.

Lewis, W. M., Jr., Hamilton, S. K., Jones, S. L., and Runnels, D. D. (1987). Major element chemistry, weathering and element yields for the Caura River drainage, Venezuela. *Biogeochemistry* **4**, 159–181.

Lindberg, S. E., Lovett, G. M., Richter, D. D., and Johnson, D. W. (1986). Atmospheric deposition and canopy interactions of major ions in a forest. *Science* **231**, 141–145.

Lloyd, J. (1999a). The CO₂ dependence of photosynthesis, plant growth responses to atmospheric CO₂ and their interactions with soil nutrient status II. Temperate and boreal forest productivity and the combined effects of increasing CO₂ concentration and increased nitrogen deposition at a global scale. *Functional Ecol.* **13**, 439–459.

Lloyd, J. (1999b). Current perspectives on the terrestrial carbon cycle. *Tellus* **51B**, 336–342.

Lloyd, J., and Farquhar, G. D. (1996). The CO₂ dependence of photosynthesis, plant growth responses to atmospheric CO₂ and their interactions with soil nutrient status I. General principles and forest ecosystems. *Functional Ecol.* **10**, 4–32.

Lloyd, J., and Farquhar, G. D. (2000). A clarification of some issues raised by Poorter (1988). *Global Change Biol.* **6**, 000–000

Lloyd, J., Grace, J., Miranda, A. C., Mier, P., Wong, S. C., Miranda, H., Wright, I., Gash, J. H. C., and McIntyre, J. (1995). A simple calibrated model of Amazon rain forest productivity based on leaf biochemical properties. *Plant Cell and Environ.* **18**, 1129–1145.

Lodge, D. J., McDowell, W. H., and McSwiney, C. P. (1994). The importance of nutrient pulses in tropical forests. *Trends Ecol. Evol.* **9**, 384–397.

Lopez-Hernandez, D., Siegert, G., and Rodriguez, J. V. (1986). Competitive adsorption of phosphate with malate and oxalate by tropical soils. *Soil Sci. Soc. Am. J.* **50**, 1460–1462.

Lovelock, C. E., Kyllo, D., Popp, M., Isopp, H., Virgo, A., and Winter, K. (1997). Symbiotic vesicular-arbuscular mycorrhizae influence maximum rates of photosynthesis in tropical tree seedlings grown under elevated CO_2. *Aust. J. Plant Physiol.* **24**, 185–194.

Lovelock, C. E., Kyllo, D., and Winter, K. (1996). Growth responses to vesicular-arbuscular mycorrhizae and elevated CO_2 in seedlings of a tropical tree, *Beilschmieda pendula. Functional Ecol.* **10**, 662–667.

Lovelock, C. F., Winter, K., Mersits, R., and Popp, M. (1998). Responses of communities of tropical tree species to elevated CO_2 in a forest clearing. *Oecologia* **116**, 207–218.

Magid, J., Tiessen, H., and Condron, L. M. (1995).. Dynamics of organic phosphorus in soils under natural and agricultural ecosystems. In *"Humic Substances in Terrestrial Ecosystems"* (A. Piccolo, Ed.). Elsevier, Amsterdam.

Malhi, Y., Baldocchi, D. D., and Jarvis, P. J. (1999). The carbon balance of tropical, temperate and boreal forests. *Plant Cell and Environ.* **22**, 715–740.

Malhi, Y., Nobre, A. D., Grace, J., Kruijt, B., Pereira, M. G. P., Culf, A., and Scott, S. (1998). Carbon dioxide transfer above a central Amazonian rain forest. *J. Geophys. Res.* **103**, 31593–31612.

Marschner, H. (1995). *"Mineral Nutrition of Higher Plants,"* 2nd ed. Academic Press, London.

Martinelli, L. A., Piccolo, M. C., Townsend, A. R., Vitousek, P. M., Cuevas, E., McDowell, W., Robertson, G. P., Santos, O. C., and Treseder, K. (1999). Nitrogen stable isotopic composition of leaves and soil: Tropical versus temperate forests. *Biogeochemistry* **46**, 45–65.

Martínez-Yrízar, A. (1995). Biomass distribution and primary productivity of tropical dry forests. In *"Seasonally Dry Tropical Forests"* (S. H. Bullock, H. A. Mooney, and E. Medina, Ed.), pp. 326–345. Cambridge University Press, Cambridge.

Matson, P. A. and Vitousek, P. M. (1987). Cross-system comparison of soil nitrogen transformations and nitrous oxide fluxes in tropical forest soils. *Global Biogeochem. Cycles* **1**, 163–170.

Mattingly, G. E. C. (1975). Labile phosphate in soils. *Soil Sci.* **119**, 369–375.

McDowell, W. H. (1998). Internal nutrient fluxes in a Puerto Rican rain forest. *J. Tropical Ecol.* **14**, 521–536.

McGill, W. B., and Cole, C. V. (1981). Comparative aspects of cycling of organic C, N, S and P through soil organic matter. *Geoderma* **26**, 267–286.

McGuire, A. D., Melillo, J. M., and Joyce, L. A. (1995). The role of nitrogen in the response of net primary productivity to elevated carbon dioxide. *Annu. Rev. Ecol. Systematics* **26**, 473–503.

McKane, R. B., Rastetter, E. B., Melillo, J. M., Shaver, G. R., Hopkinson, C. S., Fernandes, D. N., Skole, D. L., and Chomentowski, W. H. (1995). Effects of global change on carbon storage in tropical forests of South America. *Global Biogeochem. Cycles* **9**, 329–350.

McLaren, R. G., and Cameron, K. C. (1996). *"Soil Science. Sustainable Production and Environmental Protection."* Oxford University Press.

McPharlin, I. R., and Bieleski, R. L. (1989). P_i efflux and influx in P-adequate and P-deficient *Spirodela* and *Lemna. Aust. J. Plant Physiol.* **16**, 391–399.

Medina, E. (1984). Nutrient balance and physiological processes at the leaf level. In *"Physiological Ecology of Plants in the Wet Tropics"* (E. Medina, H. Mooney and C. Vásquez-Yánes, Ed.), pp. 139–154. Dr. W. Junk, The Hague.

Medina, E., and Cuevas, E. (1989). Patterns of nutrient accumulation and release in Amazonian forests of the Rio Negro Basin. In *"Mineral Nutrients in Tropical Forest and Savanna Ecosystems"* (J. Proctor, Ed.), pp. 217–240. Blackwell Scientific Publications, Oxford.

Medina, E., and Klinge, H. (1983). Productivity of tropical forests and tropical woodlands. In *"Encyclopedia of Plant Physiology 12D."* (O. L. Lange, P. S. Nobel, C. B. Osmond, and H. Ziegler, Ed.), pp. 281–303. Springer-Verlag, Berlin, Heidelberg, New York.

Meir, P. (1996). *"The Exchange of Carbon Dioxide in Tropical Forest."* Ph.D thesis, University of Edinburgh, Edinburgh.

Meuller-Harvey, I., Juo, A. S. R., and Wild, A. (1985). Soil organic C, N, S and P after forest clearance in Nigeria: mineralisation rates and spatial variability. *J. Soil Sci.* **36**, 586–591.

Muljadi, D., Posner, A. M., and Quirk, J. P. (1966). The mechanism of phosphate adsorption by kaolinite, gibbsite and pseudoboehmite. Part I. The isotherms and the effect of pH an adsorption. *J. Soil Sci.* **17**, 212–229.

Neill, C., Piccolo, M. C., Steudler, P. A., Melillo, J. M., Feigl, B. J., and Cerri, C. C. (1995). Nitrogen dynamics in soils and active pastures in the western Brazilian Amazon Basin. *Soil Biol. Biochem.* **27**, 1167–1175.

Newberry, D. M. C., Alexander, I. J., and Rother, J. A. (1997). Phosphorus dynamics in a lowland African rainforest: the influence of ectomycorhizal trees. *Ecol. Monographs* **67**, 367–409.

Nye, P. H. (1961). Organic matter and nutrient cycles under moist tropical forest. *Plant and Soil* **13**, 333–346.

Nye, P. H. and Bertheux, M. H. (1957). The distribution and significance of phosphorus in forest and savanna soils of the Gold Coast and its agricultural significance. *J. Agri. Sci.* **49**, 141–149.

Oberbauer S. F., Strain B. R., and Fetcher N. (1985). Effect of CO_2-enrichment on seedling physiology and growth of two tropical tree species. *Physiologia Plantarum* **65**, 352–356.

Ognalaga, M., Frossard, E, and Thomas, E. (1994). Glucose-1-phosphate and myo-inositol hexaphosphate adsorption mechanism on geothite. *Soil Sci. Soc. Am. J.* **58**, 332–337.

Olander, L. P., and Vitousek, P. M. (2000). Regulation of soil phosphatase and chitinase activity by N and P availability. *Biogeochemistry* **49**, 175–190.

Parker, G. G. (1983). Throughfall and stemflow in the forest nutrient cycle. *Adv. Ecol. Res.* **13**, 57–133.

Parton, W. J., Stewart, J. W. B., and Cole, C. V. (1988). Dynamics of C, N, P and S in grassland soils: A model. *Biogeochemistry* **5**, 109–131.

Peng, S., Eissenstat, D. M., Graham, J. H., Williams, K., and Hodge, N. C. (1993). Growth depression in mycorrhizal citrus at high phosphorus supply: Analysis of carbon costs. *Plant Physiol.* **101**, 1063–1071.

Phillips, O. L., Malhi, Y., Higuchi, N., Laurance, L. F., Nuñez, V. P. Vasqueth, M. R. Laurance, S. G., Ferreira, L. V., Stern, M., Brown, S., and Grace, J. (1998). Changes in the carbon balance of tropical forests: Evidence from long-term plots. *Science* **282**, 439–442.

Poorter, H. (1993). Interspecific variation in the growth-response of plants to an elevated ambient CO_2 concentration. *Vegetatio* **104/105**, 77–97.

Poorter, H. (1998). Do slow-growing species and nutrient stressed plants respond relatively strongly to elevated CO_2 ? *Global Change Biol.* **4**, 693–697.

Proctor, J. (1987). Nutrient cycling in primary and secondary rainforests. *Appl. Geogr.* **7**, 135–152.

Phillips. O. L., Hall, P., Gentry, A. H., Sawyer, S. A., and Vásquez, R. (1994). Dynamics and species richness of tropical rainforests. *Proc. Natl. Acad. Sci. U.S.A.* **91**, 2805–2809.

Raaimakers, D., and Lambers, H. (1996). Response to phosphorus supply of tropical tree seedlings: a comparison between a pioneer species *Tapirira obusta* and a climax species *Lecythis corrugata*. *New Phytol.* **132**, 97–102.

Raaimakers, D., Boot, R. G. A., Dijkstra, P., Pot, S., and Pons, T. (1995). Photosynthetic rates in relation to leaf phosphorus content in pioneer versus climax tropical species. *Oecologia* **102**, 120–125.

Raghothama, K. G. (1999). Phosphate acquisition. *Annu. Rev. Plant Physiol. Plant Mol. Biol.* **50**, 665–694.

Raich, J. W., Russell, A. E., Crews, T. E., Farrington, H., and Vitousek, P. M. (1996). Both nitrogen and phosphorus limit plant production on young Hawaiian lava flows. *Biogeochemistry* **32**, 1–14.

Rayner, P. J., Enting, I. G., Francey, R. J., and Langenfeld, R. L. (1999). Reconstructing the recent carbon cycle from atmospheric CO_2, $\delta^{13}C$ and O_2/N_2 observations. *Tellus* **51B**, 213–232.

Reekie E. G., and Bazzaz F. A. (1989). Competition and patterns of resource use among seedlings of five tropical trees grown at ambient and elevated CO_2. *Oecologia* **79**, 212–222.

Reich, P. B., Ellsworth, D. S., and Uhl, C. (1995). Leaf carbon and nutrient assimilation and conservation in species of differing successional status in an oligotrophic Amazonian forest. *Functional Ecol.* **9**, 65–76.

Richards, P. W. (1996). *"The Tropical Rain Forest: An Ecological Study,"* 2nd ed. Cambridge University Press, Cambridge.

Richter, D. D., and Babbar, L. I. (1991). Soil diversity in the tropics. *Adv. Ecol. Res.* **21**, 315–389.

Rincón, E., and Huante, P. (1994). Influence of mineral nutrient availability on growth of tree seedlings from the tropical deciduous forest. *Trees* **9**, 93–97.

Sample, E. C., Soper, R. J., and Racz, G. C. (1980). Reaction of phosphorus fertilizers in soils. In *"The Role of Phosphorus in Agriculture."* (M. Stelly, Ed.), pp. 263–310. Soil Science Society of America, Madison, Wisconsin.

Sanchez, P. A. (1976). *"Properties and Management of Soils in the Tropics."* Wiley, New York.

Sanyal, S. K., and DeDatta, S. K. (1991). Chemistry of phosphorus transformations in soil. *Adv. Soil Sci.* **16**, 1–120.

Schachtman, D. P., Reid, R. J., and Ayling, S. M. (1998). Phosphorus uptake by plants: from soils to cell. *Plant Physiol.* **116**, 447–453.

Sibanda, M. M., and Young, S. D. (1989). The effect of humus acids and soil heating on the availability of phosphate in oxide-rich tropical soils. In *"Mineral Nutrients in Tropical Forest and Savanna Ecosystems."* (J. Proctor, Ed.), pp. 71–84. Blackwell Scientific Publications, Oxford.

Silver, W. (1994). Is tropical nutrient availability related to plant nutrient use in humid tropical forests. *Oecologia* **98**, 336–343.

Silver, W. L., Lugo, A. E., and Keller, M. (1999). Soil oxygen availability and biogeochemistry along rainfall and topographic gradients in upland wet tropical forest soils. *Biogeochemistry* **44**, 301–328.

Singh, J. S., Raghubanshi, A. S., Singh, R. S., and Srivastava, S. C. (1989). Microbial biomass acts as a source of plant nutrients in dry tropical forest and savanna. *Nature* **338**, 499–500.

Smith, S. E., and Read, D. J. (1997). *"Mycorrhizal Symbiosis."* Academic Press, San Diego, CA.

Sollins, P., Robertson, G. P., and Uehara, G. (1988). Nutrient mobility in variable- and permanent-charge soils. *Biogeochemistry* **6**, 181–199.

Staddon, P. L., and Fitter, A. H. (1998). Does elevated atmospheric carbon dioxide affect arbuscular mycorrhizas? *Trends Ecol. Evolution* **13**, 455–458.

Stark, N. M., and Jordan, C. F. (1978). Nutrient retention by the root mat of an Amazonian rain forest. *Ecology* **59**, 434–437

Stoorvogel, J. J., Janssen, B. H., and Van Breemen, N. (1997). The nutrient budgets of a watershed and its forest ecosystem in the Taï National Park in Côte d'Ivoire. *Biogeochemistry* **37**, 159–172.

Strauss, R., Brmmer, G. W., and Barrow, N. J. (1997). Effects of crystallinity of geothite. II. Rates of sorption and desorption of phosphate. *Eur. J. Soil Sci.* **48**, 101–114.

Swap, R., Garstang, M., and Greco, S. (1992). Saharan dust in the Amazon Basin. *Tellus* **44B**, 133–149.

Tanner, E. V. J., Vitousek, P. M., and Cuevas, E. (1998). Experimental investigation of nutrient limitations of forest growth on wet tropical mountains. *Ecology* **79**, 10–22.

Thompson, B. D., Clarkson, D. T., and Brain, P. (1990). Kinetics of phosphorus uptake by germ-tubes of the arbuscular fungus *Gigaspora marginata*. *New Phytol.* **116**, 647–653.

Tiessen, H., Chacon, P., and Cuevas, E. (1994a). Phosphorus and nitrogen status in soil and vegetation along a toposequence of dystrophic rainforests on the upper Rio Negro. *Oecologia* **99**, 145–150.

Tiessen, H., Cuevas, E., and Chacon, P. (1994b). The role of soil organic matter in sustaining soil fertility. *Nature* **371**, 783–785.

Tiessen, H., Salcedo, I. H., and Sampaio, E. V. S. B. (1992). Nutrient and soil organic matter dynamics under shifting cultivation in semi-arid northeastern Brazil. *Agri. Ecosystems Environ.* **38**, 139–151.

Uehara, G., and Gillman, G. P. (1981). *"The Mineralogy, Chemistry and Physics of Tropical Soils with Variable Charge Clays."* Westview, Boulder, Colorado.

Uhl, C., and Jordan, C. F. (1984). Succession and nutrient dynamics following forest cutting and burning in Amazonia. *Ecology* **65**, 1476–1490.

Veenendaal, E. M., Swaine, M. D., Lecha, R. T., Walsh, M. F., Abebrese, I. K., and Owusu-Afriyie, K. (1996). Responses of West African forest tree seedlings to irradiance and soil fertility. *Functional Ecol.* **10**, 501–511.

Veneklass, E. J. (1990). Nutrient fluxes in bulk precipitation and throughfall in two montane tropical forests, Columbia. *J. Ecol.* **78**, 974–992.

Vitousek, P. (1984). Litterfall, nutrient cycling an nutrient limitation in tropical forests. *Ecology* **65**, 285–298.

Vitousek, P. M., Fahey, T., Johnson, D. W., and Swift, M. J. (1988). Element interactions in forest ecosystems: Succession, allometry and input-output budgets. *Biogeochemistry* **5**, 7–34.

Vitousek, P. M., and Farrington, H. (1997). Nutrient limitation and soil development: Experimental test of a biogeochemical theory. *Biogeochemistry* **37**, 63–75.

Vitousek, P. M., and Sanford, R. L. Jr. (1986). Nutrient cycling in moist tropical forest. *Annu. Rev. Ecol. Systematics* **17**, 137–167.

Vitousek, P., Walker, L. R., Whitekar, L. D., and Matson, P. A. (1993). Nutrient limitations to plant growth during primary succession in Hawaii Volcanoes National Park. *Biogeochemistry* **23**, 197–215.

Walker, T. W., and Syers, J. K. (1976). The fate of phosphorus during pedogenesis. *Geoderma* **15**, 1–19.

Walter, H. (1936). Nährstoffgehalt des Bodens und natürliche Waldbestände. *Forstliche Wochenschrift Silva* **24**, 201–205, 209–213.

Walter, H. (1971). *"Ecology of Tropical and Subtropical Vegetation."* Oliver & Boyd, Edinburgh.

Watt, M. and Evans, J. R. (1999). Linking development and determinacy with organic acid efflux from proteoid roots of white lupin grown with low phosphorus and ambient or elevated atmospheric CO_2 concentration. *Plant Physiol.* **120**, 705–716.

Westin, F. C., and de Brito, J. G. (1969). Phosphorus fractions of some Venezuelan soils in relation to weathering. *Soil Sci.* **107**, 124–202.

Whipps, J. (1985). Effect of CO_2 concentration on growth, carbon distribution and loss of carbon from the roots of maize. *J. Exp. Botany* **36**, 644–651.

Whitmore, T. C. (1989). Tropical forest nutrients, where do we stand ? *A tour de horizon*. In *"Mineral Nutrients in Tropical Forest and Savanna*

Ecosystems" (J. Proctor, Ed.), pp. 1–14. Blackwell Scientific Publications, Oxford.

Williams, M. R., Fisher, T., and Melack, J. M. (1997). Chemical composition and deposition of rain in the central Amazon, Brazil. *Atmos. Environ.* **31,** 207–217.

Winter, K., and Lovelock, C. E. (1999). Growth responses of seedlings of early and late successional tropical forest trees to elevated atmospheric CO_2. *Flora* **194,** 221–227.

Würth, M. K. R., Winter, K., and Körner, C. (1998). In situ responses to elevated CO_2 in tropical forest understory plants. *Functional Ecol.* **12,** 886–895.

Younge, O. R. and Plucknett, D. L. (1966). Quenching the high phosphorus fixation of Hawaiian latosols. *Soil Sci. Soc. Am. Proc.* **30,** 653–655.

Ziska, L. H., Hogan, K. P., Smith, A. P., and Drake, B. G. (1991). Growth and photosynthetic response of nine tropical species with long-term exposure to elevated carbon-dioxide. *Oecologia* **86,** 383–389.

Trees in Grasslands: Biogeochemical Consequences of Woody Plant Expansion

Steve Archer,
Thomas W. Boutton,
and K. A. Hibbard
*Texas A&M University
College Station,
Texas*

Key Words: carbon, disturbance, grazing, fire, global change, hydrocarbons, land cover, modeling, nitrogen, NOx, savanna, sequestration, soil respiration, succession, tree-grass interactions, vegetation dynamics, woody plant encroachment

1. Introduction

The term "savanna" typically denotes plant communities or landscapes having a continuous grass layer with scattered woody plants. Although savannas are not the only vegetation type where contrasting plant life forms codominate, they are one of the most striking, geographically extensive (ca. 20% of global land surface; Scholes and Hall, 1996) and socioeconomically important examples in tropical (Tothill and Mott, 1985; Young and Solbrig, 1993) and temperate (Burgess, 1995; McPherson, 1997; Anderson *et al.*, 1999) regions. Tropical savannas cover about 1600 million ha of the terrestrial surface (Scholes and Hall, 1996), including more than half the area of Africa and Australia, 45% of South America, and 10% of India and southeastern Asia (Werner, 1991). Temperate savannas in North America occupy an estimated 50 million ha (McPherson, 1997). More importantly, savannas contain a large and rapidly growing proportion of the world's human population and a majority of its rangelands and domesticated animals. As such, they have received substantial and ever-increasing anthropogenic land-use pressure.

Many savannas are dynamic ecotones between woody plant (shrub-steppe, desert scrub, woodland or forest) and grassland formations. Savannas vary substantially with respect to the stature (shrub vs tree), canopy cover (e.g., 5–80%), functional form (evergreen vs deciduous; broad-leaved vs needle-leaved; shallow vs deeply rooted), and spatial arrangement (random, regular, or clumped) of the woody elements that compose them. Similarly, the grass layer may consist of short versus tall-statured species, bunch versus rhizomatous growth forms, and C_3, C_4, or mixed C_3/C_4 photosynthetic pathway assemblages. This variation in structural/functional characteristics reflects a rich array of interactions between climate (especially the amount and seasonality of rainfall), soils (notably, depth and texture), and disturbance (particularly grazing, browsing, and fire), as shown in Figure 1 (Walker, 1987; Backéus, 1992).

Much of the literature on savanna ecology has been devoted to describing and classifying vegetation structure. Static classification schemes minimize the importance of temporal change and divert attention from functional processes that might explain dynamic spatiotemporal variation. Grasses and woody plants may coexist in a dynamic equilibrium when climatic, edaphic, and disturbance factors interact temporally such that neither life form can exclude the other. However, a directional change in one or more of these primary controlling factors may shift the balance in favor of one life form over the other and move the system toward either grassland or shrubland/woodland. The probability, rate, and extent of such a shift may depend on local topoedaphic factors and the life-history traits and autecology of the growth forms or species involved.

Human population growth and widespread Anglo-European settlement during the 18th and 19th centuries have influenced the balance of grass–woody plant interactions worldwide. For example, extensive clearing of trees for fuel, lumber, and cropland has fragmented forests and produced anthropogenic or degraded savannas (Gadgill and Meher-Homji, 1985; Sinclair and Fryxell, 1985; Cline-Cole *et al.*, 1990; Schüle, 1990; Young and Solbrig,

[1] Current Address: Climate Change Research Center, GAIM Task Force, Institute for the Study of Earth, Oceans, and Space (EOS), Morse Hall, 39 College Road, University of New Hampshire, Durham, NH 03824-3525

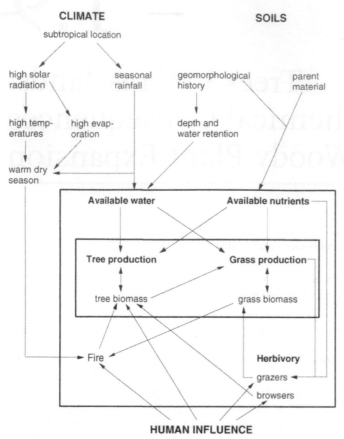

CLIMATE   SOILS

FIGURE 1 Numerous factors interact to affect the abundance of grasses and woody vegetation in drylands (from Scholes and Walker, 1993). The balance between trees and grasses (innermost level) is affected by determinants of structure and function (water, nutrients, fire, and herbivory). The outermost level contains the factors that give the determinants their characteristics. Over the past century, human influences have shifted the balance to favor woody plants through selective utilization of grasses by livestock maintained at high concentrations, elimination of browsers, and fire suppression (see Archer, 1994).

1993; Mearns, 1995). Following forest clearing, pyrophytic grasses may establish and restrict woody colonization by accelerating fire cycles and maintaining low-fertility soils (Hopkins, 1983; Mueller-Dombois and Goldammer, 1990; D'Antonio and Vitousek, 1992). In the Brazilian Cerrado, rates of agricultural expansion and clearing of savanna and woodland trees rival those reported for Amazon rain forest (Klink *et al.*, 1993). In other areas, fire suppression, eradication of indigenous savanna browsers, and the introduction of grazing livestock and exotic trees and shrubs have caused a progressive increase in woody plant density, known as bush or brush encroachment (Adamoli *et al.*, 1990; Archer, 1994; Gardener *et al.*, 1990; Miller and Wigand, 1994; Noble, 1997). As a result, areas that were once forest may become savanna-like, while areas that were once grassland or open savanna may progress toward a shrubland or woodland physiognomy. The biogeochemical consequences of this latter phenomenon are the focus of this chapter.

2. Woody Plant Encroachment in Grasslands and Savannas

Woody plant encroachment has been widespread in grassland and savanna ecosystems of North and South America, Australia, Africa, and southeast Asia over the past century (Table 1). This encroachment, typically by unpalatable trees and shrubs, has gone to completion on some landscapes and is in progress on others. It jeopardizes grassland biodiversity and threatens the sustainability of pastoral, subsistence, and commercial livestock grazing (Rappole *et al.*, 1986; Noble, 1997). As such, it may adversely impact ~ 20% of the world's population (Turner *et al.*, 1990). The proximate causes of displacement of perennial grasses by woody plants are subjects of debate. Land-use practices such as heavy grazing and reductions in fire frequency have often been implicated. However, climate change, historic atmospheric CO_2 enrichment, and exotic species introductions are potentially important contributing factors (Idso, 1995; Archer *et al.*, 1995; Polley, 1997; Polley *et al.*, 1997). Current trends in atmospheric CO_2 enrichment may exacerbate the shifts from grass to woody plant domination, especially where the invasive trees/shrubs are capable of symbiotic N_2 fixation. Expansion of woody plants into grasslands may also be favored by recent increases in atmospheric N deposition (Köchy, 1999). In addition to influencing vegetation composition, changes in each of these factors would have the potential to alter the storage and dynamics of C and N in savanna ecosystems. The net outcome of such interactions over the recent past is poorly understood and has not been well documented.

Although woody plant encroachment has long been a concern of land managers in grassland and savanna regions (e.g., Fisher, 1950, 1977), research on this problem has been primarily "applied" and focused on the effects of woody plants on grass production and the development of chemical or mechanical methods to reduce the abundance of established trees and shrubs. Despite the long-standing recognition of woody plant encroachment as a worldwide dryland management problem, little is known of the rates and dynamics of the phenomenon or its impact on fundamental ecological processes related to energy flow, nutrient cycling, and biodiversity. Grassland/savanna systems account for 30–35% of global terrestrial net primary production (Field *et al.*, 1998). Hence, when woody species increase in abundance and transform shrublands into woodlands, grasslands into savannas, or savannas into shrublands and woodlands, the potential to alter C and N sequestration and cycling at regional and global scales may be significant. Consequently, this type of land cover change has the potential to contribute significantly to the terrestrial global carbon sink (cf. Ciais *et al.*, 1995). Savanna landforms may have a larger impact on the global carbon cycle than previously appreciated (Hall *et al.*, 1995; Ojima *et al.*, 1993; Scholes and Hall, 1996; Scholes and Bailey, 1996; Scholes and van der Merwe, 1996). Indeed, recent assessments suggest that savanna ecosystems have among the highest potential C gain and loss rates of the world's biomes (ORNL, 1998). In addition, emissions of radiatively active

TABLE 1 Survey of Studies Describing or Quantifying Woody Plant Encroachment into Grassland, Tree/Shrub Proliferation in Savannas, and Tree Encroachment into Shrubland.

Arizona
Arnold, 1950
Bahre, 1991
Bahre and Shelton, 1993
Brown, 1950
Brown, 1997
Cooper, 1960
Covington and Moore, 1994
Glendening, 1952
Humphrey and Mehrhoff, 1958
Hastings and Turner, 1965
Johnsen, 1962
Kenney et al., 1986
Martin, 1975
Martin and Turner, 1977
McClaran and McPherson, 1995
McPherson et al., 1993
Miller, 1921
Reynolds and Glendening, 1949
Savage and Swetnam, 1990
Smith and Schmutz, 1975
Vivrette and Muller, 1977
Williams et al., 1987
Young and Evans, 1981

California
Bossard, 1991
Bossard and Rejmanek, 1994
Callahan and Davis, 1993
McBride and Heady, 1968
Hobbs and Mooney, 1986

Colorado
Baker and Weisberg, 1997
Mast et al., 1997
Mast et al., 1998
Veblen and Lorenz, 1991

Idaho
Anderson and Holte, 1981
Burkhardt and Tisdale, 1976
Zimmerman and Neunschwander, 1984

Iowa
Wang et al., 1993

Kansas
Abrams, 1986
Bragg and Hulbert, 1976
Briggs and Gibson, 1992
Knapp and Seastedt, 1986
Knight et al., 1994
Loehle et al., 1996
Owensby et al., 1973

Minnesota
Grimm, 1983
Johnston et al., 1996

Montana
Arno and Gruell, 1986
Arno et al., 1995

Nebraska
Johnson, 1994
Steinauer and Bragg, 1987
Steuter et al., 1990

New Mexico
Branscomb, 1958
Buffington and Herbel, 1965
Connin et al., 1997
Dick-Peddie, 1993
Gibbens et al., 1992
Hennessy et al., 1983
McCraw, 1985
York and Dick-Peddie, 1969

Nevada
Blackburn and Tueller, 1970

North Dakota
Potter and Green, 1964

Oregon
Knapp and Soule, 1996
Knapp and Soule, 1998
Miller and Rose, 1995
Miller and Halpern, 1998
Miller and Rose, 1999
Skovlin and Thomas, 1995
Soule and Knapp, 1999

Oklahoma
Engle et al., 1996
Snook, 1985

South Dakota
Bock and Bock, 1984
Progulske, 1974

Texas
Ansley et al., 1995
Archer et al., 1988
Archer, 1989
Boutton et al., 1998
Bogusch, 1952
Bray, 1901
Bruce et al., 1995
Ellis and Schuster, 1968
Foster, 1917
Inglis, 1964
Johnston, 1963
McKinney, 1996
McPherson et al., 1988
Nelson and Beres, 1987
Smeins et al., 1974
Scanlan and Archer, 1991
Weltzin et al., 1997
Wondzell and Ludwig, 1995

Utah
Madany and West, 1983
Yorks et al., 1992

Washington
Rummell, 1951

Wyoming
Fisher et al., 1987

Regional Assessments
Glendening and Paulsen, 1955
Gruell, 1983
Hart and Laycock, 1996
Humphrey, 1958
Humphrey, 1987
Johnson, 1987
Leopold, 1951
Milchunas and Lauenroth, 1993
Miller and Wigand, 1994
McClaran and McPherson, 1995
Reichard and Hamilton, 1997
Rogers, 1982
Robinson, 1965
Tieszen and Archer, 1990
Tieszen and Pfau, 1995
Wall, 1999
West, 1988
Young et al., 1979

AFRICA
Acocks, 1964
Ambrose and Sikes, 1991
Ben-Shaher, 1991
Bews, 1917
Bond et al., 1994
Friedel, 1987
Grossman and Gandar, 1989
Höchberg et al., 1994
Holmes and Cowling, 1997
Jeltsch et al., 1997
Le Roux, 1997
Menaut et al., 1990
Norton-Griffiths, 1979
O'Connor and Roux, 1995
Palmer and van Rooyen, 1998
Ramsay and Rose Innes, 1963
Reid and Ellis, 1995
Ringose et al., 1996
Sabiiti, 1988
Schwartz et al., 1996
Scott, 1966
Shantz and Turner, 1958
Skarpe, 1990
Skarpe, 1991
Thomas and Pratt, 1967
Trollope, 1982
Van Vegten, 1983
West, 1947

AUSTRALIA
Booth and Barker, 1981
Bowman and Panton, 1995
Bren, 1992
Brown and Carter, 1998
Burrows et al., 1985
Burrows et al., 1998
Burrows et al., 1990
Cook et al., 1996
Cunningham and Walker, 1973
Gardiner and Gardiner, 1996
Grice, 1996
Grice, 1997
Harrington et al., 1979
Harrington and Hodgkinson, 1986
Hodgkin, 1984
Lonsdale, 1993
Noble, 1997
Panetta and McKee, 1997

CANADA
Archibold and Wilson, 1980
Brown, 1994
Köchy, 1999

SOUTH AMERICA
Adamoli et al., 1990
Bücher, 1982
Bücher, 1987
Distel and Boo, 1996
Dussart et al., 1998
San José and Fariñas, 1983
San José and Fariñas, 1991
San José et al., 1991
San José and Montes, 1997
San José et al., 1998
Schofield and Bucher, 1986

OTHERS
Backéus, 1992
Binggeli, 1996
Walker et al., 1981
Skarpe, 1992

Note. Documentation includes historical observations, long-term monitoring, repeat ground or aerial photography, stable carbon isotope analysis, dendrochronology, and, in some cases, simulation modeling. The focus is on arid and semi-arid "rangelands." Hence, studies documenting tree/shrub invasion of abandoned agricultural fields (cf, Smith, 1975; Johnston et al., 1996; De Steven, 1991) or regeneration following forest clearing are not included here. Studies discussing or reviewing causes or consequences of woody encroachment into grasslands/savannas (cf, Humphrey, 1953; Fisher, 1977; Smeins, 1983; Rappole et al., 1986; Grover and Musick, 1990; Schlesinger et al., 1990; Archer, 1994; Archer et al., 1995; Idso, 1995; Polley et al., 1997) are also excluded. Some papers in the list reference other papers which have documented woody plant increases in historical times (Backéus, 1992; Noble, 1997). An updated and more extensive version of this table, including a list of woody genera, can be found at http://cnrit.tamu.edu/rlem/faculty/archer/.

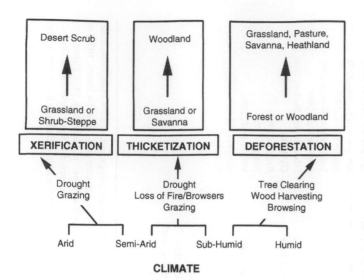

FIGURE 2 Xerification/desertification (West, 1986) and deforestation have received much attention. Although increases in woody plant abundance in drylands are geographically widespread and well documented (Table 1), little is known of the ecological consequences of this vegetation change (adapted from Archer and Stokes, 2000).

trace gases, NO_x and aerosols from savanna fires may contribute significantly to global emissions and influence climate and atmospheric chemistry (Crutzen and Andreae, 1990; Hao *et al.*, 1990; Crutzen and Goldammer, 1993).

Desertification has long been a topic of concern to land managers and ecologists (Moat and Hutchinson, 1995; Arnalds and Archer, 2000). More recently, changes in the storage and dynamics of C and N in the terrestrial biosphere have been evaluated with respect to deforestation, intensive agricultural practices, succession on abandoned agricultural lands, and afforestation/reforestation (Fig. 2) (Houghton *et al.*, 1987; Post, 1993). Increased abundance of woody plants in drylands has the potential to alter land surface–atmosphere interactions and atmospheric chemistry by affecting biophysical processes, and C and N storage and dynamics (e.g., Schlesinger *et al.*, 1990; Graetz, 1991; Bonan, 1997). Even so, its significance has yet to be thoroughly evaluated or quantified. Here, we review results from a case study of a subtropical dryland landscape which has been undergoing a transformation from grassland to savanna to woodland. Some of our recent work has explored the implications of this change in vegetation on the hydrological cycle (Brown and Archer, 1990; Midwood *et al.*, 1998; Boutton *et al.*, 1999). Here, our emphasis is on the rates of change in soil and plant carbon and nitrogen pools and fluxes.

3. The La Copita Case Study

3.1 Biogeographical and Historical Contexts

The La Copita Research Area (27° 40′N; 98° 12′W; elevation = 75–90 m ASL) is situated in the northeastern portion of the North American Tamaulipan Biotic Province (Blair, 1950) in the Rio Grande Plains of southern Texas. The potential natural vegetation of this region has been classified as *Prosopis–Acacia–Andropogon–Setaria* savanna (Küchler, 1964). However, the contemporary vegetation is subtropical thorn woodland (McLendon, 1991) and occupies about 12 million ha in Texas alone (Jones, 1975). The shrubs and small trees at the study site are characteristic of dry tropical and subtropical zones in Mexico, Central America, South America (Chaco, Caatinga, Caldenal), Africa, Australia, India, and southeast Asia. In many instances, it is believed that these vegetation types have replaced grasslands over large areas since the 1800s (Table 1). Current vegetation at the La Copita site, which has been grazed by domestic livestock since the late 1800s, consists of savanna parklands in sandy loam uplands that grade into closed-canopy woodlands in clay loam lowland drainages. All wooded landscape elements (upland shrub clusters and groves; lowland playa and drainage woodlands) are typically dominated by the leguminous tree *Prosopis glandulosa* in the overstory, with an understory mixture of evergreen, winter-deciduous, and summer-deciduous shrubs. Climate of the region is subtropical (mean annual temperature 22.4 °C) with warm, moist winters and hot, dry summers. Mean annual rainfall is 720 mm and highly variable (CV = 35%).

Reports from settlers indicate that much of southern Texas was grassland or open savanna in the mid-1800s (Inglis, 1964). Historical aerial photography demonstrates that woody plant cover on La Copita increased from 10% in 1941 to 40% in 1983 (Archer *et al.*, 1988). $\delta^{13}C$ and radiocarbon analyses of soil organic carbon have confirmed that C_3 trees and shrubs have displaced C_4 grasses in upland and lowland portions of the landscape in the past 100 years (Boutton *et al.*, 1998). Plant growth (Archer, 1989) and transition probability models (Scanlan and Archer, 1991), substantiated by tree ring analysis (Boutton *et al.*, 1998), indicate that most trees on the site have established over the past 100 years. The successional processes involved in woody plant community development and topoedaphic controls over spatial patterns of tree/shrub expansion have been elucidated (Archer, 1995b). Armed with information from these prior studies, we are now poised to ascertain the biogeochemical consequences of succession from grassland to woodland.

3.2 Herbaceous Retrogression and Soil Carbon Losses

Simulations with the CENTURY biogeochemistry model (Parton *et al.*, 1987, 1988, 1993) parameterized for assumed presettlement conditions ("light" grazing, fire at 10-year intervals), soil texture, and climate of the La Copita projected that soil organic carbon (SOC) would have been on the order of 2500 g m^{-2} to a depth of 0–20 cm (Hibbard, 1995). The SOC values from this assessment were then used as a baseline against which historic effects of heavy, continuous livestock grazing were evaluated. In a subsequent model run, intensification of grazing and removal of fire were initiated in 1850, a date approximating the advent of widespread, unregulated livestock grazing in southern Texas (Lehman,

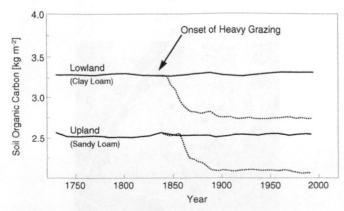

FIGURE 3 Simulation model reconstruction of changes in soil organic carbon (0–20 cm) on an upland, sandy loam soil and a lowland, clay loam soil at the La Copita Research Area in southern Texas, USA (from Hibbard, 1995). Solid lines depict steady-state SOC values expected for the climate of the site under light grazing and fire every 10 years (presettlement conditions). Dashed lines depict changes in SOC predicted to occur on two major soil types after the onset of heavy, continuous livestock grazing and cessation of fire. Steady-state values for heavy, continuous grazing and cessation of fire are within 5% of field measurements on present-day grasslands.

1969). This simulation produced a 16–29% reduction in SOC of sandy loam upland and clay loam lowland soils, respectively (Fig. 3). These results appear reasonable in that the model-generated steady-state SOC levels for sandy loam uplands (2062 g m^{-2}) approximated the average pool sizes measured in present-day grassland communities on these soils (2087 g m^{-2}). Further, the grazing-induced reductions in SOC predicted by the simulation were comparable to field measurements reported for other grazed grasslands (Bauer *et al.*, 1987; Frank *et al.*, 1995; but see also Milchunas and Lauenroth 1993). CENTURY simulations did not explicitly include potential erosion losses. The fact that simulated historic changes in SOC approximated those currently observed at the site therefore suggests that such losses may have been minimal. This inference seems reasonable, since the La Copita landscapes have relatively little topographic relief (1–3% slopes) and show no obvious physical signs of erosion (pedestals, rills, gulleys). Furthermore, soil profile structure in low-lying portions of the landscape shows no pedogenic evidence of significant translocation of soils from uplands. Elevated C and N pools in soils of developing woody communities (summarized later) thus appear to be the result of *in situ* accumulations induced by trees and shrubs rather than losses from grazed grasslands.

The present-day herbaceous vegetation is dominated by a low cover of ephemeral dicots and short-statured, weakly perennial grasses. In contrast, herbaceous vegetation on relict, protected grasslands in the region is characterized by mid- to tall-statured perennial grasses whose potential productivity (500–600 g m^{-2}; SCS, 1979) is two to three times that which has been recorded at La Copita (<270 g m^{-2}; Vega, 1991; Hibbard, 1995). Thus, it is reasonable to conclude that soil C and N storage has declined in

herbaceous communities over the past century as a result of changes in species composition, microclimate, and biomass production attributable to heavy, continuous livestock grazing on this site.

3.3 Woody Plant Encroachment and Ecosystem Biogeochemistry

Changes in soils and microclimate accompanying long-term heavy grazing may have shifted the balance in favor of N$_2$-fixing or evergreen woody plants which are better adapted than grasses to nutrient-poor soils and warmer, drier microenvironments. The establishment of trees and shrubs would have been further augmented by grazing-induced reductions in herbaceous competition and fire (Archer, 1995a). In addition, the woody plants at La Copita are highly unpalatable, and browsing by wildlife or cattle is minimal. However, fruits of the dominant tree invader (*Prosopis glandulosa*) are readily consumed by livestock, which disperse large numbers of viable seeds into grasslands (Brown and Archer, 1987). Thus, heavy, continuous, and preferential grazing of grasses by livestock has promoted woody plant encroachment via numerous direct and indirect effects (Archer, 1994). As woody communities develop in grazed grasslands, plant and soil C and N pool sizes and flux rates change as described in the following sections.

3.3.1 Plant Carbon Pools

Quantitative changes in woody plant cover at La Copita are depicted in Fig. 4 and 5. To ascertain the effects of these vegetation changes on plant carbon stocks, we linked CENTURY with a plant succession model developed for La Copita (Scanlan and Archer, 1991). We initiated woody plant encroachment in the late 1800s on a heavily grazed, fire-free landscape in which SOC content had been reduced by grazing (Fig. 3). The landscape, consisting of a

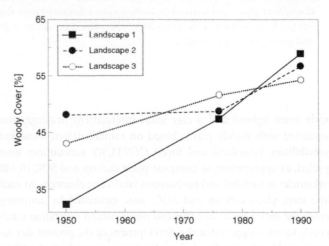

FIGURE 4 Changes in total woody plant cover on three replicated landscapes at the La Copita site in southern Texas. See Figure 5 for spatial pattern of change in various patch types on Landscape 1 (Archer and Boutton, unpublished).

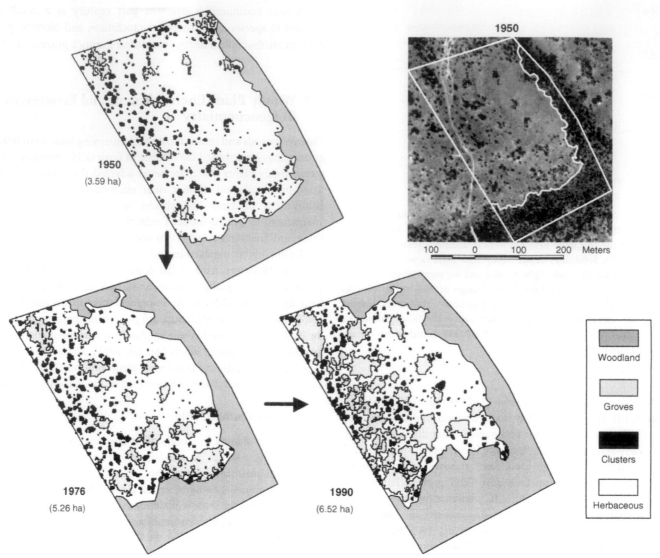

FIGURE 5 Landscape-scale changes in herbaceous and woody plant community cover from 1950 to 1990 in upland (herbaceous, discrete cluster, and grove) and lowland (woodland) plant communities at the La Copita site in southern Texas (Archer and Boutton, unpublished). Values given below dates are total hectares of woody cover (cluster + grove + woodland) for the 11.06 ha "pixel." See Figure 4 (Landscape 1) for changes in percentage woody cover.

sandy loam upland and a clay loam intermittent drainage, was populated with woody plants based on rainfall-driven transition probabilities. Grassland and forest CENTURY subroutines were applied, as appropriate, to compute plant carbon and SOC (0–20 cm) stocks in wooded and herbaceous landscape elements. At each time step, plant carbon and SOC were estimated by summing across the entire landscape (upland plus lowland vegetation patch types). Results suggest that the development of the present-day savanna parkland–woodland complex has increased plant carbon stocks 10-fold over that which would be present had the "pristine" grassland vegetation been maintained on the site (Fig. 6). Part of that increase is attributable to an increase in aboveground net pri-

mary productivity (Table 2) and part of it represents the decline in tissue turnover which occurs when herbaceous vegetation is replaced by woody vegetation. These results are conservative in that CENTURY simulations include root C mass only in the top 20 cm of soil. Biomass distributions of woody plant roots at La Copita (Watts, 1993; Boutton *et al.*, 1998; 1999; Midwood *et al.*, 1998; Gill and Burke, 1999) are typical of those of other dryland tree/shrub systems (Jackson *et al.*, 1996; Canadell *et al.*, 1996) where, relative to grasslands, there is substantially greater mass at deeper depths where turnover and decomposition are likely to be reduced. The fact that fluctuations in monthly woody plant root biomass in upper soil horizons exceeded monthly foliar litter inputs by one to

TABLE 2 Contrasts in Aboveground Net Primary Production (ANPP), Soil Physical Properties, Organic Carbon and Total Nitrogen Pools (0- to 10-cm depth), and Fluxes in Soils Associated with Woody Plant and Grazed Grassland Communities in a Sandy Loam Upland Landscape at the La Copita Research Area in Southern Texas

	Community Type	
Parameter	Herbaceous	Woody Plant
ANPP (Mg ha^{-1} year^{-1})[a]	1.9 –3.4	5.1 – 6.0
Bulk density (g cm^{-3})[a]	1.4 ± 0.01	1.1 ± 0.04
% Clay[a]	20 ± 0.7	20 ± 1
Fine roots (g m^{-2})[a]	100 –175	400 – 700
Coarse roots (g m^{-2})[a]	100 – 400	400 – 1,100
Organic C[a], %	0.84 – 0.05	2.2 ± 0.23
g m^{-2}	1165 ± 67	2352 ± 276
Potential C mineralization		
(mg C kg^{-1} soil day^{-1})[b]	7.3 ± 5.7	15.5 ± 6.8
Soil respiration (mg CO$_2$ m^{-2} year^{-1})[b]	611 ± 83	730 ± 67
Q_{10} values for *in situ* soil respiration[b]	1.2	1.4 – 2.7
Total N[a], %	0.07 ± 0.00	0.18 ± 0.02
g m^{-2}	91 ± 6	192 ± 20
N mineralization[a], g N m^{-2} year^{-1}	6 ± 1	22 ± 2
μg N g^{-1} year^{-1}	42 ± 5	200 ± 18
NO flux (ng NO-N cm^{-2} h^{-1})[c]		
Dry soil	0.2 ± 0.07	2.8 ± 0.25
Wet soil	1.1 ± 0.11	16.2 ± 2.03

Note. Maximum and minimum monthly values for samples obtained over an annual cycle are shown for root-standing crop (coarse roots ≥0.1-mm diameter); a range is presented for ANPP and Q_{10}. All other values are means ± SE.

[a] Hibbard, 1995.
[b] McCulley, 1998.
[c] Cole et al. 1996.

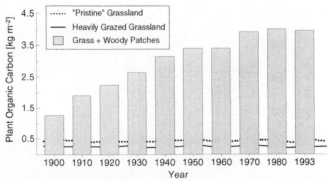

FIGURE 6 Modeled changes in whole-landscape (upland + lowland and all patch types therein) plant carbon density (aboveground + roots to 20 cm) accompanying succession from grassland to savanna parkland/woodland (from Hibbard, 1995). Dashed line depicts steady-state SOC expected for a lightly grazed grassland landscape (upland + lowland communities pooled) with fire at 10-year intervals and no woody plants; solid line depicts steady-state plant carbon density for heavily, continuously grazed grassland landscape with no fire and no woody plants. Changes in woody plant abundance on each soil type were directed by a succession model (Scanlan and Archer, 1991); subsequent changes in plant carbon stocks were then assessed with a biogeochemistry model (CENTURY; Parton *et al.*, 1994). See Figure 7 for validation results.

two orders of magnitude (Table 2) suggests that belowground inputs of organic matter drive changes in soil physical and chemical properties subsequent to woody plant establishment in grasslands. These substantial fluctuations in woody plant root biomass suggest a high turnover, which is consistent with detailed observations on woody plants in other systems (Eissenstat and Yanai, 1997; Hendricks *et al.*, 1997). In addition, turnover of grass roots may be slower than has been generally assumed (Milchunas *et al.*, 1992). Thus, increases in aboveground and belowground net primary productivities may accompany woody plant encroachment into grasslands and foster C and N accumulation.

As an independent test of the reconstruction in Figure 6, we quantified aboveground plant carbon density in patches representing the dominant community types at La Copita. This was accomplished using allometric relationships and belt-transect surveys (Northup *et al.*, 1996). Plant carbon density was then multiplied by community area measured on aerial photographs (1950, 1976, 1990) to obtain a community-level estimate. Estimates for each community type were then summed to obtain landscape-scale estimates. For patches representing various woody and herbaceous community types, CENTURY estimates of aboveground carbon density were lower than field-based estimates (Table 3), further suggesting that model estimates were conservative. Aboveground carbon density differed substantially among

the three landscapes inventoried in 1950 (Fig. 7), primarily reflecting differences in woody plant cover on this date (Fig. 4). By 1990, woody cover and carbon density were comparable on the three landscapes. The CENTURY–succession model estimates of aboveground carbon density for an "average" landscape in 1950, 1976, and 1990 closely approximated those obtained from the field-historical aerial photo approach.

3.3.2. Nonmethane Hydrocarbon Fluxes

On a regional basis, shifts from grass to woody plant domination have the potential to influence biophysical aspects of land–atmosphere interactions, such as albedo, evapotranspiration, boundary layer conditions, and dust loading (e.g., Bryant *et al.*,

TABLE 3 Observed and Predicted Aboveground Carbon Density in Patches Representing Tree/Shrub and Grassland Communities at La Copita, Texas

		Carbon Density (kg m^{-2})	
Topoedaphic Setting	Patch Type	Field Estimate	Model Estimate
Sandy loam upland	Cluster	2.9 ± 0.4	2.2
	Grove	6.3 ± 0.8	4.0
	Grassland[a]	0.05 ± 0.00	0.04
Clay loam lowland	Woodland	5.8 ± 0.8	4.5

Note. Observed data (means ± SE) are based on belt transects and plant size–biomass relationships for woody communities (Archer and Boutton, unpublished) and on clipped plots in grasslands (Archer, unpublished). Predicted values are CENTURY estimates for 100-year-old patches (Hibbard, 1995).

[a] Peak aboveground biomass.

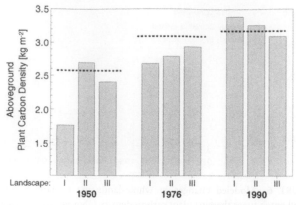

FIGURE 7 Changes in aboveground plant carbon density on three landscapes at the La Copita from 1950 to 1990 (Archer and Boutton, unpublished). Patch/soil-specific field estimates of plant carbon density (Northup *et al.*, 1996, McMurtry, Nelson and Archer, unpublished) were multiplied by patch area as measured in aerial photographs to generate whole-landscape estimates. Dashed lines denote predictions from linked CENTURY–succession model (Hibbard, 1995). See Figure 4 for changes in woody cover on the three landscapes.

1990; Pilke and Avissar, 1990; Graetz, 1991). These changes in vegetation may also influence atmospheric oxidizing capacity, aerosol burden, and radiative properties by affecting emissions of nonmethane hydrocarbons (NMHCs) such as terpenes, isoprene, and other aromatics (Fehsenfeld *et al.*, 1992). There are many sources of atmospheric NMHCs, but >90% of the global annual emission is from vegetation (Guenther *et al.*, 1995). NMHC emissions are therefore highly dependent on species composition as constrained by environmental conditions which influence plant physiology and production. The high temperatures and solar radiation fluxes associated with subtropical and tropical grasslands and savannas make these geographically extensive bioclimatic regions large potential sources of biogenic NMHC emissions. However, grasses are typically low emitters of NMHCs, whereas emissions from trees and shrubs in forest systems are highly variable, with some species being low emitters and other species being high emitters.

We hypothesized that foliar emissions of NMHCs in woody plants would be positively correlated with leaf longevity and inversely related to photosynthetic capacity. Plants characterized by low photosynthetic capacities and slow growth rates (e.g., evergreens) depend on extended leaf longevities to achieve a positive carbon balance. Preferential allocation to secondary compounds such as terpenes would help ensure foliage longevity by reducing levels of herbivory. Species with low photosynthetic capacities and high levels of secondary compounds should also dominate understory environments where low light levels preclude high growth rates and where plants are more accessible to browsers. In contrast, species selected for competitive ability would have high photosynthetic rates, high growth rates, and high rates of tissue turnover (e.g., deciduous shrubs). Allocation to secondary compounds that deter herbivory would be of lower priority since leaf

longevity is less critical to realizing a positive return on foliar investments. Such plants would be expected to preferentially allocate resources such as nitrogen to the carboxylating enzyme and productive tissues rather than to structural tissues or secondary compounds such as terpenes, and would thus be low NMHC emitters (or isoprene emitters, since isoprene is not known to be associated with defense; Coley *et al.*, 1985).

To test these hypotheses, we screened plant species representing the major growth forms at La Copita for NMHC emissions. As expected, grasses had low NMHC emission rates and several common woody species had high emission rates (Guenther *et al.*, 1999). However, there was little evidence of emissions being consistently related to woody plant taxonomy, growth form, or functional group. As a result, generalizations regarding NMHC emissions spectra for tree/shrub species assemblages in other systems do not appear feasible.

To determine if biogenic NMHC emissions have been altered as a result of the change in land cover from grass to woody plant domination at La Copita, a vegetation change model (Scanlan and Archer, 1991) was then linked with a model which predicted NMHC emissions as a function of foliar density, leaf temperature, and photosynthetic photon flux density as modulated by ambient temperature, cloud cover, precipitation, relative humidity, and wind speed (Guenther *et al.*, 1995; Guenther, 1997).

Linkage of the biogenic emissions model with the plant succession model indicated that land cover change since the early 1800s has elicited a threefold increase in isoprene emissions (Fig. 8). This increase reflected changes in vegetation composition and increases in foliar density. Model predictions of current NMHC emissions were within 20% of those measured by a tower flux system. Detailed field measurements on two common shrub species indicated that isoprene emission increased exponentially with increases in leaf temperature from 20 to 40°C and were not suppressed by drought stress. Accordingly, the model predicted that under a projected 2X-CO_2 climate, present-day biogenic NMHC emissions would double.

These estimates of changes in NMHC emissions associated with the conversion of grassland to woodland are in accordance with estimates in other ecosystems. For example, Klinger *et al.* (1998) documented a fourfold increase in total terpenoid emissions per unit foliar mass along a savanna to woodland transect in Central Africa. These changes in NMHC emissions associated with vegetation change in subtropical Texas and tropical Africa also mirror those reported for temperate forest (Martin and Guenther, 1995). Together, these results indicate the magnitude of change in NMHC emissions that could occur when climate and vegetation composition are altered. The importance of these increases in NMHC emissions is magnified at La Copita, as they occur in conjunction with elevated nitric oxide (NO) emissions from shrub-modified soils (Table 2; see Sec. 3.3.4 for elaboration).

Why are vegetation-induced increases in NMHC of concern? Biogenic hydrocarbons play an important role in generating pollutants such as O_3, CO, and organic peroxides, while influencing hydroxyl radical (OH^-) chemistry to reduce atmospheric oxida-

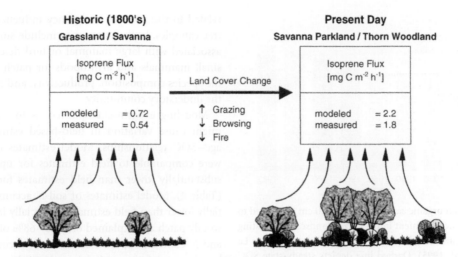

FIGURE 8 Changes in nonmethane hydrocarbon (isoprene) emissions predicted to accompany a shift from savanna grassland to a savanna woodland at the La Copita site in southern Texas (based on Guenther *et al.*, 1999). Predictions from a coupled succession–NMHC emission model are compared with values measured from flux towers. The "measured" values shown for the historic landscape are from a tower located in a savanna grassland landscape with low woody cover.

tion capacity and increase the residence time of greenhouse gases. It has been estimated that to meet current air quality standards for tropospheric ozone, anthropogenic hydrocarbon emissions would have to be reduced by only 30% in the absence of natural isoprene emissions, but by 70% in the presence of them (Monson *et al.*, 1991). Changes in NMHC–NO emissions associated with regional conversion of grassland to shrubland may therefore constitute a "moving baseline" from which to gauge tropospheric ozone production triggered by emissions from automobiles or industrial sources.

3.3.3 Soil C and N Pools

Once established, woody plants alter soils and microclimate in their immediate vicinity to affect both pool sizes and flux rates of nutrients. The result is the formation of "islands of fertility," a phenomenon which has been widely quantified in drylands (see Charley and West, 1975; Schlesinger *et al.*, 1990; Scholes and Archer, 1997; special issue of *Biogeochemistry* 42 (1/2) 1998). Three general mechanisms have been proposed to account for this (e.g., Virginia, 1986): (1) woody plants act as nutrient pumps, drawing nutrients from deep soil horizons and laterally from areas beyond the canopy, depositing them beneath the canopy via stem flow, litterfall, and canopy leaching; (2) tall, aerodynamically rough woody plant canopies trap nutrient-laden atmospheric dust that rain washes off the leaves and into the subcanopy soil; and (3) woody plants may serve as focal points attracting roosting birds, insects, and mammals seeking food, shade, or cover. These animals may enrich the soil via defecation and burrowing. For these reasons, soil carbon and nitrogen pools should increase subsequent to woody plant colonization in grazed grasslands.

At La Copita, surficial (0–10 cm) soils associated with woody plants known to have encroached over the past century have a lower bulk density, contain more root biomass, have higher concentrations of SOC and total N, and have greater rates of respiration and N mineralization than soils associated with the remaining grazed grassland communities (Table 2). As the continuity of woody plant cover increases through time, the landscape-scale soil nutrient pools and fluxes would be expected to increase and become more homogeneously distributed. Accordingly, the linked CENTURY–succession model exercise (see Section 3.3.1) predicted that by 1950, landscape-scale SOC had returned to levels which would have occurred had the "pristine" grasslands been maintained on the site (Fig. 9). By the early 1990s, landscape-scale SOC levels were about 10% higher than those expected for the "pristine" grassland, and about 30% higher than those for a heavily grazed grassland not experiencing woody plant encroachment. Forward model projections suggest SOC aggradation will continue for several hundred years, reaching equilibrium levels three times those of the present-day grazed grassland communities.

While the "island of fertility" phenomenon has been widely recognized, little is known of the rates of nutrient enrichment in tree-dominated patches. Total C and N in soil under *Acacia senegal* and *Balanites aegyptiaca* tree canopies were positively correlated with tree girth ($r^2 = 0.62$ and 0.71, respectively; Bernhard-Reversat, 1982), indicating net accumulation with time of woody plant occupancy of a patch. In temperate old fields undergoing forest succession, carbon storage increased 40% in plant + soil pools over 40 years (Johnston *et al.*, 1996). At La Copita, soil C and N were quantified under *Prosopis glandulosa* trees whose age was determined by annual ring counts. Soil organic carbon storage (top 20 cm of soil) increased linearly with tree stem age at

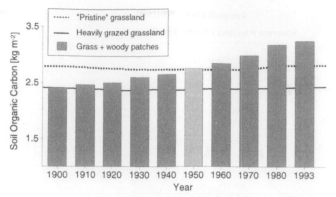

FIGURE 9 Changes in soil organic carbon (SOC; 0–20 cm) predicted to accompany woody plant encroachment into a grazed landscape consisting of a sandy loam uplands and clay loam intermittent drainages at the La Copita site (from Hibbard, 1995). Dashed line depicts steady-state SOC expected for a lightly grazed grassland landscape (upland + lowland communities pooled) with fire at 10-year intervals and no woody plants; solid line depicts steady-state SOC for heavily, continuously grazed grassland landscape with no fire and no woody plants (see Fig. 3). Bars denote SOC summed across the entire landscape and include both grassland and woody plant communities. Note that by 1950, SOC levels had increased to a level comparable to that of the "pristine" grassland (cross-hatched bar). Changes in woody plant abundance on each soil type were directed by a succession model (Scanlan and Archer, 1991); subsequent changes in soil carbon were then assessed with a biogeochemistry model (CENTURY; Parton *et al.*, 1994).

rates ranging from 11.8 to 21.5 g C m^{-2} $year^{-1}$ in sandy loam uplands woody patch types to 47.2 g C m^{-2} $year^{-1}$ in moister, clay loam woodland patches (Table 4). Rates of total N accumulation (top 20 cm of soil) ranged from 1.9 to 2.7 g N m^{-2} $year^{-1}$ in sandy uplands and averaged 4.6 g N m^{-2} $year^{-1}$ in clay loam lowlands. However, woody plant age explained only 21–68% of the variation in soil C and N sequestration rates. These low r^2 values may indicate that tree stem ages do not accurately reflect plant ages, possibly due to past disturbance and subsequent vegetative regeneration of woody cover. Low r^2's may also indicate that factors un-

related to time of tree occupancy influence soil C and N under tree canopies. Such factors may include small-scale heterogeneity associated with large mammal or bird defecation, soil mixing by small mammals and arthropods, or patch-specific differences in the species composition, productivity, and rate of development of the understory community.

Modeling experiments allowed us to control for factors that might cause variation in field-based estimates of woody plant age–SOC relationships. Model estimates of SOC accumulation were comparable to field estimates for upland patch types and substantially lower than field estimates for lowland patch types (Table 4). Model estimates of soil N accumulation were substantially lower than field estimates, especially in lowlands. Given that woody patch age explained only 26–68% of the variance in soil C and N content, our field estimates of accumulation rates cannot be taken as definitive. Model results underestimated field observations, especially for N. Reliability of model estimates of soil carbon could likely be improved with a better understanding of how turnover of the substantial root mass (Table 2) might differ among patch types. Model estimates of soil N are likely constrained by lack of information on inputs associated with N_2 fixation, atmospheric N deposition, translocation between uplands and lowlands, and root turnover.

3.3.4 Soil C and N Dynamics

Increases in the C and N pools of soils associated with woody plant communities developing on grazed grasslands at La Copita have been accompanied by increases in soil respiration, N mineralization, and nitric oxide (NO) emissions (Table 2). The increase in NO fluxes accompanying expansion of woody plants into grasslands at La Copita is noteworthy. Nitric oxide plays several critical roles in atmospheric chemistry by contributing to acid rain and by catalyzing the formation of photochemical smog and tropospheric ozone. The latter is potentially accentuated in the La Copita setting, since NO and hydrocarbon emissions (see Sec. 3.3.2) are concomitantly elevated subsequent to woody plant establishment.

The quality and quantity of organic matter inputs interact to drive soil metabolic activity (Zak *et al.*, 1994). Hence, annual soil

TABLE 4 Estimated Rates of Organic Carbon and Total Nitrogen Accumulation in Soils (0- to 20-cm Depth) Developing Beneath Woody Plants Establishing on a Former Grassland

| Location | Soil Texture | Patch Type | g C m^{-2} $year^{-1}$ | | g N m^{-2} $year^{-1}$ | |
			Field	Modeled	Field	Modeled
Upland	Sandy loam	Shrub Cluster	21.5 ($r^2 = 0.26$)	18.5	2.67 ($r^2 = 0.45$)	1.15
		Grove	11.8 ($r^2 = 0.21$)	10.5	1.90 ($r^2 = 0.51$)	0.87
Lowland	Clay loam	Woodland	47.2 ($r^2 = 0.57$)	13.1	4.64 ($r^2 = 0.68$)	0.58

Note. Field data are from linear correlations between patch age (determined by dendrochronology) and soil C and N mass (Boutton and Archer, unpublished). Model estimates are from CENTURY simulations (Hibbard, 1995). Descriptions of contrasting woody patch types can be found in Archer (1995b).

respiration rates are positively correlated with net primary productivity (Raich and Schlesinger, 1992). The elevated carbon fluxes observed with the development of woody communities in semiarid La Copita grasslands may reflect increased root (Table 2) and leaf biomass inputs and enhancement of soil moisture beneath woody plant canopies (via concentration of rainfall from stem flow, hydraulic lift, and/or reduced evaporation). Together, these biotic and abiotic factors may interact to stimulate microbial activity relative to that in grass-dominated soils. In fact, microbial biomass in woody communities is comparable to or higher than that in grassland communities at La Copita (McCulley, 1998). However, experimental irrigation, which alleviated plant water stress, enhanced photosynthesis (McMurtry, 1997), and increased soil respiration, elicited a decrease in soil microbial biomass. This suggests the elevated soil respiration observed in woody plant communities at La Copita may be a consequence more of changes in root biomass (Table 2) and respiration than of changes in microbial biomass and activity.

To estimate landscape-scale changes in soil CO_2 flux, we multiplied patch/soil-specific estimates of annual soil respiration (McCulley, 1998) by patch area. We then computed changes in patch area with a succession model (Scanlan and Archer, 1991). Landscape-scale soil respiration (kg C ha^{-1} year^{-1}) is projected to have increased from 6687 (200 YBP) to 7377 (1990s) to 7602 (200 years in future)(Table 5). This represents a 10.3% increase with the transition from historic grassland savanna to the present-day savanna parkland–thorn woodland complex, with an additional 3% increase occurring if the present savanna parkland progresses

to woodland. If mean annual temperatures increase as projected in general circulation models, further increases in soil respiration would be expected (all other factors being equal). Indeed, Q_{10} values for soil respiration in woody plant communities (1.4, 2.7, and 2.3 in cluster, grove, and woodland types, respectively) exceed those of grazed grasslands (1.2) at La Copita (McCulley, 1998). This suggests that if future temperature changes occur, the importance of recent and projected future vegetation changes on soil respiration will be further magnified. For example, the magnitude of increase in soil respiration from past grassland savanna with MAT of 22.4°C to future woodland with MAT of 28.4°C would be 22.5% (= 6687–8197 kg C ha^{-1} year^{-1}; based on Raich and Schlesinger, 1992) to 99.3% (= 6687–13,328 kg C ha^{-1} year^{-1}; based on McCulley, 1998)(Table 5). Potential changes in the amount, seasonality, and effectiveness of rainfall would have important, but as yet unknown, effects on these projections.

3.3.5 Soils as Sources and Sinks

Elevated fluxes of C and N from plants and soils following grassland-to-woodland conversion at La Copita suggest a potential for augmenting greenhouse gas accumulation and altering tropospheric chemistry, particularly if woody plant encroachment has been geographically widespread (as suggested in Table 1). However, as noted in Sec. 3.3.3, organic C and N have accumulated in soils of developing woody plant communities at La Copita, despite elevated fluxes and higher turnover rates. This indicates that inputs have exceeded outputs and that soils and vegetation at La

TABLE 5 Projected Landscape-Scale Changes in Annual Soil Respiration (SR; kg C ha^{-1} year^{-1}) Accompanying Succession from an Open Savanna/Grassland to Woodland and Potential Changes in Mean Annual Temperature

	Landscape-Scale Soil Respiration (kg C ha^{-1} year^{-1})		
Mean Annual Temperature (°C)	Past Grassland (200 YBP)[a]	Present Savanna Parkland/Woodland Complex	Future Thorn Woodland (200 YAP)[b]
A. Based on MAT/MAP regression in Raich and Schlesinger (1992)			
22.4	6687	7377	7602
25.4	6948	7666	7899
28.4	7209	7954	8197
B. Based on Q_{10} values from McCulley (1998)			
25.4	8083	9938	10,465
28.4	9480	12,499	13,328

Note. Patch- (grass and various woody communities) and soil-specific SR rates measured monthly over an annual cycle at La Copita (McCulley, 1998) were multiplied by the area of respective community types (Scanlan and Archer, 1991). Effects of mean annual temperature change (MAT, °C) on SR were estimated from (A) equations in Raich and Schlesinger (1992); for La Copita (MAT = 22.4°C and MAP = 720 mm) a 3 and 6°C increase in MAT would produce a 3.9 and 7.8% increase, respectively, in soil respiration; and (B) Q_{10} values of *in situ*, community-specific soil respiration from McCulley (1998). Estimates are probably conservative, as respiration rates used in computations were measured during a below-normal rainfall year.

[a] YBP, years before present.

[b] YAP, years after present.

Copita have been functioning as C and N sinks over the past century. A variety of factors might interact to account for the observed increases in soil C and N pools:

- The trees and shrubs which have displaced grasses may be more productive aboveground and belowground and hence deliver more organic matter into soils (see root biomass and ANPP in Table 2).

- Leaves of leguminous and nonleguminous woody plants at La Copita have higher [N] than grasses (2–4% vs <1%; Archer, unpublished). However, woody plants in these landscapes are seldom browsed by livestock or wildlife, suggesting high concentrations of secondary compounds. This could result in a significant litter quality × quantity interaction, whereby
  - •• a large fraction of the foliar biomass produced by trees and shrubs goes into the soil pool and directly as litter rather than through the herbivory pathway, and
  - •• a larger fraction of foliar biomass inputs from woody plants may be resistant to decomposition.

- Woody litter inputs and the coarser, more lignified roots of shrubs would promote C and N accumulation compared to that of grass roots and shoots.

- Shading by tree/shrub canopies reduces soil temperatures relative to those in grassland (Archer, 1995b), thus constraining potential mineralization (Q_{10} effect).

- Nitrogen accumulation is potentially a consequence of N_2 fixation by leguminous shrubs common to the site (*P. glandulosa* and several *Acacia* spp.) and/or the uptake and lateral translocation of N from grassland patches. While nodulation has been induced in controlled environments and observed under field conditions at the La Copita site (Zitzer *et al.*, 1996) and elsewhere (Virginia *et al.*, 1986; Johnson and Mayeux, 1990), methodological constraints have prevented quantification of N_2 fixation (Handley and Scrimgeour, 1997; Liao *et al.*, 1999). Root distribution studies (Watts, 1993) discount the lateral foraging hypothesis.

- La Copita is within ca. 70 km of a major oil refinery center (Corpus Christi, TX) and atmospheric N deposition has likely been significant over the past 50–75 years (e.g., Holland *et al.*, 1999). Increased N availability may have promoted woody plant expansion (e.g., Köchy, 1999) by alleviating grass–woody plant competition for soil N and by promoting growth of woody plants more than that of grazed grasses. This, in turn, may have translated into greater organic C and N inputs into soils associated with woody plants.

3.3.6 An Uncertain Future

Prosopis glandulosa currently dominates the overstory in upland and lowland woody plant communities. Depending on patch type, it constitutes 40–90% of the above-ground biomass (Archer and Boutton, unpublished) and 30–70% of the coarse root (>1-mm diameter) biomass (Watts, 1993). As such, the dynamics of *P. glan-*

dulosa must be a primary driver of changes in plant and soil C and N stocks at La Copita. Future increases in landscape nutrient pools and fluxes will reflect a combination of (a) continued growth of *P. glandulosa* and associated shrubs in existing woody plant communities and (b) expansion of woody plants into the remaining grasslands.

How likely is continued expansion? That may depend on land management practices. Relaxation of grazing pressure could enable grass biomass to accumulate and fire (prescribed or natural) to occur. Together, these could retard expansion and growth of woody plants. However, the La Copita appears to have crossed a threshold, whereby soils, seed banks, and vegetative regenerative characteristics are such that reductions in grazing pressure may be of little consequence (Archer, 1996). Relaxation of grazing would influence woody plant establishment in grassland primarily through its influences on the fire regime (Archer, 1995a; Brown and Archer, 1999). However, the remaining herbaceous clearings are small and discontinuously distributed. Hence, even if fine fuels were to accumulate, fires would be highly localized. Such fires might prevent future encroachment into remaining grassland clearings but would not likely convert woody plant communities to grassland, since the trees and shrubs at La Copita quickly regenerate by sprouting after disturbance (Scanlan, 1988; Flinn *et al.*, 1992). Expenses for clearing woody vegetation via mechanical or chemical treatments are prohibitive and generally not cost-effective, especially since the effects of the treatments are relatively short-lived. Thus, the likelihood of continued woody plant dominance is high, even with aggressive land management practices which might favor grasses.

The succession model which simulates the expansion of woody plants into remaining grasslands (Scanlan and Archer, 1991) projects that with heavy grazing and no fire, woody cover will continue to increase until the landscape goes to nearly complete canopy closure. This assumption has been substantiated by field data which indicate extension of lateral roots beyond woody canopies is minimal (Watts, 1993). Hence, there is little opportunity for between-cluster root competition and density-dependent regulation. As a result, tree/shrub densities may continue to increase until all available herbaceous clearings have been occupied and canopy cover is nearly continuous. Accordingly, woody patches on contrasting upland soils and woody patches on uplands that border woody communities of lowlands have grown and coalesced from the 1940s through the 1990s (Archer *et al.*, 1988; Stokes, 1999). However, recent studies suggest that La Copita landscapes may be reaching their carrying capacity for woody plants, owing to topoedaphic constraints (Stroh, 1995; Stokes, 1999). If this is the case, future changes in C and N pools will occur only with growth of plants in existing woody communities. Only time will tell if this is indeed the case.

As the current population of the dominant *P. glandulosa* ages, growth and biomass accumulation rates should slow, unless other woody species compensate. The understory shrubs that colonize beneath the *Prosopis* canopy subsequent to its establishment in grasslands slow *Prosopis* growth and seed production, hasten its mortality (Barnes and Archer, 1998) and prevent its reestablishment (Archer, 1995b). Thus, it appears that *P. glandulosa* will not be a component of future woodlands on this landscape. Assessments to date suggest that over the short-term, loss of *Prosopis* will not adversely affect understory shrub productivity or soil C and N pools (Hibbard, 1995; Barnes and Archer, 1996). However, none of the associated woody species appears to have the genetic potential to achieve the size of mature *Prosopis* plants, either above-or below-ground. Thus, there may be less potential for carbon storage once *Prosopis* is lost from the system, unless the remaining understory species compensate by increasing their productivity. In addition, the carbon currently stored in *Prosopis* biomass would be lost via death and decomposition, albeit rather slowly. It would be interesting to explore these scenarios with a linked CENTURY–succession model. Unfortunately, we know little of the productivity of the understory shrubs. Further, the maximum age of *P. glandulosa* is unknown and we have little basis on which to prescribe mortality from the present-day population.

4. Degradation: Ecological Versus Socioeconomic

Degradation associated with "desertification" or "xerification" in arid environments (West, 1986; Rapport and Whitford, 1999) or "deforestation" in humid environments is in sharp contrast to that associated with "thicketization" of grasslands and savannas in mesic environments (Fig. 2). Desertification and deforestation typically have negative consequences both ecologically and socioeconomically. "Thicketization" has some adverse socioeconomic implications, as it reduces the capacity of rangelands for subsistence or commercial livestock production. However, it does not necessarily represent a degraded system with respect to biodiversity, productivity, nutrient cycling, and other important ecological characteristics.

Today's La Copita landscape is clearly different from that of 100–200 YBP, but is it "degraded"? The conceptual model in Figure 10, based on the La Copita case study, proposes a degradation phase (Fig. 3) followed by an aggradation phase (Figs. 6 and 9, Table 2) that begins when unpalatable woody plants establish, grow, modify microclimate, and enrich soil nutrients. Present-day landscapes at La Copita are a rich mosaic of productive woodlands and tree-shrub patches interspersed with remnant grass-dominated patches. Current plant and soil C and N mass is substantially greater than that which occurred under "pristine" conditions. In addition, these landscapes are highly resilient following disturbance (Scanlan, 1988; Flinn *et al.*, 1992) and provide habitat for numerous wildlife species, both game and nongame. So, in this case, the system that has developed following an initial degradation phase is now ecologically diverse, productive, and functional. It would seem that it is "degraded" or "dysfunctional" (Tongway and Ludwig, 1997) only with respect to its socioeconomic value for cattle grazing. However, it has other potential socioeconomic values whose realization would necessitate a change from traditional land

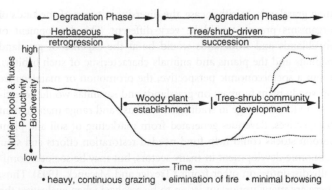

FIGURE 10 Conceptual model of ecosystem changes accompanying grazing-induced succession from grassland to woodland based on the La Copita case study. Dashed lines depict hypothesized upper and lower bounds of Y values; (i.e., lower bound is that which might occur in absence of woody plant encroachment (Fig. 3)); upper bound is that which might occur when woody plant communities mature (e.g., forward projection of Figs. 6 and 9). Values for biodiversity will vary substantially from system to system; in some cases (e.g., *Juniperus* systems) low-diversity monocultures of woody vegetation may develop. At the La Copita, this degradation–aggradation cycle has occurred over ca. 100–150 years.

uses. These include alternative classes of livestock (e.g., goats), lease hunting, charcoal production, and ecotourism.

Given the demonstrated potential for nutrient sequestration in the conversion from grassland or savanna to woodland (Johnston *et al.*, 1996; Scholes and van der Merwe, 1996; Scholes and Bailey, 1996; San José *et al.*, 1998), these lands may also have "carbon credit" value to society (e.g., Glenn *et al.*, 1992). Government or industry subsidies and payments for management practices that promote or maintain woody plant cover on rangelands would stand in sharp contrast to past rangeland management practices that have sought to eliminate or reduce woody vegetation cover using costly and often short-lived chemical or mechanical treatments that may not produce desired results (Belsky, 1996) and may convert landscapes from sinks to sources of greenhouse gases (De Castro and Kauffman, 1998). Thus, the perspective on woody plants in rangelands may shift from negative (an expensive management problem) to positive (a potential commodity).

From a biogeochemical perspective, potential benefits of C and N sequestration should be weighed against the potentially undesirable absolute increases in NO and NMHC fluxes that may accompany increases in woody plant biomass. The effects of vegetation change on the hydrological cycle must also be considered. The extent to which shifts from herbaceous to woody plant domination might reduce stem flow and groundwater/aquifer recharge remains controversial. In addition, potential increases in ecosystem transpiration associated with woody communities with high LAI and deep root systems might increase atmospheric water vapor and either offset (due to radiative properties of water) or augment (via cloud formation) benefits of C sequestration on greenhouse gas budgets. From a biodiversity perspective, shrubland/woodland communities may be more (La Copita scenario) or less diverse (many *Juniperus* communities) than grasslands. In either case, the diversity in terms of the kinds of organisms present would be very different. The development of shrublands and woodlands would be at the expense of grassland habitats and the plants and animals characteristic of such habitats. From a socioeconomic perspective, the promotion or maintenance of woody plant biomass on grasslands and savannas would necessitate a radical change in traditional land use and range management perspectives. Revenues generated from marketing of soil and plant carbon stocks could help fund needed restoration efforts and spur economic development in many sectors, but may be socioeconomically disruptive on other fronts (Trexler and Meganck, 1993). Thus, there are many important issues to be resolved when evaluating the merits of C and N sequestration associated with vegetation change.

5. Implications for Ecosystem and Natural Resources Management

Woody plant encroachment has been and continues to be a major problem in grasslands and savannas worldwide (e.g., Grossman and Gandar, 1989). Because of its direct effects on livestock production,

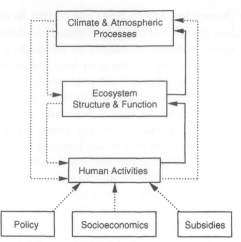

FIGURE 11 Feedbacks between climate and atmospheric processes, ecosystem structure and function, and human activities. Dashed lines depict traditional research and public awareness; solid lines denote areas requiring increased research emphasis. Understanding global change will ultimately hinge on understanding how socioeconomics, policy, and government subsidies influence human activities and land use (see Figs. 12 and 13).

FIGURE 12 LANDSAT image (1978) of the Canada (Saskatchewan/Alberta)–U.S. (Montana) border in the vicinity of the Milk River (from Knight, 1991). Subsequent to the drought and Dust Bowl of the 1930s, farmlands in Canada were repossessed by provincial or federal governments, withdrawn from cultivation, and underwent secondary succession. Intensive agriculture was maintained in the United States via elaborate farm subsidy programs. Striking contrasts in regional land cover were thus a direct result of changes in government policy. See also color insert.

encroachment of woody vegetation into grasslands has been one of the most important problems facing the ranching industry in the western United States and graziers and pastoralists in arid/semiarid regions throughout the world. This structural change in vegetation also has profound effects on the functional properties of ecosystems. Since woody plant encroachment into grasslands is occurring over large areas worldwide, ecosystem–level changes in nutrient pool sizes and fluxes will likely have important ramifications for regional and global biogeochemistry and climate. Thus, the replacement of grassland/savanna ecosystems by woodlands should be viewed not only as a local problem with economic impacts on livestock husbandry, but also in the context of longer–term, regional impacts on biogeochemistry and climate that will influence future land use options in arid and semiarid ecosystems worldwide.

Since the dawn of time, humans have been cognizant of the direct (e.g., catastrophic floods, wind storms, hail) and indirect (e.g., drought effects on food availability) effects of climate on their well-being. Ecosystem science in the 1960s and 70s focused on climatic and abiotic controls over ecosystem structure and function. During this same period, it became increasingly clear that human activities were directly responsible for significant changes in atmospheric chemistry which could feed back to affect ecosystem processes (e.g., acid rain) and human health (e.g., smog). While ecologists and natural resource managers have long been concerned with impacts of humans on ecosystems, we have only recently begun to assess how alterations of ecosystem structure and function might induce changes in climate and atmospheric chemistry, as shown in Figure 11 (Graetz, 1991; Bryant *et al.*, 1990; Pielke *et al.*, 1993, 1998). The case study presented here explicitly documents how human activities (specifically alteration of grazing and fire regimes) have modified the structure and function of a subtropical savanna grassland system in ways that may have significant impacts on climate and atmospheric processes. Anticipating future changes will largely depend on anticipating how human populations and land use will change. Land-use practices will be governed largely by socioeconomic conditions mediated by government policy and subsidies (Figs. 12 (see also color insert) and 13). Thus, the human dimension of global change is paramount and ecologists are challenged to interface ecosystem science with social science (Turner *et al.*, 1990; Walker, 1993a, b; Walker and Steffen, 1993; Vitousek, 1994).

Changes in tree–shrub–grass ratios in drylands have policy implications for federal agencies grappling with designing and implementing carbon sequestration programs. The success of such endeavors will hinge on the ability to quantitatively monitor and inventory "carbon credits" associated with various land management practices. In arid and semiarid ecosystems, this means tracking changes in woody versus herbaceous cover and understanding how shifts between these growth forms influences aboveground and belowground C and N pools and fluxes. However, landscape-scale and regional quantification of grass–woody plant mass is challenging, because woody encroachment occurs relatively slowly (decadal time scales) in a nonlinear fashion, across large and often remote areas, and in a heterogeneous manner determined by topoedaphic constraints, climate, land-use, and disturbance regimes. In addition, reductions in woody biomass also occur in

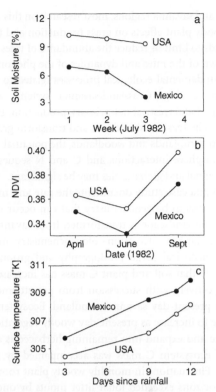

FIGURE 13 Differences in biophysical properties associated with contrasting land management policies in adjacent portions of the United States and Mexico (adapted from Bryant *et al.*, 1990). Relaxation of grazing and range improvement programs occurred in the United States subsequent to the implementation of the Taylor Grazing Act in 1934. Changes in landscape cover resulting from a change in federal policy has had biophysical consequences.

drylands where trees and shrubs are cleared using fire, herbicides, or mechanical means (e.g., roller chopping, chaining). As a result, landscapes within a region may be a mosaic of variable-strength C and N sources and sinks. At present, we lack comprehensive information on the rate of change, areal extent, and pattern of woody plant abundance in the world's drylands. Hence, it is difficult to objectively assess the role of savannas in regional/global C and N cycling. Recent advances in remote sensing show promise for quantifying changes in grass and woody plant biomass in drylands (Asner *et al.*, 1998b). These remote sensing tools, when used in conjunction with simulation modeling (Asner *et al.*, 1998a), will potentially enable functional monitoring of land-use impacts on regional biogeochemistry in savanna regions.

6. Summary

Woody plant encroachment has been widespread in grassland and savanna ecosystems over the past century. This phenomenon jeopardizes grassland biodiversity and threatens the sustainability of pastoral, subsistence and commercial livestock grazing. As such, it may adversely impact ~20% of the world's population. Although woody plant expansion has long been a concern of land managers in grassland and savanna regions, most research on this issue has focused on woody plant effects on grass production and the development methods to limit or reduce the abundance of trees and shrubs. Little is known of the rates and dynamics of the phenomenon or its impact on fundamental ecological processes related to energy flow and nutrient cycling. Grassland/savanna systems account for 30–35% of global terrestrial net primary production. Hence, when woody species increase in abundance and transform grasslands and savannas into shrublands and woodlands, the potential to alter land surface–atmosphere interactions and C and N sequestration and cycling at regional and global scales may be significant.

The La Copita case study documents the rate and magnitude of change in ecosystem biogeochemistry that can occur when a subtropical dryland landscape is transformed from savanna grassland to woodland. Linked succession–biogeochemistry models, confirmed with historical aerial photography and ground measurements, indicate that soil and plant C mass has increased 10% and 10-fold, respectively, with succession from presettlement savanna grassland to present-day savanna woodland. Ecosystem C storage will continue to increase as present-day woody vegetation communities mature and expand into remaining herbaceous areas. Accumulation of ecosystem C mass was accompanied by increases in soil N pools. Fluctuations in monthly woody plant root biomass in upper soil horizons exceeded foliar litter inputs by one to two orders of magnitude, suggesting that belowground inputs of organic matter drive changes in soil physical and chemical properties subsequent to woody plant establishment in grasslands. The deep root systems of woody plants have also increased C mass throughout the soil profile relative to that of grasslands. Increases in C and N pools have occurred in spite of increases in N mineralization, NO flux, soil respiration, and nonmethane hydrocarbon emissions.

These results are of potential global significance, given that large areas of Africa, South America, North America, and Australia have been undergoing similar land cover changes over the past century. The demonstrated capacity for carbon sequestration in this semiarid system suggests a need to reevaluate traditional perspectives on woody plants in rangelands as governments and industries seek ways to mitigate greenhouse gas emissions. However, sequestration of C by woody plants in drylands may come at the expense of elevated NO_x, NMHC, and groundwater fluxes. Regional assessments of the potential consequences of global change are hampered by a lack of quantitative information on the geographic balance between woody plant expansion and clearing in the world's extensive and often remote drylands. Recent developments in linked remote-sensing ecosystem modeling approaches show promise for alleviating these monitoring constraints.

Acknowledgments

We thank Chad McMurtry for assisting with figures and with computation of field estimates in Figure 7. Xiaolian Ren assisted with the image processing and GIS used to generate the data in Figures 4 and 5. This research was supported by grants from NASA-EOS (NAGW-2662), NASA-LCLUC (NAG5-6134), the NASA Graduate Student Research Program, DOE-NIGEC (DE-F03-90ER61010), USDA-NRICGP (96-00842), and the Texas ATRP (999902126).

References

Abrams, M. D. (1986). Historical development of gallery forests in northeast Kansas. *Vegetatio* **65**, 29–37.

Acocks, J. P. H. (1964). Karroo vegetation in relation to the development of deserts. In "Ecological Studies of Southern Africa" (D. H. S. Davis, Ed.), pp. 100–112. Dr. Junk Publishers, The Hague.

Adamoli, J., Sennhauser, E., Acero, J. M., and Rescia, A. (1990). Stress and disturbance: Vegetation dynamics in the dry Chaco region of Argentina. *J. Biogeogr.* **17**, 491–500.

Ambrose, S. H., and Sikes, N. E. (1991). Soil carbon isotope evidence for Holocene habitat change in the Kenya Rift Valley. *Science* **253**, 1402–1405.

Anderson, J. E., and Holte, K. E. (1981). Vegetation development over 25 years without grazing on sagebrush-dominated rangeland in southeastern Idaho. *J. Range Manage.* **31**, 25–29.

Anderson, R. C., Fralish, J. S., and Baskin, J. M. (Eds). (1999). "Savanna, barren and rock outcrop plant communities of North America." Cambridge University Press, New York.

Ansley, J. A., Pinchak, W. E., and Ueckert, D. N. (1995). Changes in redberry juniper distribution in northwest Texas (1948–1982). *Rangelands* **17**, 49–53.

Archer, S. (1989). Have southern Texas savannas been converted to woodlands in recent history? *Am. Nat.* **134**, 545–561.

Archer, S. (1994). Woody plant encroachment into southwestern grasslands and savannas: rates, patterns and proximate causes. In "Ecological Implications of Livestock Herbivory in the West" (M. Vavra, W. Laycock, and R. Pieper, Eds.), pp. 13–68. Society for Range Management, Denver, Colorado.

Archer, S. (1995a). Herbivore mediation of grass-woody plant interactions. *Trop. Grassl.* **29**, 218–235.

Archer, S. (1995b). Tree-grass dynamics in a *Prosopis*-thornscrub savanna parkland: reconstructing the past and predicting the future. *Ecoscience* **2**, 83–99.

Archer, S. (1996). Assessing and interpreting grass-woody plant dynamics. In "The Ecology and Management of Grazing Systems" (J. Hodgson and A. Illius, Eds.), pp. 101–134. CAB International, Wallingford, Oxon, United Kingdom.

Archer, S., Schimel, D. S., and Holland, E. A. (1995). Mechanisms of shrubland expansion: land use, climate or CO_2? *Climatic Change* **29**, 91–99.

Archer, S., Scifres, C. J., Bassham, C. R., and Maggio, R. (1988). Autogenic succession in a subtropical savanna: conversion of grassland to thorn woodland. *Ecol. Monogr.* **58**, 111–127.

Archer, S. and Stokes, C. J. (2000). Stress, disturbance and change in rangeland ecosystems. In "Rangeland Desertification" (O. Arnalds and S. Archer, Eds.), pp. 17–38. Advances in Vegetation Science Vol. **19**, Kluwer Publishing Company, Dordrevht, Netherlands.

Archibold, O. W., and Wilson, M. R. (1980). The natural vegetation of Saskatchewan prior to agricultural settlement. *Can. J. Bot.* **58**, 2031–2042.

Arnalds, O. and Archer, S. (Eds). (2000). "Rangeland Desertification." Advances in Vegetation Science Vol. 19. Kluwer Publishing Company, Dordrevht, Netherlands.

Arno, S. F. and Gruell, G. E. (1986). Douglas fir encroachment into mountain grasslands in southwestern Montana. *J. Range Manage.* **39**, 272–276.

Arno, S. F., Harrington, M. G., Fiedler, C. E., and Carlson, C. E. (1995). Restoring fire-dependent ponderosa pine forests in western Montana. *Restoration and Management Notes* **13**, 32–36.

Arnold, J. F. (1950). Changes in ponderosa pine bunchgrass ranges in northern Arizona resulting from pine regeneration and grazing. *J. Forestry* **48**, 118–126.

Asner, G. P., Bateson, C. A., Privette, J. L., El Saleous, N., and Wessman, C. A. (1998a). Estimating vegetation structural effects on carbon uptake using satellite data fusion and inverse modeling. *J. Geophys. Res.* **103**, 28,839–28,853.

Asner, G. P., Wessman, C. A., and Schimel, D. S. (1998b). Heterogeneity of savanna canopy structure and function from imaging spectrometry and inverse modeling. *Ecol. Appl.* **8**, 1022–1036.

Backéus, I. (1992). Distribution and vegetation dynamics of humid savannas in Africa and Asia. *J. Veg. Sci.* **3**, 345–356.

Bahre, C. J. (1991). "A Legacy of Change: Historic Human Impact on Vegetation of the Arizona Borderlands." University of Arizona Press, Tucson.

Bahre, C. J. and Shelton, M. L. (1993). Historic vegetation change, mesquite increases, and climate in southeastern Arizona. *J. Biogeogr.* **20**, 489–504.

Baker, W. L. and Weisberg, P. J. (1997). Using GIS to model tree population parameters in the Rocky Mountain National Park forest-tundra ecotone. *J. Biogeogr.* **24**, 513–526.

Barnes, P. W. and Archer, S. R. (1996). Influence of an overstorey tree (*Prosopis glandulosa*) on associated shrubs in a savanna parkland: implications for patch dynamics. *Oecologia* **105**, 493–500.

Barnes, P. W. and Archer, S. R. (1998). Tree-shrub interactions in a subtropical savanna parkland: Competition or facilitation? *J. Veg. Sci.* **10**, 525–536.

Bauer, A., Cole, C. V., and Black, A. L. (1987). Soil property comparisons in virgin grasslands between grazed and nongrazed management systems. *Soil Sci. Soc. Am. J.* **51**, 176–182.

Belsky, A. J. (1996). Western juniper expansion: is it a threat to arid northwestern ecosystems? *J. Range Manage.* **49**, 53–59.

Ben-Shaher, R. (1991). Successional patterns of woody plants in catchment areas in a semi-arid region. *Vegetatio* **93**, 19–27.

Bernhard-Reversat, F. (1982). Biogeochemical cycle of nitrogen in a semi-arid savanna. *Oikos* **38**, 321–332.

Bews, J. W. (1917). Plant succession in the thorn veld. *S. Afr. J. Sci.* **4**, 153–172.

Binggeli, P. (1996). A taxonomic, biogeographical and ecological overview of invasive woody plants. *J. Veg. Sci.* **7**, 121–124.

Blackburn, W. H. and Tueller, P. T. (1970). Pinyon and juniper invasion in black sagebrush communities in east-central Nevada. *Ecology* **51**, 841–848.

Blair, W. F. (1950). The biotic provinces of Texas. *Texas J. Sci.* **2**, 93–117.

Bock, J. H. and Bock, C. E. (1984). Effect of fires on woody vegetation in the pine-grassland ecotone of the southern Black Hills. *Am. Midl. Nat.* **112**, 35–42.

Bogusch, E. R. (1952). Brush invasion of the Rio Grande Plains of Texas. *Texas J. Sci.* **4**, 85–91.

Bonan, G. B. (1997). Effects of land use on the climate of the United States. *Climatic Change* **37**, 449–486.

Bond, W. J., Stock, W. D., and Hoffman, M. T. (1994). Has the Karoo spread? A test for desertification using carbon isotopes from soils. *S. Afr. J. Sci.* **90**, 391–397.

Booth, C. A. and Barker, P. J. (1981). Shrub invasion on sandplain country west of Wanaaring, New South Wales. *Soil Conservation Service of New South Wales* **37**, 65–70.

Bossard, C. C. (1991). The role of habitat disturbance, seed predation and ant dispersal on establishment of the exotic shrub *Cytisus scoparus* in California. *Am. Midl. Nat.* **126**, 1–13.

Bossard, C. C. and Rejmanek, M. (1994). Herbivory, growth, seed production, and resprouting of an exotic invasive shrub. *Biol. Conserv.* **67**, 193–200.

Boutton, T., Archer, S., and Midwood, A. (1999). Stable isotopes in ecosystem science: structure, function and dynamics of a subtropical savanna. *Rapid Commun. Mass Spectrom.* **13**, 1–15.

Boutton, T. W., Archer, S. R., Midwood, A. J., Zitzer, S. F., and Bol, R. (1998). $\delta^{13}C$ values of soil organic carbon and their use in documenting vegetation change in a subtropical savanna ecosystem. *Geoderma* **82**, 5–41.

Bowman, D. M. J. S. and Panton, W. J. (1995). Munmarlary revisited: response of a north Australian *Eucalyptus tetrodonta* savanna protected from fire for 20 years. *Aust. J. Ecol.* **20**, 526–531.

Bragg, T. B. and Hulbert, L. C. (1976). Woody plant invasion of unburned Kansas bluestem prairie. *J. Range Manage.* **29**, 19–24.

Branscomb, B. L. (1958). Shrub invasion of a southern New Mexico desert grassland range. *J. Range Manage.* **11**, 129–132.

Bray, W. L. (1901). The ecological relations of the vegetation of western Texas. *Bot. Gaz.* **32**, 99–123; 195–217; 262–291.

Bren, L. J. (1992). Tree invasion of an intermittent wetland in relation to changes in the flooding frequency of the Murray River, Australia. *Aust. J. Ecol.* **17**, 395–408.

Briggs, J. M. and Gibson, D. G. (1992). Effects of burning on tree spatial patterns in a tallgrass prairie landscape. *Bull. Torr. Bot. Club* **119**, 300–307.

Brown, A. L. (1950). Shrub invasions of southern Arizona desert grasslands. *J. Range Manage.* **3**, 172–177.

Brown, D. (1994). The impact of species introduced to control tree invasion on the vegetation of an electrical utility right-of-way. *Can. J. Bot.* **73**, 1217–1228.

Brown, J. H., Valone, T. J., and Curtin, C. G. (1997). Reorganization of an arid ecosystem in response to recent climate change. *Proc. Natl. Acad. Sci.* **94**, 9729–9733.

Brown, J. R. and Archer, S. (1987). Woody plant seed dispersal and gap formation in a North American subtropical savanna woodland: the role of domestic herbivores. *Vegetatio* **73**, 73–80.

Brown, J. R. and Archer, S. (1990). Water relations of a perennial grass and seedling versus adult woody plants in a subtropical savanna, Texas. *Oikos* **57**, 366–374.

Brown, J. R. and Archer, S. (1999). Shrub invasion of grassland: recruitment is continuous and not regulated by herbaceous biomass or density. *Ecology* **80**, 2385–2396.

Brown, J. R. and Carter, J. (1998). Spatial and temporal patterns of exotic shrub invasion in an Australian tropical grassland. *Landscape Ecol.* **13**, 93–103.

Bruce, K., Cameron, G., and Harcombe, P. (1995). Initiation of a new woodland type on the Texas coastal prairie by the chinese tallow tree (*Sapium sebiferum* (l) Roxb). *Bull. Torr. Bot. Club* **122**, 215–225.

Bryant, N. A., Johnson, L. F., Brazel, A. J., Balling, R. C., Hutchinson, C. F., and Beck, L. R. (1990). Measuring the effect of overgrazing in the Sonoran Desert. *Climatic Change* **17**, 243–264.

Bücher, E. H. (1982). Chaco and Caatinga-South American arid savannas, woodlands, and thickets. *In* "Ecology of Tropical Savannas" (B. J.

Huntley and B. H. Walker, Eds.), pp. 48–79. Springer-Verlag, New York.

Bücher, E. H. (1987). Herbivory in arid and semi-arid regions of Argentina. *Revista Chilena de Historia Natural* **60**, 265–273.

Buffington, L. D. and Herbel, C. H. (1965). Vegetational changes on a semidesert grassland range from 1958 to 1963. *Ecol. Monogr.* **35**, 139–164.

Burgess, T. L. (1995). Desert grassland, mixed-shrub savanna, shrub-steppe, or semi-desert scrub? The dilemma of coexisting growth forms. In "The Desert Grasslands" (M. P. McClaran, and T. R. Van Devender, Eds.), pp. 31–67. University of Arizona Press, Tucson.

Burkhardt, J. and Tisdale, E. W. (1976). Causes of juniper invasion in southwestern Idaho. *Ecology* **57**, 472–484.

Burrows, W. H., Beale, I. F., Silcock, R. G., and Pressland, A. J. (1985). Prediction of tree and shrub population changes in a semi-arid woodland. In "Ecolcogy and Management of the World's Savannas" (J. C. Tothill and J. J. Mott, Eds.), pp. 207–211. Australian Academy of Science, Canberra.

Burrows, W. H., Carter, J. O., Scanlan, J. C., and Anderson, E. R. (1990). Management of savannas for livestock production in north-east Australia: contrasts across the tree-grass continuum. *J. Biogeogr.* **17**, 503–512.

Burrows, W. H., Compton, J. F., and Hoffmann, M. B. (1998). Vegetation thickening and carbon sinks in the grazed woodlands of north-east Australia. In "Proceedings, Australian Forest Growers Conference." pp. 305–316. Lismore.

Callaway, R. M. and Davis, F. W. (1993). Vegetation dynamics, fire, and the physical environment in coastal Central California. *Ecology* **74**, 1567–1578.

Canadell, J., Jackson, R. B., Ehleringer, J. R., Mooney, H. A., Sala, O. E., and Schulze, E.-D. (1996). Maximum rooting depth of vegetation types at the global scale. *Oecologia* **108**, 583–595.

Charley, J. L. and West, N. E. (1975). Plant-induced soil chemical patterns in shrub-dominated semi-desert ecosystems of Utah. *J. Ecol.* **63**, 945–964.

Ciais, P., Tans, P. P., Trolier, M., White, J. W. C., and Francey, R. J. (1995). A large northern hemisphere terrestrial CO_2 sink indicated by the $^{13}C/^{12}C$ ratio of atmospheric CO_2. *Science* **269**, 1098–1102.

Cline-Cole, R. A., Main, H. A. C., and Nichol, J. E. (1990). On fuelwood consumption, population dynamics and deforestation in Africa. *World Dev.* **18**, 513–527.

Cole, J. K., Martin, R. E., Holland, E. A., Archer, S. R., Hibbard, K., and Scholes, M. (1996). Nitric oxide fluxes from a subtropical savanna. La Copita Research Area 1996 Consolidated Progress Report CPR-5047, Texas Agricultural Experiment Station.

Coley, P. D., Bryant, J. P., and Chapin, F. S. III. (1985). Resource availability and plant antiherbivore defense. *Science* **230**, 895–899.

Connin, S. L., Virginia, R. A., and Chamberlain, C. P. (1997). Carbon isotopes reveal soil organic matter dynamics following arid land shrub expansion. *Oecologia* **110**, 374–386.

Cook, G. D., Setterfield, S. A., and Maddison, J. P. (1996). Shrub invasion of a tropical wetland: implications for weed management. *Ecol. Appl.* **6**, 531–537.

Cooper, C. F. (1960). Changes in vegetation, structure, growth of southwestern pine forests since white settlement. *Ecol. Monogr.* **30**, 129–164.

Covington, W. W. and Moore, M. M. (1994). Southwestern ponderosa forest structure: changes since Euro-American settlement. *J. Forestry* **92**, 39–47.

Crutzen, P. J. and Andreae, M. O. (1990). Biomass burning in the tropics: impacts on atmospheric chemistry and biogeochemical cycles. *Science* **250**, 1669–1678.

Crutzen, P. J. and Goldammer, J. G. (1993). "Fire in the Environment: The Ecological, Atmospheric, and Climatic Importance of Vegetation Fires." John Wiley, New York.

Cunningham, G. M. and Walker, P. J. (1973). Growth and survival of mulga (*Acacia aneura* F. Muell. ex Benth) in western New South Wales. *Trop. Grassl.* **7**, 69–77.

D'Antonio, C. M. and Vitousek, P. M. (1992). Biological invasions by exotic grasses, the grass/fire cycle, and global change. *Annu. Rev. Ecol. System.* **23**, 63–87.

De Castro, E. A. and Kauffman, J. B. (1998). Ecosystem structure in the Brazilian Cerrado: a vegetation gradient of aboveground biomass, root mass and consumption by fire. *J. Trop. Ecol.* **14**, 263–283.

De Steven, D. (1991). Experiments on mechanisms of tree establishment in old-field succession: seeding survival and growth. *Ecology* **72**, 1076–1088.

Dick-Peddie, W. A. (1993). "New Mexico vegetation: past, present, and future." University of New Mexico Press, Albuquerque, New Mexico.

Distel, R. A. and Boo, R. M. (1996). Vegetation states and transitions in temperate semi-arid rangelands of Argentina. In "Proceedings, Vth International Rangeland Congress, Vol. II" (N. E. West, Ed.), pp. 117–118. Society for Range Management, Salt Lake City, Utah.

Dussart, E., Lerner, P., and Peinetti, R. (1998). Long-term dynamics of two populations of *Prosopis caldenia* Burkart. *J. Range Manage.* **51**, 685–691.

Eissenstat, D. M. and Yanai, R. D. (1997). The ecology of root life spans. *Adv. Ecol. Res.* **27**, 1–60.

Ellis, D. and Schuster, J. L. (1968). Juniper age and distribution on an isolated butte in Garza County. *Southwest. Nat.* **13**, 343–348.

Engle, D. M., Bidwell, T.G., and Moseley, M. E. (1996). Invasion of Oklahoma rangelands and forests by eastern redcedar and ashe juniper. Circular E-947, Coop. Extension Service, Oklahoma State University, Stillwater, Oklahoma.

Fehsenfeld, F., Calvert, J., Fall, R., Goldan, P., Guenther, A. B., Hewitt, C. N., Lamb, B., Liu, S., Trainer, M., Westberg, H., and Zimmerman, P. (1992). Emissions of volatile organic compounds from vegetation and the implications for atmospheric chemistry. *Global Biogeochem. Cycles* **6**, 389–430.

Field, C. B., Behrenfeld, M. J., Randerson, J. T., and Falkowski, P. (1998). Primary production of the biosphere: integrating terrestrial and oceanic components. *Science* **281**, 237–240.

Fisher, C. E. (1950). The mesquite problem in the Southwest. *J. Range Manage.* **3**, 60–70.

Fisher, C. E. (1977). Mesquite and modern man in southwestern North America. In "Mesquite: Its Biology in Two Desert Ecosystems" (B. B. Simpson, Ed.), pp. 177–188. US/IBP Synthesis Series Vol. 4, Dowden, Hutchinson & Ross, New York.

Flinn, R. C., Scifres, C. J., and Archer, S. R. (1992). Variation in basal sprouting in co-occurring shrubs: implications for stand dynamics. *J. Veg. Sci.* **3**, 125–128.

Foster, J. H. (1917). The spread of timbered areas in central Texas. *J. Forestry* **15**, 442–445.

Frank, A. B., Tanaka, D. L., Hofmann, L., and Follett, R. F. (1995). Soil carbon and nitrogen of northern Great Plains grasslands as influenced by long-term grazing. *J. Range Manage.* **48**, 470–474.

Friedel, M. H. (1987). A preliminary investigation of woody plant increase in the western Transvaal and implications for veld assessment. *J. Grassl. Soc. S. Afr.* **4**, 25–30.

Gadgill, M. and Meher-Homji, V. M. (1985). Land use and productive potential of Indian savannas. In "Ecology and Management of the World's

Savannas" (J. C. Tothill and J. J. Mott, Eds.), pp. 107–113. Australian Academy of Science, Canberra.

Gardener, G. J., McIvor, J. G., and Williams, J. (1990). Dry tropical rangelands: solving one problem and creating another. In "Australian Ecosystems: 200 Years of Utilization, Degradation and Reconstruction" (D. A. Saunders, A. J. M. Hopkins, and R. A. How, Eds.), pp. 279–286. Ecological Society of Australia, Geraldton, Western Australia.

Gardiner, C. P. and Gardiner, S. P. (1996). The dissemination of chinee apple (*Ziziphus mauritania*): a woody weed of the tropical subhumid savanna and urban fringe of north Queensland. *Trop. Grassl.* **30**, 174–174.

Gibbens, R. P., Beck, R. F., McNeely, R. P., and Herbel, C. H. (1992). Recent rates of mesquite establishment on the northern Chihuahuan Desert. *J. Range Manage.* **45**, 585–588.

Gill, R. A. and Burke, I. C. (1999). Ecosystem consequences of plant life form changes at three sties in the semiarid United States. *Oecologia* **1221**, 551–563.

Glendening, G. E. (1952). Some quantitative data on the increase of mesquite and cactus on a desert grassland range in southern Arizona. *Ecology* **33**, 319–328.

Glendening, G. E. and Paulsen, H. A. (1955). Reproduction and establishment of velvet mesquite as related to invasion of semi-desert grasslands. Technical Bulletin 1127, U.S. Department of Agriculture, Washington, DC.

Glenn, E. P., Pitelka, L. F., and Olsen, M. W. (1992). The use of halophytes to sequester carbon. *Water, Air and Soil Pollution* **64**, 251–263.

Graetz, R. D. (1991). The nature and significance of the feedback of changes in terrestrial vegetation on global atmospheric and climatic change. *Climatic Change* **18**, 147–173.

Grice, A. C. (1996). Seed production, dispersal and germination in *Cryptostegia grandiflora* and *Ziziphus mauritiana*, two invasive shrubs in tropical woodlands of northern Australia. *Aust. J. Ecol.* **21**, 324–331.

Grice, A. C. (1997). Post-fire regrowth and survival of the invasive tropical shrubs *Cryptostegia grandiflora* and *Ziziphus mauritiana*. *Aust. J. Ecol.* **22**, 49–55.

Grimm, E. C. (1983). Chronology and dynamics of vegetation change on the prairie-woodland region of southern Minnesota, USA. *New Phytol.* **93**, 311–350.

Grossman, D. and Gandar, M. V. (1989). Land transformation in South African savannah regions. *S. Afr. Geogr. J.* **71**, 38–45.

Grover, H. D. and Musick, H. B. (1990). Shrubland encroachment in southern New Mexico, USA: an analysis of desertification processes in the American southwest. *Climatic Change* **17**, 305–330.

Gruell, G. E. (1983). Fire and vegetative trends in the northern Rockies: interpretations from 1871–1982 photographs. INT-158, Ogden, Utah, USDA Forest Service Intermountain Research Station General Technical Report.

Guenther, A. (1997). Seasonal and spatial variations in natural volatile organic compound emissions. *Ecol. Appl.* **7**, 34–45.

Guenther, A., Archer, S., Greenberg, J., Harley, P., Helmig, D., Klinger, L., Vierling, L., Wildermuth, M., Zimmerman, P., and Zitzer, S. (1999). Biogenic hydrocarbon emissions and landcover/climate change in a subtropical savanna. *Phy. Chem. Earth (B)* **24**, 659–667.

Guenther, A., Hewitt, C., Erickson, D., Fall, R., Geron, C., Graedel, T., Harley, P., Klinger, L., Lerdau, M., McKay, W., Pierce, T., Scholes, B., Steinbrecher, R., Tallamraju, R., Taylor, J., and Zimmerman, P. (1995). A global model of natural volatile organic compound emissions. *J. Geophys. Res.* **100**, 8873–8892.

Hall, D. O., Ojima, D. S., Parton, W. J., and Scurlock, J. M. O. (1995). Response of temperate and tropical grasslands to CO_2 and climate change. *J. Biogeogr.* **22**, 537–547.

Handley, L. L. and Scrimgeour, C. M. (1997). Terrestrial plant ecology and ^{15}N abundance: the present limits to interpretation for uncultivated systems with orignial data from a Scottish old field. *Adv. Ecol. Res.* **27**, 133–213.

Hao, W. M., Liu, M. H., and Crutzen, P. J. (1990). Estimates of annual and regional releases of CO_2 and other trace gases to the atmosphere from fires in the tropics, based on the FAO statistics for the period 1975–1980. In "Fire in the Tropical Biota" (J. G. Goldammer, Ed.), pp. 440–462. Springer-Verlag, New York.

Harrington, G. N. and Hodgkinson, K. C. (1986). Shrub-grass dynamics in mulga communities of eastern Australia. In "Rangelands: A Resource Under Siege" (P. J. Joss, P. W. Lynch, and O. B. Williams, Eds.), pp. 26–28. Australian Academy of Science, Canberra.

Harrington, G. N., Oxley, R. E., and Tongway, D. J. (1979). The effects of European settlement and domestic livestock on the biological system in poplar box (*Eucalyptus populnea*) lands. *Aust. Rangel. J.* **1**, 271–279.

Hart, R. H. and Laycock, W. A. (1996). Repeat photography on range and forest lands in the western United States. *J. Range Manage.* **49**, 60–67.

Hastings, J. R. and Turner, R. L. (1965). "The Changing Mile: An Ecological Study of Vegetation Change with Time in the Lower Mile of an Arid and Semi-Arid Region," University of Arizona Press, Tucson.

Hendricks, J. J., Nadelhoffer, K. J., and Aber, J. D. (1997). A N−15 tracer technique for assessing fine root production and mortality. *Oecologia* **112**, 300–304.

Hennessy, J. T., Gibbens, R. P., and Tromble, J. M. A. C. M. (1983). Vegetation changes from 1935 to 1980 in mesquite dunelands and former grasslands of southern New Mexico. *J. Range Manage.* **36**, 370–374.

Hibbard, K. A. (1995). Landscape patterns of carbon and nitrogen dynamics in a subtropical savanna: observations and models. Ph.D. dissertation, Texas A&M University, College Station.

Hobbs, R. J. and Mooney, H. A. (1986). Community changes following shrub invasion of grassland. *Oecologia* **70**, 508–513.

Höchberg, M. E., Menaut, J.-C., and Gignoux, J. (1994). The influences of tree biology and fire in the spatial structure of a West African savannah. *J. Ecol.* **82**, 217–226.

Hodgkin, S. E. (1984). Shrub encroachment and its effects on soil fertility on Newborough Warren, Anglesey, Wales. *Biol. Conserv.* **29**, 99–119.

Holland, E. A., Dentener, F. J., Braswell, B. H., and Sulzman, J. M. (1999). Contemporary and pre-industrial global reactive nitrogen budgets. *Biogeochemistry* **46**, 7–43.

Holmes, P. M. and Cowling, R. M. (1997). The effects of invasion by *Acacia saligna* on the guild structure and regeneration capabilities of South African fynbos shrublands. *J. Appl. Ecol.* **34**, 317–332.

Hopkins, B. (1983). Successional processes. In "Tropical Savannas" (F. Bourliere, Ed.), pp. 605–616. Ecosystems of the World Vol. 13, Elsevier, Amsterdam.

Houghton, R. A., Boone, R. D., Fruci, J. R., Hobbie, J. E., Melillo, J. M., Palm, C. A., Peterson, B. J., Shaver, G. R., Woodwell, G. M., Moore, B., Skoles, D. L., and Myers, N. (1987). The flux of carbon from terrestrial ecosystems to the atmosphere in 1980 due to changes in land use: geographic distribution of the global flux. *Tellus* **39B**, 122–139.

Humphrey, R. R. (1953). The desert grassland, past and present. *J. Range Manage.* **6**, 159–164.

Humphrey, R. R. (1958). The desert grasslands: a history of vegetation change and an analysis of causes. *Bot. Rev.* **24**, 193–252.

Humphrey, R. R. (1987). "90 Years and 535 Miles: Vegetation Changes Along the Mexican Border." University of New Mexico Press, Albuquerque.

Humphrey, R. R. and Mehrhoff, L. A. (1958). Vegetation change on a southern Arizona grassland range. *Ecology* **39**, 720–726.

Idso, S. B. (1995). "CO_2 and the Biosphere: the Incredible Legacy of the Industrial Revolution." Special Publication, Department of Soil, Water & Climate, University of Minnesota, St. Paul.

Inglis, J. M. (1964). "A History of Vegetation on the Rio Grande Plains." Texas Parks and Wildlife Department, Austin.

Jackson, R. B., Canadell, J., Ehleringer, J. R., Mooney, H. A., Sala, O. E., and Schulze, E. D. (1996). A global analysis of root distributions for terrestrial biomes. *Oecologia* **108**, 389–411.

Jeltsch, F., Milton, S. J., Dean, W. R. J., and van Rooyen, N. (1997). Simulated pattern formation around artificial waterholes in the semi-arid Kalahari. *J. Veg. Sci.* **8**, 177–188.

Johnsen, T. N. (1962). One-seed juniper invasion of northern Arizona grasslands. *Ecol. Monogr.* **32**, 187–207.

Johnson, H. B. and Mayeux, H. S. (1990). *Prosopis glandulosa* and the nitrogen balance of rangelands: extent and occurrence of nodulation. *Oecologia* **84**, 176–185.

Johnson, K. L. (1987). Sagebrush over time: a photographic study of rangeland change. In "Biology of *Artemisia* and *Chrysothamnus*" (E. D. McArthur and B. L. Welch, Eds.), pp. 223–252. USDA Forest Service General Technical Report INT-200, Ogden, Utah.

Johnson, W. C. (1994). Woodland expansion in the Platte River, Nebraska: Patterns and causes. *Ecol. Monogr.* **64**, 45–84.

Johnston, M. C. (1963). Past and present grasslands of southern Texas and northeastern Mexico. *Ecology* **44**, 456–466.

Johnston, M. H., Homann, P. S., Engstrom, J. K., and Grigal, D. F. (1996). Changes in ecosystem carbon storage over 40 years on an old-field/forest landscape in east-central Minnesota. *Forest Ecol. Manage.* **83**, 17–26.

Jones, F. B. (1975). "Flora of the Texas Coastal Bend." Mission Press, Corpus Christi, Texas.

Kenney, W. R., Bock, J. H., and Bock, C. E. (1986). Responses of the shrub, *Baccharis pteronioides*, to livestock enclosure in southeastern Arizona. *Am. Midl. Nat.* **116**, 429–431.

Klinger, L., Greenberg, J., Guenther, A., Tyndall, G., Zimmerman, P., Bangui, M., Moutsambote, J.-M., and Kenfack, D. (1998). Patterns in volatile organic compound emissions along a savanna-rainforest gradient in Central Africa. *J. Geophys. Res.* **102**, 1443–1454.

Klink, C. A., Moreira, A. G. and Solbrig, O. T. (1993). Ecological impact of agricultural development in the Brazilian *cerrados*. In "The World's Savannas: Economic Driving Forces, Ecological Constraints and Policy Options for Sustainable Land Use" (M. D. Young and O. T. Solbrig, Eds.), pp. 259–282. Man and the Biosphere Vol. 12, Parthenon Publishing Group, Carnforth, United Kingdom.

Knapp, A. K. and Seastedt, T. R. (1986). Detritus accumulation limits productivity of tallgrass prairie. *Bioscience* **36**, 662–668.

Knapp, P. A. and Soule, P. T. (1996). Vegetation change and the role of atmospheric CO_2 enrichment on a relict site in central Oregon: 1960–1994. *Ann. Assn. Am. Geog.* **86**, 387–411.

Knapp, P. A. and Soule, P. T. (1998). Recent *Juniperus occidentalis* (western juniper) expansion on a protected site in central Oregon. *Global Change Biol.* **4**, 347–357.

Knight, C. L., Briggs, J. M., and Nelis, M. D. (1994). Expansion of gallery forest on Konza Prairie Research Natural Area, Kansas, USA. *Landscape Ecol.* **9**, 117–125.

Knight, D. H. (1991). Congressional incentives for landscape research. *Ecol. Soc. Am. Bull.* **72**, 195–203.

Köchy, M. (1999). Grass-tree interactions in western Canada. Ph.D. Dissertation, University of Regina, Regina, Canada.

Küchler, A. W. (1964). "The Potential Natural Vegetation of the Conterminous United States." American Geographical Society, New York.

Le Roux, I. G. (1997). Patterns and rate of woody vegetation cluster development in a semi-arid savanna, Kwazulu-Natal, South Africa. M.S. thesis, Department of Botany, University of Natal.

Lehman, V. W. (1969). "Forgotten Legions: Sheep in the Rio Grande Plains of Texas." Texas Western University Press, University of Texas at El Paso,

Leopold, L. B. (1951). Vegetation of southwestern watersheds in the nineteenth century. *Geog. Rev.* **41**, 295–316.

Liao, J., Boutton, T., Hoskisson, A., McCulley, R., and Archer, S. (1999). Woodland development and the N-cycle of a subtropical savanna parkland: insights from $\delta^{15}N$ of plants and soils. *Abstracts, Ecol. Soc. Am.* 1999, 274.

Loehle, C., Li, B. L., and Sundell, R. C. (1996). Forest spread and phase transitions at forest-prairie ecotones in Kansas, U.S.A. *Landscape Ecol.* **11**, 225–235.

Lonsdale, W. M. (1993). Rates of spread of an invading species—*Mimosa pigra* in northern Australia. *J. Ecol.* **81**, 513–521.

Madany, M. H. and West, N. E. (1983). Livestock grazing-fire regime interactions within montane forests of Zion National Park, Utah. *Ecology* **64**, 661–667.

Martin, P. H. and Guenther, A. B. (1995). Insights into the dynamics of forest succession and non-methane hydrocarbon trace gas emissions. *J. Biogeogr.* **22**, 493–499.

Martin, S. C. (1975). Ecology and management of southwestern semi-desert grass-shrub ranges: the status of our knowledge. RM-156, Ft. Collins, Colorado, USDA/Forest Service Research Paper.

Martin, S. C. and Turner, R. M. (1977). Vegetation change in the Sonoran Desert region, Arizona and Sonora. *Arizona Acad. Sci.* **12**, 59–69.

Mast, J. N., Veblem, T. T., and Hodgson, M. E. (1997). Tree invasion within a pine/grassland ecotone: an approach with historic aerial photography and GIS modeling. *Forest Ecol. Manage.* **93**, 181–194.

Mast, J. N., Veblem, T. T., and Linhart, Y. B. (1998). Disturbance and climatic influences on age structure of ponderosa pine at the pine/grassland ecotone, Colorado Front Range. *J. Biogeogr.* **25**, 743–755.

McBride, J. R. and Heady, H. F. (1968). Invasion of grassland by *Baccharis pilularis* DC. *J. Range Manage.* **21**, 106–108.

McClaran, M. P. and McPherson, G. R. (1995). Can soil organic carbon isotopes be used to describe grass-tree dynamics at a savanna-grassland ecotone and within the savanna? *J. Veg. Sci.* **6**, 857–862.

McCraw, D. J. (1985). Phytogeographyic history of *Larrea* in southwestern New Mexico: illustrating the historical expansion of the Chihuahuan Desert. M.A. thesis, University of New Mexico, Albuquerque.

McCulley, R. (1998). Soil respiration and microbial biomass in a savanna parkland landscape: spatio-temporal variation and environmental controls. M.S. thesis, Texas A&M University, College Station, Texas.

McKinney, L. B. J. (1996). Forty years of landscape change in Attwater's prairie chicken habitat within the Coastal Prairie of Texas. M.S. thesis, Texas A&M University, College Station, Texas.

McLendon, T. (1991). Preliminary description of the vegetation of South Texas exclusive of coastal saline zones. *Texas J. Sci.* **43**, 13–32.

McMurtry, C. R. (1997). Gas exchange physiology and water relations of co-occurring woody plant species in a Texas subtropical savanna. M.S. thesis, Southwest Texas State University, San Marcos.

McPherson, G. (1997). "Ecology and Management of North American Savannas." University of Arizona Press, Tucson, Arizona.

McPherson, G. R., Boutton, T. W., and Midwood, A. J. (1993). Stable carbon isotope analysis of soil organic matter illustrates vegetation change at the grassland/woodland boundary in southeastern Arizona, USA. *Oecologia* **93**, 95–101.

McPherson, G. R., Wright, H. A., and Wester, D. B. (1988). Patterns of shrub invasion in semiarid Texas grasslands. *Am. Midl. Natur.* **120**, 391–397.

Mearns, R. (1995). Institutions and natural resource management: access to and control over woodfuel in East Africa. In "People and Environment in Africa" (T. Binns, Ed.), pp. 103–114. John Wiley, Chichester, United Kingdom.

Menaut, J. C., Gignoux, J., Prado, C., and Clobert, J. (1990). Tree community dynamics in a humid savanna of the Côte-d'Ivoire: the effects of fire and competition with grass and neighbours. *J. Biogeogr.* **17**, 471–481.

Midwood, A., Boutton, T., Archer, S., and Watts, S. (1998). Water use by woody plants on contrasting soils in a savanna parkland: assessment with δ^2H and $\delta^{18}O$. *Plant Soil* **205**, 13–24.

Milchunas, D. G. and Lauenroth, W. K. (1993). Quantitative effects of grazing on vegetation and soils over a global range of environments. *Ecol. Monogr.* **63**, 327–366.

Milchunas, D. G., Lee, C. A., Lauenroth, W. K., and Coffin, D. P. (1992). A comparison of C-14, Rb-86, and total excavation for determination of root distributions of individual plants. *Plant Soil* **144**, 125–132.

Miller, E. A. and Halpern, C. B. (1998). Effects of environment and grazing disturbance on tree establishment in meadows of the central Cascade Range, Oregon, USA. *J. Veg. Sci.* **9**, 265–282.

Miller, F. H. (1921). Reclamation of grasslands by Utah juniper in the Tusayan National Forest, Arizona. *J. Forestry* **19**, 647–657.

Miller, R. F. and Rose, J. A. (1999). Fire history and western juniper encroachment in sagebrush steppe. *J. Range Manage.* **52**, 550–559.

Miller, R. F. and Rose, J. R. (1995). Historic expansion of *Juniperus occidentalis* southeastern Oregon. *Great Basin Nat.* **55**, 37–45.

Miller, R. F. and Wigand, P. E. (1994). Holocene changes in semiarid pinyon-juniper woodlands. *Bioscience* **44**, 465–474.

Moat, D. A. and Hutchinson, C. F. (Eds). (1995). "Desertification in Developed Countries." Kluwer Academic Publishers, London.

Monson, R. K., Guenther, A. B., and Fall, R. (1991). Physiological reality in relation to ecosystem- and global-level estimates of isoprene emission. In "Trace Gas Emissions by Plants" (T. D. Sharkey, E. A. Holland, and H. A. Mooney, Eds.), pp. 185–207. Academic Press, New York.

Mueller-Dombois, D. and Goldammer, J. B. (1990). Fire in tropical ecosytems and global environmental change: an introduction. In "Fire in the Tropical Biota: Ecosystem Processes and Global Challenges" (J. G. Goldammer, Ed.), pp. 1–10. Springer-Verlag, Berlin.

Nelson, J. T. and Beres, P. L. (1987). Was it grassland? A look at vegetation in Brewster County, Texas through the eyes of a photographer in 1899. *Texas J. Agri. Nat. Resources* **1**, 34–37.

Noble, J. C. (1997). "The Delicate and Noxious Scrub: Studies on Native Tree and Shrub Proliferation in Semi-Arid Woodlands of Australia." CSIRO Division of Wildlife and Ecology, Canberra.

Northup, B. K., Zitzer, S. F., Archer, S., and Boutton, T. W. (1996). Direct and indirect estimates of aboveground woody plant biomass. In "La Copita Research Area 1996 Consolidated Progress Report CPR-5047" (J. W. Stuth and S. Dudash, Eds.), pp. 51–53. Texas Agricultural Experiment Station, College Station.

Norton-Griffiths, M. (1979). The influence of grazing, browsing, and fire on the vegetation dynamics of the Serengeti, Tanzania, Kenya. In "Serengeti: Dynamics of an Ecosystem" (A. R. E. Sinclair and M. Norton-Griffiths, Eds.), pp. 310–352. University of Chicago Press, Chicago, Illinois.

O'Connor, T. G. and Roux, P. W. (1995). Vegetation changes (1949–71) in a semi-arid, grassy dwarf shrubland in the Karoo, South Africa: influence of rainfall variability and grazing by sheep. *J. Appl. Ecol.* **32**, 612–626.

Ojima, D. S., Bjorn, O. M. D., Glenn, E. P., Owensby, C. E., and Scurlock, J. O. (1993). Assessment of C budget for grasslands and drylands of the world. *Water, Air, and Soil Pollution* **70**, 95–109.

ORNL. (1998). Terrestrial ecosytem responses to global change: a research strategy. Environmental Sciences Division Publication No. 4821, Ecosytems Working Group, Oak Ridge National Laboratories, Oak Ridge, Tennessee.

Owensby, C. E., Blan, K. R., Eaton, B. J., and Russ, O. G. (1973). Evaluation of eastern redcedar infestations in the northern Kansas Flint Hills. *J. Range Manage.* **26**, 256–260.

Palmer, A. R. and van Rooyen, A. F. (1998). Detecting vegetation change in the southern Kalahari using Landsat TM data. *J. Arid Environ.* **39**, 143–153.

Panetta, F. D. and McKee, J. (1997). Recruitment of the invasive ornamental, *Schinus terebinthifolius*, is dependent upon frugivores. *Aust. J. Ecol.* **22**, 432–438.

Parton, W. J., Ojima, D. S., Cole, C. V., and Schimel, D. S. (1994). A general model for soil organic matter dynamics: sensitivity to litter chemistry, texture, and management. In "Quantitative Modeling of Soil Forming Processes" (R. B. Bryant, Ed.), pp. 147–167. Soil Science Society of America, Madison, Wisconsin.

Parton, W. J., Schimel, D. S., Cole, C. V., and Ojima, D. S. (1987). Analysis of factors controlling soil organic matter levels in great plains grasslands. *Soil Sci. Soc. Am. J.* **51**, 1173–1179.

Parton, W. J., Scurlock, J. M. O., Ojima, D. S., Gilmanov, T. G., Scholes, R. J., Schimel, D. S., Kirchner, T., Menaut, J.-C., Seastedt, T., Gardia Moya, E., Kamnalrut, A., and Kinyamario, J. I. (1993). Observations and modeling of biomass and soil organic matter dynamics for the grassland biome worldwide. *Global Biogeochem. Cycles* **7**, 785–809.

Parton, W. J., Stewart, J. W. B., and Cole, V. C. (1988). Dynamics of C, N, P and S in grassland soils: a model. *Biogeochemistry* **5**, 109–131.

Pielke, R. A. and Avissar, R. (1990). Influence of landscape structure on local and regional climate. *Landscape Ecol.* **4**, 133–155.

Pielke, R. A., Avissar, R., Raupach, M., Dolman, A. J., Zeng, X. B., and Denning, A. S. (1998). Interactions between the atmosphere and terrestrial ecosystems: influence on weather and climate. *Global Change Biol.* **4**, 461–475.

Pielke, R. A., Schimel, D. S., Kittel, T. G. F., Lee, T. J., and Zeng, X. (1993). Atmosphere-terrestrial ecosystem interactions: implications for coupled modeling. *Ecol. Model.* **67**, 5–18.

Polley, H. W. (1997). Implications of rising atmospheric carbon dioxide concentration for rangelands. *J. Range Manage.* **50**, 562–577.

Polley, H. W., Mayeux, H. S., Johnson, H. B., and Tischler, C. R. (1997). Atmospheric CO_2, soil water, and shrub/grass ratios on rangelands. *J. Range Manage.* **50**, 278–284.

Post, W. M. (1993). Uncertainties in the terrestrial carbon cycle. In "Vegetation Dynamics and Global Change" (A. M. Solomon and H. H. Shugart, Eds.), pp. 116–132. Chapman and Hall, New York.

Potter, L. D. and Green, D. L. (1964). Ecology of pondersoa pine in western North Dakota. *Ecology* **45**, 10–23.

Progulske, D. R. (1974). "Yellow Ore, Yellow Hair and Yellow Pine: A Photographic Study of a Century of Forest Ecology," South Dakota State University Extension Service, Brookings.

Raich, J. W. and Schlesinger, W. H. (1992). The global carbon dioxide flux in soil respiration and its relationship to vegetation and climate. *Tellus* **44B**, 81–89.

Ramsay, J. M. and Rose Innes, R. (1963). Some quantitative observations on the effects of fire on the Guinea savanna vegetation of northern Ghana over a period of eleven years. *Afr. Soils* **8**, 41–85.

Rappole, J. H., Russell, C. E., and Fulbright, T. E. (1986). Anthropogenic pressures and impacts on marginal, neotropical, semiarid ecosystems: the case of south Texas. *Sci. Total Environ.* **55**, 91–99.

Rapport, D. J. and Whitford, W. G. (1999). How ecosystems respond to stress: common properties of arid and aquatic systems. *Bioscience* **49**, 193–203.

Reichard, S. H. and Hamilton, C. W. (1997). Predicting invasions of woody plants introduced into North America. *Conserv. Biol.* **11**, 193–203.

Reid, R. S. and Ellis, J. E. (1995). Impacts of pastoralists on woodlands in South Turkana, Kenya: livestock-mediated tree recruitment. *Ecol. Appl.* **5**, 978–992.

Reynolds, H. G. and Glendening, G. E. (1949). Merriam kangaroo rat as a factor in mesquite propagation on southern Arizona rangelands. *J. Range Manage.* **2**, 193–197.

Ringose, S., Chanda, R., Musini, N., and Sefe, F. (1996). Environmental change in the mid-Boteti area of north-central Botswana: biophysical proceses and human perceptions. *Environ. Manage.* **20**, 397–410.

Robinson, T. W. (1965). Introduction, spread and areal extent of saltcedar (*Tamarix*) in the western States. Professional. Paper 491-A, US Geological Survey.

Rogers, G. F. (1982). "Then and Now: A Photographic History of Vegetation Change in the Central Great Basin Desert." Univesity of Utah Press, Salt Lake City.

Rummell, R. S. (1951). Some effects of livestock grazing on ponderosa pine forest and range in central Washington. *Ecology* **32**, 594–607.

Sabiiti, E. N. (1988). Fire behaviour and the invasion of *Acacia sieberiana* into savanna grassland openings. *Afr. J. Ecol.* **26**, 301–313.

San José, J. J. and Fariñas, M. R. (1983). Changes in tree density and species composition in a protected *Trachypogon* savanna, Venezuela. *Ecology* **64**, 447–453.

San José, J. J. and Fariñas, M. R. (1991). Temporal changes in the structure of a *Trachypogon* savanna protected for 25 years. *Acta Oecologia* **12**, 237–247.

San José, J. J., Fariñas, M. R., and Rosales, J. (1991). Spatial patterns of trees and structuring factors in a *Trachypogon* savanna of the Orinoco Llanos. *Biotropica* **23**, 114–123.

San José, J. J. and Montes, R. A. (1997). Fire effects on the coexistence of trees and grasses in savannas and the resulting outcome on organic matter budget. *Interciencia* **22**, 289–298.

San José, J. J., Montes, R. A., and Farinas, M. R. (1998). Carbon stocks and fluxes in a temporal scaling from a savanna to a semi-deciduous forest. *Forest Ecol. Manage.* **105**, 251–262.

Savage, M. and Swetnam, T. W. (1990). Early 19th-century fire decline following sheep pasturing in a Navajo ponderosa pine forest. *Ecology* **71**, 2374–2378.

Scanlan, J. C. (1988). Spatial and temporal vegetation patterns in a subtropical *Prosopis* savanna woodland, Texas. Ph.D. Dissertation, Texas A&M University, College Station.

Scanlan, J. C. and Archer, S. (1991). Simulated dynamics of succession in a North American subtropical *Prosopis* savanna. *J. Veg. Sci.* **2**, 625–634.

Schlesinger, W. H., Reynolds, J. F., Cunningham, G. L., Huenneke, L. F., Jarrell, W. M., Virginia, R. A., and Whitford, W. G. (1990). Biological feedbacks in global desertification. *Science* **247**, 1043–1048.

Schofield, C. J. and Bucher, E. H. (1986). Industrial contributions to desertification in South America. *Trends Ecol. Evol.* **1**, 78–80.

Scholes, R. C. and Bailey, C. L. (1996). Can savannas help balance the South African greenhouse gas budget. *Afr. J. Sci.* **92**, 60–61.

Scholes, R. J. and Archer, S. R. (1997). Tree-grass interactions in savannas. *Annu. Rev. Ecol. Syst.* **28**, 517–544.

Scholes, R. J. and Hall, D. O. (1996). The carbon budget of tropical savannas, woodlands and grasslands. In "Global Change: Effects on Conifer-

ous Forests and Grasslands" (A. I. Breymeyer, D. O. Hall, J. M. Melillo, and G. I. Agren, Eds.), pp. 69–100. SCOPE Vol. 56, John Wiley, Chichester, United Kingdom.

Scholes, R. J. and van der Merwe, M. R. (1996). Sequestration of carbon in savannas and woodlands. *Environ. Professional* **18**, 96–103.

Scholes, R. J. and Walker, B. H. (1993). "An African Savanna: Synthesis of the Nylsvley Study." Cambridge University Press, Cambridge.

Schüle, W. (1990). Landscapes and climate in prehistory: interaction of wildlife, man, and fire. In "Fire in the Tropical Biota" (J. G. Goldammer, Ed.), pp. 273–315. Springer-Verlag, New York.

Schwartz, D., de Foresta, H., Mariotti, A., Balesdent, J., Massimba, J. P., and Girardin, C. (1996). Present dynamics of the savanna-forest boundary in the Congolese Mayombe: a pedological, botanical and isotopic (^{13}C and ^{14}C) study. *Oecologia* **106**, 516–524.

Scott, J. D. (1966). Bush encroachment in South Africa. *S. Afr. J. Sci.* **63**, 311–314.

SCS. (1979). Soil Survey of Jim Wells County, Texas. USDA/Soil Conservation Service & Texas Agricultural Experiment Station, College Station, Texas.

Shantz, H. L. and Turner, B. L. (1958). Photographic documentation of vegetational changes in Africa over a third of a century. College of Agriculture, University of Arizona, Tucson.

Sinclair, A. R. E. and Fryxell, J. M. (1985). The Sahel of Africa: ecology of a disaster. *Can. J. Zool.* **63**, 987–994.

Skarpe, C. (1990). Structure of the woody vegetation in disturbed and undisturbed arid savanna, Botswana. *Vegetatio* **87**, 11–18.

Skarpe, C. (1991). Spatial patterns and dynamics of woody vegetation in an arid savanna. *J. Veg. Sci.* **2**, 565–572.

Skarpe, C. (1992). Dynamics of savanna ecosystems. *J. Veg. Sci.* **3**, 293–300.

Skovlin, J. M. and Thomas, J. W. (1995). Interpreting long-term trends in Blue Mountain ecosystems from repeat photography. General Technical Report PNW-315, USDA Forest Service Pacific Northwest Research Station.

Smeins, F. E. (1983). Origin of the brush problem-a geographical and ecological perspective of contemporary distributions. In "Proceedings, Brush Management Symposium" (K. W. McDaniel, Ed.), pp. 5–16. Texas Tech University Press, Lubbock.

Smeins, F. E., Taylor, T. W., and Merrill, L. B. (1974). Vegetation of a 25-year exclosure on the Edwards Plateau, Texas. *J. Range Manage.* **29**, 24–29.

Smith, A. J. (1975). Invasion and ecesis of bird-disseminated woody plants in a temperate forest sere. *Ecology* **56**, 19–34.

Smith, D. A. and Schmutz, E. M. (1975). Vegetative changes on protected versus grazed desert grassland range in Arizona. *J. Range Manage.* **28**, 453–457.

Snook, E. C. (1985). Distribution of eastern redcedar on Oklahoma rangelands. In "Eastern Redcedar in Oklahoma" (R. F. Wittwer and D. M. Engle, Eds.), pp. 45–52. Publication E-849, Coop. Extension Serv, Division of Agriculture, Oklahoma State University, Stillwater.

Soule, P. T. and Knapp, P. A. (1999). Western juniper expansion on adjacent disturbed and near-relict sites. *J. Range Manage.* **52**, 525–533.

Steinauer, E. M. and Bragg, T. B. (1987). Ponderosa pine (*Pinus ponderosa*) invasion of Nebraska sandhills prairie. *Am. Midl. Nat.* **118**, 358–365.

Steuter, A. A., Jasch, B., Ihnen, J., and Tieszen, L. L. (1990). Woodland/grassland boundary changes in the middle Niobrara valley of Nebraska identified by $^{13}C/^{12}C$ values of soil organic matter. *Am. Midl. Nat.* **124**, 301–308.

Stokes, C. J. (1999). Woody plant dynamics in a south Texas savanna: pattern and process. Ph.D. Dissertation, Texas A&M University, College Station.

Stroh, J. C. (1995). Landscape development and dynamics of a subtropical savanna parkland, 1941–1990. Ph.D. Dissertation, Texas A&M University, College Station.

Thomas, D. B. and Pratt, D. J. (1967). Bush control studies in the drier areas of Kenya. IV. Effects of controlled burning on secondary thicket in upland *Acacia* woodland. *J. Appl. Ecol.* **4**, 325–335.

Tieszen, L. L. and Archer, S. R. (1990). Isotopic assessment of vegetation changes in grassland and woodland systems. In "Plant Biology of the Basin and Range" (C. B. Osmond, L. F. Pitelka, and G. M. Hidy, Eds.), pp. 293–321. Ecology Studies 80 Springer-Verlag, New York.

Tieszen, L. L. and Pfau, M. W. (1995). Isotopic evidence for the replacement of prairie forest in the Loess Hills of eastern South Dakota. In "Proceedings, 14th North American Prairie Conference: Prairie Biodiversity" (D. C. Hartnett, Ed.), pp. 153–165.

Tongway, D. and Ludwig, J. (1997). The nature of landscape dysfunction in rangelands. In "Landscape Ecology, Function and Management: Principles from Australia's Rangelands" (J. Ludwig, D. Tongway, D. Freudenberger, J. Noble, and K. Hodgkinson, Eds.), pp. 49–61. CSIRO Publishing, Melbourne.

Tothill, J. C. and Mott, J. J. (Eds). (1985). "Ecology and Management of the Worlds' Savannas." Australian Academy of Science, Canberra, ACT.

Trexler, M. C. and Meganck, R. (1993). Biotic carbon offset progams: sponsors of or impediment to economic development? *Climate Res.* **3**, 29–136.

Trollope, W. S. W. (1982). Ecological effects of fire in South African savannas. In "Ecology of Tropical Savannas" (B. J. Huntley and B. H. Walker, Eds.), pp. 292–306. Ecological Studies, No. 42, Springer-Verlag, Berlin.

Turner, B. L. I., Clark, W. C., Kates, R. W., Richards, J. F., Mathews, J. T., and Meyers, W. B. (1990). "The Earth as Transformed by Human Action." Cambridge University Press, New York.

Van Vegten, J. A. (1983). Thornbush invasion in a savanna ecosystem in eastern Botswana. *Vegetatio* **56**, 3–7.

Veblem, T. T. and Lorenz, D. C. (1991). "The Colorado Front Range: A Century of Ecological Change." University of Utah Press, Salt Lake City.

Vega, A. J. (1991). Simulating the hydrology and plant growth of south Texas rangelands. Ph.D. Dissertation, Texas A&M University, College Station.

Virginia, R. A. (1986). Soil development under legume tree canopies. *Forest Ecol. Manage.* **16**, 69–79.

Virginia, R. A., Jenkins, M. B., and Jarrell, W. M. (1986). Depth of root symbiont occurrence in soil. *Biol. Fertil. Soils* **2**, 127–130.

Vitousek, P. M. (1994). Beyond global warming: Ecology and global change. *Ecology* **75**, 1861–1876.

Vivrette, N. J. and Muller, C. H. (1977). Mechanism of invasion and dominance of coastal grassland by *Mesembryanthemum crystallinum*. *Ecol. Monogr.* **47**, 301–318.

Walker, B. H. (Ed.) (1987). "Determinants of Tropical Savannas." IRL Press, Oxford, United Kingdom.

Walker, B. H. (1993a). Rangeland ecology—understanding and managing change. *Ambio* **22**, 80–87.

Walker, B. H. (1993b). Stability in rangelands: ecology and economics. In "Proceedings, XVII International Grassland Congress" (M. J. Baker, J. R. Crush, and L. R. Humphreys, Eds.), pp. 1885–1890. New Zealand/Australia, Keeling & Monday, Palmerston North.

Walker, B. H., Ludwig, D., Holling, C. S., and Peterman, R. M. (1981). Stability of semi-arid savanna grazing systems. *J. Ecol.* **69**, 473–498.

Walker, B. H. and Steffen, W. L. (1993). Rangelands and global change. *Rangel. J.* **15**, 95–103.

Wall, T. (1999). Western juniper encroachment into aspen communities in the northwest Great Basin. M.S. Thesis, Oregon State University, Corvallis.

Wang, Y., Cerling, T. E., and Effland, W. R. (1993). Stable isotope ratios of soil carbonate and soil organic matter as indicators of forest invasion of prairie near Ames, Iowa. *Oecologia* **95**, 365–369.

Watts, S. E. (1993). Rooting patterns of co-occurring woody plants on contrasting soils in a subtropical savanna. M.S. thesis, Texas A&M University, College Station.

Weltzin, J. F., Archer, S., and Heitschmidt, R. K. (1997). Small-mammal regulation of vegetation structure in a temperate savanna. *Ecology* **78**, 751–763.

Werner, P. A. (1991). "Savanna Ecology and Management: Australian Perspectives and Intercontinental Comparisons." Blackwell Science, London.

West, N. E. (1986). Desertification or xerification? *Nature* **321**, 562–563.

West, N. E. (1988). Inter-mountain deserts, shrubsteppes and woodlands. In "North American Terrestrial Vegetation" (M. G. Barbour and W. D. Billings, Eds.), pp. 209–230. Cambridge University Press, New York.

West, O. (1947). Thorn bush encroachment in relation to the management of veld grazing. *Rhodesia Agri. J.* **44**, 488–497.

Williams, K., Hobbs, R. J., and Hamburg, S. P. (1987). Invasion of an annual grassland in Northern California by *Baccharis pulularis* sp. *consanguinea. Oecologia* **72**, 461–465.

Wondzell, S. and Ludwig, J. A. (1995). Community dynamics of desert grasslands: influences of climate, landforms and soils. *J. Veg. Sci.* **6**, 377–390.

York, J. C. and Dick-Peddie, W. A. (1969). Vegetation changes in southern New Mexico during the past one-hundred years. In "Arid Lands in Perspective" (W. G. McGinnes and B. J. Goldman, Eds.), pp. 155–166. University of Arizona Press, Tucson.

Yorks, T. P., West, N. E., and Capels, K. M. (1992). Vegetation differences in desert shrublands of western Utah's Pine Valley between 1933 and 1989. *J. Range Manage.* **45**, 569–578.

Young, J. A., Eckert, R. E., and Evans, R. A. (1979). Historical perspectives regarding the sagebrush ecosystem. In "The Sagebrush Ecosystem: A Symposium." pp. 1–13. College of Natural Resources, Utah State University, Logan.

Young, J. A. and Evans, R. A. (1981). Demography and fire history of a western juniper stand. *J. Range Manage.* **34**, 501–506.

Young, M. D. and Solbrig, O. T. (1993). "The World's Savannas: Economic Driving Forces, Ecological Constraints and Policy Options for Sustainable Land Use." Parthenon Publishing Group, Carnforth, United Kingdom.

Zak, D. R., Tilman, D., Parmenter, R. R., Rice, C. W., Fisher, F. M., Vose, J., Milchunas, D., and Martin, C. W. (1994). Plant production and soil microorganisms in late-successional ecosystems: a continental-scale study. *Ecology* **75**, 2333–2347.

Zimmerman, G. T. and Neunschwander, L. F. (1984). Livestock grazing influences on community structure, fire intensity and fire frequency within the douglas-fir/ninebark habitat type. *J. Range Manage.* **37**, 104–110.

Zitzer, S. F., Archer, S. R., and Boutton, T. W. (1996). Spatial variability in the potential for symbiotic N_2 fixation by woody plants in a subtropical savanna ecosystem. *J. Appl. Ecol.* **33**, 1125–1136.

1.10

Biogeochemistry in the Arctic: Patterns, Processes, and Controls

S. Jonasson
Botanical Institute,
Physiological Ecology
Research Group,
University of Copenhagen,
Copenhagen K, Denmark

F. S. Chapin, III
Institute of Arctic Biology,
University of Alaska,
Fairbanks, Alaska

G.R. Shaver
The Ecosystems Center,
Marine Biological
Laboratory, Woods Hole,
Massachusetts

1. Introduction

Biogeochemical cycles of carbon (C) and other elements depend on biological processes, which operate at different rates depending on environmental conditions. In the Arctic, the cycles are generally slower than in most other ecosystems because of climatic constraints and strong seasonality. However, within the Arctic there is also a pronounced variation in ecosystem structure and function on different scales. Some of the biological variability is associated with easily interpreted latitudinal and climatic gradients of progressively shorter growing season and lower temperatures toward the north. For instance, differences of 1 or 2 weeks in the time of snow-melt or a growing-season mean temperature difference of less than 1°C are much more significant in the high Arctic than at lower latitudes with longer and milder growing seasons. This creates latitudinal gradients of, e.g., decreasing stocks of plant biomass and decreasing ecosystem nutrient content and productivity from the southern toward the northern Arctic. The latitudinal constraints, at the same time, make life and ecosystem processes increasingly dependent on stochastic, between-year variation in temperature from lower to higher latitudes.

However, local variations in environment can constrain ecosystem function just as much as large-scale regional differences can, due to the narrow range within which the biota must operate. Such local heterogeneity can translate into gradients over short distances that are just as pronounced as the large-scale climatic constraints (Shaver *et al.*, 1996; Shaver and Jonasson, in press).

In the following, we give a brief biogeochemical characterization of arctic ecosystems, which we contrast with neighboring ecosystems further south. We also discuss the biogeochemical variability among arctic ecosystems, the controls that have led to their formation, and shorter-term controls on their stability. We emphasize those processes and interactions among ecosystem components that are least well understood and that would most greatly improve our understanding of the controls over ecosystem structure and function. Finally, we give a brief summary of responses in organisms to environmental manipulations across a range of contrasting arctic ecosystems. On the basis of these responses, we seek to identify the sources of major ecosystem controls at various levels of complexity and discuss common and distinctive controls of the biogeochemical functioning of the ecosystem types.

2. Tundra Organic Matter

2.1 Distribution of Organic Matter

Compared to other ecosystem types, arctic and alpine vegetation have low plant biomass and stocks of carbon, mainly because of the lack of a tree stratum and the often spotty vegetation. For

TABLE 1 Comparisons of Carbon Pools in Arctic-Alpine Tundra, in the Neighboring Boreal Zone, and the World's Total.

	Area (10^6 km)	Soil (g m^{-2})	Vegetation (g m^{-2})	Soil/Vegetation	Total carbon (10^{12} kg)		
					Soil	Vegetation	Soil + Vegetation
Arctic and Alpine tundra	10.5	9200	550	17	96	5.7	102
Boreal woodlands	6.5	11,750	4150	2.8	76	27	103
Boreal forest	12.5	11,000	9450	1.2	138	118	256
Terrestrial total	130.3	5900	7150	0.8	772	930	1702

The soil pools do not include the most recalcitrant fractions. (After McGuire *et al.*, 1997).

instance, in the Terrestrial Ecosystem Model (TEM), which has been used to simulate regional and global C fluxes, the stock of fixed C per unit area in the arctic and alpine vegetation is assumed to be about 550 g/m^2. This corresponds to 6–7% of the amount per unit area in the neighboring boreal forest and 8% of the global average (McGuire *et al.*, 1997; Table 1). In contrast, excluding the most recalcitrant fractions with turnover times of a millennium or more, the amounts of "reactive" soil C per unit area approach those of the boreal forest and of the transition zone of open woodlands between the forest and the tundra and are about 50% higher than the terrestrial average. These differences in distribution give a soil to plant C ratio of approximately 17 in the arctic/alpine regions, compared with 1.2 in the neighboring boreal forest and a terrestrial average of 0.8 (Table1).

The C content in the soils of the Arctic is usually estimated at about 14% of the total global soil carbon (Post *et al.*, 1982). However, estimates of both soil and vegetation carbon in the northern ecosystems vary considerably, both in estimates across the entire region and in estimates of the content in major vegetation types (Bliss and Matveyeva, 1992; Oechel and Billings, 1992; Gilmanov and Oechel, 1995; Shaver and Jonasson, in press). In spite of the variability, all estimates agree that there is a general trend from the southern to the northern Arctic of decreasing amounts of organic matter incorporated into both soils and plant biomass (Table 2). It also appears that the soil organic matter (SOM) pool decreases from oceanic toward continental regions. For instance, Chris-

tensen *et al.* (1995) found much lower C pools both in wet coastal polygonal tundra and in mesic sedge-grass tundra in the region east of the Yamal Peninsula than in western Siberia. The much lower C pool in east than in west Siberia coincides with lower summer temperature and precipitation, reflecting the higher continentality of the climate. Also, the ^{14}C gradients in soil profiles were steeper in the east, indicating lower C accumulation rates there (Christensen *et al.*, 1999).

Despite these patterns, there is generally more variation among ecosystem types within a given region than within a single ecosystem type across the entire latitudinal range of the Arctic (Shaver and Jonasson, in press; Table 2). For instance, organic matter content and primary production in both vegetation and soil can vary by more than three orders of magnitude across ecosystem types within the same region while the between-region variation in single ecosystem types is less than 10-fold (Shaver and Jonasson, in press). Similarly, the estimated ratio of organic matter content in soil and vegetation between similar (and dominating) ecosystems across the regions varies 2- to 5-fold, while the within-region variation among ecosystem types is at least 200-fold (from less than unity to 40), as in the low Arctic. It is reasonable to assume that the ratio of soil to plant carbon reflects the "end result" of various processes that control atmospheric C sequestration through photosynthesis and net primary production (NPP) on one hand, and organic matter turnover and respiratory C losses in soils and plants on the other. This can be

TABLE 2 Estimates (g/m^2) of Organic Matter Mass in Soil, Plant Biomass, and Net Primary Production (NPP) in Main Arctic Ecosystem Types.

	Soil	Vegetation	NPP	Soil/Vegetation	Soil/NPP	Vegetation/NPP	Area (% of total)
Low arctic							
Tall shrub	400	2600	1000	0.15	0.4	2.6	3
Low shrub	3800	770	375	4.9	10	2.1	23
Tussock/sedge							
dwarf shrub	29000	3330	225	8.7	129	15.8	17
Wet sedge/mire	38750	959	220	40	176	4.3	16
Semidesert	9200	290	45	32	204	6.4	6
High arctic							
Wet sedge/mire	21000	750	140	28	150	5.4	2
Semidesert	1030	250	35	4.1	29	7.1	18
Polar desert	20	2	1	10	20	2.0	15

After Shaver and Jonasson (in press) based on data from Bliss and Matveyeva (1992) and Oechel and Billings (1992).

exemplified by almost identical C fixation and respiratory C loss measured in a high arctic wet sedge ecosystem in northeast Greenland (Christensen *et al.*, 2000) and in a similar ecosystem type in southern low arctic Alaska (Shaver *et al.*, 1998). The relatively small latitudinal differences relative to the local variability suggest that landscape-scale variations in environment and disturbance are stronger than the control associated with large-scale variations in climate.

Part of the large variation in estimates of biomass and NPP is, however, methodological. There is a 5-fold variation among studies and years in estimates of aboveground biomass and NPP at a single site (Toolik Lake tussock tundra)(Epstein *et al.*, 2000). The largest variation reflects whether nonvascular plants were included or excluded and the definition of "live moss." Moss biomass varies 10-fold among studies at that site. However, even aboveground vascular biomass is more variable than can be readily explained by spatial or annual variability and probably depends on where the ground surface is defined by different investigators—a detail seldom described in published methods. Belowground biomass is even more variable among studies, and estimates are highly sensitive to methods and to the care with which roots are separated from organic matter.

There appears to be a bimodal pattern of soil-to-plant-C ratios across the entire Arctic, with large areas of both low and high ratios. About 75% of the Arctic consists of ecosystems with a maximum ratio of 10 and the remaining 25% has a ratio of over 20 (Table 2). Hence, the overall "Arctic mean" of about 13 (Table 2) or cold regions mean of 17 with alpine regions included (Table 1) has limited biological relevance, assuming that the ratios actually reflect biological processes. The differences become even more evident in the ratios of accumulated soil organic matter and net primary production, ranging from less than unity to about 200 (Table 2). Basically, this indicates a pronounced difference in organic matter incorporation and turnover among the different arctic ecosystem types. The differences also indicate that cold regions have a large potential capacity to both sequester and lose C if the climatic conditions change and that these changes affect the controls of C accumulation and loss.

The variations in the ratio of soil organic matter and NPP are closely coupled to variations in snow accumulation patterns and hydrologic regime, which often depends strongly on topography. Wet ecosystem types always have high content of soil organic matter and high SOM/NPP ratio, regardless of latitudinal position. In contrast, wind-exposed and dry ecosystems have lower amounts of soil organic matter and low SOM-to-NPP ratios, except for low arctic semideserts with a very high ratio (Table 2). This suggests that hydrological conditions play an overriding role for the ecosystem processes and that the hydrological regime is the main determinant shaping the ecosystems. The topographic variation in SOM/NPP ratio also indicates that water exerts its effects on carbon storage primarily by restricting decomposition in wet sites rather than by restricting production in dry sites. This coupling between topography and hydrological features creates a mosaic of ecosystem types with distinctly different structure and function

within short distances in the arctic landscape (Shaver and Jonasson, in press).

2.2 Patterns and Controls of Organic Matter Turnover between Ecosystem Types

The present magnitude of soil organic matter accumulation is a function of the balance between organic matter production and decomposition. The decomposition rate generally increases with increased temperature but decreases with soil water saturation at a level where permanent or periodic anoxic conditions are created. On the other hand, net primary production appears to be less affected by anoxia because most plant species in the wettest ecosystems are adapted to anoxic conditions. For instance, the vascular plants have aerenchymatous tissues, which lead atmospheric air from the canopy to the belowground, inundated plant parts, and the plants thereby avoid oxygen depletion (Chapin *et al.*, 1996). The strong constraints on decomposition, but lower constraints on plant productivity in wet tundra, together with high proportions of sphagna of low decomposability, appears to be the main factors explaining the accumulation of soil C in, e.g., wet sedge tundra and arctic mires.

In contrast, it appears that the rates of C incorporation and C loss are much more balanced in the drier ecosystems, leading to low C sequestration, except in low arctic semideserts which have an even higher SOM-to-NPP ratio than the wet ecosystems (Table 2). This shift to high accumulation in the semideserts could be because of drought-limited decomposition, as decomposition rates generally decline below about 200% soil water content (Heal *et al.*, 1981). Furthermore, dry tundra is dominated by slowly growing, evergreen dwarf shrubs with sclerophyllous tissues, which may increase the constraints on decomposition further and lead to the high organic matter accumulation. Indeed, the most productive ecosystem types in tundra are in areas with surface or subsurface flowing water, such as low arctic, riverside willow scrubs in which the tall shrubs have high biomass and high net primary production rate and the soil is rapidly turned over by the decomposers (see Table 2). Hence, the rate of the C cycle in this part of the hydrological gradient is high through a combination of favorable hydrological conditions and lifeform properties of the vegetation, which interact to keep both plant production and microbial decomposition rates high.

This strong hydrologic control on ecosystem structure and function coincides with observations in laboratory experiments of increased soil C turnover with decreasing soil water content and increased water table depth in wet and moist tundra (Billings *et al.*, 1982; Johnson *et al.*, 1996). Indeed, microcosm experiments have shown that water conditions, rather than temperature, exert the main control on C exchange, and that it is ecosystem respiration rather than photosynthesis that is affected (Johnson *et al.*, 1996). However, if the depth to the water table increases and the soil dries, temperature becomes increasingly important for the C balance with large net C losses with increasing temperatures (Billings *et al.*, 1982; Shaver *et al.*, 1998). Also, reported net losses of C from tussock tundra during a series of dry and warm years (Oechel *et al.*, 1993) indicate that the C exchange in moist tundra is controlled most strongly by respiration. For this reason, a wetter climate may increase C sequestration in present mesic and wet systems depending on the balance between precipitation and evapotranspiration. The greater C accumulation rate in the western Siberian tundra then in the east (Christensen *et al.*, 1999) suggests that a combination of moister and warmer conditions may lead to substantially increased C accumulation, particularly in the coldest tundra regions.

Dry tundra may be regulated differently because the low soil water content might limit decomposition (Heal *et al.*, 1981; Oberbauer *et al.* 1996). However, it is uncertain to what extent observed low rates of CO_2 fluxes in dry tundra are due to reduced respiration in microorganisms versus that in plants (Illeris and Jonasson, 1999). In spite of this uncertainty, it appears that predicted changed climatic conditions in the Arctic (IPCC, 1996) can lead to both decreased and increased C sequestration by different arctic ecosystems. It also appears that the same change of environmental conditions may have different effects across ecosystem types and could even lead to different directions of the changes in C balance between neighboring systems.

3. Tundra Nutrients

3.1 Nutrient Distribution and Controls of Nutrient Cycling

The ecosystem pools of mineral nutrients are divided between the above- and belowground biota, the dead organic matter, and an inorganic pool, which probably constitutes the major source of plant-available nutrients. The difference in distribution of nutrients between soil and vegetation is still more pronounced than that in the distribution of C because the soil organic matter generally is enriched in nutrients compared with the vegetation (Jonasson and Michelsen, 1996). For instance, the ratio of C to nitrogen (N) is typically about 20 for soils and 60 for arctic vegetation (McGuire *et al.*, 1995). Hence, the soil N concentration is about three times higher than the concentration in the vegetation. Thus, given a soil-to-vegetation C ratio of 17 (Table 1), the soil-to-vegetation ratio of N in the arctic and alpine regions triples in comparison to the C ratio and reaches about 50.

In spite of the large soil nutrient stores and low requirements of nutrient absorption by the low plant biomass, arctic ecosystems are still sensitive to nutrient inputs. This is because the nutrient mineralization rate, i.e., the transformation of nutrients from organic, plant-unavailable form to inorganic, available form, is low and usually sets the limits to primary production (Nadelhoffer *et al.*, 1992). Any changes in the ecosystems that trigger increased nutrient supply generally lead to increased primary production (Kielland and Chapin, 1992).

The major limiting elements are N, phosphorus (P), or both together, usually with N limitation in dry and mesic ecosystem types and P limitation in wet ecosystems (Shaver and Chapin, 1995).

Hence, while variation in hydrology seems to be the proximate source of large-scale variability in ecosystem structure and function across landscapes, variation in supply rate of inorganic N and P seems to regulate the more detailed structure and function within each ecosystem type.

3.2 Nutrient Mineralization and Plant Nutrient Uptake

The net mineralization of nutrients is about an order of magnitude lower in arctic ecosystems than in the boreal region, largely because of constrained microbial activity (Nadelhoffer *et al.*, 1992). There is also a pronounced seasonal variation in net mineralization. Several studies of both litter and soil organic matter mineralization have shown that winter mineralization is higher than summer mineralization (Giblin *et al.*, 1991; Hobbie and Chapin, 1996; Shaver *et al.*, 1998), which is surprising given the expected decrease of microbial activity rate with decreasing temperature. Furthermore, net N mineralization may even be negative during the growing season; i.e., inorganic N is immobilized instead of being released. For instance, Hobbie and Chapin (1996) found high winter release of N in litter contrasting with immobilization of N during summer in spite of mass loss (Fig. 1). In fact, the total amounts of N in the litter even increased during summer, showing that N was transported into the litter mass. Similarly, five of six ecosystem types in an Alaskan tundra showed negative summer net N mineralization in the SOM, while annual mineralization was positive because of high non-growing-season mineralization rates (Giblin *et al.*, 1991).

A possible reason for this annual pattern of net mineralization could be that the decomposing microorganisms themselves absorbed and immobilized the nutrients they mineralized from the litter and soil organic matter during the summer. In contrast, nutrients could have been released passively from the microbes during winter when their activity was lower and part of the populations probably died, or during repeated freezing and thawing in autumn and spring (Giblin *et al.*, 1991; Schimel *et al.*, 1996). However, most studies have shown a discrepancy between estimated low annual net nutrient mineralization of, e.g., N and a much higher annual plant nutrient uptake (Schimel and Chapin, 1996), which needs to be explained. There are at least three possible explanations: (1) Net mineralization measured in summer only may underestimate annual mineralization, as described above. (2) Nutrient mineralization measured by the "buried bag method," which excludes plant roots, allows microbes to monopolize and immobilize the nutrients, of which a part otherwise would have been taken up by the plants if they had not been denied access. (3) Much of the nitrogen may be acquired by plants in the organic form and bypass the mineralization step. The relative importance of these issues in explaining the discrepancy between measured nutrient mineralization and nutrient uptake is currently unknown.

Strong regulation of plant nutrient availability by the population dynamics of soil microorganisms seems logical because the microbial biomass contains large amounts of nutrients, even in

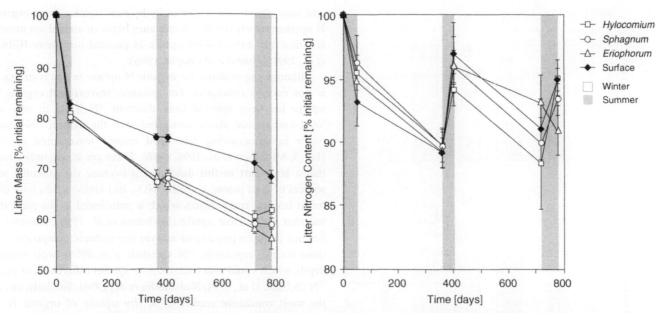

FIGURE 1 Comparison of changes in mass (left) and nitrogen content (right) in *Betula papyrifera* leaf litter in litterbags deposited on the soil surface, in *Hylocomium* or *Sphagnum* moss mats or in *Eriophorum* tussocks during three growing seasons and two nongrowing seasons. (From Hobbie and Chapin, 1996, with kind permisson from Kluwer Academic Publishers).

comparison to the vegetation. For instance, in a subarctic heath, the plant–microbial–SOM pools (to 15 cm depth) contained C in the proportions 19:2.5:78.5, N in the proportions 10:6.5:83.5, and P in the proportions 11:30:59. The proportions of soil inorganic N and P were below 1 (Jonasson *et al.*, 1999a). Hence, while the microbial C pool was much smaller than the plant C pool, the microbial N content approached the N content in plants and microbial P exceeded plant P content. Indeed, the microbial N pool in Alaskan tundra approximately equalled the amount in plant roots (Hobbie and Chapin, 1998).

Microbes contained a relatively constant proportion (2.5–2.7%) of total ecosystem C across seven Alaskan tundra sites, with the quantity of C being determined by two independent methods (Cheng and Virginia, 1993). The N incorporated in the soil microorganisms was about 7% of the total soil N, i.e., a proportion almost identical to that estimated in the subarctic heath (Jonasson *et al.*, 1999a, calculated from Cheng and Virginia, 1993).

The quantities of nutrients in microbes are large compared with the annual plant nutrient uptake, suggesting that even relatively limited dieback of the microbial populations can lead to release of an appreciable proportion of the plants' annual nutrient requirement. Indeed, it is known that the annual uptake of P by wet tundra vegetation can be almost entirely accounted for by P released through nutrient flushes from the microbial biomass (Chapin *et al.*, 1978). It is possible, therefore, that the supply rate of nutrients to the soil inorganic pool varies depending on the conditions for microbial population growth or decline and that plant nutrient availability fluctuates inversely to microbial nutrient demand.

If the annual pattern of net mineralization is regulated mainly by microbial immobilization–mobilization cycles, microbes may

be more effective than plants as competitors for nutrients during periods of the growing season (Harte and Kinzig, 1993; Jonasson *et al.*, 1996; Schimel and Chapin 1996; Schimel *et al.*, 1996). Indeed, a laboratory experiment showed that stimulated microbial activity after addition of a labile C source to the soil increased microbial nutrient uptake to the extent of causing strong limitation of plant growth due to nutrient deficiency (Schmidt *et al.*, 1997a, b). Hence, the stimulated microbial activity led to increased competition for nutrients between soil microbes and plants and not to increased decomposition and release of inorganic nutrients. These observations suggest that labile C regulates the microbial nutrient mineralization by increasing immobilization as the amount of available carbon increases, while net mineralization increases under conditions of microbial C limitation. This indeed corresponds to field observations. A higher rate of net N and P mineralization of low-quality tundra moss litter than of higher quality vascular plant litter has been reported from arctic tundra (Hobbie, 1996) and other comparable nutrient-limited ecosystems (Verhoeven *et al.*, 1990; Updegraff *et al.*, 1995). In contrast, the decomposition rate positively correlated with litter quality and, hence, was negatively correlated with the rate of mineralization, showing that differences in decomposability between litter types do not always result in differences in mineralization rates.

3.3 Are there Unaccounted Plant Sources of Limiting Nutrients?

The assumption hitherto has been that plants only, or almost exclusively, take up N in inorganic form as NH_4^+ or NO_3^-. This assumption has been questioned recently, and there are strong

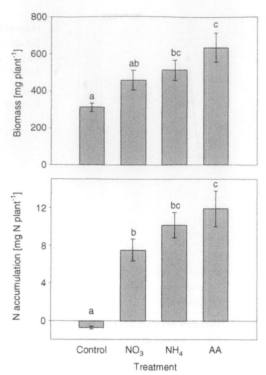

FIGURE 2 Biomass and nitrogen accumulation (means ± SE) in _Eriophorum vaginatum_ grown in NO_3, NH_4, and amino acids (AA) during 24 days. Treatment responses with the same superscript letter above the bars are not significantly different at $p \leq 0.05$. (Redrawn from Chapin _et al._, 1993, with kind permission from Nature).

indications that tundra plants of different life-forms take up organic N either directly or through their mycorrhiza. It has long been known from laboratory experiments that plants having ecto- or ericoid mycorrhiza can access organic N through their fungal partner (Abuzinadah and Read, 1988; Read _et al._, 1989). Chapin _et al._ (1993) and Kielland (1994) showed more recently that some plants can take up amino acids more rapidly than inorganic nitrogen from hydroponic culture in the absence of mycorrhizae (Fig. 2). Many dry and mesic vegetation types in the Arctic are dominated either by ericaceous species with ericoid mycorrhiza, e.g., dry heathlike vegetation types, or by _Salix_ and _Betula_ shrubs that host ectotrophic mycorrhiza and dominate large areas of mesic tundra in the low Arctic. Hence, if organic N uptake also occurs _in situ_, the vegetation over large areas of tundra has the potential capability to access organic N directly, without being dependent on mineralization and possible competition with soil microbes. However, even if plants avoid the mineralization step by absorbing organic N, they are still dependent on soil microbes for protease activity and the solubilization of organic N. Organic N uptake has also been reported in a number of dominant species of graminoid-dominated tundra, ranging from wet _Carex_ sedge meadows, through mesic _Eriophorum vaginatum_ tussock tundra to _Kobresia_ heaths (Chapin _et al._, 1993; Schimel and Chapin, 1996; Raab _et al._, 1996). These species generally lack ectomycorrhizae

and may absorb this N primarily by root uptake. Plant organic N uptake may explain the discrepancy between annual net mineralization and annual plant uptake as pointed out above (Giblin _et al._, 1991; Schimel and Chapin, 1996).

Although the evidence for organic N uptake _in situ_ is strong, it is not entirely conclusive. For instance, mycorrhizal organic N uptake has been inferred from different ^{15}N levels in non- or VAM-mycorrhizal plants, compared with the natural ^{15}N abundance in co-occurring ecto- and ericoid mycorrhizal species (Fig. 3; Michelsen _et al._, 1996, 1998). There are strong indications that a large part of this difference is because the potential soil sources of N to plants, i.e., NH_4, NO_3, and amino acids, have different isotopic composition, which is manifested in the plant tissue after the nutrient uptake (Michelsen _et al._, 1998). However, it has not yet been possible to analyze the isotopic composition of these sources separately. ^{15}N signature also differs with rooting depth, which could also contribute to species differences in plant ^{15}N (Schulze _et al._, 1994; Nadelhoffer _et al._, 1996). Similarly, one of the most conclusive studies of _in situ_ uptake of organic N in graminoids of mesic and wet tundra (Schimel and Chapin, 1996) could not entirely eliminate the possibility that the organic N taken up was first mineralized by microbes within or outside the rhizosphere. However, evidence of organic N uptake is also being reported from other ecosystems, so it appears that this alternative pathway for N uptake is progressively being confirmed (Näsholm _et al._, 1998).

The importance of organic N uptake for plant production in the Arctic is evident because most functional plant groups and species with potential N uptake capacity are those that dominate the vegetation. For instance, _Empetrum hermaphroditum_ or _E. nigrum_ having ericoid mycorrhiza often dominate southern dry arctic vegetation. In more mesic areas _Cassiope tetragona_ with ericoid mycorrhiza, dwarf _Salix_ with ectomycorrhiza or _Eriophorum_

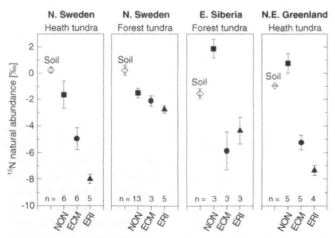

FIGURE 3 Mean $\delta^{15}N$ (± SE) of the bulk soil organic matter and plant species without mycorrhiza (NON), with ectomycorrhizal (ECM) or ericoid mycorrhizal (ERI) fungi at four different heath or forest tundra sites. n is number of species analyzed. (From Michelsen _et al._, 1998, with kind permission from Springer-Verlag).

vaginatum with ability to absorb organic N without mycorrhiza are prominent species across large areas. Similarly, wet tundra sites are dominated by other graminoids which, like *E. vaginatum*, have been shown to absorb amino acids. The effect on the N cycle of these species that are known to absorb organic N needs to be more closely evaluated and quantified.

4. Biogeochemical Responses to Experimental Ecosystem Manipulations

4.1 Applicability of Experimental Manipulations

Experimental manipulations of the environment have been a common way of exploring how tundra ecosystems and their components react to environmental changes. Such experiments have, indeed, given valuable information on environmental controls at various levels of resolution, although of necessity the information is limited by observation series over relatively short time spans. Hence, the time constraints of experimentation set a limit to the observations of long-term responses, which must be sought from other sources and by other means, such as modeling. This may lead to a risk of confounding short-term (years to decade), transient changes with long term processes, and careful consideration must be taken in extrapolations to longer time-scales and in parameterization of the models.

Experimental manipulations are, hence, most valuable for detection of short-term responses within ecosystems, but less useful for detection of larger scale processes and those acting over long time-spans, which structure the tundra landscape. For instance, we are not aware of any experiment in the Arctic, that has been conducted over a time-span long enough to detect changes in soil carbon pools or large-scale hydrological changes.

In spite of these shortcomings, a large part of our knowledge of processes in tundra ecosystems rests on results from manipulations undertaken during the last 20 years, mostly involving water or nutrient additions or changes of light and temperature within well-defined ecosystem types.

4.2 Responses to Water Applications

Water has been applied to several arctic ecosystem types to simulate increased precipitation, with the expectation that the additions would enhance plant production. The production could be stimulated either directly as a response to decreased drought, most likely to occur in dry polar deserts or semideserts (Aleksandrova, 1988; Bliss *et al.*, 1984), or indirectly by enhancement of nutrient supply to plants. For instance, increase in soil moisture facilitates the transport of nutrients toward the plants' roots (Chapin *et al.*, 1988) and creates favorable environments for N fixation (Gold and Bliss, 1995).

Water additions alone have generally not led to any detectable changes in plant productivity or nutrient cycling, even in dry, high

arctic ecosystem types (Press *et al.*, 1998b). The generally low response contrasts with the control of water on community structure, distribution of organic matter, and the turnover of C and mineral nutrients. These differences indicate that water exerts an overall long-term control on ecosystem development and function and that the tundra is buffered against short-term fluctuations in moisture conditions within the levels of the additions, which usually have been within the range of natural between-year variation in precipitation.

However, water addition has in some cases interacted with temperature enhancement or fertilizer addition and increased the productivity of single plant species (Press *et al.*, 1998a). Surprisingly, water even when applied in "moderate" amounts can have negative effect on dry plant communities. Robinson *et al.* (1998) found that combined water and fertilizer addition to a high arctic semidesert caused high winter injury of plants in some years, probably because winter hardening was delayed (Press *et al.*, 1998b).

4.3 Response to Nutrient Addition and Warming

Because productivity in most arctic ecosystems is constrained by low nutrient availability, the most common responses to nutrient addition are increases in nutrient uptake and plant nutrient mass followed by increased plant production and biomass. Similar effects on growth can be expected in response to warming through direct responses of increased productivity in a warmer environment and through enhanced nutrient uptake as a result of increased nutrient supply rate, as mineralization is likely to increase in the warmed soils (Nadelhoffer *et al.*, 1992).

As expected, addition of fertilizer has almost always led to increase in nutrient uptake, tissue nutrient mass, and net primary production. Tissue turnover rates generally have increased because of community changes toward increased proportion of species with short leaf-longevity. In several cases, fertilizer addition has also led to transient responses in biomass. For instance, while the biomass increased during the first two years of fertilizer addition to subarctic, northern Swedish forest-floor vegetation (Parsons *et al.*, 1994), the response did not continue after five years (Press *et al.*, 1998a) because the grass *Calamagrostis lapponica* expanded strongly and affected growth of dwarf shrubs and mosses negatively (Potter *et al.*, 1995; Press *et al.*, 1998a). Similarly, the canopy density and mass of the deciduous *Betula nana* increased in Alaskan tussock tundra over nine years of fertilizer application and reduced the biomass of most other species (Chapin *et al.*, 1995). As a result of the increase of *B. nana* and decline of other species, the mass of vegetation C underwent no, or only small changes. In contrast, NPP increased because species with long leaf-longevity were replaced by the deciduous *B. nana*, implying that the tissue turnover rate also increased. However, aboveground biomass increased strongly after another six years, i.e., after 15 years of treatment, because *B. nana* continued to accumulate biomass (Shaver *et al.*, unpublished data). Much of this increase was due to wood formation resulting in a decline of tissue turnover rate in comparison to the response after nine years. Hence, it appears that the transient responses of the vegetation are coupled to changes in the dominance of single species and particularly of those that form a dense or elevated canopy. Indeed, in vegetation types without any pronounced change in relative proportions of dominant species or life forms following fertilizer addition such as in Swedish treeline and high-altitude heaths and in Alaskan wet sedge tundra, most dominant life forms increased. This resulted in up to a doubling of biomass after 5–9 years of treatment (Jonasson *et al.*, 1999b; Shaver *et al.*, 1998).

The nutrient content in the fertilized vegetation increased strongly in all vegetation types where nutrient analyses were done (no analyses have been done in the forest floor vegetation). In the strongly responding Swedish treeline and high-altitude heaths, the increase in N and P was only slightly higher than the proportional increase in biomass due to relatively small effects on vegetation nutrient concentration, except for a strong increase in nutrient concentration of mosses (Jonasson *et al.*, 1999b). In contrast, the nutrient incorporation in the tussock tundra doubled (N) or tripled (P) after nine years of fertilizer addition, in spite of unchanged biomass, due to an increase of tissue nutrient concentration in the vegetation. The increase of nutrient concentration was particularly high in the mosses, as at the Swedish sites (Chapin *et al.*, 1995). Also the nutrient requirement, i.e., the uptake into the new growth, increased strongly and tripled (N) or increased seven- to eight fold (P) in graminoids and deciduous shrubs, indicating that at least the transient changes in turnover were much more pronounced than the changes in standing stocks.

A different response pattern was found at a polar semidesert. Fertilizer addition increased plant coverage strongly during the first years of treatment but was set back after an exceptionally warm and wet autumn and winter with strong winter injury and high mortality of the dominant *Dryas octopetala*. This effect was probably because winter hardening was delayed in plants that had received extra nutrients (Robinson *et al.*, 1998), and it highlights the importance to the vegetation of "unusual" climatic events. However, the moss cover increased strongly, contrasting with the usually decreased coverage in fertilized Alaskan tundra sites and in the Swedish sites, particularly in those where the canopy density of the vascular plants increased. Indeed, it appears that the cryptogams generally increase in coverage and biomass after fertilizer addition until a point is reached where the negative effects of increasing vascular plant cover and litter override the positive effect of fertilizer addition (Jonasson, 1992; Jonasson *et al.*, 1999b; Molau and Alatalo, 1998; Chapin *et al.*, 1995). This nonlinear response in mosses and lichens has a particular relevance because the cryptogams are important regulators of heat and water exchange between the soil and the atmosphere (Tenhunen *et al.*, 1992; McFadden *et al.*, 1998). At the same time, they affect the N cycle through the N-fixation ability of many dominant lichen species and blue-green algae associated with mosses. They also affect the turnover rates of organic matter because of low decomposability of their tissues (Hobbie, 1996).

The response to fertilizer addition, indeed, shows a generally strong sensitivity of arctic tundra to any change that leads to increased availability of production-limiting nutrients, for instance in N deposition. Furthermore, local disruption of the organic horizon has led to strong increase of soil nutrient mineralization, plant nutrient uptake, and in many plant species a doubling of tissue nutrient concentration in heavily exploited tundra (B. Forbes, unpublished data), mirroring the effects of fertilizer addition.

Warming of tundra vegetation within the range of predicted temperature enhancement of 2–4°C for the next century has generally led to smaller changes than those induced by fertilizer addition and always to greater responses than those after water addition (Shaver and Jonasson, 1999). For instance, temperature enhancement in the high-arctic semidesert increased plant cover within the growing seasons but the effect did not persist from year to year (Robinson *et al.*, 1998). The strongest effect was on sexual and asexual reproduction and seed germinability, which increased strongly (Wookey *et al.*, 1993, 1994). The demonstrated enhanced reproductive success is likely to increase colonization of the present large areas of bare soil surfaces and, hence, increase plant coverage and carbon sequestration.

In the low Arctic, community biomass and nutrient mass changed little in response to warming in two Alaskan tussock sites (Chapin *et al.*, 1995; Hobbie and Chapin, 1998) and in two wet sedge tundra sites (Shaver *et al.*, 1998), coincident with relatively low changes in soil nutrient pools and net mineralization. In the tussock tundra the lack of response basically was because some species increased in abundance and others decreased (Chapin and Shaver, 1985; Chapin *et al.*, 1995), similar to a pattern observed in the subarctic Swedish forest floor vegetation (Press *et al.*, 1998a). In the Alaskan tundra, where nutrients were analyzed, this led to redistribution of nutrients within the vegetation, with increased proportions allocated to the vascular plants and decreasing proportions allocated to the cryptogams.

However, the responses to warming were much stronger in the Swedish tree-line heath and in the fellfield (Jonasson *et al.*, 1999b). The biomass in the low-altitude heath increased by about 60% after air warming by about 2.5°C with little additional effect after a further warming by about 2°C. In contrast, the biomass approximately doubled after the low-temperature enhancement and tripled in the higher temperature enhancement treatments at the colder fellfield. Hence, the growth response increased from the climatically relatively mild forest understory through the treeline heath to the cold, high-altitude fellfield where the response to warming was of the same magnitude as the response to fertilizer addition. Along with the increase in tissue nutrient mass, the nutrient concentration in individual species either remained unchanged, increased or decreased. In some species, particularly at the cold fellfield site, the nutrient concentration declined strongly coincident with increased productivity, suggesting that they responded strongly to the direct effect of warming, and possibly that their nutrient stress increased due to the temperature-induced growth (Jonasson *et al.*, 1999b; Graglia *et al.*, 1997).

After combined warming and fertilizer addition the biomass and vegetation N and P mass increased additively or synergistically at the Swedish sites, but there was a negative temperature × fertilizer interaction in both the Alaskan tussock and wet sedge tundra sites. The negative interaction occurred because plant respiration increased in the combined treatment and led to decreased biomass and nutrient incorporation in the vegetation (Shaver *et al.*, 1998).

In the warming treatments, net nutrient mineralization increased only slightly at the Swedish sites (Schmidt *et al.*, 1999) and cannot explain the increase of tissue N and P mass and illustrate the discrepancy between net nutrient mineralization and nutrient uptake discussed above. Furthermore, the microbial biomass was largely unaffected by both nutrient addition and warming and the microbial nutrient content increased only when there was also a marked increase of soil inorganic nutrients. That is, the microbes absorbed extra nutrients only when the nutrient sink strength declined in plants (Jonasson *et al.*, 1999b). This speaks against strong plant–microbe competition and suggests that the plants, indeed, are able to sequester nutrients even when there is a substantial microbial sink. On the other hand, the nematode density also increased and the proportion of fungal feeders increased with warming (Ruess *et al.*, 1999). Nematodes are the main predators on the soil microflora, so their increased population density and the changes in their trophic structure suggest that the microbial productivity and activity may have increased but that the biomass was kept at a constant level due to predation. This uncertainty highlights the potential biogeochemical regulation of soil processes by the soil fauna, which so far is almost entirely unknown in the Arctic.

4.4 Responses in Ecosystem Carbon Balance

Ecosystem C exchange has been measured in a few experiments, mostly in Alaskan wet and moist tundra (Fig. 4). In the wet tundra both gross ecosystem production, i.e., the photosynthetic gains, and respiratory C losses increased with nutrient addition. The increases were particularly pronounced in P and NP addition treatments with a strong N × P interaction, which was similar to the response pattern in the biomass (see above). Warming, in contrast, had smaller effects on CO_2 fluxes but still increased or tended to increase the fluxes. The net ecosystem productivity, which is the difference between the C fluxes into and out of the ecosystem, increased strongly after fertilizer addition and also tended to increase after warming. Hence, the increase in photosynthetic carbon sequestration was more pronounced than the increase in respiration with warming only. However, as with the biomass response to combined warming and fertilizer addition, there was a negative interaction with decreased net ecosystem production due to pronounced respiratory C losses when the two treatments were combined.

Increased gross C fixation was found early in the growing season in tussock tundra after 3.5 years of warming, but the net ecosystem production still was negative because of a higher growing season respiratory C loss (Hobbie and Chapin, 1998). Also, measurements in the Swedish treeline heath after seven years of treatment showed a mid-season C loss in warmed plots relative to

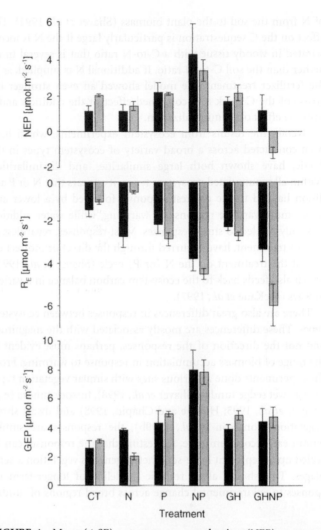

FIGURE 4 Mean (± SE) net ecosystem production (NEP), ecosystem respiration R_E and gross ecosystem production (GEP), in two Alaskan wet sedge tundra ecosystems subjected to N, P, and NP addition, greenhouse warming (GH) and combined NP and warming (GHNP). CT is untreated controls. (Redrawn from Shaver *et al.*, 1998, with kind permission from Ecological Society of America).

controls. However, the methods did not allow the inclusion of more than very low-growing vascular plants into the chambers for the flux measurement. Because the vascular plant biomass had increased strongly, it is likely therefore that the net ecosystem C balance for the system as a whole was positive, as after fertilizer addition (Christensen *et al.*, 1997).

Modeling of the C balance for tussock tundra based on responses to experimental treatments (McKane *et al.*, 1997) showed that warming first is likely to decrease the ecosystem C pool as a consequence of increased respiration. However, as soil N is mineralized and taken up by the vegetation, growth increases and offsets the respiratory losses and the model predicts a slight long-term increase in ecosystem C stock. Part of the increased sequestering of C is because the C-to-N ratio in the vegetation is higher than the ratio in the soil. Hence, the system as a whole can increase its C content without any increase in the N content by redistribution of N from the soil to the plant biomass (Shaver *et al.*, 1992). The effect on the C sequestration is particularly large if the N is incorporated in woody tissue with a C-to-N ratio that is several times higher than the soil C-to-N ratio. If additional N is supplied, as in the fertilizer treatment, the model showed an even stronger increase of the C stock as a combined effect of the addition and a priming effect on N mineralization.

Overall, the results from ecosystem experiments, which have been conducted across a broad variety of ecosystem types in the Arctic, have shown both large similarities and dissimilarities. Within all manipulated ecosystem types, it appears that N or P addition has led to the greatest response, followed by a lower and much more variable response to warming, while water addition generally has led to small responses. Most responses, regardless of type of treatment, have occurred through the direct or indirect effect of the treatment on the N (or P) cycle (Shaver *et al.*, 1992), which also feeds back to the ecosystem carbon balance in a variety of ways (McKane *et al.*, 1997).

There are also great differences in responses between ecosystem types. These differences are mostly associated with the magnitude and not the direction of the responses, perhaps most evident in the range of biomass accumulation in response to warming. From the experiments done at various sites with similar vegetation types of, e.g., wet sedge tundra (Shaver *et al.*, 1998), tussock tundra (e.g., Chapin *et al.*, 1995; Hobbie and Chapin, 1998) and dwarf shrub vegetation (Jonasson *et al.*, 1999b), the responses are similar within each ecosystem type, suggesting that the responses can be scaled up to represent large-scale heterogeneous vegetation assemblages. This should allow realistic modeling of longer-term responses to environmental change across broad regions of tundra.

from microbes by absorbing low-molecular-weight organic N compounds directly or indirectly through ecto- and ericoid mycorrhiza. Organic N uptake can probably explain part of the great discrepancies between measured low annual N mineralization and much higher annual plant N uptake.

Experimental manipulations of various vegetation types across the Arctic have given much information on the controls of ecosystem processes. Almost all vegetation types have responded strongly and consistently to fertilizer application by, e.g., increased net primary production and plant N and P mass, and in most cases by increase of standing biomass. The responses to warming have been more variable, ranging from no to pronounced increase in biomass, whereas the response to water addition has been small. Modeling and summer gas exchange studies suggest that fertilizer addition leads to enhanced sequestering of C by the ecosystems and it is likely that the ecosystems that have shown pronounced biomass increase after warming also have increased their C pool sizes. In ecosystems with low response in biomass accumulation, warming generally leads to strongly increased respiration, which results in a short-term carbon loss from the system. However, a model based on the experimental data from one such ecosystem type, the tussock tundra, showed that the respiratory C loss levels off over a longer time period as the N supply rate from the soil to the vegetation increases, resulting in enhanced plant growth and carbon sequestration.

The strength of the environmental response by vegetation is variable among sites but shows consistent patterns within similar vegetation types. This suggests that the responses can be scaled up to the regional scale for realistic future modeling of longer-term responses to environmental change.

5. Summary

The Arctic as a whole is characterized by high content of soil organic matter and low plant biomass. The organic matter accumulation, biomass, and net primary production generally decrease from south to north, but the variability is even greater among neighboring ecosystem types. Local variation is strongly related to topography, which creates gradients of snow depth and water availability that exert landscape-scale controls over ecosystem structure and function. Permanent wet or moist ecosystem types generally have the largest stocks of soil organic matter as a consequence of constrained microbial decomposition.

In spite of large soil stores of organic matter and plant nutrients, net primary production within almost all arctic ecosystem types is limited by low availability of plant-available nutrients, particularly N, and of P in wet ecosystem types. This is due to slow microbial mineralization rates associated with low temperature and often combined with extreme wet or dry conditions in several ecosystem types. Furthermore, soil microorganisms immobilize nutrients and may even act as competitors with the plants for nutrients during the growing season, when the nutrient demand by both microbes and plants is high. Recent research has, however, shown that plants can partially circumvent possible competition

Acknowledgments

This contribution is based largely on research supported by the US National Science Foundation, the Danish and Swedish Research Councils, and the Swedish Environmental Protection Board.

References

Abuzinadah, R. A., and Read, D. J. (1988). Amino acids as nitrogen sources for ectomycorrhizal fungi. Utilization of individual amino acids. *Trans. Br. Mycol. Soc.* **91**, 473–479.

Aleksandrova, V. D. (1988) "Vegetation of the Soviet Polar Deserts." Cambridge University Press, UK.

Billings, W. D., Luken, J. O., Mortensen, D. A., and Peterson, K. M. (1982). Arctic tundra: A source or sink for atmospheric carbon dioxide in a changing environment? *Oecologia* **53**, 7–11.

Bliss, L. C., and Matveyeva, N. V. (1992). Circumpolar arctic vegetation. In "Arctic Ecosystems in a Changing Climate, an Ecophysiological Perspective." (F. S. Chapin III, J. L. Jefferies, J. F. Reynolds, G. R. Shaver, and J. Svoboda, Eds.), pp. 59–89. Academic Press, San Diego.

Bliss, L. C., Svoboda, J., and Bliss, D. I. (1984). Polar deserts, their plant cover and plant production in the Canadian High Arctic. *Holarct. Ecol.* **7**, 305–324.

Chapin, F. S., III, and Shaver, G. R. (1985). Individualistic growth response of tundra plant species to manipulation of light, temperature, and nutrients in a field experiment. *Ecology* **66**, 564–576.

Chapin, F. S., III, Barsdate, R. J., and Barel, D. (1978). Phosphorus cycling in Alaskan coastal tundra: A hypothesis for the regulation of nutrient cycling. *Oikos* **31**, 189–199.

Chapin, F. S., III, Bret-Harte, M. S., Hobbie, S. E., and Zong, H. (1996). Plant functional types as predictors of transient responses of arctic vegetation to global change. *J. Veg. Sci.* **7**, 347–358.

Chapin, F. S., III, Fetcher, N., Kielland, K., Everett, K. R., and Linkins, A. R. (1988). Productivity and nutrient cycling of Alaskan tundra: Enhancement by flowing water. *Ecology* **69**, 693–702.

Chapin, F. S., III, Moilanen, L., and Kielland, K. (1993). Preferential use of organic nitrogen for growth by a nonmycorrhizal arctic sedge. *Nature* **361**, 150–153.

Chapin, F. S., III, Shaver, G. R., Giblin, A. E., Nadelhoffer, K. G., and Laundre, J. A. (1995). Responses of arctic tundra to experimental and observed changes in climate. *Ecology* **76**, 694–711.

Cheng, W., and Virginia, R. A. (1993). Measurements of microbial biomass in arctic tundra soils using fumigation-extraction and substrate-induced respiration procedures. *Soil Biol. Biochem.* **25**, 135–141.

Christensen, T. R., Jonasson, S., Callaghan, T. V., and Havström, M. (1995). Spatial variation in high latitude methane flux—A transect across tundra environments in Siberia and the European arctic. *J. Geophys. Res.* **100(D20)**, 21035–21045.

Christensen, T. R., Michelsen, A., Jonasson, S., and Schmidt, I. K. (1997). Carbon dioxide and methane exchange of a subarctic heath in response to climate change related environmental manipulations. *Oikos* **79**, 34–44.

Christensen, T. R. Jonasson, S., Callaghan, T. V., and Havström, M. (1999). Carbon cycling and methane exchange in Eurasian tundra ecosystems. *Ambio* **28**, 239–244.

Christensen, T. R., Friborg T., Sommerkorn, M., Kaplan, J., Illeris, L., Soegaard, H., Nordstroem, C., and. Jonasson, S. (2000). Trace gas exchange in a high-arctic valley 1: Variations in CO₂ and CH₄ flux between tundra vegetation types. *Global Biogeochem. Cycles.* **14**, 701–713

Epstein, H. E., Walker, M. D., Chapin, F. S. III, and Starfield, A. M. (2000). A transient, nutrient-based model of arctic plant community response to climatic warming. *Ecol. Appl.* **10**, 824–841

Giblin, A. E., Nadelhoffer, K. J., Shaver, G. R., Laundre, J. A., McKerrow, A. J. (1991). Biogeochemical diversity along a riverside toposequence in arctic Alaska. *Ecol. Monogr.* **61**, 415–436.

Gilmanov, T. G. and Oechel, W. C. (1995). New estimates of organic matter reserves and net primary productivity of the North American tundra ecosystems. *J. Biogeogr.* **22**, 723–741.

Gold, W. G. and Bliss, L. C. (1995). Water limitations and plant community development in a polar desert. *Ecology* **76**, 1558–1568.

Graglia, E., Jonasson, S., Michelsen, A., and Schmidt, I. K. (1997). Effects of shading, nutrient application and warming on leaf growth and shoot densities of dwarf shrubs in two arctic/alpine plant communities. *Ecoscience* **4**, 191–198.

Harte, J. and Kinzig, A. P. (1993). Mutualism and competition between plants and decomposers: Implications for nutrient allocation in ecosystems. *Am. Nat.* **141**, 829–846.

Heal, O. W., Flanagan, P. W., French, D. D., and MacLean S. F., Jr. (1981). Decomposition and accumulation of organic matter in tundra. In "Tundra Ecosystems: A Comparative Analysis." (L. C. Bliss, O. W. Heal, and J. J. Moore, Eds.), pp. 587–633. Cambridge University Press, UK.

Hobbie, S. E. (1996). Temperature and plant species control over litter decomposition in Alaskan tundra. *Ecol. Monogr.* **66**, 503–522.

Hobbie, S. E. and Chapin, F. S. (1996). Winter regulation of tundra litter carbon and nitrogen dynamics. *Biogeochemistry.* **35**, 327–338.

Hobbie, S. E. and Chapin, F. S. (1998). The response of tundra plant biomass, aboveground production, nitrogen, and CO₂ flux to experimental warming. *Ecology* **79**, 1526–1544.

Illeris, L. and Jonasson, S. (1999). Soil and plant CO₂ production in response to variations in soil moisture and temperature and to amendment with N, P and C. *Arct. Antarct. Alp. Res.* **31**, 264–271.

IPCC. (1996). Technical summary. In "Climate Change 1995: The Science of Climate Change." (J. T. Houghton, L. G. Meira Filho, B. A. Callandar, N. Harris, A. Kattenberg, and K. Maskell, Eds.), pp. 9–49. Cambridge University Press, Cambridge, United Kingdom.

Johnson, L. C., Shaver, G. R., Giblin, A. E., Nadelhoffer, K. J., Rastetter, E. R., Laundre, J. A., and Murray, G. L. (1996). Effects of drainage and temperature on carbon balance of tussock tundra microcosm. *Oecologia* **108**, 737–748.

Jonasson, S. (1992). Growth responses to fertilization and species removal in tundra related to community structure and clonality. *Oikos* **63**, 420–429.

Jonasson, S. and Michelsen, A. (1996). Nutrient cycling in subarctic and arctic ecosystems, with special reference to the Abisko and Torneträsk region. *Ecol. Bull.* **45**, 45–52.

Jonasson, S., Michelsen, A., and Schmidt, I. K. (1999a). Coupling of nutrient cycling and carbon dynamics in the Arctic, integration of soil microbial and plant processes. *Appl. Soil Ecol.* **11**, 135–146.

Jonasson, S., Michelsen, A., Schmidt, I. K., Nielsen, E. V. (1999b). Responses in microbes and plants to changed temperature, nutrient and light regimes in the arctic. *Ecology* **80**, 1828–1843.

Jonasson, S., Michelsen, A., Schmidt, I. K., Nielsen, E. V., and Callaghan, T. V. (1996). Microbial biomass C, N, and P in two arctic soils and responses to addition of NPK fertilizer and sugar: implications for plant nutrient uptake. *Oecologia* **106**, 507–515.

Kielland, K. (1994). Amino acid absorption by arctic plants: Implications for plant nutrition and nitrogen cycling. *Ecology* **75**, 2373–2383.

Kielland, K., and Chapin, F. S. III. (1992). Nutrient absorption and accumulation in arctic plants. In "Arctic Ecosystems in a Changing Climate: An Ecophysiological Perspective." (F. S. Chapin III, R. L. Jefferies, J. F. Reynolds, G. R. Shaver, and J. Svoboda, Eds.), pp. 321–335. Academic Press, San Diego.

McGuire, A. D., Melillo, J. M., Kicklighter, D. W., and Joyce, L. A. (1995). Equilibrium responses of soil carbon to climate change: Empirical and process-based estimates. *J Biogeogr.* **22**, 785–796.

McGuire, A. D., Melillo, J. M., Kicklighter, D. W., Pan, Y., Xiao, X., Helfrich, J., Moore, B. M., III, Vorosmarty, C. J., and Schloss, A. L. (1997). Equilibrium responses of global net primary production and carbon storage to doubled atmospheric carbon dioxide: Sensitivity to changes in vegetation nitrogen concentration. *Global Biochem. Cycles* **11**, 173–189.

McFadden, J. P., Chapin, F. S., III, and Hollinger, D. Y. (1998). Subgrid-scale variability in the surface energy balance of arctic tundra. *J. Geophys. Res.* **103**, 28947–28961.

McKane, R. B., Rastetter, E. B., Shaver, G. R., Nadelhoffer, K. J., Giblin, A. E., Laundre, J. A., and Chapin, F. S., III. (1997). Climatic effects on tundra carbon storage inferred from experimental data and a model. *Ecology* **78**, 1170–1187.

Michelsen, A., Schmidt, I. K., Jonasson, S., Quarmby, C., and Sleep, D. (1996). Leaf ¹⁵N abundance of subarctic plants provides field evidence that ericoid, ectomycorrhizal and non- and arbuscular mycorrhizal species access different sources of soil nitrogen. *Oecologia* **105**, 53–63.

Michelsen, A., Quarmby, C., Sleep, D., and Jonasson, S. (1998). Vascular plant ¹⁵N abundance in heath and forest tundra ecosystems is closely correlated with presence and type of mycorrhizal fungi in roots. *Oecologia* **115**, 406–418.

Molau, U. and Alatalo, J. M. (1998). Responses of subarctic–alpine plant communities to simulated environmental change: Biodiversity of bryophytes, lichens and vascular plants. *Ambio* **4**, 322–329.

Nadelhoffer, K. J., Giblin, A. E., Shaver, G. R., and Linkins, A. E. (1992). Microbial processes and plant nutrient availability in arctic soils. In "Arctic Ecosystems in a Changing Climate, an Ecophysiological Perspective." (F. S. Chapin III, J. L. Jefferies, J. F. Reynolds, G. R. Shaver, and J. Svoboda, Eds.), pp. 281–300. Academic Press, San Diego.

Nadelhoffer, K. J., Shaver, G. R., Fry, B., Giblin, A. E., Johnson, L. C., and McKane, R. B. (1996). ^{15}N natural abundance and N use by tundra plants. *Oecologia* **107**, 386–394.

Näsholm, T., Ekblad, A., Nordin, A., Giesler, R., Högberg, M., and Högberg, P. (1998). Boreal forest plants take up organic nitrogen. *Nature* **392**, 914–916.

Oberbauer, S. F., Cheng, W., Gillespie, C. T., Ostendorf, B., Sala, A., Gebauer, R., Virginia, R. A., and Tenhunen, J. D. (1996). Landscape patterns of carbon dioxide exchange in tundra ecosystems. In "Landscape Function and Disturbance in Arctic Tundra." (J. F. Reynolds and J. D. Tenhunen, Eds.), *Ecological Studies series* **Vol. 120**, pp. 223–256. Springer-Verlag, Berlin, Heidelberg.

Oechel, W. C. and Billings, W. D. (1992). Effects of global change on the carbon balance of arctic plants and ecosystems. In "Arctic Ecosystems in a Changing Climate, an Ecophysiological Perspective." (F. S. Chapin III, J. L. Jefferies, J. F. Reynolds, G. R. Shaver, and J. Svoboda, Eds.), pp. 139–168. Academic Press, San Diego.

Oechel, W. C., Hastings, S. J., Jenkins, M., Reichers, G., Grulke, N., and Vourlitis, G. (1993). Recent change of arctic tundra ecosystems from a carbon sink to a source. *Nature* **361**, 520–526.

Parsons, A. N., Welker, J. M., Wookey, P. A., Press, M. C., Callaghan, T. V. and Lee, J. A. (1994). Growth responses of four sub-arctic dwarf shrubs to simulated environmental change. *J. Ecol.* **82**, 307–318.

Post, W. M., Emanuel, W. R., Zinke, P. J., and Stangenberger, A. G. (1982). Soil carbon pools and world life zones. *Nature* **298**, 156–159.

Potter, J. A., Press, M. C., Callaghan, T. V., and Lee, J. A. (1995). Growth responses of *Polytrichum commune* Hedw. and *Hylocomium splendens* (Hedw.) Br. Eur. to simulated environmental change in the Subarctic. *New Phytol.* **131**, 533–541.

Press, M. C., Potter, J. A., Burke, M. J. W., Callaghan, T. V., and Lee, J. A. (1998a). Response of a subarctic dwarf shrub heath community to simulated environmental change. *J. Ecol.* **86**, 315–327.

Press, M. C., Callaghan, T. V., and Lee, J. A. (1998b). How will European arctic ecosystems respond to projected global environmental change? *Ambio* **27**, 306–311.

Raab, T. K., Lipson, D. A., and Monson, R. K. (1996). Non-mycorrhizal uptake of amino acids by roots of the alpine sedge *Kobresia myosurides*: Implications for the alpine nitrogen cycle. *Oecologia* **108**, 488–494.

Read, D. J., Leake, J. R., and Langdale, A. R. (1989). The nitrogen nutrition of mycorrhizal fungi and their host plants. In "Nitrogen, Phosphorus and Sulphur Utilization by Fungi." (L. Boddy, R. Marchant, and D. J. Read, Eds.), pp. 181–204. Cambridge University Press, UK.

Robinson, C. H., Wookey, P. A., Lee, J. A., Callaghan, T. V., and Press, M. C. (1998). Plant community responses to simulated environmental change at a high arctic polar semi-desert. *Ecology* **79**, 856–866.

Ruess, L., Michelsen, A., Schmidt, I. K., and Jonasson, S. (1999). Simulated climate change affecting microorganisms, nematode density and biodiversity in subarctic soils. *Plant and Soil* **212**, 63–73.

Schimel, J. P. and Chapin, F. S. III. (1996). Tundra plant uptake of amino acid and NH_4^+ nitrogen in situ: Plants compete well for amino acid N. *Ecology* **77**, 2142–2147.

Schimel, J. P., Kielland, K., and Chapin, F. S. III. (1996). Nutrient availability and uptake by tundra plants. In "Landscape Function and Disturbance in Arctic Tundra." (J. F. Reynolds and J. D. Tenhunen, Eds.), *Ecological Studies series* Vol. **120**, 203–221. Springer-Verlag, Berlin, Heidelberg.

Schmidt, I. K., Michelsen, A., and Jonasson, S. (1997a). Effects of soil labile carbon on nutrient partitioning between an arctic graminoid and microbes. *Oecologia* **112**, 557–565.

Schmidt, I. K., Michelsen, A., and Jonasson, S. (1997b). Effects on plant production after addition of labile carbon to arctic/alpine soils. *Oecologia* **112**, 305–313.

Schmidt, I. K., Jonasson, S., and Michelsen, A. (1999). Mineralization and microbial immobilization of N and P in arctic soils in relation to season, temperature and nutrient amendment. *Appl. Soil Ecol.* **11**, 147–160.

Schulze, E-D., Chapin, F. S. III, and Gebauer, R. (1994). Nitrogen nutrition and isotope differences among life forms at the northern treeline of Alaska. *Oecologia* **100**, 406–412.

Shaver, G. R., and Chapin, F. S. III. (1995). Long-term responses to factorial NP fertilizer treatment by Alaskan wet and moist tundra sedge species. *Ecography* **18**, 259–275.

Shaver, G. R. and Jonasson, S. (1999). Response of arctic ecosystems to climate change: Results of long-term field experiments in Sweden and Alaska. *Polar Res.* **18**, 245–252.

Shaver, G. R. and Jonasson. S. (in press). Productivity of arctic ecosystems. In "Terrestrial Global Productivity." (J. Roy, B. Saugier, and H. A. Mooney, Eds.). Academic Press, San Diego.

Shaver, G. R., Billings, W. D., Chapin, F. S., III, Giblin, A. E., Nadelhoffer, K. J., Oechel, W. C., and Rastetter, E. B. (1992). Global change and the carbon balance of arctic ecosystems. *BioScience* **42**, 433–441.

Shaver, G. R., Johnson, L. C., Cades, D. H., Murray, G., Laundre, J. A., Rastetter, E. R., Nadelhoffer, K. J., and Giblin, A. E. (1998). Biomass and CO_2 flux in wet sedge tundras: Responses to nutrients, temperature, and light. *Ecol. Monogr.* **68**, 75–97.

Shaver, G. R., Laundre, J. A., Giblin, A. E., and Nadelhoffer, K. J. (1996). Changes in vegetation biomass, primary production, and species composition along a riverside toposequence in arctic Alaska. *Arc. Alp. Res.* **28**, 363–379.

Tenhunen, J. D., Lange, O. L., Hahn, S., Siegwolf, R., and Oberbauer, S. F. (1992). In "Arctic Ecosystems in a Changing Climate, an Ecophysiological Perspective." (F. S. Chapin III, J. L. Jefferies, J. F. Reynolds, G. R. Shaver, and J. Svoboda, Eds.), pp. 213–237. Academic Press, San Diego.

Updegraff, K., Pastor, J., Bridgham, S. D., and Johnston, C. A. (1995). Environmental and substrate controls over carbon and nitrogen mineralization in northern wetlands. *Ecol. Applic.* **5**, 151–163.

Verhoeven, J. T. A., Maltby, E., and Schmitz, M. B. (1990). Nitrogen and phosphorus mineralization in fens and bogs. *J. Ecol* **78**, 713–726.

Wookey, P. A., Welker, J. M., Parsons, A. N., Press, M. C., Callaghan, T. V., and Lee, J. A. (1994). Differential growth, allocation and photosynthetic responses of *Polygonum viviparum* to simulated environmental change at a high arctic polar semi-desert. *Oikos* **70**, 131–139.

Wookey, P. A., Parsons, A. N., Welker, J. M., Potter, J. A., Callaghan, T. V., Lee, J. A., and Press, M. C. (1993). Comparative responses of phenology and reproductive development to simulated environmental change in sub-arctic and high arctic plants. *Oikos* **67**, 490–502.

1.11

Evaporation in the Boreal Zone During Summer—Physics and Vegetation

Francis M. Kelliher
Manaaki Whenua—Landcare Research,
Lincoln, New Zealand

Jon Lloyd,
Corinna Rebmann,
Christian Wirth, and
Ernst Detlef Schulze
Max Planck Institute for Biogeochemistry,
Jena, Germany

Dennis D. Baldocchi
University of California,
Berkeley, California

1. Introduction

The boreal zone is a Northern Hemisphere, circumpolar annulus ranging from around 50° to as far as 80°, including land and the Atlantic, Pacific, and Arctic Oceans. To the north, in summer and including much of the Arctic Ocean, associated archipelago, and 80% of Greenland, ice covers on the order of 10 Tm^2 (10 million km^2; Barry, 1995; Fig. 1). Low temperatures are thus one feature of the boreal zone. Vast areas of the zone sometimes become covered by ice, which happened most recently only ca. 10,000 years ago. This influenced the soil, some still underlain by permafrost, and the biodiversity of the vegetation (McGlone, 1996). Vegetation covers most of the boreal zone's land today and its physical features grade from south to north; notably, height and cover decrease significantly. The vegetation and terrestrial ecosystem evaporation rates during summer, the subjects of this chapter, are profoundly influenced by climate near the earth's surface. Because of the thermal properties of water, proximity to the open ocean moderates climate. In summer, the maritime climate is wetter (i.e., more water supply) and cooler (i.e., less atmospheric demand). A maritime influence can also delay the ends of cold and warm seasons. The magnitude of this effect varies throughout the boreal zone depending on the timing and extent of sea ice formation, if present nearby, and the range of sea-surface temperatures (Hertzman, 1997). Energy is carried to and from the ocean by the wind

in a process known as advection. A vortex of westerly winds tends to encircle the north pole although it is significantly *interrupted* by a perpendicular cordillera in western North America (Hare, 1997) and the Ural Mountains in western Siberia. Orographic effects greatly increase the precipitation rate on the windward side. Leeward, in the absence of water vapor advection from elsewhere, precipitation decreases toward the east until another ocean is approached.

The boreal zone's, remoteness and harsh winter climate have led to much of it being sparsely populated by people, especially in Siberia. Although not completely isolated from anthropogenic influence, of which fire control may be most significant, the boreal zone is relatively pristine. This attribute is significant for a baseline study of natural processes in the terrestrial biosphere, an opportunity that is becoming increasingly difficult to realize in much of the world. Evaporation is an important natural process because of its role in the surface-energy balance and the close linkages between plant water use and carbon gain. Beyond climate, elucidating the physics of evaporation is a necessary prerequisite to its study. An example is the interpretation of the recently decreasing evaporation rates during summer from pans of water located in European Russia, Siberia, and some other places. Peterson *et al.* (1995) attributed this trend to increasing cloudiness and, paradoxically, suggested that it indicated a decreasing terrestrial ecosystem evaporation; apparently failing to see how more (not

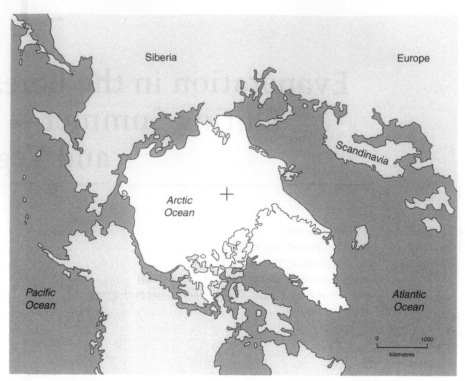

FIGURE 1 Sketch map of the boreal zone and polar ice cap (a cross marks the North Pole) with land, open ocean and ice illustrated with light and heavy shaded patterns and white, respectively. The outer limit of the map is approximately 50°N and a 1000 km scale is given. The ice extends to an average southern limit during summer after Barry (1995).

less) evaporation is required to satisfy the converse of their argument. Because Russia and most terrestrial ecosystems are not well supplied with water in summer, pan and terrestrial ecosystem evaporation rates tend to be inversely, not proportionally, related (Brutsaert and Parlange, 1998). Hence, decreasing pan evaporation rates indicate increasing terrestrial ecosystem evaporation in accordance with the recent trend of increasing precipitation (Brutsaert and Parlange, 1998). This example also illustrates how the short but hot and dry summers in the boreal zone involve significant surface–atmosphere energy exchange. Our purpose is to examine the regulation of this exchange in terms of evaporation from vegetation and soil by utilizing available field data, about half of which has been published in the past two years, with some new information from Siberia. As a synopsis, we distill the relevant physics, including the physics of the atmosphere and soil, and connect with the boreal zone's hydrology and its vegetation via plant physiology and the availability of nitrogen, a commonly limiting nutrient.

2. Climate and Soil Water

Precipitation or water supply is fundamental to terrestrial ecosystem evaporation. In the boreal zone, it is relatively sparse. Precipitation cannot generally be exceeded by evaporation in the long term. Ground water collection areas are exceptional in that water

supply for evaporation is greater than the precipitation. An example is the type of wetland known as a fen. However, precipitation is the primary source of water in the boreal zone and besides the orographic effect mentioned earlier, long-term records suggest that average annual rates (computed as annual totals divided by 365 days) decrease in Siberia and North America from around 2 mm day^{-1} in the south down to only 0.3 mm day^{-1} in the north (Table 1). In Scandinavia, there seems to be little or no latitudinal gradient in the 1.5 mm day^{-1} annual average precipitation rate, probably reflecting its general proximity to the sea. The boreal zone's short summers are generally wetter than other times of the year with rainfall averaging 0.6–3.5 mm day^{-1} for June–August, except for northern Scandinavia.

To examine the summer rainfall gradient more closely, we utilize a scaled-up 1° latitude by 1° longtitude grid-cell data set of average monthly rainfall (Leemans and Cramer, 1991). The data are divided into four boreal regions: North America (48.5–68.5°N, 58.5–163.5°W, 776 cells), European Russia (52.5–68.5°N, 22.5–59.5°E, 161 cells), Western Siberia (48.5–69.5°N, 60.5–89.5°E, 331 cells), and Eastern Siberia (49.5–72.5°N, 90.5–178.5°E, 1168 cells). For North America, which is the driest of the regions except on the windward/western side of the cordillera, regression shows that the inverse linear relation between daily summer rainfall and latitude has a slope of −0.09 mm per degree and offset of 7.2 mm accounting for 76% of the variation (Fig. 2A). Results obtained for European Russia are −0.05 mm per degree,

TABLE 1 Latitudinal Transects of Climate Data from Müller (1982) for the Boreal Zone

Site	Latitude/Longitude (degrees)	Average Temperature (°C)		Average Precipitation Rate (mm day^{-1})		Drought Coefficient
		Jun–Aug	Year	Jun–Aug	Year	Jun–Aug
Western North America						
Barrow	71.3 N/156.8 W	2.6	−12.4	0.6	0.3	0.24
Whitehorse	60.7 N/135.0 W	13.1	−0.7	1.1	0.7	0.33
Fort Nelson	58.8 N/122.6 W	15.3	−1.1	2.0	1.2	0.38
Prince George	53.9 N/122.7 W	13.8	3.3	2.1	1.7	0.45
Central Siberia						
Dudinka	69.4 N/86.2 E	8.7	−10.7	1.2	0.7	0.48
Yeniseysk	58.3 N/92.2 E	15.6	−2.2	1.9	1.3	0.52
Minusinsk	53.7 N/91.7 E	17.8	−0.1	1.8	0.9	0.42
Eastern Siberia						
Bulun	70.8 N/127.8 E	8.1	−14.5	0.8	0.3	0.36
Yakutsk	62.0 N/129.8 E	16.3	−10.2	1.2	0.6	0.32
Bomnak	54.7 N/129.0 E	16.2	−4.9	3.5	1.5	0.42
Scandanavia						
Vardo	70.4 N/31.1 E	8.4	1.6	1.4	1.5	0.47
Stockhom	59.4 N/18.0 E	16.4	6.6	2.0	1.5	0.43

The drought coefficient (ζ) for a period is equal to the number of days with precipitation divided by the total number of days. It is used to estimate the probability of a day being dry (ϕ_d) for a given time (t) since rainfall in the integrated Poisson Process model $\phi_d = 1 - e^{-\zeta t}$.

5.3 mm, and 42%, respectively, suggesting a more variable and wetter regime (Fig. 2B). For Western Siberia at −0.11 mm per degree, 9.0 mm, and 80%, respectively, the latitudinal gradient is similar and generally as tight as in North America but, as in European Russia, rainfall intensity is 0.5–0.9 mm day^{-1} greater (Fig.

2C). North and south of 60°N, there are different rainfall regimes in Eastern Siberia (Fig. 2D). The north is much drier. Rainfall is also less variable with a latitudinal gradient and offset nearly equal to those of European Russia although regression only accounts for 14% of the variation. South of 60°N the latitudinal gradient and

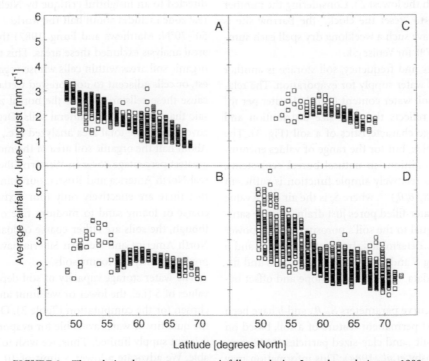

FIGURE 2 The relation between average rainfall rate during June through August 1988 (mm day^{-1}) computed as total rainfall for the 3 months (mm) divided by 92 days) and latitude (degrees North) in boreal North America (A), European Russia (B), Western Siberia (C) and Eastern Siberia (D). The data and regions are described in the text.

offset: were greater, -0.15 mm per degree and 11.3 mm, respectively ($r^2 = 0.40$).

Longitudinal variation in Siberian rain during summer includes a generally decreasing fall going eastward from the Ural Mountains, as noted by Schulze et al. (1999), with some influence of orography and proximity to the major rivers and the Sea of Okhotsk at the eastern frontier. For example, at latitude 66°N, average rainfall declines by nearly a factor of 2 from 2.5 to 1.3 mm day^{-1} going from 62 to 168°E. The relation is described by a line of slope -0.007 mm per degree and offset 2.65 mm; regression accounts for 42% of the variation (data not shown).

Rainfall frequency can be important because surface drying during fine weather greatly reduces the ground evaporation rate, which can be a significant fraction of the total evaporation from boreal ecosystems with sparse vegetation (e.g., Kelliher *et al.*, 1998). For the summer months of June–August, rainfall frequency may be considered a Poisson process with the probability of each event being independent (although, in a continental climate, terrestrial surface-energy partitioning may contribute to rainfall occurrence.) It is represented by a simple exponential function of time since the last event (t) (e.g., $\phi_d = 1 - e^{-\zeta t}$ where ϕ_d and ζ are the probability of a day being dry and a drought coefficient, respectively; Kelliher *et al.*, 1997). The coefficient ζ is computed simply as an inverse of the average time between rainfalls. In drier eastern Siberia and western North America, like rainfall, ζ tends to decrease significantly from south to north, but it is relatively constant throughout central Siberia and Scandinavia (Table 1). Table 1 shows that the probability of a weeklong dry period after rainfall varies from 0.03 (Yeniseysk with the highest ζ tabulated) to 0.19 (Barrow with the lowest ζ). Considering the number of rainy days, or chances to "reset the clock," the Barrow site is thus relatively certain to have such a weeklong dry spell each summer, while the chance is 57% for Yeniseysk.

Besides rainfall intensity and frequency, soil storage is another important determinant of water supply for evaporation. The relation between volumetric soil water content (θ, m^3 of water per m^3 of soil) and suction (S) reflects the pore-size distribution and quantifies the water storage characteristics of a soil (Fig. 3). The curve's shape can be complex, but for the range of values encountered under most field conditions in natural boreal ecosystems, where S is rarely near 0, a relatively simple function is sufficient (Campbell, 1985); $S(\theta) = S_e(\theta/\theta_s)^{-b}$, where S_e is the air entry value of S at which the largest water-filled pores just drain, θ_s is the saturation value of θ that is equal to the soil's porosity, and the power coefficient b is empirically determined. In practice, both S_e and b are determined by plotting S and θ on logarithmic scales and fitting a straight line to the data, the regression slope and offset being b and S_e, respectively.

Values of water-release curve parameters S_e, θ_s, and b have been related to texture, the most permanent feature of a soil, based on the percentages of sand-, silt-, and clay-sized particles (1448 samples from 35 locations; Cosby *et al.*, 1984). This information may be combined with the 1° latitude by 1° longitude grid-cell texture database of Zobler (1986) to estimate soil water storage capacities

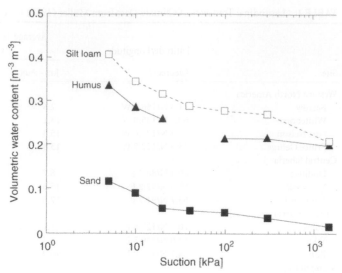

FIGURE 3 The relation between volumetric water content (θ, m^3 m^{-3}) and suction (S, kPa) for sand and silt loam mineral soils (solid squares and line and open squares and dashed line, respectively) and humus (i.e., organic matter; solid triangles and line). The sand came from the upper 0.1-m depth of soil beneath a stand of *Pinus sylvestris* trees located in central Siberia (Kelliher *et al.*, 1998), the silt loam from the 0.2–0.3-m depth of soil beneath a stand of *Larix gmelinii* trees in eastern Siberia (Kelliher *et al.*, 1997), and the humus from depth 0.14 m in a 0.3-m- deep forest floor beneath a stand of *Tsuga heterophylla* and *Thuja plicata* trees near Vancouver, Canada (Plamandon *et al.*, 1975).

throughout the world. Caution is thus required and the reader is directed to an insightful critique by Nielsen *et al.* (1996). The boreal zone includes about half the world's wetland area (2.6 Tm^2 for 50–70°N; Matthews and Fung, 1987) that has organic soils. Our areal analysis excluded these areas. This necessitated estimation of organic soil areas within cells whose vegetation is classified as forest, or cells adjacent to nonforested wetlands that are forested, because these cells are part of the boreal zone's forest/wetland mosaic that also contains mineral soils. Organic soils of this mosaic cover 9% of the total area analyzed (i.e., 1.3 of 14.2 million km^2). About half the organic soil area of the mosaic is in Siberia with the greatest percentage cover in Western Siberia at 24%. For all of boreal North America and Russia, our mineral soil analysis suggests that there are effectively only four textural classes ranging from coarse or loamy sand to medium fine or loam (Table 2). Mostly though, the soils are either coarse or loamy. Large areas of boreal North America and western Siberia have coarse soils, but Russia predominantly has loamy soils.

The water storage capacity of soil depends variably on the two values of S (i.e., the lower or wet limit and the upper or dry limit) chosen for the computation (Table 2). Our purpose is to estimate the quantity of water available for evaporation over a range where it is not supply limited. Thus, we wish to quantify the supply variable. We advocate a lower S limit of 10 kPa. This value is analogous to the so-called field capacity. In terms of the associated θ, it is generally the wettest a soil can become because drainage then

TABLE 2 Boreal Zone Regional Soil Water Storage Capacities for 1° Latitude by 1° Longitude Grid Cells Based on the Global Soil Texture Database of Zobler (1986) with Hydraulic Parameter Values from Cosby *et al.* (1984)

Region (degrees of latitude (N)/longitude)	Total Area(Tm^2)	LS	SCL	SL	L
		Fraction of total area minus wetland area in each mineral soil texture class			
North America (48.5−68.5/58.5−163.5 W)	4.55	0.40	0.14	0.08	0.38
European Russia (52.5−68.5/22.5−59.5 E)	0.83	0.16	0.02	0.14	0.68
Western Siberia (48.5−69.5/60.5−89.5 E)	1.52	0.29	0.06	0.00	0.65
Eastern Siberia (49.5−72.5/90.5−178.5 E)	7.25	0.01	0.09	0.00	0.90
Suctions		Water storage capacity, mm water per 0.1 m depth of soil			
10 and 100 kPa		8	9	11	13
10 and 1500 kPa		13	16	19	22

Wetland areas with organic soils were excluded using the global database of Matthews (1989) as described by Matthews and Fung (1987) and in the text. Water storage capacities were derived from the difference between water contents determined in the laboratory on mineral soil cores subject to 10 and 100 or 10 and 1500 kPa of applied suction. From left to right, the mineral soil texture classes of Zobler (1986)/Cosby *et al.* (1984)(abbreviation, percentages of sand:silt:clay) are coarse/loamy sand (LS, 82:12:6), medium fine/sandy clay loam (SCL, 58:15:27), medium coarse/sandy loam (SL, 58:32:10), and medium/loam (L, 43:39:18).

becomes minimal. In exceptional cases, lower values of *S* (and higher values of θ) result from impeded drainage, rainfall intensity in excess of the soil's infiltration capacity, and groundwater intrusion. Next, we turn to the upper limit of *S*. Kelliher *et al.* (1998) synthesized the scarce forest literature identifying how evaporation rate was limited by the available energy (see later discussion) from when *S* and θ were equal to the field capacity until a drier critical value of θ was reached. This critical value seems to be surprisingly conservative and it is equal to θ when about half the root zone depth of soil water is depleted. (See also Choudhury and DiGirolamo (1998) who argue that the fraction is smaller at 0.4 based on a review of all available data, with about half from agricultural studies.) Thereafter, as the soil further dries, the forest evaporation rate declines sharply in response to the declining water supply. In the coarse soils they examined, the critical value of θ corresponded to $S \approx 100$ kPa. We adopt this value as our upper limit of *S*. However, for comparative purposes, we also report another, much higher value of *S* (and drier soil): 1500 kPa. This represents the so-called permanent wilting point, a long-advocated concept based on experiments conducted with potted sunflowers. It is essentially the driest a soil can become in terms of *S* and θ.

The mineral soils analyzed in Table 2 can store 8 to 13 mm of water per 0.1 m depth for *S* between 10 and 100 kPa. For *S* between 10 and 1500 kPa, these soils store about twice as much, or 13 to 22 mm of water per 0.1 m depth. Water storage capacity grades with texture from the lowest value for coarse soils to the greatest for loams. Boreal zone water-release data are available for sand and silt loam mineral soils beneath Siberian *Pinus sylvestris* and *Larix gmelinii* forests, respectively (Fig. 3; Kelliher *et al.*, 1997; 1998). The well-sorted sand (i.e., the particles are mostly of a similar size) holds only 4 and 8 mm of water per 0.1 m depth for *S* of 10−100 and 10−1500 kPa, respectively. For the silt loam, the respective storage capacities are 7 and 14 mm. These soils are thus at the low end of the data given in Table 2. Comparable organic

matter data are available for humus from the floor of a *Tsuga heterophylla* and *Thuja plicata* forest near Vancouver, Canada (such measurements were not made for humus from the Siberian forests because of its very shallow depth at those two sites). The respective storage capacities of the humus are 7 and 9 mm (note how the humus water-release curve flattens for *S* between 100 and 1500 kPa; Plamandon *et al.*, 1975). This illustrates the limited nature of organic matter or peaty soil water storage capacity for $S > 10$ and especially > 100 kPa. This reflects the predominance of larger pores that are mostly emptied by little or low suction in humified organic matter.

The nagging question emerges of what depth of soil is relevant to evaporation. For example, a dry surface-layer of relatively shallow depth, about 1−10 mm, dramatically reduces soil evaporation compared to the rate obtained when the surface is wet and the weather fine, especially for coarse-textured material (e.g., Kelliher *et al.*, 1998). In terms of vegetation transpiration, plant root density declines strongly with depth as quantified by a recent global synthesis of available data (Jackson *et al.*, 1997). Jackson *et al.* represent the vertical distribution of fine root biomass by the power function ($Y = 1 - c^d$), where *Y* is the cumulative fraction varying between 0 and 1, *d* the depth (cm), and *c* an extinction coefficient. For boreal forest and tundra vegetation, *c* is 0.943 and 0.909, respectively. This suggests that forest and tundra have 44 and 61%, respectively, of their fine roots in the upper 0.1 m of soil. The percentages are 69 and 85, and 83 and 93% for the upper 0.2 and 0.3 m of soil, respectively. For a depth of 0.3 m and *S* values of 10 and 100 kPa, the mineral soils shown in Table 2 can store 25−40 mm of water. For comparison, the corresponding value for the so-called average soil of the world is 41 mm (Nielsen *et al.*, 1983).

Related to this limited water storage capacity of soils in the boreal zone is the climatological feature, especially in Siberia, of an abrupt transition from frozen winter to warm summer. Snow melt thus occurs relatively quickly and it involves relatively large quantities of water. Much of this water is generally lost, however, as a pulse contributor to extreme river flow rates in spring. For example, peak flow rate at the northern mouth of the Yenesei River in central Siberia was once a remarkable 70,000 m^3 s^{-1} (Beckinsale, 1969). This loss of winter precipitation in spring may result from surplus water unable to seep into the still-frozen soil (Walter, 1985). If the soil is not frozen, high-volume drainage may bypass the matrix through macropores (Clothier et al. (1998) review the pertinent physics involved.)

Summer days in the boreal zone are long, the daily period of illumination increasing virtually to 24 h in the north. Radiation is a primary driving variable of earth surface–atmosphere energy exchange including evaporation. On a clear day, for an unpolluted atmosphere, shortwave irradiance at the earth's surface obtains a maximum of around 70–80% of the extraterrestrial value due to

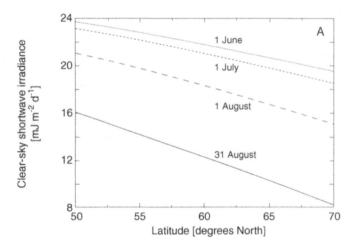

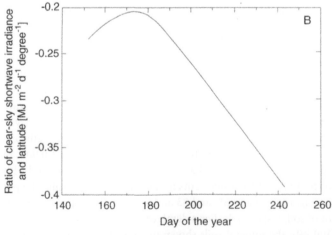

FIGURE 4 (A) Relations between daily clear-sky shortwave irradiance at the earth's surface, computed with the atmospheric transmissivity = 0.7, and latitude from 50 to 70°N on 1 June (dotted line), 1 July (short dashes line), 1 August (long dashes line) and 31 August (solid line). (B) Relation between the ratio of daily clear-sky shortwave irradiance and latitude (i.e., slopes of lines like those shown in (A)) and day of the year from 1 June (day 152) through 31 August (day 243).

attenuation by water vapor and dust (i.e., atmospheric transmissivity = 0.7–0.8). For a clear sky during summer (June–August) in the boreal zone, daily shortwave irradiance varies significantly with a peak on 21 June and a minimum on 31 August (e.g., 22.0 and 12.3 MJ m^{-2} day^{-1} at 60°N, respectively, for these two days with atmospheric transmissivity = 0.7.) Irradiance decreases with increasing latitude but the rate of change is not constant in the boreal zone (Fig. 4A). Differences in irradiance across the boreal latitudes are least in June around the summer solstice and greatest in terms of decline during August. From linear regression of irradiance and latitude, the offset is relatively constant at 34–36 MJ m^{-2} day^{-1} but the slope is least on 21 June at -0.21 MJ m^{-2} day^{-1} degree^{-1}, declining linearly thereafter to -0.39 MJ m^{-2} day^{-1} degree^{-1} on 31 August (Fig. 4B). The boreal zone, of course, does not always have a completely clear sky. During fine summer weather in Northern Canada, an additional 15% attenuation of shortwave radiation by smoke aerosols from forest fires has been reported (Miller and O'Neill, 1997). On a completely overcast day, the atmospheric transmissivity would typically be only around 0.25. As an example of the integrated effect of atmospheric radiation attenuation, for June–August 1997 at a relatively sunny aspen forest site in Northern Canada (53.7°N, 106.2°W), the measured shortwave irradiance averaged 80% of the clear-sky value computed for an atmospheric transmissivity of 0.8 (T.A. Black, personal communication).

3. Evaporation Theory

The evaporation rate (E) obtained for an extensive, wet surface in dynamic equilibrium with the atmosphere and in the absence of advection (E_{eq}) may be written (Slatyer and McIlroy, 1961; McNaughton, 1976) as

$$E_{eq} = [\epsilon/\lambda(\epsilon + 1)]R_a, \qquad (1)$$

where ϵ is the rate of change of latent heat content of saturated air with change in sensible heat content, λ the latent heat of vaporization, and R_a the available energy flux density. The partitioning of R_a into E according to Eq. (1) is strongly temperature-dependent, especially through term ϵ (Fig. 1). All of the available energy may thus be dissipated by evaporation if $E = E_{eq}$ at the so-called partitioning temperature of 33°C (Priestley, 1966; Priestley and Taylor, 1972; Calder et al., 1986) including consideration of the entrainment of dry air from aloft into the convective boundary layer (De Bruin, 1983). In addition to temperature and R_a, it also takes time for $E \rightarrow E_{eq}$ (McNaughton and Jarvis, 1983; Finnigan and Raupach, 1987), but daily or longer periods are sufficient.

To illustrate the application of Eq. (1), we can conduct a global average annual computation. The global average surface temperature is 15°C according to Graedel and Crutzen (1993). At this temperature, $\lambda = 2.465$ J g^{-1} and $\epsilon/(\epsilon + 1) = 0.63$. For the earth's continents, Baumgartner and Reichel (1975) estimate $R_a/\lambda = 850$ mm year^{-1} so that $E_{eq} = 535$ mm year^{-1}. This may be compared to their continental $E = 480$ mm year^{-1} independently determined and constrained by conservation of mass according to a global water balance including the oceans. Consequently, for the terrestrial biosphere as a whole, E_{eq} is within 10% of E and therefore not significantly different, given a reasonable error associated with the computations.

Equation (1) may seem a purely meteorological model but it does not ignore the evaporating surface, particularly in its application to the terrestrial biosphere including vegetation and soil (Priestley and Taylor, 1972). First and foremost, the surface affects R_a by determining the shortwave radiation reflection coefficient or albedo, which a recent review showed varied by a factor of 2 in the boreal zone (Baldocchi *et al.*, 2000). Albedo depends also on ground-surface wetness (i.e., darkness), which is particularly relevant to the boreal zone because of the commonly sparse nature of the vegetation, and on the solar zenith angle because of vegetation architecture effects. For the boreal zone, it is thus relevant to note how fire can alter vegetation structure (Wirth *et al.*, 2000). Second, the surface temperature, also used in Eq. (1) to determine ϵ and λ, largely determines the outgoing longwave radiation. Surface temperature in turn depends on R_a and its partitioning into λE and sensible heat (H), heat that you can sense because it warms the air, via the surface energy balance ($R_a = \lambda E + H$). Aerodynamic roughness of vegetation or of ground affects the surface temperature as well. In addition, for outgoing longwave radiation, a dry mineral or plant-litter surface will have a significantly lower emissivity than vegetation or wet ground (soil or litter) (Monteith and Unsworth, 1990). Vegetation density determines the amount of net all-wave irradiance that is absorbed. Ground surface wetness (thermal admittance) also affects the fraction of the remaining energy dissipated by conduction into the ground. The rate of conduction can be variably, and overwhelmingly, affected by underlying ice during summer in the boreal zone (e.g., Fitzjarrald and Moore, 1994).

For plant leaves, the surface may be considered explicitly by a model of E or transpiration rate (E_t);

$$E_t = D_0 g_{st}, \qquad (2)$$

where D_0 is air saturation deficit, expressed as a dimensionless specific-humidity deficit, at the leaf surface and g_{st} is the stomatal conductance for water vapor transfer. It is g_{st}, rather than E_t, that has been the focus of most research, although some careful studies including D_0 have also been conducted (Mott and Parkhurst, 1991; Alphalo and Jarvis, 1991; 1993; Monteith, 1995). Using Eq. (2) and rearranging Eq. (1), we may write an equation for the terrestrial surface conductance for water vapor transfer (g_S) as

$$g_S = [\epsilon/\lambda(\epsilon + 1)] \, R_a/D_0. \qquad (3)$$

Written similarly, g_S is also called the climatological conductance (Monteith and Unsworth, 1990). In any case, like g_{st}, g_S represents the surface control of E by balancing radiant energy supply and atmospheric demand (D_0). Consequently, as shown below, evaluating terrestrial E in relation to E_{eq} includes an assessment of the underlying surface as well as the meteorology and water balance.

The value of E in relation to E_{eq}, like the value of g_S, indicates the evaporative nature of the surface so $E < E_{eq}$ or $E/E_{eq} < 1$ reflects surface dryness or stomatal closure as well as the balance of energy exchange between the atmosphere and the underlying surface. By definition, $E > E_{eq}$ can be caused only by advection. As implied above with respect to the partitioning of temperature, this may also result from the entrainment of dry air from above the convective boundary layer that develops daily over the earth surface. To further illustrate the relation between E and E_{eq} in terms of surface characteristics, it is helpful to write the Penman–Monteith equation (Monteith and Unsworth, 1990),

$$E = (\epsilon R_a + \rho D g_A)/[\lambda(\epsilon + 1 + g_A/g_S)], \qquad (4)$$

where ρ is the density of air, D is the air saturation deficit above the evaporating surface expressed as a dimensionless specific-humidity deficit, and g_A is the total aerodynamic conductance assuming similarity between heat and water vapor transfer processes in the atmosphere. Some limits of this equation suggest that there are at least three surface-related conditions leading to $E = E_{eq}$ (Raupach, 2001), namely:

(i) $g_S \rightarrow \infty$ by definition for a completely wet surface with the corollary $D \rightarrow 0$ (Slatyer and McIllroy, 1961), although this requires no entrainment of dry air from above the near-surface boundary layer (i.e., the closed box model of McNaughton and Jarvis, 1983),

(ii) $g_A \rightarrow 0$ by definition for a surface beneath a completely calm atmosphere (Thom, 1975), and

(iii) E is not sensitive to g_A (Thom, 1975) for a completely smooth surface or a surface completely isolated from the influence of D on E, a derivative of (ii).

Condition (i) depends mostly on precipitation frequency although precipitation interception, hydraulic conductance characteristics of soil, ground surface (e.g., litter), and plants, and ground water storage capacity also contribute. Conditions (ii) and (iii) depend on the surface roughness, mostly reflecting vegetation height or a lack of it. For all conditions, larger-scale meteorology is relevant too, as are the physical and physiological feedback processes critically analyzed recently by Raupach (1998).

Returning to plant leaves, through stomata, there is an intrinsic connection between E_t and the net rate of carbon assimilation (A) that may be written as

$$A = (C_a - C_0)(g_{st}/1.6), \qquad (5)$$

where C_a and C_0 are carbon dioxide concentrations in the atmosphere and substomatal cavity, respectively. Division of g_{st} by 1.6 accounts for the difference in diffusion coefficients for water vapor and carbon dioxide (Massmann, 1998). Further, the maximum A or carbon assimilation capacity of the leaf has been found to be proportional to the leaf nitrogen content (Field and Mooney,

1986; Evans, 1989). According to Eq. (5) and when a wide range of plants are compared, maximum g_{st} is also proportional to leaf nitrogen content [Schulze *et al.*, 1994; g_{st} correlates with carbon assimilation capacity (Wong *et al.*, 1979)]. In this way, stomata link water, carbon, and nutrient cycles. It is no wonder then that a voluminous literature reflects a virtually universal interest in stomatal behavior among plant physiologists and biometeorologists (Körner, 1994).

4. Evaporation during Summer and Rainfall

The first boreal ecosystem we consider is that most analogous to water, namely, the wetland. During summer, P averaged 0.8 mm day^{-1} less than E for wetlands (Table 3). However, by definition, this comparison is incomplete for fens because water supply to this system exceeds rainfall by virtue of ground-water intrusion, although this is often difficult to quantify with certainty. In Table 3, the Saskatchewan, Zotino, and Schefferville sites were fens but it

is not clear if ground water supplemented water supply in the other wetlands studied. Generally and on average, $E = E_{eq}$ at 2.6 mm day^{-1}. Dry air inevitably entrains into the convective boundary layer (i.e., advection from aloft) on fine summer days in the boreal zone at a rate proportional to the surface sensible heat flux density. Consequently, the attainment of $E = E_{eq}$ suggests either that one of the aforementioned limits of the Penman– Monteith equation has been reached or that the surface was relatively dry, as can happen even for a wetland. Indeed, for the two cases where $E < E_{eq}$, conditions during the measurements were at least sometimes considered relatively dry. Although advection enhanced E at Southern James Bay, the similarity between P and E there is striking (Rouse *et al.*, 1987). The topography and hydrology involved in the formation and maintenance of the wetlands studied may be such that these ecosystems are often not sufficiently extensive (i.e., the wetland landscape is patchy) to avoid the influence of horizontal advection.

For tundra sites at $60°–80°$N, P has generally been observed to be nearly identical to E in summer, averaging 1.5 mm day^{-1}. There is no clear trend of E or P changing with latitude for these sites.

TABLE 3 Average Rates of Precipitation (P) and Evaporation (E) and the Ratio of E and the Equilibrium Evaporation Rate (E/E_{eq}) during Summer from Boreal Zone Vegetation

Surface Site	Lattitude/Longitude	P	E (mm day^{-1})	E/E_{eq}	Source
Wetland					
Southern James Bay	51.0 N/80.0 W	3.1	3.5	1.4	Rouse *et al.*, 1987
Central Hudson Bay	58.7 N/94.1 W	0.9	2.6	1.0	Rouse *et al.*, 1987
Kinosheo Lake	51.6 N/81.8 W	0.9	1.4	0.6	den Hartog *et al.*, 1994
Schefferville	54.9 N/66.7 W	1.4	2.8	1.0	Moore *et al.*, 1994
Thompson	55.9 N/98.4 W	2.2	2.4	1.0	Lafleur *et al.*, 1997
Zotino	61.0 N/89.0 E	1.9[a]	2.6	0.8	Schulze *et al.*, 1999
Saskatchewan	54.0 N/105.0 W	2.2	2.9	1.0	Suyker and Verma, 1998
Tundra					
Hardangervidda	60.3 N/7.7 E	2.0	2.0	1.0	Skartveit *et al.*, 1975
King Christian Island	77.8 N/101.2 W	0.7	0.9	0.6	Addison and Bliss, 1980
Axel Heiberg Island	79.5 N/90.8 W	1.4	1.5	1.0[a]	Ohmura, 1982
Imnavait Creek	68.6 N/149.3 W	1.9	1.8	1.0	Kane *et al.*, 1990
U-pad	70.3N/148.9 W	1.0	1.3	n.d.	Vourlitis and Oechel, 1997
Happy Valley	69.1N/148.8 W	2.1	1.5–2.7	0.6–0.9	McFadden *et al.*, 1998
Deciduous broad-leaved forest (full leaf)					
Betula/Populus, Moscow	55.8 N/37.5 E	2.3[a]	4.7	1.1	Rauner, 1976
Populus/Corylus, Prince Albert	53.7 N/106.2 W	2.2	2.2	1.0	Black *et al.*, 1996
Deciduous needle-leaved forest (full leaf)					
Larix, Churchill	58.8 N/94.0 W	1.4	2.1	0.8	Lafleur, 1992
Larix, Yakutsk	61.0 N/128.0 E	1.2[a]	1.9	0.6	Kelliher *et al.*, 1997
Evergreen needle-leaved forest					
Pinus, Jadraas	60.8/16.5 E	2.0	3.4	0.9	Lindroth, 1985
Pinus, Lac du Bonnet	50.2/95.9 W	2.0	1.4	0.8	Amiro and Wuschke, 1987
Pinus, Norunda	60.3 N/17.3 E	1.5	1.9	0.9	Grelle *et al.*, 1997
Pinus, Nipawin	53.0 N/104.0 W	1.7[b]	1.5	0.6	Baldocchi *et al.*, 1997
Pinus, Zotino	61.0 N/89.0 E	1.9[a]	1.3	0.4	Kelliher *et al.*, 1998
Picea, Schefferville	54.9 N/66.7 W	1.8	1.8	0.5	Fitzjarrald and Moore, 1994
Picea, Candle Lake	54.0 N/105.1 W	1.7[b]	2.0	0.7	Jarvis *et al.*, 1997

[a] average for June through August from Müller (1982).

[b] from Baldocchi *et al.* (1997) but excludes a 110 mm storm, on a day halfway through the 117 day study and following a week of 23.1 mm rainfall, that was assumed to drain beyond the root zone of the sandy soil.

This suggests that local regimes can be influenced by other factors such as proximity to the sea and elevation/orography (see Vourlitis and Oechel, 1997). In one far-north study where only 9% of the site was covered by vascular plants, $E \ll E_{eq}$ although $E > P$ (Addison and Bliss, 1980). The tundra landscape is also patchy (see Skartveit *et al.*, 1975). This leads to some variability in E although soil and sensible heat flux densities are much more spatially variable (McFadden *et al.*, 1998). The relatively lower net available energy and consequently the sensible heat flux densities in the northern part of the boreal zone during summer probably limit entrainment there. The tight water balance and nutrient cycles and generally lower temperatures seem to have an overwhelming influence on the attained evaporation rate of tundra. Consequently, in general for tundra, we also find $E = E_{eq}$.

Boreal zone forest stands are also part of a patchy landscape at least in terms of tree age, size, and density, which determine stand structure. More commonly, though, the stands of trees are interspersed with herbaceous and shrubby vegetation and wetlands. On average, P is found to be 0.5 mm day^{-1} less than E (= 2.3 mm day^{-1}) for forests in summer. Rainfall interception has been included in this comparison because E comprised wet and dry canopy evaporation rates except for the Yakutsk study by Kelliher *et al.* (1997). The difference between rainfall and evaporation rates is up to 2 mm day^{-1} for the two oldest studies. These studies also report by far the highest evaporation rates. These E measurements were made using the Bowen ratio technique, which is notoriously difficult to employ over forests because of the small gradients in temperature and humidity there. Except for Lafleur (1992), who also used this technique, the other measurements were made directly by eddy covariance.

For the two broad-leaved forests studied, E was found to be significantly greater than or equal to P but $E \approx E_{eq}$ on average, while E/E_{eq} averaged only 0.7 for the needle-leaved forest studies. The two *Larix* forests had remarkably similar P and E, but E_{eq} was 0.6 mm day^{-1} higher at the eastern Siberian site mostly reflecting the effect of warmer temperatures there. Nevertheless, $E/E_{eq} < 1$ for both *Larix* and *Pinus* forests. These data reflect the well-known degree of surface control of E by needle-leaved forests that follows from examination of Eq. (4) as g_S/g_A declines (McNaughton and

Jarvis, 1983). For nearly 30 years, beginning with the seminal *Pinus sylvestris* forest study of Stewart and Thom (1973), this has been interpreted to indicate or demonstrate the importance of stomatal control of forest evaporation. This is generally correct for relatively well-watered needle-leaved forests of the temperate zone (as noted by Kelliher *et al.* (1998), precious few studies of forest E have been done in stands subject to soil water deficit) that possess large leaf areas, but broad-leaved forests are not so straightforward (Baldocchi and Vogel, 1996). Moreover, extension of this temperate zone knowledge to the much drier and thus usually sparse forests of the boreal zone may be perilous. This is because forests of the boreal zone generally have two significant sources of E namely, the tree canopy as implied above and the understorey including vegetation and soil (Table 4). Consequently, quantitative analysis of E in terms of g_S requires caution in terms of a basic tenet of Eq. (4) (i.e., commonality of height/location of the momentum sink and heat and water vapor sources). This is one of the many reasons we mostly examine boreal zone E in terms of E_{eq}.

Looking more closely at the boreal *Pinus* forest data, while P is relatively similar for the five sites, E from the three *Pinus sylvestris* stands varies by a factor of 2.6. This is mostly attributable to data from the two Swedish stands that are not far apart and yet E at Jadraas is nearly twice that at Norunda. However, E/E_{eq} is the same and relatively large for both sites, suggesting that the difference in E may be attributed to vagaries of the weather during the two studies. This may explain the large defiance of mass/water conservation during Lindroth's study, where $E \gg P$. In central Siberia $P < E$, reflecting a greater contribution of ground evaporation that was much more closely coupled to rainfall frequency limited to a major storm at the beginning of the 18-day study (Table 4). Tree transpiration does not vary so dynamically and, at first glance including Eqs. (2) and (4), its contribution to E seems simply proportional to the overstorey leaf area index regardless of the plant lifeform, genus, or species. Although this may seem an intuitive conclusion, it only accounts for the physics involved in radiation interception and energy balance and the quantity of leaves. The recently published contrary data of Zimmermann *et al.* (2000) are unique. They found that the stand-level transpiration rates of *Pinus sylvestris* trees in a central Siberian chronosequence (aged

TABLE 4 Measured Daily Values of the Percentage of Forest Evaporation (Average with Range in Parentheses) Emanating from the Understorey during Summer in Six Boreal Forests

Overstorey	Overstorey Leaf Area Index	Understorey	% Evaporation from Understorey	Source
Deciduous broad-leaved forest (full leaf)				
Populus	5.1	*Corylus*	22	Black *et al.*, 1996
Deciduous needle-leaved forest (full leaf)				
Larix	1.5	*Betula, Salix, Ledum* /lichen, moss, water	65 (45–87)	Lafleur and Schreader, 1994
Larix	1.5	*Vaccinium, Arctostaphylos*	50	Kelliher *et al.*, 1997
Evergreen needle-leaved forest				
Pinus	3–4	*Vaccinium*, moss	17	Grelle *et al.*, 1997
Pinus	2	*Vaccinium, Arctostaphylos*/lichen	25 (10–40)	Baldocchi *et al.*, 1997
Pinus	1.5	lichen	54 (33–92)	Kelliher *et al.*, 1998

TABLE 5 Maximum Half-Hourly and Daily Forest Evaporation Rates (E_{max}) Measured by the Eddy Covariance Technique above Each of Four *Pinus sylvestris* Stands during July 1996

Tree Age (years)	Tree Leaf Area Index[a]	Half-Hourly E_{max} (mm h^{-1})	Daily E_{max} (mm d^{-1})	Daily E_{max}/E_{eq}	Tree Leaf Nitrogen Content (mg g^{-1})
7	0.2[a]	0.3	1.3	0.5	13.0
53	2.5[b]	0.4	3.2	0.8	11.6
215	1.5[c]	0.3[c]	2.3[c]	0.7	9.5
Open woodland	0.04[a]	0.3	1.7	0.5	n.d.

[a] from Rebmann *et al.*, 1999.
[b] computed from leaf biomass and leaf area density data of Christian Wirth (personal communication).
[c] from Kelliher *et al.*, 1998.

The chronosequence of stands, located 40 km from the village of Zotino in central Siberia, includes an open woodland with a low density of relatively large trees of undetermined age. The values of E_{max} were measured during fine weather when the tree canopy was dry. Also shown are the ratio of daily E_{max} to the equilibrium rate (E_{eq}) and tree leaf nitrogen content.

28–430 years) during July 1995 were proportional to sapwood area and not leaf area index or tree age.

Forests are long-lived vegetation. Although fire probably always limits the lifespan of trees in the boreal zone, they can live for hundreds of years. We are aware of only one study of forest E in a chronosequence of stands. This was also conducted near Zotino, central Siberia, in July 1996; eddy covariance measurement systems were employed simultaneously at four sites within a *ca.* 20-km^2 area (Rebmann *et al.*, 1999; Table 5). The trees varied in age by over 200 years and all were growing on well-drained sand except for a 53-year-old stand located at a drainage collection area where the ground water was <1 m beneath the surface throughout the measurements. This seemed to account for the relatively higher tree leaf area index there (Table 5). However, stand disturbance by fire and the time elapsed therafter are generally considered more influential in determining tree leaf area index in the region (Wirth *et al.*, 2000). For comparison of the data, we begin with the capacity for evaporation by examining maximum half-hourly E during fine weather when the canopy was dry. It is essentially indistinguishable throughout the four stands. This suggests no effect of tree aging, in agreement with Zimmermann *et al.* (2000), although the dominance of ground evaporation in the well-drained stands when E is at a maximum after rainfall was also important. This leads to examination of the maximum daily E during fine weather and the corresponding E_{eq}. The daily data from the 7- and 215-year-old stands and the open woodland were from the first fine day after a significant 12 mm of rain. Nevertheless, the daily data vary across the chronosequence, as noted by Schulze *et al.* (1999) for average values. The highest values of daily maximum E and E/E_{eq} were obtained in stands with the highest tree leaf area index. These stands had the highest tree canopy transpiration rates. The 53-year-old stand had a relatively high tree leaf nitrogen content and the highest leaf area index and E, emphasizing the contribution of tree transpiration. By contrast, owing to the relatively short time since disturbance, the 7-year-old stand had the highest tree leaf nitrogen content but the lowest leaf area index and significantly lower E and E/E_{eq}. Soil evaporation composed half of E in the 215-year-old stand (Kelliher *et al.*,

1998) and the fraction was probably greater in the youngest stand and open woodland. Even on the day after rainfall, atmospheric demand evidently can exceed water supply in these younger stands so that soil evaporation is limited by a drying surface layer.

5. Forest Evaporation, Tree Life Form and Nitrogen

Among boreal zone vegetation, forest E is unique in its variable relation with E_{eq} (Table 3). According to three of these studies that are comparable, the differences in E/E_{eq} are associated with the three tree life-forms found in the boreal zone, namely, deciduous broad-leaved, deciduous needle-leaved, and evergreen needle-leaved (Table 6). During summer, the former two life-forms mostly bear fully grown leaves (see Black *et al.*, 1996). However, life-form is not the only cause of difference in E among these three forests. Leaf area index was identically small in the two sparse needle-leaved forests and greater by more than a factor of 3 in the relatively closed-canopy, broad-leaved forest. Because of this, half of E emanated from the soil of the *Pinus* forest and from the understorey vegetation (leaf area index = 0.5) and soil of the *Larix* forest. The corresponding value was 22% for the understorey (leaf area index = 3.3) of the *Populus* forest, including only 5% from the soil.

The large contribution of soil evaporation in the two Siberian forests meant E was closely coupled to rainfall frequency. Rapid surface drying occurs in the boreal zone during summer because of the generally high D. Thereafter, during fine weather, E is greatly reduced from the turbulence-driven, wet-surface, energy-limited rate to a much lower rate limited by diffusion through the dry surface layer. An illustrative example comes from measurements made during and after the 12-mm fall of rain on 12 July 1996 [two 1.5-h long showers that ended at 1130 hours (1.5 mm) and 1900 hours (10.5 mm)] at the central Siberian *Pinus sylvestris* forest site. The soil in that forest was mostly (65%) covered with a 30-mm-thick carpet of lichen (the rest being forest floor beneath the tree crowns) that had a surface area index of 6 and (dry) bio-

TABLE 6 The Ratio of Forest Evaporation (E) to the Equilibrium Rate (E_{eq}) and Overstorey (Tree) Leaf Nitrogen Content and Area Density (Leaf Area, Expressed on a One-Sided Basis, Produced per kg of Carbon Assimilated)(Normalized Values in Parentheses) during Summer in Three Life Forms of Boreal Forest

Life-form	Summer E/E_{eq}	Overstorey Leaf Nitrogen Content (mg g^{-1})	Overstorey Leaf Area Density (m^2 kg^{-1})	Sources
Deciduous broad-leaved forest (full leaf)	1.0	30 (1)	20.0 (1)	Black *et al.*, 1996; Dang *et al.*, 1997
Deciduous needle-leaved forest (full leaf)	0.6	16 (0.53)	7.2 (0.36)	Kelliher *et al.*, 1997; Vygodskaya *et al.*, 1997
Evergreen needle-leaved forest	0.4	10 (0.33)	1.8 (0.09)	Kelliher *et al.*, 1998; Wirth, 1998 (personal communication).

The three tree species compared, with their leaf area index and age in years, respectively in parentheses, are *Populus tremuloides* (5.1, 70), *Larix gmelinii* (1.5, 130), and *Pinus sylvestris* (1.5, 215), respectively.

mass density of 0.8 kg m^{-2}. Throughfall, captured in five 150-mm diameter lysimeters at ground level, indicated that the tree and lichen canopies intercepted 5.1 (42%) and 1.5 mm of rain, respectively. The lichen canopy thus intercepted 22% of the tree canopy throughfall, a percentage identical to that of feather moss covering the floor of a boreal *Picea mariana* forest in northern Canada (Price *et al.*, 1997), and 78% or only 5.4 mm actually reached the sandy soil. At 0900 hours on 13 July, the lichen water content was 1.7 g g^{-1}, which is within the range of values obtained for boreal forest mosses in equilibrium over distilled water (1.5–2.5 g g^{-1}; Busby and Whitfield, 1978). For Busby and Whitfield, these values were only 10–20% of the saturated values (determined by immersion followed by several minutes of drainage before weighing), suggesting that 80–90% of the water was retained as capillary films on (i.e., not within) their moss samples. Our field evaporation data support this conclusion because on the fine day after the rain, the lysimeters averaged a weight loss equivalent to 1.6 mm day^{-1}, suggesting that soil evaporation contributed only 0.1 mm (see also Skre *et al.*, 1983). On the following two days, soil evaporation was 1.2 and 0.6 mm day^{-1}, respectively, and it remained at about 0.5 mm day^{-1} for the next nine fine days until the next rainfall (Kelliher *et al.*, 1998). Rainfall intercepted by the lichen canopy was thus "gone" within one day. Nearly a quarter of the rainfall that reached the soil was evaporated away on the second day and, as stated above, the asymptotic lower limit of soil evaporation was reached by the third day.

Beyond the leaf area index and fractional understorey E differences, overstorey leaf nitrogen contents (N) are also significantly different among the three life-forms of forest. Recalling Eq. (2), for simplicity, we consider N in light of its proportionality to the maximum value of g_S ($g_{S_{max}}$ (mm s^{-1}) = 0.9 N (mg g^{-1}) for vegetation; Schulze *et al.*, 1994). For the *Populus*, *Larix*, and *Pinus* forests, we thus estimate $g_{S_{max}}$ = 27, 14, and 9 mm s^{-1}, respectively. Reasonably corroborating the Siberian estimates are $g_{S_{max}}$ values derived from micrometeorological measurements using the Penman–Monteith equation (Monteith and Unsworth, 1990): 10 and 8 mm s^{-1} for the *Larix* and *Pinus* forests, respectively (Kelliher *et al.*, 1997; Schulze *et al.*, 1999). The proportionality between $g_{S_{max}}$ and N originates at the scale of a leaf with the

maximum value of stomatal conductance ($g_{st_{max}}$) (Schulze *et al.*, 1994). The evergreen needle-leaved boreal forest data of Roberntz and Stockfors (1998) are illustrative here. Fertilized *Picea abies* trees growing in northern Sweden (64°N) gave values of $g_{st_{max}}$ and N, both of which were 30% larger than for their unfertilized counterparts.

Although g_S in Eq. (2) does not constantly equal $g_{S_{max}}$, it is an important parameter determining the capacity of vegetation for evaporation governed by atmospheric demand or D_0 in Eq. (2). Because E cannot always meet this demand, it is also instructive to examine the relation between g_S and D_0. The Lohammer function fits most field data of this relation and it is simply written as $g_S = g_{S_{max}} [1/\{1 + (D_0/D_{50})\}]$, where D_{50} is the value of D_0 when $g_S = g_{S_{max}}/2$. For the Siberian *Larix* and *Pinus* forest data, Schulze *et al.* (1999) found that a single relation emerged and we find that the Lohammer function fits it sufficiently with $D_{50} = 10$ mmol mol^{-1} and a slightly different $g_{S_{max}}$ = 11 mm s^{-1}.

There are thus a number of similarities and differences between the deciduous and evergreen needle-leaved Siberian forests that can affect E/E_{eq} during summer. The two forests had the same sparse overstorey leaf area indices (1.5), response of g_S to D_0, and understorey contributions to E. On the other hand, in terms of the vegetation, overstorey N and $g_{S_{max}}$ were greater for *Larix* than for *Pinus*, in agreement with the wider ranging study of Kloeppel *et al.* (1998). Leaf area density (m^2 of leaf area produced per kg of carbon assimilated) is larger by a factor of 4 for *Larix* than *Pinus*, but the *Pinus* leaves live up to 6 years in the Zotino stand. The evergreen *Pinus* leaves also contain higher concentrations of secondary compounds such as lignin that deter herbivores (Kloeppel *et al.*, 1998) and waxes which, along with stomatal closure, minimize winter desiccation when soil water is frozen and unavailable. These additional constituents, not used in carbon assimilation and thus not determining stomatal conductance (Wong *et al.*, 1979), which is a component of g_S, may also effectively dilute N (Kloeppel *et al.*, 1998). Another relevant climatological difference is the warmer and drier summer conditions, leading to a generally higher D_0, in the eastern Siberian *Larix* forest (Table 1). Higher $g_{S_{max}}$ and D_0 in the *Larix* forest, and Eq. (2), thus explain the higher E/E_{eq} there compared to the central Siberian *Pinus* forest.

Equation (5) states how stomata connect evaporation, carbon assimilation, and nutrient cycles. Distilled in the CANVEG model, the coordination of low leaf N, leaf area, photosynthetic capacity, g_S, and E/E_{eq} was demonstrated for the boreal forest by Baldocchi *et al.* (2000). Experimentally, the fertilized *Picea abies* trees of Roberntz and Stockfors (1998) obtained a maximum value of A (A_{max})(Wong *et al.*, 1979) that was, like $g_{st_{max}}$ and N, 30% larger than for their unfertilized counterparts. Furthermore, over a four-year period, stemwood growth rate of the fertilized trees was greater by a factor of 4.3 (Linder *et al.*, 1996). For co-occurring *Larix* and *Pinus* trees, including two central Siberian sites, leaf carbon isotope discrimination was 1–3‰ greater for *Larix* (Kloeppel *et al.*, 1998). Kloeppel *et al.* interpreted their data as indicating a greater water loss per unit of carbon assimilated for *Larix* because of its higher stomatal conductance.

The deciduous broad-leaved forest evaporates at the equilibrium rate during summer, like the boreal zone's other vegetation. As stated earlier, overstorey N and leaf area index in the *Populus* forest are greater by as much as a factor of 3 than in the needle-leaved Siberian forests. This is not coincidental. The higher N of *Populus* leaves reflects its earlier successional position following a disturbance (i.e., fire) that releases a significant quantity of nitrogen. As time goes on following disturbance, the later successional coniferous forest leaves "lose" much of this N in woody tissue (Schulze *et al.*, 1999), although subsequent nonlethal fires may release more nitrogen. Photosynthetic capacity is proportional to N (Schulze *et al.*, 1994), so *Populus* forest is relatively fast-growing, including a large leaf area index that intercepts virtually all of the above-forest irradiance (recall how soil E was only 5% of the total in the *Populus* forest).

Higher N also means $g_{S_{max}}$ is much larger in the *Populus* forest than in the needle-leaved Siberian forests. Using Eq. (2) to further explore the effects of life-form on E will be partly dependent on wind speed because the larger *Populus* leaves may be isolated from the effects of D during relatively calm conditions when essentially $D \neq D_0$. This is incorporated into Eq. (4) but, because of the generally similar and tall heights of the three forests (the *Larix*, *Pinus*, and *Populus* trees were up to 20, 22, and 21 m tall, respectively), it will generally not significantly affect the nature of our comparison and so we shall continue with Eq. (2). Thus, if the *Populus* forest D_{50} in the Lohammer function is similar to the *Larix* and *Pinus* forest value of 10 mmol mol^{-1} and the D_0 climate is similar to that in Siberia, although we know it varies there, the higher E/E_{eq} of the *Populus* forest could be explained solely by the higher N and $g_{S_{max}}$. In northern Canada, leaf carbon isotope discrimination values were around 3‰ greater for *Populus* than for *Pinus*, indicating a much more conservative water use efficiency for the latter (Brooks *et al.*, 1997). Brooks *et al.* obtained consistent results in years with above- and below-average precipitation, suggesting that environmental changes did not alter the life-form ranking. More generally, in a global study of leaf carbon isotope discrimination, Lloyd and Farquhar (1994) showed that deciduous species use water less efficiently than evergreen conifers.

In terms of deciduous versus evergreen habit, a boreal tree's life-form also reflects air bubble production inside its conduits during the freezing of xylem sap because dissolved air in the sap is insoluble in ice (i.e., winter embolism; Sperry, 1995). Larger xylem conduits tend to be more vulnerable to cavitation by freezing and thawing than smaller ones. *Populus* can avoid this problem because of its deciduous habit, whereas *Larix* is an exception, although its habit is reflected in its higher N compared to *Pinus*, as discussed earlier. For *Populus tremuloides* trees in Alaska (*ca.* 65°N, 148°W), vessel diameter ranged from 5 to 50 μm with around half being between 20 and 30 μm (Sperry *et al.*, 1994). For the xylem of conifers, lumen diameters in tracheids (i.e., the radial hole) are generally around an order of magnitude smaller and every few millimeters, the sap has to pass through the extremely fine pores of bordered pit membranes (Whitehead and Jarvis, 1981; Vysotskaya and Vaganov (1991) report distributions of radial tracheid diameters for co-occurring central Siberian *Pinus* and *Larix* trees that are similar to the range of values given by Sperry *et al.*, but one cell-wall thickness was included with the tracheid lumens.) In accordance with Poisueille's equation, increased conduit size should lead to increased conducting efficiency of the xylem. Sperry *et al.* found that the branch xylem hydraulic conductance was nearly an order of magnitude larger for *Populus tremuloides* than for *Larix laricina*. This would also confer an evaporative advantage on *Populus*, compared to *Larix* and *Pinus*, which would be enhanced by its relatively higher leaf area index to yield a greater forest E relative to E_{eq}.

6. Conclusions

In the boreal zone, summer rainfall frequency and quantity decline significantly from south to north, except in maritime Scandinavia. Correspondingly, the vegetation changes from forest to tundra with wetland being relatively ubiquitous. Average rainfall rate and soil water storage capacity range from ca. 0.5–2.0 mm day^{-1} and 8–13 mm water per 0.1 m soil, respectively. Snowmelt generally ensures a relatively full store of soil water at the beginning of summer. Wetland can receive a supplemental supply of water that accounts for its evaporation (averaging 2.6 mm day^{-1}), sometimes exceeding rainfall. Tundra and forest evaporation (averaging 1.5 and 2.2 mm day^{-1}, respectively) and rainfall rates are almost equal, illustrating the dominant effect of summer precipitation on terrestrial ecosystem evaporation rate. For wetland, tundra, and broad-leaved deciduous forest, seasonal average evaporation obtains the theoretically expected equilibrium rate. Given patchiness of the boreal landscape and entrainment of dry air into the convective boundary layer on fine summer days, we do not consider these surfaces to be completely devoid of influence on E. However, a variety of sometimes unrelated factors apparently compensates over the course of a summer. For deciduous and evergreen needle-leaved forests, evaporation is about 70 and 50%, respectively, of the equilibrium rate, indicating an overwhelming degree of surface control. Among the three tree life-forms found in the boreal zone, forest evaporation rate is physiologically related to overstorey leaf habit, xylem anatomy, and especially successional position following disturbances such as fire. For the forests compared, this determines leaf nitrogen content and in turn the maximum stomatal and surface conductances and the leaf area index via the effects on photosynthetic and growth rates. The leaf area index affects understorey evaporation rate which is half the total in the needle-leaved forests, and is governed largely by rainfall frequency. These conclusions about tree life-form are supported by data on leaf carbon isotope discrimination reflecting close linkages between nutrients, water use, and carbon gain in boreal forests.

Acknowledgments

FMK is grateful to the Max Planck Institute for Biogeochemistry for funding his attendance at the meeting that led to this chapter. His support for preparation of the essay came from a long-term grant for atmospheric research from the New Zealand Foundation for Research, Science and Technology. Michael Raupach and Frank Dunin generously, persistently, and patiently gave FMK many valuable lessons about the equilibrium evaporation rate. In New Zealand, writing a paper about the boreal zone could not have been done without the excellent interloan support of librarians Izabella Kruger and Chris Powell. Brian Amiro, Andy Black, Achim Grelle, Andy Suyker, and Sashi Verma kindly contributed by correspondence to the preparation of Table 3. John Hunt and David Whitehead did helpful critiques of a draft manuscript. Finally, we are indebted to Nina Wagner for her expert translation of the Vysotskaya and Vaganov paper from Russian to English.

References

Addison, P. A., and Bliss, L. C. (1980). Summer climate, microclimate, and energy budget of a polar semidesert on King Christian Island, N. W. T., Canada. *Arctic Alpine Res.* **12**, 161–170.

Alphalo, P. J., and Jarvis, P. G. (1991). Do stomata respond to relative humidity? *Plant Cell Environ.* **14**, 127–132.

Alphalo, P. J., and Jarvis, P. G. (1993). The boundary layer and the apparent responses of stomatal conductance to wind speed and to the mole fractions of CO_2 and water vapour in the air. *Plant Cell Environ.* **16**, 771–783.

Amiro, B. D., and Wuschke, E. E. (1987). Evapotranspiration from a boreal forest drainage basin using an energy balance/eddy covariance technique. *Boundary Layer Meteorol.* **38**, 125–139.

Baldocchi, D. D., and Vogel, C. A. (1996). Energy and CO_2 flux densities above and below a temperate broad-leaved forest and boreal pine forest. *Tree Physiol.* **16**, 5–16.

Baldocchi, D. D., Vogel, C. A., and Hall, B. (1997). Seasonal variation of energy and water vapor exchange rates above and below a boreal jack pine forest. *J. Geophys. Res.* **102**, 28,939–28,951.

Baldocchi, D. D., Kelliher, F. M., Black, T. A., and Jarvis, P. G. (2000). Climate and vegetation controls on boreal zone energy exchange. *Global Change Biol.* **6**, 69–83.

Barry, R. G. (1995). Land of the midnight sun. *In* "Polar Regions", pp. 28–39, RD Press, Surrey Hills, New South Wales, Australia.

Baumgartner, A., and Reichel, E. (1975). "The World Water Balance". Elsevier, Amsterdam.

Beckinsale, R. P. (1969). River regimes. *In* "Water, Earth and Man", (R. J. Chorley, Ed.), pp. 455–471. Methuen, London.

Black, T. A., den Hartog, G., Neumann, H., Blanken, P., Yang, P., Nesic, Z., Chen, S., Russel, C., Voroney, P., and Stabler, R. (1996). Annual cycles of CO_2 and water vapor fluxes above and within a boreal aspen stand. *Global Change Biol.* **2**, 219–230.

Brooks, J. R., Flanagan, L. B., Buchmann, N., and Ehleringer, J. R. (1997). Carbon isotope composition of boreal plants: functional grouping of life forms. *Oecologia* **110**, 301–311.

Brutsaert, W., and Parlange, M. B. (1998). Hydrologic cycle explains the evaporation paradox. *Nature* **396**, 30.

Busby, J. R., and Whitfield, D. W. A. (1978). Water potential, water content, and net assimilation of some boreal forest mosses. *Can. J. Bot.* **56**, 1551–1558.

Calder, I. R., Wright, I. R., and Murdiyarso, D. (1986). A study of evaporation from tropical rain forest—West Java. *J. Hydrol.* **89**, 13–31.

Campbell, G. S. (1985). "Soil Physics with BASIC." Elsevier, Amsterdam.

Choudhury, B. J., and DiGirolamo, N. E. (1998). A biophysical model of global land surface evaporation using satellite and ancillary data. I. Model description and comparison with observations. *J. Hydrol.* **205**, 164–185.

Clothier, B. E., Vogeler, I., Green, S. R., and Scotter, D. R. (1998). Transport in unsaturated soil: Aggregates, macropores, and exchange. *In* "Physical Nonequilibrium in Soils: Modelling and Application" (H. Magdi Salam and L. Ma, Eds.), pp. 273–295. Ann Arbor Press, Chelsea, Michigan.

Cosby, B. J., Hornberger, G. M., Clapp, R. B., and Ginn, T. R. (1984). A statistical exploration of the relationships of soil moisture characteristics to the physical properties of soils. *Water Resour. Res.* **20**, 682–690.

Dang, Q. L., Margolis, H. A., Sy, M., Coyea, M. R., Collatz, G. J., and Waltham, C. L. (1997). Profiles of photosynthetically active radiation, nitrogen and photosynthetic capacity. *J. Geophys. Res. D* **102**, 28,845–28,860.

De Bruin, H. A. R. (1983). Evapotranspiration in humid tropical regions. *In* "Hydrology of humid tropical regions." (R. Keller, Ed.), pp. 299–311. IAHS Publication No. 140.

den Hartog, G., Neumann, H. H., King, K. M., and Chipanshi, A. C. (1994). Energy budget measurements using eddy correlation and Bowen ratio techniques at the Kinosheo Lake tower site during the Northern Wetlands Study. *J. Geophys. Res. D* **99**, 1539–1549.

Evans, J. R. (1989). Photosynthesis and nitrogen relationships in leaves of C_3 plants. *Oecologia* **78**, 9–19.

Field, C. B., and Mooney, H. A. (1986). The photosynthetic–nitrogen relationship in wild plants. *In* "On the Economy of Plant Form and Function" (T. J. Givnish, Ed.), pp. 25–56. Cambridge Univ. Press, Cambridge, United Kingdom.

Finnigan, J. J., and Raupach, M. R. (1987). Transfer processes in plant canopies in relation to stomatal characteristics. *In* "Stomatal Function." (E. Zieger, G. D. Farquhar, and I. R. Cowan, Eds.), pp. 385–429. Stanford University Press, Stanford, California.

Fitzjarrald, D. R., and Moore, K. E. (1994). Growing season boundary layer climate and surface exchanges in a subarctic lichen woodland. *J. Geophys. Res. A* **99**, 1899–1917.

Graedel, T. E., and Crutzen, P. J. (1993). "Atmospheric Change: A Global System Perspective." Freeman, New York.

Grelle, A., Lundberg, A., Lindroth, A., Moran, A.-S., and Cienciala, E. (1997). Evaporation components of a boreal forest: Variation during the growing season. *J. Hydrol.* **197**, 70–87.

Hare, F. K. (1997). Canada's climate: An overall perspective. *In* "The Surface Climates of Canada" (W. G. Bailey, T. R. Oke, and W. R. Rouse, Eds.), pp. 3–20, McGill–Queen's Univ. Press, Montreal.

Hertzman, O. 1997. Oceans and the coastal zone. *In* "The Surface Climates of Canada" (W. G. Bailey, T. R. Oke, and W. R. Rouse, Eds.), pp. 101–123. McGill-Queen's University Press, Montreal.

Jackson, R. B., Mooney, H. A., and Schulze, E.-D. (1997). A global budget for fine root biomass, surface area, and nutrient contents. *Proc. Natl. Acad. Sci.* USA **94**, 7362–7366.

Jarvis, P. G., Massheder, J. M., Hale, S. E., Moncrieff, J. B., Rayment, M., and Scott, S. L. (1997). Seasonal variation of carbon dioxide, water vapor, and energy exchanges of a boreal black spruce forest. *J. Geophys. Res.* **D 102**, 28,953–28,966.

Kane, D. L., Gieck, R. E., and Hinzman, L. D. (1990). Evapotranspiration from a small Alaskan arctic watershed. *Nordic Hydrol.* **21**, 253–272.

Kelliher, F. M., Hollinger, D. Y., Schulze, E.-D., Vygodskaya, N. N., Byers, J. N., Hunt, J. E., McSeveny, T. M., Milukova, I., Sogatchev, A., Varlargin, A., Ziegler, W., Arneth, A., and Bauer, G. (1997). Evaporation from an eastern Siberian larch forest. *Agric. For. Meteorol.* **85**, 135–147.

Kelliher F. M., Lloyd, J., Arneth, A., Byers, J. N., McSeveny, T. M., Milukova, I., Griegoriev, S., Panfyrov, M., Sogatchev, A., Varlargin, A., Ziegler, W., Bauer G., and Schulze, E.-D. (1998). Evaporation from a central Siberian pine forest. *J. Hydrol.* **205**, 279–296.

Kloeppel, B. D., Gower, S. T., Treichel, I. W., and Kharuk, S. (1998). Foliar carbon isotope discrimination in *Larix* species and sympatric evergreen conifers: A global comparison. *Oecologia* **114**, 153–159.

Körner, Ch. (1994). Leaf diffusive conductances in the major vegetation types of the globe. *In* "Ecophysiology of Photosynthesis" (E.-D. Schulze and M. M. Caldwell, Eds.), Ecological Studies , Vol. 100, pp. 463–490. Springer-Verlag, Berlin.

Lafleur, P. M. (1992). Energy balance and evapotranspiration from a subarctic forest. *Agric. For. Meteorol.* **58**, 163–175.

Lafleur, P. M., McCaughey, J. H., Joiner, D. W., Bartlett, P. A., and Jelinski, D. E. (1997). Seasonal trends in energy, water, and carbon dioxide fluxes at a northern boreal wetland. *J. Geophys. Res.* **D 102**, 29,009–29,020.

Lafleur, P. M., and Schreader, C. P. (1994). Water loss from the floor of a subarctic forest. *Arctic Alpine Res.* **26**, 152–158.

Leemans, R., and Cramer, W. (1991). The IIASA climate database for mean monthly values of temperature, precipitation and cloudiness on a terrestrial grid (RR–9118). IIASA, Laxenberg.

Linder, S., McMurtrie, R. E., and Landsberg, J. J. (1996). Global change impacts on managed forests. *In* "Global Change and Terrestrial Ecosystems." (B. Walker and W. Steffen, Eds.), pp. 275–290. Cambridge University Press, Cambridge, United Kingdom.

Lindroth, A. (1985). Seasonal and diurnal variation of energy budget components in coniferous forests. *J. Hydrol.* **82**, 1–15.

Lloyd, J., and Farquhar, G. D. (1994). ^{13}C discrimination during CO_2 assimilation by the terrestrial biosphere. *Oecologia* **99**, 201–215.

McFadden, J. P., Chapin III F. S., and Hollinger, D. Y. (1998). Subgrid-scale variability in the surface energy balance of arctic tundra. *J. Geophys. Res.* **D 103**, 28,947–28,961.

McGlone, M. S. (1996). When history matters: Scale, time, climate and tree diversity. *Global Ecol. Biogeogr. Lett.* **5**, 309–314.

McNaughton, K. G. (1976). Evaporation and advection I: Evaporation from extensive homogeneous surfaces. *Quart J. R. Meteorol. Soc.* **102**, 181–191.

McNaughton, K. G., and Jarvis, P. G. (1983). Predicting effects of vegetation changes on transpiration and evaporation. *In* "Water Deficits and Plant Growth" (T. T. Kozlowski, Ed.), Vol. VII, pp. 1–47, Academic Press, New York.

Massmann, W. J. (1998). A review of the molecular diffusivities of H_2O, CO_2, CH_4, CO, O_3, SO_2, NH_3, N_2O, NO, and NO_2 in air, O_2 and N_2 near STP. *Atmos. Environ.* **32**, 1111–1127.

Matthews, E. (1989). Global data bases on distribution, characteristics and methane emissions of natural wetlands: Documentation of archived data tape. NASA Technical Memorandum 4253.

Matthews, E., and Fung, I. (1987). Methane emission from natural wetlands: Global distribution, area, and environmental characteristics of sources. *Global Biogeochem. Cycles* **1**, 61–86.

Miller, J. R., and O'Neill, N. T. (1997). Multialtitude airborne observations of insolation effects of forest fire smoke aerosols at BOREAS: Estimates of aerosol optical parameters. *J. Geophys. Res.* **D 102**, 29,729–29,736.

Monteith, J. L. (1995). Accommodation between transpiring vegetation and the convective boundary layer. *J. Hydrol.* **166**, 251–263.

Monteith, J. L., and Unsworth, M. H. (1990). "Principles of Environmental Physics.", Edward Arnold, London.

Moore, K. E., Fitzjarrald, D. R., Wofsy, S. C., Daube, B. C., Munger, J. W., Bakwin, P. S., and Crill, P. (1994). A season of heat, water vapor, total hydrocarbon, and ozone fluxes at a subarctic fen. *J. Geophys. Res.* **D 99**, 1937–1952.

Mott, K., and Parkhurst, D. F. (1991). Stomatal responses to humidity in air and helox. *Plant Cell Environ.* **14**, 509–515.

Müller, M. J. (1982). "Selected Climatic Data for a Global Set of Standard Stations for Vegetation Science." Dr. W. Junk Publishers, The Hague.

Nielsen, D. R., Kutilek, M., and Parlange, M. B. (1996). Surface soil water content regimes: Opportunities in soil science. *J. Hydrol.* **184**, 35–55.

Nielsen, D. R., Reichardt, K., and Wierenga, P. J. (1983). Characterization of field-measured soil-water properties. *In* "Isotope and Radiation Techniques in Soil Physics and Irrigation Studies 1983." pp. 55–78. Report IAEA–SM–267/40, International Atomic Energy Agency, Vienna.

Ohmura, A. (1982). Evaporation from the surface of the arctic tundra on Axel Heiberg Island. *Water Resour. Res.* **18**, 291–300.

Peterson, T. C., Golubev, V. S., and Groisman, P. Y. (1995). Evaporation losing its strength. *Nature* **377**, 687–688.

Plamandon, A. P., Black, T. A., and Goodell, B. C. (1975). The role of hydrologic properties of the forest floor in watershed hydrology. *In* "Watersheds in Transition." pp. 341–348. American Water Resources Association, Colorado Springs, Colorado.

Priestley, C. H. B. (1966). The limitation of temperature by evaporation in hot climates. *Agric. Meteorol.* **3**, 241–246.

Priestley, C. H. B., and Taylor, R. J. (1972). On the assessment of the surface heat flux and evaporation using large-scale parameters. *Mon. Weather Rev.* **100**, 81–92.

Price, A. G., Dunham, K., Carleton, T., and Band, L. (1997). Variability of water fluxes through the black spruce (*Picea mariana*) canopy and feather moss (*Pleurozium schreberi*) carpet in the boreal forest of Northern Manitoba. *J. Hydrol.* **196**, 310–323.

Rauner, J. U. L. (1976). Deciduous forests. *In* "Vegetation and the Atmosphere." (J. L. Monteith, Ed.), pp. 241–264. Academic Press, London.

Raupach, M. R. (1998). Influences of local feedbacks on land–air exchanges of energy and carbon. *Global Change Biol.* **4**, 477–494.

Raupach, M. R. (2001). Combination theory and equiluibrium evaporation. *Quart. J. R. Meteorol. Soc.* (in press).

Rebmann, C., Kolle, O., Ziegler, W., Panyorov, M., Varlargin, A., Kelliher, F. M., Milyokova, I., Wirth, C., Lühker, B., Lloyd, J., Valentini, R., Dore, S., Marchi, G., Schulze, E.-D. (1999). Exchange of carbon dioxide and water vapour between the atmosphere and three central Siberian pine forest stands with different aged trees, land-use and fire history. *Agric. For. Meteorol.* (submitted).

Roberntz, P., and Stockfors, J. (1998). Effects of elevated CO_2 concentration and nutrition on net photosynthesis, stomatal conductance and needle respiration of field-grown Norway spruce trees. *Tree Physiol.* **18**, 233–241.

Rouse, W. R. (1990). The regional energy balance. *In* "Northern Hydrology: Canadian Perspectives" (T. D. Prowse and C. S. L. Ommanney, Eds.), pp. 187–206. National Hydrology Research Institute Report 1, Environment Canada, Ottawa.

Rouse, W. R., Hardhill, S. G., and Lafleur, P. (1987). The energy balance in the coastal environment of James Bay and Hudson Bay during the growing season. *J. Climatol.* **7**, 165–179.

Schulze, E.-D., Kelliher, F. M., Körner, Ch., Lloyd, J., and Leuning, R. (1994). Relationships between maximum stomatal conductance, ecosystem surface conductance, carbon assimilation rate and plant nutrition: A global ecology scaling exercise. *Annu. Rev. Ecol. Systematics* **25**, 629–660.

Schulze, E.-D., Lloyd, J., Kelliher, F. M., Wirth, C., Rebmann, C., Lühker, B., Mund, M., Milykova, I., Schulze, W., Ziegler, W., Varlargin, A., Valentini, R., Dore, S., Grigoriev, S., Kolle, O., and Vygodskaya, N. N. (1999). Productivity of forests in the Eurosiberian region and their potential to act as a carbon sink–A synthesis. *Global Change Biol.* **5**, 703–722.

Skartveit, A., Rydén, B. E., and Kärenlampi, L. (1975). Climate and hydrology of some Fennoscandian tundra ecosystems. *In* "Fennoscandian Tundra Ecosystems. Part 1, Plants and Microorganisms" (F. E. Wielgolaski, Ed.), pp. 41–53, Springer-Verlag, Berlin.

Skre, O., Oechel, W. C., and Miller, P. M. (1983). Moss leaf water content and solar radiation at the moss surface in a mature black spruce forest in central Alaska. *Can. J. Forest Res.* **13**, 860–868.

Slatyer, R. O., and McIlroy, I. C. (1961). "Practical Microclimatology with Special Reference to the Water Factor in Soil–Plant–Atmosphere Relations." UNESCO.

Sperry, J. S. (1995). Limitations on stem water transport and their consequences. *In* "Plant Stems: Physiology and Functional Morphology" (B. L. Gartner, Ed.), pp. 105–149. Academic Press, San Diego.

Sperry, J. S., Nichols, K. L., Sullivan, J. E. M., and Eastlack, S. E. (1994). Xylem embolism in ring-porous, diffuse-porous, and coniferous trees of Northern Utah and interior Alaska. *Ecology* **75**, 1736–1752.

Stewart, J. B., and Thom, A. S. (1973). Energy budgets in pine forest. *Quart. J. R. Meteorol. Soc.* **99**, 154–170.

Suyker, A. E., and Verma, S. B. (1998). Surface energy fluxes in a boreal wetland. *In* "Proceedings of the 23rd American Meteorological Society Conference on Agricultural and Forest Meteorology, 2–6 November 1998," p. 271. Albuquerque, New Mexico.

Thom, A. S. (1975). Momentum, mass and heat exchange of plant canopies. *In* "Vegetation and the Atmosphere" (J. L. Monteith, Ed.), Volume 1, pp. 57–109. Academic Press, London.

Vourlitis, G. L., and Oechel, W. C. (1997). Landscape-scale CO_2, H_2O and energy flux of moist–wet coastal tundra ecosystems over two growing seasons. *J. Ecol.* **85**, 575–590.

Vygodskaya, N. N., Milukova, I., Varlargin, A., Tatarinov, F., Sorgachov, A., Bauer, G., Hollinger, D. Y., Kelliher, F. M., and Schulze, E.-D. (1997). Leaf conductance and CO_2 assimilation of *Larix gmelinii* under natural conditions of Eastern Siberian boreal forest. *Tree Physiol.* **17**, 607–615.

Vysotskaya, L. G., and Vaganov, E. A. (1991). Variability of radial tracheid sizes in the annual rings of some conifers. *Bot. Zh.* **76**, 564–571.

Walter, H. (1985). "Vegetation of the Earth," 3rd ed. translated from German by O. Muise, Springer-Verlag, Berlin.

Whitehead, D., and Jarvis, P. G. (1981). Coniferous forests and plantations. *In* "Plant Water Deficits and Growth." (T. T. Kozlowski, Ed.), Vol. VII, pp. 49–152. Academic Press, NY.

Wirth, C., Schulze, E.-D., Schulze, W., von Stünzner-Karbe, D., Ziegler, Milukova, I., Sogatchov, A., Varlargin, A., Panfyrov, M., Griegoriev, S., Kusnetova, W., Siry, M., Hardes, G., Zimmerman, R., and Vygodskaya, N. N. (2000). Above-ground biomass in pristine Siberian Scots pine forest as controlled by competition and fire. *Oecologia* **121**, 66–80.

Wong, C.-S., Cowan, I. R., Farquhar, G. D. (1979). Stomatal conductance correlates with photosynthetic capacity. *Nature* **282**, 424–426.

Zimmermann, R., Schulze, E.-D., Wirth, C., Schulze, E. E., McDonald, K. C., Vygodskaya, N. N., and Ziegler, W. (2000). Canopy transpiration in a chronosequence of central Siberian pine forests. *Global Change Biol.* **6**, 25–37.

Zobler, L. (1986). "A world Soil File for Global Climate Modelling." NASA Technical Memo 87802.

Roberntz, P., and Stockfors, J. (1998). Effects of elevated CO_2 concentration and nutrition on net photosynthesis, stomatal conductance and needle respiration of field-grown Norway spruce trees. *Tree Physiol.* **18**, 233–241.

Rouse, W. R. (1990). The regional energy balance. In "Northern Hydrology, Canadian Perspectives" (T. D. Prowse and C. S. L. Ommanney, Eds.), pp. 187–206. National Hydrology Research Institute, Report 1, Environment Canada, Ottawa.

Rouse, W. R., Hardill, S. G., and Lafleur, P. (1987). The energy balance in the coastal environment of James Bay and Hudson Bay during the growing season. *J. Climatol.* **7**, 165–179.

Schulze, E. D., Kelliher, F. M., Korner, Ch., Lloyd, J., and Leuning, R. (1994). Relationships between maximum stomatal conductance, ecosystem surface conductance, carbon assimilation rate and plant nutrition: A global ecology scaling exercise. *Annu. Rev. Ecol. Systematics* **25**, 625–660.

Schulze, E.-D., Lloyd, J., Kelliher, F. M., Wirth, C., Rebmann, C., Lühker, B., Mund, M., Milyukova, I., Schulze, W., Ziegler, W., Varlagin, A., Valentini, R., Dore, S., Grigoriev, S., Kolle, O., and Vygodskaya, N. N. (1999). Productivity of forests in the Eurosiberian boreal region and their potential to act as a carbon sink – A synthesis. *Global Change Biol.* **5**, 703–722.

Sharratt, B. S., and Kirschbaum, J. (1972). Climate and hydrology of some peat Permafrost tundra ecosystems. In "Permacausedzan Ecosystems" (E. E. Weber, Eds.) pp. 212–223. Springer-Verlag, Berlin.

Sheila, D. L., and Blake, S. M. (1980). Mass leaf water content and snow reduction at the moss surface ?? a mature black spruce stand.

Sleger, D. J., and McNaughton, D. J. (1981). Dynamics of the relationship with respect to the Water Factor ... In "Plant-Atmosphere Relation..."

Smith, J. J. (1981). Distribution on earth water ... Ecol. Syst. ...

Smith, K. L., Cameron, R. L., and Nuttall, R. L. (1994). Distribution of mass potato tubers potato, and continuous transmission from forest in boreal forests. *Ecology* **75**, 1736–1772.

Stewart, J. B., and Thom, A. S. (1973). Energy budgets in pine forest. *Quart. J. R. Meteorol. Soc.* **99**, 154–170.

Stylar, A. D., and Verma, S. B. (1994). Surface energy fluxes in a boreal wetland. In "Proceedings of the 23rd American Meteorological Society Conference on Agricultural and Forest Meteorology, 2–6 November 1998," p. 174. Albuquerque, New Mexico.

Thom, A. S. (1975). Momentum, mass, and heat exchange of plant canopies. In "Vegetation and the Atmosphere" (J. L. Monteith, Ed.), Vol. 1, pp. 57–109. Academic Press, London.

Verhulst, G. L., and Oechel, W. C. (1991). Landscape scale CO_2, H_2O and energy flux of moist-wet coastal tundra ecosystems over two growing seasons. *J. Ecol.* **85**, 575–590.

Vygodskaya, N. N., Milukova, I., Varlagin, A., Tatarinov, F., Sogatchev, A., Bauer, G., Hollinger, D. Y., Kelliher, F. M., and Schulze, E.-D. (1997). Leaf conductance and CO_2 assimilation of *Larix gmelinii* under natural conditions of Eastern Siberian boreal forest. *Tree Physiol.* **17**, 607–615.

Vyarosxkovi, I. G., and Nigunova, I. A. (2001). Imbalance of boundary layer in the annual river in some centers. *Res. J. Bot. Sci.* **...**

Walter, H. (1988a). "Vegetation of the Earth and its related from Climate." G. Mohler Springer-verlag, Berlin.

Whitehead, D., and Jarvis, P. G. (1981). Coniferous forest and plantations. In "Tissue Water Deficits and ..." (T. Kozlowski, Ed.), Vol. VI, pp. 49–152. Academic Press, NY.

Wirth, C., Schulze, E.-D., Schulze, W., von Stünzner-Karbe, D., Ziegler, W., Miljukowa, I. M., Sogatchev, A., Varlagin, A., Panvyörov, M., Grigoriev, S. V., Kusnetzova, W., Zimmermann, R., and Vygodskaya, N. N. (2000). Above ground biomass and structure of pristine Siberian Scots pine forests as controlled by competition and fire. *Oecologia* **121**, 66–80.

Woods, G. A., Cowan, I. R., Farquhar, G. D. (1979). Stomatal and cuticular transpiration with photosynthesis in plants. *Photosynth.* **53**, 636–670.

Wring, R. L., Richardson, I. R., Russell, G. D., Jorol, J., and ... Sartz, S. J., Schulz, H. (...). ... mass exchange with the atmosphere in a spruce-feathermoss boreal forest. *J. Geophys. Res.* ...

Wring, R. A. (...). The BOREAS mesonet in the southern study area ... NASA Technical Memo (...).

1.12

Past and Future Forest Response to Rapid Climate Change

Margaret B. Davis
University of Minnesota,
St. Paul, Minnesota

In response to large changes of climate during the Holocene, geographical ranges of tree species shifted northward in eastern North America, with range extensions occurring at rates of 10–100 km per century. Long-distance dispersal of seeds, an important mechanism for rapid range extension, is documented by fossil evidence for colonies established well in front of the continuous population. Average jump distance as trees moved into the Great Lakes region was 80–100 km for eastern hemlock (*Tsuga canadensis*), with wind-dispersed seed, and 40 km for beech (*Fagus grandifolia*), which has animal-dispersed seed. Jump dispersal distances estimated from range maps, by measuring distances between outlying colonies and the continuous population, are again larger for hemlock than for beech—40 km versus 8 km. Interactions with resident vegetation were constraints on migration rates. Invasions of individual forest stands by hemlock were restricted to stands dominated by white pine (*Pinus strobus*). Stands that were dominated by hardwoods at the time of invasion were not invaded by large numbers of hemlock and are now dominated by sugar maple (*Acer saccharum*) and basswood (*Tilia americana*). Fine-scale studies of fossil records from hemlock stands by T. E. Parshall show that several centuries elapsed after the first hemlock trees were established before hemlock became dominant in the stand, replacing resident white pine. Stand simulations suggest that delays of this length could be caused by competition from resident canopy trees.

If future climate changes caused by doubled greenhouse gas concentrations occur within a century, seed dispersal is inadequate to accomplish significant changes in ranges. On this time scale, interactions with resident vegetation become important. Resident vegetation will constrain colonization of microhabitats that become more favorable as the climate changes, and resident

canopy trees will inhibit population expansion of minor species that are better adapted to the new climatic regime.

1. Introduction

A rise in global temperature of about 2°C is expected as greenhouse gases reach doubled concentrations in the atmosphere (Houghton *et al.* 1996). How will forests respond?

The Holocene fossil record provides the only observations we have of forest responses to climate warming of this magnitude. In many regions that were not actually covered by ice, plant species expanded from local refuges as the climate warmed in the early Holocene (Thompson, 1988; Tsukada, 1988; Markgraf *et al.*, 1995; McGlone, 1997). But in eastern North America and western Europe, a major response to Holocene warming was a latitudinal shift of species ranges as tree populations "migrated" northward to populate the newly deglaciated landscape. Analyses of climate changes using global circulation models suggest that tree migrations tracked the movements of climate-spaces to which particular species were adapted (Prentice *et al.*, 1991). Migrations involved movements of species boundaries at rates averaging 10–100 km per century (Davis, 1981; Huntley and Birks, 1982; Huntley and Webb, 1988; Delcourt and Delcourt, 1987; Pitelka *et al.*, 1997). If greenhouse gases double in the coming century (Houghton *et al.*, 1996), future changes may involve shifting of species boundaries an order of magnitude more rapidly. Is such rapid migration possible?

This chapter examines the Holocene fossil record of the Great Lakes region for information on the process of tree migration. Seed dispersal and interactions with resident vegetation are considered because both factors could have constrained the rate of advance of trees. Finally, I will discuss the results in the context of global warming, in order to identify the constraints that are likely to be important in the coming century.

2. Long-Distance Dispersal

Recent discussions of tree migration have emphasized the importance of seed dispersal (Pitelka *et al.*, 1997; Clark *et al.*, 1998). Dispersal appears to be limiting, because in forests seed travels only a few meters from the source tree. Diffusion models that use observed dispersal parameters are unable to simulate the rates of migration observed in the Holocene record (Clark, 1998). Models that are able to simulate Holocene migration rates invoke seed dispersal hundreds or even thousands of meters farther than observed in forests (Clark, 1998; Shigesada *et al.*, 1995; S. Sugita, Ehime University, personal communication). Yet despite the importance of long-distance dispersal, we know little about it. Greene and Johnson (1995) have developed models of seed dispersal, testing them against seed trap data from very large open areas. Appreciable numbers of seeds, 1–10% of their densities in forests, were found in traps 1000–1600 m from the forest edge. Direct observation of seeds released from towers indicated that 1–2% were caught in updrafts, dispersing quite differently than predicted assuming a con-

stant wind speed (Greene and Johnson, 1995). Johnson and Adkisson (1985) observed bluejays (*Cyanocitta cristata*) transporting beechnuts (*Fagus grandifolia*) 4 km to cache them near their nesting areas. These rare observations are important because they demonstrate that a small proportion of seeds is available for dispersal across distances very much greater than observed in seed traps in forests. However, direct observations do not provide information on maximum dispersal distances, nor on the frequency of long-distance dispersal events.

Additional information about seed dispersal is contained in the fossil record. Dispersal events can be inferred wherever there is evidence that populations were founded at large distances from the source population. It is, of course, technically difficult to detect a small population using fossil pollen or macrofossils, and even more difficult to demonstrate that the small population was isolated from the main population (Davis *et al.*, 1991). Pollen studies in Sweden, however, record the establishment of individual colonies of beech (*Fagus sylvatica*) in the late Holocene (Bjorkman, 1996) and macrofossils demonstrate that populations of spruce (*Picea abies*) grew far in advance of the expanding species front for thousands of years (Kullman, 1996). East of James Bay, Canada, small colonies of larch (*Larix laricina*) have become established in patches during the past 1500 years as the population has expanded. Some of these colonies have fused into a continuous distribution,

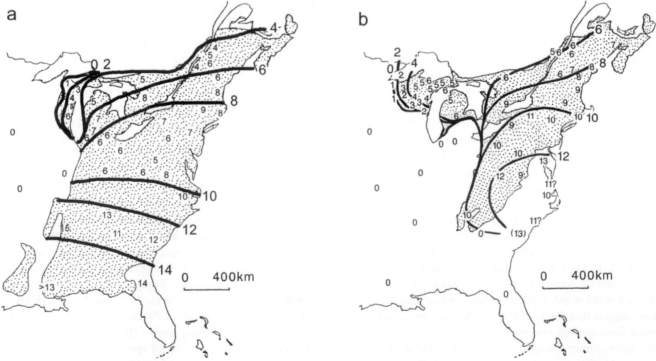

FIGURE 1 Migration maps for beech (a) and for hemlock (b). Heavy black lines indicate the approximate position of the migrating front at 2000-year intervals. Small numbers indicate the C-14 age (in 10^3 year) of a steep 10-fold increase in pollen accumulation rates (grains cm^2 year^{-1}) and/or pollen percentages at an individual fossil site (maps modified from Davis, 1981, Davis et al.; 1986). Within the Great Lakes region, small numbers indicate establishment dates for populations of beech using the criteria established by Webb (1987) and Woods and Davis (1989). Stippled areas indicate present range.

but along the western species limit small populations remain isolated from one another (Penalba and Payette, 1997).

More detailed information is available for American beech (*Fagus grandifolia*) and eastern hemlock (*Tsuga canadensis*), because their migration into the Great Lakes region of North America was studied for the purpose of contrasting migration patterns of an animal-dispersed species (beech) with those of a wind-dispersed species (hemlock) (Davis, 1987). Outlying populations were relatively easy to document, because pollen evidence for colonies founded on the far side of any of the Great Lakes provides convincing evidence of disjunction between the new colonies and the parent population. In this manner long-distance dispersal was demonstrated for both species (Webb, 1987; Davis *et al.*, 1986; Davis, 1987; Woods and Davis, 1991).

Beech expanded northward from Florida during the early Holocene, arriving in southernmost Michigan 7000–8000 years ago (Davis, 1981; Bennett, 1987; Webb, 1988)(Fig. 1a). Detailed studies of beech migration into the Great Lakes region reveal that while beech was spreading northward on the eastern side of Lake Michigan, colonies were established on the western shore. Colonies were established on the western shore of the lake as early

as 6000 years ago. Beech had not yet colonized the northern shore of the lake, so the disjunct populations must have been established by long-distance dispersal. Seeds were transported directly across the lake, a distance of about 100 km, or by many jumps between hypothetical islands of favorable habitat within the prairie vegetation at the southern end of the lake (Webb, 1987)(Fig. 1a). Webb (1986) speculates that the extinct passenger pigeon could have been the vector for dispersal. Four thousand years ago, additional colonies were established farther north along the western shore of Lake Michigan. About 2500 years ago, the northern shore of the lake was colonized and all populations expanded, coalescing into a continuous population by 2000 years B. P. Within the past millennium, as beech reached its western limit, a large disjunct population was established west of Marquette, MI, 40 km beyond the species boundary (Woods and Davis, 1989).

Hemlock moved northward up the Appalachians and along the eastern seaboard before moving westward into the upper Great Lakes region (Fig. 1b). Between 6000 and 5000 years ago, there was a sudden increase of pollen from trace quantities to 10–20% of tree pollen in sediment cores from many lakes in Michigan (Fig. 2). The sudden increase occurred throughout hemlock's

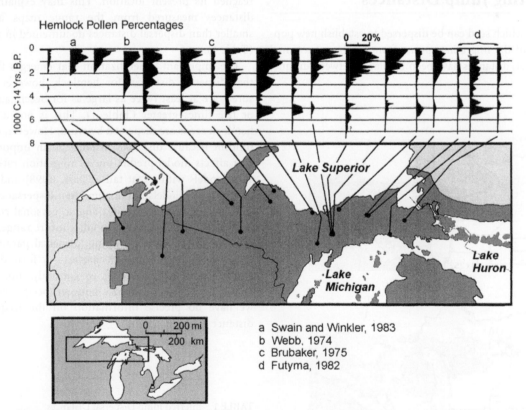

FIGURE 2 Pollen percentages of hemlock (as percent tree pollen) in sediment from 17 lakes and bogs in northern Michigan. Radiocarbon ages (indicated on ordinate) are based on bulk sediment; hard-water errors are large at the nine lakes at the eastern end of the transect, a region of calcareous bedrock. All the eastern lakes show trace quantities of pollen starting about 7000 years ago, and a sharp increase of hemlock pollen between 6000 and 5000 years ago. The western lakes document rapid westward migration between 5500 and 4500 years ago, and much slower westward migration 3000–1500 years ago (Davis *et al.*, unpublished data, and references cited in figure).

range in lower Michigan, and in the eastern half of upper Michigan (Fig. 1b). The sudden invasion of such a wide area suggests that many previously established colonies were expanding rapidly in response to climate changes that favored hemlock. Populations coalesced and the species frontier migrated westward as a continuous front (Fig. 2)(Davis *et al.*, 1986; Davis, 1987; Davis *et al.*, unpublished). The source of seed for the colonies established more than 6000 years ago must have been east of Lake Huron—southern and central Ontario, which had been invaded by hemlock 8000 years ago (Kapp, 1977; Bennett, 1987; Fuller, 1998). Although the precise trajectory of dispersal is unknown, the distances required are large–over 100 km. Dispersal across such great distances is believable, however, because after hemlock had spread across Michigan and into Wisconsin, reaching its present western limit 1500 years ago, colonies became established at several locations in Minnesota. The nearest colonies are 100 km from the species front, and the farthest locations are an additional 110 km beyond them. Pollen records from two of the outlying colonies in Minnesota establish their origin as 1200 years ago (Calcote, 1986).

3. Estimating Jump Distances

The distance to which seed can be dispersed to establish new populations ahead of the migrating species front has been measured in two ways. First, distances have been tabulated between disjunct

colonies demonstrated in the fossil record and the species frontier as it existed at that time. The most conservative estimate was chosen in all cases. For example, islands in Lakes Michigan or Huron were presumed to have been "stepping-stones," and dispersal distances were measured over stretches of open water. The results are illustrated in Figure 3, and the data are summarized in Table 1. Jump-dispersal distances (Pielou, 1979) are at least twice as great for hemlock as for beech. There are several instances of leaps of 100 km or more for hemlock. The largest leap for beech is 100 km across Lake Michigan, but other jumps are smaller, in general 10–40 km.

The second method for measuring dispersal uses detailed range maps prepared for Wisconsin and a part of upper Michigan using witness tree data collected before settlement in the early 19th century (Davis *et al.*, 1991). In the range maps shown in Figure 4, distances were measured between outlying colonies and the continuous species limit. Again, if there were colonies between the main species limit and more distant colonies, the assumption was made that intermediate colonies acted as stepping stones. Estimates of dispersal distances measured in this way are conservative, because some of the outlying colonies may have been established well before the continuous population reached its present location. This may explain why dispersal distances measured from the range maps are consistently smaller than dispersal distances documented in the fossil record (Table 1).

The important generalization that emerges from the data is that dispersal distances for hemlock, however they are measured, are at least twice as large as for beech, and possibly four or five times greater (Table 1). This is not unexpected since hemlock seeds are dispersed by wind. However, the result is significant because dispersal distance is an important parameter that affects model predictions of migration rates—both diffusion models with a "fat tail" (Clark, 1998) and modified scattered colony models that assume that dispersal can occur out to some maximum distance (S. Sugita, personal communication). If these models are used to predict future ranges of tree species resulting from global warming, dispersal parameters will have to be determined for each species—a formidable task. Even with so much data available, we can only approximate the difference in dispersal distance between beech and hemlock, and we have no precise information on the frequency of long-distance dispersal events.

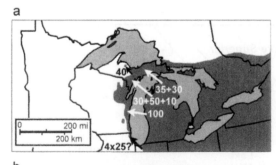

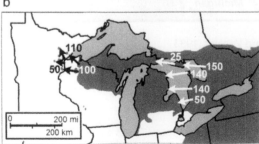

FIGURE 3 Maps showing distances between outlying colonies and the main population at the time the colonies were established, as indicated by the fossil pollen records of beech (a) and hemlock (b) (S. Webb, 1987; Woods and Davis, 1989; Davis *et al.*, 1986; Calcote, 1986; Davis *et al.*, unpublished data).

TABLE 1   Inferred Jump Dispersal Distances

Jump distances implied by fossil records of establishment of outlying colonies	
Beech (*Fagus grandifolia*)	40 km (n = 7)
Hemlock (*Tsuga canadensis*)	80–100 km (n = 7,8)
Average distance to outlying colonies on presettlement range map	
Beech (*Fagus grandifolia*)	8 km (n = 16)
Hemlock (*Tsuga canadensis*)	39 km (n = 46)

a - Beech

b - Hemlock

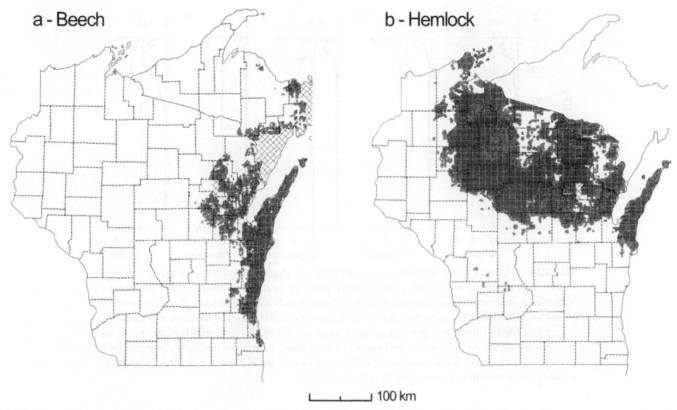

└────────┴────────┘ 100 km

FIGURE 4 Maps showing the distribution of beech and hemlock in Wisconsin at the time of the Federal Land Office Survey in the early 19th century (redrawn from Davis *et al.*, 1991).

4. Interactions with Resident Vegetation—Constraints on Establishment

The migration of hemlock has been studied at a fine spatial scale, using sediment from small forest hollows that provide a pollen record of the history of individual forest stands a few hectares in size (Sugita, 1994; Calcote, 1995; 1998). These fine-scale studies record the invasion of individual forest stands. Did resident vegetation influence the pattern of invasion? This question has been considered at length in the literature on invasions by exotic species, with evidence cited by several authors that resident vegetation, or its absence on disturbed sites, can influence invasion success (Crawley, 1987; Drake, 1990; Lodge, 1993). For the present discussion we are interested in the effect this phenomenon could have on overall migration rate.

Fossil pollen in a series of small forest hollows about 10 m in diameter provides a record of hemlock invasion of individual forest stands along a 10-km transect in northern Michigan. The distribution of species within the present-day forest, which has never been clearcut, is patchy—a mosaic of stands dominated by hemlock interspersed with mixed stands and large patches dominated by sugar maple (*Acer saccharum*). Pollen diagrams from four

hemlock stands and four maple stands extend back to the time hemlock invaded the forest about 3000 years ago. Prior to hemlock invasion, all the hemlock stands had been dominated by white pine (*Pinus strobus*). After hemlock invaded, it coexisted with pine for a thousand or more years, until hemlock displaced white pine in three of the four stands. White pine had also been abundant in one of the stands now dominated by hardwoods. This stand was also invaded by hemlock, but following a windstorm, hemlock was eliminated and maple became dominant. In contrast, other stands now dominated by maple were already dominated by hardwoods at the time hemlock was invading pine stands. They were never invaded by large numbers of hemlock. Establishment was probably prevented by the same factors that discourage the establishment of hemlock seedlings in maple stands today. Hardwood litter provides a poor seedbed for hemlock, and the light and nutrient regimes favor the growth of maple seedlings that shade hemlock seedlings (Ferrari, 1993; Davis *et al.*, 1994; 1998).

In this example, a portion of the landscape was occupied by resident vegetation that inhibited establishment of a migrating species. At present maple stands make up 12% of the upland landscape. Lakes and wetlands compose another 34% of the area (Pastor and Broschart, 1990), leaving only about half of the landscape available for colonization by hemlock. Was the rate of hemlock

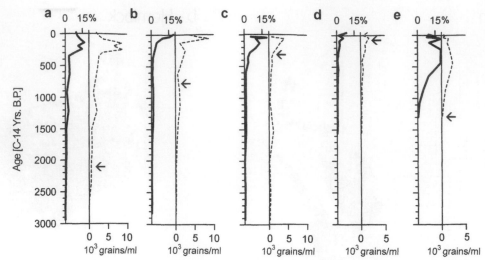

FIGURE 5 Hemlock pollen percentages and concentrations plotted against the C-14 age of sediment in five small forest hollows located within hemlock stands in western Wisconsin. Arrows indicate the oldest sediment in which fossil hemlock stomata are found at each hollow. Stomata indicate the presence of one or more hemlock trees within 20 m of the hollow (Parshall, 1999). The increases in pollen concentrations (dashed lines) and percentages (solid lines) in the last 200–300 years indicate increasing hemlock population densities within the nearest 1–3 ha of forest. Most records indicate a long establishment phase between the initial colonization of the stand and the population increase. [Modified figure reprinted, with the author's permission, from Parshall (1998)].

migration slowed because establishment was restricted? Unfortunately, the answer is ambiguous. Migration was indeed slow in this part of Michigan 3000 years ago, but we cannot separate the effects of climate from the constraint on establishment (Davis, 1987; Davis *et al.*, 1994; 1998).

5. Interactions with Resident Vegetation—Competition for Light and Resulting Constraints on Population Growth

Many introduced species show an "establishment phase," years or decades following introduction when the invading plants or animals are not seen in their new environment. Then suddenly the invading organisms seem to be everywhere, in a rapidly expanding population. Some believe the establishment phase represents a period when the density of colonists is so low that they are undetectable, while others believe it represents a period when genetic adaptation to a new environment is taking place (Ewel, 1986; Baker, 1965).

Direct observation of the establishment phase has not been possible because the invading organisms are so rare. A retrospective record provides more information, obtainable from small hollows in forest stands that currently include the invading species. Parshall (1998) studied invasion by eastern hemlock of five hemlock stands in western Wisconsin, using fossil pollen, conifer needles, and stomates from conifer needles. Fossil pollen in the forest hollows reflects hemlock density within 50–80 m (Sugita, 1994; Calcote, 1995), while the fossil needles, or stomates from needles, demonstrate the presence of one or more hemlock trees within 20 m (Parshall 1998). A remarkable feature of Parshall's data (Fig. 5) is the long lag at four of the five sites between the first appearance of hemlock (shown by fossil stomates: arrows in Fig. 5) and the population expansion indicated by increased amounts of hemlock pollen. At one stand, the lag—i.e., the establishment phase—lasts 1000 years, at others several hundred years (Parshall, 1998). Thus hemlock, although present in the forest, was unable to increase for several centuries in most of these stands (Parshall, 1998.)

Population growth may have been delayed by competition from resident canopy trees. Competition for light is the mechanism suggested by gap model simulations. In a simulated sugar maple forest subjected to a sudden climate cooling of 2°C, the change in canopy dominants from sugar maple to spruce (*Picea rubens*) took 200 years. In the simulations, sugar maple saplings were replaced by spruce, but canopy maple continued to shade the better-adapted spruce in the understory until the canopy trees reached the end of their normal lifespan (Davis and Botkin, 1985). Disturbance can speed up replacement (Davis and Botkin, 1985; Overpeck *et al.*, 1990), but natural disturbance rates in hardwood forests of the Great Lakes region are quite low, resulting in canopy lifetimes of 150–200 years (Frelich and Lorimer,

1991; Frelich and Graumlich, 1994; Parshall, 1995; Parshall *et al.*, in review).

6. Discussion

If future warming were to occur slowly, with temperature increases associated with CO_2 doubling spread out over 500 years, migrations could occur as they did during the Holocene. The rates of northward range extension would have to be the maximum recorded, however, 50–100 km per century. In this unlikely scenario, seed dispersal would be an important variable limiting the rate of advance. We have shown that seed dispersal can occur over long distances—a few tens of kilometers to over a hundred kilometers—but the frequency of recorded dispersal events is not high. Long-distance dispersal that resulted in successful colonies is recorded for each species about once per millennium during the Holocene, whereas a much higher frequency will be required for future change. Dispersal by humans, however, is likely to occur, making natural dispersal mechanisms less important. But for non-commercial trees, as well as the herbs, shrubs, mosses, and fungi that make up forest ecosystems, natural dispersal will remain an important limitation to adjustment to climate change. Many future forests will doubtless lack species that we now consider important components of the ecosystem.

If climate change occurs rapidly, with 2°C warming by 2100 A.D. (Houghton *et al.*, 1996), then seed dispersal will be too slow to accomplish significant vegetation adjustment. Range extensions could occur in this time frame only for trees that already had outlying colonies to the north. Hemlock, for example, had a fringe of outlying populations beyond its range limit (Little, 1971). In Wisconsin the colonies are 40 km from the range boundary on average (Fig. 4). If outlying populations like these have survived logging along the northern range limit, they can serve as centers of infection for the surrounding landscape. Expanding these preexisting populations could allow hemlock to extend its range by an average of 40 km within the coming century. Outlying colonies of beech are only 10 km from the range limit in Wisconsin, suggesting that this species could extend its range northward only by 10 km (Fig. 4, Table 1). These range extensions are small relative to the displacements of potential ranges by hundreds of kilometers that are likely with climate changes accompanying doubled CO_2 (Davis and Zabinski, 1992; Sykes *et al.*, 1996).

Under a rapid climate change scenario, factors that constrain establishment and population growth become much more important than seed dispersal. On this time scale the likely responses are changes in species abundances and distributions within regions where the tree is already growing. Fossil records and simulations show that competition from canopy trees can delay the population expansion of tree species that are better adapted than resident dominants to new climate conditions (Davis and Botkin, 1985). Another likely response to a rapidly changing climate is redistribution of tree species on the landscape, involving, for example, dispersal from drier substrates to more mesic sites. In this case possible inhibition of establishment by local vegetation could be important in delaying adjustment to changing climate.

Recent literature on forest response to future climate change emphasizes dispersal limitations (Pitelka *et al.*, 1997; Clark *et al.*, 1998). The review I have presented here suggests that natural dispersal is unlikely to accomplish adaptation to future climate. Competition and stand dynamics are much more important constraints on the decadal scale we need to consider if greenhouse gases continue to accumulate at present rates.

7. Conclusions

1. Range shifts in response to climate changes over the past 11,000 years were slow compared to the rapid range adjustments that will be necessary in the coming century as greenhouse gases double in concentration.

2. Past range shifts were accomplished by seed dispersal 10–100 km beyond the species range limit. The frequency of long-distance dispersal events has not been adequately measured, but data show clearly that the average dispersal distance differs between species. Migration models will have to include species-specific dispersal parameters.

3. Establishment of new populations of migrating trees was limited by resident vegetation.

4. Population expansion by newly established colonies of migrating trees was delayed for decades or centuries by competition from resident canopy trees.

5. The fossil record of forest tree response to changing climate suggests that in the coming century seed dispersal will not be adequate to accomplish range shifts rapidly enough to track future climate. On the time scale of decades, resident vegetation that inhibits establishment and disturbance regimes that control forest stand dynamics will be important limitations to the rate of forest adaptation to changing climate.

Acknowledgments

I appreciate the opportunity to participate in the celebration of the founding of the Max-Planck Institute for Biogeochemie at Jena. I congratulate the Institute and wish it success as it seeks to understand the changing global ecosystem. The research reviewed here was supported by the National Science Foundation, Grants DEB8012159, DEB8407943, BSR8615196, BSR8916503, DEB9221371, and by the Mellon Foundation. I gratefully acknowledge the generosity of T.E. Parshall and S. Sugita, who allowed me to present their research results, and I thank Holly Ewing, David Lytle, Christine Douglas, Randy Calcote, and Shinya Sugita for helpful comments on an earlier version of the manuscript.

References

Baker, H. G. 1965. Characteristics and modes of origin of weeds. In "The Genetics of Colonizing Species." (H. G. Baker and G. L. Stebbins, Eds.), p. 147-173. Academic Press, New York.

Bennett, K. D. 1987. Holocene history of forest trees in southern Ontario. *Can. J. Bot.* **65,** 1792–1801.

Bjorkman, L. 1996. Long-term population dynamics of *Fagus sylvatica* at the northern limits of its distribution in southern Sweden: a paleoecological study. *Holocene* **6,** 225–234.

Brubaker, L. B. 1975. Postglacial forest patterns associated with till and outwash in north central upper Michigan. *Q. Res.* **5,** 499–527.

Calcote, R. R. 1986. Hemlock in Minnesota: 1200 years as a rare species. M.S. Thesis. University of Minnesota, Minneapolis, Minnesota.

Calcote, R. R. 1995. Pollen source area and pollen productivity: evidence from forest hollows. *J. of Ecol.* **83,** 391–602.

Calcote, R. 1998. Identifying forest stand types using pollen from forest hollows. *Holocene* **8,** 423–432.

Clark, J. S. 1998. Why trees migrate so fast: confronting theory with dispersal biology and the paleorecord. *Am. Nat.* **152,** 204–224.

Clark, J. S., Fastie, C., Hurtt, G., Jackson, S. T., Johnson, C., King, G. A., Lewis, M., Lynch, J., Pacala, S., Prentbice, I. C., Schupp, E. W., Webb, T. III, and Wyckoff, P. 1998. Reid's Paradox of rapid plant migration. *Bioscience* **48 (1),** 13–24.

Crawley, M. J. 1987. What makes a community invasible? In "Colonization, Succession and Stability." (A. J. Gray, M. J. Crawley, and P. J. Edwards, Eds.), pp.1–29. Blackwell Scientific Publishers, Oxford, United Kingdom.

Davis, M. B. 1981. Quaternary history and the stability of forest communities. In "Forest Succession: Concepts and Application." (D. C. West, H. H. Shugart, and D. B. Botkin, Eds.), pp. 132–153. Springer-Verlag, New York.

Davis, M. B. 1987. Invasions of forest commmunities during the Holocene: beech and hemlock in the Great Lakes region. In "Colonization, Succession and Stability." A. J. Gray, M. J. Crawley, and P. J. Edwards, Eds.), pp. 373–393. Blackwell Sci., Oxford, United Kingdom.

Davis, M. B. and Botkin, D. B. 1985. Sensitivity of cool-temperate forests and their fossil pollen record to rapid temperature change. *Q. Res.* **23,** 327–340.

Davis, M. B., Woods, K. D., Webb, S. L., and Futyma, R. P. 1986. Dispersal versus climate: expansion of *Fagus* and *Tsuga* into the upper Great Lakes region. *Vegetatio* **67,** 93–103.

Davis, M. B., Schwartz, M. N., and Woods, K. 1991. Detecting a species limit from pollen in sediments. *J. of Biogeogr.* **18,** 653–668.

Davis, M. B. and Zabinski, C. 1992. Changes in geographical range resulting from greenhouse warming: effects on biodiversity in forests. In "Global Warming and Biological Diversity." (R. L. Peters and T. E. Lovejoy, Eds.), pp. 297–308. Yale University Press, New Haven, Connectient.

Davis, M. B., Sugita, S., Calcote, R. R., Ferrari, J. B., and Frelich, L. E. 1994. Historical development of alternate communities in a hemlock-hardwood forest in northern Michigan, USA. "Large Scale Ecology and Conservation Biology." In (P. J. Edwards, R. May, and N. R. Webb, Eds.), pp. 19–39 Blackwell Scientific Publications, Oxford, United Kingdom.

Davis, M. B., Calcote, R. R., Sugita, S., and Takahara, H. 1998. Patchy invasion and the origin of a hemlock-hardwoods mosaic. *Ecology* **78,** 2641–2659.

Delcourt, P. A. and Delcourt, H. R. 1987. "Long-Term Forest Dynamics of the temperate zone." Springer-Verlag, New York.

Drake, J. 1990. Communities as assembled structures: do rules govern pattern? *Tree* **5,** 159–164.

Ewel, J. J. 1986. Invasibility: lessons from south Florida. In "Ecology of Biological Invasions of North America and Hawaii." (H. A. Mooney and J. A. Drake, Eds), pp. 214–230. Springer-Verlag, New York.

Ferrari, J. B., 1999. Fine-scale patterns of leaf litterfall and nitrogen cycling in an old-growth forest. *Can. J. Forestry Res.* **29,** 291–302

Frelich, L. E. and Lorimer, C. G. 1991. Natural disturbance regimes in hemlock-hardwood forests of the Upper Great Lakes Region. *Ecol. Monogr.* **61,** 145–164.

Frelich L. E. and Graumlich, L. J. 1994. Age-class distribution and spatial patterns in an old-growth hemlock-hardwood forest. *Can. J. Forestry Res.* **24,** 1939–1947.

Fuller, J. L. 1998. Ecological impact of the mid-Holocene hemlock decline in southern Ontariuo, Canada. *Ecology* **74,** 2337–3351.

Futyma, R. P. 1982. Postglacial vegetation of eastern Upper Michigan. Ph.D. Thesis, University of Michigan, Ann Arbor, Mchigan.

Greene, D. F. and Johnson, E. A. 1995. Long-distance dispersal of tree seeds. *Can J. Bot.* **73,** 1036–1045.

Huntley, B. and Birks, H. J. B. 1981. "An Atlas of Past and Present Pollen Maps for Europe: 0–13,000 Years." Cambridge University Press, Cambridge, United Kingdom.

Huntley, B. and Webb, T. III, 1988. "Vegetation History." Kluwer Sci. Dordrecht, Netherlands.

Houghton, J. T., Filho, L. G. M., Callander, B. A. Harris, N., Kattenberg, A., and Maskell, K., Eds. 1996. "Climate Change 1995: The Science of Climate Change. IPPC Report." Cambridge University Press, Cambridge, United Kingdom.

Johnson, W. C. and Adkisson, C. S. 1985. Dispersal of beech nuts by blue jays in fragmented landscapes. *Am. Midland Nat.* **113,** 319–324.

Kapp, R. O. 1977. Late Pleistocene and postglacial plant communities of the Great Lakes Region. In "Geobotany." (R.C. Romans.), pp. 1–27 Ed. Plenum, New York.

Kullman, K. 1996. Norway spruce present in the Scandes Mountains, Sweden at 8000 B.P.: new light on Holocene tree spread. *Global Ecol. Biogeogr. Lett.* **5,** 94–101.

Little, E. L. Jr. 1971. "Atlas of United States Trees. Vol. 1. Conifers and Important Hardwoods." Misc. Publ. 1146. USDA Forest Service, U.S. Government Printing Office, Washington D.C.

Lodge, D. M. 1993. Biological invasions: lessons for ecology. *Tree* **8,** 133–137.

Markgraf, V., McGlone, M., and Hope, G. 1995. Neogene paleoenvironmental and paleoclimatic change in southern temperate ecosystems—a southern perspective. *Tree* **10,** 143–147.

McGlone, M. S. 1997. The response of New Zealand forest diversity to Quaternary climates. PP. 73–80. In Past and Future Rapid Environmental Changes: The spatial and Evolutionary Responses of Terrestrial Biota. (B. Huntley, W. Cramer, A. V. Morgan, H. C. Prentice, and J. R. M. Allen Eds.) Springer-Verlag, Berlin.

Overpeck, J. T., Rind, D., and Goldberg, R. 1990. Climate-induced changes in forest disturbance and vegetation. *Nature* **343,** 51–53.

Parshall, T. E. 1995. Canopy mortality and stand-scale change in a northern hemlock-hardwood forest. *Can. J. Forestry Res.* **25,** 1466–1478.

Parshall, T. E. 1998. Hemlock invasion and population increase near the species limit in Wisconsin. PhD Thesis, University of Minnesota, St. Paul, Minnesota.

Parshall, T. E., Calcote, R., Davis, M. B., Walker, K. and Douglas, C. Tree mortality in an old-growth hemlock-hardwood forest. *J. of Vegeta. Sci.* (in review).

Pastor, J. and Post, W. M. 1988. Response of northern forests to CO_2-induced climate change. *Nature* **334,** 55–58.

Penalba, M. C. and Payette, S. 1997. Late-Holocene expansion of eastern larch (*Larix laricina* {DuRoi] K.Koch) in northwestern Quebec. *Q. Res.* **48**, 114–121.

Pielou, E. C. 1979. "Biogeography." Wiley-Interscience, New York.

Pitelka, L. F. and Plant Migration Workshop Group. 1997. Plant migration and climate change. *Am. Scientist* **85**, 464–423.

Shigesada, N., Kawasaki, K, and Takeda, Y. 1995. Modeling stratified diffusion in biological invasions. *Am. Nat.* **146**, 229–251.

Sugita, S. 1994. Pollen representation in Quaternary sediments: theory and methods in patchy vegetation. *J. Eco.* **82**, 879–898.

Swain, A. and Winkler, M. 1983. Forest and disturbance history at Apostle Islands National Lakeshore. *Park Sci.* **3**, 3–5.

Thompson, R. S. 1988. Western North America. In "Vegetation History." (B. Huntley and T. Webb III, Eds.), pp. 454–458. Kluwer, Dordrecht, Netherlands.

Tsukada, M. 1988. Japan. In "Vegetation History." (B. Huntley and T.Webb III, Eds.), pp. 459–518. Kluwer, Dordrecht, Netherlands.

Webb, T III. 1974. A vegetational history from northern Wisconsin. Evidence from modern and fossil pollen. *Am. Midland Nat* **92**, 12–34.

Webb, T, III. 1988. Eastern North America. In "Vegetation History." (B. Huntley and T. Webb III, Eds.), pp. 387–414. Kluwer Acad. Publ., Dordrecht, Netherlands.

Webb, S. L. 1986. Potential roles of passenger pigeons and other vertebrates in the rapid Holocene migrations of nut trees. *Q. Res.* **26**, 367–375.

Webb, S. 1987. Beech range extension and vegetation history: pollen stratigraphy of two Wisconsin Lakes. *Ecology* **68**, 1993–2005.

Woods, K. D. and Davis, M. B. 1989. Paleoecology of range limits: beech in the upper peninsula of Michigan. *Ecology* **70**, 681–696.

Biogeochemical Models: Implicit versus Explicit Microbiology

Joshua Schimel

University of California,
Santa Barbara
Santa Barbara, California

1. Introduction

Microbiology examines the smallest level of the organization of life, while biogeochemistry considers the largest. What then, is the appropriate relationship between these fields? The point could be argued several ways. On the one hand, much of biogeochemistry is simply microbial physiology writ large. On the other hand, the vast gap in scale could mean that information about microbial physiology and community dynamics has limited direct utility in large-scale biogeochemical studies. The present chapter will consider this issue and will discuss some aspects of how microbiological understanding is (or is not) currently incorporated into biogeochemistry. I will also identify some directions for future research that could enhance the linkage between the two fields. As simulation models are a primary tool for linking fields and scales, much of this discussion will be targeted at how biogeochemical models handle microbiological processes and how this might change in the future.

Microorganisms (including bacteria, fungi, single-celled algae, and protozoa; Madigan *et al.*, 1997) are ubiquitous on Earth. They include the greatest diversity of all living things and they are dominant players in almost all global biogeochemical processes. In the C cycle, terrestrial plants may carry out slightly more than half the total global primary productivity, but single-celled algae in the ocean account for most of the rest (Schlessinger, 1997). The vast bulk of decomposition is carried out by fungi and bacteria (though mediated in some cases by faunal food webs). In the nitrogen cycle, essentially all the important transformations are carried out by microorganisms, including mineralization, N_2-fixation, nitrification, and denitrification. The same is true of the cycles of sulfur, phosphorus, and many other elements as well; the dominant biological transformations are microbial, often bacterial.

2. Microbiology in Biogeochemical Models

Although there is great diversity within microbial groups in terms of physiology and environmental responses, the populations of some microbial groups can vary dramatically over time, and microbial biomass can be an important reservoir of labile nutrients, microbial physiologists and community ecologists rarely interact directly with biogeochemists. Biogeochemical models of ecosystem C and N cycling rarely include microbiology explicitly. Models typically use simple response functions for processes and almost never include microbial population size as an active control on specific process rates. A number of soil organic matter models include a pool labeled "microbial biomass." However, to quote McGill (1996):

> Further, although biomass was a frequent "pool" in these models, its treatment was often indistinguishable from active forms of SOM. One might consider inclusion of soil biomass in this way to be tokenism.

Thus, one can make an argument that biogeochemical models tend to ignore microbiology or at least simplify it beyond recognition. This conclusion would suggest that there is little real, direct application of microbiological information to larger scale biogeochemical modeling. I believe, however, that conclusion would be wrong. First, several models do go into a reasonable amount of physiological detail for some processes; e.g., denitrification in the model DNDC (Li, 1996; Leffelaar *et al.*, 1988). More importantly and more generally, though, most biogeochemical models try to go directly from inputs of environmental parameters directly to outputs of process rates. Thus the microbiology, rather than being nonexistent, is "implicit." It is buried in the equation structure of a model as kinetic constants and response functions. Implicit microbiology is quite different from no microbiology. A lot of basic microbial physiology and process study went into developing the response functions in most biogeochemical models. In fact, to do implicit microbiology, incorporating important mechanistic accuracy, requires a sound understanding of the processes and their control.

While making microbiology implicit in models has the advantage of simplicity, it also, however, has limitations. A model converts assumptions into predictions. By testing those predictions against reality, it allows the investigator to test the strength and accuracy of the assumptions, and how they interact with each other. That is easier to do when the assumptions are made explicit and incorporated into a model as a mechanism. When processes are made implicit, rather than explicit, it is harder to test the validity of the assumptions about those processes. To quote McGill (1996) again:

> To restrict microbial biomass to a measurable SOM component, however, renders models nonmechanistic and removes the possibility that a model might simulate changes in SOM dynamics as a result of changes in activity or characteristics of soil organisms.

This raises the question: is there a need to become more mechanistic about the role of microorganisms in biogeochemical models? If so, what aspects of microbial processes and community dynamics should be considered for more focused study? In the remainder of this chapter I would like to consider these questions.

3. Dealing with Microbial Diversity in Models

There are two common core assumptions among biogeochemical models (and therefore among biogeochemists) that are worth evaluating from the microbial perspective. The first is that microbial physiologies are global. That is, microbial processes can be modeled across a range of conditions by using a single equation such as the following (Parton et al., 1987):

$$dC/dt = K \times M_d \times T_d \times C \qquad (1)$$

In this equation, C is the size of a soil carbon pool, K is a first-order rate constant, and M_d and T_d are reducing functions based on temperature and moisture. Each process has a single K value and a single reducing functions for each environmental driver. The assumption is that the fundamental response functions do not change with environmental conditions or the composition of the microbial community. To say this in another way, "microbial diversity has no discrete 'role' to play with respect to ecosystem function." (Finlay *et al.*, 1997). But is this true? Or alternatively, can changes in microbial communities change the nature of the response functions? I

and others have written rather extensively on this question (Schimel, 1995; Schimel and Gulledge, 1998; Brussard et al., 1997; Groffman and Bohlen, 1999; Wall and Moore, 1999). The general conclusion of these reviews has been that for processes that are carried out by physiologies that are broadly distributed across the microbial world (e.g., glucose metabolism), or for processes that we measure as single processes but that are really an aggregates of many specific processes (e.g., soil respiration), the composition of the microbial community is not often a major control on process dynamics. For processes that are physiologically "narrow" (i.e., carried out by physiologically/phylogenetically limited groups of organisms), such as nitrification and CH_4 production and consumption, the composition of the microbial community is sometimes a substantial control on process dynamics.

Most of the good existing case studies illustrating these points have been discussed in the review papers mentioned above. However, one new study is worth mentioning. Bodelier et al. (2000) examined the effects of rice plants and N fertilization on the dynamics of methanotrophs in rice soil. They showed that while Type II methanotrophs dominated the bulk soil, rice plants selected for populations of Type I methanotrophs in the rhizosphere. They also showed that these Type I methanotrophs are N-limited and that with fertilization, Type I populations increase dramatically, significantly reducing net CH_4 efflux from the system. Type II methanotrophs, while also CH_4-saturated, did not show a strong increase on N fertilization. This stimulation by added NH_4^+ is the direct opposite of the inhibition commonly found in upland soils (Gulledge and Schimel, 1998). Thus, the types and activities of methanotrophs present became a significant control on the overall methane flux from the rice ecosystem.

The conclusion from these studies is that there are cases where microbial community composition significantly affects the nature and environmental responses of biogeochemical processes. In these cases, assumptions of global physiologies and unitary response functions fail. Evaluating how diverse and significant these effects are, and then finding effective ways to integrate them into biogeochemical models is one area where microbial ecologists and biogeochemists should collaborate.

4. Kinetic Effects of Microbial Population Size

The second key assumption that models make is that microbial processes are never limited by the size of the microbial population. This assumption is clearly stated by Chertov and Komarov (1996) in their discussion of the SOMM model:

> The number and species composition of decomposing organisms is dependent on the biochemical properties of organic debris and on hydrological and thermal conditions. We postulate that there are no barriers for a rapid invasion of new organisms. Thus, it is possible to calculate the decompo-

sition coefficients for the communities as a function of the biochemical properties of litter, temperature and moisture.

Finlay et al. (1997) state the same idea even more bluntly:

(2) Microbial diversity in an ecosystem is never so impoverished that the microbial community cannot play its full part in biogeochemical cycling. The species complement of the microbial community quickly adapts, even to momentous changes in the local environment.

The assumption that microbial communities will always rapidly adapt to the available environment and substrate supply is a fundamental assumption in using first order kinetics, as in Eq. (1). Most biogeochemical processes are catalyzed, however, and catalyzed reactions invariably show Michaelis–Menten kinetics,

$$dX/dt = k \times E \times S/(K_m + S), \tag{2}$$

where X is the product concentration, k is a reaction constant, E is the catalyst concentration, S is substrate concentration, and K_m is the half saturation constant. $k \times E$ is usually expressed as V_{max} (Roberts, 1977), the maximum velocity possible for the reaction, but it is worth making clear that V_{max} is a linear function of the catalyst concentration. If most microbial processes actually follow Michaelis–Menten type kinetics, how can biogeochemical models represent them as 1st order? For modeling reaction kinetics across a wide range of concentrations, no 1st order model will work adequately (Fig. 1a). However, if substrate concentrations are moderate ($\sim K_m$) but do not vary over an excessively wide range, it is possible to fit a line to the kinetic response, even if it is not strictly 1st order (Fig. 1b). Alternatively, if substrate concentration is very low ($S \ll K_m$), Eq. (2) reduces to

$$dX/dt = V_{max} \times S/K_m \tag{3}$$

In this equation, kinetics become 1st order with V_{max}/K_m as the rate constant (Fig. 1c). Note that this constant still includes catalyst concentration (i.e., population size of the active microbes) within the V_{max} term. This raises two questions: first, which conditions that allow a 1st order approximation occur in nature, and second, if all the rate constant terms are actually linear functions of the microbial population, why do models leave population size out?

To address the first of these questions, Table 1 presents data on the ratio of basal/maximal rates of soil respiration, nitrification, and denitrification. If the ratio is very small, then one can conclude that the process is naturally occurring at very low substrate concentrations, whereas a ratio close to 1 would indicate that the process is close to becoming substrate-saturated. A ratio of 0.5 would imply that the process was occurring at a substrate concentration close to K_m. Table 1 is far from exhaustive, but represents the range of behaviors that occur for these processes. Respiration usually operates at between 20 and 65% of its maximum rate. This may cover a narrow enough range of concentrations so that using a 1st order approximation for a portion of the Michaelis–Menten curve (Fig. 1b) would not introduce large errors. Nitrification,

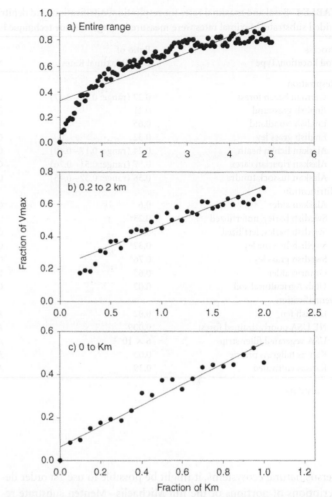

FIGURE 1 Fitting 1st order curves to catalyzed reaction kinetics. The individual data points were generated in a spreadsheet model of Michaelis–Menten kinetics with a random $\pm 5\%$ error introduced to the individual values. The straight lines were fit to these data. The three panels all show the same data but over different ranges: (a) over the entire range from zero to close to substrate saturation; (b) over the range from 0.2 to 2 K_m; (c) over the range from 0 to K_m.

however, appears to operate over a wide range, from 3 to 75%. However, the higher values appear to occur in natural ecosystems, while the lower values occur in agricultural systems. Regular fertilization may produce very large populations. To cover the entire range of systems, any 1st order assumption would fail, but if systems are divided into agricultural and nonagricultural, the ranges may be small enough to allow 1st order fits. Denitrification, however, behaves still differently, commonly proceeding at less than 1% of its potential rate, a range in which a 1st order assumption should work well. I hypothesize that denitrification operates so much below potential because it is carried out by aerobic organisms that switch to denitrification as a "back up" physiology when soils go anaerobic (Zumft, 1992). Thus, it should be possible to grow a large population of organisms under aerobic conditions; this would provide overcapacity when soils go anaerobic. So it appears that with some limited reparameterizing for agricultural

TABLE 1 Basal and maximal rates for respiration, nitrification, and denitrification across a range of ecosystems. Basal rates were measured without added substrate. Maximal rates were measured using the same technique but with saturating amounts of substrate added.

Process Soil Location/Type	Ratio of Basal/Maximal Rates	Notes on Approach	Reference
Respiration			
German beech forest	0.27 (range: 0.12–0.47)	Glucose amended soils	Anderson and Joergensen (1997)
English grassland	0.21	Glucose amended soils	Lin and Brookes, 1999
English woodland	0.65	Glucose amended soils	Lin and Brookes, 1999
English grass ley	0.32	Glucose amended soils	Lin and Brookes, 1999
Alaskan lichen heath	0.24 (range: 0.17–0.40)	Glucose amended soils	Cheng *et al.*, 1998
Alaskan riparian carex	0.57 (range: 0.51–0.76)	Glucose amended soils	Cheng *et al.*, 1998
Alaskan tussock tundra	0.38 (range: 0.32–0.45)	Glucose amended soils	Cheng *et al.*, 1998
Nitrification			
Alaskan alder	0.6	Gross nitrification/chlorate slurry	Clein and Schimel, 1995
Swedish barley, unfertilized	0.23[a]	Chlorate amended core/slurry	Berg and Rosswall, 1987
Swedish barley, fertilized	0.25[a]	Chlorate amended core/slurry	Berg and Rosswall, 1987
Swedish lucerne ley	0.32[a]	Chlorate amended core/slurry	Berg and Rosswall, 1987
Swedish grass ley	0.76[a]	Chlorate amended core/slurry	Berg and Rosswall, 1987
Ontario alder	0.65	Unamended/amended chlorate slurry	Hendrickson and Chatarpaul, 1984
Utah Agricultural soil	0.03	Gross nitrification/slurry	Shi and Norton, 2000
Dentrification			
Danish fen	0.02	Amended anaerobic cores	Ambus and Christensen, 1993
NE USA poorly drained forest	0.009	Amended anaerobic cores	Groffman *et al.*, 1991
USA vegetated filter strip	6×10^{-5}	Amended anaerobic cores	Groffman *et al.*, 1991
Kansas tallgrass prairie	0.03	Amended aerobic cores	Groffman, 1991
Kansas cultivated	0.39	Amended aerobic cores	Groffman, 1991

[a] July data.

versus natural ecosystems, it might be possible to use 1st order descriptions of portions of the full Michaelis–Menten substrate response curves for most of these processes. However, biomass size would still be a part of the effective rate constant, and so should still be a measurable control on the actual rate of the process in the field. This again raises the question: how can biogeochemical models exclude biomass as a factor in kinetic responses?

The effects of biomass size on process kinetics would not necessarily be apparent under some conditions. If the microbial population size is constant, then biomass can be incorporated as part of the rate constant. If population sizes vary linearly with specific environmental conditions, then biomass can be incorporated into the appropriate response function. However, neither situation is actually true. Microbial population sizes are not constant. Total microbial biomass can vary over time by a factor of 2 or more, sometimes with little obvious correlation to season or weather (Wardle, 1998). More specific populations, such as nitrifiers, denitrifiers, and methanotrophs, can vary more than that (e.g., Acea and Carballas, 1996; Berg and Rosswall, 1987; Both et al., 1992; Saad and Conrad, 1993). There are also stresses that can rapidly reduce biomass by as much as a factor of 2 such as rewetting a dry soil (Bottner, 1985; Kieft et al., 1987) and freeze–thaw (Morley et al., 1983). Thus, microbial studies suggest that biomass can vary enough to have large impacts on overall process rates. This once again raises the question: why do models almost invariably exclude biomass as

an active control? It is because of the second assumption implicit in using first order kinetics: microbes grow quickly enough so that they can rapidly recover from any stress, as quoted from Finlay et al. (1997) above. For models operating at the ecosystem scale and above, as long as populations can recover from stress over periods of days to weeks that assumption would probably be valid. As *Escherichia coli* can double in 20 min, and *Penicilium* spp. can cover a piece of bread in days, this probably does not seem like an unreasonable assumption. However, it is not necessarily valid.

5. Microbial Recovery from Stress

Many microorganisms (particularly some fungi) grow slowly. Even those that regrow quickly may have to recolonize habitats after stress-induced local extinction. Recolonization and regrowth dynamics in soils are not well understood. However, there are some data that suggest that they may be important in controlling ecological processes.

Clein and Schimel (1994) found that a single one-day drying–rewetting could reduce microbial respiration in birch litter for more than 60 days in a lab assay, causing a 25% reduction in total C respired. Schimel et al. (1998) did a field study to expand on that work. They placed bags containing birch litter in the field and used watering and drought shelters to establish treat-

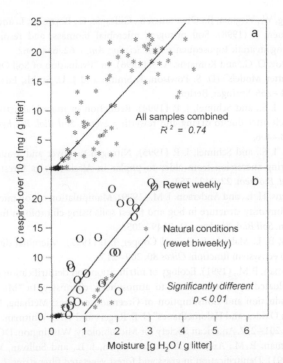

FIGURE 2 Carbon respired over a 10-day incubation in the lab on litter samples that had been incubated under different treatments. Panel (a) shows data from all treatments pooled into a single analysis, while (b) shows only data from the rewet weekly and natural conditions treatments. The continuously moist treatment was not significantly different from the rewet weekly treatment. Data are from Schimel *et al.*, 1999.

ments including: (a) continually moist, (b) watered weekly, and (c) natural conditions (which was actually biweekly rain). Samples were harvested every week over a month. Respiration rates at field moisture were measured over a 10-day period in the lab to establish the potential activity of the extant community. When C respired over the 10-day lab incubation was expressed as a function of moisture, the R^2 for all the data pooled was 0.74 (Fig. 2a), which is often considered quite adequate for modeling ecological data. However, when the different treatments were analyzed separately, there were tighter responses and significant differences between treatments, even after accounting for the moisture of the sample (Fig. 2b). Samples that had experienced longer drought in the natural conditions had lower respiration, biomass, and specific activity at any given water content than samples that were wet more frequently. Thus, this provided clear evidence that the stress history substantially affected the size and functioning of the microbial community and that it could not recover over at least the 10-day incubation following harvest. A study examining multiple freeze–thaw cycles showed similar results (Schimel and Clein, 1996). Each stress cycle reduced the ability of the surviving microbial community to process organic matter and respire, without evidence of recovery over the course of a one-month experiment.

Another kind of stress that is important in trace gas dynamics is shifting aeration/anaerobiosis. In wetland systems the aeration history of the system is a significant controller of CH_4 efflux and

the lag in the development of microbial communities is long. Moore and Roulet (1993) equilibrated microcosms for 20 days saturated, then spent 25 days dropping the water table to 50 cm, left them drained for 15 days, and then reflooded them over an additional 25 days. Methane fluxes were between 9 and 116-fold greater as the water table was dropping than as it was rising. This difference was a combination of the release of pore-water CH_4 and the inability of methanogens and methanotrophs to adapt to the changes in aeration state. Temperature variations can also produce significant hysteresis in process rates, mediated through changes in microbial communities (Updegraff et al., 1998).

Thus, I believe that episodic stress events can reduce the size of even the bulk respiring community substantially enough to have measurable effects on total process rates at the ecosystem scale. These studies and others (e.g., Dickens and Anderson, 1999; Yavitt and Lang, 1990) also suggest that recovery from such stresses may not be as rapid as many have assumed. As many ecosystems experience such episodic pulse stresses with some regularity (e.g., freezing–thawing in northern and alpine systems, drying–rewetting in arid and semiarid systems) it seems that these effects may have ecosystem consequences. The actual importance of such variations in specific microbial populations on larger and longer scales has really not been well explored. We know very little about historical legacy effects that are mediated through microbial communities. The few studies that have actually looked for such effects often find them (e.g., Updegraff et al., 1998 and others cited above), suggesting that the assumption in biogeochemical models that microbial population size never controls process rates, and that microbial processes therefore have no history may be wrong. This all suggests that studying such legacy effects and microbial community dynamics at biogeochemistry-relevant scales is a fruitful area for collaboration among microbial ecologists and biogeochemists.

6. Conclusions: Integrating Biogeochemistry and Microbiology

While there are few good case studies for the two issues that I have raised (different physiological responses and biomass limitation of process rates), I believe that that is less because the cases are rare than because few researchers have designed studies to test these possibilities. Thus, as a message to a new Institute for Biogeochemistry, I believe that the important point is that the microbiology that has been incorporated implicitly into most biogeochemical models is microbial physiology. I have argued here and in other papers (Schimel, 1995; Schimel and Gulledge, 1997) that these models may need to consider aspects of microbial community ecology as well. Ignoring these effects will probably only very rarely produce order-of-magnitude errors, but I believe that there will be cases where 25–50% errors may be likely. Additionally, much of the unexplained error and surprises in current biogeochemical studies may be due to unaccounted-for microbial community dynamics (Schimel, 2000). Incorporating microbial community effects into biogeochemical models will require efforts at both the microbial and biogeochemical modeling ends:

1. From the microbial side, we need more research targeted at understanding when and where microbial community effects, either through variations in physiology or through changes in populations sizes, have large-scale impacts. Many short-term lab studies have been done, but few have attempted to extrapolate to larger spatial and temporal scales.
2. From the biogeochemical modeling side, we need to experiment with models where the microbiology is less implicit by incorporating microbial community effects. Models should consider incorporating response functions that may vary over time with environmental conditions, as proposed by Updegraff et al. (1998). They should also consider incorporating the size of the active microbial population as a control on the rate of a process.

Through collaborative efforts on these two points, we can determine how best to deal with the relevant microbial dynamics in biogeochemical models, and how to predict them. From this, I hope we will be able to develop models that simply, yet effectively incorporate microbial community dynamics. These models will clearly not incorporate a lot of detailed and explicit microbiology. Rather, we must develop a solid enough understanding of the relevant microbial community dynamics so that we can model past the microbes, going directly from environmental drivers to process rates, thus making the microbiology, once again, implicit.

References

Acea, M. J., and Carballas T. (1996). Changes in physiological groups of microorganisms in soil following wildfire. *FEMS Microbiol. Ecol.* **20**, 33–39.

Ambus, P., and Christensen, S. (1993). Denitrification variability and control in a riparian fen irrigated with agricultural drainage water. *Soil Biol. Biochem.* **25**, 915–923.

Anderson, T. -H., and Joergensen, R. G. (1997). Relationship between SIR and FE estimates of microbial biomass C in deciduous forest soils at different pH. *Soil Biol. Biochem.* **29**, 1033–1042.

Berg, P., and Rosswall, T. (1987). Seasonal variations in abundance and activity of nitrifiers in four arable cropping systems. *Microb. Ecol.* **13**, 75–87.

Bodelier, P. L. E., Roslev, P., Henckel, T., and Frenzel, P. (2000). Ammonium stimulates methane oxidation in rice soil. *Nature* **403**, 421–424.

Both, G. J., Gerards, S., and Laanbroek, H. J. (1992). Temporal and spatial variation in the nitrite-oxidizing bacterial community of a grassland soil. *FEMS Microbiol. Ecol.* **101**, 99–112.

Bottner, P. (1985) Response of microbial biomass to alternate moist and dry conditions in a soil incubated with ^{14}C- and $^{15>}N$-labeled plant material. *Soil Biol. Biochem.* **17**, 329–337.

Brussaard, L., BehanPelletier, V. M., Bignell, D. E., Brown, V. K., Didden, W., Folgarait, P., Fragoso, C., Freckman, D. W., Gupta, V. V. S. R., Hattori, T., Hawksworth, D. L., Klopatek, C., Lavelle, P., Malloch, D. W., Rusek, J., Soderstrom, B., Tiedje, J. M., and Virginia, R. A. (1997). Biodiversity and ecosystem functioning in soil. *Ambio* **26**, 563–570.

Cheng, W., Virginia, R., Oberbauer, S., Gillespie, C., Reynolds, J., and Tenhunen, J. (1998). Soil nitrogen, microbial biomass, and respiration along an arctic toposequence. *Soil Sci. Soc. Am. J.* **62**, 654–662.

Chertov. O. G., and Komarov, A. S. (1996). In "Evaluation of Soil Organic Matter Models" (D. S. Powlson, P. Smith, and J. U. Smith, Eds.), pp. 231–236. Springer, Berlin.

Clein, J. S., and Schimel, J. P. (1994). Reduction in microbial activity in birch litter due to drying and rewetting events. *Soil Biol. Biochem.* **26**, 403–406.

Clein. J. S., and Schimel, J. P. (1995). Nitrogen turnover and availability during succession from alder to poplar in Alaskan taiga forests. *Soil Biol. Biochem.* **27**, 742–752.

Dickens, H. E., and Anderson, J. M. (1999). Manipulation of soil microbial community structure in bog and forest soils using chloroform fumigation. *Soil Biol. Biochem.* **31**, 2049–2058.

Finlay, B. J., Maberly, S. C., and Cooper, J. I. (1997). Microbial diversity and ecosystem function. *Oikos* **80**, 209–213.

Groffman, P. M., (1991). Ecology of nitrification and denitrification in soil evaluated at scales relevant to atmospheric chemistry. In "Microbial Production and Consumption of Greenhouse Gases: Methane, Nitrogen Oxides, and Halomethanes" (J. E. Rogers and W. B. Whitman, Eds.), pp. 201–217. American Society for Microbiology, Washington, DC.

Groffman, P. M., Axelrod, E. A., Lemunyon, J. L., and Sullivan, W. M. (1991). Denitrification in grass and forest vegetated filter strips. *J. Environ. Qual.* **20**, 671–674.

Groffman, P. M. and Bohlen, P. J. (1999). Soil and sediment biodiversity—Cross-system comparisons and large-scale effects. *BioScience* **49**, 139–148.

Gulledge, J., and Schimel, J. P. (1998). Low-concentration kinetics of atmospheric CH_4 oxidation in soil and the mechanism of NH_4^+ inhibition. *Appl. Environ. Microbiol.* **64**, 4291–4298.

Hendrickson, O., and Chatarpaul, L. (1984). Nitrification potential in an alder plantation. *Can. J. For. Res.* **14**, 543–546.

Kieft L. T., Soroker, E., and Firestone, M. K. (1987). Microbial biomass response to a rapid increase in water potential when a dry soil is wetted. *Soil Biol. Biochem.* **19**, 119–126.

Leffelar, P. A., and Wessel, W. W. (1988). Denitrification in a homogeneous, closed system: Experiment and simulation. *Soil Sci.* **146**, 335–349.

Li, C. (1996). The DNDC model. In "Evaluation of Soil Organic Matter Models" (D. S. Powlson, P. Smith, and J. U. Smith, Eds.), pp. 263–268 Springer, Berlin.

Lin, Q., and Brookes, P. C. (1999). An evaluation of the substrate induced respiration method. *Soil Biol. Biochem.* **31**, 1969–1983.

Madigan, M. T., Martinko, J. M., and Parker, J. (1997). "Brock Biology of Microbiology" 8th. Prentice Hall, Upper Saddle River, New Jersey.

McGill, W. B. (1996). Review and classification of ten soil organic matter (SOM) models. In "Evaluation of Soil Organic Matter Models" (D. S. Powlson, P. Smith, and J. U. Smith, Eds.), pp. 111–132. Springer, Berlin.

Moore, T. R., and Roulet, N. T. (1993). Methane flux: water table relations in northern wetlands. *Geophys. Res. Lett.* **20**, 587–590.

Morley C. R., Trofymow, J. A., Coleman, D. C., and Cambardella, C. (1983). Effects of freeze–thaw stress on bacterial populations in soil microcosms. *Microb. Ecol.* **9**, 329–340.

Parton, W. J., Schimel, D. S., Cole, C. V., and Ojima, D. S. (1987). Analysis of factors controlling soil organic matter levels in Great Plains grasslands. *Soil Sci. Soc. Am. J.* **51**, 1173–1179.

Roberts, D. V. (1977). "Enzyme Kinetics." Cambridge University Press. Cambridge.

Saad, O., and Conrad, R. (1993). Temperature dependence of nitrification, denitrification, and turnover of nitric oxide in different soils. *Biol. Fertil. Soil.* **15**, 21–27.

Schimel, J. (1995). Ecosystem consequences of microbial diversity and community structure. In "Arctic and Alpine Biodiversity: Patterns, Causes, and Ecosystem Consequences" (F. S. Chapin and C. Korner, Eds.), pp. 239–254. Springer-Verlag, Berlin.

Schimel, J. (2000). Rice, microbes, and methane. *Nature* **403**, 375–377.

Schimel, J. P., and Gulledge, J. (1998). Microbial community structure and global trace gases. *Global Change Biol.* **4**, 745–758.

Schimel, J. P., Gulledge, J. M., Clein-Curley, J. S., Lindstrom, J. E., and Braddock, J. F. (1999). Moisture effects on microbial activity and community structure in decomposing birch litter in the Alaskan taiga. *Soil Biol. Biochem.* **31**, 831–838.

Schimel, J. P., and Clein, J. S. (1996). Microbial response to freeze–thaw cycles in tundra and taiga soils. *Soil Biol. Biochem.* **28**, 1061–1066.

Schlessinger, W. H. (1997). "Biogeochemistry," 2nd ed.. Academic Press, San Diego.

Shi, W., and Norton, J. M. (2000). Microbial control of nitrate concentrations in an agricultural soil treated with dairy waste compost or ammonium fertilizer. Soil Biol. Biochem. in press.

Updegraff, K., Bridgham, S. D., Pastor, J., and Weishampel, P. (1998). Hysteresis in the temperature response of carbon dioxide and methane production in peat soils. *Biogeochemistry* **43**, 253–272.

Wall, D. H., and Moore, J. C. (1999). Interactions underground—Soil biodiversity, mutualism, and ecosystem processes. *BioScience* **49**, 109–117.

Wardle, D. (1998). Controls of temporal variability of the soil microbial biomass: a global-scale synthesis. *Soil Biol. Biochem.* **30**, 1627–1637.

Yavitt, J. B., and Lang, G. E. (1990). Methane production in contrasting wetland sites: Response to organo-chemical components of peat to sulfate reduction. *Geomicrobiol. J.* **8**, 27–46.

Zumft, W. G. (1992). The denitrifying prokaryotes. In "The Prokaryotes" (A. Balows, H. G. Tršper, M. Dworkin, W. Harder, and K. H. Schleifer, Eds.), pp. 554–582. Springer-Verlag, New York.

1.14

Global Soil Organic Carbon Pool

Michael Bird
Australian National University,
  Canberra,
Australia

Hana Santrůcková
University of South Bohemia,
  Ceské Budíjovice,
Czech Republic

John Lloyd
Max Planck Institute for
  Biogeochemistry,
Jena, Germany

Elmar Veenendaal
Harry Oppenheimer Okavango
  Research Center,
Maun, Botswana

1. Introduction: the Soil Carbon Pool and Global Change

Estimates of the size of the global soil organic carbon (SOC) pool have ranged between 700 Pg (Bolin, 1970) and 2946 Pg (Bohn, 1976), with a value of around 1500 Gt now generally accepted as the most appropriate (Table 1). This value is considered to be between one-half (e.g., Townsend *et al.*, 1995) and two-thirds (e.g., Trumbore *et al.*, 1996) of the total terrestrial carbon pool.

The SOC pool plays an important role in modulating anthropogenic changes to the global carbon cycle. On the one hand, human activities such as land clearing, agriculture, and biomass burning lead to large emissions of CO_2 from the SOC pool, which tend to continue long after the initial perturbation to standing biomass has ceased. Conversely, the terrestrial biosphere is thought to be sequestering $\sim 1-2$ Pg/year of anthropogenic CO_2 (the "missing sink") as a result of enhanced photosynthetic carbon fixation (e.g., Dixon *et al.*, 1994). While standing biomass is thought to be responsible for the enhanced uptake required to balance the global anthropogenic CO_2 budget, the SOC pool is thought to provide the longer term transient sink for much of this carbon (Smith and Shuggart, 1993). This is due to the comparatively long time required for the SOC pool to establish a new equilibrium with the enhanced rates of delivery of carbon to the soil from standing biomass.

Despite the major role of the SOC pool in the global carbon cycle, the dynamics of carbon exchange both within the SOC pool and between the SOC pool and the other major global carbon reservoirs are poorly constrained (Townsend *et al.*, 1995; Tans *et al.*, 1990; Trumbore, 1993; Fung *et al.*, 1997). Major uncertainties still surround the size of the SOC pool, the capacity of the SOC pool to store additional carbon sequestered by living biomass, and the response of the SOC pool to changes in climate. Reducing these uncertainties will require more robust estimates of the size of the soil carbon pool and rates and fluxes through the soil carbon pool, as well as the development of additional constraints given by $\delta^{13}C$ and $\Delta^{14}C$ measurements (Fung *et al.*, 1997).

Implicit in the above statement is the need to be able to better predict variations in the SOC pool spatially in terms of soil substrate type, geomorphology, and climate, and assess the accuracy of such predictions against a consistent global observational database. The binding provisions of the Kyoto Protocol and the future possibility of carbon trading between countries introduce further urgency into efforts to establish verifiable inventories of carbon stocks and fluxes of carbon into and out of the SOC pool—the largest but least understood terrestrial carbon reservoir (IGBP Terrestrial Carbon Working Group, 1998).

This chapter examines the status of observational studies of the SOC pool, global trends in the soil carbon pool, and the relationship between observational studies and efforts to model the dynamics of carbon exchange in the SOC pool and suggests how

GLOBAL BIOGEOCHEMICAL CYCLES IN THE CLIMATE SYSTEM

TABLE 1 Estimates of the Size of the Global SOC Pool to 100 cm

Study	*Soil* C (Pg)
Bolin (1970)	700
Bohn (1982)	2946
Baes *et al.* (1977)	1080
Bazilevich (1974)	1392
Schlesinger (1977)	1456
Aitjay *et al.* (1979)	2070
Post *et al.* (1982)	1395
Eswaran *et al.* (1993)	1576
Batjes (1996)	1500

TABLE 2 Factors Affecting the SOC Pool

➤ Temperature
➤ Precipitation/evaporation
➤ Soil texture (parent material/time)
➤ Geomorphology
  Drainage/elevation
  Slope
  Microenvironment
➤ Nutrient status (parent material/time)
➤ Natural disturbance
  Fire
  Drought
  Insects/disease
  Windthrow
  Grazing pressure
➤ Anthropogenic disturbance
  Clearing/afforestation
  Biomass burning
  Agriculture
  Grazing pressure
  Climate change
  CO_2/N fertilization

observational and modeling studies might be better integrated to improve the major uncertainties still surrounding the dynamics of the SOC pool.

2. Factors Affecting the Distribution of Soil Organic Carbon

Jenny (1941) first elucidated the factors likely to affect the SOC content of soil, although the broad relationships between soil, vegetation, and climate had been identified previously. Also utilizing data from later studies, Table 2 lists the major factors controlling SOC inventories.

Climate (temperature and precipitation) exerts a major influence on SOC at the global scale by controlling the levels of input from live biomass into the soil. Climate also influences the rate at which

carbon delivered to the soil is cycled through the SOC pool and ultimately respired back to the atmosphere by microbial biomass, or is lost from the profile as dissolved organic carbon (Fig. 1). Climate, in combination with other factors, controls initial litter quality (nitrogen content, lignin content etc.; Melillo *et al.*, 1982) and processes that modify the nature of organic carbon while in the SOC pool. Climate influences SOC distribution through the soil profile by influencing the efficiency and depth of illuviation and effective bioturbation (e.g., Holt and Coventry, 1990) and is a key fac-

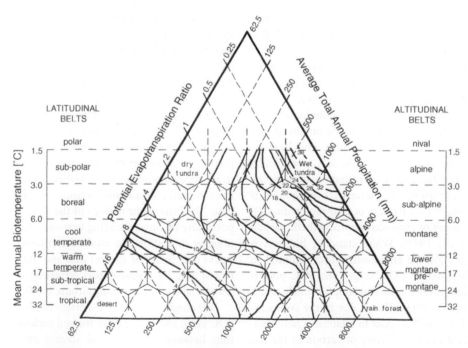

FIGURE 1 SOC distribution in kg/m³ among Holdridge life-zones (Post *et al.*, 1982).

tor affecting the rate of production and the mineralogy of the soil substrate (e.g., Goh *et al.*, 1976). It should also be noted that time plays an important role in determining the nature of the soil substrate, as soil-forming processes operate on time scales from years in the case of pedogenesis on recent alluvial sediments to millions of years in the case of the deep weathering of continental cratons.

At the local scale (i.e., for a given climate) several other factors modulate the distribution of SOC across the landscape. Of primary importance at this scale is soil texture (Parton *et al.*, 1987), a variable that is closely linked to other parameters such as bedrock type, nutrient status (cation exchange capacity), water holding capacity, illuviation and bioturbation rates, root penetration resistance, and the availability of oxygen to support aerobic microbial respiration. It is convenient that these variables tend to be coupled in such a way that soil texture becomes a useful proxy for all of them, with SOC levels generally increasing with decreasing particle size of the soil substrate.

Geomorphology exerts control on soil carbon levels by determining erosion/accretion rates of sediment and SOC and access to water/nutrients and through the provision of local microenvironments (e.g., climatic or protection from fire) that allow the development of vegetation types that sequester and cycle carbon at different rates. Local microenvironments can also modify microbial respiration rates.

The role of a variety of natural and anthropogenic disturbances in modifying SOC inventories has received increased attention in recent decades owing to the large role that land-use change plays in determining the magnitude of transfer between the terrestrial carbon source/sink and the atmospheric CO_2 reservoir. Some disturbances such as deforestation/logging, agricultural, and grazing practices are clearly anthropogenic while others such as windthrow, climatic extremes (drought, etc.), insect plagues, and diseases are more directly attributable to natural causes. In the case of disturbances such as biomass burning, it is often more difficult to ascribe a uniquely anthropogenic or natural cause, and the possibility of anthropogenic climate change means that all the above disturbances might now in part be related to human activities. In this sense, CO_2/nitrogen fertilization effects and afforestation can also be classed as anthropogenic disturbances affecting the SOC pool.

The most immediate effect of the above disturbances on the SOC pool is to modify the rate at which carbon is delivered to the soil, either by changing the rate at which carbon is sequestered by living biomass, or by diverting carbon sequestered by live biomass directly or indirectly back to the atmosphere. Disturbance can also modify the rate at which carbon is returned to the atmosphere by microbial respiration.

In many cases disturbance can lead to long-term changes in local vegetation and soil structure which means that during the period over which the disturbance is maintained, and over which a new equilibrium is established following the cessation of disturbance, the local SOC pool can act as either a source to, or a sink from, the atmosphere. Thus, disturbance can lead to permanent changes in SOC inventories and to transient changes in carbon fluxes from the disturbed area. Hence time is an important vari-

able when considering the response of the SOC pool during and following disturbance.

The complex interactions possible between all of the above variables are one of the root causes of the many uncertainties surrounding the dynamics of carbon exchange through and within the soil carbon pool.

3. Global Variations in the SOC Pool

Predictable variations in the SOC pool have been observed in either comparative studies between soils from contrasting climates (Jenny, 1961; Trumbore, 1993) or contrasting soil textural types (Parton *et al.*, 1987), transects along climatic gradients (Townsend *et al.*, 1995), transects across soil chrono/topo sequences (Goh *et al.*, 1976;), and comparisons between disturbed and undisturbed areas (Townsend *et al.*, 1995; Desjardin *et al.*, 1993). In addition, some studies have attempted to look at SOC trends on the continental (e.g., Spain *et al.*, 1983; Moraes *et al.*, 1995) or global (e.g., Post *et al.*, 1982) scale using data compiled from literature sources and have been able to deduce major trends in SOC inventories with respect to climate and soil type, but generally with much scatter in the data.

From this understanding of the behavior of the SOC pool, models such as Rothamsted (Jenkinson and Rayner, 1977) and Century (Parton *et al.*, 1993) have been developed that allow results from regional validation studies to be extrapolated to the global scale (Schimel *et al.*, 1994). Such models divide the SOC pool up into three to five pools with turnover times ranging from years to thousands of years, and the sizes of these pools for a given soil texture are determined by climate-driven interactions between plant carbon inputs, nutrients, microbial respiration, and leaching of DOC (Fig. 2). In some cases, these models have been tested

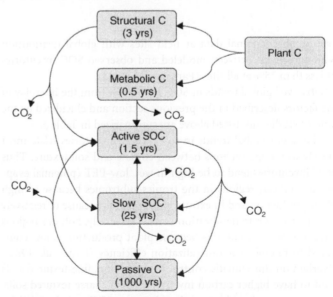

FIGURE 2 Carbon flows in the Century model (Parton *et al.*, 1987).

TABLE 3 Observational Trends in Soil Carbon Storage

Trend	Reference
Climate-specific trends	
SOC turnover rates decrease with decreasing temperature (T) and rainfall (ppt)	Trumbore *et al.* (1996)
	Trumbore (1993)
Proportion of total SOC in upper soil increases with decreasing T (and ppt?)	Zinke *et al.* (1986)
	Spain *et al.* (1983)
Microbial respiration rates decrease with decreasing T (and ppt?)	Lloyd and Taylor (1994)
	Raich and Schlesinger, (1992)
SOC stocks decrease with increasing T (at constant ppt)	Post *et al.* (1982)
	Jenny (1980)
SOC stocks increase with increasing ppt (at constant T)	Post *et al.* (1982)
	Jenny (1980)
$\delta^{13}C$ in surface SOC in C_3 biomes increases with decreasing T	Bird *et al.* (1996)
$\delta^{13}C$ in surface SOC in grass-dominated biomes decreases with decreasing T	Bird and Pousai (1997)
Location-specific trends	
SOC stocks are generally lower in coarse-textured than fine-textured soils	Parton *et al.* (1987)
	Schimel *et al.* (1994)
SOC stocks are generally higher in topographically low positions (valleys)	Malo *et al.* (1974)
	Spain *et al.* (1983)
Disturbance generally leads to changed SOC stocks and fluxes	Dalal and Meyer (1986)
	Harrison *et al.* (1995)
SOC stocks (per cm³) generally decrease with increasing depth in the profile	Spain *et al.* (1983)
	Desjardin *et al.* (1993)
The apparent "age" of SOC generally increases with depth in the profile	Townsend *et al.* (1995)
	Desjardin *et al.* (1993)
SOC stocks are generally higher under trees in mixed tree–grass systems	Kellman (1979)
	Bird and Pousai (1997)
SOC from woody tissues has a longer turnover time than nonwoody tissues	Parton *et al.* (1987)
	Bird and Pousai (1997)
In well-drained soils, SOC $\delta^{13}C$ values generally increase with depth	Agren *et al.* (1996)
	Krull and Retallack (in press)
In poorly drained soils, SOC $\delta^{13}C$ values generally decrease with depth	Agren *et al.* (1996)
	Krull and Retallack (in press)
In episodically poorly drained soils, SOC $\delta^{13}C$ values are constant with depth	Agren *et al.* (1996)
	Krull and Retallack (in press)
Fine clay-associated SOC has a higher $\delta^{13}C$ than coarse, particulate SOC	Desjardin *et al.* (1993)
	Bird and Pousai (1997)
Fine SOC has a longer residence time in soil than coarse SOC	Buyanovsky *et al.* (1994)
In mixed C_3/C_4 systems, SOC $\delta^{13}C$ increases with decreasing particle size	Bird and Pousai (1997)

against observational data at field sites with global distribution with agreement between modeled and observed SOC inventories of less than 25% at all sites (Parton *et al.*, 1993).

Observed global trends in SOC that result from the interplay of the factors described in the previous section and elucidated by the kind of studies discussed above are summarized in Table 3.

The major global trends in soil carbon inventories relate most closely to the interactions between climate and soil texture. Thus SOC inventories tend to be high in wet, low-PET (potential evapotranspiration) regions in the tropics/subtropics because of high plant production and in extratropical regions because of relatively low microbial remineralization rates. Conversely, hot, dry regions have low SOC inventories because plant production is low compared to microbial remineralization efficiency (Post *et al.*, 1982). Overlain on this climatic control in any region, fine-textured soils tend to have higher carbon inventories than coarse-textured soils, due to organomineral interactions between SOC and clay particles

that lead to the physical protection or chemical stabilization of a higher proportion of SOC (Mayer, 1994; Skjemstad *et al.*, 1993; 1996). The broad relationship between SOC, climate, and soil texture is illustrated in Fig. 3 (see also color insert).

Local geomorphic effects tend to lead to higher carbon densities at locations that are lower in the landscape, due to downslope movement of carbon and nutrients and more reliable access to ground and surface water. Carbon inventories tend to be higher along watercourses, and in the case of fire-prone regions carbon inventories tend to be higher in local topographic depressions that are protected from fire. The extreme case of geomorphic localization of SOC is in the development of peat bogs.

The distribution of trees in a landscape has a significant effect on carbon inventories, with higher inventories generally present under tree canopies (e.g., Liski, 1996). The effect of tree distribution is particularly pronounced in savanna ecosystems (Kellman, 1979; Bird *et al.*, in press; Bird and Pousai, 1997). The distribution

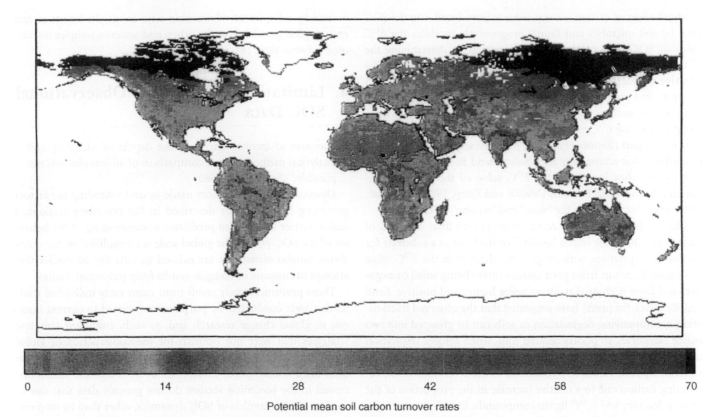

| 0 | 14 | 28 | 42 | 58 | 70 |

Potential mean soil carbon turnover rates

FIGURE 3 Potential mean soil carbon turnover rates extrapolated to the global scale using the temperature and soil texture relationships from the Century model (Schimel *et al.*, 1994). See also color insert.

of trees within a savanna and the location of savanna/woodland/forest boundaries are themselves the results of complex interactions between climate, fire frequency, soil type, geomorphology, and grazing pressure (e.g., Archer, 1990).

At the local scale several general trends in SOC can typically be identified within a single soil profile. As a result of most input of carbon to a soil profile being introduced from the overlying standing biomass, SOC generally decreases down the profile, approximated by a log-log function (Zinke *et al.*, 1986). As much as 50% of the total SOC inventory to 1 m may be present in the top 20 cm of the profile, with the surface here being taken as the top of the mineral soil horizon. The degree to which carbon is concentrated in the upper soil layers is a function of soil type, rooting depth, and climate. Intensive, deep bioturbation such as characterizes most tropical regions leads to a more gradual decline in carbon stocks with depth compared with colder regions where bioturbation is minimal and permafrost may limit downward movement of SOC. Grassland soils tend to have a higher proportion of SOC in deeper soil layers than in comparable forested regions, possibly related to deeper rooting in grassland ecosystems.

There is generally an increase in the average turnover time for carbon in soil (as indicated by ^{14}C) with decreasing temperature and or precipitation (Trumbore, 1993), reflected in a slowing of microbial respiration rates (Raich and Schlesinger, 1992). In an individual soil profile, the ^{14}C age of soil carbon tends to increase with depth in a soil profile (Townsend *et al.*, 1995; Desjardin *et al.*, 1993), and also with decreasing particle size (Buyanovsky *et al.*, 1994), reflecting an increase in the relative proportion of old refractory carbon in fine particles and the deep soil.

Many studies have examined the controls of $\delta^{13}C$ distribution in soils. The primary control on the $\delta^{13}C$ composition of SOC is the $\delta^{13}C$ value of the carbon being delivered to the SOC pool from live biomass. Large differences are controlled by the distribution of C_3 and C_4 vegetation in terrestrial ecosystems, which in turn are determined by climate (Lloyd and Farquhar, 1994). Smaller differences are determined by factors such as altitude (Bird *et al.*, 1994), soil water availability (Stewart *et al.*, 1995), irradiance (Ehrlinger *et al.*, 1986), and the degree of reutilization of respired CO_2 in closed canopies (Van der Merwe and Medina, 1989).

Carbon can remain in the SOC pool for a long period after its assimilation by vegetation. This means that not all carbon in the SOC pool is in equilibrium with the isotopic composition of the modern atmosphere due to a decrease in the $\delta^{13}C$ value of the atmosphere as a result of fossil fuel burning since industrialization (Freidli *et al.*, 1986; Bird *et al.*, 1996). "Old" carbon will therefore be enriched by up to 1.5‰ compared to recent carbon. This "terrestrial Seuss effect" needs to be considered when calculating the $\delta^{13}C$ value of CO_2 respired to the atmosphere from the SOC pool (Fung *et al.*, 1997).

Metabolism of carbon, once it enter is introduced into the SOC pool by soil microbes and fauna, progressively modifies the $\delta^{13}C$ value of SOC. Two major processes compete in determining the $\delta^{13}C$ value of the carbon remaining from metabolic processes. The selective utilization of nutrient- and energy-rich compounds such as sugars and proteins tends to increase the relative proportions of components such as lignin in the remaining carbon, and these compounds have $\delta^{13}C$ values lower than the bulk biomass of which they were a part (Benner *et al.*, 1987). Conversely, kinetic fractionation effects that accompany metabolism tend to favor the respiration of ^{12}C, thus increasing the $\delta^{13}C$ value of the carbon partitioned into microbial biomass (Macko and Estep, 1984; Blair *et al.*, 1985) and ultimately into the "slow" and "passive" SOC pools.

The model proposed by Agren *et al.* (1996) links the $\delta^{13}C$ of carbon in SOC to the initial "quality" of the litter as a substrate for microbial respiration, with progressive changes in the $\delta^{13}C$ value of degraded carbon from poor quality litters being small or negative, and from high quality litters being higher and positive. Krull and Retallack (in press) have suggested that the observed fractionation accompanying degradation in soils can be grouped into two major categories. In poorly aerated soils, when aerobic respiration is retarded, organic matter accumulates and selective utilization of some compounds leads to a reduction in the $\delta^{13}C$ value of the remaining carbon due to a relative increase in the proportion of the more refractory low $-^{13}C$ lignin compounds. In well-aerated soils, kinetic isotopic fractionation dominates and the remaining carbon is enriched in ^{13}C.

A further mechanism that has yet to be considered is the role of macrofauna in the decomposition of organic matter. In the seasonally dry tropics, for example, it has been estimated that 20% of organic matter decomposition results from the action of termites (Holt and Coventry, 1990). Since methane is an abundant product of the decomposition of organic matter in termite nests, it is possible that the remaining SOC residue is substantially enriched in ^{13}C.

The concentration of the partly "stabilized" products of microbial metabolism in the fine fraction of the soil coupled with the decreasing atmospheric CO_2 $\delta^{13}C$ value since industrialization means that old, refractory, clay-associated carbon has a higher $\delta^{13}C$ value than coarser particulate carbon (Kracht and Bird, in review). In tropical savannas, this trend of increasing $\delta^{13}C$ values with decreasing particle size is augmented by the preferential accumulation of C_4-derived carbon in the fine-particle-size fractions (Bird and Pousai, 1997).

At the regional/global level, there have been relatively few observational studies of variations in SOC $\delta^{13}C$ value. Bird *et al.* (1996) demonstrated the existence of a latitudinal gradient in the $\delta^{13}C$ value of surface SOC in forest soils, with tropical forest soils having approximately $1-1.5‰$ lower $\delta^{13}C$ values than those of high-latitude forest soils. This was attributed to a temperature gradient in the turnover times for carbon in the soil (the "terrestrial Seuss effect") and possibly to a $\sim 0.5‰$ temperature effect (Bird and Pousai, 1997).

Bird and Pousai (1997) found that $\delta^{13}C$ values and carbon content on a rainfall transect through northern Australia could be de-

scribed by a log-linear relationship with monsoon forests at one end, tropical grasslands at the other, and savanna samples distributed between these two end-members.

4. Limitations of Available Observational SOC Data

"Because of inconsistencies in the depths of sampling and analytical methods, direct comparison of all samples was not possible." (Spain *et al.*, 1983)

Despite the many advances made in understanding the factors governing SOC behavior described in the preceding sections, a major barrier to a refined predictive understanding of the behavior of the SOC pool at the global scale is exemplified by the quote above. Similar statements are echoed in virtually all studies that attempt to compare or compile results from published studies.

These problems largely result from many early individual studies that were conducted for purposes not related to current interests in global change research and, as such, employed sampling strategies that were not optimal for the construction of global SOC inventories or for the modeling of SOC dynamics. In addition, the available global inventories of SOC and the trends discussed in the preceding section do not provide data that can be used to validate models of SOC dynamics, other than by confirming gross trends in carbon storage.

These problems can be grouped into three "types"—definition, sampling, and analysis. A significant problem exists with respect to the definition of what constitutes soil carbon. Many studies make a strong distinction between "mineral soil," the organic "O-horizon," and "litter" and most estimates of the size of the SOC pool refer only to the size of the SOC pool in the mineral soil. However, in cool temperate regions, a significant portion of the total SOC is present in the litter and O-horizon of many soils. This carbon may have been "dead" for a century or more. Much of this material can be considered SOC in the sense that it is dead organic carbon in a variable state of decomposition, which provides a source of nutrients for the living rooted biomass and a substrate for microbial metabolism.

A more appropriate definition of SOC for global change requirements might be that SOC represents all dead carbon from the land surface down (i.e., from the top of the conventional litter horizon) with no dimension greater than 2 mm, including such carbon present in the litter and O-horizons. All material greater than 2 mm is thus considered litter, a size that is commonly used in soil science and a distinction that is easily made and readily quantifiable. The present distinction between litter and the O-horizon is ambiguous in that in reality a continuum exists between carbon in litter, O-horizon, and mineral soil. The present distinction also cannot account for the presence of subsurface litter derived from recently dead roots, the turnover dynamics of which may be more similar to surface litter than the mineral soil.

An additional definition problem relates to the description of sampling localities in many studies simply in terms of soil type.

This again may be of use in pedogenic studies, and does allow some inferences to be drawn regarding the factors likely to control SOC dynamics. However, it does not provide the quantifiable textural and geomorphic information that is crucial in determining SOC dynamics and required as input for modeling studies (Parton *et al.*, 1987). Many studies are also not conducted with a view to separately quantifying carbon inputs from different sources. In forests, local carbon inventories have been shown to vary widely depending on the distribution of trees (Liski, 1996). In mixed savanna ecosystems, the carbon inputs from tree and grass sources are not necessarily equivalent, and will vary spatially with the proportion of each source (this problem is particularly acute in terms of the $\delta^{13}C$ value of tropical savanna SOC).

Lack of an agreed sampling protocol presents a major problem in comparing the results from different studies. Individual studies sample over different depth increments to different depths (or by horizons) and in many cases do not incorporate soil density data. Where no soil density data are available, authors have attempted to use a variety of inferential techniques to estimate carbon density from %C data. While this can provide a crude estimate of carbon densities, Bird (1998) for example, found that one of the effects of grazing on similar basalt-derived soils was to increase soil density in the 0- to 5-cm interval by 25–35% and any attempt to estimate carbon density from a single approximated soil density estimate would introduce a similar error into the calculated carbon densities.

The general use of soil pits means that samples and bulk density measurements are usually made horizontally into the pit wall at the mid-point of the sample interval, rather than by collecting the whole depth interval; this can potentially introduce (smaller) biases if the distribution of carbon with depth is not uniform. In addition, there is little uniformity in the literature with regard to whether samples are sieved or not, and whether/how soil carbonate is considered.

Further large uncertainties are introduced into previously published data by the use of different analytical techniques. Combustion in oxygen at high temperature (in a variety of forms) is now the preferred method of determining carbon abundances in soil samples. However, much of the literature data on which many of the regional and global SOC inventory studies have been based also include data generated using either Walkley–Black wet oxidation or "loss on ignition" techniques (e.g., Moraes *et al.*, 1995).

It has long been known that the Walkley–Black technique underestimates total organic carbon and if any attempt is made to compensate for this fact, it is common to multiply the measured values by a factor of 1.3. This factor seems to derive from the original work of Walkley and Black (1934) and represents the average under-estimate of carbon from the analysis of seven soil samples, where the factor calculated for the individual samples ranged between 1.16 and 1.66. The value of 1.3 has been widely applied (e.g., Little *et al.*, 1962) but because Walkley–Black oxidation does not measure "recalcitrant" carbon, it can be expected that the proportion of such recalcitrant carbon in soils will be highly variable depending on the nature of the carbon delivered to the soil from standing biomass, fire frequency, and other soil parameters.

Loss on ignition measures "organic matter," and a single invariant factor (generally 0.58, the van Bremmelen factor) is usually multiplied by the observed organic-matter content to calculate the percentage of carbon in the sample. However, it is known, for example, that this factor is variable and can be as low as 0.45 in the tropics (Burringh, 1984). Uncertainties in this factor will flow directly into uncertainties in the calculated carbon densities.

The above discussion suggests that depending on the sampling and analytical protocols used to estimate carbon inventories at the same single site, the results might easily differ by 10–20%. To this uncertainty must be added the inherently high variability in the SOC pool at all spatial scales in response to the factors outlined in preceding sections.

Unlike live vegetation, there is currently no means to remotely sense SOC stocks. SOC stocks can be estimated from SOC models coupled to NPP estimates derived from remotely sensed data, but observational SOC data are not usually collected in such a way as to be able to provide a means of validating model results.

The situation is even more difficult with respect to current knowledge of the carbon-isotope composition of the SOC pool. Work in this area with few exceptions (e.g., Bird *et al.*, 1996; Bird and Pousai, 1997) has focused on using carbon isotopes as a tracer of carbon dynamics and vegetation change in local regions. There is an urgent need for techniques that can be used to provide observational estimates of the carbon-isotope composition of the SOC pool and of carbon fluxes from the SOC pool to the atmosphere.

Several models are now available that can predict isotope fractionation by the terrestrial biosphere (e.g., Lloyd and Farquhar, 1994) and this knowledge provides important constraints on source/sink distributions when coupled with measurements of the isotopic composition of the atmosphere (Cias *et al.*, 1995; Fung *et al.*, 1997; Bakwin *et al.*, 1998).

There is currently no way of better constraining or testing these model results with regional observational terrestrial isotopic data. Because the SOC pool integrates the isotopic signature of local vegetation over several to many years, it potentially provides the best integrated measure of the carbon isotope composition of regional biomass, if this signature can be adequately isolated from the isotopic effects of degradation (e.g., Agren *et al.*, 1996) and the terrestrial "Seuss effect" (Bird *et al.*, 1996; Fung *et al.*, 1997).

There is a need for a new class of SOC data, collected globally in a consistent fashion, which allows the direct comparison of results across a wide range of climatic conditions, which can be better integrated with remote-sensed vegetation indices, and which is more amenable to the validation of models of global carbon cycle dynamics and SOC dynamics.

5. A Stratified Sampling Approach

Discussion in the previous section has highlighted the problems inherent in attempting to utilize previously published data to obtain global observational trends in the SOC pool at sufficient resolution to be of use in resolving major problems in global change research, such as the location of the "missing sink." It is also likely that the political requirements of the Kyoto Protocol will require more accurate and verifiable estimates of the size of the SOC pool and fluxes through the SOC pool than are currently possible.

The approach of Parton *et al.* (1993) represents a useful step in this direction, holding biome constant (using 11 grassland sites with global distribution) and allowing climate and soil texture to vary. However, to define global trends better, a more instructive approach might be to hold soil texture constant and allow climate (and thereby vegetation) to vary. Thus, undisturbed vegetation on sandy soils would provide a "low" end-member, and vegetation on basalt-derived (or similar) soils would provide a "high" end-member for SOC inventories and fluxes (including isotopes) under any given climate. These two estimates could then be mixed as appropriate to provide estimates for regional storage and fluxes in a given area based on the knowledge of soil texture distribution.

Vegetated aeolian sand bodies from the last glacial maximum are common in many parts of the world, as are sandy sediments in palaeochannels and glacial deposits. Likewise, fine-textured soils derived from basalts (or intrusive equivalents) are common and

widely distributed across the globe. It should be noted that sandy soils have a significant advantage over fine-textured soils in that they are usually freely drained and there is little scope for the physical or chemical protection of SOC which can complicate interpretation of results, once the soil is disturbed by sampling.

Such an approach neglects secondary topographic effects, but these can be avoided in the sample set by restricting sampling to locations high in the local topography. The approach therefore cannot cope, for example, with local effects such as the occurrence of peats or different SOC dynamics around watercourses, but neither can any other current approach.

Holdridge (1947) provided a simple climatic classification of vegetation in terms of rainfall and precipitation, dividing the world into 30 life zones. While this classification has been superceded by more recent work, it can serve as a guide to the number of sample regions that might be needed to cover the globe, that is, 120 locations (fine- and coarse-textured sites duplicated in each life zone).

A consistent set of sampling depths must be applied at each site that attempt to partly separate mostly recent SOC (0–5 cm) from deeper soil layers (5–30 cm and 30–100 cm). In temperate areas,

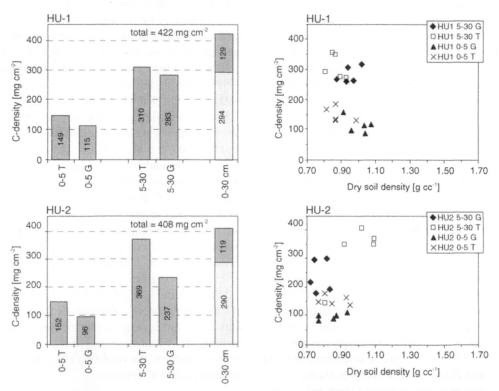

FIGURE 4 Relationship between soil carbon density and soil bulk density, as well as inventories calculated from averaging of this data. The two sites are on ungrazed basalt-derived soils in the Hughenden–Charters Towers region of North Queensland (Bird, 1998), and the samples were collected along transects of approximately 1000 m in each case. Samples were collected as soil cores from the 0–5cm and 5–30cm depth intervals. Because the vegetation is a tree–grass mosaic, one-half of the total samples was collected at half-crown-distance from trees (−T) and the other half was collected at approximately equal distances from local trees (−G). In the construction of the site inventories, the total from the −T and −G samples was weighted according to the estimated percent crown (−T) cover at the site.

where thick surface-organic layers are present, it may be necessary to quantify these separately, and begin depth sampling at the top of the mineral soil. All discussions below define the soil surface as the top of the conventional "litter" horizon.

Where trees are widely separated, or both trees and grass are present in the biome, a separate suite of samples must be collected from each category as tree distribution has a major impact on local variability in the SOC pool as discussed in preceding sections. All samples must be subject to particle-size fractionation to provide the textural information required for the interpretation of observed variations and as model input.

Even when an attempt is made to control variables in the manner proposed above, there will still be local variability that cannot be encompassed if sampling is restricted to a few soil pits. Liski (1996) has suggested, for a boreal forest on a sandy substrate in Finland, that a minimum of 30 samples are required for a 10% confidence interval on the mean value obtained for the carbon inventory, and the number of samples required is likely to be higher than this for more heterogeneous tree–grass ecosystems. Carter *et al.* (1998) found that 15 random soil samples were required from $1° \times 1°$ grid cells in Queensland (Australia) to define the average SOC content of a cell to within 10% of the "true" mean, while about 40 samples were required for an estimate to within 5% of the mean.

Based on the number of sample regions and the number of samples per region suggested above, a global sampling program with 3 or 4 depths and 3 or 4 size fractions per sample would require between 200,000 and 300,000 individual analyses of carbon

content. The analysis of this number of samples would be prohibitive. However, a stratifed sampling approach provides a mechanism for "bulking" many equivalent individual samples into a single "stratified" sample, encompassing local variability and reducing the required analytical effort to approximately 12,000 samples.

For such an approach to be viable, it must be able to reproduce results obtained from the averaging of many results from individual samples from the same area. Figure 4 shows results for individual samples taken from two transect sites (~1000 m length) in ungrazed savanna areas on basalt-derived soils in north Queensland (Bird, 1998). It also shows the carbon inventories calculated from averaging these data for areas beneath tree canopies and areas remote from trees and the calculated average carbon density for the sites weighted according to the percentage crown cover at the site. There is considerable scatter among the individual samples but averaging 10 samples at each site yields closely comparable total inventories (408 vs 422 mg/cm^2 0–30 cm).

A comparison between these results (40 analyses per site) and the results obtained using the stratified sample approach (four analyses per site) both for basalt soils and for a sandy granite-derived soil from the same region is provided in Fig. 5. The inventories obtained using each approach give essentially identical results and highlight the large texture-controlled differences in carbon inventories between the granite and basalt soils. The concordance between the results from the single sample and stratified sample approaches is to be expected as mixing between carbon inventories (i.e., per unit volume) can be expected to be linear.

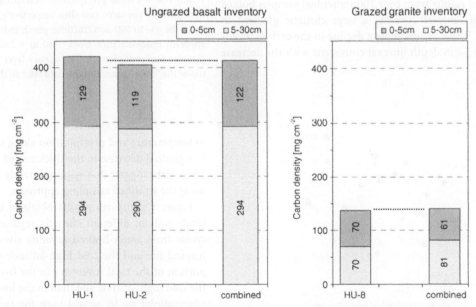

FIGURE 5 Comparison of the results from Figure 4 with the stratified approach, whereby a volumetric fraction of each sample type was added to a bulk sample, producing four site-averaged samples (0–5T, 0–5G, 5–30T, and 5–30G) from 40 individual samples. These samples were analyzed as for the individual samples and the inventories weighted according to percent crown cover as for the individual samples. Results are shown for the basalt-derived soils from Figure 4 and for a site on sandy granite-derived soil.

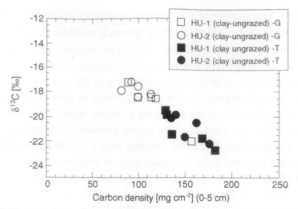

FIGURE 6 Relationship between the δ^{13}C value of SOC and soil carbon density for the samples shown in Figure 4.

The δ^{13}C results from these samples (Fig. 6) provide support for the conclusion that mixing between carbon sources (trees and grass) is linear and that therefore the analysis of stratified samples provides a representative value for the regions sampled. The relationship is further confirmed by similar results for both sandy and clay soils from Zimbabwe (Bird *et al.*, 2000) as shown in Fig. 7. The regular decrease in δ^{13}C value with increasing carbon density relates to higher carbon per unit area in those areas beneath C$_3$ trees and/or the longer residence time of C$_3$-derived carbon once it enters the SOC pool. Below 5 cm depth, the relationship is less clear owing to the effects of degradation on the isotopic composition of SOC and to less tight coupling between SOC at this depth and the vegetation currently overlying the site.

Figure 8 shows results from over 700 individual samples bulked into six regional samples covering a large climatic gradient in western Canada. These results show a decline in the carbon inventory of the 0- to 5-cm depth interval consistent with the decrease

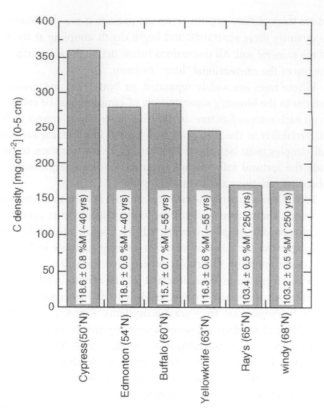

FIGURE 8 Carbon densities for the 0–5 cm interval of stratified samples (approximately 75 samples bulked into one sample per region) on a transect from the U.S.–Canada border to the Canadian Arctic, through Alberta and the Northwest Territories. "Biotemperatures" (as defined by Holdridge, 1947) range from 10.5°C for Cypress (50°N) to 1.5°C for Windy (68°N), while precipitation (excluding snowfall) ranges from 750 to 60 mm for the same two sites, respectively. Also shown is the ^{14}C activity of the 63- to 500-μm fraction, which indicates a consistent increase in apparent residence time from ~40 to ~250 years with decreasing temperature and precipitation. The surface layer in all cases is taken to be the top of the "litter" layer, and the data refer to the <2000-μm fraction.

in temperature and precipitation along the gradient, which results in a gradual decrease in the thickness of the litter and O-horizons. These data suggest that regional trends in SOC can be elucidated using the stratified sampling approach.

Figure 9 compares results obtained using this approach from sandy soils in different climatic regions. Carbon inventories increase from water-limited savanna sites toward both the humid tropical site and the cold high-latitude site. In addition, the proportion of the total inventory in the 0–5 cm interval is higher in the cold high-latitude site than in the low-latitude sites. Both these observations are in accord with the trends to be expected from discussion in earlier sections, with the development of thick litter and O-horizons in high-latitude soils and their absence in lower-latitude soils.

The total range of inventories between these sites is a factor of ~3, but it should be noted that no humid high-latitude sites are

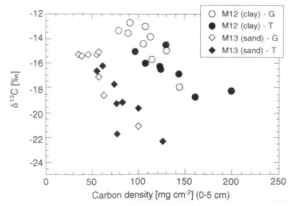

FIGURE 7 Similar relationships between δ^{13}C and SOC density pertain to both sandy and clay soils at the Matopos Research Station, Zimbabwe (Bird *et al.*, 2000). The relationship appears to be independent of the fire regime imposed, from an annual burn to complete protection from fire since the experiment began (1947–48).

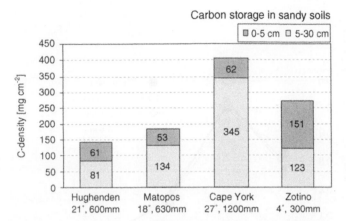

FIGURE 9 Comparison of results produced using the stratified sampling approach from several different climates, all on sandy soils. The surface layer in all cases is taken to be the top of the "litter" layer and refers to the <2000-μm fraction. Data are from the following sources: Matopos, Zimbabwe: Bird *et al.*, (2000); Hugheneden, Australia: Bird (1998); Cape York, Australia: Kracht and Bird (unpublished data); Zotino, Siberia: Bird (unpubl. data)

represented and, for example, the 0- to 5-cm inventory for the Edmonton sample (Fig. 5) is approximately the same as the total 0- to 30-cm interval inventories at the Cape York and Zotino sites, and thus the 0- to 30-cm inventory for the Edmonton site can be expected to be considerably greater than at either of these sites.

An additional advantage of the stratified sampling approach is that it enables a suite of more time-consuming analyses to be performed on a greatly reduced sample set. For example, Fig. 10 shows results for the particle-size distribution, carbon distribution, and carbon-isotope composition of the 0- to 5-cm interval. Despite the similar mineral particle-size distributions in the two data sets, climate, vegetation, and fire regime have imparted very different characteristics to the distribution of carbon and carbon-isotopes between the different size fractions. Again, these differences are readily explicable in terms of the processes described in preceding sections (see figure caption).

Stratified samples are also amenable to radiocarbon analysis and can potentially provide better "average" turnover times for particular climatic/soil-texture conditions than site-specific studies, using any of the available methodologies (e.g., Bird *et al.*, 1996; Trumbore, 1993; Harrison *et al.*, 1995). ^{14}C results for the 63- to 500-μm fraction for the stratified Canadian samples discussed above are provided in Fig. 8. Again, the apparent turnover times for carbon in this size fraction increases with decreasing temperature and precipitation in keeping with expectations based on discussion in preceding sections.

An as yet little explored possibility is using the ^{14}C activity of microbial carbon (Ladd and Amato, 1988) or microbially respired CO_2 to obtain an integrated measure of the average turnover time of SOC in a soil sample. The rationale behind this

statement is that while the microbial carbon pool turns over rapidly, the ^{14}C activity of the microbial carbon pool will be determined by the amount-weighted ^{14}C activity of the substrate being metabolized. Thus if recalcitrant, degraded carbon is not being utilized for microbial metabolism (and thus is not relevant to the global cycle on decadal–centennial timescales) it will not be recorded in the ^{14}C activity of the microbial carbon or microbially respired CO_2.

The measurement of the ^{14}C activity of CO_2 microbially respired from root-free soil in the laboartory is technically straightforward (Santruckova *et al.*, in press), but the possibility exists that in disturbing the soil from its field location may make "protected" carbon available for metabolism. The physical separation of microbial carbon from a soil sample requires the preparation of two samples (Jenkinson, 1988). The first sample is extracted with K_2SO_4, while the second is fumigated with chloroform and then extracted with K_2SO_4. The chloroform fumigation renders a proportion of the microbial carbon extractable in K_2SO_4. The proportion and ^{14}C activity of the microbial carbon are then determined by mass balance. While this approach has the advantage of providing an instantaneous "snapshot" of microbial carbon at the time of collection (assuming that microbial activity is halted by drying or freezing soon after collection), the neccessity for a mass balance calculation introduces additional uncertainty into the calculation of the microbial ^{14}C activity, and it is also unclear whether the fumigation step only liberates microbial carbon (Badalucco *et al.*, 1992).

As an initial test of this approach, the samples representing the "warmest" (Edmonton) and "coldest" (Windy) locations on the Canadian transect were analyzed using both techniques. Figure 11 shows the results obtained using both techniques. The microbially respired CO_2 results suggest reasonable turnover times slightly shorter than the 63–500 μm fraction of the same soils. The results obtained from the fumigation–extraction technique yields much shorter apparent turnover times with much larger errors. One explanation for the discrepancy between the two results is that microbes do not partition all carbon equally between assimilative and respiratory processes. It is possible that younger more energy-/nutrient-rich carbon is more likely to be partitioned into biomass while older carbon is used to support respiration (J. Schimel, personal communication). The fact that the microbial carbon ^{14}C activity of the Edmonton sample lies above the predicted relationship between ^{14}C activity and turnover time for 1996 further suggests that a carbon atom can be cycled within the microbial carbon pool for an average of ~10 year before it is respired or passes to the "slow/passive" pool in this sample.

These preliminary results suggest that the respired CO_2 technique might provide more reliable estimates of SOC turnover time than the fumigation–extraction technique. In sandy soils where there is little chance for carbon to be physically protected, the potential problems identified above relating to the physical disturbance of the sample might be avoidable.

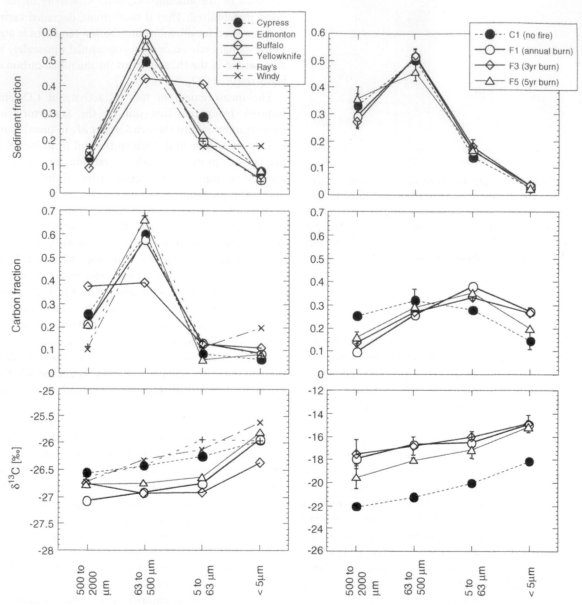

FIGURE 10 Size distribution of clastic particles as well as size distribution and isotopic composition of carbon in stratified samples from the Canadian transect (Fig. 8) and the Matopos fire trials (Fig. 7). While the clastic size distribution is similarly sandy for most samples, there are dramatic differences in the distribution of carbon between the same size fractions. Carbon is uniformly present in the coarser size fractions of the Canadian soils, but is variably enriched in the finer fractions of the Matopos samples. The distribution of carbon between size fractions at Matopos is controlled by fire frequency, with a regular increase in carbon in the finer fractions with increasing fire frequency, leading to the combustion of coarse material. The $\delta^{13}C$ value of carbon increases with decreasing particle size in the Canadian samples by $\sim 1\text{‰}$ due to the terrestrial Seuss effect and fractionation associated with microbial degradation (Bird and Pousai, 1997). The $\delta^{13}C$ value of carbon in the Matopos samples increases by $2-3\text{‰}$ with decreasing particle size owing to the differential input of C_4-derived carbon into the fine fractions or the selective preservation of C_3-derived carbon in the coarse fractions (Bird and Pousai, 1997). The $\delta^{13}C$ values of the Canadian samples are typical of high-latitude C_3 forests (Bird *et al.*, 1996), while the Matopos values are indicative of a variable input of C_4-derived carbon. It should be noted that the Matopos results were obtained by duplicating the stratified sampling at two sites subjected to each of the burning regimes in each case. The comparability of the results (indicated by the error bars) from these independent duplicate samplings provides further evidence that the stratified sampling approach can produce reliable results.

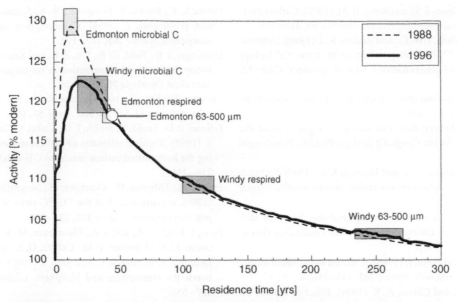

FIGURE 11 ^{14}C activity of microbially respired CO_2, microbial carbon (by fumigation–extraction), and the 63- to 500-μm particle-size fraction of the Edmonton and Windy stratified samples (see Fig. 8). Also shown is the relationship between residence time and ^{14}C activity expected for 1996 when the samples were collected and 1988, the last time that values as high as observed for the Windy microbial carbon sample could have occurred in the SOC pool.

6. Conclusions: Sandworld and Clayworld

Discussion in the previous sections has demonstrated that the mechanisms underlying observed trends in the SOC pool are understood to the extent that models have been developed that can adequately describe observed patterns of carbon distribution at well-documented field sites. Further discussion has also demonstrated that the number of such sites with consistency in sampling and analytical protocols sufficient to provide further constraints for global models is small.

While it is not intended to supplant further process-oriented studies at individual sites, it has been argued that there is a need for a new class of SOC data, aimed at refining our understanding of how the determinants of SOC behavior interact at the global scale. This style of data must use consistent methodologies from globally distributed sites to produce information on SOC stocks, fluxes, potential respiration, and isotopes. This information must be consistently coupled to textural and depth-distribution data that take local variability into account and can provide regional-scale estimates in a format suitable for comparison with model-derived data. Some effort in this direction is already being made (Paustian *et al.*, 1995; Falloon *et al.*, 1998).

Sandworld and clayworld, that is, observational estimates of SOC parameters from coarse- and fine-textured substrates distributed across the full spectrum of global climatic zones, are achievable technically and logistically. They represent a natural extension of process-based site-specific studies and will be required to further refine our understanding of the SOC pool at the global scale. Such a refinement will be required to address the major current questions surrounding the functioning of the global carbon cycle and to answer the political questions that will arise as a result of the ratification of the Kyoto Protocol.

Acknowledgment

The completion of this manuscript was greatly facilitated by a visiting fellowship provided by the Max Planck Institute for Biogeochemistry.

References

Aitjay, G. L., Ketner, L. P., and Duvingneaud, P. (1979). Terrestrial primary production and phytomass. In "The Global Carbon Cycle—Scope." (B. Bolin, E. T. Degens, S. Kemper, and P. Ketner, Eds.), pp. 129–181. Wiley, New York. Vol. 13.

Agren, G. I., Busatta, E. and Balesdent, J. (1996). Isotope discrimination during decomposition of organic mater: a theoretical analysis. *Soil Sci. Am. J.* **60**, 1121–1126.

Archer, S. (1990). Development and stability of grass/woody mosaics in a subtropical savanna parkland, Texas, U.S.A. *J. Biogeogr.* **17**, 453–462.

Badalucco, L., Gelsomino, A., Dell'Orco, S., Grego, S., and Nannipieri, P. (1992). Biochemical characterization of soil organic compounds extracted by 0.5 M K_2SO_4 before and after chloriform fumigation. *Soil Biol. Biochem.* **24**, 569–578.

Baes, C. F., Goeller, H. E., Olson, J. S., and Rotty, R. M. (1977). Carbon dioxide and climate: The uncontrolled experiment. *Am. Sci.* **65**, 310–320.

Bakwin, P. S., Tans, P. P., White, J. W. C., and Andres, R. J. (1998). Determination of the isotopic ($^{13}C/^{12}C$) discrimination by terrestrial biology from a global network of observations. *Global Biogeochem. Cycles* **12**, 555–562.

Batjes, N. H. (1996). Total carbon and nitrogen inventories in soils of the world. *Eur. J. Soil Sci.* **47**, 151–163.

Bazilevich, N. L. (1974). Energy flow and biological regularities of the world ecosystems. *Proc. 1st Int. Congr. Of Ecology*, Puddoc, Wageningen 468–470.

Benner, R., Fogel, M. L., Sprague, E. K. and Hodson, R. E. (1987) depletion of ^{13}C in lignin and its implication for stable–isotope studies. *Nature* **329**, 708–710.

Bird, M. I. (1998). Changes in soil C following land-use change in Australia. Report prepared for Environment Australia (Australian Greenhouse Office), unpublished, 58 p.

Bird, M. I., Chivas, A. R., and Head, J. (1996). A latitudinal gradient in carbon turnover times in forest soils. *Nature* **381**, 143–146.

Bird, M. I., Haberle, S. G., and Chivas, A. R. (1994). Effect of altitude on the carbon-isotope composition of forest and grassland soils from Papua New Guinea. *Global Biogeochem. Cycles* **8**, 13–22.

Bird, M. I. and Pousai, P. (1997). $\delta^{13}C$ Variations in the surface soil organic carbon pool. *Global Biogeochem. Cycles*, **11**, 313–322.

Bird, M. I., Santruckova, H., and Lloyd, J. (in review). Soil carbon storage and dynamics on a latitudinal transect in Canada. *Eur. J. Soil Sci.*

Bird, M. I., Veenendaal, E., Moyo, C., Lloyd, J., and Frost, P. (2000). Effect of fire and soil texture on soil carbon dynamics in a sub-humid savanna, Matopos, Zimbabwe. *Geoderma* **94**, 71–90.

Blair, N., Leu, A., Munos, E., Olsen, J., Kwong, E., and des Marais, D. (1985). Carbon isotopic fractionation in heterotrophic microbial metabolism. *Appl. Environ. Microbiol.* **50**, 996–1001.

Bohn, H. L. (1982). Estimate of organic carbon in world soils. *Soil Sci. Soc. Am. J.* **4**, 1118–1119.

Bolin. (1970). The carbon cycle. *Sci. Am.* **223**, 124–130.

Burringh. (1984). Organic carbon in soils of the world. In "The Role of Terrestrial Vegetation in the Global Carbon Cycle: Measurement by Remote Sensing." (G. M. Woodwell, Ed.), pp. 91–109. Wiley, New York.

Buyanovsky, G. A., Aslam, M., and Wagner, G. H. (1994). Carbon turnover in physical soil fractions. *Soil Sci. Soc. Am. J.* **58**, 1167–1173.

Cais, P., Tans, P. P., White, J. C. W., Trolier, M., Francey, R. J., Berry, J. A., Randall, D. R., Sellers, P. J., Collatz, J. G. and Schimel, D. S. (1995). Partitioning of land and ocean uptake of CO_2 as inferred by $\delta^{13}C$ measurements from the NOAA Climate Monitoring and Diagnostics Laboratory Global Air Sampling Network *J. Geophys. Res.* **100**, 5051–5070.

Carter, J. O., Howden, S. M., Day, K. A., and McKeon, G. M. (1998). Soil carbon, nitrogen, phosphorus and biodiversity in relation to climate change. In "Evaluation of the Impact of Climate Change on Northern Australian Grazing Industries." *Final Report, Rural Industries Research and Development Corp. Project* **139A**, 185–280.

Dalal, R. C. and Mayer, R. J. (1986). Long-term trends in fertility of soils under continuous cultivation and cereal cropping in southern Queensland. II. Total organic carbon and its rate of loss from the soil profile. *Aust. J. Soil Res.* **24**, 281–292.

Desjardin, T., Andreux, F., Volkoff, B., and Cerri, C. C. (1993). Organic carbon and ^{13}C contents in soils and soil size fractions, and their changes due to deforestation and pasture installation in eastern Amazonia. *Geoderma* **61**, 103–118.

Dixon, R. K., Brown, S., Houghton, R. A., Solomon, A. M., Trexler, M. C., and Wisniewski, J. (1994). Carbon pools and flux of global forest ecosystems. *Science* **263**, 185–190.

Ehleringer, J. R., Field, C. B., Lin, Z-F., and Kuo, C-Y. (1986). Leaf carbon isotope and mineral composition in subtropical plants along an irradiance cline. *Oecologia* **70**, 520–526.

Eswaran, H., Van den berg, E., and Reich, P. (1993). Organic carbon in soils of the world. *Soil Sci. Soc. Am. J.* **57**, 192–194.

Falloon, P. D., Smith, P., Smith, J. U., Szabo, J., Coleman, K., and Marshall, S. (1998). Regional estimates of carbon sequestration potential—linking the Rothamsted carbon model to GIS databases. *Biol. Fertil. Soil* **27**, 236–241.

Freidli, H., Lötscher, H., Oeschger, H., Siegenthaler, U., and Stauffer, B. (1986). Ice core record of the $^{13}C/^{12}C$ ratio of atmospheric CO_2 in the past two centuries. *Nature* **324**, 237–238.

Fung, I. Field, C. B., Berry, A., Thompson, M. V., Randerson, J. T., Malmström, C. M., Vitousek, P. M., Collatz, G. J., Sellers, P. J., Randall, D. A., Denning, A. S., Badeck, F., and John, J. (1997). Carbon 13 exchanges between the atmosphere and biosphere. *Global Biogeochem. Cycles* **11**, 507–533.

Fung, I., Field, C. B., Berry, J. A., Thompson, M. V., Randerson, J. T., Malmström, C. M., Vitousek, P. M., Collatz, G. J., Sellers, P. J., Randall, D. A., Denning, A. S., Badeck, F., and John, J. (1997). Carbon 13 exchanges between the atmosphere and biosphere. *Global Biogeochem. Cycles* **11**, 507–533.

Goh, K. M., Rafter, T. A., Stout, J. D., and Walker, T. W. (1976). The accumulation of soil organic matter and its carbon isotope composition in a chronosequence of soils developed on aeolian sand in New Zealand. *Soil Science* **27**, 89–100.

Harrison, K. G., Post, W. M., and Richter, D. D. (1995). Soil carbon turnover in a recovering temperate forest. *Global Biogeochem. Cycles* **9**, 449–454.

Holdridge, L. R. (1947). Determination of world plant formations from simple climatic data. *Science* **105**, 367–368.

Holt, J. A. and Coventry, R. J. (1990). Nutrient cycling in Australian savannas. *J. Biogeogr.* **17**, 427–432.

IGBP Terrestrial Carbon Working Group. (1998). The terrestrial carbon cycle: Implications for the Kyoto Protocol. *Science* **280**, 1393–1394.

Jenkinson, D. S. (1988). Determination of microbial biomass in soil: Measurement and turnover. In "Soil Biochemistry." (E. A. Paul and J. N. Ladd, Eds.), Vol. 5, pp. 415–471 Dekker, New York.

Jenkinson, D. S. and Rayner, J. H. (1977). The turnover of soil organic matter in some Rothamsted classical experiments. *Soil Sci.* 123, 298–305.

Jenny, H. (1941). "Factors of Soil Formation." McGraw–Hill, New York.

Jenny, H. (1961). Derivation of state factor equations of soils and ecosystems. *Soil Sci. SOC. Am. Proc.* **25**, 385–388.

Kellman, M. (1979). Soil enrichment by neotropical savanna trees. *J. Ecol.* **67**, 565–577.

Kracht, O. and Bird, M. I. (in review). Effect of texture on the isotopic composition of soil organic carbon. *Aust. J. Soil Res.*

Krull, E. S. and Retallack, G. J. (in press). Stable carbon isotope depth profiles from palaeosols across the Permian–Triassic boundary. *Geol. Soc. Am. Bull.*

Ladd, J. N. and Amato, M. (1988). Relationahips between biomass ^{14}C and soluble organic ^{14}C in a range of fumigated soils. *Soil Biol. Biochem.* **20**, 115–116.

Liski, J. (1996). Variations in soil organic carbon and thickness of soil horizons within a boreal forest stand—effect of trees and implications for sampling. *Silva Fennica* **29**, 255–266.

Little, I. P., Haydock, K. P., and Reeve, R. (1962). The correlation of Walkley–Black organic carbon with the value obtained by total combustion in Queensland soils. *CSIRO Aust. Div. Rep.* No.3/62.

Lloyd, J. and Farquhar, G. D. (1994). $\delta^{13}C$ discrimination during CO_2 assimilation by the terrestrial biosphere. *Oecologia* **99**, 201–215.

Lloyd, J. and Taylor, J. A. (1994). On the temperature dependence of soil respiration. *Funct. Ecol.* **8**, 315–323.

Macko, S. A. and Estep, M. L. F. (1984). Microbial alteration of stable nitrogen and carbon isotopic composition of organic matter. *Org. Geochem.* **6**, 787–790.

Malo, D. D., Wercester, B. K., Cassel, D. K., and Mazdorf, K.D. (1974). Soil–landscape relationships in a closed drainage system. *Soil Sci. Soc. Am. Proc.* **38**, 813–818.

Mayer, L. M. (1994). Relationships between mineral surfaces and organic carbon concentrations in soils and sediments. *Chem. Geol.* **114**, 347–363.

Melillo, J. M., Aber, J. D., and Muratore, J. F. (1982). Nitrogen and lignin control of hardwood leaf litter decomposition dynamics. *Ecology* **63**, 621–626.

Moraes, J. L., Cerri, C. C., Melillo, J. M., Kicklighter, D., Neill, C., Skole, D. L., and Steudler, P. A. (1995). Carbon stocks of the Brazilian Amazon Basin. *Soil Sci. Soc. Am. J.* **59**, 244–247.

Parton, W. J., Schimel, D. S., Cole, C. V., and Ojima, D. S. (1987). Analysis of factors controlling soil organic matter levels in Great Plains Grasslands. *Soil Sci. Soc. Am. J.* **51**, 1173–1179.

Parton, W. J., Scurlock, M. O., Ojima, D. S., Gilmanov, T. G., Scholes, R. J., Schimel, D. S., Kirchner, T., Menaut, J-C., Seastedt, T., Garcia Moya, E., Kamnalrut, A., and Kinyamario, J. I. (1993). Observations of biomass and soil organic carbon dynamics for the grassland biome worldwide. *Global Biogeochem. Cycles* **7**, 785–809.

Paustian, K., Elliott, E. T., Collins, H. P., Cole, C. V., and Paul, E. A. (1995). Use of a network of long-term experiments for analysis of soil carbon dynamics and global change—The North American model. *Aust. J. Exp. Agric.* **35**, 929–939.

Post, A. M., Emanuel, W. R., Zinke, P. J., and Strangenberger, A. G. (1982). Soil carbon pools and world life zones. *Nature* **298**, 156–159.

Raich, J. W. and Schlesinger, W. H. (1992). The global carbon dioxide flux in soil respiration and its relationship to vegetation and climate. *Tellus* **44B**, 81–99.

Santruckova, H., Bird, M. I., and Lloyd, J. (in press). ^{13}C fractionation associated with heterotrophic metabolism in grassland soils. *Funct. Ecol.*

Schimel, D. S., Braswell, B. H., Holland, E. A., McKeown, R., Ojima, D. S., Painter, T. H., Parton, W. J., and Townsend, A. R. (1994). Climatic, edaphic and biotic controls over storage and turnover of carbon in soils. *Global Biogeochem. Cycles* **8**, 279–293.

Schlesinger, W. H. (1977). Carbon balance in terrestrial detritus. *Annu. Rev. Ecol. Syst.* **8**, 51–81.

Skjemstad, J. O., Clarke, P., Taylor, J. A., Oades, J. M., and McClure, S. G. (1996). The chemistry and nature of protected carbon in soil. *Aust. J. Soil Res.* **34**, 251–277.

Skjemstad, J. O., Janik, L. J., Head, M. J., and McClure, S. G. (1993). High energy ultraviolet photo-oxidation: A novel technique for studying physically protected matter in clay- and silt-sized aggregates. *J. Soil Sci.* **44**, 485–499.

Smith, T. M. and Shuggart, H. H. (1993). The transient response of terrestrial carbon storage to a perturbed climate. *Nature* **361**, 523–526.

Spain, A. V., Isbell, R. F., and Probert, M. E. (1983). Soil organic matter. In "Soils : An Australian Viewpoint." pp. 551–563. Division of Soils, CSIRO, Melbourne.

Stewart, G. R., Turnbull, M. H., Schmidt, S., and Erskine, P. D. (1995). ^{13}C abundance in plant communities along a rainfall gradient: A biological indicator of water availability. *Aust. J. Plant Physiol.* **22**, 51–55.

Tans, P., Fung, I. Y., and Takahashi, T. (1990). Observational constraints on the atmospheric CO_2 budget. *Science* **247**, 1431–1438.

Townsend, A. R., Vitousek, P. M., and Trumbore, S. E. (1995). Soil organic matter dynamics along gradients in temperature and land-use on the island of Hawaii. *Ecology* **76**, 721–733.

Trumbore, S. E. (1993). Comparison of carbon dynamics in tropical and temperate soils using radiocarbon measurements. *Global Biogeochem. Cycles* **7**, 275–290.

Trumbore, S. E., Chadwick, O. A., and Amundsen, R. (1996). Rapid exchange between soil carbon and atmospheric carbon dioxide driven by temperature change. *Nature* **272**, 393–396.

Van der Merwe, N. J. and Medina, E. (1989). Photosynthesis and $^{13}C/^{12}C$ ratios in Amazonian rain forests. *Geochim. Cosmochim. Acta.* **53**, 1091–1094.

Walkley, A. and Black, I. A. (1934). An examination of the Degtjareff method for determining soil organic matter, and a proposed modification of the chromic acid titration method. *Soil Sci.* **37**, 29–38.

Zinke, P. J., Strangenberger, A. G., Post, W. M., Emanuel, W. R., and Olson, J. S. (1986). Worldwide organic soil carbon and nitrogen data. NDP-018, Carbon Dioxide Information Centre, Oak Ridge National Laboratory, Oak Ridge, Tennessee 136p.

1.15

Plant Compounds and Their Turnover and Stabilization as Soil Organic Matter

Gerd Gleixner,
Claudia J. Czimczik,
Christiane Kramer,
Barbara Lühker, and
Michael W.I. Schmidt
Max Planck Institut for
  Biogeochemistry,
Jena, Germany

1. Introduction

The increase in atmospheric CO_2 because of fossil fuel emissions has been identified as a major driving force for global climate change. Soil organic matter (SOM) is expected to be an important sink for this carbon (Ciais *et al.*, 1995; Schimel, 1995; Steffen *et al.*, 1998). However, at higher mean temperatures, this sink may act as additional source for CO_2 if it is accessible to microbial decomposition. To understand these complex interactions between stabilization and decomposition of SOM, it is crucial to investigate not only the turnover and stability, but also the chemical nature of soil organic matter.

Plant biomass, formed by photosynthesis from atmospheric CO_2, is the first organic substrate in the terrestrial carbon cycle (Fig. 1). The net biomass formation rate is estimated as up to $1.7 \, GT(10^{15} \, g)$ carbon per year, while the global pools of living biomass and atmospheric carbon amount to 620 and 720 Gt C, respectively. However, plants can store this carbon only temporarily. During decay, biomass is rapidly mineralized by microorganisms and less than 1% of photosynthetically assimilated CO_2 enters the more stable SOM pool. Despite this low rate, this pool has accumulated 1580 Gt carbon over centuries and millennia. This is more than the sum of the atmospheric and biological carbon pools. So far, the mechanisms and factors controlling the accumulation and remobilization of carbon in soils are only marginally understood.

The following chapter will provide basic biogeochemical knowledge of the formation and decomposition of primary plant biomass initiating SOM formation. Better awareness of these phytochemical and microbial processes is the basis for understanding soil organic matter chemistry and consequently stability. Addition-

ally, nonbiotic factors and processes, e.g., oxygen partial pressure, water, radiation, and fire, are involved in the formation of SOM. Of particular interest is the formation of black carbon, e.g., charred material remaining from biomass burning and soot, as these compounds are thought to be the most stable fractions of carbon in soils. This present chapter will review current knowledge on the stabilization of organic compounds. The focus will be on the chemical stability of molecules, the interactions of organic molecules with clay or metal (Fe or Al) oxides and hydroxides, and the possibility of biological carbon stabilization. Finally, current knowledge of turnover of SOM is presented.

2. Pathways of Soil Organic Matter Formation

2.1 Formation and Decomposition of Biomass

Carbon turnover in terrestrial ecosystems is mostly linked to biochemical reactions of three types of organisms. Primary biomass is produced by autotrophic organisms, mainly plants. Their biomass is transformed into new but chemically similar secondary biomass of consumers. These are connected by trophic relations in food chains and carbon recycling systems. Nonliving biomass is again mineralized by decomposers to carbon dioxide, water, and minerals. The basic biochemical pathways such as glycolysis, the pentose-phosphate cycle (Calvin cycle), and the Krebs cycle are for all organisms nearly identical. Only a few main biochemical pathways produce metabolites for biomass production, in particular cell walls.

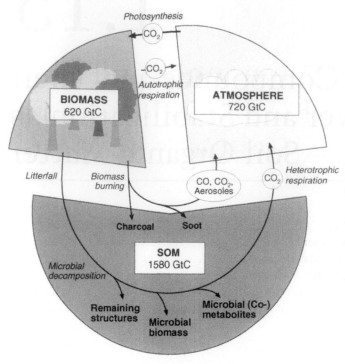

FIGURE 1 Major processes, pools, and fluxes involved in the formation of soil organic matter.

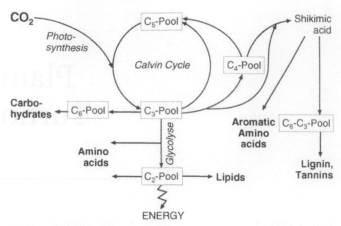

FIGURE 2 Scheme of biochemical pathways and pools leading to carbohydrates, lignin, lipids, and other metabolites.

Most important for all organisms is the carbohydrate metabolism, which provides metabolic energy for reproduction and growth. The central part of carbohydrate metabolism is the intermediate C_3 pool (Fig. 2), where primary assimilates enter and the glycolytic breakdown to energy and CO_2 starts. This pool also provides precursors for the polymerization of structural (cellulose) and storage (starch) compounds via the C_6 pool and for the regeneration of the photosynthetic CO_2 acceptors, the C_5 pool of the Calvin cycle. Other intermediates from the Calvin cycle, e.g., from the C_4 pool, and intermediates from the C_3 pool generate the C_6–C_3 pool (phenylpropanes). This pool is the starting point for the production of aromatic and phenolic compounds, e.g., lignin.

The C_2 pool, which is also part of glycolytic breakdown, is the starting point for lipid synthesis. In contrast, amino acids have several precursors and they are connected to a range of pools and metabolic pathways. To understand the structural and chemical similarity and possible differences between organisms, the following biochemical groups are described more in detail: carbohydrates, phenylpropanes and their associated derivatives, amino acids, lipids, and the major cell wall constituents.

2.1.1 Carbohydrates

Carbohydrates are the initial carbon and energy source for metabolism and therefore the most important metabolites for biological life. They cover a broad range of molecules consisting of mainly five (pentose) or six (hexose) carbon atoms, which form oxygen-containing ring structures (Fig. 3). Their degree of polymerization is linked to different cellular and biological functions. Monosaccharides, such as glucose, are soluble sugars of the cell that are directly involved in metabolic reactions. Disaccharides, e.g., sucrose, are often involved in the transport of carbohydrates

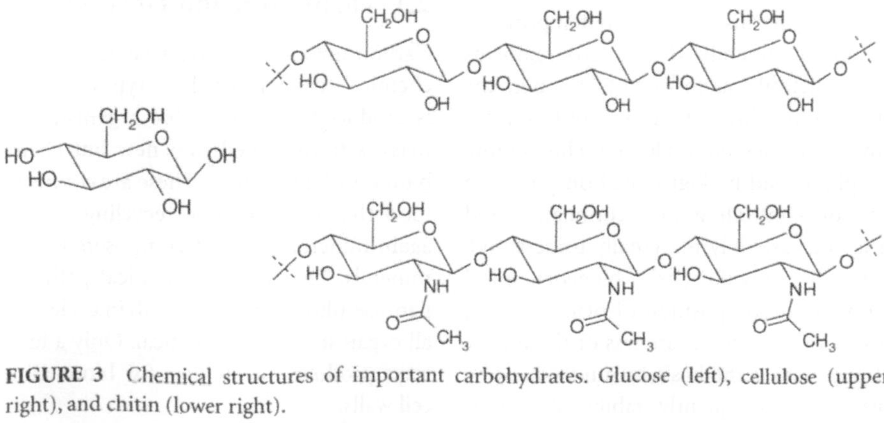

FIGURE 3 Chemical structures of important carbohydrates. Glucose (left), cellulose (upper right), and chitin (lower right).

in plants. Most abundant in nature are polysaccharides (Fig. 3), such as cellulose starch, hemicelluloses, and chitin. Cellulose and starch are polymers of glucose, hemicelluloses are a mixture of polymers from other hexoses and pentose units, and chitin is formed from a nitrogen-containing derivative of glucose (Fig. 3). Most of these polymers either form the cellular structure or are used as storage compounds. Polysaccharides are the major structural part of plant and microbial cell walls; in microorganisms they are associated with lipids and proteins. Some carbohydrates are preferentially found in microorganisms, e.g., the hexose fucose, while pentoses such as arabinose or xylose are typical constituents of plants.

Generally, carbohydrates are rapidly decomposed, as they are part of energy metabolism. Therefore, in plants cellulose is protected by other compounds against breakdown. Cellulose fibers are surrounded by hemicelluloses (Barton *et al.*, 1999), which are additionally crusted with lignin, which is highly resistant to metabolic breakdown (Paul and Clark, 1996). Additionally the nonenzymatic browning reaction (Maillard reaction) stabilizes carbohydrates forming hydroxymethylfurfurals from sugars and amino acids.

2.1.2 Phenylpropanes

Derivatives of the C_6-C_3 pool are the most important secondary products of organisms. They are involved in the stabilization of tissues, especially lignins, in the chemical communication of plants and in important electron transport processes. Most abundant are lignins in woody plants and derivatives of gallic acids (tannins).

2.1.2.1 Lignin

Besides cellulose, lignin is the most abundant constituent of wood (Killops and Killops, 1993). The production of lignin is specific to terrestrial life, stabilizing plant tissues during growth. It consists of three different alcohols from the C_6-C_3 pool, namely coumaryl alcohol, coniferyl alcohol, and sinapyl alcohol (Fig. 4). Their relative abundance in lignin indicates for different

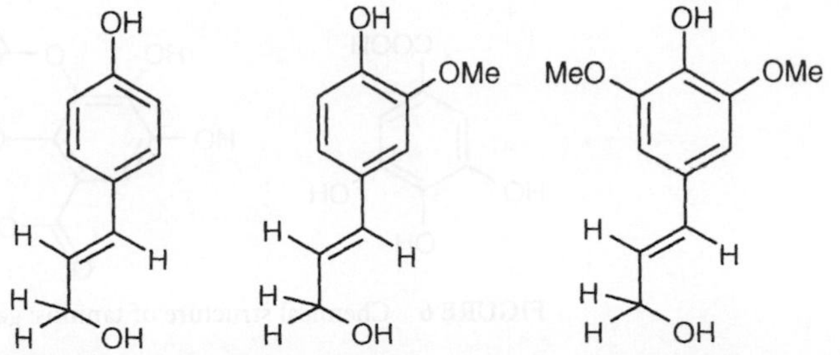

FIGURE 5 Partial structure of lignin, which is a polyphenol built up from units of phenylpropane derivatives (Fig. 4) by condensation and dehydrogenation within the plant. Condensation points, lignin precursors via ether, and C–C bonds are indicated. Modified after Killops and Killops (1993).

plant types; e.g., coniferyl alcohol units dominate conifers and sinapyl alcohol is only found in the lignin of deciduous trees. The three-dimensional network of lignin is formed by polymerization of free radicals of the monomers. Mainly ether links and the C_3-groups form covalent cross linkages, but additionally more stable C–C-bonds are formed in this nonspecific reaction (Fig. 5).

Lignin is only decomposed by highly specialized organisms, e.g., white rot fungi, called after the residue, which is white cellulose. Lignin decomposers have specific enzymes, namely lignin peroxidase, manganese peroxidase, and laccase, which catalyze the strongest biological oxidations (free oxygen radicals). They degrade the phenol structure to CO_2, but the carbon is not used for metabolic reactions (cometabolic breakdown)(Fritsche, 1998). Once lignin is broken into monomers, microorganisms gain access to the protected carbohydrates. Since the use of oxidases for breakdown requires molecular oxygen, lignin is consequently decomposed mainly in terrestrial environments. In marine systems lignin remains undecomposed. It is therefore a biomarker for terrestrial input.

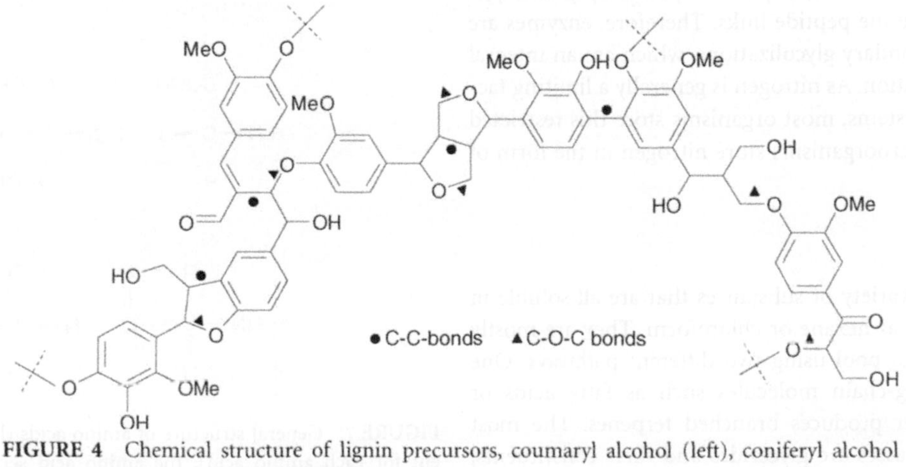

● C–C-bonds ▲ C–O–C bonds

FIGURE 4 Chemical structure of lignin precursors, coumaryl alcohol (left), coniferyl alcohol (middle), and sinapyl alcohol (right).

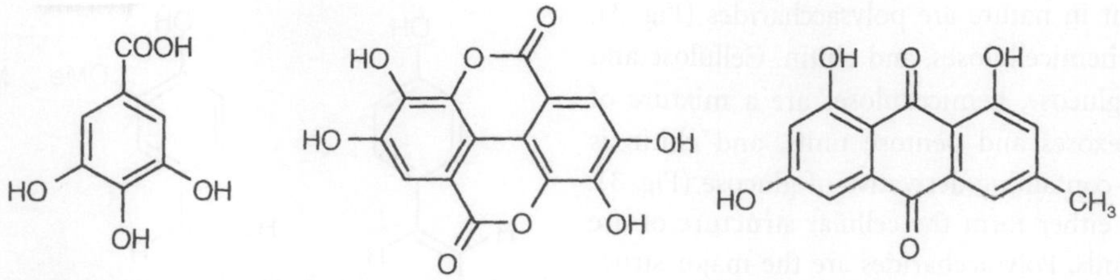

FIGURE 6 Chemical structure of tannins: gallic acid (left), ellagic acid (middle), emoldin (right).

2.1.2.2 Tannins and related compounds

Tannins are widespread in nature but they are less abundant than lignin. They are part of the chemical defense and attractant system of plants, which make them less palatable to herbivores. Their chemical composition is used for chemotaxonomic classifications. Tannins are polyhydroxyaromatic acids, especially gallic acid or ellagic acid, which are, like lignin, produced via the C_6–C_3 pool (Fig. 6). In general, these compounds are resistant to microbial attack.

Similar structures and functions are found in anthraquinones, e.g., emoldin. They are found in higher plant tissues, particularly bark, heartwood, and roots, but also in a range of organisms including fungi, lichens, vascular plants, and insects.

2.1.3 Amino Acids

Amino acids are important elements of organisms, because they are substrates for protein synthesis and enzymes. Microorganisms also liberate amino acids as exoenzymes to degrade complex organic matter outside their cells to smaller digestible monomers. Amino acids with the common α-amino-acid structure originate from various metabolic pathways (Fig. 7). Most nitrogen in organisms and in soil organic matter is found as amino groups. In contrast to plant and animal cell walls, amino acids are the major constituents of microbial cell walls. Here they are linked to a carbohydrate structure and form glycoproteins, proteoglycans, and peptidoglycans.

Proteins and enzymes are readily decomposed by proteolytic enzymes that hydrolyze the peptide links. Therefore, enzymes are often protected by secondary glycolizations, which are an integral part of cell communication. As nitrogen is generally a limiting factor for terrestrial ecosystems, most organisms store this restricted element; e.g., some microorganisms store nitrogen in the form of γ-amino butyric acid.

2.1.4 Lipids

Lipids include a great variety of substances that are all soluble in nonpolar solvents such as hexane or chloroform. They are mostly synthesized from the C_2 pool using two different pathways. One pathway produces long-chain molecules such as fatty acids or alcohols, and the other produces branched terpenes. The most important substance classes are glycerides and their constituents, e.g., fatty acids, waxes and related compounds, e.g., cutin and

suberin, various kinds of terpenoids, e.g., steroids or hopanoids, and tetrapyrrole pigments, e.g., chlorophyll. Lipids are often highly specific biomarkers that are used in taxonomic classifications.

2.1.4.1 Glycerides

Glycerides consist of glycerin, an alcohol from the C_3 pool, which is esterified with three fatty acids (Fig. 8) to form fats as an energy store. In phospholipids one fatty acid is replaced by phosphoric acid. Phospholipids form membranes that isolate the inner part of cells from the surrounding environment because of their arrangement as a bilayer. The hydrophobic alkyl chains of the fatty acids are directed toward the inner side of the bilayer and the hydrophilic phosphate ends form the surface of the membrane. Membranes are most important for cellular function and therefore are part of all organisms. The composition of fatty acids in membranes is specific to source organisms and hence is used to describe microbial community structures (Olsson, 1999).

2.1.4.2 Terpenoids

The branched isoprene unit, which is also synthesized from the C_2 pool, is the basic structure of terpenoids. Less condensed structures are used as volatile pheromones, e.g., jasmonic acid, menthol, or camphor, or as natural rubber material. More condensed structures such as steroids and hopanoids are part of membranes, influencing their fluidity. They are also highly specific to their source organisms. Best known are cholesterol (in animals and plants), ergosterol (in fungi), and brassicasterol (in diatoms). Besides cellulose, hopanoids are the most abundant biomolecules.

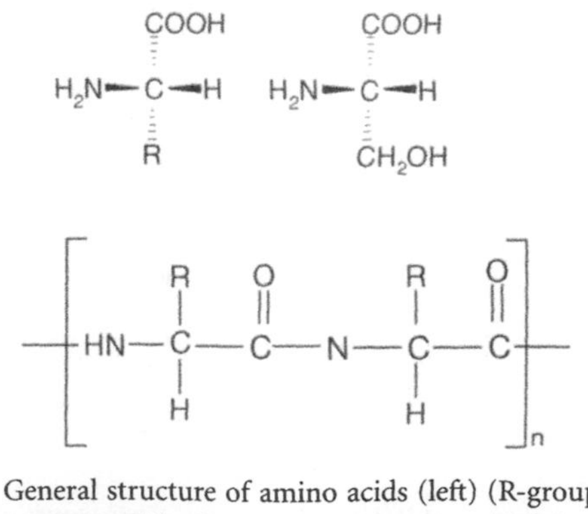

FIGURE 7 General structure of amino acids (left) (R-groups are different for each amino acid), the amino acid serine (right), amino acids linked to a peptide chain (beyond).

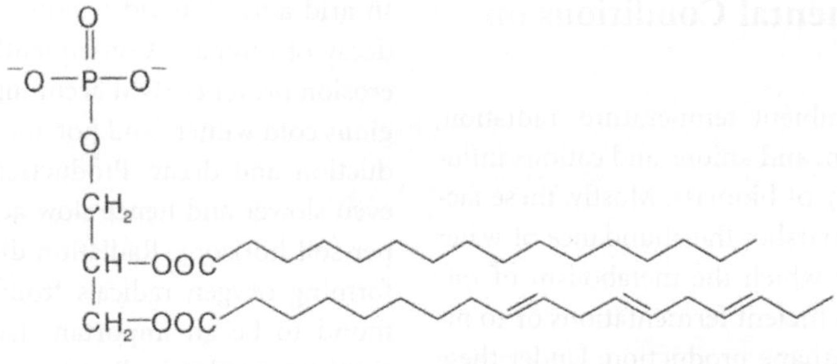

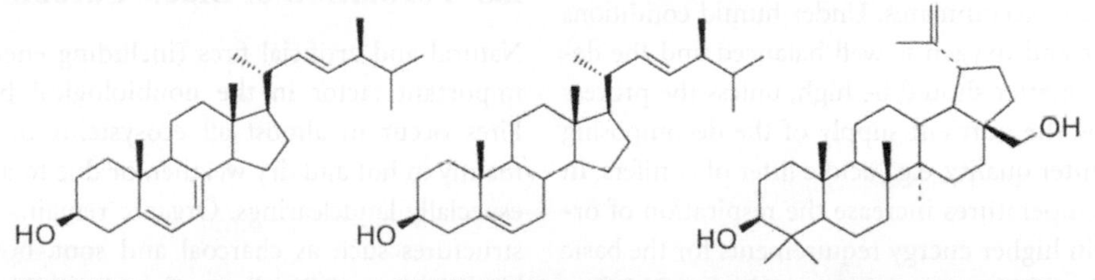

FIGURE 8 Chemical structure of lipids: Triglyceride (upper, glycerin esterified with one phosphate group, and a saturated and an unsaturated fatty acid unit) and terpenoids (lower, ergosterol from fungi (left), brassicasterol from diatoms (middle), and a hopanoid from plants (right)).

They were discovered in the late 60s in geological samples (Albrecht and Ourisson, 1969), but are present in low concentrations in almost all organisms.

2.1.5 Cellular Components of Terrestrial Plants

The development of terrestrial life required specialized cellular components to resist atmospheric influences like drought, high oxygen concentrations, or wind. In the last case, the three-dimensional lignin network was built to stabilize cell walls of terrestrial plants. To resist drought, two different strategies evolved. One strategy, used by plants, is to protect the exposed part of outer cells with less permeable, hydrophobic compounds, such as waxes. The other strategy, used by microorganisms, is to use gel-like substances as cell walls and extracellular polymeric substances to retain water.

2.1.5.1 Waxes

Waxes, in particular cutin and suberin, are polymerized and cross linked structures of hydroxy fatty acids that are resistant to oxidation and to microbial and enzymatic attack. Cutin is found on the outer surface of plant tissue while suberin is mainly associated with roots and bark of plants. Both contain an even number of carbons in the range from C_{16} to C_{26}. Cutan and suberan are also highly aliphatic polymers lacking ester cross linkages. They are linked by carbohydrate structures to form glycolipids that are integral parts of microbial cell walls (De Leeuw and Largeau, 1993).

2.1.5.2 Microbial cell walls and extracellular polymeric substances

Biosynthetic efforts of organisms, reproduction and growth, are connected to the synthesis of new cell walls. Outside the cell wall often extracellular polymeric substances similar to these cell wall

components produce a "diffusion space" that anchors exoenzymes. While fungi use chitin, glucan, or even cellulose to form their cell walls, bacteria use more complex materials, such as glycolipids, peptidoglycans, proteoglycans, and glycoproteins. Glycolipids consist of carbohydrates and lipids, whereas peptidoglycans, proteoglycans, and glycoprotein consist of amino acid polymers and carbohydrates or chitin. The latter three differ only in their relative composition and cross linkage. They are high-molecular-weight compounds with a rigid, gel-like structure stabilizing the extracellular and intracellular reaction space.

To summarize this biochemical and structural diversity, it becomes obvious that primary biomass of plants is dominated by carbohydrates, e.g., hexoses and pentoses, and lignin. Lipids and amino acids are also present but they are generally less abundant. In contrast, secondary biomass of microorganisms is dominated by carbohydrates, e.g., hexoses and chitin, in combination with lipids and proteins. Living biomass is protected from decay by cellular defense mechanisms using, for example, tannin-like structures, whereas nonliving biomass is metabolized rapidly and similar biochemical compounds are formed at different trophic levels. Only less palatable molecules, e.g., hopanoids, tannins, or antibiotics, are resistant to decay. They are less abundant and no organism is adapted to feed on them because the energetic cost of metabolic breakdown is too high. Often these compounds are mineralized in a cometabolic way using different exoenzymes of various organisms. No energy is provided from this process. Basically, in principle, all organic substances can be broken down by microorganisms. Thus additional processes are needed to stabilize carbon in the SOM pool. These are (a) environmental conditions, (b) fires, and (c) the direct interaction of organic matter with mineral particles in soil.

2.2 Influence of Environmental Conditions on SOM Formation

Environmental factors such as ambient temperature, radiation, and the availability of water, oxygen, and anions and cations influence directly or indirectly the decay of biomass. Mostly, these factors are coupled. In peatland and marshes the abundance of water forms anaerobic conditions under which the metabolism of microorganisms shifts to less energy-efficient fermentations or to nitrate and sulfate reduction and methane production. Under these conditions, the breakdown of aromatic and phenolic substances such as lignin, which requires molecular oxygen, is not possible and nondegraded biomass accumulates. Under humid conditions the availability of water and oxygen is well balanced and the decomposition of organic matter should be high, unless the process of biomass is limited by the nutrient supply of the decomposing microorganisms or by litter quality, e.g., acidic litter of conifers. In tropical regions high temperatures increase the respiration of organisms, which results in higher energy requirements for the basic metabolism. Thus, as the SOM turnover rates are high, SOM does not generally accumulate in tropical ecosystems. Moreover, high amounts of rainfall in combination with deeply weathered soil profiles enable the transport of dissolved organic matter (DOM) to deeper soil horizons from which DOM is transported to rivers.

In arid and semiarid regions, water restricts the production and decay of biomass. Consequently, the turnover of SOM is low but erosion prevents SOM accumulation in these regions. In boreal regions cold winters and hot and dry summers restrict biomass production and decay. Production of biomass is slow but decay is even slower and hence slow accumulation of SOM occurs in upper soil horizons. Radiation directly affects the decay of biomass, forming oxygen radicals from water. This mechanism has been found to be an important factor in the oxidation of DOM in northern peatlands (Bertilsson *et al.*, 1999).

2.3 Formation of Black Carbon

Natural and artificial fires (including energy production) are an important factor in the nonbiological breakdown of biomass. Fires occur in almost all ecosystems due to natural lightning, mainly in hot and dry weather, or due to anthropogenic activities, especially landclearings. Organic remains of fires are recalcitrant structures such as charcoal and soot, both often referred to as black carbon (BC). Charcoal is the solid residue of the biomass burned, whereas soot is generated in the gas phase of a fire. The global BC production for the 1980s is estimated at 0.04–0.6 Gt per year from vegetation fires and 0.007–0.024 Gt per year from fossil fuel combustion (Kuhlbusch and Crutzen, 1995). Hence, BC

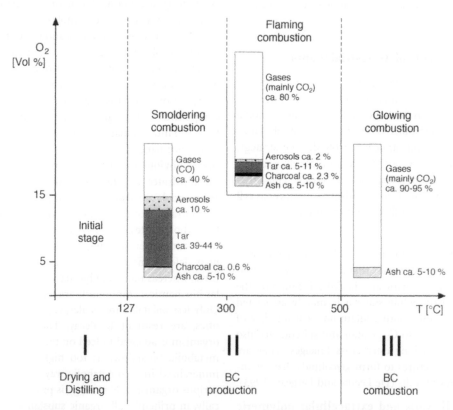

FIGURE 9 Biomass-C partitioning in different combustion stages of natural fires with an average combustion efficiency of 80wt-%. Values for gas and tar are ased on estimations. (Fearnside *et al.*, 1993; Kuhlbusch *et al.*, 1996; Laursen *et al.*, 1992; Lobert and Warnatz, 1993; Simoneit, 1999.)

TABLE 1 Emissions from Biomass Combustion*

Initial Drying and Distilling	Smoldering	Flaming	Glowing
Water (H_2O)	Carbon Monoxide (CO)	Carbon dioxide (CO_2)	Carbon dioxide (CO_2)
	Methane (CH_4)	Ethyne (C_2H_2)	
Alcohols	Non-methane hydrocarbons		Carbon monoxide (CO)
	(NMHC, mainly monounsaturated C_{1-7})	Nitric oxide (NO)	
Aldehydes	Polycyclic aromatic hydrocarbons (PAHs)	Nitrous oxide (N_2O)	
Terpenes	Ammonia (NH_3)	Nitrogen (N_2)	
	Hydrogen cyanide (HCN)	Cyanogen (NCCN)	
	Acetonitrile (CH_3CN)	Sulfur dioxide (SO_2)	
	Cyanogen (NCCN)	Aerosols	
	Amines, heterocycles, amino acids	(40 wt % Soot)	
	Methyl chloride (CH_3Cl)		
	Sulfur compounds		
	(H_2S, COS, DMS, DMDS)		
	Aerosols (4 wt % Soot)		

*The highest number of compounds is produced within the smoldering stage, whereas the highest amount of (CHNS-) emissions derives from the flaming stage of a fire (modified after Lobert and Warnatz, 1993 #5967).

is found in the soil of almost all ecosystems (Goldberg, 1985; Schmidt and Noack, 2000; Skjemstad *et al.*, 1996).

The combustion process itself is rather complex, yielding mainly volatile products (50–100%), e.g., gases and aerosols; to a lower extent solids remain (0–50%), e.g., ash, charcoal, and tar. The composition and yield of these products is determined by fuel properties and the combustion process itself, e.g., temperature, oxygen concentration, and stage of combustion. The whole process can be divided into three main stages (Fig. 9). During the initial stage (up to 127°C) the fuel is dried by distillation. Water and other high-volatile compounds, mainly lipids, are lost (Table 1). In the second stage, BC is produced under smoldering or flaming conditions depending on temperature and oxygen concentration. In the final stage at temperatures above 500°C and high oxygen concentrations, BC is burned again to CO_2 and CO.

Smoldering conditions (below 300°C, 5–15 vol. % O_2) produce mainly aerosols and gases. The latter are a mixture of molecules of low molecular weight at various oxidation stages (Table 1), whereas aerosols are more complex molecules that are often highly specific to the type of fuel (Simoneit, 1999)(Table 2). They can be unaltered fuel constituents that are released by steam-stripping or thermodesorption (mainly from lipids) or thermally less altered pyrolysis products (mainly from carbohydrates and lignin). The corresponding processes are dehydration and oxidation leading to depolymerization and fragmentation reactions. Additionally molecules, e.g., less condensed polycyclic aromatic hydrocarbons (PAH), dioxins, or soot are *de novo* synthesized (Lobert and Warnatz, 1993; Simoneit, 1999).

Flaming conditions (>300°C, >15 vol. % O_2) form fully oxidized gases (Table 1). High-molecular-weight substances of the fuel

TABLE 2 Organic Aerosols from Biomass Burning*

Compound Group	Plant Source	Product properties
Monosaccharides (e.g., levoglucosan)	Cellulose	Thermally altered
Methoxyphenols	Lignin	Thermally altered
Amino acids, amines, heterocycles (1/3 of fuel N is emitted as particle)	Proteins	Thermally altered
n-Alkanes	Epicuticular waxes	Natural
n-Alkenes	Epicuticular waxes/lipids	Thermally altered
n-Alkanoic acids	Internal lipid substances	Natural
n-Alkanols	Epicuticular waxes	Natural
Diterpenoids	Gymnosperm resins, waxes	Natural/thermally altered
Triterpenoids	Angiosperm waxes, gums	Natural/thermally altered
Steroids	Internal lipid substances	Natural/thermally altered
Wax esters	Lipid membranes, waxes	Natural
Triterpenoid esters	Internal lipid substances	Natural
Polycyclic aromatic Hydrocarbons (e.g., retene)	Multiple sources (Gymnosperms)	Thermally altered
Soot	Multiple sources	Thermally altered

*Lobert and Warnatz, 1993; Simoneit, 1999.

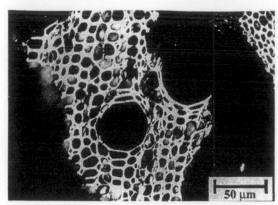

FIGURE 10 Cross section of partially burned particles from the Cretaceous–Tertiary boundary layer, as seen by reflected light microscopy (Kruge *et al.*, 1994). Particles show typical plant xylem cell structures.

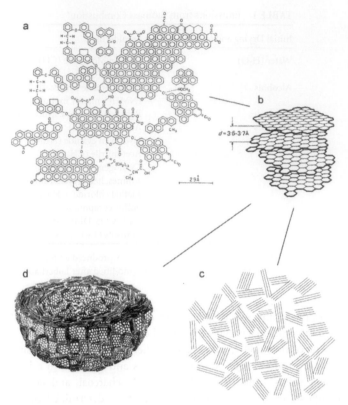

FIGURE 11 Soot structure as (a) produced in the laboratory (Sergides *et al.*, 1987), forming (b) basic structural units of 3–4 layers (Heidenreich *et al.*, 1968), (c) randomly oriented basic structural units shown as a 2-dimensional schematic diagram, (d) onion-type particle with several condensation seeds (Heidenreich *et al.*, 1968).

are progressively broken down to intermediate-molecular-weight tar products by free-radical-controlled pyrolysis. The tar fraction provides the energy for the fire. It is a mixture of low-volatile pyrolysis products mainly from lignin and lipids. Depending on temperature and oxygen supply, they are further cracked, become volatilized, and are fully oxidized. Also, thermodesorption of thermally nonaltered tar products continues. In the flame, molecular rearrangements with free radicals forming soot are maximized.

The solid residue from the second stage of vegetation fires is black carbon, e.g., charcoal and soot. Charcoal is the remains of the solid fuel phase and often still holds the morphological properties of the biomass burned (Fig. 10). Its yield depends mainly on the lignin content of the fuel. From 26% to 39% of lignin char is produced. Initially higher condensed lignins from hardwood lead to higher charcoal production (Jakab *et al.*, 1997; Wiedemann *et al.*, 1988). Charcoal is mainly produced under flaming conditions (Kuhlbusch and Crutzen, 1995).

In contrast, soot is synthesized *de novo* within the flame. Basic soot structures are multilayers of highly condensed PAHs (Fig. 11 a, b). These multilayers are either randomly oriented, or well ordered, forming three-dimensional "onion" structures (Figs 11c, 11d). The initial reaction forming aromatic soot structures involves free CH and CH_2 radicals and intermediate reactive C_3H_3 molecules. Reactions leading to further growth of the soot molecule are still rather speculative (Lobert and Warnatz, 1993). The same building reactions are described for less condensed PAHs and fullerenes. However, like charcoal, formation, soot formation occurs mainly under flaming combustion, while smoldering leads to the production of smaller, less condensed PAHs.

BC was assumed to be stable on geological time scales, as charcoal particles of similar particle size were found at various depths of 65×10^6-year-old marine sediments (Herring, 1985). Moreover, BC was found to resist various oxidation procedures, e.g., wet-chemical or thermal treatment (Kuhlbusch, 1995). However, recent carbon and oxygen isotopic studies suggest that BC degrades in soils and well-oxygenated marine sediments takes less than a century (Bird *et al.*, 1999; Middleburg *et al.*, 1999). Free-radical mechanisms, e.g., photochemical (Ogren and Charlson, 1983) or microbial cometabolic breakdown (Shneour, 1966; Winkler, 1985), are proposed to be responsible for BC degradation. Corresponding breakdown products such as benzenepolycarboxylic acids have been found in various soils (Glaser *et al.*, 1998; Hayatsu *et al.*, 1979). However, both the observed long-term stability of BC and the proposed degradation reactions are still poorly understood.

3. Stabilization of Soil Organic Matter

Formation and degradation of plant biomass, through biological and thermal processes, produce molecules differing in their intrinsic or chemical stability. More stable compounds are potentially preserved in the genesis of SOM, whereas others are transformed into biomass again. Additionally, the interaction of SOM with the soil matrix, e.g., mineral particles and metal oxides and hydroxides, may stabilize carbon in this pool. The formation of stable aggregates forming closed environments, or the adsorption of molecules on inner and outer surfaces of clay minerals may reduce the effect of exoenzymes. However, adsorption of organic matter will not be in the focus of this chapter. However, the current knowledge of the importance of individual stabilization mechanisms are still limited.

2.4 Chemical Stability of Molecules

Chemical stability of molecules is often determined by physical and biological parameters. Physically, molecules are only destabilized when the activation energy needed for bond breaking is available. As a rough estimate, this energy can be derived from the heat of combustion or corresponding bond energy, indicating that double and triple bonds are most stable, followed by homopolar C-C and C-H bonds. Heteropolar C-O and C-N bonds are most unstable. Aromatic and phenolic systems are further stabilized by resonance phenomena of translocated electrons. In biological systems the required activation energy is lowered by specific enzymes that catalyze the breakdown of molecules. For this purpose, the "active center" of the enzymes is formed in the geometry of the "transition state" between the two reaction stages. Further reduction in the activation energy for biological bond breaking is reached using sequences of enzymatic steps for bond breaking. However, this procedure needs a whole set of enzymes that are usually available only for the most common natural products. Additionally, a specific molecular environment is needed for enzymes to catalyze reactions. This implies that enzymes need to get access to their substrates. Consequently, the decay rate for polymeric substances is often slower than that for single molecules.

These interactions can be illustrated using ^{14}C-labeled monomeric and polymeric compounds (Fig. 12)(Azam *et al.*, 1985). Carbohydrates, lipids, and proteins are degraded rapidly. Most organisms have the complete set of enzymes to degrade these major metabolic products completely and to produce metabolic energy and metabolites. Also, monomeric lignin constituents mineralize rapidly. Even if carbon was labeled in the more stable aromatic ring systems, most carbon was recovered as respired CO_2. Only small amounts of radioactivity were found in the microbial biomass, indicating cometabolic breakdown of ring systems by exoenzymes (Azam *et al.*, 1985). In contrast, the stability of lignin dramatically increased when polymeric substances were

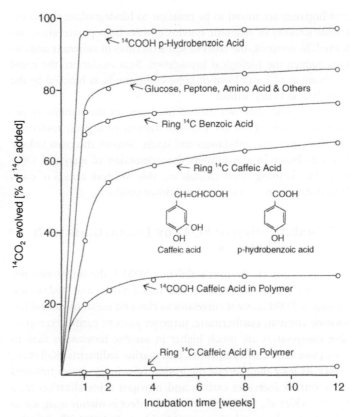

FIGURE 12 Decomposition of ^{14}C-labeled lignin analog. From Haider and Martin (1975) in Paul and Clark (1996).

used (Fig. 12). Obviously, the polymeric structure prevents degradation by exoenzymes.

Similar results were obtained for the degradation of BC. Here it is suggested that the resistance to microbial or photochemical degradation depends on the condensed and disordered molecular structure (Almendros and Dorado, 1999). Mainly, the high degree of internal cross linkages stabilizes black carbon.

The long-term stability of natural polymers can be assessed from geological samples (Table 3). Besides aromatic ring structures of lignin and tannin, the polyaromatic systems of steranes

TABLE 3 Occurrence of Presently Known Biomacromolecules and Their Potential for Survival during Sedimentation and Diagenesis*

Biomacromolecules	Occurrence	Preservation potential
Cellulose	Vascular plants, some fungi	−/+
Chitin	Arthropods, copepods, crustacea, fungi, algae	+
Lignins	Vascular plants	+ + + +
Tannins	Vascular plants, algae	+ + +/+ + +
Suberans/cutans	Vascular plants	+ + + +
Suberins/cutins	Vascular plants	+/+ +
Proteins	All organisms	−/+
Glycolipids	Plants, algae, eubacteria	+/+ +
Lipopolysaccharides	Gram-negative eubacteria	+ +

*Modified after Tegelaar (1989) in de Leeuw and Largeau, (1993)

and hopanes are found to be resistant to biodegradation. The intrinsic stability of aromatic systems enables their preservation potential. In contrast, the cross linked structures of suberans and cutans reduce the biological breakdown. Beta oxidation, the usual mechanism of breakdown of these compounds, is blocked by the cross link in beta position.

It is obvious that two main factors control the stability of organic molecules. First, the intrinsic stability of organic molecules stabilizes aromatic substances and lipids. Second, the cross linkage between biomolecules inhibits the interaction of enzymes. Only reactions forming small radicals are able to break bonds of cross linked structures and release breakdown products.

3.2 Stabilization of SOM by Interactions with the Soil Matrix

Another major mechanism stabilizing SOM is the interaction between SOM and clay particles and metal oxides and hydroxides. Evidently, SOM content correlates to clay and metal oxide and hydroxide content. Furthermore, turnover rates of easily decomposable compounds are much higher in aerobic fermenters than in soils (Van Veen and Paul, 1981), and marine sediments (Keil *et al.*, 1994). At least the thermal disruption of soil aggregates followed by rewetting increases carbon and nitrogen mineralization rates (Gregorich *et al.*, 1989). However, this effect is mainly assigned to microbial carbon and nitrogen (Magid *et al.*, 1999). Thus, stabilization of SOM may occur via formation of closed environments (aggregates) and via sorption of SOM to the mineral matrix (primary particles).

The interaction of organic matter with free mineral particles (sand, silt, clay) may form micro and macroaggregates. Microaggregates (<250 μm in diameter) are the basic structural units in soils that are neither disrupted by water nor affected by agricultural practices. Microaggregates may form macroaggregates, larger than 250 μm. Both micro and macroaggregates contain primary particles, organic matter, and pores of different sizes (Tisdall and Oades, 1982). The mechanical stability of aggregates is determined mainly by their contents of microbial biomass and water-extractable carbohydrates (Haynes, 2000).

Primary particles can be separated using ultrasonic dispersion followed by either gravity or density separation (Amelung and Zech, 1999; Schmidt *et al.*, 1999; Turchenek and Oades, 1979). Large, light particles are assumed to represent remaining plant biomass, whereas small, dense particles would represent highly degraded material and microbial remains (Christensen, 1996; Turchenek and Oades, 1979). This is supported by several independent observations on content and composition of SOM by particle size fractions, mainly of A horizons (Fig. 13). Concentrations of C and N increase with decreasing particle size. More than 50% of total soil C and N are found in the clay fractions and more than 90% in the combined clay-and-silt fraction (Christensen, 1996). Concurrently, C/N ratios decrease from values typical of plants (C/N $\sim$ 40) in the sand fraction to values typical of microorganisms (C/N $\sim$ 10) in the clay fraction (Gregorich *et al.*, 1989). The amount of

hydrolyzeable N (10–40% of total N) can mainly be attributed to amino acids and amino sugars (Stevenson, 1982) whereas in the insoluble remainder, in addition to amino functions (Knicker *et al.*, 1997), heterocyclic N compounds were detected (Leinweber and Schulten, 1998). However, the stabilization and the sources of N-containing compounds in soils are only poorly understood. Sand-size particles are dominated by polysaccharides and nondegraded lignins from plant residues, confirmed by bulk chemical carbon functionality (Baldock *et al.*, 1997; Mahieu *et al.*, 1999) and molecular markers (Guggenberger *et al.*, 1994; 1995; Hedges *et al.*, 1988; Oades *et al.*, 1987; Schulten and Leinweber, 1991; Turchenek and Oades, 1979), whereas lipids are scarcely detected in this fraction. Silt-size fractions are dominated by degraded lignins, whereas plant waxes, microbial lipids, and carbohydrates dominate clay fractions (Fig. 13).

Microbial availability of organic matter for decomposition can be limited by organomineral interactions such as adsorption onto clay particles or complexation with polyvalent cations (Oades *et al.*, 1988; Sollins *et al.*, 1996). The incorporation of cationic amides into interlayers of clay minerals (Huang and Schnitzer, 1986), or the formation of highly persistent microbial spores may be involved in this stabilization (Danielson *et al.*, 2000; Kanzawa *et al.*, 1995). In alisols, SOM forms organominerals associated with clay minerals, whereas in podzols organic matter is complexed by iron. Generally, clay contents are positively correlated with SOM concentrations when other factors such as vegetation, climate, and hydrology are similar (Davidson, 1995). Recent research, however, seems to indicate the existence of a distinct protective capacity, characteristic of individual soils (Hassink *et al.*, 1997; Hassink and Whitmore, 1997c). Some volcanic soils may have a greater stabilizing influence on organic matter than predicted from their clay contents (Parfitt *et al.*, 1997). These observations may be explained by the presence of allophane and ferrihydrite, both of which have a large specific surface capable of adsorbing organic molecules.

3.3 Biological Stabilization of Organic Matter in Soils

Summarizing the presented results on the genesis and stabilization of SOM, it is possible to develop a conceptual model of SOM turnover including the microbial lifecycle (Fig. 14). Coarse N-depleted litter added to soils will be broken down by shredders, e.g., woodlice or earthworms, into smaller particles. The main result of the process is an increase of litter surface for inoculation with microorganisms, which transform cellulose and lignin into easily decomposable and N-containing microbial biomass. The inoculation takes place in the guts of these animals. In nature these inoculated feces are often "eaten" a second time to get access to the transformed food. Termites and ants, for example, have "fungal gardens" to digest biomass. Microorganisms are not able to incorporate particles into their cells directly. Only small molecules such as amino acids or sugars can diffuse into their cells. This implies that macromolecules are digested outside their cells using exoenzymes

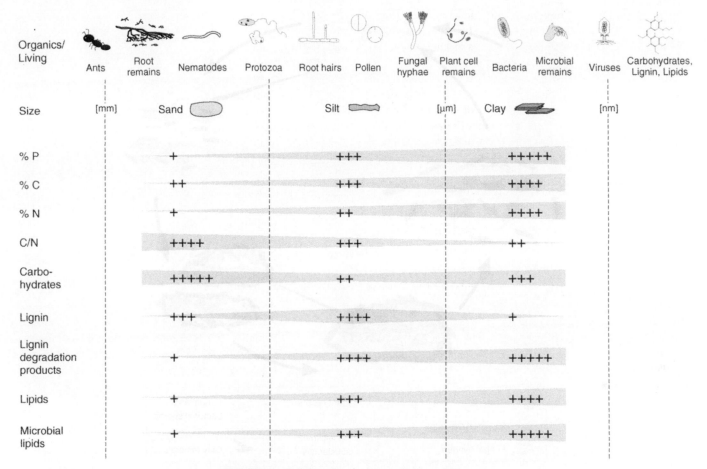

FIGURE 13 Stabilization of carbon in biological life cycles.

and that these enzymes stay within diffusion distance. Therefore, after substrate contact, microorganisms produce sticky carbohydrates (alginates, extracellular polymeric substances) to allow close contact between exoenzymes and substrate within this diffusion space (Fig. 14). Additionally this "glue" forms stable aggregates with soil minerals which exclude other microorganisms from this environment (Fig. 14). Inside the aggregates organic matter will be digested by the aggregate-forming organisms using a set of exoenzymes, e.g., cellulases, proteases, lyases. Under oxidative conditions, nonspecific oxidases are additionally able to degrade most compounds using small oxygen radicals. After substrate depletion, the carbohydrates of the diffusion space are again incorporated into the cell and highly persistent spores are formed. After spore formation, aggregates are destabilized due to changing geometry. The constituents of the aggregates are rebound as free primary particles, adsorbed spores, and recalcitrant organic matter (Fig. 14).

This conceptual model explains the existing experimental evidence. Coarse particles (sand size) from disrupted aggregates are mainly nondegraded plant remains, e.g., cellulose and lignin. As cellulose degrades more rapidly than lignin, smaller sand-size particles are relatively enriched in partially degraded lignin. In clay-size particles mainly microbial cells and spores (lipid, protein, and carbohydrate) are found. Moreover, this would suggest that enrichments of nitrogen in smaller particles are microbial remains. Only highly crosslinked structures and intrinsically stable substances have the potential to survive this process. These results suggest that SOM can be stabilized biologically: active protection of carbon from decay by cellular defense mechanism in combination with storage of carbon in soil food webs are the suggested mechanisms. In order to understand and proove the underlying processes, the turnover of carbon at the molecular level using applicable tracers has to be studied.

4. Turnover of Soil Organic Matter

Appropriate tracers to investigate SOM turnover rates are ^{14}C and ^{13}C, the two naturally occurring isotopes of [12]C. The radioactive ^{14}C atom is continuously formed in the atmosphere by solar radiation and from the remaining ^{14}C in organic compounds their age can be estimated. The mean natural abundance of ^{13}C is constant; however, small variations in the ^{13}C/^{12}C ratio identify sources and processes involved in the formation of organic molecules. The best-known examples are the isotopic difference between "heavy" C_4 plants and "light" C_3 plants and the isotopic enrichment of food chains.

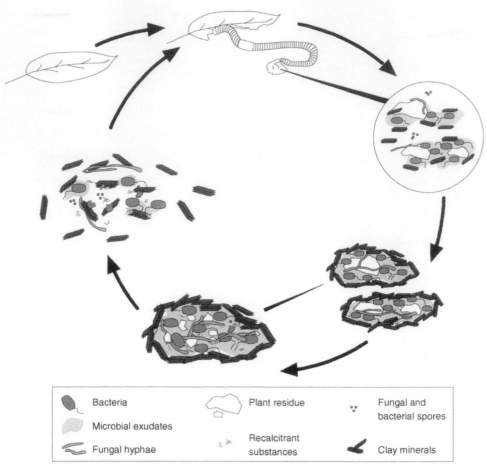

	Bacteria		Plant residue		Fungal and bacterial spores
	Microbial exudates				
	Fungal hyphae		Recalcitrant substances		Clay minerals

FIGURE 14 Elemental and molecular characteristics of different particle-size fractions of soil organic matter and size relation to organic matter.

The ¹⁴C age of SOM in depth profiles of different soil types indicates that SOM in deeper horizons can reach mean ages between 1000 and 15,000 years (Bol *et al.*, 1999; Jenkinson *et al.*, 1999). In peat even ¹⁴C ages of 40,000 years were determined (Zimov *et al.*, 1997). However, these ages are only mean ages as in SOM an unknown proportion of old and continuously added new carbon is measured simultaneously (Wang *et al.*, 1996) and consequently, even in upper horizons, recalcitrant matter can be found. So far neither in bulk chemical nor in physical fractions have substances substantially older than this mean value been identified (Balesdent and Guillet, 1987; Wang *et al.*, 1996). Recently, compound-specific ¹⁴C ages indicated for the first time that terrestrial biomarkers (lipids) are ten times older than bulk organic matter in marine environments (Eglinton *et al.*, 1997). In general, mean ages of SOM are highest in both wet-and-cold and dry-and-hot ecosystems (see above) having high or low carbon accumulation rates, respectively, and low turnover rates. In contrast, tropical rainforests with high turnover rates have the lowest mean ages.

Additionally, the ¹⁴C signal, introduced by atmospheric thermonuclear bomb tests at the end of the 60s, can be used to investigate turnover rates of SOM. This signal is often found in the upper 5 cm of wet-and-cold soils or dry-and-hot soils, indicating that the carbon is still present after 30 years. In ecosystems with high turnover rates this peak often appears in deeper horizons. Using modeling approaches, the distribution of the ¹⁴C signal over the profile suggests that tropical soil consists mainly of SOM with a mean residence time below 10 years (Trumbore, 1993), which is in good agreement with mean residence times of 6 years determined for dead trees in rainforests (Chambers *et al.*, 2000). For soils of temperate climates, SOM pools of different mean residence times (10, 100, 1000 years) are used to model the ¹⁴C distribution.

Carbon turnover rates can be estimated using natural labeling techniques with ¹³C in combination with vegetation shifts from "light" C₃ plants to "heavy" C₄ plants. (Balesdent and Guillet, 1987). Coarse particles that are mainly fresh litter have mean residence times between 0.5 and 20 years, whereas the carbon in the clay fraction has mean residence times of about 60–80 years (Balesdent, 1996). Recently, the direct determination of molecular turnover rates using pyrolysis products was applied to SOM after vegetation change (Gleixner *et al.*, 1999). This technique indicated mean residence times between 9 and 220 years for individual pyrolysis products for the first time. Most intriguing was the fact that some of the more resistant pyrolysis products were derived from proteins. This in fact supports the possibility of SOM stabilization in the form of microbial carbon, either actively protected or adsorbed on metal oxides.

5. Conclusion

The turnover and stability of SOM depends mainly on environmental and biological parameters. Either biomass production or decomposition rates are affected. Additionally, soil matrix and litter quality and fire frequencies stabilize carbon in soils. From the presented results it is obvious that ecosystems have different mechanisms for stabilizing SOM, which lead to different chemistries of the stable compounds. For a better understanding of SOM in the terrestrial carbon cycle and to identify the "missing carbon sink," some major points have to be considered:

1. The content of SOM depends mainly on four functions: (a) biomass input, (b) decomposition rate, (c) retention capacity, and (d) carbon output. All these functions are controlled by environmental and biological parameters.
2. The chemical type of stable carbon is specific to each ecosystem. Therefore, isotopic tracers are more appropriate to understand turnover and stability of SOM.
3. Retention of carbon as microbial biomass in combination with "active" protection as "biological carbon stabilization" may be an important factor controlling carbon accumulation in soils.

To identify the corresponding processes and mechanisms we will need:

1. to investigate compound specific mean residence times of stable compounds and biomarkers,
2. to develop new soil carbon models that are able to model the molecular turnover of ^{13}C and ^{14}C.

The combined information will give new insight into soil carbon turnover and will help to understand and to quantify ecosystem-specific retention mechanisms for carbon. Additionally, this information may identify the carbon sink capacities of soils.

References

Albrecht, P. and Ourisson, G. (1969). Terpene alcohol isolation from oil shale. *Science* **163**, 1192–1193.

Almendros, G. and Dorado, J. (1999). Molecular characteristics related to the biodegradability of humic acid preparations. *Eur. J. Soil Sci.* **50**, 227–236.

Amelung, W. and Zech, W. (1999). Minimisation of organic matter disruption during particle-size fractionation of grassland epipedons. *Geoderma* **92**, 73–85.

Azam, F., Haider, K. and Malik, K.A. (1985). Transformation of carbon-14-labeled plant components in soil in relation to immobilization and remineralization of nitrogen-15 fertilizer. *Plant and Soil* **86**, 15–26.

Baldock, J. A., Oades, J. M., Nelson, P. N., Skene, T. M., Golchin, A., and Clarke, P. (1997). Assessing the extent of decomposition of natural organic materials using solid-state ^{13}C NMR spectroscopy. *Aust. J. Soil Res.* **35**, 1061–1083.

Balesdent, J. (1996). The significance of organic separates to carbon dynamics and its modelling in some cultivated soils. *Eur. J. Soil Sci.* **47**, 485–493.

Balesdent, J. A. M. and Guillet, B. (1987). Natural ^{13}C abundance as a tracer for studies of soil organic matter dynamics. *Soil Biol. Biochem.* **19**, 25–30.

Barton, S. D., Nakanishi, K., and Meth-Cohn, O. (1999). "Comprehensive Natural Products Chemistry," 1st ed. Elsevier, Amsterdam.

Bertilsson, S., Stepanauskas, R., Cuadros-Hansson, R., Graneli, W., Wikner, J., and Tranvik, L. (1999). Photochemically induced changes in bioavailable carbon and nitrogen pools in a boreal watershed. *Aquatic Microbial Ecol.* **19**, 47–56.

Bird, M. I., Moyo, C., Veenendaal, E. M., Lloyd, J., and Frost, P. (1999). Stability of elemental carbon in a savanna soil. *Global Biogeochem. Cycles* **13**, 923–932.

Bol, R. A., Harkness, D. D., Huang, Y., and Howard, D. M. (1999). The influence of soil processes on carbon isotope distribution and turnover in the British uplands. *Eur. J. Soil Sci.* **50**, 41–51.

Chambers, J. Q., Higuchi, N., Schimel, J. P., Ferreira, L. V., and Melack, J. M. (2000). Decomposition and carbon cycling of dead trees in tropical forests of central Amazon. *Oecologia* **122**, 380–388.

Christensen, B. T. (1996). "Carbon in Primary and Secondary Organomineral Complexes". pp. 97–165. Springer, New York.

Ciais, P., Tans, P. P., Trolier, M., White, J. W. C., and Francey, R. J. (1995). A large Northern Hemisphere terrestrial CO₂ sink indicated by the C-13/C-12 ratio of atmospheric CO_2. *Science* **269**, 1098–1102.

Danielson, J. W., Zuroski, K. E., Twohy, C., Thompson, R. D., Bell, E., and McClure, F. (2000). Recovery and sporicidal resistance of various *B. subtilis* spore preparations on porcelain penicylinders compared with results from AOAC test methods. *J. AOAC International* **83**, 145–155.

Davidson, E. R. (1995). What are the physical, chemical and biological process that control the formation and degradation of nonliving organic matter? In "Role of Nonliving Organic Matter in the Earth's Carbon Cycle, Dahlem Workshop Report 16." (R. G. Zepp and C. Sonntag, eds.), pp. 305–324. Wiley, Chichester.

De Leeuw, J. R. and Largeau, C. (1993). A review of macromolecular organic compounds that comprise living organisms and their role in kerogen, coal, and petroleum formation. In "Organic Geochemistry" (M. H. Engel and S. A. Macko, Eds.), pp. 23–72. Plenum Press, New York.

Eglinton, T. I., Benitez-Nelson, B. C., Pearson, A., McNichol, A. P., Bauer, J. E., and Druffel, E. R. M. (1997). Variability in radiocarbon ages of individual organic compounds from marine sediments. *Science* **277**, 796–799.

Fearnside, P. M., Leal, N. J., and Fernandes, F. M. (1993). Rainforest burning and the global carbon budget: Biomass, combustion efficiency, and charcoal formation in the Brazilian Amazon. *J. Geophys. Res.* **98**, 16733–16743.

Fritsche, W. (1998). "Umwelt-Mikrobiologie." Fischer, Jena. Ulm

Glaser, B., Haumaier, L., Guggenberger, G., and Zech, W. (1998). Black carbon in soils: The use of benzenecarboxylic acids as specific markers. *Org. Geochem.* **29**, 811–819.

Gleixner, G., Bol, R., and Balesdent, J. (1999). Molecular insight into soil carbon turnover. *Rapid Commun. Mass Spectrom.* **13**, 1278–1283.

Goldberg, E. D. 1985. "Black Carbon in the Environment." Wiley, New York.

Gregorich, E. G., Kachanoski, R. G., and Voroney, R. P. (1989). Carbon mineralization in soil size fractions after various amounts of aggregate disruption. *J. Soil Sci.* **40,** 649–659.

Guggenberger, G., Christensen, B. T., and Zech, W. (1994). Land-use effects on the composition of organic matter in particle-size separates of soil. I. Lignin and carbohydrate signature. *Eur. J. Soil Sci.* **45,** 449–458.

Guggenberger, G., Zech, W., and Thomas, R. J. (1995). Lignin and carbohydrate alteration in particle-size separates of an oxisol under tropical pastures following native savanna. *Soil Biol. Biochem.* **27,** 1629–1638.

Haider, K. and Martin, J. P. (1975). Decomposition of specifically carbon-14 labelled benzoic and cinnamic acid derivatives in soil. *Soil Sci. Soc. Am. Proc.* **39,** 657–662.

Hassink, J., Matus, F. J., Chenu, C., and Dalenberg, J. W. (1997). Interactions between soil biota, soil organic matter, and soil structure. In: "Soil Ecology in Sustainable Agricultural Systems." (L. Brussard and R. Ferrera-Cerrato, Eds.), pp. 15–35. CRC Press, Boca Raton, FL.

Hassink, J. and Whitmore, A. P. (1997). A Model of the physical protection of soils. *Soil Sci. Soc. Am. J.,* **61(1),** 131–139.

Hayatsu, R., Winans, R. E., McBeth, R. L., Scott, R. G., Moore, L. P., and Studier, M. H. (1979). Lignin-like polymers in coals. *Nature* **278,** 41–43.

Haynes, R. J. (2000). Interactions between soil organic matter status, cropping history, method of quantification and sample pretreatment and their effects on measured aggregate stability. *Biol. Fertil. Soils* **30,** 270–275.

Hedges, J. I., Blanchette, R. A., Weliky, K., and Devol, A. H. (1988). Effects of fungal degradation on the CuO oxidation products of lignin: A controlled laboratory study. *Geochim. Cosmochim. Acta* **52,** 2717–2726.

Heidenreich, R. D., Hess, W. M. and Ban, L. L. (1968). A test object and criteria for high resolution electron microscopy. *J. Appl. Crystollogr.* **1,** 1–19.

Herring, J. R. (1985). Charcoal fluxes into sediments of the North Pacific Ocean: The Cenozoic record of burning. *Geophys. Monogr.* **32,** 419–442.

Huang, P. M. and Schnitzer, M. (1986). "Interaction of Soil Minerals with Natural Organics and Microbes. Soil Science Society of America, Madison, WI.

Jakab, E., Faix, O., and Till, F. (1997). Thermal decomposition of milled wood lignins studied by thermogravimetry/mass spectroscopy. *J. Anal. Appl. Pyroly.* **40–41,** 171–186.

Jenkinson, D. S., Meredith, J., Kinyamario, J. I., Warren, G. P., Wong, M. T. F., Harkness, D. D., Bol, R., and Coleman, K. (1999). Estimating net primary production from measurements made on soil organic-matter. *Ecology* **80,** 2762–2773.

Kanzawa, Y., Harada, A., Takeuchi, M., Yokota, A., and Harada, T. (1995). *Bacillus curdlanolyticus* sp. Nov and *Bacillus kobensis* sp. Nov, which hydrolyze resistant curdlan. *Int. J. Systematic Bacteriol.* **45,** 515–521.

Keil, R. G., Tsamakis, E., Fuh, C. B., Giddings, J. C., and Hedges, J. I. (1994). Mineralogical and textural controls on the organic composition of coastal marine sediments: Hydrodynamic separation using SPLITT-fractionation. *Geochim. Cosmochim. Acta* **58,** 879–893.

Killops, S. D. and Killops, V. J. (1993). "An Introduction to Organic Geochemistry." Longman Scientific & Technical, London

Knicker, H., Lüdemann, H.-D., and Haider, K. (1997). Incorporation studies of NH_4^+ during incubation of organic residues by ^{15}N-CPMAS-NMR-spectroscopy. *Eur. J. Soil Sci.* **48(3)** 431–442.

Kruge, M. A., Stankiewicz, A. B., Crelling, J. C., Montanari, A., and Bensley, D. F. (1994). Fossil charcoal in Cretaceous–Tertiary boundary strata: Evidence for catastrophic firestorm and megawave. *Geochim. Cosmochim. Acta* **58,** 1393–1397.

Kuhlbusch, T. A. J. (1995). Method for determining black carbon in residues of vegetation fires. *Environ. Sci. Technol.* **29,** 2695–2702.

Kuhlbusch, T. A. J., Andreae, M. O., Cachier, H., Goldammer, J. G., Lacaux, J.-P., Shea, R., and Crutzen, P. J. (1996). Black carbon formation by savanna fires: Measurements and implications for the global carbon cycle. *J. Geophys. Res.* **101,** 23,651–23,665.

Kuhlbusch, T. A. J. and Crutzen, P. J. (1995). Toward a global estimate of black carbon in residues of vegetation fires representing a sink of atmospheric CO_2 and a source of O_2. *Global Biogeochem. Cycles* **9,** 491–501.

Laursen, H. K., Ferek, R. J., and Hobbs, P. V. (1992). Emission factors for particles, elemental carbon, and trace gases from the Kuwait oil fires. *J. Geophys. Res.* **97,** 14491–14497.

Leinweber, P. and Schulten, H. R. (1998). Nonhydrolyzable organic nitrogen in soil size separates from long-term agricultural experiments. *Soil Sci. Soc. Am. J.* **62,** 383–393.

Lobert, J. M. and Warnatz, J. (1993). Emissions from the combustion process in vegetation. In "Fire in the Environment" (Crutzen, P. J., Goldammer, J. G., Eds.), pp. 15-37. Wiley, Chichester.

Magid, J., Kjaergaard, C., Gorissen, A., and Kuikman, P. J. (1999). Drying and rewetting of a loamy sand soil did not increase the turnover of native organic matter, but retarded the decomposition of added C-14-labelled plant material. *Soil Biol. Biochem.* **31,** 595–602.

Mahieu, N., Powlson, D. S., and Randall, E. W. (1999). Statistical analysis of of published carbon-13 CPMAS NMR spectra of soil organic matter. *Soil Sci. Soc. Am. J.* **63,** 307–319.

Middleburg, J. J., Nieuwenhuize, J., and Van Breugel, P. (1999). Black carbon in marine sediments. *Marine Chem.* **65,** 245–252.

Oades, J. M., Vassallo, A. M., Waters, A. G., and Wilson, M. A. (1987). Characterization of organic matter in particle size and density fractions from a red-brown earth by solid-state carbon-13 NMR. *Aust. J. Soil Res.* **25,** 71–82.

Ogren, J. A. and Charlson, R. J. (1983). Elemental carbon in the atmosphere: Cycle and lifetime. *Tellus* **35,** 241–254.

Olsson, P. A. (1999). Signature fatty acids provide tools for determination of the distribution and interaction of mycorrhizal fungi in soil. *FEMS Microbiol. Ecol.* **29,** 303–310.

Parfitt, R. L., Theng, B. K. G., Whitton, J. S., and Sheperd, T. G. (1997). Effects of clay minerals and land use on organic matter pools. *Geoderma* **75,** 1–12.

Paul, E. A., and Clark, F. E. (1996). "Soil Microbiology and Biochemistry," 2nd ed. Academic Press, San Diego.

Schimel, D. S. (1995). Terrestrial ecosystems and the carbon cycle. *Global Change Biol.* **1,** 77–91.

Schmidt, M. W. I. and Noack, A. G. (2000). Black carbon in soils and sediments: Analysis, distribution, implications, and current challenges. *Global Biogeochem. Cycles.* **14(3),** 777–793.

Schmidt, M. W. I., Rumpel, C., and Kögel-Knabner, I. (1999). Evaluation of an ultrasonic dispersion method to isolate primary organomineral complexes from soils. *Eur. J. Soil Sci.* **50,** 1–8.

Schulten, H.-R. and Leinweber, P. (1991). Influence of long-term fertilization with farmyard manure on soil organic matter: Characteristics of particle size fractions. *Biol. Fertil. Soils* **12,** 81–88.

Sergides, C. A., Jassim, J. A., Chughtai, A. R., and Smith, D.M. (1987). The structure of hexane soot. III. Ozonation studies. *Appl. Spectrosc.* **41,** 482–492.

Shneour, E. A. (1966). Oxidation of graphitic carbon in certain soils. *Science* **151,** 991–992.

Simoneit, B. R. T. 1999. A review of biomarker compounds as source indicators and tracers for air pollution. *Environ. Sci. Pollut. Res.* **6,** 159–169.

Skjemstad, J. O., Clarke, P., Taylor, J. A., Oades, J. M., and McClure, S. G. (1996). The chemistry and nature of protected carbon in soil. *Aust. J. Soil Res.* **34**, 251–271.

Steffen, W., Noble, I., Canadell, J., Apps, M., Schulze, E. D., Jarvis, P. G., Baldocchi, D., Ciais, P., Cramer, W., Ehleringer, J., Farquhar, G., Field, C. B., Ghazi, A., Gifford, R., Heimann, M., Houghton, R., Kabat, P., Korner, C., Lambin, E., Linder, S., Mooney, H. A., Murdiyarso, D., Post, W. M., Prentice, I. C., Raupach, M. R., et al. (1998). The Terrestrial Carbon Cycle—Implications For the Kyoto Protocol. *Science* **280**, 1393–1394.

Stevenson, F. J. (1982). "Humus Chemistry." Wiley, New York.

Tegelaar, E. W., Derenne, S., Largeau, C., and de Leeuw, J. W. (1989). A reappraisal of kerogen formation. *Geochim. Cosmochim. Acta* 3, 3103–3107.

Tisdall, J. M. and Oades, J. M. (1982). Organic matter and water stable aggregates in soils. *J. Soil Sci.* **33**, 141–163.

Trumbore, S. E. (1993). Comparison of carbon dynamics in tropical and temperate soil using radicarbon measurements. *Global Biogeochem. Cycles* **7**, 275–290.

Turchenek, L. W. and Oades, J. M. (1979). Fractionation of organo-mineral complexes by sedimentation and density techniques. *Geoderma* **21**, 311–343.

Van Veen, J.A. and Paul, E.A. (1981). Organic carbon dynamics in grassland soils. *Can. J. Soil Sci.* **61**, 185–201.

Wang, Y., Amundson, R., and Trumbore, S. (1996). Radiocarbon dating of soil organic matter. *Q. Res.* **45**, 282–288.

Wiedemann, H. G., Riesen, R., Boller, A., and Bayer, G. (1988). From wood to coal: A compositional thermogravimetric analysis. In "Compositional Analysis by Thermogravimetry" (C. M. Earnest, Ed.), pp. 227–244. American Society for Testing and Materials, Philadelphia.

Winkler, M. G. (1985). Charcoal analysis for paleoenvironmental interpretation: A chemical assay. *Q. Res.* **23**, 313–326.

Zimov, S. A., Voropaev, Y. V., Semiletov, I. P., Davidov, S. P., Prosiannikov, S. F., Chapin, F. S., Chapin, M. C., Trumbore, S., and Tyler, S. (1997). North Siberian lakes—A methane source fueled by Pleistocene carbon. *Science* **277**, 800–802.

1.16

Input/Output Balances and Nitrogen Limitation in Terrestrial Ecosystems

Peter Vitousek
Stanford University
Stanford, California

Christopher B. Field
Carnegie Institution of
*   Washington*
Stanford, California

Why does the availability of N often limit net primary production (NPP) and other processes in terrestrial ecosystems? For N to limit NPP in the long term, two conditions must be met: N must be lost from terrestrial ecosystems by pathways that cannot be prevented fully by N-demanding organisms, and the power of N_2 fixation to add new N to N-limited ecosystems must be constrained. We utilize a simple model to explore the consequences of (a) losses by dissolved organic nitrogen, transformation dependent trace gas fluxes, and spatial/temporal variation in the supply versus demand for N, and (b) constraints on N_2 fixation caused by disproportionately severe effects of P limitation, grazing, and shade intolerance on symbiotic N_2 fixers. The results of these analyses suggest that the pervasiveness of N limitation in terrestrial ecosystems is strongly shaped by processes that are not well understood.

1. Introduction

The biological availability of nitrogen—its pattern, dynamics, and regulation—has attracted a great deal of research for several decades. Why has there been such intense focus on just one of the 13 or so essential elements that higher plants obtain from soil? There are a number of reasons that the nitrogen cycle has been and remains particularly interesting to terrestrial ecologists:

1. The supply of biologically available N demonstrably limits ecosystem properties and processes over much of the earth. It controls yield in most intensive agricultural systems and controls plant growth, net primary productivity (NPP), species composition and chemistry, and trophic structure in many managed and natural systems (e.g., Tilman, 1987; Berendse et al. 1993).

2. The global cycle of N has been altered to an astonishing degree by human activity. Humanity more than doubles the quantity of N_2 fixed annually on land, and greatly increases fluxes of the N-containing trace gases from land to the atmosphere, and those of nitrate from land to aquatic systems (Galloway et al. 1995; Howarth et al., 1996; Vitousek et al., 1997a).

3. On the one hand, human activities that increase the availability of N in N-limited systems can cause net storage of C; this may be an important component of the "missing sink" for anthropogenic CO_2 (Schimel et al., 1995; Townsend et al., 1996). On the other hand, N limitation may be an important constraint on the ability of terrestrial ecosystems to store C in response to anthropogenic CO_2 enrichment (Melillo et al., 1996).

4. Anthropogenic fixed N causes or contributes to a wide range of environmental problems, from forest dieback and the loss of biological diversity on land to acidification and unhealthy concentrations of nitrate in streamwater and groundwater to eutrophication of estuaries and ocean margins to increasing concentrations of the reactive gas nitric oxide regionally and of the greenhouse gas nitrous oxide globally (Schulze, 1989; Nixon et al., 1996; Vitousek et al., 1997a; Aber et al., 1998).

5. N limitation is economically important; humanity spends tens of billions of dollars annually on N fertilizer and its application.

6. There is a fascinating intellectual puzzle concerning N limitation. Given the ubiquitous distribution of N_2-fixing

organisms that can draw upon the essentially unlimited supply of atmospheric N_2, how can N limitation be anything more than a marginal or transient phenomenon (Vitousek and Howarth, 1991)?

The nature of this puzzle is perhaps best appreciated by examining lake ecosystems. Twenty-five years ago, there was an intense controversy in the United States and Canada concerning nutrient limitation—specifically, concerning what controls the anthropogenic eutrophication of lakes. C, N, and P all had their proponents, until experimental studies with whole lakes demonstrated unambiguously that while the supply of C and N may affect photosynthesis and other processes in lakes, the longer-term accumulation of algal biomass is driven by P enrichment. P supply is controlling in the long term because lake surface water is an open system with respect to C and N; the concentration of CO_2 in surface water can be drawn down by biological uptake, but then more will enter in by diffusion from the atmosphere. Similarly, the supply of fixed N can be drawn down by biological uptake, but then N_2-fixing cyanobacteria will have a substantial competetive advantage over other phytoplankton, dominate the producer community, and bring the quantity of fixed N more or less into equilibrium with that of P, at the N:P ratio required by phytoplankton (Schindler, 1977). There is still excellent work being done on the interactions between C, N, and P in aquatic ecosystems (e.g., Elser *et al.*, 1996), but the fact that eutrophication generally is controlled by P supply, and the reasons for that control, are not in dispute.

Why aren't terrestrial ecosystems more lake-like? Their N cycle is open, at least potentially, so why is NPP in many terrestrial systems N-limited? Before addressing this question directly, we should note several points. First, the question is explicitly comparative across elements. Why is N more important than P, Ca, K, or B in controlling NPP, net ecosystem production (NEP), and other processes in many terrestrial ecosystems?

Second, by saying that NPP is N-limited, we do not assert that only N is limiting; multiple resource limitation is the rule in ecosytems (Bloom *et al.*, 1985; Field *et al.*, 1992). Biomass accumulation in lakes is limited by light as well as P; NPP in terrestrial systems can be limited by water, CO_2, light, and one or more soil-derived nutrients simultaneously. However, it would be surprising to find that N, P, Ca, Mg, K, B, and all other soil-derived elements were equally limiting in any ecosystem; in practice the supply of one or two elements is controlling at any given time. Soil-derived elements can neither be obtained independently nor readily traded off for each other (Rastetter and Shaver, 1992; Gleeson and Tilman, 1992).

Finally, our analysis will be focused on inputs and outputs of elements at the ecosystem level and on their controls. In the short term, nutrients can limit NPP or other processes when organisms' demands for an element exceed the supply of that element; for N and P, that generally means that potential uptake exceeds mineralization. A particular element may be limiting because it cycles more slowly than another; for example, biochemical mineraliza-

tion of P by extracellular enzymes can allow P to cycle more rapidly than N, driving ecosystems toward limitation by N (McGill and Cole, 1981). However, in the longer term (centuries to millennia), the balance between inputs to and outputs from ecosystems determines the quantity of elements that can cycle within ecosystems.

2. Long-Term Nutrient Limitation

Ecosystems are open systems, with the potential for inputs and outputs of all biologically essential elements. In the long term, for any element to limit NPP or other ecosystem processes, one essential condition must be met:

1. The element must be lost from the system by some pathway(s) in addition to the loss of "excess available nutrients" (defined below); these additional losses must be large enough to balance element inputs at a point where the supply of that element (within the system) remains limiting to NPP.

For N, a second condition also is essential:

2. Some process(es) must constrain rates of biological N_2 fixation to the extent that N_2 fixers cannot respond to N deficiency sufficiently to eliminate it.

For the first condition, many ecosystem models, conceptual and others, assume that losses of elements occur from a pool of excess available nutrients—nutrients that remain in the soil when plants and microbes have taken up all the nutrients that they can use—and that this pool (of a particular nutrient) is vanishingly small when that nutrient limits NPP or other ecosystem processes (Vitousek and Reiners, 1975). Simple models of N saturation and its consequences are based on this approach (Ågren and Bosatta, 1988; Aber, 1992). If this conceptual model were accurate, however, and if nutrient supply and demand were relatively constant in space and time, then no nutrient could remain limiting indefinitely. In any real ecosystem, losses of a limiting nutrient would be near zero, inputs from outside the system would accumulate, and eventually the pool of that limiting nutrient within the system would increase to the point where it no longer limited NPP (or did so only marginally) (Hedin *et al.*, 1995; Vitousek *et al.*, 1998). Nutrient limitation can be sustained in the long run by loss pathways that are independent of excess available nutrients, or where spatial or temporal variation allows losses that are not wholly preventable. What are these pathways? (Note that this necessary condition implicitly includes rates of nutrient input; alternative pathways of element loss may be sufficient to sustain nutrient limitation when inputs are low, but not in the face of high rates of input.)

While the first condition applies to all essential elements, the second is specific to N. Biological N_2 fixation is capable of adding tens to hundreds of kg ha^{-1} year^{-1} to ecosystems (Sprent and Sprent, 1990), more than enough to meet plant and microbial demand for N in a short time, and to overwhelm the capacity of alternative N loss pathways, and so rapidly offset N limitation. It is this potential to respond to deficiency with large, biologically-controlled inputs that makes the widespread nature of N limitation such a puzzle. Biological processes within ecosystems can affect inputs of other elements—for example, plant and microbial activity can enhance rates of rock weathering (Davis *et al.*, 1985; Cochran and Berner, 1997)—but not in a regulatory way, not with the ability to enhance inputs of a particular element when it is deficient within the system. What processes constrain N_2 fixation in N-limited ecosystems, and so sustain N limitation?

With these conditions and questions in mind, why is P more often limiting in lakes than on land, at least in the temperate zone? P limits lake productivity because (a) unlike C and N, there are no mechanisms that can increase inputs of P when it is in short supply, as discussed above; (b) P is relatively immobile within and through terrestrial ecosystems, so inputs of P to lakes are small; and (c) lakes have an uncontrollable loss of P, in the sinking of particulate organic matter out of the euphotic zone.

In contrast, terrestrial ecosystems include soils that develop from parent material containing large quantities of P (and Ca, Mg, K, and other elements). The supply of P and other elements via weathering of parent material is large relative to the requirements of organisms, for thousands to hundreds of thousands of years after unweathered parent material begins its development into soil (Walker and Syers, 1976).

Once the weathering source of P and other elements is depleted, limitation by P or another rock-derived element becomes possible (Walker and Syers, 1976, Vitousek *et al.*, 1997b); atmospheric inputs of P in particular are very small (Newman, 1995). Accordingly, while sustained P limitation to NPP is unlikely in ecosystems of the north temperate or boreal zones, where the frequency of glaciation should suffice to maintain weathering as a source of minerals within soil, P (and base cation) limitation could be more frequent on geologically old substrates in the tropics or subtropics (Vitousek *et al.*, 1997b, Kennedy *et al.*, 1998, Chadwick *et al.*, 1999).

3. A Simple Model

Why does N supply limit NPP and other ecosystem processes in many terrestrial ecosystems? In contrast to P, N is absent from most parent material (not all—see Dahlgren, 1994); it must be accumulated from the atmosphere. Nevertheless, even low inputs of N over thousands of years should more than account for the quantities of N we observe in most ecosystems (Peterjohn and Schlesinger, 1990); still less time is required where N fixers are abundant.

We evaluated alternative pathways of N loss, constraints to inputs via N_2 fixation, and their interactions and consequences using a simple model (Vitousek and Field, 1999). This model is not intended to represent ecosytem dynamics in detail, but it is useful in examining the logical consequences of plausible assumptions about N inputs and outputs and their controls and consequences.

In its simplest form, the model includes two types of primary producers: nonfixers and symbiotic N_2 fixers. Nitrogen becomes available in the soil through N mineralization and (secondarily) atmospheric inputs; N mineralization occurs when decomposition reduces the soil C:N ratio below a threshold. In effect this gives microbes priority over plants for available N. The nonfixer is assumed to take up all available N, up to a ceiling set by light availability. If available N remains in the soil above that ceiling, it is lost from the system—implicitly by nitrate leaching or denitrification. If all of the available N is taken up by the nonfixer and light remains available, then (and only then) the N_2 fixer can grow and fix N_2, up to the ceiling set by light availability. In effect, this gives nonfixers priority for available N and for light in proportion to available N. This assumption is too strong, in that symbiotic N fixers can make use of already-fixed N in the soil (Pate, 1986; McKey, 1994). However, it is conservative in that it tends to eliminate symbiotic N fixers from simulated ecosystems, and so downplay N fixation—and yet (as we will show) it is insufficient to maintain N limitation on NPP (Vitousek and Field, 1999).

We assume that 10% of plant C and N is lost annually as litter; accordingly, plant biomass is close to a 10-year running mean of NPP. A mass balance for N in the system is maintained, so that biologically fixed N ultimately increases the quantity and availability of N in the system.

Results of a long-term run of the model, starting with no plants, C, or N in the system, are summarized in Fig. 1. If N fixation is excluded (set to zero), then the system must depend on a low rate of atmospheric deposition to accumulate N. It takes millennia to accumulate sufficient N to the point where it scarcely limits biomass accumulation (Fig. 1a), but as long as N can only be lost when it is in excess, N will accumulate to this point. Allowing N_2 fixation causes N to accumulate and biomass to equilibrate much more rapidly (Fig. 1c), but the equilibrium N accumulation and NPP are the same with or without fixation.

4. Pathways of N Loss

We can identify three pathways that could remove N from terrestrial ecosystems, even though it limits NPP therein. These are loss of dissolved organic N (DON), loss of N trace gases by transformation-dependent pathways, and losses of N as a consequence of temporal or spatial heterogeneity in the supply versus demand for available N within ecosystems.

4.1 Dissolved organic N

Hedin *et al.* (1995) suggested that losses of DON could represent an uncontrollable leak of fixed N from ecosystems, one that could balance the very low atmospheric N inputs in the low-input Chilean temperate forest they studied. While the controls of DON flux are not well understood, DON loss appears to be much less dependent on the N status of ecosystems than is nitrate leaching (e.g., Currie *et al.*, 1996). Where DON loss is substantial (and inputs are small),

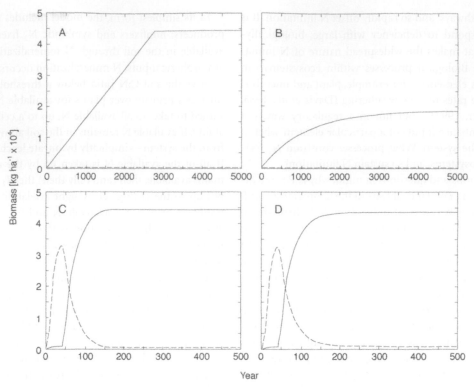

FIGURE 1 Biomass of a nonfixer (solid line) and a symbiotic N_2 fixer (dashed line) as function of time, starting with no plant or soil C or N. (A) No N_2 fixation, and losses of N occur only from the pool of excess available N. (B) No N_2 fixation; N losses (as DON or transformation-dependent trace gases) can also occur in proportion to the quantity of N cycling in the system. (C) N_2 fixation can occur; N losses only as excess available N. Note change in *x*-axis. (D) With N_2 fixation and the additional pathways of N loss. Revised from Vitousek and Field (1999).

this pathway of removal of N could keep N from accumulating to the point where it no longer limits NPP (Hedin *et al.*, 1995).

4.2 Transformation-Dependent Trace Gas Flux

In a sense similar to DON, N trace gases that are produced and lost in the course of nitrification could be regarded as leaks of potentially available N (Firestone and Davidson, 1989), whereas losses via denitrification could be more analogous to leaching of nitrate. The nitrification process is internal to the N cycle of many ecosystems, while denitrification utilizes a pool of N that can accumulate when N is available in excess (Vitousek *et al.*, 1998).

Vitousek and Field (1999) evaluated the effects of N losses through DON and transformation-dependent trace-gas fluxes by modifying the model described above so that a constant fraction (5%) of net N mineralization is lost. With that additional loss pathway in place, and with N fixation turned off, N accumulation, NPP, and biomass in the simulated system equilibrated to much lower levels (e.g., Fig 1b for biomass) than in the case where only excess N was lost. Vitousek *et al.* (1998) reported similar results from a simpler model; further, they showed that boosting simulated atmospheric deposition of N from 2 to 10 kg ha^{-1} year^{-1} was sufficient to overwhelm the additional loss pathways of N and to

take the system to the original, non-N-limited equilibrium. Clearly, alternative loss pathways that are dependent on N transformations rather than excess available N are sufficient to cause sustained and substantial limitation by N in low–input systems. Results are similar if these additional losses are made dependent on the total quantity of soil organic N.

Given the assumptions about inputs, outputs, and their controls, these conclusions are robust despite the simplicity of the model. For N to limit NPP and biomass accumulation in the long term, there must be N losses from pathways other than excess available N, and there must be additional constraints on N fixation even stronger than our (assumed) priority of the nonfixers for fixed N and light, in the model as well as in the world. These same processes cause simulated N limitation in the more complex Century model. Century calculates several pathways of loss of N, including losses by leaching and denitrification, from the pool of available N that is left over after biological uptake; losses of N as DON (calculated as a complex function of decomposition and water flux); and losses of N as nitrification-dependent trace gas flux (calculated as a constant fraction of gross N mineralization (Metherell *et al.*, 1993, Parton *et al.*, 1996)). Schimel *et al.* (1997) used Century to show that globally the cycles of C, N, and water equilibrate with each other in the long run, with N always in relatively short supply (so that it generally limits NPP within Century). Vitousek *et al.* (1998) ran a tropical forest version of Century with DON and nitrification-dependent losses turned off; when N losses can only occur via excess available N, Century simulates a system with greater N pools, greater productivity, and no N limitation at equilibrium. The world according to Century is limited by N in part because (a) it includes substantial losses of N by pathways that are independent of excess available N and (b) it does not allow for substantial N fixation.

4.3 Temporal/Spatial Variation in N Supply versus Demand

This third pathway involves loss of available N when it is in temporary or local surplus, even though N supply limits NPP most of the time or over most of the area. Temporary excesses of supply over demand can occur on time scales from centuries (disturbance/regeneration cycles in forests) to seasonal or even day-to-day. Large-scale disturbance can cause a short-term excess of N supply over demand, leading to losses (Vitousek and Reiners, 1975; Bormann and Likens, 1979). More importantly, fire and harvest themselves cause substantial losses of N; where these are the important agents of disturbance, ecosystem N budgets are characterized by long periods of accumulation (and potentially limitation) punctuated by brief periods of large losses.

Year-to-year variations in climate can also drive temporary imbalances in N supply and demand, particularly in water-limited systems. We evaluated this process by modifying the model above (Vitousek and Field, 1999) to include water as a resource—making production, decomposition, and losses all constrained equivalently by water supply. We ran the model with constant but low

water availability, no N_2 fixation, and only excess N lost; NPP equilibrated at a lower level than in Fig. 1a, in direct proportion to water supply, but N did not limit production or biomass accumulation at that equilibrium value. We then introduced random year-to-year variation in precipitation. As a consequence of this variation, available N was in excess in some years (e.g., wet years following several dry years) and could be lost. However, N was in short supply in most years, to the point that it limited NPP—and this N limitation was exacerbated by N losses during the years when N was in excess. We quantified the extent of N limitations by simulating additions of N fertilizer each year. Without year-to-year variation in precipitation, added N had little effect on NPP or biomass; when N was added to a system with year-to-year variation, NPP and biomass accumulation were enhanced by 20% on average (Fig. 2).

The model used here is relatively simple, but the results make sense given our understanding of controls on decomposition and mineralization in ecosystems. Also, more complex models yield similar results. Year-to-year variations in NPP in Century are controlled by interactions between precipitation and precipitation-induced variations in N mineralization (Burke *et al.*, 1997); these year-to-year variations can drive losses of N even when it limits NPP in most years. The Pnet model also predicts year-to-year variations in nitrate leaching from deciduous forest watersheds as a consequence of variations in precipitation; these predictions are strongly supported by watershed-level observations (Aber and Driscoll, 1997). This mechanism could help to explain observations that the N cycle appears to be more open in semiarid areas than in mesic forest ecosystems, in the sense that both inputs and outputs of N are larger relative to N pools within ecosystems (Austin and Vitousek, 1998).

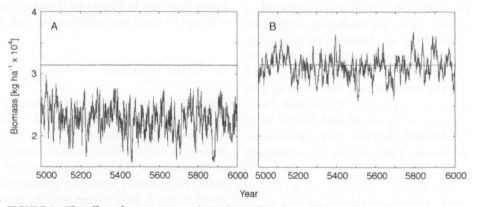

FIGURE 2 The effect of year-to-year variation in precipitation on N limitation to biomass accumulation in a simulated system without N_2 fixation or alternative pathways of N loss. The straight lines represent simulated plant biomass without year-to-year variation; simulated additions of N fertilizer (in part B) increase biomass and production by ~2%. With year-to-year variation, biomass and production are lower and variable. Simulated additions of N fertilizer increase biomass and production by ~25% on average, to the level of the system without year-to-year variation—demonstrating that year-to-year variation in precipitation induces N limitation. (A) No fertilization. (B) N fertilizer added.

5. Constraints on N Fixation

Pathways of loss that are independent of excess available nutrient pools exist—and where they are quantitatively important, they are sufficient to explain nutrient limitation by elements other than N. Moreover, given the greater mobility of N relative to P and (as nitrate) relative to most other elements, and given the importance of N trace-gas fluxes, it is reasonable to speculate that these pathways of loss would make N more likely than P to limit NPP in many terrestrial ecosystems, in the long term—were it not for N_2 fixation. However, a system dominated by N fixers has the capacity to add N at least as fast as it can be lost, by all of these pathways. How can N_2 fixation be sufficiently constrained so that N_2 fixers do not respond to N deficiency with increased growth and activity?

N_2 fixation in the model already appears to be strongly constrained, ultimately by a higher cost of N acquisition for fixers, proximately by an absolute priority for fixed N (and the light and water equivalent to that fixed N) given to nonfixers (Vitousek and Field, 1999). Nevertheless, even where alternative pathways of N loss are important, the model simulates sufficient N_2 fixation to overwhelm N limitation in a very short time (Fig. 1d), and to sustain an equilibrium biomass that is barely limited by N. To the extent that N supply limits real terrestrial ecosystems, N_2 fixation in the world must be constrained more and/or differently than is N_2 fixation in the model.

Vitousek and Field (1999) explored three additional constraints on N_2 fixation: P availability, differential herbivory on N_2 fixers, and a lower shade-tolerance of N_2 fixers.

5.1 P Limitation

P limitation on N_2 fixation is widely observed in aquatic systems; in terrestrial ecosystems there is good evidence for it from agricultural, pastoral, and some natural systems (Eisele *et al.*, 1989; Smith, 1992; Crews, 1993). The model estimates P availability within a mass-balanced P cycle, with inputs via weathering, outputs via leaching, and a labile adsorbed fraction. Nonfixers are given priority for P, in proportion to the amount of fixed N available. If not enough available P is present to match available N, nonfixers are P limited. If P remains available after nonfixers have taken up what they can, N_2 fixers can use it—at a lower C:P ratio than that of nonfixers (Pate, 1986), and up to the overall limit to NPP set by light or water availability (Vitousek and Field, 1999). Limitation by other elements (e.g., Mo; Silvester, 1989) could be treated similarly.

5.2 Grazing

Differential grazing on N_2 fixers in comparison to nonfixers is often observed (e.g., Hulme, 1994; 1996; Ritchie and Tilman, 1995; Ritchie *et al.*, 1998); it can occur because N_2 fixers generally have higher concentrations of N and protein than do non-fixers. Indeed, McKey (1994) suggested that legumes could have evolved the rhizobial symbiosis in part because of their N-demanding lifestyle. High levels of chemical defense can reduce the amount of grazing on fixers, but this in effect further raises the energetic cost

of N_2 fixation. There is good evidence that preferential grazing by deer on legumes virtually eliminates N_2 fixers and is responsible for maintaining N limitation of production and biomass accumulation at Cedar Creek, Minnesota (Ritchie and Tilman, 1995; Ritchie *et al.*, 1998). The model includes this preferential grazing by removing more biomass from fixers than nonfixers, effectively reducing production by N_2 fixers early in soil development (Vitousek and Field, 1999). A more realistic demographic analysis of the effects of grazing on N_2 fixers could yield a more sustained effect.

5.3 Shade Tolerance

It could be difficult for fixers to colonize under an established canopy of nonfixers, due to their greater cost for N acquisition (Gutschick, 1987; Vitousek and Howarth, 1991). If N_2 fixers have systematically lower shade tolerance, they would be unable to respond to N deficiency in a closed-canopy ecosystem, even where N is limiting to NPP and biomass accumulation. The model simulates this effect by suppressing growth of N_2 fixers in proportion to a sigmoidal function of the biomass of nonfixers.

A comparison of the initial model (with N fixers, and "with versus without" alternative pathways of N loss) with a revised version of the model that includes these three additional constraints to N_2 fixation is displayed in Figure 3. Neither the additional constraints on N_2 fixation alone (Fig. 3c) nor the alternative pathways of N-loss alone are sufficient to drive more than a marginal N limitation on biomass, at equilibrium (Fig. 3b). However, the combination (Fig. 3d) of both alternative loss pathways and additional constraints on N_2 fixation yields a system that is strongly limited by available N at equilibrium. For completeness, one would need to carry out a similar analysis of nonsymbiotic N_2 function. Some of the same mechanisms (e.g., energy cost of N_2 fixation, P limitation to N_2 fixers) could be important, and some others (e.g., decomposers might not be limited by N supply, even in sites where NPP is N limited) could also contribute Vitousek and Hobbie 2000.

6. Conclusions

Overall, we think that Figure 3a displays the logical consequences of the ways that many ecologists think of the N cycle in terrestrial ecosystems—with losses of N occurring primarily when N is available in excess, with N_2 fixation constrained to some extent by its energetic cost. If this were a reasonable representation of the world, N limitation would be a transient phenomenon, there would be very little C storage resulting from human alteration of the N cycle, and any stimulation in NPP and/or C storage resulting from increased CO_2 would not be constrained for long by N availability.

The pervasiveness of terrestrial ecosytems where N availability demonstrably limits NPP and other ecosystem processes suggests that alternative pathways of N loss and additional constraints on N_2 fixation should be fundamental parts of our view of the N cycle. To the extent that these are important, anthropogenic

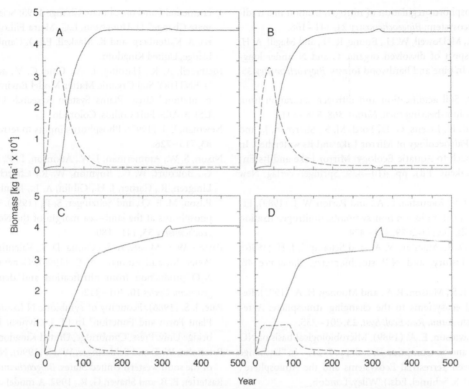

FIGURE 3 The effects on biomass and the extent of N limitation of adding constraints on N_2 fixation. The constraints include differentially severe effects of grazing, P limitation, and shade intolerance on N_2 fixers. In each case, a pulse of simulated N fertilizer was added beginning in year 300 (50 kg N ha^{-1} year^{-1}, continued for 20 years) to illustrate the extent of N limitation. (A) No additional constraints to N_2 fixation, losses of excess available N only. (B) No additional constraints; losses of N by additional pathways. (C) N_2 fixation constrained by grazing, P limitation, and shade intolerance; losses of excess available N only. (D) N_2 fixation constrained, and losses of N by additional pathways. Revised from Vitousek and Field (1999).

changes in the N cycle can have fundamental effects on terrestrial ecosystems; anthropogenic N could increase C storage on land at equilibrium; and the long-term effects of elevated CO_2 on N availability will depend on how CO_2 interacts with pathways of N loss and with the processes constraining N_2 fixation (Vitousek and Field, 1999). These are mechanisms and interactions that we ought to try to understand.

Acknowledgments

This research was supported by grants from the National Science Foundation and the Andrew Mellon Foundation.

References

Aber, J. D.(1992). Nitrogen cycling and nitrogen saturation in temperate forest ecosystems. *Trends Ecol. Evolut.* 7, 220–223.

Aber, J. D., and Driscoll, C. T. (1997). Effects of land use, climate variation, and N deposition on N cycling and C storage in northern hardwood forests. *Global Biogeochem. Cycles* 11, 639–648.

Aber, J. D., McDowell, W. H., Nedelhoffer, K., Magill, A., Berntson, G., Ka-

makea, M., McNulty, S., Currie, W., Rustad, L., and Fernandez, I. (1998). Nitrogen saturation in temperate forest ecosystems. *BioScience* 48, 921–934.

Ågren, G. I., and Bosatta, E. (1988). Nitrogen saturation of terrestrial ecosystems. *Environ Pollut.* 54, 185–197.

Austin, A. T. and Vitousek, P. M. (1998). Nutrient dynamics on a precipitation gradient in Hawai'i. *Oecologia* 113, 519–529.

Berendse, F., Aerts, R., and Bobbink, R. (1993). Atmospheric nitrogen deposition and its impact on terrestrial ecosystems. In "Landscape Ecology of a Stressed Environment" C. C. Vos and P. Opdam, (Eds.), pp. 104–121. Chapman & Hall, London.

Bloom, A. J., Chapin III, F. S., and Mooney, H. A. (1985). Resource limitation in plants—An economic analogy. *Annu. Rev. Ecol. Syst.* 16, 363–393.

Bormann, F. H., and Likens, G. E. (1979). "Pattern and Processes in a Forested Ecosystem". Springer-Verlag, New York.

Burke, I. C., Lauenroth, W. K., and Parton, W. J. (1997). Regional and temporal variation in net primary production and nitrogen mineralization in grasslands. *Ecology* 78, 1330–1340.

Chadwick, O. A., Derry, L. A., Vitousek, P. M., Huebert, B. J., and Hedin, L. O. (1999). Changing sources of nutrients during four million years of ecosystem development. *Nature* 397, 491–497.

Cochran, M. F., and Berner, R. A. (1997). Promotion of chemical weathering by higher plants: Field observations on Hawaiian basalts. *Chem.l Geol.* 132, 71–85.

Crews, T. E. (1993). Phosphorus regulation of nitrogen fixation in a traditional Mexican agroecosystem. *Biogeochemistry* **21**, 141–166.

Currie, W. S., Aber, J. D., McDowell, W. H., Boone, R. D., and Magill, A. H. (1996). Vertical transport of dissolved organic C and N under long-term N amendments in pine and hardwood forests. *Biogeochemistry* **35**, 471–505.

Dahlgren, R. A. (1994). Soil acidification and nitrogen saturation from weathering of ammonium-bearing rock. *Nature* **368**, 838–841.

Davis, M. B., Moeller, R. E., Likens, G. E., Ford, M. S., Sherman, J., and Goulden, C. (1985). Paleoecology of Mirror Lake and its watershed. In An Ecosystem Approach to Aquatic Ecology: Mirror Lake and Its Environment." (G. E. Likens, Ed), pp. 410–429. Springer-Verlag, New York.

Eisele, K. A., Schimel, D. S., Kapustka, L. A., and Parton W. J. (1989). Effects of available P and N:P ratios on non-symbiotic dinitrogen fixation in tall grass prairie soils. *Oecologia* **79**, 471–474.

Elser, J. J., Dobberfuhl, D. R., MacKay, N. A., and Schampel, J. H. (1996). Organism size, life history, and N:P stoichiometry. *Bioscience* **46**, 674–684.

Field, C. B., Chapin III, F. S., Matson, P. A., and Mooney, H. A. (1992). Responses of terrestrial ecosystems to the changing atmosphere: A resource-based approach. *Annu. Rev. Ecol. Syst.* **23**, 201–235.

Firestone, M. K., and Davidson, E. A. (1989). Microbiological basis of NO and NO_2 production and consumption in soil. pp. 7–21. In "Exchange of Trace Gases Between Terrestrial Ecosystems and the Atmosphere". (M. O. Andreae and D. S. Schimel, Eds.), Wiley, London.

Galloway, J. N., Schlesinger, W. H., Levy II, H., Michaels, A., and Schnoor, J. L. (1995). Nitrogen fixation: Atmospheric enhancement—Environmental response. Global *Biogeochem. Cycles* **9**, 235–252.

Gleeson, S. K., and Tilman, D. (1992). Plant allocation and the multiple limitation hypothesis. *Am. Nat.* **139**, 1322–1343.

Gutschick, V. P. (1987). "A Functional Biology of Crop Plants." Timber Press, Portland.

Hedin, L. O., Armesto, J. J., and Johnson, A. H. (1995). Patterns of nutrient loss from unpolluted, old-growth temperate forests: evaluation of biogeochemical theory. *Ecology* **76**, 493–509.

Howarth, R. W., Billen, G., Swaney, D., Townsend, A., Jaworski, N., Lajtha, K., Downing, J. A., Elmgren, R., Caraco, N., Jordan, T., Berendse, F., Freney. J., Kudeyarov, V., Murdoch, P., and Zhu Zhao-liang. (1996). Regional nitrogen budgets and riverine N & P fluxes for the drainages to the North Atlantic Ocean: Natural and human influences. *Biogeochemistry* **35**, 181–226.

Hulme, P. E. (1994). Seedling herbivory in grassland: Relative impact of vertebrate and invetebrate herbivores. *J. Ecol.* **82**, 873–880.

Hulme, P. E. (1996). Herbivores and the performance of grassland plants: A comparison of arthropod, mollusc and rodent herbivory. *J. Ecol.* **84**, 43–51.

Kennedy, M. J., Chaduick, O. A., Vitousek, P. M., Derey, L. A., and Hendeicks, D. M., (1998). Changing Sources of base cations during ecosystem development, Hawaian Islands. *Geology* **26**, 1015–1018.

McGill, W. B. and Cole, C. V. (1981). Comparative aspects of cycling of organic C, N, S, and P through soil organic matter. *Geoderma* **26**, 267–286.

McKey, D. (1994). Legumes and nitrogen: The evolutionary ecology of a nitrogen-demanding lifestyle. "Advances in Legume Systematics: Part 5—The Nitrogen Factor." J. L. Sprent and D. McKey, (Eds.), pp 211–228. Royal Botanic Gardens, Kew, England.

Melillo, J. M., Prentice, I. C., Farquhar, G. D., Schulze, E.-D., and Sala, O. E. (1996). Terrestrial ecosystems: Biotic feedbacks to climate. In "Intergovernmental Panel on Climate Change 1995: Scientific Assessment of Climate Change". (J. Houghton, L.G. Meira Filho, B. A. Callander, N. Harris, A. Kattenberg, and K. Maskell, Eds.), Cambridge Univ. Press, Cambridge, United Kingdom.

Metherell, A. K., Harding, L. A., Cole, C. V., and Parton, W. J. (1993). "CENTURY Soil Organic Matter Model Environment: Technical Documentation." Great Plains System Research Unit Technical Report 4, USDA-ARS, Fort Collins, Colorado.

Newman, E. I. (1995). Phosphorus inputs to terrestrial ecosystems. *J. Ecol.* **83**, 713–726.

Nixon, S. W., Ammerman, J. W., Atkinson, L. P., Berounsky, V. M., Billen, G., Boicourt, W. C., Boynton, W. R., Church, T. M., Ditoro, D. M., Elmgren, R., Garber, J. H., Giblin, A. E., Jahnke, R. A., Owens, N. P. J., Pilson, M. E. Q., and Seitzinger, S. P., (1996). The fate of nitrogen and phosphorus at the land–sea margin of the North Atlantic Ocean. *Biogeochemistry* **35**, 141–180.

Parton, W. J., Mosier, A. R., Ojima, D. S., Valentine, D. W., Schimel, D. S., Weier, K., and Kulmala, A. E. (1996). Generalized model for N_2 and N_2O production from nitrification and denitrification. *Global Biogeochem. Cycles* **10**, 401–412.

Pate, J. S. (1986). Economy of symbiotic N fixation. "On the Economy of Plant Form and Function" In (T. J. Givnish, Ed.), pp. 299–325. Cambridge Univ. Press, Cambridge, United Kingdom.

Peterjohn, W. T. and Schlesinger, W .H. (1990). Nitrogen loss from deserts of the southwestern United States. *Biogeochemistry* **10**, 67–79.

Rastetter, E. B. and Shaver, G. R., 1992. A model of multiple element limitation for acclimating vegetation. *Ecology* **73**, 1157–1174.

Ritchie, M. E. and Tilman, D., (1995). Responses of legumes to herbivores and nutrients during succession on a nitrogen-poor soil. *Ecology* **76**, 2648–2655.

Ritchie, M. E., Tilman, D., and Knops, J. M. H. (1998). Herbivore effects on plant and nitrogen dynamics in oak savanna. *Ecology* **79**, 165–177.

Schimel, D. S., Brassell, B. H., and Parton, W. J. (1997). Equilibration of the terrestrial water, nitrogen, and carbon cycles. *Proc. Natl. Acad. Sci. USA* **94**, 8280–8283.

Schimel, D. S., Enting, I. G., Heimann, M., Wigley, T. M. L., Raynaud, D., Alves, D., and Siegenthaler U., (1995). CO_2 and the carbon cycle. In "Climate Change 1994: Radiative Forcing of Climate Change". (J. T. Houghton, L. G. Meira Filho, and K. Maskell Eds.), pp. 39–71. Cambridge Univ. Press, Cambridge, United Kingdom.

Schindler, D. W. (1977). Evolution of phosphorus limitation in lakes. *Science* **195**, 260–262.

Schulze, E.-D. (1989). Air pollution and forest decline in a spruce (*Picea abies*) forest. *Science* **244**, 776–783.

Silvester, W. B. (1989). Molybdenum limitation of asymbiotic nitrogen fixation in forests of Pacific Northwest America. *Soil Biol. Biochem.* **21**, 283–289.

Smith, V. H. (1992). Effects of nitrogen:phosphorus supply ratios in nitrogen fixation in agricultural and pastoral systems. *Biogeochemistry* **18**, 19–35.

Sprent, J. I., and Sprent, P. (1990). "Nitrogen Fixing Organisms: Pure and Applied Aspects." Chapman & Hall, London.

Tilman, D., (1987). Secondary succession and the pattern of plant dominance along experimental nitrogen gradients. *Ecological Monogzaphs* **57**, 189–214.

Townsend, A. R., Braswell, B. H., Holland, E. A., and Penner, J. E. (1996). Spatial and temporal patterns in terrestrial carbon storage due to deposition of fossil fuel nitrogen. *Ecol. Appl.* **6**, 806–814.

Vitousek, P. M., Aber, J. D., Howarth, R. W., Likens, G. E., Matson, P. A., Schindler, D. W., Schlesinger, W. H. and Tilman, D. (1997a). Human alteration of the global nitrogen cycle: Sources and consequences. *Ecol. Appl.* **7**, 737–750.

Vitousek, P. M., Chadwick, O. A., Crews, T., Fownes, J., Hendricks, D., and Herbert, D. (1997b). Soil and ecosystem development across the Hawaiian Islands. *GSA Today* **7**(9), 1–8.

Vitousek, P. M., and Field, C. B. (1999). Ecosystem constraints to symbiotic nitrogen fixers: A simple model and its implications. *Biogeochemistry* **46**, 179–202.

Vitousek, P. M., Hedin, L. O., Matson, P. A., Fownes, J. H., and Neff, J., (1998). Within-system element cycles, input–output budgets, and nutrient limitation. "Successes, Limitations, and Frontiers in Ecosystem Science". (M. Pace and P. Groffman, Eds.), pp. 432–452. Springer-Verlag, Berlin.

Vitousek, P. M., and Hobbie, S. (2000). Heterotrophic nitrogen Fixation in decomposing litter: Patterns and regulation. *Ecology* **81**, 2366–2376.

Vitousek, P. M., and Howarth, R. W. (1991). Nitrogen limitation on land and in the sea: How can it occur? *Biogeochemistry* **13**, 87–115.

Vitousek, P. M., and Reiners, W. A. (1975). Ecosystem succession and nutrient retention: A hypothesis. *Bioscience* **25**, 376–381.

Walker, T. W. and Syers, J. K. (1976). The fate of phosphorus during pedogenesis. *Geoderma* **15**, 1–19.

1.17

Interactions between Hillslope Hydrochemistry, Nitrogen Dynamics, and Plants in Fennoscandian Boreal Forest

Peter Högberg

Section of Soil Science, Department of Forest Ecology SLU, Umeå, Sweden

1. Introduction

The title of this chapter may sound odd to a chemist, as one may rightly ask if N is not a chemical element like others in the context of hillslope hydrochemistry. I will argue here, however, that N is different from other nutrient elements in many ways, and especially so because (a) it is in general not derived by weathering of minerals, (b) its supply and dynamics are under particularly strong biological control, and (c) its availability often exerts a strong and direct control on primary productivity. This, in turn, means that we should also ask what factors (or factor) control(s), and interact(s) with, the availability of N (e.g., Cole and Heil, 1981; Vitousek and Howarth, 1991).

This chapter focuses on the controls on plant productivty in Fennoscandian boreal forests, but the discussion may well be applicable to larger areas of temperate forests developed on young glacial till soils and sediment soils.

2. A Historical Perspective

That N can limit production in temperate forests was first revealed through experiments conducted by Mitchell and Chandler in the U.S. (1939) and Hesselman and Romell (Fig. 1) in Sweden (reviewed by Tamm, 1991). Meanwhile, many forest soil scientists tended to focus on the role of other elements, notably the so-called base cations, possibly because benefits of liming had been recognized in agriculture. The first generation of liming trials in

forests were, thus, initiated to see if the treatment could increase forest growth by increasing the biological turnover of organic matter, rather than to counteract soil acidification (Hüttl and Zöttl, 1993). Implications of the fact that N, unlike Ca, Mg, K, etc., is not supplied by weathering of minerals (but see Holloway et al., 1998) went largely unnoticed for some time.

Viro (1951; 1955) conducted an extensive survey of links between soil chemistry and forest productivity in Finland. He observed a correlation between exchangeable Ca and forest productivity. Most interestingly, Dahl et al. (1961), in a reexamination of his material, commented that the correlation was not firm evidence of a direct limitation of forest growth by Ca supply. Rather, they thought there must be an influence of the supply of Ca on the turnover of N in the soil. Subsequently, Dahl et al. (1967), in a regional survey in Norway, demonstrated a strong correlation between %N and base saturation of the mor layer (Fig. 2). They also showed that plant community composition and forest productivity changed continuously along the regression of %N versus base saturation. A similar pattern was found in a large survey comprising 921 forests in southern Finland (Lahti and Väisänen, 1987), where exchangeable Ca (and pH) in the soil was correlated with %N, and where there was also a correlation between these variables and plant community composition and forest productivity (Fig. 2).

Early in the 20th century, Cajander (1909; 1926) described relations between forest vegetation type (the composition of the understorey or "field-layer" plant community) and forest productivity. Forest productivity and the distribution of forest types are

FIGURE 1 Lars-Gunnar Romell in action trying to pull nutrients through the soil–plant system. This drawing may be used to illustrate the fact that N supply is linked to the supply of other elements. It was made by the Norwegian Dyring in 1948 on wall paper in the Flaka hut, 60 km from Umeå in northern Sweden, where more systematic research on forest biogeochemistry started in the 1920s.

closely intimately linked with landscape topography, i.e., more productive forests with tall herbs in the understorey are regularly found in toe slope areas (e.g., Hägglund and Lundmark, 1977). In fact, one can reasonably well predict forest productivity based on topographic maps alone (Holmgren, 1994).

After the alarming reports of forest decline in Central Europe in the late 1970s, the focus of discussions on forest nutrition was on

the negative effects of acidity, especially of Al^{3+}, on roots (e.g., Hütterman and Ulrich, 1984). A dominating idea was that N in soil was already, or would soon be, in excess of plant demand because of high levels of N deposition (e.g., Nihlgård, 1985; Aber et al., 1989), and this is no doubt true for some areas of Central Europe, in particular. A contemporary model of acid deposition effects on tree growth (Sverdrup et al., 1992) treated N as a nonlimiting nutrient and the base cations Ca, Mg, and K as limiting. Hence, liming of forests was advocated as a means to counteract the suspected negative effects of soil acidification on tree growth even in Sweden (Sverdrup et al., 1994) despite the comparatively low load of acid deposition there. However, there was substantial experimental evidence to suggest that (a) forests in Sweden (as elsewhere in Fennoscandia) were still, in general, strongly N-limited, and (b) within reasonable limits, further acidification should not pose an immediate threat to forest productivity in Sweden (Binkley and Högberg, 1997). During much of the discussions about soil acidity effects on tree growth, some basic interactions between other elements and N supply were overlooked. A lower forest productivity on more acidic soils was to some extent interpreted as a growth decline caused by soil acidification, and not as a natural condition related to a low N supply and not necessarily involving any effects of acid rain (see the example from Betsele below).

It should be pointed out here that there is a major difference between the deeper forest soils that predominate in Fennoscandia and the shallow soils that are regionally important in Norway, in particular. The latter have limited buffer capacity and are particularly sensitive to acid rain.

3. Nitrogen Supply and Forest Productivity in a Landscape Perspective: Hypotheses

There may be several explanations for the increase in forest productivity (and the correlation between base cations and N) down slope major slopes in Fennoscandian forest landscapes (Tamm, 1991; Giesler et al., 1998; Table 1). One category of these explanations (Fig. 3) refers to transport of N toward toe slope areas. As N is lost by leaching from large recharge areas, there should be a concentrated flux of N in the smaller groundwater discharge areas. Rohde (1987) reported that groundwater discharge areas on average comprise approximately 10% of the landscape in Swedish forests, which would lead to a ninefold increase per unit area of the N flux in the groundwater discharge areas as compared to that in recharge areas. Transport of N should increase as a result of disturbances, e.g., after a forest fire, the flux of NO_3^- should increase (Reiners, 1981). It may well be that the generally wetter ecosystems in toe slope positions acted as comparatively intact biological nutrient traps, while forests upslope were more damaged by fire (as fires usually climb upslope), and hence became more leaky. Also, according to this kind of explanation, the correlation between N and base cations occurs because they are solutes in the same water flow.

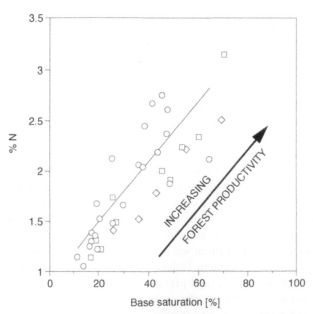

FIGURE 2 Correlations between %N and base saturation of the mor-layer: (a) ○, in a regional survey in Hedmark county, Norway (Dahl et al., 1967); (b)◇, in a larger regional survey of southern Finland (Lahti and Väisänen, 1987); and (c) □, along the 90 m long gradient at Betsele, northern Sweden (Giesler et al., 1998). The regression line is from Dahl et al. (1967).

TABLE 1 Theoretical N Balance over a Period of 100 Years in a Discharge Area and Its Associated Recharge Area in Boreal Forest in Northern Sweden*

	Recharge area	Discharge area
Deposition[a]	+200	+200
N$_2$ fixation[b]	+50	+150
Fire[c]	−100	0
Leaching[d]	−100	−200 (−1200)
Lateral inflow	0	+900
Balance	+50	+1050 (+50)

*It is assumed that the recharge area covers 90% of the total area. All figures are in kg N ha^{-1} 100 year^{-1}.

[a]Has increased from a low preindustrial (and pre-intensive-agricultural) background to 2 kg N ha^{-1} year^{-1} (Lövblad et al., 1995).

[b]Estimates based on Nohrstedt (1985).

[c]Assumes a fire return interval of 100 years (Zackrisson, 1977), and that fires consume mainly the upper part of the mor-layer and some woody debris on the ground.

[d]Leaching data from Degermark (1985; 1987; 1988; 1989). The high value given in brackets for the discharge area was calculated to bring its N-sequestration rate down to the level of that in the recharge area.

However, stream water and groundwater originating in recharge areas in unpolluted parts of Fennoscandia have never been reported to contain appreciable amounts of N (e.g., > 2 kg N ha^{-1} year^{-1}), except in connection with disturbance, e.g., forest N fertilization and clearfelling (Tamm et al., 1974; Wiklander, 1981). After clearfelling, forests downslope of the felled area may show increased growth, which, given the strong N limitation, may be due to a flux of N from the clearfelling to the forest downslope (Lundell and Albrektsson, 1997). In contrast to the situation with N, the concentrations of base cations, notably of Ca, is higher than the N concentration in groundwater, in particular when the water has passed a long way through the soil and the bedrock. Data from a typical recharge area on acidic bedrock in northern Sweden show that the concentration of Ca is an order of magnitude higher than that of total N in groundwater (Degermark, 1985; 1987; 1988; 1989). Back-of-the-envelope calculations suggest that losses of N from such recharge areas commonly are ∼1 kg N ha^{-1} year^{-1}, potentially contributing to a flux of ∼9 kg N ha^{-1} year^{-1} in discharge areas, given that the latter compose 10% of the area (see above and Table 1).

Another category of explanations (Fig. 3) refers to *in situ* processes in the groundwater discharge areas. This category of explanations of the higher N supply in discharge areas includes (Tamm, 1991; Chapin et al., 1988; Giesler et al., 1998):

1. conditions more conducive for N$_2$-fixation, at present or earlier in site history,
2. higher rates of autotrophic nitrification because of higher soil pH,
3. higher rates of N mineralization and higher *in situ* flux of solutes to roots and mycorrhizae because of wetter conditions, and

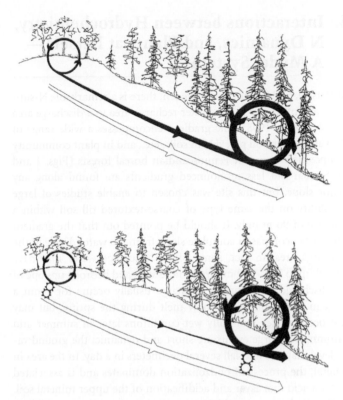

FIGURE 3 Possible interpretations of mechanisms underlying the increase in N supply that regularly occurs down slopes in Fennoscandian boreal forests. In the upper graph, the importance of the flux of N (and base cations) from recharge to discharge areas is stressed. In the lower graph, it is emphasized that fluxes of N down slopes are small, and that there must be processes promoting the high *in situ* N turnover in discharge areas, and that these processes are probably linked to the high pH (which is maintained by the high flux of base cations) in such areas.

4. smaller losses of N during fires because of wetter conditions (and lower position in the landscape).

There is evidence from a study of 20 sites in central Sweden that N$_2$ fixation by free-living microorganisms is positively correlated with pH and extractable Ca (Nohrstedt, 1985), which commonly are higher in discharge as compared to recharge areas (but see the example from Betsele below). As regards 2 and 3, it is often held that N turnover (and especially autotrophic nitrification) in acid soils should be stimulated by an increase in pH (e.g., Alexander, 1977; Kreutzer, 1995). Water limitation is a complex issue, as it may affect activity of soil organisms or flux of solutes to roots (Chapin et al., 1988), as well as it might imply a direct water limitation on photosynthesis. Hence, it is difficult, even experimentally, to determine if water supply or its effects on N supply limit plant growth.

It is likely that processes of both categories described above (Fig. 3) are highly relevant. It appears complicated, but of great theoretical interest, to determine the relative importance of inflow of N versus increased *in situ* N turnover as components of the higher N availability in groundwater discharge areas.

4. Interactions between Hydrochemistry, N Dynamics, and Plants at Betsele — A Model System

At Betsele, 64°N in northern Sweden, there is a remarkable N-supply gradient from a groundwater recharge area to a discharge area (Giesler *et al.,* 1998). The gradient encompasses a wide range of the variation in soil pH, %N in soils, etc., and in plant community composition found in Fennoscandian boreal forests (Figs. 2 and 4). Similar but less pronounced gradients are found along any major slope, but this site was chosen to enable studies of large variability on the same type of coarse-textured till soil within a distance of 90 m only. It should be pointed out that the gradient does not encompass a complete slope from the water divide to the discharge area; in fact, some of the discharge water may have flowed up to 700–800 m through the soils and bedrock upslope of the discharge area. Surface discharge usually occurs for about a week in connection with snow melt during the spring, but may also occur under unusually wet conditions later in summer and autumn. Discharge events are short and dynamic; the groundwater level may rise and fall several decimeters in a day. In the area in general, the process of podzolization dominates and is associated with an acid mor layer and acidification of the upper mineral soil. In the discharge area, this process is, however, sometimes interrupted by "titration" events, bringing in base-rich water (and no doubt some N).

Along the 90 m transect, forest productivity increases by a factor of 3 in the direction of the discharge area. The forest stand is about 125 years old throughout, but changes from a dominance of *Pinus sylvestris* L. to a dominance of *Picea abies* (Karst.) L. The tallest *P. abies* in the discharge area are 36 m tall, compared with 22 m tall *P. sylvestris* in the recharge area. Also, there is a remarkable shift in the understory from mainly 1- to 2-dm-tall dwarf shrubs of *Vaccinium* spp. through a zone of dwarf shrubs and low herbs to a luxurious stand of tall herbs, e.g., 1- to 2-m-tall *Aconitum septentrionale* L., in the discharge area.

Soil solution pH in the mor-layer increases from about 3.5 in the recharge area to about 6.5 in the discharge area (Fig. 4). In the upper mineral soil, Al^{3+} dominates the exchange complex in the recharge area, but is gradually replaced by Ca^{2+} toward the discharge area. Moreover, in mor-layer of the recharge area, levels of inorganic N are low (< 30 μmol L^{-1}), but as one approaches the discharge area, levels of NH_4^+ rise (>100 μmol L^{-1}), and finally NO_3^- becomes the dominant inorganic N species in the discharge area, where levels of inorganic N are highest (180 μmol L^{-1}). Levels of total P rise gradually at first, but then increase more rapidly toward the discharge area. In contrast, levels of PO_4 in the soil solution fall remarkably toward nil in the discharge area.

Thus, analyses of soil chemistry confirm the suspected increase in N supply in the direction of the discharge area. Foliar analysis and plant growth bioassays also confirm this increase in N supply. The low availability of PO_4 in the discharge area is confirmed by plant growth bioassays as well as ^{32}P root uptake bioassays on

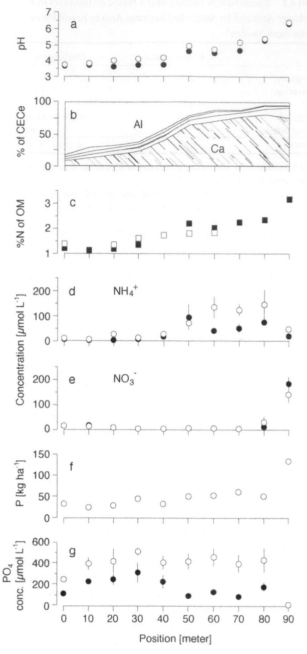

FIGURE 4 Variations in soil chemistry (Giesler *et al.,* 1998) along the 90-m long gradient at Betsele (which starts at 0 m in a recharge area and ends at 90 m in a discharge area) along which forest productivity increases threefold. (a), pH in the mor-layer (open symbols, the upper O1 horizon; closed symbols, the lower O2 horizon); (b), exchangeable cations in the mineral E horizon (CECe = effective cation exchange capacity). The same trend in Al and Ca saturation is found in the mor-layer and the mineral Bs horizon, but in the mor-layer Al is organically complexed, while it occurs as Al^{3+} in the mineral E and Bs horizons; (c), %N in the mor-layer (symbols as in a); (d), concentration of NH_4^+ in soil solution in the mor-layer (symbols as in a); (e), concentration of NO_3^- in soil solution in the mor-layer (symbols as in a); (f), amount of total P in the mor-layer; (g), concentration of PO_4 in the soil solution in the mor-layer (symbols as in a).

roots collected in the field. Apparently, the low availability of P in the discharge area is caused by high levels of organically complexed Fe and Fe-oxyhydroxides with a high capacity to bind P, e.g., into Fe phosphates.

The study at Betsele confirms that the supply of N is strongly correlated with soil pH, and especially the supply of Ca (Giesler *et al.,* 1998; Fig. 4). This should partly reflect the fact that both N and Ca are components of the groundwater, but it is noteworthy that NO_3^- is a dominant inorganic N species in the discharge area, which indicates that also *in situ* processes are important, as NO_3^- only occurs at trace levels in the soil solution elsewhere. Also, preliminary data suggest that calculated rates of N mineralization increase with the increase in pH (P. Högberg, A. Nordgren and M, Högberg, unpublished). It is also of interest that P, rather than N, is limiting in the discharge area; hence this is a naturally N-saturated system. As regards discussions about ratios between base cations (BC) and Al as predictors of effects of acid rain on forest growth, it is clear here, that in an area with very low levels of deposition of N and S, there is a correlation between the Ca/Al ratio and productivity (Fig. 4). Some authors have argued that the correlations between such ratios and forest growth are evidence of the impact of acid rain on forest growth, and that ratios of BC/Al < 1, or of Ca/Al < 1, in the mineral soil, represent situations when soil acidity is detrimental to root function and forest growth (Sverdrup *et al.,* 1992; Cronan and Grigal, 1995). Any effects of such ratios on forest growth in Fennoscandia are confounded, if not totally obscured, by effects of the variability in N supply, which is correlated with pH (which, in turn, correlates positively with BC/Al or Ca/Al, cf. Fig. 4).

Interestingly, the relations between soil pH and N supply found along the short transect at Betsele are very much the same as those proposed by Read (1986; 1991) as typical for long latitudinal and altitudinal gradients from polar/Alpine conditions through temperate coniferous forests and nemoral deciduous forests to dry and warm temperate "steppe" conditions. Read proposed that under cold and wet conditions, decomposition of organic matter would be slow and incomplete and result in the formation of acid, peaty soils or thick mor-layers with slow mineralization of organic N. Dominant plants are typically ericaceous dwarf shrubs with ercoid mycorrhiza and a high potential capacity to use organic N sources, e.g., amino acids. Under slightly warmer (and drier) conditions, in temperate forests, soil organic matter is more readily decomposed, and along with organic N sources, plants can use more NH_4^+ produced by mineralization. Dominant plants are ectomycorrhizal trees with a high potential capacity to use both simple organic N sources and NH_4^+. Under warmer, drier conditions, in temperate grasslands ("steppe"), mineralization is rapid, soil pH is high, and NO_3^- becomes an important N source. Dominant plants have arbuscular mycorrhiza. At Betsele, the sequence in terms of changes in soil pH, C/N ratio, concentration and species of inorganic N, and type of mycorrhiza, is, indeed, very much the same (Giesler *et al.,* 1998). It has not yet been demonstrated in the field that plants along the transect differ very much in their use of the different

potential N sources. Preliminary analyses have, however, shown that amino acids are the dominant N species in the soil solution of the mor-layer in the recharge, but not in the discharge area (A. Nordin, P. Högberg, and T. Näsholm, unpublished). This is most interesting, since plants with all three major types of mycorrhiza, ericoid, ecto-, and arbuscular, are capable of taking up at least glycine, as was recently demonstrated in the field (Näsholm *et al.,* 1998). Also, nonmycorrhizal sedge has been demonstrated to use simple organic N sources (Chapin *et al.,* 1993). However, despite many years of research on N cycling and plant N uptake, we do not know more exactly the fractional contribution of different potential N sources to plant uptake in the field and how this varies latitudinally and across landscapes.

5. Experimental Evidence

Since the 1940s, experiments with additions of nutrients on hundreds of plots in the field have shown a more or less strong N limitation on forest growth in Fennoscandian forests (Tamm, 1991; Binkley and Högberg, 1997). In such trials, additions of NH_4NO_3 increase forest production, but they frequently acidify the mineral soil (Fig. 5; Tamm, 1991; Binkley and Högberg, 1997). This means that the BC/Al (or Ca/Al) ratio is lowered, while forest production increases, and definitely implies that the N supply has proximal control of forest productivity.

Furthermore, trees have a low demand for Ca in relation to the supply. Extensive laboratory tests, as well as field evidence, suggest that the demand for N is more than an order of magnitude higher than the demand for Ca in the two major conifer species, *Picea abies* and *Pinus sylvestris* (Ingestad, 1979; Linder, 1995), while in boreal forests, foliar analysis frequently shows that the concentration of N is at most only twice the concentration of Ca (e.g., Edfast *et al.,* 1990).

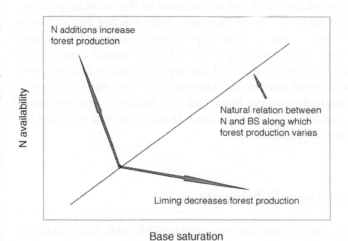

FIGURE 5 The basic correlation between the supply of N (and forest production) and soil base saturation (cf. Fig. 2), and the effects (arrows) of experimental treatments on forest growth.

Any positive effects of Ca on tree growth, as suggested by the correlation between Ca supply and forest growth, cannot thus be direct, but can potentially occur in a longer perspective, provided increases in soil pH increase soil N turnover. However, in typical boreal forest soils with a C/N ratio > 30 in the mor-layer, the reverse, i.e., increased N immobilization, commonly occurs after liming (Persson and Wirén, 1996) and is most likely the reason why forest growth often declines over a period of several decades after liming (Fig. 5; Derome *et al.*, 1986).

It is noteworthy that experimental additions of elements do not, in the shorter term, necessarily mimic a higher natural level of supply. Microorganism communities likely evolve in relation to specific site conditions, to which sudden changes in nutrient supply rate, or pH, are a perturbation. Possibly, as suggested by the liming trials, it may take decades before the microorganism community is in balance with a new chemical regime. This contrasts to natural conditions, e.g., at a site like Betsele, where the discharge area has, because of its position in the landscape, maintained a relatively high and stable pH since the last deglaciation 9200 year ago, while the surrounding soils in recharge areas have been acidified gradually through podzolization (but with forest fires as major intermittent disturbances leading to transient increases in soil pH).

6. Conclusions

There is strong experimental evidence that the supply of N exerts proximal control on forest growth in Fennoscandian boreal forests and that any positive effect of Ca is likely indirect and related to the long-term influence of soil pH on microbes turning over N in soils. The strong correlation between soil pH (and Ca supply) and forest growth occurs partly also because Ca and N are components of the same water fluxes in landscapes. The relative contribution of *in situ* N turnover processes versus inflow of N to the greater N supply (and hence productivity) in groundwater discharge as compared to that in recharge areas is not known. A more thorough understanding of the biogeochemistry and controls on productivity in these forest ecosystems requires that important links between soil chemistry and soil biology are identified and explored. Such studies will probably reveal a discrepancy between short-term effects of chemical manipulations on biota and the relations between natural chemical variability and biota. Hence, there is a need for long-term experiments.

Acknowledgments

My research has been sponsored by SJFR, NFR, SNV, and the EC. This chapter has benefitted from my introduction to this area of research by C.-O. Tamm and many discussions with D. Binkley, R. Giesler, H. Grip, M. Högberg, H.-Ö. Nohrstedt, and T. Näsholm. I thank E.-D. Schulze for the kind invitation to write this chapter.

References

Alexander, M. (1977) "Introduction to Soil Microbiology," 2nd ed. John Wiley, New York.

Aber, J. D., Nadelhoffer, K. J., Steudler, P., and Melillo, J. (1989). Nitrogen saturation in northern forest ecosystems. *BioScience* **39**, 378–386.

Binkley, D. and Högberg, P. (1997). Does atmospheric deposition of nitrogen threaten Swedish forests? *For. Ecol. Manage.* **92**, 119–152.

Cajander, A. K. (1909). Über Waldtypen. *Acta For. Fenn.* **1**, 1–175.

Cajander, A. K. (1926). The theory of forest types. *Acta For. Fenn.* **29(3)**, 1–108.

Chapin, F. S. III, Fetcher, N., Kielland, K., Everett, K. R., and Linkins, A. E. (1988). Productivity and nutrient cycling of Alaskan tundra: Enhancement by flowing soil water. *Ecology* **69**, 693–702.

Chapin, F. S. III, Moilanen, L., and Kielland, K. (1993). Preferential use of organic nitrogen for growth by a non-mycorrhizal arctic sedge. *Nature* **361**, 150–153.

Cole, C. V. and Heil, R. D. (1981). Phosphorus effects on terrestrial nitrogen cycling. *Ecol. Bull.* **33**, 363–374.

Cronan, C. S. and Grigal, D. F. (1995). Use of calcium/aluminium ratios as indicators of stress in forest ecosystems. *J. Environ. Qual.* **24**, 209–226.

Dahl, E., Gjems, O., and Kjelland-Lund, J. Jr. (1967). On the vegetation of Norwegian conifer forest in relation to the chemical properties of the humus layer. *Medd. Norsk Skogsforsoksv.* **85**, 501–531.

Dahl, E., Selmer-Anderssen, C., and Saether, R. (1961). Soil factors and the growth of Scotch pine: A statistical re-interpretation of data presented by Viro (1955). *Soil Sci.* **92**, 367–371.

Degermark, C. (1985). "Climate and Chemistry of Water at Svartberget. Reference Measurements 1984." Vindeln Exp. For. Stat., Vindeln.

Degermark, C. (1987). "Climate and Chemistry of Water at Svartberget. Reference Measurements 1986." Vindeln Exp. For Stat., Vindeln.

Degermark, C. (1988). "Climate and Chemistry of Water at Svartberget. Reference measurements 1987." Vindeln Exp. For. Stat., Vindeln.

Degermark, C. (1989). "Climate and Chemistry of Water at Svartberget. Reference Measurements 1988." Vindeln Exp. For. Stat., Vindeln.

Derome, J., Kukkola, M., and Mälkönen, E. (1986). "Forest Liming on Mineral Soils. Results from Finnish Experiments." Report 3084, Swedish Environmental Protection Agency, Solna.

Edfast, A.-B., Näsholm, T., and Ericsson, A. (1990). Free amino acid concentrations in needles of Norway spruce and Scots pine on different sites in areas with two levels of nitrogen deposition. *Can. J. For. Res.* **20**, 1132–1136.

Giesler, R., Högberg, M., and Högberg, P. (1998). Soil chemistry and plants in Fennoscandian boreal forest as exemplified by a local gradient. *Ecology* **79**, 119–137.

Hägglund, B. and Lundmark, J.-E. (1977). Site index estimation by means of site properties. Scots pine and Norway spruce in Sweden. *Stud. For. Suec.* **138**, 1–38.

Holloway, J. M., Dahlgren, R. A., Hansen, B., and Casey, W. H. (1998). Contribution of bedrock nitrogen to high nitrate concentrations in stream water. *Nature* **395**, 785–788.

Holmgren, P. (1994). Topographic and geochemical influence on the forest site quality, with respect to *Pinus sylvestris* and *Picea abies* in Sweden. *Scand. J. For. Res.* **9**, 75–82.

Hütterman, A. and Ulrich, B. (1984). Solid phase–solution–root interaction in soils subjected to acid deposition. *Philos. Trans. R. Soc. London Ser. B* **305**, 352–368.

Hüttl, R. F. and Zöttl, H. W. (1993). Liming as a mitigation tool in Germany's declining forests—Reviewing results from former and recent trials. *For. Ecol. Manage.* **61**, 325–338.

Ingestad, T. (1979). Mineral nutrient requirements of *Pinus sylvestris* and *Picea abies* seedlings. *Physiol. Plant.* **45**, 373–380.

Kreutzer, K. (1995). Effects of forest liming on soil processes. *Plant Soil* 168–169, 447–470.

Lahti, T. and Väisänen, R. A. (1987). Ecological gradients of boreal forests in South Finland: An ordination test of Cajander's forest type theory. *Vegetatio* **68**, 145–156.

Linder, S. (1995). Foliar analysis for detecting and correcting nutrient imbalances in Norway spruce. *Ecol. Bull.* **44**, 178–190.

Lövblad, G., Kindbom, K., Grennfelt, P., Hultberg, H., and Westling, O. (1995). Deposition of acidifying substances in Sweden. *Ecol. Bull.* **44**, 17–34.

Lundell, Y. and Albrektsson, A. (1997). Downslope effects of clear-cutting in Sweden on diameter increment of *Picea abies* and *Pinus sylvestris*. *Scand. J. For. Res.* **12**, 241–247.

Mitchell, H. L. and Chandler, R. F. (1939). The nitrogen nutrition and growth of certain deciduous trees of northeastern United States. With a discussion of the principles and practice of leaf analysis as applied to forest trees. *Black Rock For. Bull.* **11**, 1–94.

Näsholm, T., Ekblad, A., Nordin, A., Giesler, G., Högberg, M., and Högberg, P. (1998). Boreal forest plants take up organic nitrogen. *Nature* **392**, 914–916.

Nihlgård, B. (1985). The ammonium hypothesis—An additional explanation for the forest dieback in Europe. *Ambio* **14**, 2–8.

Nohrstedt, H.-Ö. (1985). Nonsymbiotic nitrogen fixation in the topsoil of some forest stands in central Sweden. *Can. J. For. Res.* **15**, 715–722.

Persson, T. and Wirén, A. (1996). Effekter av skogsmarkskalkning på kväveomsättningen. In "Skogsmarkskalkning." (H. Staaf, T. Persson, and U. Bertills, Eds.), pp. 70–91. Report 4559, Swedish Environmental Protection Agency, Stockholm.

Read, D. J. (1986). Non-nutritional effects of mycorrhizal infection. *In* "Physiological and Genetical Aspects of Mycorrhizae." (V. Gianinazzi-Pearson and S. Gianinazzi, Eds.), pp. 169–176. INRA, Paris.

Read, D. J. (1991). Mycorrhizas in ecosystems. *Experientia* 47, 376–391.

Reiners, W. A. (1981). Nitrogen cycling in relation to ecosystem succession. *Ecol. Bull.* **33**, 507–528.

Rohde, A. (1987). "The Origin of Streamwater Traced by Oxygen-18". Doctoral thesis, University of Uppsala.

Sverdrup, H., Warfvinge, P., and Rosén, K. (1992). A model for the impact of soil solution Ca:Al ratio, soil moisture, and temperature on tree base cation uptake. *Water Air Soil Pollut.* **61**, 365–383.

Sverdrup, H., Warfvinge, P., and Nihlgård, B. (1994). Assessment of soil acidification effects on forest growth in Sweden. *Water Air Soil Pollut.*. **78**, 1–36.

Tamm, C.-O. (1991). Nitrogen in terrestrial ecosystems. *Ecol. Stud.* **81**, 1–115.

Tamm, C.-O., Holmen, H., Popovic, B., and Wiklander, G. (1974). Leaching of plant nutrients from forest soils as a consequence of forestry operations. *Ambio* **3**, 211–221.

Viro, P. J. (1951). Nutrient status and fertility of forest soil. I. Pine stands. *Metsätiet. Tutkimuslait. Julk.* **39**, 1–47.

Viro, P. J. (1955). Use of ethylendiaminetetraacetic acid in soil analysis. II. Determination of soil fertility. *Soil Sci.* **80**, 69–74.

Vitousek, P. M. and Howarth, R. W. (1991). Nitrogen limitation on land and in the sea: How can it occur? *Biogeochemistry*. 87–115.

Wiklander, G. (1981) Clear-cutting and the nitrogen cycle. Heterogenous nitrogen leaching after clear-cutting. *Ecol. Bull.* **33**, 642–647.

Zackrisson, O. (1977). Influence of forest fires on the north Swedish boreal forest. *Oikos* **29**, 22–32.

1.18

The Cycle of Atmospheric Molecular Oxygen and Its Isotopes

Martin Heimann

Max Planck Institute for Biogeochemistry, Jena, Germany

1. Introduction

In the history of the earth, the cycles of carbon and atmospheric molecular oxygen are closely coupled to the development of life because of the fundamental biochemical reactions occurring during photosynthesis and respiration. However, the role of the two cycles in the earth system, at least on time scales up to 10^6 years, is distinctly different. The major atmospheric branch of the global carbon cycle, i.e., carbon dioxide, constitutes a potent greenhouse gas with the potential to control the climate of the earth. On the other hand, changes in the abundant atmospheric oxygen, at least on time scales less than 10^6 years, are too small to significantly impact the radiative balance of the atmosphere and have been proven very difficult to measure directly. Therefore, up to now the cycle of atmospheric oxygen has not received much attention in global change science. However, because of the tight coupling of the carbon and oxygen cycles, variations in atmospheric oxygen reflect also important processes in the carbon cycle. Now, with recently developed analytical techniques to accurately measure the variations in atmospheric O_2, the global cycle of oxygen as a diagnostic tool has drawn much interest.

Oxygen has three stable oxygen isotopes: ^{16}O, ^{17}O, and ^{18}O. Most biological, chemical, and physical processes in which oxygen is involved are affected by the different masses of the oxygen atoms, leading to fractionation processes that induce varying isotopic ratios of the different oxygen-containing molecules in the earth system. Most of these fractionation processes are relatively well understood and/or empirically measured, which makes observations of the oxygen isotope ratios an additional important diagnostic tool.

Here I briefly review some recent applications of using oxygen and its isotopes as diagnostic tracers of the global carbon cycle. The focus is on atmospheric O_2 and CO_2 and the oxygen isotope ratios in each of these molecules. Because of the vigorous mixing in the atmosphere, spatiotemporal variations of these species in atmospheric air reflect large-scale surface processes. The unraveling and quantification of this information necessitates a combination of a model of atmospheric transport and a model of the surface processes. This review outlines some components of the latter, though it does not intend to be comprehensive. A more extensive review on some of the topics addressed here, albeit with a different point of view, may be found in Keeling (1995).

2. Atmospheric Molecular Oxygen

2.1 Overview

Oxygen is one of the most abundant elements on earth. It is contained in most rocks and it is a fundamental constituent of the water molecule. Oxygen exists in the atmosphere in the form of molecular O_2; with a content of 20.95%, it is the second most abundant atmospheric gas after molecular nitrogen. Atmospheric molecular oxygen is produced during photosynthesis by terrestrial vegetation and marine phytoplankton, and it is consumed during autotrophic respiration by plants and respiration of organic carbon by heterotrophic organisms on land and in the sea. Both of these processes also involve the transformation of carbon from inorganic to organic forms and vice versa; hence they form the fundamental linkage points between the biogeochemical cycles of carbon and oxygen. Since both processes involve water, they also constitute important linkage points between molecular oxygen and the hydrological cycle, which are important for isotopic exchanges as described further below.

Figure 1 shows in the upper two panels a simplified scheme of the natural global cycles of carbon and atmospheric molecular oxygen. The scheme depicts only the fundamental flows between

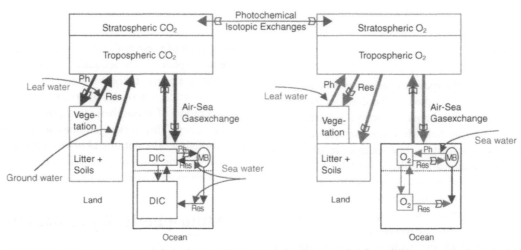

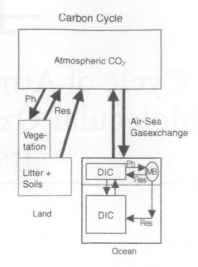

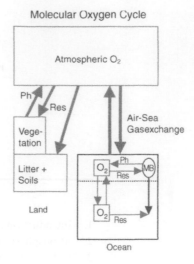

FIGURE 1 Upper panels: simplified scheme of the natural global cycles of carbon (left) and atmospheric molecular oxygen (right). Ph, photosynthesis; Res, respiration; MB, marine biota; DIC, dissolved inorganic carbon (H_2CO_3, HCO_3^-, CO_3^{2-}). Pool sizes are not shown to scale. Lower panels: corresponding schemes of the cycles of the oxygen isotopes in CO_2 (lower left panel) and in O_2 (lower right panel). Dark blue arrows indicate links to oxygen isotopes in the hydrological cycle. Red whiskers on arrows indicate exchanges during which fractionation processes occur. See also color insert.

the atmosphere, the terrestrial biosphere, and the oceans. Minor exchanges with the geosphere (e.g., volcanism, weathering) and by river flows are neglected. Also ignored are the pools and reactions with minor atmospheric-carbon containing constituents: CO, CH_4, hydrocarbons, etc. At first sight, the carbon and oxygen cycles seen almost reciprocal, with O_2 and CO_2 being produced and consumed during photosynthesis and respiration in clearly defined stoichiometric ratios. However, there is a fundamental difference between the two cycles in the oceans. Atmospheric CO_2 is buffered by the large oceanic carbonate system (dissolved inorganic carbon: $DIC = H_2CO_3 + HCO_3^- + CO_3^{2-}$), which comprises more than 50 times the carbon contained in the atmosphere. Depending on the time scale, any perturbation to

atmospheric CO_2 is diluted to a considerable extent by this large oceanic carbon reservoir. On the other hand, molecular oxygen is dissolved in the ocean only in minute amounts; hence perturbations to the oxygen cycle are not damped by exchanges with the ocean. Therefore, the dynamics of the two cycles are quite different, and this forms the basis of the diagnostic approaches described in the sections below.

2.2 Measurement Techniques

In photosynthesis and respiration processes, about 1.1 mol of O_2 is exchanged for 1 mol of CO_2, which implies that the induced atmospheric variations in O_2 are similar in magnitude to those in

CO_2. Hence, to be useful, atmospheric O_2 concentrations have to be measured with an accuracy of at least 0.1 ppmv. Considering the background atmospheric O_2 concentration of about 20%, this implies a measurement sensitivity of better than 10^{-6}. Keeling and Shertz (1992) reported the first accurate measurements of variations in atmospheric O_2 using an interferometric technique. Since these pioneering measurements, O_2 variations have also been measured by means of mass spectroscopy (Bender *et al.*, 1996). Both of these techniques determine the ratio of O_2/N_2 of an air sample relative to a laboratory gas standard. Based on a rough quantitative assessment of the atmospheric N_2 cycle, it is easy to see that atmospheric variations of N_2 are expected to be on the order of 10^{-8} or smaller, implying that changes in the O_2/N_2 ratio primarily reflect changes in O_2.

The long-term maintenance of the constancy of the gas standards constitutes one of the big challenges in oxygen measurement work, because a technique accurate enough to determine the absolute oxygen content of air samples has not existed as yet. Recently, the group of R. Keeling has developed two new approaches to measure continuously O_2/N_2 ratios in atmospheric air, one by ultraviolet spectroscopy (Stephens, 1999) and the other by measuring the paramagnetic susceptibility of oxygen (Manning *et al.*, 1999), both of which also demonstrate the required sensitivity. These new continuous measurement techniques have opened up the possibility of much more extensive global monitoring of O_2/N_2 ratios than has been possible so far.

O_2/N_2 ratios are commonly expressed in "permeg," a unit that is defined as the relative deviation of the measured O_2/N_2 ratio from the standard multiplied by 10^6. Because the atmosphere contains 20.95 vol % O_2, a variation of 1 ppmv of O_2 corresponds to a shift in the O_2/N_2 ratio of 4.773 permeg.

2.3 Global Atmospheric Trends in CO_2 and O_2

The most significant information from oxygen measurements to date has been the separation of the net terrestrial uptake from the oceanic uptake of anthropogenic excess CO_2. The burning of fossil fuels and the emissions from cement production induce an in-

creasing trend in atmospheric CO_2 concentration, and, because the burning of fossil fuels requires oxygen, a decreasing trend in atmospheric O_2. The oceans and the land take up a sizeable fraction of the excess CO_2. But, as discussed above, while terrestrial uptake by photosynthesis involves the generation of O_2, the ocean does not affect the O_2 balance. Hence global budget equations for CO_2 and O_2 may be formulated as

$$\frac{d}{dt} N_{CO_2,a} = Q_{foss} - (S_{ocean} + S_{land}) \tag{1}$$

$$\frac{d}{dt} N_{O_2,a} = f_{foss} Q_{foss} - f_{land} S_{land} + Q_{ocean}. \tag{2}$$

Here $N_{CO_2,a}$ and $N_{O_2,a}$ a are the global atmospheric contents of CO_2 and O_2, while Q_{foss} denotes the carbon emissions from the burning of fossil fuels and from cement production. S_{ocean} and S_{land} denote the CO_2 sinks on land and in the ocean. f_{foss} is the stoichiometric factor for the industrial emissions (mol of O_2 consumed per mol of CO_2 generated) and f_{land} is the stoichiometric factor for terrestrial carbon uptake. The term Q_{ocean} denotes potential outgassing of dissolved oxygen from the ocean, for example, that one induced by global warming. The size of this term is believed to be small although not entirely negligible.

The two equations are readily solved for the two unknowns S_{land} and S_{ocean}. Since they constitute two linear equations in two unknowns, their solution can also be represented in graphical form (Keeling *et al.*, 1996; see also Fig. 2 below). Direct atmospheric measurements of O_2/N_2 started in 1989 (Keeling and Shertz, 1992) in La Jolla and at several stations in the early 1990s (Bender *et al.*, 1996; Keeling *et al.*, 1996; Battle *et al.*, 2000). Using analyses of O_2 in archived air from Cape Grim, Langenfelds *et al.* (1999) were able to extend the record back to 1979. Oxygen measurements have also been reported from air extracted from Antarctic firn, dating back to the late 1970s (Battle *et al.*, 1996).

An update of the global budget representing 1990–1997 is presented in Table 1. The atmospheric O_2 trends averaged over

TABLE 1 Global O_2 and CO_2 Budgets Averaged over 1990–1997: Numerical Values of Terms in Eqs. (1) and (2)*

Variable	Value	Standard Error	Variance Fraction to Error of Land Uptake (%)	Variance Fraction to Error of Ocean Uptake
$\frac{d}{dt} N_{O_2,a}$	−15.6 permeg year^{-1}	± 0.87 permeg year^{-1}	44	66
$\frac{d}{dt} N_{CO_2,a}$	1.34 ppmv year^{-1}	± 0.02 ppmv year^{-1}	0	4
Q_{foss}	6.26 GtC year^{-1}	± 0.38 GtC year^{-1}	38	3
f_{foss}	−1.39	± 0.04	8	12
f_{land}	−1.1	± 0.05	1	1
Q_{ocean}	0.3 permeg year^{-1}	± 0.6 permeg year^{-1}	9	14
S_{ocean}	1.94 GtC year^{-1}	± 0.65 GtC year^{-1}		
S_{land}	1.47 GtC year^{-1}	± 0.80 GtC year^{-1}		

*For the explanation of the last two columns see text.

this period have been determined by merging the data from Alert, Canada and La Jolla, California (Keeling *et al.*, 1996) with observations from Point Barrow, Alaska and Cape Grim, Tasmania (Battle *et al.*, 2000). In being merged, the records, were first deseasonalized by fitting functions consisting of a seasonal cycle represented by four harmonics and a linear trend to the individual records. Subsequently, the records were merged based on the offsets determined by fitting linear trends to overlapping parts of the records. Then, annual averages overlapping by 6 months were formed. The global inventory change over 1990–1997 was determined from the difference between the annual averages of 1997 and 1990. Since observations in the early record were somewhat sparse, the value for 1990 was computed as the average of the five annual means centered at 1989.5, 1990.0, 1990.5, 1991.0, and 1991.5. The corresponding global average CO_2 trend was determined with a similar procedure from monthly CO_2 observations of the Point Barrow and the Cape Grim monitoring stations reported by the Climate and Monitoring Diagnostics Laboratory of the NOAA (Conway *et al.*, 1994).

Table 1 includes also the error analysis and the fraction of the error variance of the land and ocean uptake estimates generated by the uncertainties in the individual terms. Interestingly, the uncertainty in the land uptake is dominated about equally by errors in the fossil fuel emissions and the atmospheric O_2/N_2 trend, while the uncertainty of the ocean uptake is dominated by the error of the global O_2/N_2 trend only. This behavior results from canceling effects in the solutions of Eqs. (1) and (2) for the two sink terms. Thus, although the oxygen budget [Eq. (2)] in principle only determines the land uptake term, it is the fact that we know the atmospheric CO_2 trend very well, which tightly couples the two equations and leads to a smaller overall error estimate of the ocean uptake term. It is readily seen that a further reduction of the error in the global O_2/N_2 trend will primarily reduce the error of the ocean uptake estimate. Table 1 also shows that the errors contributed by uncertainties in the stoichiometric factors (f_{land} and f_{foss}) are at present relatively minor.

There is also a significant uncertainty induced by the largely unknown ocean outgassing term (Q_{ocean}). The value 0.3 permeg year^{-1} adopted for this term in Table 1 reflects an ocean warming rate of the order of 1 W m^{-2} as inferred from oceanographic data (Levitus *et al.*, 2000) and from global warming simulations (Roeckner *et al.*, 1999). The conversion of warming to outgassing of O_2 is computed with a ratio of approximately 1.5×10^{-9} mol O_2 per J of heat uptake (see Keeling *et al.*, 1993). Thereby the effect on the atmospheric O_2/N_2 ratio of the corresponding outgassing of N_2 has also to be taken into account. Although the corresponding conversion factor of 2.2×10^{-9} mol N_2 per J of heat uptake is slightly larger than that for O_2, molecular nitrogen is four times more abundant in the atmosphere, hence the effect of thermally driven N_2 outgassing on the atmospheric O_2/N_2 ratio is only half as large as for O_2 and of opposite sign. There is a considerable uncertainty in the value for the ocean outgassing component. Furthermore, a warming ocean may also affect the natural cycling of O_2 between the atmosphere and the sea: the increased

stratification of the ocean may prevent deeper, oxygen-undersaturated waters to come into contact with the atmosphere. This may result in an enhanced O_2 outgassing as compared to the purely thermal outgassing effect described above. The magnitude of this enhancement, however, is very difficult to assess. In the present analysis we only include the direct thermal effect based on an ocean warming rate of 1 W m^{-2} and include the potential enhancement due to increases in stratification in the uncertainty estimate of this term. A graphical representation of the global budget equations (1) and (2) in the form of an arrow diagram is shown in Figure 2.

2.4 Seasonal Cycles and Mean Annual Spatial Gradients

The seasonal signal in atmospheric CO_2 in the northern hemisphere is mostly generated by the terrestrial biosphere (Heimann *et al.*, 1986; 1989; 1998; Fung *et al.*, 1987; Knorr and Heimann, 1995), oceanic seasonal fluxes being largely buffered by the ocean chemistry and the slow sea–air gas exchange of CO_2. This is not true for O_2 which in the Northern Hemisphere, at least in oceanic

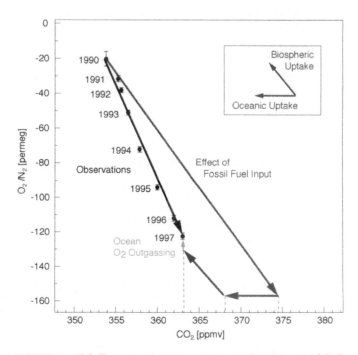

FIGURE 2 Globally averaged O_2/N_2 ratio (vertical axis) versus globally averaged CO_2 mixing ratio (horizontal axis). The annually averaged observations have been determined from the station records as described in the text. The black arrow shows the observed trend for 1990–1997 determined by the fitting procedure as described in the text. The red arrows depicts the expected change due to the fossil emissions during 1990–1997. The effects of the ocean and the land biosphere is shown with the blue and the green arrow, whereby their slopes are determined by their respective O_2 versus CO_2 contributions (see inset). The purple vertical arrow reflects an estimate of the oceanic O_2 outgassing induced by ocean warming. See also color insert.

areas, exhibits a seasonal signal about twice as large as the corresponding seasonal cycle in atmospheric CO_2. Hence the magnitude of the seasonal oceanic component of O_2 is similar to that of the terrestrial component. In the Southern Hemisphere, with small land areas the oceanic O_2 signal dominates largely the seasonal signal (Keeling and Shertz, 1992).

Since terrestrial exchanges in O_2 occur in relatively fixed stoichiometric ratios (Severinghaus, 1995), it is the oceanic O_2 signal that is of primary interest for monitoring in the atmosphere. The terrestrial component can be removed from the atmospheric measurements conveniently by introducing of the tracer atmospheric potential oxygen (APO), conveniently defined as the atmospheric oxygen signal that would indeed result if all atmospheric carbon were oxidized with the stochiometric constant of terrestrial biospheric carbon (Stephens *et al.*, 1998):

$$APO = \delta O_2 + f_{land} ([CO_2] + 2 [CH_4] + 0.5 [CO]). \quad (3)$$

Here δO_2 denotes the observed deviation of the atmospheric O_2 concentration from a standard. The atmospheric tracer APO is dominated primarily by oceanic gas exchanges in addition to a relatively small contribution from fossil fuel not accounted for by the terrestrial stoichiometric factor (i.e., the fossil fuel component scaled by the factor $f_{foss} - f_{land}$). Observations of the seasonal variation of APO in conjuction with surface-water oxygen measurements have been used to constrain the large-scale magnitude of the air–sea gas exchange coefficient (Keeling *et al.*, 1998) and of marine productivity (Six and Maier-Reimer, 1996; Balkanski *et al.*, 1999). Mean annual gradients of APO have also been shown to provide powerful constraints on biogeochemical air–sea fluxes computed by ocean-circulation models with an embedded ocean carbon cycle (Stephens *et al.*, 1998).

2.5 Continental Dilution of the Oceanic O_2/N_2 Signal

Observations of the atmospheric O_2/N_2 ratio will also be an important tool to constrain the zonal transport of air between the continents and the oceans within the Northern Hemisphere. Recently, atmospheric measurements of CO_2 from the global monitoring networks have been used to discriminate net carbon balances of different continental-scale regions by inversion studies (e.g., Rayner *et al.*, 1999; Fan *et al.*, 1998; Kaminski *et al.*, 1999; Heimann and Kaminski, 1999; see also the contribution by Rayner in this volume). While these approaches yield relatively robust estimates for the carbon balances of the whole northern and southern extra-tropics and the tropics, credible estimates for smaller regions such as Europe, Asia, or North America are difficult to establish. The reasons for this difficulty can be traced to the presently inadequate monitoring network (~ 100 monitoring stations, mostly located in oceanic areas) and to difficulties in faithfully representing the flushing of air over the continents and the oceans in the atmospheric-transport models employed in the inversion studies. Of particular concern are "rectifying" effects

(Heimann *et al.*, 1986; Keeling *et al.*, 1989; Denning *et al.*, 1995) generated by the strong seasonally varying surface sources, such as CO_2. Because of the temporal covariance between seasonal atmospheric transport (seasonal changes of the vertical stability over the continents, monsoon circulations, seasonal ITCZ movements, etc.) and a seasonal source at the surface, mean annual atmospheric concentration patterns are generated. These patterns are in principle indistinguishable from the patterns generated by net annual sources and sinks. Hence if one wants to invert atmospheric concentration patterns in terms of net surface sources and sinks, the "rectifying" patterns have to be correctly represented in the employed transport models. However, the magnitude of these effects is largely unknown; different atmospheric models yield dramatically different "rectifying" patterns when forced with the same purely seasonal surface source (Law *et al.*, 1996). Unfortunately, there does not exist a direct atmospheric tracer that can be used to evaluate the rectifying effects simulated by the various models.

As described in the previous section, the tracer APO has no significant seasonal sources over the continents, but is mostly generated by the seasonal oceanic O_2 exchanges. Hence the dilution of this signal into the interiors of the continents in principle provides a test for the transport representation in the models. Of course, the usefulness of this test depends on continental O_2 monitoring stations, which currently are not existing, but are planned for the near future.

As an example, Figure 3 shows the modeled amplitude of the seasonal signal of terrestrial CO_2 and of the oceanic O_2 (oceanic component of APO) in the lower planetary boundary layer (at about 380 m above the surface) simulated with the global TM3 transport model (updated from Heimann, 1995) by using a horizontal resolution of approximately. 4° latitude by 5° longitude and 19 layers in the vertical dimension. The model predictions of the Hamburg model of the oceanic carbon cycle (HAMOCC3; Maier-Reimer, 1993) with the phytoplankton–zooplankton model of Six and Maier-Reimer (1996) have been used to prescribe the monthly oceanic O_2 exchanges (atmospheric simulation in the upper panel). The simple diagnostic biosphere model (SDBM) of Knorr and Heimann (1995) has been used to prescribe the seasonal terrestrial sources in the CO_2 simulation (lower panel). The significant zonal structures of the oceanic O_2 amplitude field and its dilution over the continents is a feature that remains to be verified by atmospheric measurements.

Detecting the dilution of the oceanic seasonal cycle signal in O_2 over the Northern Hemisphere continents is relatively straightforward and does not involve a detailed analysis and determination of the seasonal signal. All that has to be monitored is the O_2 versus CO_2 relationship over the course of one year. Since the seasonal signals in O_2 from the ocean and the terrestrial biosphere are closely in phase, this O_2-CO_2 relationship is expected to fall approximately on one line, with, however, a slope determined by the magnitude of the oceanic signal. The principle is shown in Figure 4. If there were no oceanic contribution, the slope would merely reflect the biosphere stoichiometric factor (-1.1). A larger slope indicates a significant oceanic contribution. Figure 5 shows the

a

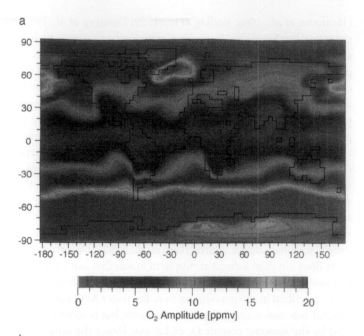

O₂ Amplitude [ppmv]

b

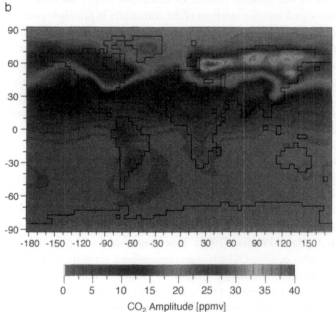

CO₂ Amplitude [ppmv]

FIGURE 3 Amplitude of the seasonal signal in the lower planetary boundary layer (at approximately 380 m above the surface) generated by the terrestrial biosphere in the CO_2 mixing ratio (lower panel) and by oceanic exchanges in the atmospheric O_2/N_2 ratio (upper panel) as simulated with the TM3 atmospheric transport model. See text for the model setup description. See also color insert.

spatial variation of the slope between O_2 and CO_2 in the lower troposphere resulting from the model simulations described above. The color code has been chosen, such that only the variations in the Northern Hemisphere are highlighted; i.e., where the aforementioned relationship between the terrestrial and oceanic seasonal cycle is expected to hold. This is no longer the case in the

tropics and in the southern hemisphere, where a more complex relationship exists between O_2 and CO_2. It is seen that the slope of the relationship over the Atlantic and Pacific oceans reaches values above 2. Over the interior of the continent this ratio is progressively reduced to values of 1.3–1.4. It is expected that this pattern will vary considerably between different models, and should therefore provide a critical check on the realism of the modeled transport.

3. Stable Isotopes of Oxygen

3.1 Overview

The three stable oxygen isotopes, ^{16}O, ^{17}O, and ^{18}O in both, CO_2 and O_2, constitute important tracers of the global carbon and oxygen cycles. Besides revealing crucial information in local process studies (see Lloyd, this volume), they may also be observed and modeled on the global scale. Unlike carbon or oxygen, however, these isotopes are not conserved in the carbon or oxygen cycle, but are constantly exchanged at a few critical connection points with the hydrological cycle (which contains much more oxygen than is present in CO_2 and O_2). Figure 1 shows the two cycles schematically in the lower two panels and indicates the locations where exchanges with water determine the isotopic composition of the flows of CO_2 and O_2 (blue arrows). On land these are:

1. The leaf water, which determines the isotopic composition of the O_2 generated by photosynthesis and the CO_2 generated by autotrophic respiration.
2. The soil/groundwater determining the isotopic composition of CO_2 generated by heterotrophic respiration.

In the oceans, the isotopic signature of seawater determines the O_2 produced by photosynthesis and the oxygen isotopic composition of DIC generated during remineralization of organic material and carbonate.

In addition, fractionation processes during phase transitions between the major carbon and oxygen reservoirs modify the isotopic composition of atmospheric CO_2 and O_2. These fractionation steps are indicated in Figure 6 with the red whisker symbols on the exchange flow arrows. In the stratosphere, there exists an exchange link between the oxygen isotope cycles of CO_2 and O_2 generated by photochemical processes (Bender et al., 1994). While this link is of minor importance for the atmospheric budgets of the oxygen isotopes, it is essential for the isotopic ^{17}O anomaly (see Section 3.4 below).

Both, (a) changes in the isotopic composition of the hydrological cycle due to, e.g., climate variations and (b) changes in the carbon and oxygen fluxes between the reservoirs therefore can induce changes in the oxygen isotopic composition of atmospheric CO_2 and O_2. Dynamically, after a perturbation a new atmospheric steady state of the oxygen isotope ratios establishes within a few years in the case of $^{18}O/^{16}O$ in CO_2 and within about 1200 years in

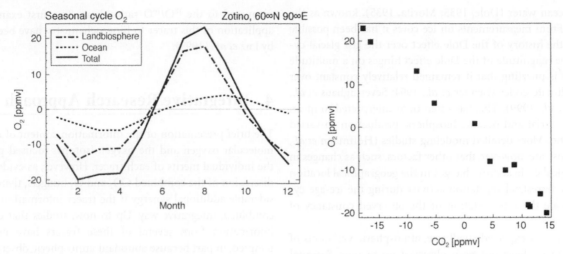

FIGURE 4 Relationship between the seasonal cycles of CO_2 and O_2 in the interior of the Eurasian continent (at Zotino, 60°N, 90°E) within the planetary boundary layer. The left diagram shows the seasonal signal components in O_2 induced from the terrestrial and oceanic seasonal sources; the right-hand panel shows the modeled relation between the seasonal cycles of O_2 and CO_2 (monthly averages).

the case of $^{18}O/^{16}O$ in O_2 (Bender *et al.*, 1994). This difference arises from the different atmospheric turnover times of CO_2 and O_2. Observations of spatiotemporal isotopic variations in the atmosphere may be related to either one or a combination of these two principal driving factors. Clearly, there exists a remarkable correspondence between the cycles of the oxygen isotopes and the global cycles of carbon and oxygen, which becomes evident on comparing the upper and lower panels in Figure 1.

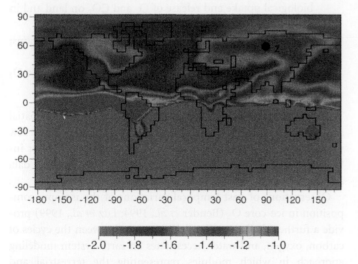

FIGURE 5. Slope of the modeled relationship between the seasonal cycles of O_2 and CO_2 in the planetary boundary layer. The color scale has been selected such that values in the Northern Hemisphere are highlighted, where the relationship between the seasonal signals of the two tracers is essentially linear. The black dot indicates the location of the Zotino (60°N, 90°E) station displayed in Figure 4. See also color insert.

3.2 ^{18}O in CO_2

The $^{18}O/^{16}O$ ratio of CO_2 is primarily controlled by exchanges with the terrestrial carbon systems. Exchanges with the surface ocean are controlled by temperature-dependent fractionation during gas exchange and the isotopic signature of surface waters, which is highly correlated with the salinity (Craig and Gordon, 1965). Marine biological processes do not significantly affect this signature.

Of interest are the atmospheric seasonal cycles, the annual mean gradients, and interannual variations as recorded at the global monitoring networks. If the $^{18}O/^{16}O$ signature of surface waters is known (i.e., ground water and the evaporatively enriched leaf water), then both, the observed seasonal cycle and the meridional gradient of $^{18}O/^{16}O$ in CO_2 provide a powerful tool to constrain on regional and global scales the gross photosynthesis (GPP) of the terrestrial biosphere. A forward modeling study to demonstrate this has been performed by Ciais *et al.* (1997a, b). The atmospheric observations may also be used in an inverse modeling approach to constrain GPP and its driving factors on a regional basis (Peylin *et al.*, 1999). Interannual variations of $^{18}O/^{16}O$ in CO_2 have been documented in observations, but are relatively difficult to interpret, as these are controlled to a large extent by changes in the driving hydrological cycle over land.

3.3 The Dole Effect

Terrestrial and oceanic biospheric processes drive with similar relative weight the $^{18}O/^{16}O$ ratio of atmospheric O_2. The most important effect is the fractionation occurring during consumption of O_2 by heterotrophic respiration. This fractionation leads to a global atmospheric $^{18}O/^{16}O$ isotope ratio enriched by about 23.5%

relative to ocean water (Dole, 1935; Morita, 1935), known as the Dole effect. From measurements on ice cores it has been possible to establish the history of the Dole effect over the last glacial cycles. Since the magnitude of the Dole effect hinges on a multitude of factors, it is puzzling that it remained relatively constant over the glacial climate cycles (Bender *et al.*, 1994; Severinghaus *et al.*, 1998; Petit *et al.*, 1999). This has often been interpreted to mean that, the terrestrial and oceanic biospheric production co-varied in a fixed ratio. More detailed modeling studies (Hoffmann *et al.*, 1998) demonstrate, however, that other factors, such as changes in the hydrological cycle and/or changes in the geographical location of the major terrestrial vegetation activity during the ice-age cycles, complicate the interpretation of the observed constancy of the Dole effect.

Because of their expected small size, atmospheric variations of the $^{18}O/^{16}O$ in O_2 have not been observed up to now. Seasonal variations and mean annual gradients are expected to be on the order of a few permeg (Seibt, 1997). Just as the combination of CO_2 and O_2 provides a very powerful constraint on the carbon cycle, a combination with measurements of the oxygen isotope ratios in both CO_2 and O_2 would constitute two further powerful constraints.

3.4 ^{17}O

Oxygen consists of three stable isotopes: ^{16}O, ^{17}O and ^{18}O. For most applications only the ratio between the more abundant isotopes, ^{18}O and ^{16}O, is measured. If all fractionation processes in the environmental system were purely mass-dependent, measurements of $^{17}O/^{16}O$, would be redundant, as they could be predicted from the $^{18}O/^{16}O$ ratio. However, it has recently been observed that photochemical exchange between O_2, O_3, and CO_2 in the stratosphere involves mass-independent fractionation among the oxygen isotopes (1990; Thiemens *et al.*, 1995a, b Thiemens, 1999). Thereby O_2 becomes anomalously depleted, while CO_2 becomes anomalously enriched. Because of this, measurements of $^{17}O/^{16}O$ in atmospheric O_2 and/or CO_2 provide an independent piece of information. Conveniently, one may define an ^{17}O anomaly tracer ($\Delta^{17}O$) (Thiemens *et al.*, 1995b)

$$\Delta^{17}O = \delta^{17}O - 0.52 \times \delta^{18}O, \tag{4}$$

where the symbol δ denotes an isotope ratio expressed as a deviation from a standard in units of ‰. The scaling factor in the definition of this anomaly tracer has been chosen so that it captures the mass-dependent fractionation. Thus, wherever at a phase transition a mass-dependent oxygen isotope fractionation occurred, $\Delta^{17}O$ would remain constant. Thus, $\Delta^{17}O$ constitutes a tracer with pathways identical to $^{18}O/^{16}O$ in the cycles of CO_2 and O_2 except that it is not fractionated. It is constantly being generated in the stratosphere and "destroyed" at the exchanges with the hydrological cycle in the land biosphere and the ocean. Because of these properties, measurements of this tracer are easier to interpret as

compared to the $^{18}O/^{16}O$ ratio alone. Two first examples of the application of this tracer to biogeochemistry have been presented by Luz *et al.*, (1999).

4. Integrative Research Approach

The brief presentation of the information content of atmospheric molecular oxygen and the oxygen isotopes focused primarily on the individual merits of each tracer. However, as evident from the discussion of the combined CO_2 and O_2 budgets, there exists considerable additional synergy if the tracer information is used in a combined, integrative way. Up to now, studies that combine the information from several of these tracers have not been attempted, in part because abundant atmospheric observations with sufficient measurement accuracy are not yet readily available. In particular, the oxygen isotope ratio measurements in atmospheric O_2 are only now becoming precise enough to reveal spatial and temporal patterns in the present-day atmosphere. If they were so precise, however, the potential of an integrative approach would be substantial: The tracers discussed above provide a total of six independent constraints (concentration of CO_2 and O_2, $^{18}O/^{16}O$ in CO_2 and O_2, and $^{17}O/^{16}O$ in CO_2 and O_2), which may be used in a combined way to quantitatively deduce the six major carbon fluxes of interest including photosynthesis, respiration on land and in the sea, together with the gross air–sea gas exchange fluxes. Of course, the application of this approach requires the knowledge of:

1. the processes occurring at linkage points between the oxygen cycle and the carbon cycle, (e.g., stoichiometry between biological uptake and release of O_2 and CO_2 on land and in the sea;

2. the fractionation processes involved at the phase transitions; and

3. the isotopic composition of the water that is imparted to O_2 and CO_2 formed during photosynthesis or respiration.

This information must be available on the temporal and spatial scales of interest. It remains a research challenge for the next few years to develop a modeling framework into which the tracer information can be integrated, possibly by means of advanced data assimilation methods.

Observations of past temporal variations of the isotopic composition in ice-core O_2 (Bender *et al.*, 1994; Luz *et al.*, 1999) provide a further challenge. The tight coupling between the cycles of carbon, oxygen, and water necessitates an earth-system modeling approach in which modules representing the terrestrial and oceanic carbon cycle are coupled into a global climate model that would also include a description of the oxygen isotopes in the hydrological cycle. Such models are currently being developed. For such a model framework the oxygen isotopes in O_2 will provide a promising model-validation tool in the future (Hoffmann *et al.*, 1998).

References

Balkanski, Y., Monfray, P., Battle, M., and Heimann, M. (1999). Ocean primary production derived from satellite data: An evaluation with atmospheric oxygen measurements. *Global Biogeochem. Cycles* 13, 257–271.

Battle, M., Bender, M., Sowers, T., Tans, P. P., Butler, J. H., Elkins, J. W., Ellis, J. T., Conway, T., Zhang, N., Lang, P., and Clarke, A. D. (1996). Atmospheric gas concentrations over the past century measured in air from firn at the South Pole. *Nature* 383, 231–235.

Battle, M., Bender, M. L., Tans, P. P., White, J. W. C., Ellis, J. T., Conway, T., and Francey, R. J. (2000). Global carbon sinks and their variability inferred from atmospheric O$_2$ and δ^{13}C. *Science* 287, 2467–2470.

Bender, M., Ellis, T., Tans, P., Francey, R., and Lowe, D. (1996). Variability in the O$_2$/N$_2$ ratio of southern hemisphere air, 1991–1994—Implications for the carbon cycle. *Global Biogeochem. Cycles* 10, 9–21.

Bender, M., Sowers, T., and Labeyrie, L. (1994). The Dole effect and its variations during the last 130,000 years as measured in the Vostok ice core. *Global Biogeochem. Cycles* 8, 363–376.

Ciais, P., Denning, A. S., Tans, P. P., Berry, J. A., Randall, D. A., Collatz, G. J., Sellers P. J., White, J. W. C., Trolier, M., Meijer, H. A. J., Francey, R. J., Monfray, P., and Heimann, M. (1997a). A three-dimensional synthesis study of δ^{18}O in atmospheric CO$_2$. 1. Surface fluxes. *J. Geophys. Res.–Atmos.* 102, 5857–5872.

Ciais, P., Tans P. P., Denning, A. S., Francey, R. J., Trolier, M., Meijer, H. A. J., White J. W. C., Berry, J. A., Randall, D. A., Collatz, G. J., Sellers, P. J., Monfray, P., and Heimann M. (1997b). A three-dimensional synthesis study of δ^{18}O in atmospheric CO$_2$. 2. Simulations with the TM2 transport model. *J. Geophys. Res.–Atmos.* 102, 5873–5883.

Conway, T. J., Tans, P. P., Waterman, L. S., and Thoning, K. W. (1994). Evidence for interannual variability of the carbon cycle from the national oceanic and atmospheric administration climate monitoring and diagnostics laboratory global air sampling network. *J. Geophys. Res.–Atmos.* 99, 22,831–22,855.

Craig, H. and Gordon, A. (1965). Deuterium and oxygen-18 variation in the ocean and marine atmosphere. *Proc. Con. Stable Isotopes in Oceanography Studies of Paleotemperature*, 9–130.

Denning, A. S., Fung, I. Y., and Randall, D. (1995). Latitudinal gradient of atmospheric CO$_2$ due to seasonal exchange with land biota. *Nature* 376, 240–243.

Dole, M. (1935). The relative atomic weight of oxygen in water and air. *J. Am. Chem. Soc.* 57, 2731.

Fan, S., Gloor, M., Mahlman, J., Pacala, S., Sarmiento, J., Takahashi, T., and Tans, P. (1998). A large terrestrial carbon sink in North America implied by atmospheric and oceanic carbon dioxide data and models. *Science* 282, 442–446.

Fung, I. Y., Tucker, C. J., and Prentice, K. C. (1987). Application of advanced very high-resolution radiometer vegetation index to study atmosphere-biosphere exchange of CO$_2$. *J. Geophys. Res.–Atmos.* 92, 2999–3015.

Heimann, M. (1995). The TM2 tracer model, model description and user manual, pp. 47. German Climate Computing Center (DKRZ).

Heimann, M., Esser, G., Haxeltine, A., Kaduk, J., Kicklighter, D. W., Knorr, W., Kohlmaier, G. H., McGuire, A. D., Melillo, J., Moore, B., Otto, R. D., Prentice, I. C., Sauf, W., Schloss, A., Sitch, S., Wittenberg, U., and Wurth, G. (1998). Evaluation of terrestrial carbon cycle models through simulations of the seasonal cycle of atmospheric CO$_2$: First results of a model intercomparison study. *Global Biogeochem. Cycles* 12, 1–24.

Heimann, M. and Kaminski, T. (1999). Inverse modelling approaches to infer surface trace gas fluxes from observed atmospheric mixing ratios. In "Approaches to Scaling of Trace Gas Fluxes in Ecosystems (A. F. Bowman, E.d.), pp. 277–295. Elsevier, Amsterdam.

Heimann, M., Keeling, C. D., and Fung, I. Y. (1986). Simulating the atmospheric carbon dioxide distribution with a three-dimensional tracer model. In "The Changing Carbon Cycle: A Global Analysis." (J. Trabalka and D. E. Reichle, Eds.), pp. 16–49. Springer Verlag.

Heimann, M., Keeling, C. D., and Tucker, C. J. (1989). A three dimensional model of atmospheric CO$_2$ transport based on observed winds: 3. Seasonal cycle and synoptic time scale variations. In "Aspects of Climate Variability in the Pacific and the Western Americas." Vol. 55 (D. H. Peterson, Ed.), pp. 277–303. American Geophysical Union.

Hoffmann G., Werner M., and Heimann M. (1998). Water isotope module of the ECHAM atmospheric general circulation model—a study on time scales from days to several years. *J. Geophys. Res.–Atmos.* 103, 16871–16896.

Kaminski, T., Heimann, M., and Giering, R. (1999). A coarse grid three-dimensional global inverse model of the atmospheric transport 2. Inversion of the transport of CO$_2$ in the 1980s. *J. Geophys. Res.–Atmos.* 104, 18555–18581.

Keeling, R. F. (1995). The atmospheric oxygen cycle—The oxygen isotopes of atmospheric CO$_2$ and O$_2$ and the O$_2$/N$_2$ ratio. *Rev. Geophys.* 33, 1253–1262.

Keeling, R. F., Najjar, R. P., Bender, M. L. and Tans, P. P. (1993). What atmospheric oxygen measurements can tell us about the global carbon cycle. *Global Biogeochem Cycles* 7, 37–67.

Keeling, R. F., Piper, S. C., and Heimann, M. (1996). Global and hemispheric CO$_2$ sinks deduced from changes in atmospheric O$_2$ concentration. *Nature* 381, 218–221.

Keeling, C. D., Piper, S. C., and Heimann, M. (1989). A three dimensional model of atmospheric CO$_2$ transport based on observed winds 4. Mean annual gradients and interannual variations. In "Aspects of Climate Variability in the Pacific and the Western Americas." Vol. 55 (D. H. Peterson, Ed.), pp. 305–363. American Geophysical Union, Washington D.C.

Keeling, R. F. and Shertz S. R. (1992). Seasonal and interannual variations in atmospheric oxygen and implications for the global carbon cycle. *Nature* 358, 723–727.

Keeling, R. F., Stephens, B. B., Najjar, R. G., Doney, S. C., Archer, D., and Heimann, M. (1998). Seasonal variations in the atmospheric O$_2$/N$_2$ ratio in relation to the kinetics of air–sea gas exchange. *Global Biogeochem. Cycles* 12, 141–163.

Knorr, W. and Heimann, M. (1995). Impact of drought stress and other factors on seasonal land biosphere CO$_2$ exchange studied through an atmospheric tracer transport model. *Tellus Ser. B–Chem. Phys. Meteorol.* 47, 471–489.

Langenfelds, R. L., Francey, R. J., Steele, L. P., Battle, M., Keeling, R. F., and Budd, W. F. (1999). Partitioning of the global fossil CO$_2$ sink using a 19-year trend in atmospheric O$_2$. *Geophys. Res. Lett.* 26, 1897–1900.

Law, R. M., Rayner, P. J., Denning, A. S., Erickson, D., Fung, I. Y., Heimann, M., Piper, S. C., Ramonet, M., Taguchi, S., Taylor, J. A., Trudinger, C. M., and Watterson, I. G. (1996). Variations in modeled atmospheric transport of carbon dioxide and the consequences for CO$_2$ inversions. *Global Biogeochem. Cycles* 10, 783–796.

Levitus, S., Antonov, J. I., Boyer, T. P., and Stephens, C. (2000). Warming of the world ocean. *Science* 287, 2225–2229.

Luz, B., Barkan, E., Bender, M. L., Thiemens, M. H., and Boering, K. A. (1999). Triple-isotope composition of atmospheric oxygen as a tracer of biosphere productivity. *Nature* 400, 547–550.

Maier-Reimer, E. (1993). Geochemical cycles in an ocean general circulation model. Preindustrial tracer distributions. *Global Biogeochem. Cycles* 7, 645–678.

Manning, A. C., Keeling, R. F., and Severinghaus, J. P. (1999). Precise atmospheric oxygen measurements with a paramagnetic oxygen analyzer. *Global Biogeochem. Cycles* **13,** 1107–1115.

Morita, N. (1935). The increased density of air oxygen relative to water oxygen. *J. Chem. Soc. Jpn* **56,** 1291.

Petit, J. R., Jouzel, J., Raynaud, D., Barkov, N. I., Barnola, J. M., Basile, I., Bender, M., Chappellaz, J., Davis, M., Delaygue, G., Delmotte, M., Kotlyakov, V. M., Legrand, M., Lipenkov, V. Y., Lorius, C., Pepin, L., Ritz, C., Saltzman, E., and Stievenard, M. (1999). Climate and atmospheric history of the past 420,000 years from the Vostok ice core, Antarctica. *Nature* **399,** 429–436.

Peylin, P., Ciais, P., Denning, A. S., Tans, P. P., Berry, J. A., and White, J. W. C. (1999). A 3-dimensional study of $\delta^{18}O$ in atmospheric CO_2: Contribution of different land ecosystems. *Tellus Ser. B–Chem. Phys. Meteorol* **51,** 642–667.

Rayner, P. J., Enting, I. G., Francey, R. J., and Langenfelds, R. (1999). Reconstructing the recent carbon cycle from atmospheric CO_2, $\delta^{13}C$ and O_2/N_2 observations. *Tellus Ser. B–Chem. Phys. Meteorol.* **51,** 213–232.

Roeckner, E., Bengtsson, L., Feichter, J., Lelieveld, J., and Rodhe, H. (1999). Transient climate change simulations with a coupled atmosphere–ocean GCM including the tropospheric sulfur cycle. *J. Climate* **12,** 3004–3032.

Seibt, U. (1997). "Simulation der $^{18}O/^{16}O$-Zusammensetzung von atmosphärischem Sauerstoff". Diploma Thesis. University of Tübingen, Germany.

Severinghaus, J. (1995). "Studies of the Terrestrial O_2 and Carbon Cycles in Sand Dune Gases and in Biosphere 2." Thesis. Columbia University, Palisades, N.Y.

Severinghaus, J. P., Sowers, T., Brook, E. J., Alley, R. B., and Bender, M. L. (1998). Timing of abrupt climate change at the end of the Younger Dryas interval from thermally fractionated gases in polar ice. *Nature* **391,** 141–146.

Six, K. D. and Maier-Reimer, E. (1996). Effects of plankton dynamics on seasonal carbon fluxes in an ocean general circulation model. *Global Biogeochem. Cycles* **10,** 559–583.

Stephens, B. B. (1999). "Field-Based Atmospheric Oxygen Measurements and the Ocean Carbon Cycle." Thesis, University of California, San Diego, CA.

Stephens, B. B., Keeling, R. F., Heimann, M., Six, K. D., Murnane, R., and Caldeira, K. (1998). Testing global ocean carbon cycle models using measurements of atmospheric O_2 and CO_2 concentration. *Global Biogeochem. Cycles* **12,** 213–230.

Thiemens, M. H., Jackson, T. L., and Brenninkmeijer, C. A. M. (1995a). Observation of a mass independent oxygen isotopic composition in terrestrial stratospheric CO_2, the link to ozone chemistry, and the possible occurrence in the martian atmosphere. *Geophy. Res. Lett.* **22,** 255–257.

Thiemens, M. H., Jackson, T., Zipf, E. C., Erdman, P. W., and Vanegmond, C. (1995b). Carbon dioxide and oxygen isotope anomalies in the mesosphere and stratosphere. *Science* **270,** 969–972.

Thiemens, M. H. (1999). Atmosphere science—mass-independent isotope effects in planetary atmospheres and the early solar system [review]. *Science* **283,** 341–345.

1.19

Constraining the Global Carbon Budget from Global to Regional Scales — The Measurement Challenge

R. J. Francey,
P. J. Rayner, and
C. E. Allison

CSIRO Atmospheric Research,
Aspendale, Victoria and
Cooperative Research,
Centre for Southern
Hemisphere Meteorology,
Clayton,
Victoria, Australia

1. Introduction

The long lifetime and rapid mixing of CO_2 in the atmosphere provide a large-scale integration of surface fluxes, while, with sufficient measurement precision, signatures of individual surface source or sink regions can still be detected. The three-dimensional Bayesian synthesis inversion technique was introduced into global carbon cycle modeling by Enting *et al.* (1993; 1995). Measurements of atmospheric CO_2 mixing ratios and stable carbon isotope ratios from globally distributed sampling sites for selected years were interpreted using an atmospheric transport model to determine regional sources and sinks of atmospheric carbon. The inversion process is inherently unstable, and requires additional constraints, in this case the spatial distribution of known sources and sinks, and prior estimates of the surface fluxes. When those prior estimates are independently and rigorously determined, the Bayesian technique provides a promising framework within which the various studies of regional carbon fluxes (and associated process information) can be reconciled with changes in the global atmospheric carbon content. A particular advantage is the potential for systematic treatment of uncertainty in the various components of the inversion. Application of the Bayesian technique to global carbon budgeting is still in the early stages of development.

2. Present Status of Global C-Models

Recently, Rayner *et al.* (1999) developed a 3D time-dependent inversion model to determine interannual variability in the regional terrestrial and oceanic uptake of fossil-fuel CO_2 over the last two decades. The Rayner *et al.* study is used here as a benchmark against which the potential for improved precision and spatial resolution of flux estimates from atmospheric composition measurements is explored.

In the Rayner study, extended records of monthly average concentrations of CO_2 in background air measured at 12 or 25 selected sites in the NOAA/CMDL (Climate Monitoring and Diagnostics Laboratory) global flask-sampling network were employed. To determine partitioning of carbon between oceanic and terrestrial reservoirs, $^{13}C/^{12}C$ in CO_2 from one site and a new determination at the same site (Cape Grim) of the trend in O_2/N_2 over two decades (Langenfelds *et al.*, 1999a, b) were used. The small number of selected sites for CO_2, can be compared to the current number of global sampling sites that approaches 100, many with records that are decadal or longer. For $^{13}C/^{12}C$, at least two global sampling networks have made measurements from several sites since the early 1980s (see Francey *et al.*, 1995 and Keeling *et al.*, 1995), yet only one site-record was used. In the case of O_2/N_2 there is no other reliable information available over this time frame. The limited site selections for the Rayner *et al.* study reflect a very real concern about the quality and intercalibration of records from different measurement laboratories.

Figure 1 is adopted from Rayner *et al.* and illustrates the uncertainty ascribed to prior flux estimates (Fig. 1a) and the modified uncertainties resulting from the inversion of the atmospheric measurements (Fig. 1b). The grid-scale of the inversion model has dimensions of 8° latitude × 10° longitude. The numbers refer to flux uncertainties (in Gt C year^{-1}) representing over 25 larger aggregated areas selected as characteristic source regions for the prior source estimates. A reduction in uncertainty in a region from Fig. 1a to 1b indicates that effective constraints are imposed by the atmospheric measurements, and it is no coincidence that the larger improvements occur in regions best represented by atmospheric sampling sites (for example, North America compared to South America). There are still regions of the globe where uncertainties are relatively large ($\sim \pm 1$ Gt C year^{-1}, compared to global fossil-fuel emissions of around 6 Gt C year^{-1}). Even where uncertainties appear to be relatively small, e.g., North America at

± 0.5 Gt C year^{-1}, this should be viewed against the net derived sink in this study of 0.3 Gt C year^{-1} and the total fossil source of ~ 1.6 Gt C year^{-1}. Uncertainties are often reduced when regions are aggregated, but even regions in Fig. 1 are too large for many policy needs.

The potential advantages of the atmospheric inversion approach compared to more conventional on-the-surface carbon accounting methods are that, first, flux estimates are firmly bounded by the global growth rate of atmospheric CO_2, perhaps the best determined of all inputs to a global carbon budget. Second, if the regional uncertainties can be reduced to levels small enough to detect important changes in net continental emissions and uptakes, then atmospheric monitoring provides an opportunity for continuous, relatively low-cost, globally consistent monitoring. The Kyoto Protocol, 1997 and more recently, COP4 of the UN Framework Convention on Climate Change, Buenos Aires, November 1998, present a new and urgent challenge to the atmospheric science community to provide and monitor regional carbon fluxes for verification and/or regulatory purposes.

Of the three broad inputs to the Bayesian synthesis inversion, namely, atmospheric transport models, surface flux constraints, and atmospheric measurements, all have experienced rapid progress over the last few years. In the atmospheric transport area, the problem of estimating GCM model error is significant. However, progress has been made (a) with increasing availability of analyzed wind fields (Trenberth, 1992; e.g., permitting an examination of the impact of interannual variation in transport on measured parameters) and (b) with identification of model differences in the on-going series of TRANSCOM model comparisons (e.g., Law *et al.*, 1996; Denning *et al.*, 1999). Considerable research effort is now focussed on "bridging the scale gap" between the typical grid cells of the transport models and the volume of atmosphere represented by the atmospheric measurement at surface sites. This research introduces boundary-layer and regional-transport models, direct flux measurement campaigns, and vertical profiling of CO_2 and related trace species. Significant uncertainties are perceived to remain, for example, in the representation of mass transport in the tropical areas.

A large volume of new information is also emerging on the interaction between terrestrial ecosystems and the atmosphere with process-oriented campaigns focused on major ecosystems such as the Amazon and Siberia. Recent perspectives and a summary of the advances in knowledge of the understanding of the role of the terrestrial biosphere in the global carbon cycle is provided by Schimel (1995) and Lloyd (1999). The situation is similar for the interaction with the world's oceans. Extensive on-going surveys of ocean parameters are elucidating air–sea gas exchange constraints on carbon uptake by the oceans (e.g., Takahashi *et al.*, 1997; Heimann and Maier-Reimer, 1996), while similar constraints are emerging from the development of ocean general circulation models, e.g., Orr (1999) and Sarmiento *et al.* (1998). The formal integration of the information on terrestrial and oceanic fluxes as additional constraints in the Bayesian inversion framework is in its infancy. Even with these various streams of information, the

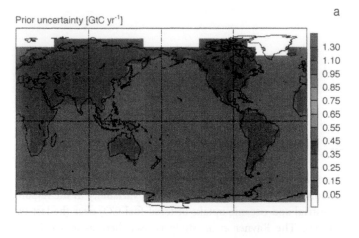

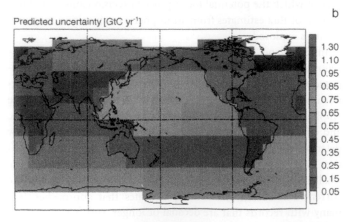

FIGURE 1 Prior and predicted estimates of uncertainty in air–surface fluxes of CO_2 as the result of a 3D Bayesian synthesis inversion of atmospheric CO_2, $\delta^{13}C$, and O_2/N_2 data from selected sites for the period 1980–1995 (Rayner *et al.*, 1999). See also color insert.

carbon cycle remains an underdetermined system that requires more and better-calibrated measurements.

3. Global CO_2-Measurements

The rest of this contribution concentrates on recent and potential progress in the measurement of atmospheric CO_2 mixing ratios and related species. The challenge for such measurement programs is to monitor, with high precision, the temporal changes and/or spatial gradients of CO_2 and related species. Conventional methodologies for monitoring atmospheric CO_2, developed over the past 40 years, show a number of shortcomings when examined in the light of the requirements for improved estimates of regional fluxes from baseline atmospheric-composition measurements.

Measurements of carbon dioxide mixing ratios are made at over 100 globally distributed "baseline" sites (i.e., fixed or mobile sites for which measurements reflect CO_2 behavior over large spatial scales). The requirement for large-scale representation has heavily influenced global sampling strategies insofar as the great majority of sampling sites is located to access marine boundary-layer air. In fact, for the smaller sampling networks, zonal representation was a common assumption. Furthermore, data are still generally selected to reinforce the marine boundary layer bias, though this is changing. Most results are now reported to one or more data banks, including the Carbon Dioxide Information Analysis Center (CDIAC) World Data Centre—A, for Atmospheric Trace Gases, established in 1982 by the Oak Ridge National Laboratory, Tennessee, and the World Meteorological Organisation (WMO) World Data Centre for Greenhouse Gases (WDCGG) in the Japan Meteorological Agency, established in 1990. In late 1995, a Co-operative Atmospheric Data Integration Project (CADIP-CO_2) was commenced in the NOAA Climate Monitoring and Diagnostics Laboratory (CMDL), U.S.A., using data from much the same sources, with the aim of providing an integrated "globally-consistent" data set, GLOBALVIEW-CO_2, for modeling studies. At the heart of GLOBALVIEW is a data extension and integration technique (Masarie and Tans, 1995) that addresses difficulties such as those related to missing data or introduction of new stations. However, interlaboratory calibration remains a problem.

Around 17 different laboratories from 12 nations are involved in the measurement and reporting of CO_2 data to these data banks. Historically, the WMO has taken responsibility for the intercalibration of measurements in different laboratories. Primary activities have involved the establishment of a Central Calibration Laboratory to maintain and provide access to "primary" CO_2-in-air standards measured with high-precision manometric techniques, and initiation of blind "round-robin" intercalibrations involving the circulation of high-pressure cylinders containing CO_2-in-air among participating laboratories. In addition, the WMO has provided a forum of "CO_2 Measurement Experts," now held once every two years to assess progress and plan future activities, with each meeting producing a WMO technical report.

Results from two recent WMO CO_2 round robins are summarized in Figure 2, adopted from WMO technical reports (Pearman, 1993; Peterson, 1997). As an example of the procedure, the most recent round-robin, (b), was proposed at the July 1995 8th WMO CO_2-Experts Meeting in Boulder, Colorado, and was completed in time for an initial assessment at the 9th Meeting of Experts on the Measurement of Carbon Dioxide Concentration and Associated Tracers (endorsed by International Atomic Energy Agency), Aspendale, Australia, 1–4 September 1997. NOAA CMDL prepared three sets of three cylinders of air with nominal CO_2 mixing ratios of 345, 360, and 375 ppm. Each set was distributed to one of three groups of around eight laboratories (in North America and the Southern Hemisphere, Asia, and Europe). A target inter-laboratory precision of 0.05 ppm was identified by this community to achieve a "network precision" of 0.1 ppm. This precision is appropriate for the merging of data from different sites to estimate regional fluxes via synthesis inversion studies (WMO, 1987). This level of precision is comparable to that of an individual measurement in the better operational systems; the "target" precision of 0.05 ppm refers more to the requirement for precise average temporal values (e.g., annual or seasonal) and for precise large-scale values (e.g., GCM grid scale to hemispheric). Note that Fig. 2 results usually represent the average of multiple measurements on a cylinder.

The most important point to be drawn from Fig. 2 is that there are significant (>0.05 ppm) and variable calibration differences between laboratories, which are not currently accounted for in the CDIAC and WDCGG data bases, or in the GLOBALVIEW data assimilation. Another general observation is that there is a significant overall improvement going from the first to second round-robin (while the actual laboratories are not identified, the identification of country is sufficient to make this inference). However, in the second, more precise intercalibration a new concern about linearity emerges, with a majority of participants measuring lower than the low-mixing-ratio tank and higher than the high-mixing-ratio tank. The fact that the degree of "nonlinearity" varies widely suggests that this is an issue for many laboratories; it also argues for an independent verification of both the manometric technique and the scale propagation, e.g., by using gravimetric dilution techniques.

The unsatisfactory situation for CO_2 mixing-ratio intercalibration is also evident for $\delta^{13}C$ of CO_2. The International Atomic Energy Agency (IAEA) conducts a co-operative research program on "Isotope-aided Studies of Atmospheric Carbon Dioxide and Other Greenhouse Gases" with an objective of providing whole-air standards for the measurement of greenhouse gas isotopes. Figure 3 shows preliminary results from the first circulation of "CLASSIC" (Circulation of Laboratory Air Standards for Stable Isotope interComparisons) standards, where the initial round-robin has been restricted to four laboratories with the longest involvement in sampling the background atmosphere from a network of stations. Here the community has set a required target precision of 0.01‰ for temporal or large-scale averages, which is even more demanding than the case with CO_2 mixing ratios since the typical precision on an individual measurement is around 0.03‰.

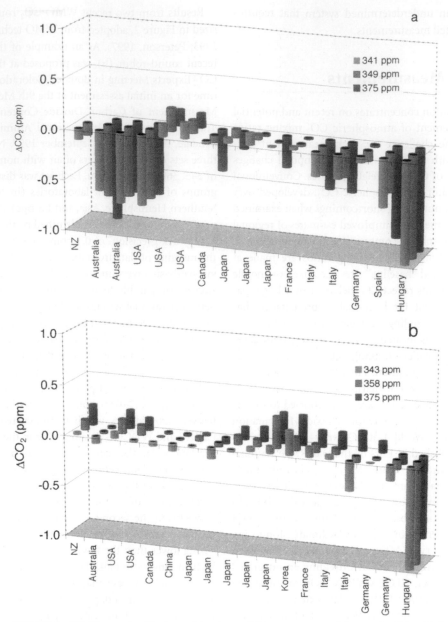

FIGURE 2 WMO round-robin intercalibrations of CO_2 measurement laboratories (identified by country only). Plotted are measured differences from mixing ratios assigned by NOAA CMDL. Data for (a) a circulation conducted between 1991–93, and (b) between 1995–97. See also color insert.

Preliminary results of this round robin are given in Fig. 3 (Allison *et al.*, in press). Measured differences are reported with respect to initial measurements conducted at CSIRO in November 1996. CSIRO(2) refers to CSIRO measurements conducted after circulation in July 1998, confirming the stability of the tank standards. Measurements on pure CO_2 samples scatter by about ± 0.02‰, outside the required target. For the analyses of the whole-air standards the situation is much worse, with reported values scattered over a range of almost ± 0.1‰, suggesting serious

differences between pretreatments to extract CO_2 from air. Furthermore, there also appears to be a linearity problem with the CSIRO measurement compared to the other three laboratories.

The situation is even more serious than indicated by the round-robin comparisons. Since 1992, with the aim of confirming our ability to merge data from two different measuring laboratories, CMDL and CSIRO commenced an "operational intercalibration" (also referred to as the ICP, Inter-Comparison Program, also the "flask-air-sharing" comparison). Both CSIRO and CMDL

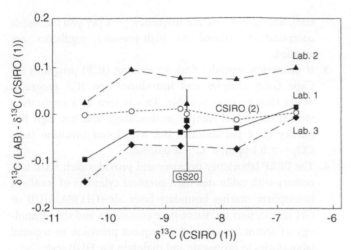

FIGURE 3 IAEA round robin intercalibrations of $\delta^{13}C$ of CO_2 using both pure CO_2 (GS20) and whole air in high-pressure cylinders (in which $\delta^{13}C$ is related to CO_2 mixing-ratio difference from ambient values by about $-0.05‰$ ppm^{-1}). USA and Japanese measurement laboratories are identified by number only. Plotted are measured differences from d$\delta^{13}C$ assigned by CSIRO prior to circulation. CSIRO(2) refers to analyses after circulation.

networks collect pairs of flasks 3 or 4 times per month, from the Cape Grim station on the northwest tip of Tasmania. Approximately twice per month, one of a pair of CMDL flasks has been routinely routed through CSIRO's GASLAB for analysis prior to analysis in CMDL and Institute for Alpine and Arctic Research, University of Colorado (INSTAAR, for the isotopic measurements on CMDL flask samples). The process is facilitated by the unusually small sample requirements for precise analysis in GASLAB (Francey *et al.*, 1996). Once per month, the results of the multi-comparisons (CO_2, CH_4, CO, N_2O, H_2, $\delta^{13}C$, $\delta^{18}O$) in both routine flask sampling of Cape Grim air from each laboratory and from the ICP flasks are automatically processed and reported via ftp in both laboratories (Masarie *et al.*, submitted).

No systematic influence of GASLAB measurements on CMDL flasks has been detected. Figure 4 shows the results of the ICP flask comparisons for CO_2 and for $\delta^{13}C$. Compared to cylinder intercomparisons, the precision on the ICP comparisons is low (individual measurements) but the frequency is high. The CO_2 results are startling. The Australian calibration scale was established to within ~0.01 ppm at ambient CO_2 mixing ratios by repeated analysis of 10 cylinders initially characterized by CMDL. Return of a subset of the cylinders after two years confirmed this agreement to within a few hundredths of a ppm, as have comparisons of other cylinders. Despite this agreement in calibration scales (see also Fig. 2b, Australia), there is a consistent mean difference (CSIRO-CMDL) in the ICP flasks of 0.17 ± 0.17 ppm.

The reason for this offset in flasks compared to high-pressure cylinders is not yet fully understood. However, development of a low-flow (15 ml min^{-1}), high-precision (~7 ppb), and highly

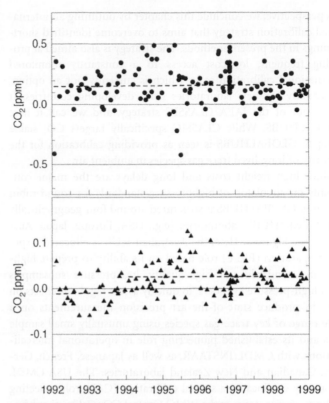

FIGURE 4 (CSIRO-INSTAAR) measured differences on Cape Grim air from the same flask as a function of flask collection dåte, for CO_2 (red circles) and $\delta^{13}C$ (blue triangles).

stable NDIR CO_2 analyzer at CSIRO (G. Da Costa and L. P. Steele, in preparation) has provided clues that high-pressure regulators are a likely contributor to such offsets.

The $\delta^{13}C$ comparison in Fig. 4 illustrates another advantage of the ICP. The (CSIRO–INSTAAR) difference begins at close to zero, or slightly negative, and early in 1994 jumps to a positive value. After 1994, the difference is consistent with highpressure cylinder intercomparisons included in Fig. 3. The discrepancy between the laboratories, if applied globally, translates into a partitioning error of around 1 Gt C year^{-1} between the two laboratories. The continuity of the ICP data has permitted detection of the onset of the problem with reasonable accuracy, and the identification of possible contributing factors that occurred around this time.

4. The Global CO_2-Measuring Network

It is clear that such unanticipated discrepancies between results from different measuring laboratories are a major obstacle for high-precision merging of data sets. The merging is highly desirable from the point of view of maintaining adequate spatial monitoring of global trends and for identification of regional source/sink changes from atmospheric inversion techniques. From this perspective, we conclude this chapter by outlining an international calibration strategy that aims to overcome identified shortcomings in the present methods. The strategy is also aimed at providing frequent, low-cost access to a constantly monitored international calibration scale, which is currently not an option, particularly for new laboratories from developing countries. It grows out of the IAEA CLASSIC strategy, and we call it here GLOBALHUBS. While CLASSIC specifically targets CO_2 stable isotopes, GLOBALHUBS is seen as providing calibration for the majority of long-lived trace gas species in ambient air.

Since high freight costs and long delays are the major constraints on circulaton of highpressure standards for round-robin exercises, GLOBALHUBS is structured around four geographically distributed "HUB" Laboratories (e.g., USA, Europe, Japan, Australia), see Figure 5. Here, the Australian HUB is allocated a special preparaticn (PREP) role based on its ability to prepare high-pressure standards and high-quality, low-pressure subsamples from high-pressure cylinders, to quickly assess regulator effects on CO_2, to produce state-of-the art precision measurements on a wide range of key trace-gas species using unusually small sample sizes and its established pioneering role in operational intercalibrations with CMDL/INSTAAR, as well as Japanese, French, German, Canadian and New Zealand laboratories. The USA CMDL laboratory is allocated a special calibration (CAL) role reflecting both its current status as the WMO Central CO_2 Calibration laboratory (with absolute manometric standards), also and its potential to implement the results of the GLOBALHUBS comparisons into the GLOBALVIEW globally- consistent trace gas data sets.

A common HUB scale is maintained by a variety of approaches:

1. An upgraded CLASSIC rotation between the HUB laboratories is conducted at least once per year. With upgraded and certified regulators, CO_2, $\delta^{13}C$, $\delta^{18}O$, CH_4, N_2O, CO, H_2, etc. can be established to high precision (e.g., $CO_2 \sim \pm 0.01$ ppm, $\delta^{13}C \sim \pm 0.01$‰) with respect to the CORE laboratory scale for air standards covering the full range of anticipated clean air values. The CLASSIC highpressure cylinders ("circulators") are accompanied by a range of pure-CO_2 standards for the isotope measurements. The CLASSIC rotation, though relatively cumbersome and expensive, provides a long standard lifetime (decades for the air standards, and many decades for CO_2 isotope standards). It also provides precise detector response information from both air and pure CO_2 standards.

2. It introduces "oscillator" exchanges between the PREP laboratory and the other three HUB laboratories. The containers are high-quality, electropolished, four-liter stainless steel filled at 1–4 bar pressure by decanting from high-pressure cylinder air standards comprising CO_2-free air (CO_2 stripped from ambient Southern Hemisphere marine boundary-layer air) plus ~ 360 ppm of GS20 (or equivalent, with near-ambient CO_2 isotopic ratios). The oscillator air standards provide moderate to high precision (e.g., $CO_2 \sim \pm 0.01$ to 0.03 ppm, $\delta^{13}C \sim \pm 0.01$ to 0.03‰, depending on required sample

size), and have moderate frequency (~ 4 per year). Possible complications related to high-pressure regulators are avoided.

3. It maintains/upgrades flask air-sharing (ICP) programs for Cape Grim samples and introduces new ICP programs where they become possible. This is seen as a verification step. It uses exact sample methodology and has high frequency (2–4 per month), but with lower precision (e.g., $CO_2 \sim \pm 0.1$ ppm, $\delta^{13}C \sim \pm 0.03$‰).

4. The PREP laboratory prepares and provides each HUB laboratory with calibrated high-pressure cylinders of southern hemisphere marine boundary-layer air (GLOBALHUB or GH tanks), plus oscillator-type containers and the technology to decant into these for frequent provision to regional laboratories to propagate and maintain the HUB scale.

5. The HUB structure permits rapid assessment and dissemination of communityapproved calibration scale adjustments (e.g., arising from new manometric or gravimetric determinations via the CAL laboratory) or of new methods (e.g., "continuous flow" technology for $\delta^{13}C$; D. Lowe, NIWA, NZ, personal communication).

Initial funding is required to establish the HUB capability in existing advanced laboratories, and to secure their long-term involvement. Once the HUB scale is established quite modest regional funding can maintain operation and access. The current strategy is to seek endorsement from WMO and IAEA, and a commitment to continue their roles for planning, assessment, and dissemination of results, with particular encouragement to laboratories from developing countries. Coordinated establishment costs and regional operating costs are being sought from international funding bodies with a charter to support atmospheric composition/climate change research.

We speculate here on the improvements to atmospheric inversion studies of air–sea and air–land carbon fluxes that might flow from more effective global calibration strategies for CO_2, $\delta^{13}C$, and O_2/N_2. A realistic ambition for the precision of year-to-year and large spatial-scale differences over the next 5 years for more than 100 station networks, i.e., using merged data from different measurement laboratories, is close to an order of magnitude improvement (0.2–0.02 ppm, 0.2–0.02‰ for CO_2, $\delta^{13}C$, respectively). For O_2/N_2, perhaps 20–50 sites might contribute to similar precision improvements. With parallel improvements in atmospheric transport and surface flux parameterization, surface fluxes on current GCM grid scales may be improved from current levels of ~ 1 Gt C year^{-1} to better than 0.1 Gt C year^{-1}.

Acknowledgments

Colleagues at CSIRO Atmospheric Research and the CRC for Southern Hemisphere Meteorology have contributed greatly to the results and perspectives aired here. Paul Steele provided valuable comments on this manuscript. We thank the IAEA and WMO

GLOBALHUBS-
Global Quality Control for Long-Lived Trace Gas Measurements

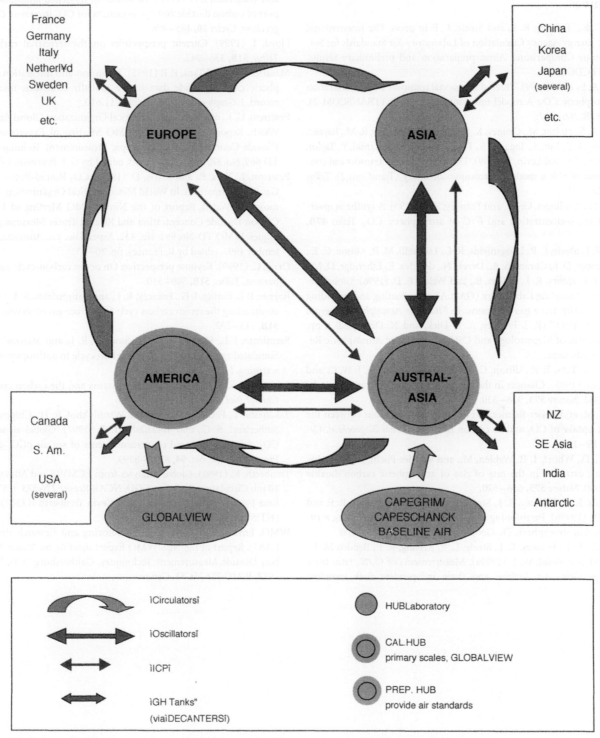

FIGURE 5 Proposed international comparision strategy for laboratories measuring long-lived atmospheric trace gases in air. Identification of laboratories is nominal only. See also color insert.

for support in developing the GLOBALHUBS strategy, the WMO for permission to include round-robin results in press, and CMDL and INSTAAR for the use of ICP results.

References

Allison, C. E., Francey, R. J., and Steele, L. P. in press. The International Atomic Energy Agency Circulation of Laboratory Air Standards for Stable Isotope Comparisons: Aims, preparation and preliminary results. *IAEA TECDOC.*

Denning, A. S., *et al.* (1999). Three-dimensional transport and concentration of atmospheric CO_2: A model intercomparison study (TRANSCOM 2). *Tellus,* **51B,** 266–297.

Denning, A. S., Holzer, M., Gruney, K. R., Heimann, M., Law, R. M., Rayner, P. J., Fung, I. Y., Fan, S., Taguchi, S., Friedlingstein, P., Balkanski, Y., Taylor, J. A., Maiss, M., and Levin, I. (1999). Three-dimensional transort and concentration of SF6: a model intercomparison study (TransCom 2). *Tellus* **51B,** 226–297

Enting, I. G., Trudinger, C. M., and Francey, R. J. (1995). A synthesis inversion of the concentration and $\delta^{13}C$ of atmospheric CO_2. *Tellus* **47B,** 35–52.

Francey, R. J., Steele, L. P., Langenfelds, R. L., Lucarelli, M. P., Allison, C. E., Beardsmore, D. J., Coram, S. A., Derek, N., de Silva, F., Etheridge, D. M., Fraser, P. J., Henry, R. J., Turner, B., and Welch, E. D. (1996). Global Atmospheric Sampling Laboratory (GASLAB): supporting and extending the Cape Grim trace gas programs. In *"Baseline Atmospheric Program (Australia) 1993."* (R. J. Francey, A. L. Dick, and N. Derek, Eds.), pp. 8–29. Bureau of Meteorology and CSIRO Division of Atmospheric Research, Melbourne.

Francey R. J., Tans, P. P., Allison, C. E., Enting, I. G., White, J. W. C. and Trolier, M. (1995). Changes in the oceanic and terrestrial carbon uptake since 1982. *Nature* **373,** 326–330.

Heimann, M. and Maier-Reimer, E. (1996). On the relations between the oceanic uptake of CO_2 and its carbon isotopes. *Global Biogeochem. Cycles* **10,** 89–110.

Keeling, C. D., Whorf, T. P., Wahlen, M., and van der Plicht, J. (1995). Interannual extremes in the rate of rise of atmospheric carbon dioxide since 1980. *Nature* **375,** 666–670.

Langenfelds, R. L., Francey, R. J., Steele, L. P., Battle, M., Keeling, R. F. and Budd, W. (1999b). Partitioning of the global fossil CO_2 sink using a 19-year trend in atmospheric O_2. Geophys. Res. Lett. **26,** 1897–1900.

Langenfelds, R. L., Francey, R. J., Steele, L. P., Keeling, R. F., Bender, M. L., Battle, M. and Budd, W. F. (1999a). Measurements of O_2/N_2 ratio from the Cape Grim Air Archive and three independent flask sampling programs. In *"Baseline Atmospheric Program (Australia), 1996."* (J. L. Gras, N. Derek, N. W. Tindale, and A. L. Dick, (Eds.), p. 57–70. Bureau of Meteorology and CSIRO Division of Atmospheric Research, Melbourne.

Law, R. M., Rayner, P. J., Denning, A. S., Erickson, D., Fung, I. Y., Heimann, M., Piper, S. C., Ramonet, M., Taguchi, S., Taylor, J. A., Trudinger, C. M., and Watterson, I. G. (1996). Variations in modelled atmospheric transport of carbon dioxide and consequences for CO_2 inversion. *Global Biogeochem. Cycles* **10,** 483–496.

Lloyd, J. (1999). Current perspectives on the terrestrial carbon cycle. *Tellus,* **51B,** 336–342.

Masarie, K. A. and Tans, P. P. (1995). Extension and integration of atmospheric carbon dioxide data into a globally consistent measurement record. *J. Geophys. Res.* **100,** 11,593–11,610.

Pearman, G. I., in World Meteorological Organization Global Atmosphere Watch, Report of the Seventh WMO Meeting of Experts on Carbon Dioxide Concentration and Isotopic Measurement Techniques, WMO TD 669, No. 88, Rome, Italy, 1993, edited by G. I. Pearman, PP 104–104.

Peterson, J., Tans, P., and Kitzis, D. (1999). CO_2 Round-Robin Reference Gas Intercomparison. In World Meteorological Organization Global Atmosphere Watch, Report of the Ninth WMO Meeting of Experts on Carbon Dioxide Concentration and Related Tracer Measurement Techniques, WMO TD-No. 952. No. 132, Aspendale, Vic. Australia, 1–4 September 1997, edited by R. Francey, pp. 30–33.

Orr, J. C. (1999). Keynote perspective: On ocean carbon-cycle model comparison. *Tellus,* **51B,** 509–510.

Rayner, P. J., Enting, I. G., Francey, R. J., and Langenfelds, R. L. (1999). Reconstructing the recent carbon cycle from trace gas observations. *Tellus,* **51B,** 213–232.

Sarmiento, J. L., Hughes, T. M. C., Stouffer, R. J., and Manabe, S. (1998). Simulated response of the ocean carbon cycle to anthropogenic climate warming. *Nature,* **393,** 245–249.

Schimel, D. S. (1995). Terrestrial ecosystems and the carbon cycle. *Global Change Biol.,* **1,** 77–91.

Takahashi, T., Feely, R. A., Weiss, R., Wanninkhof, R. H., Chipman, D. W., Sutherland, S. C., and Takahashi, T. T. (1997). Global air-sea flux of CO_2: An estimate based on measurements of sea-air pCO_2 difference. *Proc. Natl. Acad. Sci.* **94,** 8292–8299.

Trenberth, K. (1992). Global Analyses from ECMWF and Atlas of 1000 to 10 mb Circulation Statistics, CGD, *NCAR Report TN-373 + STR* 205 p., June 1992. For a copy please contact Kevin Trenberth at (303)497-1318. (NTIS # PB92 218718/AS).

WMO, Environmental Pollution Monitoring and Research Programme. (1987). Report of the NBS/WMO Expert meeting on Atmospheric Carbon Dioxide Measurement Techniques, Gaithersburg, MD, USA, June 1987, *WMO TD-No.* **51,** Geneva.

Carbon Isotope Discrimination of Terrestrial Ecosystems— How Well Do Observed and Modeled Results Match?

Nina Buchmann and
Jed O. Kaplan
*Max Planck Institut for
 Biogeochemistry,
Jena, Germany*

Terrestrial ecosystems play an important role in the global carbon cycle. Recently, δ^{13}C ratios of CO_2 in the atmosphere have been used in general circulation models to constrain the global carbon budget and imply location and magnitude of carbon sources and sinks. These models rely on scaling modeled δ^{13}C ratios of soil and plant components to the ecosystem level, but no validation with measured ecosystem level estimates has been accomplished. However, the isotopic signature of the biosphere is highly variable through space and time. We have compiled a global dataset of measurements on ecosystem carbon isotope discrimination (Δ_e) and used this dataset to validate a global terrestrial biosphere model that simulates Δ_e (BIOME3.5). Measured Δ_e values (based on ecosystem measurements) averaged 18‰ globally, while the global modeled estimate (with BIOME3.5) averaged 15.6‰. These differences between the measurements and the model may be due to site selection and lack of representative coverage of certain ecosystem types as well as to model parameterization. Field measurements in deserts, C_3 and C_4 grasslands, and savanna systems are very limited or do not exist yet. The latitudinal bands between 40° and 20° S, 20° and 30° N or >70°N are not covered. The model, which does not incorporate information about land use, simulates a mean Δ_e intermediate between those used in other modeling studies. The effects of land use may confound the global Δ_e signal. The model also shows that the ratio of ecosystem assimilation to canopy conductance is closely related to the ecosystem's Δ_e except in tropical savannas where roughly equal amounts of C_3 and C_4 vegetation coexist. Thus, Δ_e is a useful tool for investigating the global carbon cycle as it provides information not only on isotopic fractionation during terrestrial CO_2 exchange with the atmosphere but also ecophysiological information on the water status of the vegetation. Future analyses of the global carbon budget need to account for the magnitude and the heterogeneity of the terrestrial isotopic signature as a 3‰ underestimate of modeled Δ_e can cause up to a 20% reduction in the estimated strength of the terrestrial carbon sink.

1. Introduction

Terrestrial ecosystems play an important role in the global carbon (C) budget. The C amounts released by anthropogenic activities are higher than the observed increase in atmospheric CO_2 concentrations measured globally by almost a factor of 2 (IPCC, 1996). After accounting for a large oceanic sink, models predict a large C sink to be located in the Northern Hemisphere, particularly in the terrestrial biosphere (Tans *et al.*, 1990; Ciais *et al.*, 1995; Enting *et al.*, 1995; Francey *et al.*, 1995). However, the annual partitioning among different terrestrial carbon sinks is still under debate (budget for 1980 to 1990; IPCC, 1996; Keeling *et al.*, 1996; Schimel *et al.*, 2000).

Thus, ecosystem physiology, specifically the CO_2 gas exchange between terrestrial ecosystems and the atmosphere, is of primary interest for global change research (Walker and Steffen, 1996). Atmospheric CO_2 concentrations and the corresponding carbon isotope ratios of that CO_2 ($\delta^{13}C_{trop}$) fluctuate seasonally, mainly due to changes in the terrestrial C fluxes (Conway *et al.*, 1994; Trolier *et al.*, 1996). In addition, measurements starting in the early 1980s showed that atmospheric $\delta^{13}C$ ratios decreased by $-0.025‰$ during the 1980s, but that this rate of change approached almost zero between 1990 and 1993 (Trolier *et al.*, 1996). Understanding these changes in $\delta^{13}C_{trop}$ requires detailed knowledge about the ^{13}C signature of different compartments and fluxes in terrestrial ecosystems as well as about the ^{13}C fractionation taking place during the biospheric CO_2 exchange with the atmosphere.

The $^{13}CO_2$ exchange of the biosphere with the atmosphere is influenced by the interactions of the turbulence regime and ecosystem physiology (Fig. 1). The turbulence regime will influence the mixing of CO_2 with different isotopic compositions between the biosphere and the atmosphere. Ecosystem physiology will affect the signature of the biospheric flux and the magnitude of this flux. Strong feed back mechanisms exist such as the effect of high turbulence on ecosystem assimilation or of low ecosystem gas-exchange rates on CO_2 concentrations. The main ecosystem processes that alter the signature of canopy CO_2 are assimilation, autotrophic respiration, and heterotrophic respiration, each carrying ^{13}C signals integrated over different time spans. The leaf carbon-isotope ratios ($\delta^{13}C_{leaf}$) reflect current carbon isotope ratios of tropospheric CO_2 ($\delta^{13}C_{trop}$) or canopy CO_2 ($\delta^{13}C_{canopy}$) as this CO_2 is fixed during leaf photosynthesis. During this fixation and subsequent carboxylation, discrimination against the heavier $^{13}CO_2$ takes place (Δ_{leaf}). In contrast, the $\delta^{13}C$ ratios of litter ($\delta^{13}C_{litter}$) and soil organic carbon ($\delta^{13}C_{SOC}$) carry isotopic signals from past times due to the long residence times of organic matter in the soil. Thus, both reflect conditions with lower tropospheric CO_2 concentrations ($[CO_2]$) and higher tropospheric $\delta^{13}C$ ratios,

prior to the combustion of ^{13}C-depleted fossil fuel (isotopic disequilibrium; Enting *et al.*, 1995). Consequently, the $\delta^{13}C$ ratio of soil respiration ($\delta^{13}C_{Rs}$) carries this long-term "ecosystem memory," dependent on the turnover rates of soil organic matter. At the ecosystem level, two parameters integrate these various spatial and temporal scales and describe the $^{13}CO_2$ fluxes: the $\delta^{13}C$ of ecosystem respiration ($\delta^{13}C_{ER}$; Flanagan and Ehleringer, 1998) and the ecosystem carbon discrimination (Δ_e; Buchmann *et al.*, 1998).

$\delta^{13}C_{ER}$ ratios and Δ_e estimates describe the ^{13}C signature of ecosystem CO_2 fluxes and quantify the biospheric fractionation at the ecosystem level (Flanagan and Ehleringer, 1998; Buchmann *et al.*, 1998). Both parameters integrate the ^{13}C signature of ecosystem CO_2 exchange with the troposphere, weighted by both the flux rates of above and below-ground processes and the contribution of all species present (see below for necessary measurements and calculations). Furthermore, both $\delta^{13}C_{ER}$ and Δ_e values reflect land-use history, due to mixing of litter and slow turnover rates of soil carbon (Buchmann and Ehleringer, 1998). Thus, both estimates of the ^{13}C signature of terrestrial CO_2 exchange can be used to constrain general circulation models' identification and quantification of C sinks or sources. $\delta^{13}C_{trop}$ ratios, which have been measured at selected stations within international networks since 1990 (Ciais *et al.*, 1995; Enting *et al.*, 1995; Fung *et al.*, 1997), have found application in general circulation models only very recently. These models rely on scaling modeled $\delta^{13}C$ ratios of soil and plant components to the ecosystem level. To date no comparison between flask-derived and ecosystem-level estimates of ecosystem discrimination has been accomplished.

Bakwin *et al.* (1998) estimated the ^{13}C signatures of biospheric discrimination using flask data from 17 stations of the NOAA/CMDL network (National Oceanic and Atmospheric Administration/Climate Monitoring and Diagnostics Laboratory). Their global estimate of biospheric discrimination averaged 16.8 $\pm$ 0.8‰. However, the comparison of this mean with two recent model simulations showed large discrepancies (Lloyd and Farquhar, 1994; Fung *et al.*, 1997). The model of Fung *et al.* (1997) predicted a much stronger latitudinal gradient, whereas the model of Lloyd and Farquhar (1994) resulted in higher discrimination values. Differences in the distribution of C_3 versus C_4 vegetation and the smoothing effect of atmospheric transport were thought to be responsible for the disagreement of modeled versus measured data (Bakwin *et al.*, 1998).

In this paper, we compare modeled Δ_e estimates from the BIOME3.5 model and measured Δ_e estimates from ecosystem studies. Using flask data from canopy air collections within and above 50 different forest and agricultural sites might allow us to test the modeled Δ_e estimates more realistically since the smoothing effect of atmospheric transfer observed with the tropospheric air collections will be eliminated. Furthermore, this comparison will be used to identify major gaps in the spatial representation of our study sites and to test the hypothesis that Δ_e estimates are related to the water use efficiency of terrestrial vegetation as postulated by Buchmann *et al.* (1998).

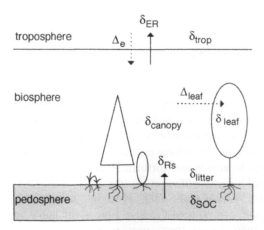

FIGURE 1 Conceptual model of ecosystem $^{13}CO_2$ exchange with the atmosphere. $\delta_{canopy} = \delta^{13}C$ of canopy air CO_2, Δ_e = ecosystem carbon isotope discrimination, $\delta_{ER} = \delta^{13}C$ of CO_2 respired by the ecosystem, $\delta_{leaf} = \delta^{13}C$ of foliage, Δ_{leaf} = leaf carbon-isotope discrimination, $\delta_{litter} = \delta^{13}C$ of litter, $\delta_{Rs} = \delta^{13}C$ of soil respired CO_2, $\delta_{SOC} = \delta^{13}C$ of soil organic carbon, $\delta_{trop} = \delta^{13}C$ of tropospheric CO_2.

2. Experimental and Analytical Methods

Flask samples of canopy air from a variety of ecosystems were generally taken from different heights within the canopy at different times during the day and the night. These measurements represent a wide range of canopy CO_2 concentrations ($[CO_2]$). Dried canopy air was drawn through glass flasks, $[CO_2]$ were measured before the flasks were closed, and flasks were brought to the lab for further analyses (For more details on how to sample and how to prepare the air samples for mass spectrometric analyses, see Buchmann *et al.*, 1998.) Carbon isotope ratios of canopy air ($\delta^{13}C_{canopy}$) were measured using an isotope ratio mass spectrometer (IRMS, precision between 0.03 and 0.3‰, depending on the IRMS used). Carbon isotope ratios ($\delta^{13}C$) were calculated as

$$\delta^{13}C = 1000\left(\frac{R_{sample}}{R_{standard}} - 1\right), \tag{1}$$

where R_{sample} and $R_{standard}$ are the $^{13}C/^{12}C$ ratios of the sample and the standard (PeeDee Belemnite), respectively (Farquhar *et al.*, 1989).

The $\delta^{13}C$ of respired CO_2 during ecosystem respiration ($\delta^{13}C_{ER}$) should represent a weighted average of all respiration processes within the ecosystem. The so-called "Keeling plot" method is based on measurements of $[CO_2]$ and $\delta^{13}C$ in air and can be used to estimate $\delta^{13}C_{ER}$. If the inverse $[CO_2]_{canopy}$ are plotted against their corresponding $\delta^{13}C_{canopy}$ values (so-called "Keeling plot"), a linear relationship is obtained. This relationship reflects the mixing of tropospheric CO_2 with an additional CO_2 source that is depleted in ^{13}C compared to the troposphere. The following linear equation describes the mixing model adequately (Keeling, 1961a, b):

$$\delta^{13}C_{canopy} = \frac{[CO_2]_{trop}}{[CO_2]_{canopy}}\left(\delta^{13}C_{trop} - \delta^{13}C_{ER}\right) + \delta^{13}C_{ER}. \tag{2}$$

The intercept of this equation has been used to identify the carbon isotope ratio of the additional CO_2 source, e.g., in forest canopies; of ecosystem respiration ($\delta^{13}C_{ER}$). This method has several advantages over scaling results from small-scale enclosure studies (e.g., of foliage, stem, and soil respiration) to estimate ecosystem respiration (Lavigne *et al.*, 1997). The Keeling plot method integrates spatially over all autotrophic and heterotrophic respiration fluxes within the ecosystem. Furthermore, it results in a flux-weighted estimate of $\delta^{13}C_{ER}$ that includes plant respiration as well as fast and slowly decomposing carbon pools and their carbon isotopic signatures.

Ecosystem carbon-isotope discrimination Δ_e (Buchmann *et al.*, 1998) was calculated using the $\delta^{13}C_{ER}$ ratio, the *y*-intercept of the regression equation (Eq. 3) and the corresponding tropospheric $\delta^{13}C$ ratio ($\delta^{13}C_{trop}$) as

$$\Delta_e = \frac{\delta^{13}C_{trop} - \delta^{13}C_{ER}}{1 + \delta^{13}C_{ER}}. \tag{3}$$

Tropospheric $\delta^{13}C$ ratios were obtained from international networks such as the NOAA/CMDL Cooperative Flask Sampling Network (National Oceanic and Atmospheric Administration/Climate Monitoring and Diagnostics Laboratory) in cooperation with INSTAAR (Stable Isotope Laboratory at the Institute of Arctic and Alpine Research), the CSIRO network (Commonwealth Scientific and Industrial Research Organization) or the SIO network (Scripps Institution of Oceanography). Within these networks, tropospheric air samples, collected generally in remote areas, are analyzed for $[CO_2]$ (all sites) and $\delta^{13}C$ ratios (at selected sites; Trolier *et al.*, 1996).

Two major sources of error should be considered for the Δ_e estimates obtained from field measurements, errors associated with the $\delta^{13}C$ of tropospheric CO_2 and the $\delta^{13}C$ of respired CO_2. The precision of the tropospheric background data, e.g., collected by NOAA/CMDL is <0.5 ppm for $[CO_2]$, and ± 0.03‰ for $\delta^{13}C$ (Conway *et al.*, 1994; Trolier *et al.*, 1996). The larger error is associated with the estimates of $\delta^{13}C_{ER}$ due to the nature of regression analysis (extrapolating to the *y*-intercept). Summarizing 49 "Keeling-plot" analyses, the standard error for $\delta^{13}C$ of respired CO_2 averaged 0.98‰ (Buchmann *et al.*, 1998).

3. Description of the Model

BIOME3.5 is an equilibrium terrestrial biosphere model based largely on the BIOME3 model of Haxeltine and Prentice (1996). BIOME3.5's main differences from its predecessor include the addition of a module to calculate isotopic discrimination during photosynthesis (Δleaf), the reparameterization of the original plant functional types (PFTs), and the addition of several new PFTs to reflect poorly represented vegetation types in the arctic and arid subtropics. Like BIOME3, BIOME3.5 is a coupled carbon and water flux model that predicts global vegetation distribution, structure, and biogeochemistry. The model is driven by an arbitrary global ambient CO_2 concentration (in this case 360 ppmv), and a globally gridded climatology dataset (for T_{min} see Bartlein, 1998, other data Leemans and Cramer, 1991). In addition, the model uses global information on soil texture and soil depth and recently available global surveys on rooting depth, frost resistance, and photosynthetic pathway (Woodward, 1987; FAO, 1995; Haxeltine and Prentice, 1996; Jackson *et al.*, 1996; Kern and Bartlein, 1998; Ehleringer *et al.*, 1997). The model is run globally at a 0.5° resolution.

Model operation is based on a suite of 13 PFTs representing broad, physiologically distinct classes of vegetation from arctic cushion forbs to tropical rainforest trees. Each PFT is assigned absolute bioclimatic limits (Table 1) that determine whether or not its net primary productivity (NPP) is calculated for a given gridcell. The core of the model is a coupled carbon and water flux scheme that determines the leaf area index (LAI) that maximizes NPP. Given a certain soil-water balance, calculated on a pseudo-daily timestep, the model iteratively calculates the LAI that yields the maximum gross photosynthetic uptake and the corresponding

TABLE 1 Absolute Bioclimatic Limits[*]

	T_c		T_{min}		GDD_5		Sd
	min	max	min	max	min	max	min
Trees							
Tropical evergreen			0				
Tropical raingreen			0				
Temperate broadleaved evergreen		20	−10	0	1200		
Temperate summergreen	−15	10	−42	0	1200		
Subtropical/temperate conifer	−19	20	−45	0	900		
Boreal evergreen	−32.5	1		0	250		
Boreal deciduous		5		0	250		
Nontrees							
Temperate grass		15			350		
Tropical grass	20		−3				
Desert woody shrub			−45		500		
Tundra woody shrub		−5		0		400	10
Cold herbaceous		−5		0		400	
Cushion forb/lichen/moss		−5		0		400	

[*]T_c, mean temperature of the coldest month in °C; T_{min}, absolute minimum temperature in °C; GDD_5, growing degree-days on a 5°C base; Sd, the minimum survivable winter snowpack in cm.

canopy conductance. NPP is then calculated as the difference between gross photosynthetic uptake and maintenance respiration. Various environmental factors including variation in soil texture with depth and seasonal patterns of precipitation as well as the ambient concentration of atmospheric CO_2 have an effect on transpiration and carbon gain. PFT-specific parameters determine the sensitivity of each PFT to environmental changes (Table 2). Photosynthetic pathway is also PFT-specific, with a C_3-type for woody plants and a C_4-type representing tropical and subtropical grasslands and C_4 desert shrubs (such as some *Atriplex* species). For computational reasons, the C_4 subtypes, NADP-ME, NAD-ME, and PCK, are not separated; CAM photosynthesis is not considered.

Monthly mean NPP is summed on an annual basis for each PFT. The woody PFT with maximum NPP is considered the dominant PFT, except in special cases where grass or mixtures of grass and trees would be expected to dominate because of an inferred disturbance regime or soil moisture constraints. The dominant and subdominant PFTs are expanded into 22 classes of terrestrial vegetation biomes. All of the biogeochemical output from the model represents the dominant PFT for a grid-cell, as there is no explicit accommodation for mixed-PFT grid-cells. However, in the case of savannas and some mixed tree–grass temperate plant communities, the output variables (including Δleaf) are given an NPP-weighted average of the grass and tree types.

BIOME3.5 has the new feature of calculating isotopic discrimination against $^{13}CO_2$ during photosynthesis at the leaf level (Δleaf) and total ecosystem discrimination ($Δ_e$). The discrimination model for Δleaf is closely related to that of Lloyd and Farquhar (1994). The main difference is that the BIOME3.5 model explicitly simulates the concentration of CO_2 in the chloroplast through optimization calculations balancing carbon gain with water loss. Only a maximum c_i/c_a ratio is prescribed for each PFT

(optratio, Table 2). The actual c_i/c_a is subsequently modeled by the optimization calculation. Maximum c_i/c_a ratios were compiled from a literature survey on laboratory studies (Kaplan, in preparation) and from maximum $δ^{13}C$ values measured for leaf material of all PFTs (Lloyd and Farquhar, 1994; Lloyd, personal communication).

Additionally, we developed a model for $Δ_e$ that is based on the theories presented by Buchmann *et al.* (1998) and Flanagan and Ehleringer (1998). Monthly $Δ_e$ values are estimated as the flux-weighted difference in discrimination against ^{13}C from NPP and heterotrophic respiration (R_h). Photosynthate, with a specific ^{13}C content determined by the Δleaf value, is incorporated into the plant on a seasonally integrated flux-weighted basis. A simple model for R_h determines the monthly flux of respired CO_2 and $^{13}CO_2$ to the atmosphere (Sitch *et al.*, 1999; Foley, 1995; Lloyd and Taylor, 1994). The source of respired CO_2 is the aggregated annual NPP for the dominant vegetation type in a grid-cell. This carbon stock is arbitrarily divided into three pools according to the scheme of Foley (1995). Each pool is subjected to a degree of isotopic fractionation during respiration based on the assumed decay rate of the pool. Because the processes underlying carbon isotope fractionation during respiration are poorly understood, fractionation in each pool is assigned a constant value. The fractionation factor increases with pool age (Buchmann *et al.*, 1997; Ciais *et al.*, 1995; Ehleringer *et al.*, 2000).

4. ^{13}C Signature of Ecosystem Respiration

Recently, an increasing number of studies have evaluated the $^{13}CO_2$ exchange of terrestrial ecosystems, applying the experimental methods outlined above. Results for 51 different forest and

TABLE 2 Plant-Specific Physiological Parameters[*]

	P	G_{min}	E_{max}	R_{30} (%)	L_m	kk	optratio	T_{pC3}	Tcurve	Rfact	Alloc	Fire (%)
Trees												
Tropical evergreen	e	0.5	10	69	18	0.7	0.95	10	1	0.8		25
Tropical raingreen[a]	r	0.5	10	70	9	0.7	0.9	10	1	0.8	1	20
Temperate broadleaved evergreen	e	0.2	4.8	67	18	0.6	0.8	5	1	1.4	1.2	40
Temperate summergreen[b]	s	0.8	10	65	7	0.6	0.8	4	1	1.6	1.2	50
Subtropical/temperate conifer	e	0.2	4.8	52	30	0.5	0.9	3	0.9	0.8	1.2	40
Boreal evergreen	e	0.5	4.5	83	24	0.5	0.8	0	0.8	4	1.2	33
Boreal deciduous[b]	s	0.8	10	83	6	0.4	0.9	0	0.8	4	1.2	33
Nontrees												
Temperate grass[c,d,e]	r	0.8	6.5	83	8	0.4	0.65	4.5	1	1.6	1	40
Tropical grass[c,e]	r	0.8	8	57	10	0.4	0.65	10	1	0.8	1	40
Desert woody shrub[e,f]	e	0.1	1	53	12	0.3	0.7	5	1	1.4	1	33
Tundra woody shrub[f]	e	0.8	1	93	8	0.5	0.9	−7	0.6	4	1	33
Cold herbaceous[g]	s	0.8	1	93	8	0.3	0.75	−7	0.6	4	1	33
Cushion forb/lichen/moss[f]	e	0.8	1	93	8	0.6	0.8	−12	0.5	4	1.5	33

[*] P, phenological type (e, evergreen, r, raingreen, s, summergreen); G_{min}, minimum canopy conductance (mm/s); E_{max}, maximum daily transpiration rate (mm/day); R_{30}, percent of roots in the top 30 cm of soil; L_m, leaf longevity (months); optratio, the maximum allowed C_i/C_a ratio; kk, the Beer's law extinction coefficient; T_{pC3}, minimum monthly temperature for C_3 photosynthesis; Tcurve, modifier to the curve response of photosynthesis to temperature; Rfact, modifier to the curve response of maintenance respiration to temperature; Alloc, modifier to the minimum allocation; Fire, the soil moisture percent threshold at which a fire day is counted.

[a] Soil moisture threshold of 60 % for leaf flushing and 50 % for leaf fall.

[b] Requirement of 200 growing degree-days on a 5°C basis to grow a full canopy.

[c] Soil moisture threshold of 30% for leaf flushing and 20% for leaf fall.

[d] Requirement of 100 growing degree-days on a 0°C basis to grow a full canopy.

[e] Presence of C_4 photosynthesis, minimum monthly temperature for C_4 photosynthesis is 10°C.

[f] Presence of sapwood respiration.

[g] Requirement of five growing degree-days on a 0°C basis to grow a full canopy.

agricultural sites were available to the authors (published and unpublished datasets) and were used in the following analysis (original data are given in the Appendix). When more than one estimate of the ^{13}C signature of ecosystem respiration was published, a growing season mean was calculated (arithmetic mean). Thus, seasonal variability of $\delta^{13}C_{ER}$ was not considered (but see Buchmann *et al.*, 1998).

Estimates of $\delta^{13}C_{ER}$ varied globally between $-29.4‰$ (Yakir and Wang, 1996; Harwood, 1997) and $-20‰$ (Yakir and Wang, 1996; Buchmann and Ehleringer, 1998), averaging -25.3 ± 2.2 (SD) ‰ (Fig. 2A). Tropical forests and agricultural stands exhibited very low $\delta^{13}C_{ER}$ values, while agricultural C_4 stands in Mediterranean and temperate regions showed the highest $\delta^{13}C_{ER}$ values. Physiological constraints of C_3 photosynthesis as well as the expression of the C_4 photosynthetic pathway resulted in this large global spread of $\delta^{13}C_{ER}$ values. The smallest variability of $\delta^{13}C_{ER}$ ratios was observed in the humid tropics ($<10°$ N or S), reflecting relatively stable microclimatic conditions as well as the restriction of study sites to forest stands. Greatest variability was found for agricultural stands, illustrating the pronounced effects of land-use change (Buchmann and Ehleringer, 1998). Under these circumstances, changes in the photosynthetic pathway of vegetation cover from C_3 to C_4 (and vice versa) result in a mixture

of organic matter in the soil that is being respired carrying a mixture of both ^{13}C signatures. No clear latitudinal trend was found, supporting results from tropospheric air measurements (Fig. 2B). Calculating $\delta^{13}C_{ER}$ ratios from the NOAA/CMDL flask sampling network and from very tall towers (up to 600 m), Bakwin *et al.* (1998) found an average $\delta^{13}C_{ER}$ of $-24.7 \pm 0.8‰$ (mean $\pm SD$) for the biogenic carbon exchange with the atmosphere, slightly higher than the average from measured $\delta^{13}C_{ER}$ estimates.

5. Modeled Ecosystem Carbon Discrimination

Model results for Δ_e estimates display several important features reflecting global trends in vegetation composition and water status (Fig. 3). Generally, ecosystems in cool, wet environments exhibit the greatest Δ_e values, whereas C_4-dominated, tropical grasslands show the lowest Δ_e estimates. However, within the area of C_3-dominated vegetation, there is a wide range of Δ_e values, representative of changing water status of the vegetation and plant adaptation to these environments. Dry deserts have characteristically low Δ_e values due to the plants' water conserving strategy. High ratios of assimilation to stomatal conductance result in low C_i/C_a ratios, and therefore in low foliar carbon discrimination (Farquhar *et al.*, 1989). Ecosystems in Mediterranean-type winter-rain climates show a higher Δ_e because most of plant production takes place in the wet winter season (i.e., in the Mediterranean basin). A study of natural vegetation and unirrigated agriculture on Crete showed that Mediterranean plants had unexpectedly depleted foliar $\delta^{13}C$ ratios and therefore high leaf-carbon discriminations (Kaplan, unpublished data). In contrast, dry savannas and woodlands in summer-rain areas have lower Δ_e values in apparent response to the higher evaporative demand (i.e., in Australia). In some temperate deserts that are dominated by C_3 grasses and C_3 shrubs, the presence of C_4 woody shrubs, which are simulated in this model, further depresses the Δ_e signal from that region (i.e., southwestern U.S.A.).

Latitudinal averages of flux-weighted Δ_e estimates display a trimodal distribution with Δ_e values being greatest in the boreal zones of both hemispheres and the humid tropics (Fig. 4). Due to large longitudinal variations within a latitudinal band of 0.5°, the moist, highly productive tropical rain forests (between 10° N and S) overshadow any signal from C_4 equatorial grasslands such as those in East Africa. In the boreal zone (between 55° and 80° N), Δ_e is generally high because of low ratios of assimilation to stomatal conductance (i.e., intrinsic water-use efficiencies) of boreal plants. In addition, waterlogged soils due to permafrost and low evaporative demand in maritime influenced areas (i.e., Chile, Alaska) tend to increase Δ_e. Boreal forests dominated by *Larix, Betula, Sorbus* and other deciduous species show even higher Δ_e values than evergreen taiga, presumably because of their physiological ability to transpire more water and hence achieve a higher C_i/C_a ratio. These model results are corroborated by field studies on leaf $\delta^{13}C$ ratios (Lloyd and Farquhar, 1994, Michelsen *et al.*, 1996).

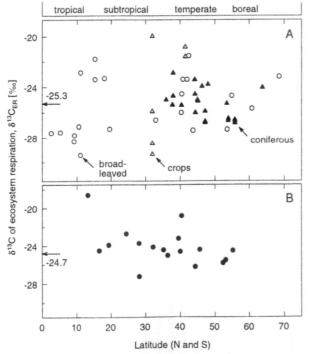

FIGURE 2 Observed carbon isotope ratios of ecosystem respiration. $\delta^{13}C_{ER}$ was determined as the intercept of "Keeling plots," i.e., the regression of inverse canopy [CO_2] against the corresponding $\delta^{13}C_{canopy}$. A: $\delta^{13}C_{ER}$ based on measurements of canopy air for broad-leaved and coniferous forests and crop stands. Original data are given in the Appendix. B: $\delta^{13}C_{ER}$ based on tropospheric measurements within the NOAA network (after Bakwin *et al.*, 1998).

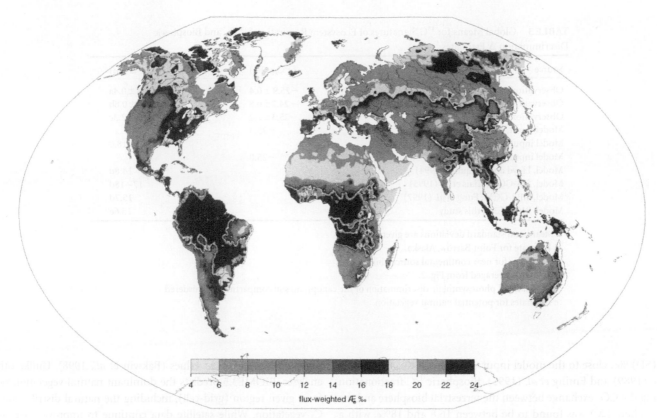

FIGURE 3 Modeled ecosystem carbon-isotope discrimination. See Sec. 3 for details on BIOME3.5.

Mean values for modeled Δ_e estimates are lowest in subtropical C_4-dominated grasslands, especially in Africa and Australia. In the temperate prairies of central North America and Eurasia, C_4 grasslands are only seasonally dominant and share latitude bands with both forests and deserts. Thus, in these regions Δ_e values are correspondingly intermediate.

6. Comparison of Observed and Modeled Δ_e Estimates

Good agreement between measured and modeled Δ_e values is found for the boreal zone, the humid tropics as well as for ecosystems in the Southern Hemisphere (Fig. 4). However, measured values for Δ_e are higher than modeled estimates in the temperate latitudes. In the mid-latitudes, there are several reasons modeled results might differ from observations by up to 5‰. We will discuss several potential explanations in the paragraphs below. The main issue is probably a result of measurement sampling strategy (see below) and the wide integration over bioclimatic space that the model makes in a 0.5° latitudinal band.

The observed global mean $\delta^{13}C_{ER}$ ratios and Δ_e estimates were surprisingly similar, independent whether free tropospheric CO_2 or canopy air CO_2 were measured (Table 3). Due to higher measurement precision and the use of only one analytical laboratory,

both estimates using free tropospheric air measurements (Trolier *et al.*, 1996; Bakwin *et al.*, 1998) showed lower standard deviations than the estimate based on canopy air measurements from 50 different ecosystem studies (0.4 and 0.8‰ vs 2.2‰). Mean $\delta^{13}C_{ER}$ values from these three independent estimates ranged between -24.7 and $-25.9‰$, with a global mean $\delta^{13}C_{ER}$ value of $-25.3 \pm$

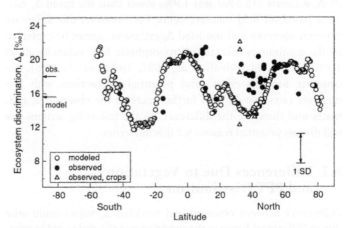

FIGURE 4 Comparison of observed and modeled ecosystem discrimination estimates for natural vegetation and crop systems. Observed Δ_e values were calculated using Eq. 3 (see Appendix); modeled Δ_e values were estimated using BIOME3.5.

TABLE 3   Global Means for ^{13}C Signatures of Ecosystem Respiration δ^{13} C$_{ER}$ and Biospheric Discrimination Δ Estimates[*]

Source	δ^{13}C$_{ER}$ (‰)	Δ(‰)
Observations, Trolier *et al.* (1996)	-25.9 ± 0.4	-17.8 ± 0.4a
Observatios, Bakwin *et al.*(1998)	-24.7 ± 0.8	-16.8 ± 0.8b
Observation, this study	-25.3 ± 2.2	-18.0 ± 2.3c
Model Input, Keeling *et al.* (1989)	-25.3	
Model input, Tans *et al.* (1993)		18.0
Model input, Enting *et al.* (1995)	-25.0	
Model, Lloyd and Farquhar (1994)		14.8d
Model, SiB-GCM, Ciais *et al.* (1995)		17–18d
Model, SiB2-GCM, Fung *et al.* (1997)		15.7d
Model, BIOME3.5, this study		15.6e

[*]Means and standard deviations are given
a Estimate for Point Barrow, Alaska.
b Estimates for near continental source/sink regions.
c Estimates averaged from Fig. 2.
d Estimates of photosynthetic discrimination of the canopy, no soil compartment conisdered.
e Estimates for potential natural vegetation.

0.6 (SD) ‰, close to the model input values for δ^{13}C$_{ER}$ by Keeling *et al.* (1989) and Enting *et al.* (1995). Biospheric ^{13}C fractionation during the CO$_2$ exchange between the terrestrial biosphere and the atmosphere (Δ_e) was found to be between 16.8 and 18‰, with a global mean for observations of 17.5‰, independent of the experimental method used.

However, modeled estimates differed by a maximum of 2.7‰ from this observed mean value for Δ_e (17.5‰). Closest agreement was found with the model estimate by Ciais and colleagues (17–18‰). Differences might arise from the fact that Δ_e values from actual measurements did naturally include the soil compartment (see above) whereas most of the models estimated only the canopy ^{13}C discrimination, excluding the soil compartment and its associated isotopic effects. However, although BIOME3.5 did include the "long-term memory" effect of the soil compartment, its Δ_e estimate (15.6‰) was 1.9‰ lower than the mean Δ_e estimate based on field measurements. Furthermore, this difference between observed and modeled Δ_e estimates cannot be explained by the continuous decrease of tropospheric δ^{13}C values by about 1.3‰ since 1744 (Friedli *et al.*, 1987, Trolier *et al.*, 1996). Accounting for this effect (and potential interactions with low turnover rates) would even further increase the observed Δ_e estimates and therefore the difference. In the following sections, we will discuss potential reasons for this difference.

6.1 Differences Due to Vegetation and PFT Distribution

Differences between observed and modeled Δ_e values could arise due to differences between the modeled and the real world in vegetation distribution or PFT distribution. Particularly, the distribution of C$_4$ plants is assumed to be one of the major factors contributing to differences between Δ_e values based on tropospheric air measure-

ments and modeled Δ_e values (Bakwin *et al.*, 1998). Unlike other analyses, BIOME3.5 predicts the dominant natural vegetation type for a given region (grid-cell), including the natural distribution of C$_4$ vegetation. While satellite data continue to improve, there is a lack of consistent data on actual vegetation type, seasonal variability, and distribution. BIOME3.5 circumvents the need for a predefined vegetation map, but includes a few other caveats.

Assumptions must be made about the predominant vegetation in several areas. One important limitation of BIOME3.5 is its inability to simulate biogeochemically the coexistence of different PFTs within a grid-cell and the variation due to habitats. For an analysis of Δ_e values, this limitation is not important in most cases because the physiological parameters of the PFTs are similar enough that under localized environmental conditions the PFTs behave similarly. Thus, flux rates and correspondingly the magnitude of isotopic fractionation are similar. However, in tropical savannas and warm-temperate grasslands, C$_3$ and C$_4$ plants may often be codominants in the same grid-cell. In this situation, the model takes an empirical approach of arbitrarily assigning a percentage of the grid-cell's NPP and consequent Δ_e to the grass and woody PFTs based on each PFTs NPP relative to the other. The NPP of any given PFT is calculated as if it was growing alone in the grid-cell. In the case of mixed C$_3$–C$_4$ biomes, NPP and Δ_e are scaled to reflect a mixed ecosystem. Since the main differences between observed and modeled Δ_e values (BIOME3.5) were observed in the mid-latitudes, natural vegetation and PFT distribution could contribute to this observed difference (see below).

6.2 Differences Due to Vegetation Change

Only a few of the new generation of computationally expensive dynamic vegetation models can accurately simulate a situation that must incorporate the transient effects of competition, disturbance, and mortality. The potential "memory effect" that decaying soil carbon may have when the dominant vegetation is in a successional or transition stage (Houghton, 1995; Neill *et al.*, 1996; Buchmann and Ehleringer, 1998) cannot be simulated by an equilibrium vegetation model used to date (this paper; Lloyd and Farquhar, 1994; Fung *et al.*, 1997). These effects, while exacerbated by anthropogenic land-use change, may also be present in natural ecosystems. Low-frequency but catastrophic disturbance regimes such as those in the arid subtropics may effect a long-term shift in Δ_e. Simultaneously, seasonal variability can cause a shift from C_3-grass or shrub-dominated ecosystems to C_4-grass-dominated ones, with expected lags in the response of $\delta^{13}C_{ER}$. Shifts in vegetation distribution due to climate change may also be a source of incongruity in the signature of $\delta^{13}C_{ER}$ and therefore Δ_e.

Because BIOME3.5 cannot simulate the dynamics of changing carbon pools and is not supplied with information on the $\delta^{13}C$ of atmospheric CO_2, it is impossible to make an estimate of isotopic disequilibrium. However, global isotopic disequilibrium of ^{13}C is estimated to be less than 0.3–0.5‰, thus within the range of uncertainty of the observed Δ_e as well as the modeled Δ_e (Enting *et al.*, 1995; Fung *et al.*, 1997). This analysis further suggests that the information from inverse modeling techniques about isotopic disequilibrium is limited by the wide spatial, and possibly temporal heterogeneity in Δ_e. The estimates presented here along with estimates of global biosphere Δ_e presented by others (Table 3) differ substantially. These differences illustrate the uncertainty in prescribing a mean global value of Δ_e, which has often been the case when Δ_e was used for constraining deconvolution analyses (Tans *et al.*, 1993).

While the current state of land use has been incorporated into some modeling studies (Fung *et al.*, 1997; Lloyd and Farquhar, 1994), no model to date has performed a sensitivity analysis on the importance of transient effects of land-use change. Transient land-use changes are especially important for addressing the question of isotopic disequilibrium. When the dominant vegetation changes to that of a different photosynthetic pathway, i.e., with the conversion of forest to C_4 cropland, the $\delta^{13}C_{ER}$ ratios would be expected to respond, albeit with a time-lag that would vary among ecosystems (Tans *et al.*, 1993). However, recent widespread conversion to C_4 crops is seen only in a few areas in the tropics (J. Lloyd, personal communication). There, a significant disequilibrium may exist between the isotopic signatures of soil-respired CO_2 and the carbon in the standing biomass.

In temperate regions of the Northern Hemisphere, widespread maize production may cause an increase in the isotopic content of the regional carbon stock (i.e., a shift to heavier ^{13}C signatures). However, this signal would result in a higher global mean $\delta^{13}C_{ER}$ and a lower Δ_e; thus it cannot explain the observed 5‰ difference in Δ_e at midlatitudes between field observations and BIOME3.5 estimates.

6.3 Differences Due to the Water Regime

The BIOME3.5 model may underestimate ecosystem discrimination in places where water stress or environmental limitations on plant productivity are present during prolonged periods of the growing season (e.g., in winter-rain areas or deserts). Other modeling studies (Lloyd and Farquhar, 1994; Fung *et al.*, 1997) take an even more empirical approach to vegetation distribution and physiology and also have difficulty to simulate Δ_e properly in dry places. Discrepancies between observed and modeled Δ_e estimates were expected for latitudes where agricultural C_3 crops replaced the natural vegetation. Thus, higher observed than modeled Δ_e could arise because crop species are mainly bred for productivity, and only to a minor extent for low water use. In addition, great efforts are generally taken to ensure high water availability to agricultural fields, thus lowering the need to conserve water through stomatal regulation of photosynthesis. For the similar reason, Δ_e estimates for C_4 crops were expected to be lower than modeled Δ_e values. These factors could contribute to both the observed differences, though the result may be confounding.

Other limitations of the model include assumptions made based on the driving data, soil hydrology model, and physiology of PFTs. BIOME3.5, using only monthly means, does not simulate the nonuniform nature of weather events. Physical parameters regarding soil structure, depth, and water-holding capacity are poorly constrained. We do not model plants' access to deep groundwater and other aquifers. Finally, various unknowns in the physiological parameters of certain PFTs, such as photosynthetic response to low temperatures, are coarsely parameterized.

6.4 Differences Due to Selection of Field Sites

Further discrepancies between observed and modeled Δ_e estimates might arise from biased site selection. Most of the study sites were located in the higher latitudes between 30° and 60° (Fig. 2). Terrestrial ecosystems within certain latitudinal bands such as 40°–20° S, 20°–30° N, or >70° N have not been studied at all (Fig. 4). Thus, the representation of global vegetation is still rather poor despite the 50 different study sites used for this comparison. This lack of field observations in deserts, C_3 or C_4 grasslands, savannas, or shrublands skews the distribution of observed Δ_e estimates. This could result in overestimation of Δ_e from field measurements. Spatial heterogeneity within a biome is often better known than differences among biomes or vegetation types (Flanagan *et al.*, 1996; Buchmann *et al.*, 1997; 1998).

In general, only limited information is available for ecosystems under naturally or anthropogenically disturbed conditions (e.g., herbivory, fire, wind-fall, logging, clear-cut, severe air pollution). However, these conditions do not cause disagreements with modeled Δ_e estimates since they are not considered in BIOME3.5 or the other models either. Thus, the lack of predominantly C_4 sites could contribute to the observed pattern of lower Δ_e values from the models.

7. Ecophysiological Information from Δ_e

It is well established that leaf carbon-isotope discrimination provides valuable information about the ratio of leaf assimilation to stomatal conductance (intrinsic water-use efficiency; Farquhar *et al.*, 1989). Based on field observations of Δ_e, Buchmann *et al.* (1998) postulated that ecosystem discrimination might increase with decreasing ratio of ecosystem assimilation to canopy conductance (A/G_c, in mol CO_2 per mol H_2O). Using BIOME3.5, we tested this hypothesis for the 22 biomes simulated by the model (Fig. 5).

We found a strong negative relationship of Δ_e with A/G_c. Within the C_4 biomes, deserts and tropical grasslands show low Δ_e values, but high A/G_c ratios. Tropical savannas have a much lower A/G_c ratio, mainly due to the mixture with tropical, drought-deciduous C_3 trees. Obviously, this new vegetation component increases the water loss from savanna regions compared to pure C_4 ecosystems. Differences in physiological regulation of the gas exchange between the two photosynthetic pathways is probably the main reason for the drop in A/G_c (Ehleringer *et al.*, 1997). However, effects due to the mixture with deciduous or semideciduous trees and shrubs might also be responsible for this pattern in savannas (Larcher, 1994). Furthermore, deciduous and semideciduous tropical forests and woodlands show lower A/G_c ratios but similar Δ_e compared to evergreen tropical forests. Broad leaf and needle leaf evergreen forests in temperate regions also exhibit higher A/G_c ratios than deciduous temperate forests, at the same Δ_e value. Thus, the advantage of the evergreen life form that is

well established at the plant and the leaf levels (Aerts, 1995) also shows at the ecosystem level. The model's simulation of leaf area index, which is optimized for A/G_c, is closely correlated to Δ_e in most cases, indicating the importance of canopy structure and canopy roughness for determining Δ_e (unpublished mss.).

Highest A/G_c within boreal and arctic biomes was modeled for steppe tundra. Its A/G_c ratio was of a similar magnitude as that of tropical grasslands, but with a higher corresponding Δ_e estimate. Tundra Δ_e values of about 12‰ indicated that the C_3 vegetation was probably growing under conditions of low water availability (i.e., in polar deserts). The highest Δ_e estimate coupled with the lowest A/G_c was found for deciduous taiga forests, clearly indicating the isotopic signature of C_3 vegetation and high water supply during the growing season. Both boreal ecosystems span almost the entire range of the observed ecosystem discrimination and the ratio of ecosystem assimilation to canopy conductance, representing the two end-members of this negative relationship between Δ_e and AG_c for C_3 vegetation.

8. Conclusions

- All three estimates of Δ_e based on measurements of free tropospheric or canopy air were close to each other (Table 3). Since the spatial representation of the flask-sampling networks is still limited (Tans *et al.*, 1996), the collection of canopy air to deduce Δ_e can be recommended.

- The strong negative relationship between Δ_e and A/G_c supported the hypothesis proposed by Buchmann *et al.* (1998). Despite the variations of Δ_e within a biome, the use of Δ_e seems promising to detect differences in the ratio of carbon to water fluxes among biomes or changes in this ratio due to climatic or environmental conditions.

- Modeling Δ_e without consideration of land-use changes and the transient dynamics of clearance, agriculture, abandonment, and succession could contribute to the observed differences between modeled and measured Δ_e values. A sensitivity analysis of the importance of land-use change at a global scale is therefore necessary. Currently dynamic models are computer-intensive and coarser in PFT-specific-ness, but only future simulations with these models will be able to assess the long-term importance of "memory effect" and isotopic disequilibrium.

- Differences between flask-derived and modeled estimates of ecosystem discrimination (up to 3‰) were due to the lack of measured Δ_e estimates in certain ecosystems and/or regions and to model parameterization. Field measurements filling the spatial gaps as well as modeling actual vegetation cover and its physiology should be of high priority in global ecology research: According to Fung *et al.* (1997), an underestimate of Δ_e in global models by 3‰ (such as the difference we observed here) translates into a carbon flux of 0.7 Gt C year[-1] or into an overestimation of the terrestrial biospheric carbon sink of 20%.

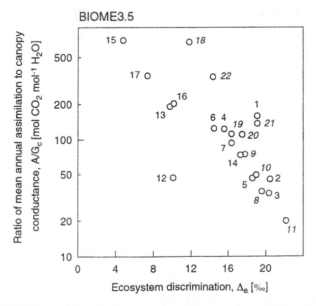

FIGURE 5 Relationship between Δ_e estimates and modeled ratio of ecosystem assimilation to canopy conductance (A/G_c; using BIOME3.5). 1, tropical evergreen forest; 2, tropical semideciduous forest; 3, tropical deciduous forest/woodland; 4, temperate broadleaf evergreen forest; 5, temperate deciduous forest; 6, temperate conifer forest; 7, warm mixed forest; 8, cool mixed forest; 9, cold mixed forest; 10, evergreen taiga/montane forest; 11, deciduous taiga/montane forest; 12, tropical savanna; 13, temperate sclerophyll woodland; 14, temperate woodland; 15, tropical grassland; 16, temperate grassland; 17, desert: shrubland and steppe; 18, steppe tundra; 19, shrub tundra; 20, dwarf shrub tundra; 21, prostrate shrub tundra; 22, cushion-forb, lichen and moss tundra.

Appendix

Site	Latitude	N/S	Longitude	E/W	Δ_e [‰]	Intercept $\delta^{13}C_{ER}$ [‰]	SE of $\delta^{13}C_{ER}$ [‰]	Time	Reference	Life form
Forests, grasslands										
Rainforest	42	S	147	W	16	−23.40			Francey et al., 1985	Broad-leaved
Cerrado	15.33	S	47.36	W	13.80	−21.80		May	Miranda et al., 1997	Broad-leaved
Cerrado	15.33	S	47.36	W	15.40	−23.40		November	Miranda et al., 1997	Broad-leaved
Rainforest	10.5	S	61.6	W	19.5	−27.1			Lloyd et al., 1996	Broad-leaved
Rainforest	2.6	S	59.6	W	20.26	−27.60			Quay et al., 1989	Broad-leaved
Rainforest	5.2	N	53	W	20.42	−27.56	0.30	Annual mean	Buchmann et al., 1997a	Broad-leaved
Rainforest	9.1	N	79.51	W	21.1	−28.30			Buchmann et al., 1997a	Broad-leaved
Rainforest	9.2	N	79.8	W	20.7	−27.8			Lancaster, 1990	Broad-leaved
Seasonal deciduous rainforest	11	N	61	W	21.42	−29.42			Harwood, 1997, thesis	Broad-leaved
Seasonal deciduous rainforest	11	N	61	W	16.6	−22.87			Broadmeadow et al., 1992	Broad-leaved
Rainforest	18	N	67	W	15.32	−23.32	1.41		Ehleringer (unpublished)	Broad-leaved
Rainforest	19.5	N	105.1	W	20.2	−27.3			Lancaster, 1990	Broad-leaved
Quercus spp.	33	N	116.6	W	19.4	−26.6			Lancaster, 1990	Broad-leaved
Populus tremuloides	40.34	N	111.2	W	18	−26.00	1.30	Annual mean	Buchmann et al., 1997b	Broad-leaved
Acer spp.	40.47	N	111.46	W	17.1	−24.55	0.50	Annual mean	Buchmann et al., 1997b	Broad-leaved
Deciduous forest	40.8	N	77.8	W	16.1	−23.4			Lancaster, 1990	Broad-leaved
Deciduous forest	42.5	N	72.2	W	13.54	−21.54	0.84		Ehleringer (unpublished)	Broad-leaved
Acer/Alnus spp.	43.7	N	72.6	W	20.3	−27.4			Lancaster, 1990	Broad-leaved
Populus tremuloides	53.63	N	106.2	W	19.66	−27.34	0.41	Annual mean	Flanagan et al., 1997	Broad-leaved
Quercus spp.	55	N	2	W	16.69	−24.69			Harwood, 1997, thesis	Broad-leaved
Populus tremuloides	55.89	N	98.68	W	19.09	−26.70	0.39	Annual mean	Flanagan et al., 1996	Broad-leaved
Tundra	60.8	N	161.88	W	18.5	−25.7			Lancaster, 1990	Broad-leaved
Tussock tundra	68.6	N	149.3	W	15.9	−23.2			Lancaster, 1990	Broad-leaved
Pinus spp.	36	N	76	W	18.3	−25			Keeling, 1961	Coniferous
Pinus spp.	37.8	N	119.7	W	18.2	−25.4			Lancaster, 1990	Coniferous
Sequoia sempervirens	38	N	122	W	18.1	−24.7			Keeling, 1961	Coniferous
Pinus/Abies spp.	38	N	120	W	16.4	−22.9			Keeling, 1961	Coniferous
Pinus contorta	40.39	N	110.54	W	18.3	−25.42	1.27	Annual mean	Buchmann et al., 1997b	Coniferous
Juniperus occidentalis	44.2	N	121.2	W		−23.42	0.49	Annual mean	Ehleringer and Cook, 1998	Coniferous
Pinus ponderosa	44.3	N	121.4	W		−24.56	0.46	Annual mean	Ehleringer and Cook, 1998	Coniferous
Pseudotsuga menziesii	44.4	N	122.4	W		−26.01	0.79	Annual mean	Ehleringer and Cook, 1998	Coniferous
Picea/Tsuga spp.	45	N	123.6	W	17.3	−25.00	0.52	Annual mean	Ehleringer and Cook, 1998	Coniferous
Pinus resinosa	45.12	N	75.37	W	17.69	−25.10	0.15	Annual mean	Berry et al., 1997	Coniferous
Pinus spp.	46.3	N	114.2	W	16.7	−23.95			Lancaster, 1990	Coniferous
Abies amabilis	47.19	N	121.35	W	18.9	−25.90	0.60	Annual mean	Buchmann et al., 1998	Coniferous
Abies amabilis	47.19	N	121.35	W	19.3	−26.80	1.10	Annual mean	Buchmann et al., 1998	Coniferous
Abies amabilis	47.19	N	121.35	W	19.1	−26.70	2.40	Annual mean	Buchmann et al., 1998	Coniferous
Pseudotsuga/Tsuga	48	N	121	W	17.5	−23.8			Keeling, 1961	Coniferous
Pinus/Picea spp.	53.5	N	118.3	W	18.1	−25.3			Lancaster, 1990	Coniferous
Pinus banksiana	53.92	N	104.69	W	18.97	−26.65	0.31	Annual mean	Flanagan et al., 1997	Coniferous
Picea mariana	53.99	N	105.12	W	18.81	−26.48	0.26	Annual mean	Flanagan et al., 1997	Coniferous
Picea mariana	55.91	N	98.52	W	18.97	−26.58	0.11	Annual mean	Flanagan et al., 1997	Coniferous

(continued)

Appendix *(Continued)*

					Annual mean				
Pinus banksiana	55.93	N	98.62	W	19.22	−26.83	0.22	Flanagan *et al.*, 1997	Coniferous
Pinus sylvestris., Picea abies	63.8	N	20.3	E		−24.06	0.48	Högberg and Ekblad, 1996	Coniferous
Crops									
Zea mays	32	N	35	E	12.25	−19.97		Yakir and Wang, 1996	Crops
Triticum aestivum	32	N	35	E	22.05	−29.36		Yakir and Wang, 1996	Crops
Triticum aestivum	32	N	35	E	21.11	−28.47		Yakir and Wang, 1996	Crops
Gossipium hirsutum	32	N	35	E	18.43	−25.91		Yakir and Wang, 1996	Crops
Medicago sativa Zea mays	41.5	N	111.5	W	13.2	−20.84		Buchmann and Ehleringer, 1998	Crops
Medicago sativa Zea mays	41.5	N	111.5	W	13.8	−21.61		Buchmann and Ehleringer, 1998	Crops
Global mean					17.95	−25.27			
SD					2.30	2.18			
SE					0.34	0.31			

References

Aerts, R. (1995). The advantages of being evergreen. *Trends Ecol. Evolut.* **10**, 402–407.

Bakwin, P. S., Tans, P. P., White, J. W. C., and Andres, R. J. (1998). Determination of the isotopic ($^{13}C/^{12}C$) discrimination by terrestrial biology from a global netweork of observations. *Global Biogeochem. Cycles* **12**, 555–562.

Bartlein, P. (1998). Personal communication.

Buchmann, N. and Ehleringer, J. R. (1998). A comparison of CO_2 concentration profiles, carbon isotopes, and oxygen isotopes in C_4 and C_3 crop canopies. *Agri. Forest Meteorol.* **89**, 45–58.

Buchmann, N., Kao, W. Y., and Ehleringer, J. R. (1997). Influence of stand structure on carbon-13 of vegetation, soils, and canopy air within deciduous and evergreen forest in Utah, United States. *Oecologia (Berlin)* **110**, 109–119.

Buchmann, N., Brooks, J. R., Flanagan, L. B., and Ehleringer, J. R. (1998). Carbon isotope discrimination of terrestrial ecosystems. *In* "Stable Isotopes and The Integration of Biological, Ecological and Geochemical Processes." (H Griffiths, Ed.), pp. 203–221. BIOS Scientific Publishers, Oxford, United Kingdom.

Ciais, P., Tans, P. P., Trolier, M., White, J. W. C., and Francey, R. J. (1995). A large northern hemisphere terrestrial CO_2 sink indicated by the $^{13}C/^{12}C$ ratio of atmospheric CO_2. *Science* **269**, 1098–1102.

Conway, T. J., Tans, P. P., Waterman, L. S., Thoning, K. W., Kitzis, D. R., Masarie, K. A., and Zhang, N. (1994). Evidence for interannual variability of the carbon cycle from the National Oceanic and Atmospheric Administration/Climate Monitoring and Diagnostics Laboratory Global Air Sampling Network. *J. Geophys. Res.* **99**, 22,831–22,855.

Ehleringer, J. R., Buchmann, N., and Flanagan, L. B. (2000). Interpreting belowground processes using carbon and oxygen isotope ratios. *Ecol. Appl.* **10**, 412–422.

Ehleringer, J. R., Cerling, T. E., and Helliker, B. R. (1997). C_4 photosynthesis, atmospheric CO_2, and climate. *Oecologia (Berlin)* **112**, 285–299.

Enting, I. G., Trudinger, C. M., and Francey, R. J. (1995). A synthesis inversion of the concentration and $\delta^{13}C$ of atmospheric CO_2. *Tellus* **47B**, 35–52.

FAO. (1995). The Digital Soil Map of the World, version 3.5, vol. CD-ROM. Food and Agriculture Organization, Rome, Italy.

Farquhar, G. D., Ehleringer, J. R., and Hubick, K. T. (1989). Carbon isotope discrimination and photosynthesis. *Annu. Rev. Plant Physiol. Plant Mol. Biol.* **40**, 503–537.

Flanagan, L. B. and Ehleringer, J. R. (1998). Ecosystem–atmosphere CO_2 exchange: Interpreting signals of change using stable isotope ratios. *Trends Ecol. Evolut.* **13**, 10–14.

Flanagan, L. B., Brooks, J. R., Varney, G. T., Berry, S. C., and Ehleringer, J. R. (1996). Carbon isotope discrimination during photosynthesis and the isotope ratio of respired CO_2 in boreal forest ecosystems. *Global Biogeochem. Cycles* **10**, 629–640.

Foley, J. A. (1995). An equilibrium model of the terrestrial carbon budget. *Tellus* **47B**, 310–319.

Francey, R. J., Tans, P. P., Allison, C. E., Enting, I. G., White, J. W. C., and Trolier, M. (1995). Changes in oceanic and terrestrial carbon uptake since 1982. *Nature* **373**, 326–330.

Friedli, H., Siegenthaler, U., Rauber, D., and Oeschger, H. (1987). Measurements of concentration, $^{13}C/^{12}C$ and $^{18}O/^{16}O$ ratios of tropospheric carbon dioxide over Switzerland. *Tellus* **39B**, 80–88.

Fung, I. Y., Field, C. B., Berry, J. A., Thompson, M. V., Randerson, J. T., Malmstrom, C. M., Vitousek, P. M., Collatz, G. J., Sellers, P. J., Randall,
D. A., Denning, A. S., Badeck, F. W., and John, J. (1997). Carbon 13 exchanges between the atmosphere and the biosphere. *Global Biogeochem. Cycles* **11**, 507–533.

Harwood, K. (1997). Variation in atmospheric canopy $^{13}CO_2$, $C^{18}O^{16}O$, and $H_2^{18}O$. Ph.D. thesis. University of Newcastle, Newcastle upon Tyne.

Haxeltine, A. and Prentice, I. C. (1996). BIOME3: An equilibrium terrestrial biosphere model based on ecophysiological constraints, resource availiability, and competition among plant functional groups. *Global Biogeochem. Cycles* **10**, 693–709.

Houghton, R. A. (1995). Land-use change and the carbon cycle. *Global Change Biol.* **1**, 275–287.

IPCC. (1996). Climate Change 1995. "The Science of Climate Change." pp. 572. Cambridge University Press, Cambridge.

Jackson, R. B., Canadell, J., Ehleringer, J. R., Mooney, H. A., Sala, O. E., and Schulze, E. D. (1996). A global analysis of root distribution for terrestrial biomes. *Oecologia (Berlin)* **108**, 389–411.

Keeling, C. D. (1961a). The concentration and isotopic abundances of carbon dioxide in rural and marine air. *Geochim. Cosmochim. Acta* **24**, 277–298.

Keeling, C. D. (1961b). A mechanism for cyclic enrichment of carbon-12 by terrestrial plants. *Geochim. Cosmochim. Acta* **24**, 299–313.

Keeling, C. D., Piper, S. C., and Heimann, M. (1989). A three-dimensional model of atmospheric CO_2 transport based on observed winds: 4. Mean annual gradients and interannual variations. *Geophys. Monogr.* **55**, 277–302.

Keeling, R. F., Piper, S. C., and Heimann, M. (1996). Global and hemispheric CO_2 sinks deduced from changes in atmospheric O_2 concentrations. *Nature* **381**, 218–221.

Kern, J. and Bartlein, P. (1998). Soil physical parameters. (unpublished).

Larcher, W. (1994). "Ökophysiologie der Pflanzen." p. 394. Verlag Eugen Ulmer GmbH & Co, Stuttgart.

Lavigne, M. B., Ryan, M. G., Anderson, D. E., Baldocchi, D. D., Crill, P. M., Fitzjarrald, D. R., Goulden, M. L., Gower, S. T., Massheder, J. M. M., McCaughey, J. H., Rayment, M. B., and Striegl, R. G. (1997). Comparing nocturnal eddy covariance measurements to estimates of ecosystem respiration made by scaling chamber measurements at six coniferous boreal sites. *J. Geophys. Res.* **102**, 28,977–28,985.

Leemans, R. and Cramer, W. P. (1991). The IIASA climate database for mean monthly values of temperature, precipitation and cloudiness on a terrestrial grid. *Int. Inst. Appl. Sys. Anal.* RR, 91–118.

Lloyd, J. and Farquhar, G. D. (1994). ^{13}C discrimination during CO_2 assimilation by the terrestrial biosphere. *Oecologia (Berlin)* **99**, 201–215.

Lloyd, J. and Taylor, J. A. (1994). On the temperature dependence of soil respiration. *Functional Ecol.* **8**, 315–323.

Michelsen, A., Jonasson, S., Sleep, D., Havstroem, M., and Callaghan, T. V. (1996). Shoot biomass, $\delta^{13}C$, nitrogen and chlorophyll responses of two artic dwarf shrubs to in situ shading, nutrient application and warming simulating climatic change. *Oecologia (Berlin)* **105**, 1–12.

Neill, C., Fry, B., Melillo, J. M., Steudler, P. A., Moraes, J. F. L., and Cerri C. C. (1996). Forest- and pasture-derived carbon contribution to carbon stocks and microbial respiration of tropical pasture soils. *Oecologia (Berlin)* **107**, 113–119.

Schimel, D., Melillo, J., Tian, H., McGuire, A. D., Kicklighter, D., Kittel, T., Rosenbloom, N., Running, S., Thornton, P., Ojima, D., Parton, W., Kelly, R., Sykes, M., Neilson, R., and Rizzo, B. (2000). Contribution of increasing CO_2 and climate to carbon storage by ecosystems in the United States. *Science* **287**, 2004–2006.

Sitch, S., McGuire, A. D., Heimann, M., and other CCMLP participants. (1999). Seasonal cycles of atmospheric CO_2: Comparison of modelled and remotely sensed foliage phenology.

Tans, P. P., Bakwin, P. S., and Guenther, D. W. (1996). A feasible Global Carbon Cycling Observing System: A plan to decipher today's carbon cycle bases on observations. *Global Change Biol.* **2,** 309–318

Tans, P. P., Berry, J. A., and Keeling, R. F. (1993). Oceanic ^{13}C data. A new window on CO_2 uptake of oceans. *Global Biogeochem Cycles* **7,** 353–368.

Tans, P. P., Fung, I. Y., and Takahashi, T. (1990). Observational constraints on the global atmospheric CO_2 budget. *Science* **247,** 1431–1438.

Trolier, M., White, J. W. C., Tans, P. P., Masarie, K. A., and Gemey, P. A. (1996). Monitoring the isotopic composition of atmospheric CO_2: Measurements from NOAA global air sampling network. *J. Geophys. Res.* **101,** 25,897–25,916.

Walker, B. and Steffen, W. (1996). "Global Change and Terrestrial Ecosystems." p. 619. Cambridge University Press, Cambridge.

Woodward, F. I. (1987). "Climate and Plant Distribution." p. 173. Cambridge University Press, Cambridge.

Yakir, D. and Wang, X. F. (1996). Fluxes of CO_2 and water between terrestrial vegetation and the atmosphere estimated from isotope measurements. *Nature* **380,** 515–517.

References for Appendix

Berry, S. C., Varney, G. T., and Flanagan, L. B. (1997). Leaf δ^{13}C in *Pinus resinosa* trees and understory plants: Variation associated with light and CO_2 gradients. *Oecologia (Berlin)* **109,** 499–506.

Broadmeadow, M. S. J., Griffiths, H., Maxwell, C., and Borland, A. M. (1992). The carbon isotope ratio of plant organic material reflects temporal and spatial variations in CO_2 within tropical forest formations in Trinidad. *Oecologia (Berlin)* **89,** 435–441.

Buchmann, N. and Ehleringer, J. R. (1998). A comparison of CO_2 concentration profiles, carbon isotopes, and oxygen isotopes in C_4 and C_3 crop canopies. *Agri. For. Meteorol.* **89,** 45–58.

Buchmann, N., Hinckley, T. M., and Ehleringer, J. R. (1998). Carbon isotopic dynamics at the leaf and the canopy level in montane *Abies amabilis* stands in the Cascades: influence of age and nutrition. *Can. J. For. Res.* **28,** 808–819.

Buchmann, N., Guehl, J. M., Barigah, T. S., and Ehleringer, J. R. (1997a). Interseasonal comparison of CO_2 concentrations, isotopic composition, and carbon dynamics in an Amazonian rainforest (French Guiana). *Oecologia (Berlin)* **110,** 120–131.

Buchmann, N., Kao, W. Y., and Ehleringer, J. R. (1997b). Influence of stand structure on carbon-13 of vegetation, soils, and canopy air within deciduous and evergreen forest in Utah, United States. *Oecologia (Berlin)* **110,** 109–120.

Ehleringer, J. R. and Cook, C. S. (1998). Carbon and oxygen isotope ratios of ecosystem respiration along an Oregon conifer transect: Preliminary observations based upon small-flask sampling. *Tree Physiol.* **18,** 513–519.

Flanagan, L. B., Brooks, J. R., and Ehleringer, J. R. (1997). Photosynthesis and carbon isotope discrimination in boreal forest ecosystems: A comparison of functional characteristics in plants from three mature forest types. *J. Geophys. Res.* **102,** 28,861–28,869.

Francey, R. J., Gifford, R. M., Sharkey, T. D., and Weir, B. (1985). Physiological influences on carbon isotope discrimination in huon pine (*Lagarostrobus franklinii*). *Oecologia (Berlin)* **66,** 211–218.

Harwood, K. (1997). Variation in atmospheric canopy $^{13}CO_2$, $C^{18}O^{16}O$, and $H_2^{18}O$. Ph.D. thesis. University of Newcastle, Newcastle upon Tyne, United Kingdom.

Högberg, P. and Ekblad, A. (1996). Substrate-induced respiration measured in situ in a C_3-plant ecosystem using additions of C_4-sucrose. *Soil. Biol. Biochem.* **28,** 1131–1138.

Keeling, C. D. (1961). The concentration and isotopic abundances of carbon dioxide in rural and marine air. *Geochim. Cosmochim. Acta* **24,** 277–298.

Lancaster, J. (1990). Carbon-13 fractionation in carbon dioxide emitting diurnally from soils and vegetation at ten sites on the North American continent. Ph.D. thesis, University of California, San Diego, California.

Lloyd, J., Kruijt, B., Hollinger, D. Y., Grace, J., Francey, R. J., Wong, C. S., Kelliher, F. M., Miranda, A. C., Gash, J. H. C., Vygodskaya, N. N., Wright, I. R., Miranda, H. S., Farquhar, G. D., and Schulze, E. D. 1996. Vegetation effects on the isotopic composition of atmospheric CO_2 at local and regional scales: Theoretical aspects and a comparison between rain forest in Amazonia and a boreal forest in Siberia. *Aust. J. Plant. Physiol.* **23,** 371–399.

Miranda, A. C., Miranda, H. S., Lloyd, J., Grace, J. C., Francey, R. J., McIntyre, J. A., Meir, P., Riggan, P., Lockwood, R., and Brass, J. (1997). Fluxes of carbon, water and energy over Brazilian Cerrado: An analysis using eddy covariance and stable isotopes. *Plant, Cell Environ.* **20,** 315–328.

Quay, P., King, S., and Wilbur, D. (1989). $^{13}C/^{12}C$ of atmospheric CO_2 in the Amazon Basin: forest and river sources. *J. Geophys. Res.* 94, **18,** 327–18,336.

Yakir, D. and Wang, X. F. (1996). Fluxes of CO_2 and water between terrestrial vegetation and the atmosphere estimated from isotope measurements. *Nature* **380,** 515–517.

1.21

Photosynthetic Pathways and Climate

James R. Ehleringer
Department of Biology
University of Utah
Salt Lake City, Utah

Thure E. Cerling
Department of Geology and
Geophysics University
of Utah
Salt Lake City, Utah

1. Introduction

Other chapters in this volume have explored carbon cycles within and among ecosystems, especially their response to the global changes that are occurring on earth today. In this chapter, the focus shifts from factors that influence carbon flux dynamics to the ways in which the composition of the atmosphere and thermal environment influence the type of photosynthetic system that predominates within a terrestrial ecosystem. In turn, the kind of photosynthetic system present has significant impacts on the distribution of the grazing animals that are dependent on primary productivity generated across the landscape, both in the short-term and over evolutionary time periods.

Three photosynthetic pathways exist in terrestrial plants: C_3, C_4, and CAM photosynthesis (Ehleringer and Monson, 1993). C_3 photosynthesis is the ancestral pathway for carbon fixation and occurs in all taxonomic plant groups. C_4 photosynthesis occurs in the more advanced plant taxa and is especially common among monocots, such as grasses and sedges, but not very common among dicots (Ehleringer *et al.*, 1997; Sage and Monson, 1999). CAM photosynthesis occurs in many epiphytes and succulents from very arid regions, but is sufficiently limited in distribution so that CAM plants are not an appreciable component of the global carbon cycle. Therefore, this chapter will focus on the factors influencing the dynamics of C_3- and C_4-dominated ecosystems.

Photosynthesis is a multistep process in which the C from CO_2 is fixed into stable organic products. In the first step, RuBP carboxylase-oxygenase (Rubisco) combines RuBP (a 5C molecule) with CO_2 to form two molecules of phosphoglycerate (PGA, 3C molecule). However, Rubisco is an enzyme capable of catalyzing two distinct reactions: one leading to the formation of two molecules of PGA when CO_2 is the substrate and the other resulting in

one molecule each of PGA and phosphoglycolate (PG, 2C molecule) when O_2 is the substrate (Lorimer, 1981). The latter oxygenase reaction results in less net carbon fixation and eventually leads to the production of CO_2 in a process known as photorespiration:

$$RuBP + CO_2 \rightarrow PGA$$
$$RuBP + O_2 \rightarrow PGA + PG$$

The proportion of the time for which Rubisco catalyzes CO_2 versus O_2 is dependent on the $[CO_2]/[O_2]$ ratio; the reaction is also temperature-dependent, with oxygenase activity increasing with temperature. This dependence of Rubisco on the $[CO_2]/[O_2]$ ratio establishes a firm link between current atmospheric conditions and photosynthetic activity. As a consequence of Rubisco sensitivity to O_2, the efficiency of the C_3 pathway decreases as atmospheric CO_2 decreases.

C_4 photosynthesis represents a biochemical and morphological modification of C_3 photosynthesis to reduce Rubisco oxygenase activity and thereby increase the photosynthetic rate in low-CO_2 environments such as we have today (Ehleringer *et al.*, 1991; Sage and Monson, 1999). In C_4 plants, the C_3 cycle of the photosynthetic pathway is restricted to interior cells within the leaf (usually the bundle-sheath cells). Surrounding the bundle-sheath cells are mesophyll cells in which a much more active enzyme, PEP carboxylase, fixes CO_2 (but as HCO_3^-) into oxaloacetate, a C_4 acid. The C_4 acid diffuses to the bundle-sheath cell, where it is decarboxylated and refixed in the normal C_3 pathway. As a result of the higher activity of PEP carboxylase, CO_2 is effectively concentrated in the regions where Rubisco is located and this results in a high CO_2/O_2 ratio and limited photorespiratory activity.

When the focus is on ecosystem processes, an appropriate question to ask might be, "Why be concerned about the fact that different photosynthetic pathways exist?" There are several important and clear answers to this question. First, C_3 and C_4 species are capable of giving quite different photosynthetic rates and primary productivity rates, even when grown under similar environmental conditions (Sage and Monson, 1999). Second, morphological and possibly defensive-compound differences between C_3 and C_4 species lead to differences in feeding preferences among herbivores (Caswell *et al.*, 1973; Ehleringer and Monson, 1993; Sage and Monson, 1999). Third, photosynthetic pathways among intensively managed ecosystems, such as pastures and agricultural crops, differ in both productivity and water-use efficiency, exhibiting strong geographical tendencies that reflect climatic variations. Last, the natural distributions of both C_3 and C_4 species exhibit strong relationships with both atmospheric CO_2 and climate, suggesting that future plant distributions need not be similar to today's distributions.

2. A Physiological Basis for C_3/C_4 Plant Distributions

Photorespiration impacts both maximum photosynthetic rate and photosynthetic light-use efficiency (Björkman, 1966; Ehleringer and Björkman, 1977; Ehleringer and Pearcy, 1983; Sage and Monson, 1999). One consequence is that light-use efficiency or quantum yield for CO_2 uptake differ between C_3 and C_4 taxa (Ehleringer and Björkman, 1977). The quantum yield is defined as the slope of the photosynthetic light-response curve at low light levels. As the total leaf area within a canopy increases, an increasing proportion of the canopy-level carbon gain is influenced by light-use efficiency because the light level that the average leaf within the canopy is exposed to reduces with increasing total leaf area. The reduced quantum yield values in C_4 taxa are temperature independent and reflect the additional ATP costs associated with operation of the C_4 cycle (Hatch, 1987; Kanai and Edwards, 1999). In contrast, the quantum yield values of C_3 taxa are reduced as temperatures increase, because Rubisco oxygenase activity increases with temperature. As a consequence, for any given set of atmospheric CO_2 and O_2 conditions, the light-use efficiency of C_3 plants will exceed that of C_4 plants at lower air temperatures and will fall below that of C_4 plants at higher temperatures.

Cerling *et al.* (1997) and Ehleringer *et al.* (1997) modeled the effects of variations in C_3/C_4 quantum yields on predicted photosynthetic carbon gain under different environmental combinations of atmospheric CO_2 and temperature. They predicted that as atmospheric CO_2 levels decreased, C_4 photosynthesis should become increasingly more common because of its higher light-use efficiency (Fig. 1).This model predicts that C_3 plants predominated during periods of the earth's history when atmospheric CO_2 levels were above ~ 500 ppmV. Plants with the C_4 pathway are predicted to have a selective advantage only in the warmest ecosystems at atmospheric CO_2 levels close to 500 ppmV. However, as atmospheric CO_2 levels decrease further, the advantage of C_4 photosynthesis and C_4 dominance are predicted to drift toward cooler habitats.

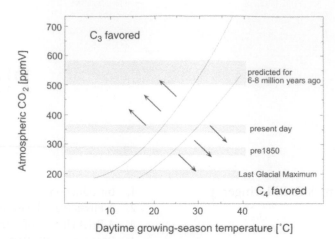

FIGURE 1 Modeled crossover temperatures of the quantum yield for CO_2 uptake for C_3 and C_4 plants as a function of atmospheric CO_2 concentrations. The boundary conditions shown are for NADP-me C_4 plants (upper boundary) and NAD-me C_4 plants (lower boundary). The crossover temperature is defined as the temperature (for a particular atmospheric CO_2 concentration) at which the quantum yields for CO_2 uptake are equivalent for both the C_3 and the C_4 plant. Figure is modified from Ehleringer *et al.* (1997).

3. A Brief History of Atmospheric Carbon Dioxide Levels

The significance of the "quantum yield" model's prediction of C_3/C_4 distributions is best viewed in the context of atmospheric CO_2 changes that have occurred over the past several hundred million years. The history of levels of atmospheric CO_2 is related to its production through volcanism relative to the losses associated with weathering, photosynthesis, and burial in the oceans (Berner, 1994, 1997). The important biogeochemical processes contributing to the change in atmospheric CO_2 are

$$CaSiO_3 + CO_2 \longrightarrow CaCO_3 + SiCO_3$$

and

$$H_2O + CO_2 \longrightarrow CH_2O + O_2,$$

where the first reaction describes weathering and the formation of carbonate sediments that are finally deposited in oceanic carbon sinks and the second reaction is an abbreviated description of the production and burial of organic matter in terrestrial sediments. The combination of these two reactions and the presence of liquid water on earth results in a long-term decline in atmospheric CO_2 values (Berner, 1991).

While there is uncertainty about the atmospheric CO_2 values prior to half a million years, most modeling and analytical approaches suggest that atmospheric CO_2 levels were substantially higher in the Cretaceous than they are today (Fig. 2, left). Modeled and experimental approaches further agree that atmospheric CO_2 levels began to decline during the late Cretaceous, eventually settling into a range of concentrations less than 500 ppmV. These

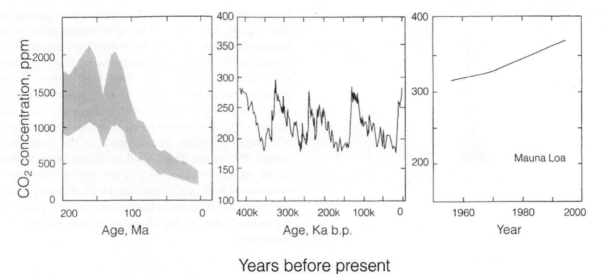

FIGURE 2   Patterns of atmospheric CO_2 concentrations through time. Left plate: reconstruction of paleo CO_2 levels between 200 Ma and present; adopted from Cerling *et al.* 1998a. Middle plate: reconstruction of atmospheric CO_2 from ice cores for the past 160,000 years; adopted from Petit *et al* (1999). Right plate: atmospheric CO_2 concentrations recorded at Mauna Loa, Hawaii since 1958; adopted from Keeling (1998). records at ORNL CDIAC

relatively low atmospheric CO_2 levels are thought to have characterized the earth's atmosphere from the late Miocene up to the dawn of the Industrial Revolution. Icecore data, particularly the lengthy Vostok ice core observations from Antarctica (Jouzel *et al.*, 1987; Petit *et al.*, 1999), indicate that over the past 420,000 years there have been oscillations in the atmospheric CO_2 from 180 to 280 ppmV, associated with glacial and interglacial periods, respectively (Fig. 2, middle). In contrast to this long-term historical pattern is an anthropogenically induced increase in atmospheric CO_2 levels, especially during the 20th century, to values well in excess of 350 ppmV in association with the continued combustion of fossil fuels (Fig. 2, right).

The answer to "why should natural selection favor the emergence of a second photosynthetic pathway?" is seen in the large decreases in atmospheric CO_2 that have occurred over the past 200 million years, particularly the changes in atmospheric CO_2 levels in the past 6–8 million year, while during the same interval atmospheric O_2 levels are thought to have remained almost constant. It is the changes in the $[CO_2]/[O_2]$ ratio that result in decreased photosynthesis by C_3 plants as photorespiration rates increase, which favors the evolution and expansion of C_4 photosynthesis. The higher activity of PEP carboxylase effectively creates a "CO_2 pump," resulting in a $[CO_2]/[O_2]$ ratio inside the bundle-sheath cells of C_4 plants that is several-fold greater than observed at sites of Rubisco activity in C_3 plants. The "quantum yield" model predicts how common C_4 photosynthesis is expected to be for any global atmospheric CO_2 level. Specifically, the model predicts the temperature ranges that should have favored C_4 over C_3 as atmospheric CO_2 declined over the last 200 million years.

The decreased atmospheric CO_2 levels have had enormous con-

sequences both for the distribution of plant communities across our planet and for animal evolution, as will be discussed below. While throughout much of history, earth had been subject to relatively high atmospheric CO_2 levels, the earth has now been in a "CO_2-starved mode" for approximately 7 Ma with periods of exceptionally low atmospheric CO_2 levels (~180 ppmV) characterizing the atmosphere during recent glacial periods.

4. Recognizing the Presence of C_3 and C_4 Ecosystems in the Paleorecord

Carbon isotope ratios can be used to identify the presence of C_3 versus C_4 photosynthesis in the fossil records. Large differences in discrimination against $^{13}CO_2$ by the initial carboxylation reactions in C_3 (RuBP carboxylase) and C_4 (PEP carboxylase) photosynthesis result in significant differences in the carbon isotope ratios ($\delta^{13}C$) of C_3 and C_4 plants (Farquhar *et al.*, 1989). Modern C_3 plants average approximately $-27‰$ and C_4 plants average approximately $-13‰$ (Fig. 3). The observed ranges of $\delta^{13}C$ values for both C_3 and C_4 plants are the result of genetic differences among taxa as well as responses to variations in environmental conditions, including light and water stress (Farquhar *et al.*, 1989; Ehleringer *et al.*, 1993; Buchmann *et al.*, 1996). Differences among C_4 photosynthetic subtypes (NADP-me, NAD-me, and PCK) contribute as much as 1–2‰ to the range of values shown in Fig. 3 (Hattersley, 1982; 1983).

Animal tissues faithfully record the isotopic composition of their food sources (Tieszen *et al.*, 1983; Hobson, 1999), but often are not preserved in the fossil records or are subject to alteration

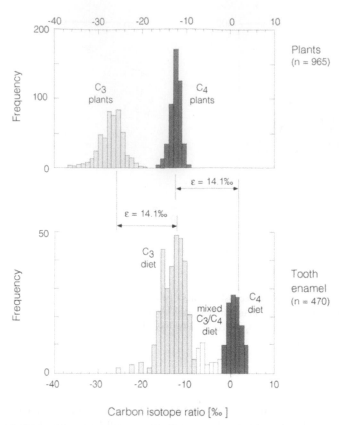

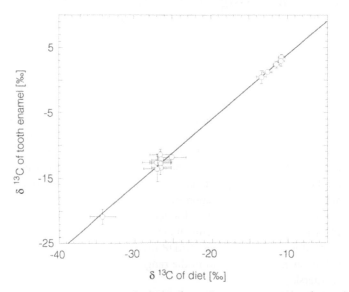

FIGURE 3 Histograms of the carbon isotope ratios of modern grasses and modern tooth enamel; adopted from Cerling *et al.* (1997).

(diagenesis) during fossilization. However, tissues such as tooth enamel are preserved without subsequent modification, thus recording the original animal diet over periods of several million years (Lee-Thorp and van der Merwe, 1987). Tooth enamel

FIGURE 4 Relationship between the carbon isotope ratio values of estimated diet and measured tooth enamel for ungulate mammals; adopted from Cerling and Harris (1999).

(bioapatite) is enriched 14.1‰ relative to a grazing mammal's diet (Fig. 4), resulting in a straightforward means of recording long-term feeding patterns by mammalian grazers (Cerling and Harris, 1999). The lower histograms in Figure 3 illustrate this offset between animals and their food sources, based on an accumulation of observations of $\delta^{13}C$ values of apatite from a wide variety of grazing species (Cerling *et al.*, 1997; Cerling and Harris, 1999). It is important to note that the variation in plant $\delta^{13}C$ values is similar in magnitude to the variation in tooth-enamel $\delta^{13}C$ values. Thus, small variations in tooth-enamel $\delta^{13}C$ values on the order of 1–2‰ are just as likely to represent variations in food quality associated with changing environmental conditions as variations in the abundances of C_3 and C_4 plants in the animal's diet or the changing carbon isotope ratio of the atmosphere.

5. Global Expansion of C_4 Ecosystems

Figure 5 shows that between 8 and 6 Ma there was a global expansion of C_4 ecosystems (Cerling *et al.*, 1997, 1998a). There is no conclusive evidence for the presence of C_4 biomass in the diets of mammals before 8 Ma (Cerling *et al.*, 1997; 1998a), although the presence of small amounts of C_4 biomass in diets is not excluded because of the uncertainty in the $\delta^{13}C$ endmember for C_3 plants. By 6 Ma there is abundant evidence for significant C_4 biomass in Asia (Cerling *et al.*, 1993; 1997; Morgan *et al.*, 1994), Africa (Morgan *et al.*, 1994; Cerling *et al.*, 1997), North America (Cerling, *et al* 1993; MacFadden and Cerling, 1999; Cerling *et al.*, 1999), and South America (MacFadden *et al.*, 1996; Cerling *et al.*, 1997), but not in Europe (Cerling *et al.*, 1997). Figure 5 documents several different ecosystem type changes as recorded in mammalian tooth enamel. While each of these regions appears to have been dominated by C_3 ecosystems earlier in the Miocene, the C_3 Pakistani ecosystem was almost completely replaced by a C_4 ecosystem; African, North American, and South American ecosystems retained both C_3 and C_4 components; European and northern portions of North American ecosystems did not show any change in the fraction of C_3 biomass, remaining at virtually 100% C_3 ecosystems. The mixture of C_3 and C_4 components within a grazing ecosystem can be achieved by one of two ways: a temporal separation with C_3 grasses active in winter–spring and C_4 grasses active in summer or a monsoonal system with C_4 grasses and C_3 woody vegetation. Without fine-scale analyses of the seasonal dynamics with tooth enamel, the isotopic record is silent as to which pattern existed.

The isotopic evidence is clear that the expansion of C_4 ecosystems was a global phenomenon, persisting until today. It was accompanied by significant faunal changes in many parts of the world. It is unlikely that the global expansion of C_4 biomass in the late Miocene was due solely to higher temperatures or to the development of arid regions. There have always been regions on earth with hot, dry climates. To explain the simultaneous global expansion of C_4 plants requires a global phenomenon. Changes in

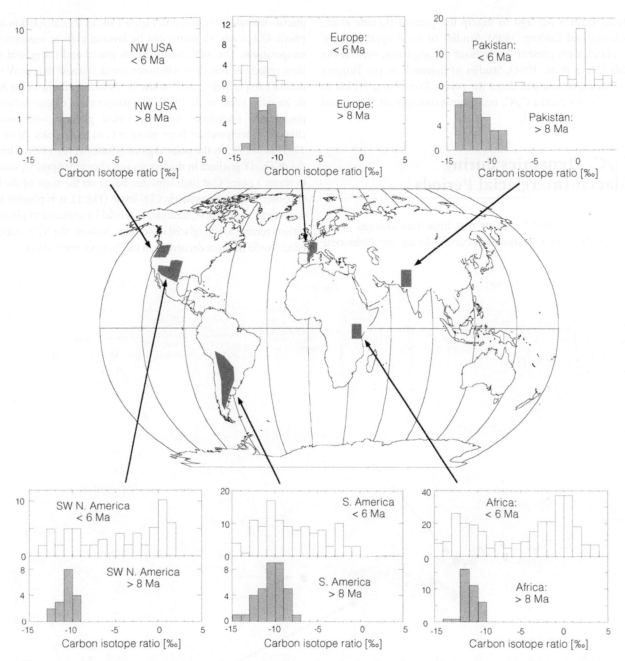

FIGURE 5 Histograms comparing the carbon isotope ratio values for fossil tooth enamel older than 8 Ma (lower charts) with those that are younger than 6 Ma for six regions of earth; adopted from Cerling *et al.* (1998a).

atmospheric CO_2 as predicted by the quantum-yield model are a strong possibility for this global mechanism. Supporting evidence indicates that the global expansion of C_4 ecosystems appears to have originated in warmer, equatorial regions and then spread to cooler regions, consistent with the temperature-sensitivity predictions of the quantum-yield model. Cerling *et al.* (1997) documented that within both modern and fossil horses (equids), the distributions of isotope ratios strongly support a decrease in the

importance of C_4 photosynthesis in moving from warm equatorial to cooler temperate latitudes.

Stable-isotope studies of paleosols from Pakistan and East Africa are in good agreement with the paleodietary studies. The Siwalik sequence in Pakistan has excellent exposures covering the last 20 Ma. $\delta^{13}C$ studies of paleosol carbonates show a virtually pure C_3 ecosystem up to about 7.5 Ma ago, a transitional period of ecosystem change lasting 1–1.5 Ma, and then C_4-dominated ecosystems from 6 Ma ago to nearly the present (Quade *et al.*, 1989; Quade and Cerling, 1995). Studies of fossil eggshell show that C$_3$ plants were present throughout the sequence, even in the last 6 Ma (Stern *et al.*, 1994). Studies of paleosols in the Turkana Basin in Africa, covering in detail the period from about 7.5 Ma to the present, show mixed C$_3$/C$_4$ ecosystems throughout this period (Cerling *et al.*, 1993; 1997).

6. C$_3$/C$_4$ Dynamics during Glacial-Interglacial Periods

The quantum-yield model predicts that important changes in the global proportions of C$_4$ biomass occurred during the Pleistocene glacial–interglacial transitions. Figure 1 shows that at very low atmospheric CO$_2$ levels, C$_4$ plants can be favored even at moderately low temperatures. The oscillation between glacial and interglacial conditions reflected an oscillation between about 180 and 280 ppmV (Fig. 2, middle), respectively, based on the CO$_2$ concentrations in the Antarctic ice cores (Petit *et al.*, 1999). The temperature change between the glacial and interglacial intervals varied globally, with estimated changes in temperature from about 5°C in the tropics (Stute *et al.*, 1995) to >15°C in the polar regions (Cuffey *et al.*, 1995). Therefore, the dCO$_2$/dT gradient in the tropics was about 20 ppm/°C, compared to about 7 ppm/°C at high latitudes. Based on the slope of the C$_3$ /C$_4$ crossover at low atmospheric CO$_2$ levels (Fig. 1), it is possible that in some regions greater C$_4$ abundance would be expected in glacial conditions relative to interglacial conditions, because the "CO$_2$ starvation" effect would be more decisive than the "temperature" effect.

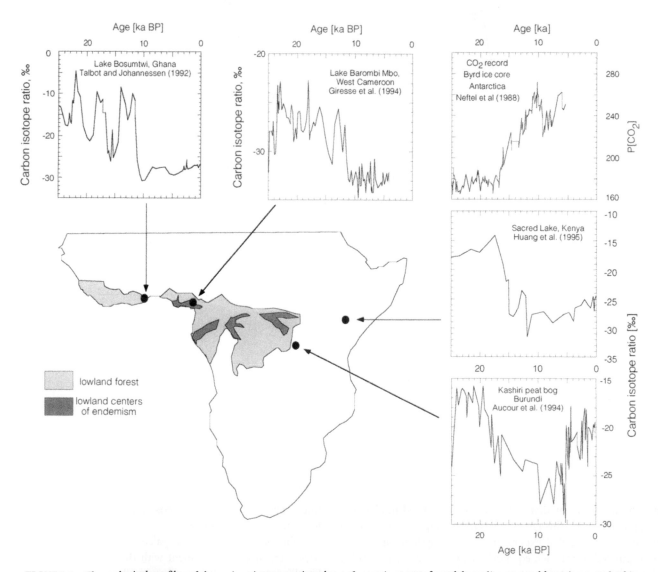

FIGURE 6 Chronological profiles of the carbon isotope ratio values of organic matter from lake sediments and bogs in central Africa. The data indicate that these areas all had more extensive C$_4$ biomass during the last glacial maximum (30–20 ka B.P. than during the Holocene (10 ka B.P. to present). Data are from Talbot and Johannessen (1992), Giresse *et al.* (1994) Aucour *et al.* (1994), and Neftel *et al.* (1988). Adopted from Cerling *et al.* (1998a).

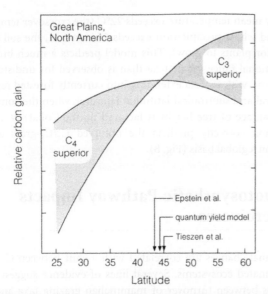

FIGURE 7 Predicted relative carbon gain by the quantum-yield model and therefore predicted competitive success by C_3- and C_4-grass canopies across the Great Plains of North America under today's atmospheric carbon dioxide levels. Noted are the predicted cross-over points from C_3- to C_4-dominance based on the quantum-yield model and the observations for soil organic matter (Tieszen *et al.*, 1997) and for aboveground harvests (Epstein *et al.*, 1997). Adopted from Ehleringer (1978).

Ehleringer *et al.* (1997) examined published reports of $\delta^{13}C$ in peat bogs and lakes from Central Africa in regions that are currently dominated by rain-forest ecosystems. The available data strongly support the hypothesis of extensive C_4 expansion during the last full glacial (Aocour *et al.*, 1993; Hillaire *et al.*, 1989)(Fig. 6). This implies extensive retreat of the African rain-forest ecosystems and has important implications for refugia during the Pleistocene which are discussed below. Farther east in Africa, sedimentary data from Sacred Lake in Kenya also show that C_4 grasses were much more common during the glacial period when C_3 vegetation would have been "CO_2 starved" (Street-Perrott *et al.*, 1995; 1997; Huang *et al.*, 1995; 1999). Following deglaciation, the C_4 abundances in the Sacred Lake region exhibited a dramatic decline correlated with the increases in atmospheric CO_2 levels.

Within North American ecosystems, there is also evidence that C_4 ecosystems were more extensive during the last glacial period than they are today. Soil carbonates from the southwestern portions of North America show that C_4 plants dominated the landscape during glacial periods, but are less abundant in these aridland ecosystems today (Cole and Monger, 1994; Liu *et al.*, 1996; Monger *et al.*, 1998). Dietary analyses of fossil herbivores from western North America also provide convincing evidence of widespread C_4 abundance in regions that have a near absence of C_4 grasses today (Connin *et al.*, 1998). While the mechanisms for the observed decline in C_4 abundance in North America require further study, the dramatic decrease in C_4 plants is correlated with the transition out of the glacial and the abrupt increases in atmospheric CO_2 levels.

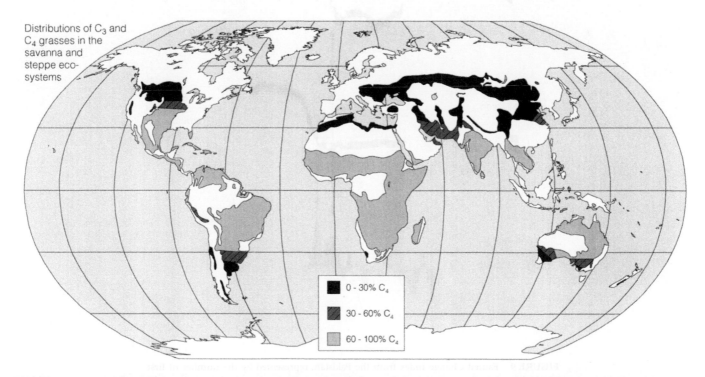

FIGURE 8 Predicted distributions of C_3 and C_4 grasses in steppe and savanna ecosystems of the world. These are the only two ecological regions where grasses are a significant fraction of the vegetation. Distribution of ecological regions is based on Bailey (1998) and the partitioning of photosynthetic pathways is based on the synthesis in Sage and Monson (1999).

7. Photosynthetic Pathway Distribution in the Modern World

The current distributions of C_4 plants within grassland ecosystems at an atmospheric CO_2 level of 350 ppmV are well predicted by the quantum-yield model (Fig. 1). Across the Great Plains of North America, the crossover between C_3- and C_4-dominated grasslands is predicted to occur at a latitude of approximately 45°N (Figure 7). Both long-term aboveground harvest studies (Epstein *et al.*, 1997) and belowground soil organic carbon studies (Tieszen *et al.*, 1997) independently indicate a C_3/C_4 transition near 45°N. In the case of C_3/C_4 grasses from the Great Plains as well as all other monocot studies, the relationships between C_3 and C_4 grass abundances were all very highly correlated with temperature (Ehleringer *et al.*, 1997). In most of these studies, >90% of the variance in C_3/C_4 abundance in today's ecosystems is explained by temperature alone.

Collatz *et al.* (1998) extended predictions of the quantum-yield model to the global scale (Fig. 8). Their model predicted that C_4 abundances are expected in all geographical regions where the monthly mean temperature exceeds 22°C (the crossover temperature) and where precipitation exceeds 25 mm (i.e., the soil must be wet for plants to grow). This model predicts a much broader distributional range for C_4 taxa than is observed for undisturbed ecosystems, with C_4 taxa extending into currently forested regions of tropical and subtropical latitudes. However, when the competitive advantage of tree height is factored in, the Collatz *et al.* extrapolation correctly predicts the observed C_3/C_4-grass abundances on a global basis (Fig. 8).

8. Photosynthetic Pathway Impacts Herbivores

Megafaunal changes are correlated with a shift between C_3- and C_4-dominated ecosystems. Several lines of evidence suggest relationships between turnover of mammalian grazing taxa and the shifts between C_3/C_4 vegetation types. Cerling *et al.* (1998a) reported abrupt changes in mammalian lineages in East Africa asso-

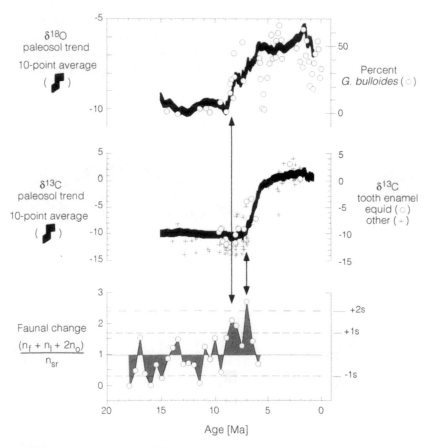

FIGURE 9 Faunal Change Index from the Pakistan, represented by the number of first (n_f) and last (n_l) occurrences, including only occurrences (n_o), normalized to species richness (n_{sr}). The Faunal Change Index is normalized to 1.0 for the total data set. Adopted from Cerling *et al.* (1998b).

ciated with the transition from C_3-dominated to C_3/C_4-dominated ecosystems. During the same time period, Cerling *et al.* (1998b) showed that abrupt changes in faunal diversity of Pakistani mammals occurred at the same time as the emergence of C_4-dominated ecosystems in Pakistan (Fig. 9). Evolutionary relationships between C_3/C_4 and horses appear to be somewhat different (Mac-Fadden and Cerling, 1996; MacFadden *et al.*, 1996). Evolution of the modern horse is associated with the transition from "browsing" to "grazing" horses, which is typically marked by the lengthening of the M1 molar, creating the high-crowned tooth (Fig. 10). However, the evolution of the M1 tooth and the increased diversity of horse taxa was not associated with the global expansion of C_4 ecosystems, because these changes occurred in a C_3 world. However, the crash in biological diversity of horses at 6 Ma is associated with the expansion of C_4-dominated ecosystems into regions that once contained only C_3 plants (Fig. 10).

Modern mammalian herbivores exhibit strong preferences for C_3 versus C_4 diets (Fig. 11), with only limited numbers of examples of mixed C_3/C_4 feeding (Figure 3). While it may be difficult to quantify the exact percentages of C_3/C_4 within the diets of some mammals, it is possible to classify the extreme grazers and browsers: hypergrazers with nearly 100% C_4 grass and hyperbrowsers with nearly 100 % C_3 browse. It is remarkable that the herbivore mammals of the savannahs and grasslands of Africa falls into such distinct C_3/C_4 dietary categories, with extreme hypergrazers such as the wildebeest standing distinct from grazers such as the oryx and zebra (Fig. 11). The modern African elephant

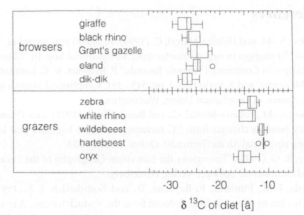

FIGURE 11 Ranges in the carbon isotope ratios of diets for African browsers and grazers. Adopted from Cerling *et al.* (1999).

(*Loxodonta*) is often regarded as a grazing animal, yet its isotopic composition strongly shows that these animals are distinctly C_3 browsers (Cerling *et al.*, 1999). In contrast, a million years ago elephants were distinctly grazers.

The selective basis for differential C_3/C_4 herbivory may be related to the differential distributions of leaf protein within C_3 and C_4 leaves (Ehleringer and Monson, 1993). In C_3 plants, relatively high protein levels are found in most mesophyll cells. These cells have relatively thin cell walls, especially when compared to the much thicker bundle-sheath cell walls (Brown, 1977). In contrast, there is relatively more protein within bundle-sheath cells of C_4 leaves than in mesophyll cells. Thus, tooth morphology in mammalian grazers would be expected to play a role in determining whether or not animals were able to extract sufficient protein from their C_3/C_4 diet. Insects such as grasshoppers show a strong preference for C_3 or C_4 food sources, but typically not for both (Isely, 1946; Caswell *et al.*, 1973; Boutton *et al.*, 1978; Ehleringer and Monson, 1993). Here it is known that there are significant differences in mandible morphology correlated with C_3/C_4 dietary preference.

9. Summary

The current distribution of C_3 and C_4 photosynthetic pathways in today's ecosystems is a strong function of temperature. Changing atmospheric CO_2 levels modify this geographical distribution. The global emergence of C_4-dominated ecosystems in the late Miocene suggests that atmospheric CO_2 levels decreased across a threshold of ~500 ppmV favoring C_4 photosynthesis over C_3 photosynthesis in warm ecosystems. More recently during glacial periods when atmospheric CO_2 levels decreased to 180 ppmV, C_4 taxa were apparently more abundant than they are today. These changes in C_3/C_4 abundances have had enormous impacts on both evolution and distribution of mammalian grazers. The mechanistic basis for this impact on mammal herbivory may be feeding preferences associated with differential digestibility of C_3 versus C_4 grasses.

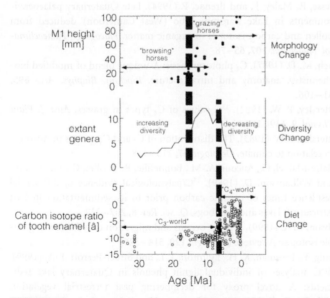

FIGURE 10 A chronology of horse evolution. Top plate: morphological changes in the height of the M1 tooth. Middle plate: diversity changes as recorded by the number of extant genera. Bottom plate: carbon isotope ratios of tooth enamel illustrating that the transition from browsing horses to grazing horses was not associated with expansion of C_4 ecosystems, but that the loss of genera was correlated in time with C_4 expansion. Adopted from Cerling (1999).

References

Aucour, A.-M. and Hillaire-Marcel, C. (1993). A 30,000 year record of ^{13}C and ^{18}O changes in organic matter from an equatorial bog. In "Climate Changes in Continental Isotopic Records." P. K. Swart, K. C. Lohmann, J. McKenzie, and S. Savin (Eds.), pp. 343–351. Geophysical Monograph 78. American Geophysical Union, Washington.

Aucour, A.-M., Hillaire-Marcel, C., and Bonnefille, R. (1994). Late Quaternary biomass changes from ^{13}C measurements in a highland peat bog from equatorial Africa (Burundi). *Q. Res.* **41**, 225–233.

Bailey, R. G. (1998). "Ecoregions the Ecosystem Geography of the Oceans and Continents." Springer-Verlag, Heidelberg.

Barnola, J.-M., Pimienta, P., Raynaud, D., and Korotkevich, Y. S. (1991.) CO_2-climate relationship as deduced from the Vostok ice core: A re-examination based on new measurements and on A re-evaluation of the air dating. *Tellus* **43B**, 83–90.

Berner, R. A. (1991). A model for atmospheric CO_2 over Phanerozoic time. *Am. J. Sci.* **291**, 339–376.

Berner, R. A. (1994). GEOCARB II: A revised model of atmospheric CO_2 over Phanerozoic time. *Am. J. Sci.* **294**, 56–91.

Berner, R. A. (1997). The rise of plants and their effect on weathering and atmospheric CO_2. *Science* **276**, 544–545.

Björkman, O. (1966). The effect of oxygen concentration on photosynthesis in higher plants. *Physiol. Plant.* **19**, 618–633.

Boutton, T. W., Cameron, G. N., and Smith, B. N. (1978). Insect herbivory on C_3 and C_4 grasses. *Oecologia* **36**, 21–32.

Brown, W. V. (1977). The Kranz syndrome and its subtypes in grass systematics. *Memoirs Torrey Bot. Club* **23**, 1–97.

Buchmann, N., Brooks, J. R., Rapp, K. D., and Ehleringer, J. R. (1996). Carbon isotope composition of C_4 grasses is influenced by light and water supply. *Plant Cell Environ.* **9**, 392–402.

Caswell, H., Reed, F., Stephenson, S. N., and Werner, P. A. (1973). Photosynthetic pathways and selective herbivory: a hypothesis. *Am. Nat.* **107**, 465–479.

Cerling, T. E. (1999). The evolution of modern grasslands and grazers. In "Paleobiology II." (D. E. Briggs, Ed.), (in press).

Cerling, T. E., Ehleringer, J. R., and Harris, J. M. (1998a). Carbon dioxide starvation, the development of C_4 ecosystems, and mammalian evolution. *Proc. R. Soc. London* **353**, 159–171.

Cerling, T. E. and Harris, J. M. (1999). Carbon isotope fractionation between diet and bioapatite in ungulate mammals and implications for ecological and paleoecological studies. *Oecologia* **120**, 347–363.

Cerling, T. E., Harris, J. M., and Leakey, M. G. (1999). Browsing and grazing in elephants: The isotope record of modern and fossil proboscideans. *Oecologia* **120**, 364–374.

Cerling, T. E., Harris, J. M., MacFadden, B. J., Leakey, M. G., Quade, J., Eisemann, V., and Ehleringer, J. R. (1997). Global vegetation change through the Miocene-Pliocene boundary. *Nature* **389**, 153–158.

Cerling, T. E., Harris, J. M., MacFadden, B. J., Quade, J., Leakey, M. G., Eisemann, V., and Ehleringer, J. R. (1998b). Miocene–Pliocene shift: One step or several? *Nature* **393**, 127.

Cerling, T. E., Wang, Y., and Quade, J. (1993). Expansion of C_4 ecosystems as an indicator of global ecological change in the late Miocene. *Nature* **361**, 344–345.

Cole, D. R. and Monger, H. C. (1994). Influence of atmospheric CO_2 on the decline of C_4 plants during the last deglaciation. *Nature* **368**, 533–536.

Collatz, G. J., Berry, J. A., and Clark, J. S. (1998). Effects of climate and atmospheric CO_2 partial pressure on the global distribution of C_4 grasses: Present, past, and future. *Oecologia* **114**, 441–454.

Connin, S. L., Betancourt, J., and Quade, J. (1998). Late Pleistocene C_4 plant dominance and summer rainfall in the southwestern United States from isotopic study of herbivore teeth. *Q. Res.* **50**, 179–193.

Cuffey, K. M., Clow, G. D., Alley, R. B., Stuiver, M., Waddington, E. D., and Saltus, R. (1995). Large Arctic temperature change at the Wisconsin–Holocene glacial transition. *Science* **270**, 455–458.

Ehleringer, J. R. (1978). Implications of quantum yield differences to the distributions of C_3 and C_4 grasses. *Oecologia* **31**, 255–267.

Ehleringer, J. and Björkman, O. (1977). Quantum yields for CO_2 uptake in C_3 and C_4 plants: dependence on temperature, CO_2 and O_2 concentrations. *Plant Physiol.* **59**, 86–90.

Ehleringer, J. R. and Cerling, T. E. (1995). Atmospheric CO_2 and the ratio of intercellular to ambient CO_2 levels in plants. *Tree Physiol.* **15**, 105–111.

Ehleringer, J. R., Cerling, T. E., and Helliker, B. R. (1997). C_4 photosynthesis, atmospheric CO_2, and climate. *Oecologia* **112**, 285–299.

Ehleringer, J. R., Hall, A. E., and Farquhar, G. D. (1993). Stable isotopes and plant carbon/water relations. Academic Press, San Diego.

Ehleringer, J. R. and Monson, R. K. (1993). Evolutionary and ecological aspects of photosynthetic pathway variation. *Annu. Rev. Ecol. Systematics* **24**, 411–439.

Ehleringer, J. R. and Pearcy, R. W. (1983). Variation in quantum yields for CO_2 uptake in C_3 and C_4 plants. *Plant Physiol.* **73**, 555–559.

Ehleringer, J. R., Sage, R. F., Flanagan, L. B., and Pearcy, R. W. (1991). Climate change and the evolution of C_4 photosynthesis. *Trends Ecol. Evolut.* **6**, 95–99.

Epstein, H. E., Lauenroth, W. K., Burke, I. C., and Coffin, D. P. (1997). Productivity patterns of C_3 and C_4 functional types in the U.S. Great Plains. *Ecology* **78**, 722–731.

Farquhar, G. D., Ehleringer, J. R., and Hubick, K. T. (1989). Carbon isotope discrimination and photosynthesis. *Annu. Rev. Plant Physiol. Mol. Biol.* **40**, 503–537.

Giresse, P., Maley, J., and Brenac, P. (1994). Late Quaternary palaeoenvironments in Lake Barombi Mbo (West Cameroon) deduced from pollen and carbon isotopes of organic matter. *Paleogeogr. Palaeoclimatol. Palaeoecol.* **107**, 65–78.

Hatch, M. D. (1987). C_4 photosynthesis: A unique blend of modified biochemistry, anatomy and ultrastructure. *Biochim. Biophys. Acta* **895**, 81–106.

Hattersley, P. W. (1982). ^{13}C values of C_4 types in grasses. *Aust. J. Plant Physiol.* **9**, 139–154.

Hattersley, P. W. (1983). The distribution of C_3 and C_4 grasses in Australia in relation to climate. *Oecologia* **57**, 113–128.

Hillaire-Marcel, C., Aucour, A.-M., Bonnefille, R., Riollet, G., Vincens, A., and Williamson, D. (1989). ^{13}C/palynological evidence of differential residence times of organic carbon prior to its sedimentation in East African rift lakes and peat bogs. *Q. Sci. Rev.* **8**, 207–212.

Hobson, K. A. (1999). Tracing origins and migration of wildlife using stable isotopes: A review. *Oecologia* **120**, 314–326.

Huang, Y., Freeman, K. H., Eglinton, T. I., and Street-Perrott, F. A. (1999). $\delta^{13}C$ analyses of individual lignin phenols in Quaternary lake sediments: A novel proxy for deciphering past terrestrial vegetation changes. *Geology* **27**, 471–474.

Huang, Y., F. A., Street-Perrott, R. A., Perrott, R. A., and Eglinton, G. (1995). Molecular and carbon isotope stratigraphy of a glacial/interglacial sediment sequence from a tropical freshwater lake: Sacred Lake, Mt. Kenya. In "Organic Geochemistry: Developments and Applications to Energy, Climate, Environment and Human History." (J. O. Grimalt and C. Dorronsoro, Eds.), pp. 826–829. AIGOA, Spain.

Isely, F. B. (1946). Differential feeding in relation to local distribution of grasshoppers. *Ecology* **27**, 128–138.

Jouzel, J., Loriu, C., Petit, J. R., Genthon, C., Barkov, N. I., Kotlyakov, V. M., and Petrov, V. M. (1987). Vostok ice core: A continuous isotope temperature record over the last climatic cycle (160,000 years). *Nature* **329**, 403–408.

Kanai, R. and Edwards, G. E. (1999). The biochemistry of C$_4$ photosynthesis. In "C$_4$ Plant Biology." R. F. Sage and R. K. Monson, eds.), pp. 49–87. Academic Press, San Diego.

Lee-Thorp, J. and van der Merwe, N. J. (1987). Carbon isotope analysis of fossil bone apatite. *S. Afr. J. Sci.* **83**, 712–715.

Liu, B., Phillips, F. M., and Campbell, A. R. (1996). Stable carbon and oxygen isotopes of pedogenic carbonates, Ajo Mountains, southern Arizona: Implications for paleoenvironmental change. *Palaeogeogr. Palaeoclimatol. Palaeoecol.* **124**, 233–246.

Lorimer, G. H. (1981). The carboxylation and oxygenation of ribulose 1,5-bisphosphate: The primary events in photosynthesis and photorespiration. *Annu. Rev. Plant Physiol.* **32**, 349–383.

MacFadden, B. J. and Cerling, T. E. (1996). Mammalian herbivore communities, ancient feeding ecology, and carbon isotopes: A 10 million-year sequence from the Neogene of Florida. *J. Verteb. Paleontol.* **16**, 103–115.

MacFadden, B. J., Cerling, T. E., and Prado, J. (1996). Cenozoic terrestrial ecosystem evolution in Argentina: Evidence from carbon isotopes of fossil mammal teeth. *Palaios* **11**, 319–327.

Monger, H. C., Cole, D. R., Gish, J. W., and Giorano, T. H. (1998). Stable carbon and oxygen isotopes in Quaternary soil carbonates as indicators of ecogeomorphic changes in the northern Chihuahuan Desert USA. *Geoderma* **82**, 137–172.

Monson, R. K. (1989). On the evolutionary pathways resulting in C$_4$ photosynthesis and Crassulacean acid metabolism. *Adv. Ecol. Res.* **19**, 57–110.

Morgan, M. E., Kingston, J. D., and Marino, B. D. (1994). Carbon isotope evidence for the emergence of C$_4$ plants in the Neogene from Pakistan and Kenya. *Nature* **367**, 162–165.

Neftel, A., Oeschger, H., Staffleback, T., and Stauffer, B. (1988). CO$_2$ record in the Byrd ice core 50,000–5,000 years BP. *Nature* **331**, 609–611.

Petit, J. R., Jouzel, J., Raynaud, D., Barkov, N. I., Barnola, J.-M., Basile, I., Benders, M., Chappellaz, J., Davis, M., Delaygue, G., Delmotte, M., Kotlyakov, V. M., Legrand, M., Lipenkov, V. Y., Lorius, C., Pépin, L., Ritz, C., Saltzman, E., and Stievenard, M. (1999). Climate and atmospheric history of the past 420,000 years from the Vostok ice core, Antarctica. *Nature* **399**, 429–436.

Quade, J. and Cerling, T. E. (1995). Expansion of C$_4$ grasses in the Late Miocene of Northern Pakistan: Evidence from stable isotopes in paleosols. *Palaeogeogr. Palaeoclimatol. Palaeoecol.* **115**, 91–116.

Quade, J., Cerling, T. E., and Bowman, J. R. (1989). Development of Asian monsoon revealed by marked ecological shift during the latest Miocene in Northern Pakistan. *Nature* **342**, 163–166.

Sage, R. F. and Monson, R. K. (1999). "C$_4$ Plant Biology." Academic Press, San Diego.

Stern, L. A., Johnson, G. D., and Chamberlain, C. P. (1994). Carbon isotope signature of environmental change found in fossil ratite eggshells from a South Asian Neogene sequence. *Geology* **22**, 419–422 .

Street-Perrott, F. A., Huang, Y., Perrott, R. A., Eglinton, G., Barker, P., Ben Khelifa, L., Harkness, D. D., and Olago, D. O. (1997). The impact of lower atmospheric CO$_2$ on tropical mountain ecosystems. *Science* **278**, 1422–1426.

Stute, M., Forster, M., Frischkorn, H., Serejo, A., Clark, J. F., Schlosser, P., Broecker, W. S., and Bonani, G. (1995). Cooling of tropical Brazil (5°C) during the last glacial maximum. *Science* **269**, 379–383.

Talbot, M. R. and Johannessen, T. (1992). A high resolution palaeoclimatic record for the last 27,500 years in tropical west Africa from the carbon and nitrogen isotopic composition of lacustrine organic matter. *Earth Planet. Sci. Lett.* **110**, 23–37.

Tieszen, L. L., Boutton, T. W., Tesdahl, K. G., and Slade, N. A. (1983). Fractionation and turnover of stable carbon isotopes in animal tissues: Implications for ^{13}C analysis of diet. *Oecologia* **57**, 32–37.

Tieszen, L. L., Reed, B. C., Bliss, N. B., Wylie, B. K., and DeJong, D. D. (1997). NDVI, C$_3$ and C$_4$ production, and distributions in Great Plains grassland land cover classes. *Ecol. Applic.* **7**, 59–78.

Biological Diversity, Evolution, and Biogeochemistry

H.A. Mooney
Stanford University,
Stanford, California

The phenomenon that I probe here is the relationship between biodiversity and biogeochemistry. To do this I take an evolutionary as well as an ecological approach. The basic question that I ask is whether one needs to take into account biodiversity in considering global biogeochemical cycles, the most important of which involve both solid and gas phases. Why is this a concern? First, we know that there are strong interactions between the terrestrial biosphere and the atmosphere—what happens with one affects the other (Mooney *et al.*, 1987). We know in a general sense that the chemistry and metabolism of organisms affect both terrestrial and atmospheric processes. The surface of the earth is covered by millions of different species of organisms, some of which have vast ranges and constitute considerable biomass. We know that biodiversity is being altered in major ways on the earth's surface. What are the effects of these modifications on biogeochemistry?

1. What Do We Have and What Are We Losing?

At present there are 1.7 million described species, nearly a million of which are insects. There are many more. UNEP (1995) gives the best guess that there are six undescribed organisms for every one that has been described. These are very large numbers. The number of these species whose existence is threatened by the activities of humans is considerable. For the better-known groups, such as mammals, birds, fish, and higher plants, we have good estimates of the numbers of threatened species. For these groups, the percentages threatened of the total known species are 18, 11, 5, and 8%, respectively. A recent survey notes that 10% of tree species are

threatened (Oldenfield *et al.*, 1998). The earth has seen the origins and demise of species through time but the current extinction rates are up to three orders of magnitude over evolutionary background values.

2. Do Species Losses Matter for Biogeochemical Cycling?

2.1 From First Principles—No

As in any comparison, one can find similarities among sets or find differences. First, I consider briefly the similarity arguments, that is, the position that species differences are not large enough to affect the kinds of processes that occur at the global level. From a functional point of view one can focus on similarities among species, as Eigen and Schuster (1977) have noted, " . . . millions of species, plants and animals, exist, while there is only one basic machinery of the cell; one universal genetic code and unique chiralities of the macromolecules." This particular observation supports the viewpoint that species differences are only frills upon the basic plan of organisms. For example, in terms of carbon fixation by plants, with the exception of very specialized organisms that use chemical energy rather than light energy, there has essentially been only one type of photosynthetic pathway, the Calvin cycle. The Hatch–Slack pathway is a relatively recent evolutionary innovation, as well as discovery by scientists (Hatch and Slack, 1966).

This similarity among species, in rather fundamental functional properties, has no doubt led to the development and use of earth system models that have little diversity content, but rather use the color of the land surface only, such as the green scum production models and the Daisy World model of Lovelock (1994). There certainly is a rationale for the use of very simple approaches in viewing the functional diversity of the world in global productivity models. The finding of Monteith (1977) that the productivity of plants is dependent on the energy absorbed rather than on species per se is an important finding that has been utilized extensively in global modeling. The CASA biogeochemistry model builds on this principle (Potter *et al.*, 1993) with corrections for the effects of environmental stress on the basic light production relationship.

Similarity can be structural as well as functional when viewed at a regional level. One of the fundamental principles of biogeography is the concept of convergent evolution. This is based on the observation that in similar climates, even though separated geographically, comparable vegetation structures will be present. It has also been shown that similarity in structure also reflects similarity in function (Mooney, 1977). It is a fact that the convergent vegetation types of the world have comparable productivities. Thus much can be said about the earth's biogeochemistry without knowing much about the millions of species that constitute the earth's biota.

2.2 From First Principles — Yes

Since no two species can coexist on the same limiting resource (Gause, 1934) there is an inexorable evolutionary drive for biotic diversification. Natural selection favors innovation in acquiring new resources in the environment be it a new way, for a given habitat, to capture light, water, or nutrients. This means that through time for a given site, a single species is not able to acquire all of the available resources to derive maximum productivity, for example. By this reasoning, removing species should impact production and hence biogeochemistry. By the same token adding species to mixtures, as is done in agroforestry, is an attempt to optimize total system productivity.

In the following I examine the apparent contradiction between these viewpoints, looking at the various kinds of diversity that are found among organisms and its significance in terms of biogeochemical functioning with the main focus, again, being on the question of whether we should be concerned about species losses (or additions).

3. Kinds of Diversity

When considering the diversity of the biological systems of the earth, we traditionally consider only species. In recent times though, there has been an additional, and important, emphasis on the other dimensions of biological diversity including genetic, population, community, ecosystem, and landscapes diversity. This has been an important change in emphasis since it points to, as one example, the very important role that landscape configuration plays in the transfer of materials in the earth system as well as showing how not accounting for changes in population sizes, within species,

obscures the real effects of the changes in biodiversity that are occurring.

However, in this paper I take a more traditional view and examine the diversity of species, discussing plants only. There is plenty of richness in this viewpoint alone. First I outline the kinds of diversity found in plants, following this with a more synthetic view of the evolution of this diversity and how this relates to biogeochemistry.

3.1 Structural Diversity

Box (1981) classified all of the different plant species into 16 different structural types (trees, small trees, etc.) and in turn into a total of 77 plant forms (e.g., evergreen tropical rainforest trees, mediterranean dwarf shrubs, etc.). This latter classification combines form, geographical distribution, and to a certain extent function (evergreen, deciduous, ephemeral). So fundamentally there are not too many different structural types of plants, as Theophrastus noted several millennia ago. These basic forms, when coalesced into communities, certainly have an influence on land surface/atmospheric models through turbulent transfer and boundary-layer effects that are often incorporated into atmospheric exchange models.

If we view structure in a broader sense there is a great array of possibilities from forms of trees, leaves, rigidity of tissues, and so forth. Many of these properties certainly affect radiation and gas exchange as well as decomposition, and hence biogeochemical cycling, both directly and indirectly. As a striking example of the importance of structure and biogeochemistry, Robinson (1990) has put forth the proposal that the evolution of high-lignin-content tissues in the Devonian, as gymnosperms and ferns evolved, caused a decomposition bottleneck, an accumulation of organic carbon and an increased atmospheric oxygen content that was not relieved until the evolution of lignolytic basidiomycetes and the evolution of angiosperms with lowered lignin contents as noted below.

3.2 Chemical Diversity

One of the arguments utilized for the protection of biodiversity is that organisms represent a depository of organic compounds of potential use to humans, such as drugs, dyes, insecticides, etc. There is indeed a large array of compounds that are found in organisms, many of which have no known function at present. These are so-called secondary compounds, or natural products. Over the past few decades research into exactly what, if any, functional roles are played by secondary compounds has been very active. The number of potential pharmaceutical extracts from tropical plants has been estimated to be about 750,000. Of the secondary compounds, the highest in number are alkaloids (10,000), flavonoids (4000), sesquiterpene lactones (3000), and diterpenoids (2000) (Harborne, 1993). The number of known secondary chemicals is increasing rapidly. For example, McGarvey and Croteau (1995) note that the number of defined terpenoids has doubled every decade since the 1970s. The number of combinations of these compounds in plants can be very high. Further, their amounts change in tissues through developmental stages (Matsuki, 1996).

3.3 Functional-Type Diversity

In recent years, classifying plants into functional types has been a very active field of research. Functional types represent characterization of groups of species performing similar functions or utilizing similar resources. This concept is equivalent to the idea of guilds of workers doing specific jobs in society. The concept of functional types is very attractive to modelers and global ecologists since it offers the potential of not having to deal with the complexity of species diversity in working on the functional questions of interest to them; biogeochemistry and land-surface–atmosphere interactions, for example. There is, however, no generally accepted way of classifying functional types. In theory, there actually could be a great number of functional types since they are usually characterized by an array of structural, chemical, and physiological characteristics modified by time and space. The possible combinations can be very great. However, Chapin *et al.* (1996) argue that the combination possibilities are not infinite since many traits are mutually related, So that you cannot have one without the other; for example high photosynthetic rate is most often linked to short-lived, herbaceous leaves with high nitrogen content.

4. Evolution of Functional Diversity

The earth has not provided a constant physical environment for the evolutionary process. This is due in part to the fact that once biotic evolution began, feedback between organisms and the environment caused changes. These modifications drove further evolutionary changes and in time, through competition among organisms, brought about the beginnings of new biotic drivers for evolution.

4.1 The Changing Atmosphere

The increase in oxygen in the atmosphere, due first to disassociation of water, and later to photosynthesis, developed slowly during the Precambrian because of initial reactions with surface materials (Lowry *et al.*, 1980). The low atmospheric oxygen content meant that no ozone was produced and thus there was no protective shield against UV radiation. With increased oxygen, respiration, rather than fermentation could proceed along with a new evolutionary tempo. It has been speculated by Berkner and Marshall (1965) that when the earth's oxygen concentration reached 0.1% of present levels, it led to sufficient UV protection for evolution on land to proceed, which correlates with the origin of land plants in the Silurian. Although the details of this hypothesis are subject to debate (Kubitzki, 1987) the general pattern is not (Lowry *et al.*, 1980).

4.2 Getting onto Land

The movement onto land and the subsequent evolutionary divergence involved a number of innovations including the development of mechanisms to enable plants to withstand atmospheric drought including cuticle, xylem, stomata, and intercellular spaces. All of these innovations occurred during the late Silurian and early Devonian around 400 million years ago (Raven, 1977). Evidently coinciding with these structural innovations came the evolution of compounds that aided both protection from UV-B and water-stress tolerance.

5. Cellulose

Although cellulose appeared 3.5 billion years ago it was not until the evolution of land plants that it became universally present in plants. Besides providing rigidity, so that cellular turgor could develop, it also provided mechanical strength due to its layered microfibril development. The tensile strength of cellulose microfibrils is over twice that of steel. The strength of cellulose is diminished, however, when it is wet. Materials embedded in the matrix of cellulose, such as lignin, hemicellulose, and pectins, add waterproofing as well as other properties. Chitin has structural properties similar to those of cellulose and could well have played a more general role in plant evolution. However, in contrast to cellulose, chitin requires reduced nitrogen, and it is thought that cellulose won the evolutionary race owing to the general nitrogen limitation in the biosphere. Organisms that utilize chitin, such as saprophytic and parasitic fungi, are found in generally nitrogen-rich environments (Duchesne and Larson, 1989). Cellulose, a polysaccharide, is one of the most abundant chemicals on earth, constituting about one-half of the earth's standing biomass.

6. Evolution of Polyphenolic Compounds

Lignin plays an enormously important role in determining the utility of cellulose as described above, by adding strength to it as well as by protecting it from pathogens. Lignin, however, also plays a number of different roles in plants and has done so through time. It is thought that the aromatic amino acids, which are the precursors to lignin, as well as tannins and flavonoids, evolved in aquatic algae as protectants against high UV-B radiation (Rozema *et al.*, 1997; Lowry *et al.*, 1980). Subsequently, derivative phenolic-acid products provided protection against microbial predation in terrestrial nonvasuclar plants and finally with the evolution of land plants, polyphenolics such as tannins and lignins provided further protection from herbivores and microorganisms, respectively, as well as mechanical strength as noted above. Also derived from the same phenolic pathway was the large class of compounds known as flavonoids, which have multiple roles, including plant–insect interactions (Rozema *et al.*, 1997).

The consequences of the evolution of lignin to the operation of the carbon cycle have been considerable. Lignin is one of the most refractory organic compounds produced by plants. It resists decay partly because it is insoluble and it has a very high C/N ratio. Lignin and lignin-degradation products inhibit the breakdown of complex carbohydrates. Complexes resulting from lignin breakdown can persist for thousands of years in aerobic soils (Robinson, 1990).

Robinson (1990) traces the evolutionary history of lignin. There was no lignin in plants in the Ordovician. Ligninlike compounds were found in some of the first vascular plants in the Silurian. She estimates that the content of lignin in these small plants was comparable to that of herbs today (10–15%). Subsequently, lignification rose to 40% in the Late Devonian, dropped to 30–35% at the end of the Mesozoic, and has subsequently declined to an average of about 20%.

7. Build-Up of Carbon and the Evolution of Decomposers

By the Robinson (1990) hypothesis, effective lignin degradation did not appear until some 200 million years after lignin evolved with the appearance of basidiomycetes in the Pennsylvanian. This could explain the build-up of organic compounds during the late Mesozoic. With increasing oxygen content of the atmosphere, the increase in basiodimycetes and the evolution of angiosperms with lower lignin-content in the Late Cretaceous, carbon stores in the soils decreased.

The events discussed so far do not address the issue of diversity; quite the contrary. The plant evolutionary path up through the Paleozoic was one of biochemical innovations that exerted generalized effects on herbivory and decomposition—universal solutions.

7.1 Evolution of Angiosperms and Insects

Within the past several million years, starting in the Cretaceous, we have seen a dramatic shift in the vegetation of the earth with a change from dominance of conifers, with their high lignin contents, to the adaptive radiation of the angiosperms. With this evolutionary event entirely new plant types have been introduced. How has this changed the diversity picture, especially in relation to biogeochemistry? Throughout the Cretaceous there was a continuing evolution of more and more complex flower types and accompanying diversification of fruit types (Friis and Crepet, 1987). With these changes have come the development of new kinds of animals involved in pollinating these flowers and dispersing the novel fruits, as well as those utilizing these specialized organs as food and habitat sources. Thus there apparently has been a concomitant diversification of animals driven by the appearance of angiosperms. This is indeed the case, as has been outlined by Crepet and Friis (1987) for insects.

The number of species of insects is quite high as noted earlier. This may be due in part to their low extinction rates over

tetrapods, for example (Lanandeira and Sepkoski, 1993). It has often been claimed that the diversity of insects has been driven by coevolution with angiosperms. However, Lanandeira and Sepkoski note that the principal trophic divisions among insects precede angiosperm evolution by a hundred million years and further the rate of evolution, at the familial level at least, was not affected by angiosperm evolution.

At the moment a debate is appearing on just how effective coevolution has been in fueling the diversity of secondary chemicals in plants. The debate is whether random variation has played a more important role than coevolutionary adaptation (see Berenbaum and Zangerl,1996 and Jones and Firn,1991, for opposing views on this issue). At the same time as these debates are going on, there also has been some questioning of the effectiveness of some compounds that have long been held to be generalized deterrents. Ayres *et al.* (1997) have shown that tannins, which bind dietary proteins and digestive enzymes and may be directly toxic, may not be universally effective. They note that "we doubt that selective pressures from folivorous insects can be the main explanation for the diversion of so much carbon, in so many plant species, into the synthesis of condensed tannins."

Although there is currently a debate on the coevolutionary interactions between insects and angiosperms, there is no doubt that the evolution of angiosperms brought about a great diversity in plant chemistry. Kubitzki and Gottlieb (1984) present an interesting case for the diversification of secondary chemicals in plants and the progressive evolution of angiosperms. They note in the course of angiosperm evolution, there was a movement away from generalized polyphenolic herbivore defensive compounds derived from the shikimic acid pathway to very specialized defensive compounds that utilize the mevalonic acid pathway such as alkaloids, iridoids, and terpenes. Lerdeau *et al.* (1997) have recently reviewed our knowledge of the production and emission of volatile organic compounds by plants. They note that monoterpenes have probably evolved as a defensive mechanism against herbivorous insects. It is interesting that not only are certain families characterized by the production of monoterpenes, such as the mint family, but that also certain regions have a high preponderance of terpene-producing plants, such as the arid fringes of mediterranean climate regions.

Kubitzki (1995) has summarized the generalized dichotomy between plants that produce secondary chemicals via the mevalonic acid pathway and those that use the shikimic acid cycle. The polyphenols produced by the shikimic acid cycle are so-called carbon-based defensive compounds. They are generalized defensive compounds of high carbon content. In contrast, those from the mevalonic acid cycle are diverse in structure, are nitrogen-based, and are highly toxic compounds. In nature, as might be expected, the distributions of these classes are different. The carbon-based defensives are found in nutrient-poor sites. The lignins in the litter, along with the ectotrophic mycorrhizal fungi associated with the plants of these sites, inhibit basidiomycetes, further impoverishing the site.

In summary, the diversification of plants and animals has also brought about a diversification of novel chemistry that in turn has had an impact on biogeochemistry. Even in fairly recent times new metabolic pathways have evolved in response to changing atmospheric conditions and have affected global biogeochemistry as discussed by Ehleringer in this volume.

8. Analysis of the Role of Diversity and Biogeochemistry

8.1 Direct Tests

To this point we have looked to the past to understand the changing relations between biodiversity and biogeochemisry. Recently, there has been considerable research into the role of species richness in ecosystem functioning, including biogeochemistry. There have been explicit experiments on this issue as well as observations on the impact of addition and deletion of species on natural ecosystems.

The experimental studies, although the most direct attack on this issue, have been somewhat controversial. Tilman *et al.*, (1996) have demonstrated through very elaborate experiments that species richness per se can influence such fundamental properties of ecosystems as nitrate retention as well as ecosystem resilience (Tilman and Downing, 1994). Through experiments that controlled not only plant species diversity but also trophic web diversity, Naeem *et al.* (1994) have demonstrated that richness can influence the carbon flux of ecosystems. Others have shown that functional types, another measure of diversity, are more important than richness per se in determining impacts on biogeochemistry (Hooper and Vitousek, 1997). The properties measured in these sorts of experiments are subject to great variability and hence subtle differences between species performance may be intrinsically more difficult to measure than among functional types.

8.2 Field Inferences

Another approach to investigating the role of species in ecosystem functioning and biogeochemistry is to view the impacts of the additions and, in some cases, the deletions of individual species on ecosystems. These studies have been quite revealing. Invasive species are altering the biotic structure of the earth's ecosystems. The breakdown of biogeographic barriers through international commerce has resulted in large numbers of species extending their natural ranges, often over many different continents. There are a great number of examples of species introductions totally altering ecosystem properties, including alterations of hydrology, nutrient cycling, and physiographic development. Additionally, invasive species can cause massive changes in ecosystem structure and such processes as fire cycles. All of these modifications directly influence biogeochemistry. The question is, at what scale? There is ample evidence of many invasive species having profound effects locally. There are also examples of invasives having large-scale regional effects on processes influencing biogeochemistry (see Drake *et al.*, 1989). The latter include the total conversion of the

intermountain west of North America by the succession of *Bromus tectorum* and the conversion of the perennial grasslands of California by mediterranean annual grasses. These conversions have respectively altered regional fire and water cycles. A striking example of an impact of an invasive species at the continental scale has been described for the effects of rinderpest on the megafauna (and human social structures) in Africa (Sinclair, 1979). The relatively large-scale plantation forestry practiced in many parts of the world, and of course agricultural conversions, have altered water, carbon, and nutrient balances of these regions, although in the latter case there has been a conversion of biotic material, as well as the resource base itself, through irrigation and fertilization. Introduction of invasive species caused major alterations of the operation of ecosystem dynamics in large water bodies, as has been evidenced by the effects of the zebra mussel on the trophic structure and water quality of the Great Lakes (Ludyanskiy *et al.*,1993) and similar effects of the Asian clam in the San Francisco Bay (Carlton, *et al.*, 1990).

Invasive species have also inhabited areas that were formally unoccupied, as in the case of *Spartina*, which is now invading the mud flats of the Pacific Northwest (Daehler and Strong, 1996), altering the physiographic development in these regions. The regional impacts of the removal of individual species, and of whole functional groups, have been described for the effects of recent as well as past megafaunal extinctions in many parts of the world (Owen-Smith, 1989). These removals have totally altered ecosystem structure and dynamics.

9. Summary

Early evolution produced organisms with generalized compounds and metabolic pathways that have had a profound effect on how the earth system operates. With the development of the angiosperms and the coincident adaptive radiation of animals, novel and diverse chemical compounds have evolved. Many of these compounds have also had impacts on biogeochemistry, in some cases only locally, but in others more generally.

We are now witnessing a major biological revolution with the mixing of biota formerly separated by oceans. Many of the new communities that are resulting have distinctive biogeochemical signatures.

References

Ayres, M. P., Claussen, T. P., Maclean, S. F. Jr., Redman, A. M., and P. B. Reichardt, (1997). Diversity of structure and antiherbivore activity in condensed tannins. *Ecology* **78**, 1696–1712.

Berenbaum, M. R. and Zangerl, A. R. (1996). Phytochemical diversity. Adaptation or random variation? In *"Phytochemical Diversity and Redundancy in Ecological Interactions."* (ed. J. T. Romeo, J. A. Saunders and P. Barbosa). pp. 1–24. Plenum Press, New York.

Berkner, L. V. and Marshall, L. C. (1965). History of the major atmospheric components. *Proc. Natl. Acad. Sci.* USA **53**, 1215–1225.

Box, E. O. (1981). "Macroclimate and Plant Forms: An Introduction to Predictive Modelling in Phytogeography." Junk, The Hague.

Carlton, J. T., Thompson, J. K., Schemel, L. E., and Nichols, F. H. (1990). Remarkable invasion of San Francisco Bay (Califonria, USA) by the Asian clam *Potamocorbula amurensis*. I. Introduction and dispersal. *Marine Ecol. Prog. Ser.* **66,** 81–94.

Chapin, F. S., BretHarte, M. S., Hobbies, S. E. & Zhong, H.L. (1996). Plant functional types as predictors of transient responses of arctic vegetation to global change. *Journal of Vegetation Science.* **7,** 347–358.

Crepet, W. L. and Friis, E. M. (1987). The evolution of insect pollination in angiosperms. In "The Origins of Angiosperms and their Biological Consequences." (E. M. Friis, W. G. Chaloner, and P. R. Crane, Eds.), pp. 108–201. Cambridge University Press, Cambridge, England.

Daehler, C. C., and Strong, D. R. (1996). Status, prediction and prevention of introduced cordgrass. Spartina spp invasions in Pacific estuaries, USA. *Biol. Conserva.* **78,** 51–58.

Drake, J. A., Mooney, H. A., di Castri, F., Groves, R. H., Kruger, F. J., Rejmanek, M., and Williamson, M. Eds. (1989). "Biological Invasions. A Global Perspective." Wiley, Chichester

Duchesne, L. C. and Larson, D. W. (1989). Cellulose and the evolution of plant life. *Bioscience* **39,** 238–241.

Friis, E. M. & Crepet, W. L. (1987). Time of appearance of floral features, In *The Origins of Angioserms and their Biological Consequences* (ed. E. M. Friis, W. G. Chaloner and P. R. Crane), PP. 145–179. Cambridge University Press, Cambridge.

Eigen, M. and Schuster, P. (1977). Hypercycle: Principle of natural self-organization. A. Emergence of hypercycle. *Naturwissenschaften* **64,** 541–565.

Gause, G. F. (1934). "The Struggle for Existence." Williams & Wilkins, Baltimore.

Harborne, J. B. (1993). "Introduction to Ecological Biochemistry." Academic Press, London.

Hatch, M. D. & Slack, C. R. (1966). Photosynthesis by sugarcane leaves. A new carboxylation reaction and the pathway of sugar formation. *Biochemical Journal* **101,** 103–111.

Hooper, D. U. and Vitousek, P. M. (1997). The effects of plant composition and diversity on ecosystem processes. *Science* **277,** 1302–1305.

Jones, C. G. and Firn, R. D. (1991). On the evolution of plant secondary chemical diversity. *Philos. Trans. R. Soc. London Ser. B* **333,** 273–280.

Kubitzki, K. (1987). Phenylpropanoid metabolism in relation to land plant origin and diversification. *J. Plant. Physiol.* **131,** 17–24.

Kubitzki, K. (1995). Plant chemistry of Amazonia in an ecological context. In "Chemistry of the Amazon." (P. R. Seidl, O. R. Gottlieb, and M. A. C. Kaplan, Eds.), pp. 126–134. American Chemical Society, Washington, D.C.

Kubitzki, K. and Gottlieb, O. R. (1984). Phytochemical aspects of angiosperm origin and evolution. *Acta Bot. Neerlandica* **33,** 457–468.

Lanandeira, C. C. and Sepkoski, J. J. Jr. (1993). Insect diversity in the fossil record. *Science* **261,** 310–315.

Lerdau, M., Guenther, A. & Monson, R. (1997). Plant production and emission of volatile organic compounds. *BioScience* **47,** 373–383.

Lowry, B., Lee, D., and Hebant, C. (1980). The origin of land plants: A new look at an old problem. *Taxon* **29,** 183–197.

Lovelock, J. E. (1994). Geophysiological aspects of biodiversity. In "Biodiversity and Global Change." (Solbrig, O. T., H. M. van Emden, and P. G. W. J. van Oordt, Eds.). CAB International: Wallingford, England, United Kingdom.

Ludyanskiy, M. L., McDonald, D., and MacNeill, D. (1993). Impact of the zebra mussel, a bivalve invader. *Bioscience* **43,** 533–544.

Matsuki, M. (1996). Regulation of plant phenolic synthesis: From biochemistry to ecology and evolution. *Aust. J. Bot.* **44,** 613–634.

McGarvey, D. and Croteau, R. (1995). Terpenoid metabolism. *Plant Cell* **7,** 1015–1026.

Monteith, J. L. (1977). Climate and efficiency of crop production in Britain. *Philos. Trans. R. Soc. London B.* **271,** 277–294.

Mooney, H. A., Ed. (1977). Convergent evolution of Chile and California-mediterranean climate ecosystems. Dowden, Hutchinson and Ross, Stroudsburg, Pennsylvania

Mooney, H. A., Vitousek, P. M., and Matson, P. A. (1987). Exchange of materials between terrestrial ecosystems and the atmosphere. *Science* **238,** 926–932.

Naeem, S., Thompson, L. J., Lawler, S. P., Lawton, J. H. & Woodfin, R. M. (1994). Declining biodiversity can alter the performance of ecosystems. *Nature* **368,** 734–737.

Oldenfield, S., Lusty, C., and MacKinven, A., Eds. (1998). "The World List of Threatened Trees." World Conservation Press, Cambridge

Owen-Smith, N. (1989). Megafaunal extinctions: The conservation message from 11,000 years B.P. *Conserv. Biol.* **3,** 405–412.

Potter, C. S., Randerson, J. T., Field, C. B., Matson, P. A., Vitousek, P. M., Mooney, H. A., and Klooster, S. A. (1993). Terrestrial ecosystem production: A process model based on global satellite and surface data. *Global Biogeochem. Cycles* **7,** 811–841.

Raven, J. A. (1977). The evolution of vascular land plants in relation to supracellular transport processes. *Adv. Bot. Res.* **5,** 153–219.

Robinson, J. M. (1990). Lignin, land plants, and fungi: Biological evolution affecting Phanerozoic oxygen balance. *Geology* **15,** 607–610.

Rozema, J., Van De Staaij, J., Bjorn, L. O., and Caldwell, M. (1997). UV-B as an environmental factor in plant life: Stress and regulation. *Trends Ecol. Evolut.* **12,** 22–28.

Sinclair, A. R. E. (1979). The Serengeti environment. In "Serengeti. Dynamics of an Ecosystem" (A. R. E. Sinclair and M. Norton-Griffiths, Eds.), pp. 31–45. University of Chicago Press, Chicago.

Tilman, D. and Downing, J. A. (1994). Biodiversity and stability in grasslands. *Nature* **367,** 363–365.

Tilman, D., Wedin, D., and Knops, J. (1996). Productivity and sustainability influenced by biodiversity in grassland ecosystems. *Nature* **379,** 718–720.

UNEP. (1995). "Global Biodiversity Assessment." Cambridge University Press, Cambridge, UK.

1.23

Atmospheric Perspectives on the Ocean Carbon Cycle

Peter J. Rayner

CRC for Southern Hemisphere
Meteorology, CIRO
Atmospheric Research
Asperdale, Australia

1. Introduction

The observed increase in atmospheric CO_2 concentration (e.g., Conway *et al.*, 1994; Keeling *et al.*, 1995), is a balance between anthropogenic inputs (such as fossil-fuel combustion and land-use change) and natural responses. The natural responses may arise directly from increasing atmospheric CO_2 concentration (e.g., extra dissolution in seawater) but may result from many other factors. An understanding of these natural responses, their sensitivities to human impacts, and their likely future trajectory forms a critical part of a well-founded projection of climate change and an attempt to manage that change. A first step is to quantify these responses via the net fluxes into various reservoirs. This has been the topic of considerable effort in the climate science and biogeochemistry communities particularly in the last decade. Summaries of the state of the science, really snapshots of evolving knowledge, are given, for example, by Schimel *et al.* (1995). I should stress at the outset that quantifying current fluxes is an early step in a process of understanding and should be considered as a measurement of the underlying behavior somewhat akin to remote sensing. However, even at this step, divergence and controversy have been the rule rather than the exception. In Sec. 2 I will give one view of the current state of the most basic question in this area: the relative roles of oceanic and terrestrial fluxes in balancing the atmospheric CO_2 budget. I will review and compare several somewhat independent lines of evidence. I will present a synthesis of these lines of evidence, although I will not attempt to draw them into one overarching framework. I will focus on the role of large-scale constraints with either gross or zero spatial information.

If the solution of the long-term budget can be regarded as understanding the basic state of the global carbon cycle, then interannual variability is a clue to the sensitivity of that state. The task is to understand the processes that control year-to-year variations in flux but even the task of estimating these fluxes is at the limits

of current capacity. In Sec. 3 I will compare estimates of this variability from models of the processes involved and atmospheric inferences to demonstrate this inconsistency. Finally, I will sketch a modeling framework that may help address this inconsistency.

2. Long-Term Mean Ocean Uptake

Here I review three largely independent lines of evidence concerning the net uptake of carbon by the global ocean. At the outset I should address some ambiguity about the word "uptake." The word should literally refer to the net storage of carbon in the ocean. However, many of the studies cited here calculate the net flux of carbon across the air–sea interface or deduce the ocean uptake from the terrestrial uptake. Carbon entering the ocean by other routes may not be counted. This becomes particularly confusing in the presence of background circulations of carbon that may predate any anthropogenic perturbation. Such a correction between fluxes and uptake (uptake is often referred to as storage to remove this ambiguity) was cited by Sarmiento and Sundquist (1992) to reconcile the flux estimates of Tans *et al.* (1990) with estimates from, for example, Sarmiento *et al.* (1992). While Sarmiento and Sundquist (1992) quoted a magnitude of 0.4–0.8 Gt C year^{-1} for the correction I believe this should contribute as much to the uncertainty as to the estimate itself.

2.1 Global and Temporal Perspective

The first strand of evidence for the global ocean uptake comes from models of the ocean carbon cycle. These generally combine descriptions of the relevant carbonate chemistry with some representation of tracer transport in the ocean. As well they need a model (usually highly simplified) of the biological processes in the ocean and finally some parameterization of the surface fluxes of

CO_2. Underlying all these is a model of the physical circulation of the ocean, usually arising from an ocean general circulation model. Three such estimates of the air–sea carbon flux are shown in Figure 1. The estimates are taken from the Ocean Carbon Model Intercomparison (OCMIP) (Orr, 1997).

There is consensus among ocean modelers, with a preliminary estimate for net uptake of anthropogenic CO_2 in the ocean of 2.0 ± 0.4 Gt C year^{-1} for 1990. Note that, since most ocean carbon cycle models do not attempt to represent the interannual variability in ocean circulation, these calculations represent an average over an ensemble of possible ocean states for 1990. Given the uncertainties in the underlying physical simulations, this convergence seems surprising. Two factors may help explain the agreement. First, several of the highly uncertain processes controlling uptake do not control the uptake on a decadal time-scale. For example, the surface and ocean are close to equilibrium, so the highly uncertain gas exchange does not force large model–model differences. Similarly, exchange with the deep ocean is not large on decadal scales. The important chemical processes, on the other hand, are well understood and hence consistent among models. Second, there are some integral or global constraints on the global ocean uptake in a model. In particular, most models used for ocean carbon studies have been checked if not tuned against the change in ^{14}C inventory arising from the above-ground nuclear tests in the 1950s and 1960s. While limited data make this check far from perfect (Heimann and Maier-Reimer, 1996), it does provide a constraint for global uptake estimates.

The range of estimates from OCMIP probably underestimates the total uncertainty in net uptake from ocean models. In Fig. 1 the local agreement is worse than the global agreement. Just as in the global case, the range is one estimate of local uncertainty in air–sea flux. Some of the differences arise from the different positioning of source or sink regions and will vanish when larger-scale integrals are considered. This cancellation will only partly offset the general behavior of uncertainties, which is that they sum quadratically. I believe that large-scale cancellation of differences, e.g., that differences in net uptake in one hemisphere are compensated in the other, is not due to compensating differences inherent in the models. Rather it arises when the models are forced to fit global constraints such as the ^{14}C inventory. I would conclude, then, that the agreement in estimates of net uptake from ocean models *does* suggest some underlying control, although the estimated uncertainty is larger than this intercomparison would suggest.

The second line of evidence comes from the time trends of some atmospheric species, mainly CO_2 and oxygen. Developing the oxygen budget of the atmosphere, which contains some problematic terms, is beyond the scope of this chapter. Briefly, the global, long-term budgets for O_2/N_2 and CO_2 can be written schematically as

$$\frac{\partial_q(CO_2)}{\partial_t} = f_F + f_B + f_O \tag{1a}$$

$$\frac{\partial_q(O_2)}{\partial_t} = S_F f_F + S_O f_O + S_B f_B, \tag{1b}$$

where q refers to atmospheric concentration, f_F, f_O, and f_B refer to fluxes to the atmosphere from fossil fuel, the ocean, and the terrestrial biosphere, respectively, and the S factors are the stoichiometric ratios of O_2 to CO_2 associated with each of these fluxes. Briefly, this assumes that the ocean makes no contribution to the atmospheric O_2/N_2 budget, which means the equation is invalid at seasonal (and probably interannual) time-scales. There are also difficulties associated with treating budgets like this over finite time intervals, particularly regarding estimates of the trend from noisy series. Readers are referred to Enting (1999) for a fuller treatment of these problems. The oxygen budget also assumes the marine biomass is in steady state, an assumption questioned by Galloway *et al.* (1995). Potential unaccounted fluxes like this should be regarded as contributing to the uncertainty in derived flux estimates. Given reasonable knowledge of f_F (Marland and Boden, 1997) Eq. 1a reduces to two algebraic equations in the unknowns f_O and f_B. All this relies on measurements of the CO_2 and O_2/N_2 trends, which were pioneered respectively by C. D. Keeling (Keeling, 1960) and R. F. Keeling (Keeling, 1988). For CO_2, the long-term trend is well characterized by the global sampling programmes in place since the 1960s, e.g., Conway *et al.* (1994) and Keeling *et al.* (1995). For O_2/N_2, two estimates of the trend have been made on decadal time-scales. One study (Battle *et al.*, 1996)

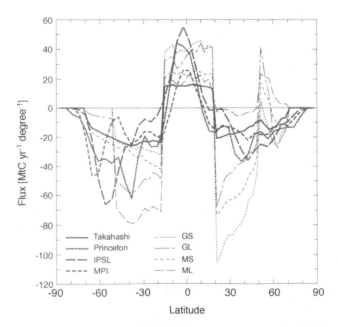

FIGURE 1 Several estimates of the zonal mean flux (Mt C year^{-1} per degree latitude) from the ocean into the atmosphere. The three blue lines are taken from OCMIP (Orr, 1997), the red line from a regionally aggregated estimate of Takahashi *et al.* (1997) and the four green lines from four atmospheric inversions as described in the text. See also color insert.

uses the record of air trapped in firn at the South Pole and some more recent atmospheric measurements. The firn measurements constrain the trend for the period 1978–1985. The other study (Langenfelds *et al.*, 1999) uses air sampled at Cape Grim, Tasmania, and archived. This record determines the trend for 1978–1997. Both records have their difficulties and are hard to compare because of the different periods they cover. Their overall trend is similar. I will use the value of -3.5 ± 0.2 ppmv year^{-1} for 1978–1997 from Langenfelds *et al.* Table 1 lists data and derived values for these terms. The contributions to the uncertainty are as follows:

1. Quoted uncertainty in both the CO_2 and O_2/N_2 trends.
2. Fossil-fuel source. Note that this has a significant impact on the terrestrial uncertainty because of its high oxygen–carbon stoichiometric ratio.
3. Uncertainties in stoichiometric ratios.
4. The potential for the ocean oxygen budget to be out of balance on any given time-scale.

Most of these terms are difficult to assign and I have used some judgment for some of them. More serious, however, is the question of just what budget is established by this simple calculation. The derived so-called terrestrial flux is in fact the net amount of carbon being reduced with carbon/oxygen ratios characteristic of photosynthetic material. According to Galloway *et al.* (1995) some of this carbon could be reduced in the ocean, resulting in changes of marine biomass, although this is probably not a major contribution on the time-scales considered here. The calculation also says nothing about the ultimate fate of the organic material formed by photosynthesis. Some is washed into the ocean by rivers to be outgassed again through the ocean surface or buried in sediments. Such a circulation probably existed in the preindustrial carbon cycle and the impact of perturbations in this budget is highly uncertain. However, Sarmiento and Sundquist [1992] estimate a maximum contribution of 0.3 Gt C year^{-1} for this term.

TABLE 1 Parameters Used in the Derivation of the Long-Term Mean Global Land and Ocean Fluxes from Trends in Atmospheric CO_2 and O_2/N_2.

Parameter	Value
$\dfrac{\partial_q(CO_2)}{\partial_t}$	1.4 ± 0.05 ppmv year^{-1}
f_F	5.73 ± 0.3 Gt C year^{-1}
$\dfrac{\partial_q(O_2)}{\partial t}$	-3.5 ± 0.2 ppmv year^{-1}
S_F	-1.38 ± 0.02
S_O	0
S_B	-1.05 ± 0.05
f_O	-2.2 ± 0.4 Gt C year^{-1}
f_B	-0.6 ± 0.5 Gt C year^{-1}

See Eq. 1a for the definition of the parameters. Uncertainties refer to one standard deviation. CO_2 growth rates are taken from the output of Rayner *et al.* (1999).

Global constraints like the O_2/N_2 trend form part of the data used by Rayner *et al.* (1999) in their synthesis inversion. They solve for the spatial distribution of surface sources and use the long-term constraint provided by the oxygen budget as a global constraint. With their large excess of degrees of freedom, they deduce a budget of -2.1 Gt C year^{-1} for ocean flux and -0.7 Gt C year^{-1} for land. The slight mismatch between this calculation and the purely global one detailed in Table 1 comes from the extra information from spatial CO_2 gradients and the use of prior estimates. Note, too, that the estimated net flux for the terrestrial biosphere includes a large positive contribution from land-use change.

2.1.1 Ice-Core Records and $\delta^{13}C$

O_2/N_2 is a useful quantity for partitioning land–ocean uptake since it is hoped to be a tracer of terrestrial and not ocean processes. The ratio of ^{13}C to ^{12}C in the atmosphere can play a similar role since photosynthesis discriminates strongly against ^{13}C while dissolution in the ocean does not. We could, in principle, use $\delta^{13}C$ in the same way as O_2/N_2 but there are several complicating factors. The carbon in fossil fuel, being a product of photosynthesis itself, is isotopically different from the carbon in the current atmosphere so that its input changes $\delta^{13}C$ in the atmosphere (the Suess effect). There are also large so-called gross fluxes between the atmosphere and underlying reservoirs. These gross fluxes influence the ^{13}C budget so we need to take them into account when using this species. The gross flux can be thought of as the number of molecules crossing the interface between two reservoirs (in one direction). In the ocean the gross flux is driven by the continual exchange between surface waters and overlying air. For the terrestrial biosphere, the important reservoir is the photosynthesized material. The amount of carbon entering this reservoir each year is known as the gross primary productivity. We can neglect the larger amount of carbon that passes in and out of the stomata of leaves without being assimilated into plants. Both these gross fluxes can be large even when there is no driver for a net flux, i.e., in steady state.

Gross fluxes can impact the isotopic composition of the atmosphere in the absence of net fluxes. If there are, for example, higher concentrations of $^{13}CO_2$ molecules in the ocean than the atmosphere then, on average, more of these are likely to leave than enter the ocean. The resulting isotopic flux is often called the *isoflux* as a convenient shorthand. The isoflux acts as a restoring term to bring the ocean and atmosphere back to equilibrium. The same holds for the terrestrial biosphere. The strength of the restoring term depends on the size of the gross flux (number of molecules crossing the interface) and the difference in the concentration of $^{13}CO_2$ molecules on each side of the interface, i.e, the magnitude of the isotopic disequilibrium. The disequilibrium, in turn, depends on the rate of change in the atmosphere and the adjustment times of underlying reservoirs. A combination of large reservoirs (hence slow adjustment) and relatively rapid change in the atmosphere has led to a large disequilibrium at present between the atmosphere and both the main reservoirs. The upshot is that the $\delta^{13}C$ budget of the atmosphere is sensitive to the magnitudes of the gross fluxes. Unfortunately, these gross fluxes are very hard to quantify from direct measurement. It is even more difficult when one considers that the disequilibrium is spatially variable so that the global isoflux must be properly flux-weighted for the effect of different regions. A side benefit of the calculations based on O_2/N_2 has been an indirect estimate of the global isoflux. To do this one first calculates the ocean and terrestrial fluxes as above, then calculates the isoflux as the residual from these fluxes, the fossil fuel input, and the atmospheric trend. Both Langenfelds *et al.* (1999) and Rayner *et al.* (1999) present such a calculation.

There is an alternative method for estimating the current isoflux. Air trapped in ice cores can provide long-term histories of CO_2 and $\delta^{13}C$ at one point over centuries or millennia. With such records, and knowledge of anthropogenic inputs, the combined CO_2 and $\delta^{13}C$ budgets can be inverted to solve for the ocean and terrestrial net fluxes. Such a calculation is presented by Joos *et al.* (1999). A similar calculation using only the CO_2 record, a model of ocean uptake and the $\delta^{13}C$ record as cross validation is presented by Trudinger *et al.* (1999). Both calculations use the ice-core data of Etheridge *et al.* (1996) and Francey *et al.* (1999) in simple box models of the atmosphere, ocean, and terrestrial biosphere. Such calculations also use simple models to track the $\delta^{13}C$ of the reservoirs and hence the disequilibrium with the atmosphere. Thus, assuming the gross fluxes are known, and subject to the accuracy of these models, this calculation will estimate the isoflux over time. Such calculations are stabilized by the role of the gross flux as a restoring term. A large gross flux will generate a small disequilibrium while a small gross flux will generate a large disequilibrium. Their product, the isoflux, may be less sensitive than either. In a set of sensitivity calculations, Trudinger *et al.* (1999) showed that the calculated isoflux for the period 1980–1990 was not very sensitive to the specified gross fluxes although a more detailed test of model parameters is still to be performed. Thus, used in this long-term role, $\delta^{13}C$ forms a valuable constraint on the current partition of uptake.

Rather than comparing estimates of net fluxes (which vary rapidly on interannual periods) Figure 2 compares the isoflux predicted from the O_2/N_2-based calculations of Rayner *et al.* (1999) and the calculations of Trudinger *et al.* (1999). There is close agreement on both magnitude and slope. Note that, when computed by an inversion, the slope of the isoflux is derived from curvature in the $\delta^{13}C$ record, which is a very subtle feature. (Indeed, the magnitude of the slope is not supported by the findings of Gruber *et al.* (1999) using the trends in ocean disequilibrium. They show no trend in disequilibrium between the mid-1970s and mid-1990s.) Their mean value is, however, consistent with estimates presented here. While the uncertainties in both methods are substantial, corresponding to a net flux of ± 0.5 Gt C year^{-1}, these two somewhat independent calculations agree that net ocean uptake in the period 1980–1990 was 2.0 ± 0.5 Gt C year^{-1}.

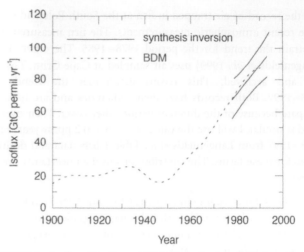

FIGURE 2 Isoflux (Gt C year^{-1}‰) from the box diffusion model of Trudinger *et al.* (1999) and the synthesis inversion of Rayner *et al.* (1999). Uncertainties for the synthesis inversion are calculated by considering the joint uncertainty in the mean value and slope.

2.2 Spatial Perspectives

In summary, then, it would appear that ocean models, the trend in O_2/N_2 and the long-term histories of CO_2 and $\delta^{13}C$ agree on an ocean uptake for the 1980s around 2 Gt C year^{-1}. The apparent consensus breaks down when spatial gradients of CO_2 and $\delta^{13}C$ in the atmosphere are considered. While an exhaustive overview of these studies is beyond us here, the general methods and results have been similar. The major forcing for the meridional structure of CO_2 in the atmosphere is the north–south gradient in fossil-fuel combustion. When this forcing is used as a flux-boundary condition for an atmospheric transport model, this produces a north–south gradient substantially larger than observed. Based on this mismatch in gradients, several authors, e.g., Enting and Mansbridge (1989), Tans *et al.* (1990) and Keeling *et al.* (1989), suggested that there should be a sink in the Northern Hemisphere to produce a countervailing negative gradient.

How this gradient was interpreted depended on ancillary information. Keeling *et al.* (1989) had already set the global ocean uptake from a calculation like that of Trudinger *et al.* (1999). They were left to decide an apportionment of the uptake among ocean basins. They posited an extra source in the southern ocean balancing an extra northern oceanic sink. This structure is consistent with the description by Broecker and Peng (1992) of the transport of carbon from formation of North Atlantic Deep Water into the southern ocean. Tans *et al.* (1990), using δpCO_2 measurements argued that the northern ocean could not account for the required sink. They therefore preferred a scenario of a substantial northern land sink. In their preferred scenario, the global ocean sink was less than 1 Gt C year^{-1}. An ocean sink in keeping with δpCO_2 measurements (with a relatively large Southern Hemisphere uptake) would exacerbate the gradient mismatch.

The carbon cycle community has sought hard the solutions to this ocean–atmosphere paradox. At global scale, Sarmiento and Sundquist (1992) adduced the skin-temperature correction from Robertson and Watson (1992) as well as the river-flux correction already mentioned. The mismatch in gradients was also reduced when the atmospheric transport of CO was considered as in Enting and Mansbridge (1991). Next, since the initial results were derived from two related models it was thought (perhaps hoped) that their transport was aberrant. If modeled transport between hemispheres was much slower than observed then the mismatch between modeled and observed meridional gradients would be purely a model artifact. A comparison of 12 atmospheric transport models was reported by Law *et al.* (1996). in the TransCom (transport comparison) study. An estimate of fossil-fuel source and seasonal biospheric exchange (annually balanced) were used as flux-boundary conditions for the contributing models. Although the response to the fossil-fuel source (embodied in the mean interhemispheric difference at the surface) varied by a factor of 2, the models from the earlier studies were at the more rapid end. Thus these models would produce a smaller north–south gradient than most, meaning the mismatch was perhaps even worse than was first thought.

While this study was under way the evidence for large northern land sinks seemed to strengthen. Two studies by Ciais *et al.* (1995a, b) included the spatial gradients in $\delta^{13}C$ as well as CO_2 in an atmospheric inversion. The results suggested northern land sinks in the range 2.5–3.5 Gt C year^{-1} for the period 1992–1994. Studies that calculated the year-to-year changes in this sink, such as Conway *et al.* (1994) and Rayner *et al.* (1999), suggest this was a period of anomalously large northern sink.

A further complication was added by Denning *et al.* (1995). Using the Colorado State University (CSU) general circulation model (GCM), they simulated a strong annual mean response to the annually balanced biospheric source. The annual mean response arose from the covariance between the seasonality of transport and the seasonality of the source. The effect had been noted earlier in Keeling *et al.* (1989) and dubbed the atmospheric rectifier by analogy with the production of a dc signal from an ac source in electrical circuits. The signal was positive in the Northern Hemisphere, strongest over land but also carried to the Northern Hemisphere observation sites. If the effect occurs in nature it would further strengthen the north–south gradient, requiring a yet larger sink in the Northern Hemisphere. The Denning *et al.* (1995) study provoked such strong interest because of the size of the effect, generating large-scale gradients from the annually averaged biosphere source roughly half those from fossil-fuel sources. Law *et al.* (1996) noted the same effect in several contributing models. In a recent sensitivity study, Law and Rayner (1999) found that the impact of the rectifier effect on atmospheric inversions was governed not only by the strength of the signal over source regions but how well it was advected to observation sites. The CSU model used by Denning *et al.* (1995) produced stronger signals at observation sites than did the NCAR-MATCH model used by Law and Rayner (1999) and hence a greater impact. The

third phase of the TransCom study will investigate these model–model differences in inversions systematically.

Finally, the terrestrial carbon cycle community has had little trouble identifying candidates for this enlarged role for land biota. Inventory studies such as Kauppi *et al.* (1992) or flux studies such as Grace *et al.* (1995) have suggested large sinks in northern and tropical forests, respectively. The possible mechanisms include the direct stimulation of net uptake by increased CO_2 concentration (so-called CO_2 fertilization), regrowth of forests on abandoned agricultural land, increased nutrient supply from other anthropogenic sources, and impacts of climate change. Many of these will be discussed elsewhere in this book.

So, 10 years after the paradox was first raised, the global carbon budget stands in a curious position. In general, the scientific community has adopted the combined evidence of the O_2/N_2 temporal trend and ocean carbon cycle models. Thus I would propose a value for inorganic fixation of 2.1 ± 0.5 Gt C year^{-1} for the decade averaged around 1990. This is probably close to the rate of change of ocean carbon inventory although the uncertainty for this may be larger. This choice reflects a belief in the simplicity particularly of the O_2/N_2 record and suggests a lack of confidence in the ability to interpret spatial gradients in the atmosphere. Given that various proposals are in place to interpret these gradients at smaller scales (almost certainly more difficult), it is important to understand the inversion of the large-scale gradient.

Investigation of this paradox between spatial and temporal information requires inversions with both types of information present. The inversion of Rayner *et al.* (1999) was one such, using both large-scale spatial gradients and the long-term O_2/N_2 constraint. Four different versions of this study, using two different observing networks and two atmospheric transport models, are shown as the four green lines in Fig. 1. The "GS" line is similar to the calculation of Rayner *et al.* (1999) except that it uses only the long-term mean sources, not the full year-to-year variations as in that calculation. Prior uncertainties are halved compared to Rayner *et al.* (1999) to reflect this lack of interannual variability. However, the prior uncertainty on the long-term mean in Rayner *et al.* (1999) is given by $\sigma/n^{1/2}$ where σ is the uncertainty for one year and n is the number of years of data in their experiment. This expression arises from the form for the standard error of the mean. Accounting for this, the uncertainties here are twice those in Rayner *et al.* (1999). The calculation uses the Goddard Institute for Space Studies (GISS) transport model and a relatively small observing network of 12 CO_2-observing sites. Unlike Rayner *et al.* (1999) it does not use $\delta^{13}C$ data. The setup of their calculation meant that $\delta^{13}C$ data had no impact on the long-term mean sources so this is not an important difference. Data are taken from GLOBALVIEW-CO2 (1999) and cover the period 1980–1995. The calculation uses data from only those months where actual observations exist. Data uncertainties are taken from Peylin *et al.* (2000) and reflect the ability of a monthly mean concentration to fit the actual flask data. They range from 0.3 ppmv at clean sites such as South Pole to over 3 ppmv at difficult observing sites near large terrestrial and industrial sources. The "GL" curve uses the same transport model but a larger CO_2-observing network, which, at its maximum, contains 65 CO_2-observing sites. The "MS" and "ML" lines are the same as the "GS" and "GL" lines except that they use the MATCH transport model as used by Law and Rayner (1999). The prior sources for all these experiments are the same balanced biospheric source from Fung *et al.* (1987) as used in Rayner *et al.* (1999) but for the ocean estimate we use the flux compilation of Takahashi *et al.* (1997). This flux is shown as the red line in Fig. 1.

The clearest feature from the various inversion calculations is the large scatter among them, and between them and the prior estimate. First, there is a substantial difference between the global ocean uptake as reported in Takahashi *et al.* (1997) and that required to match the oxygen constraint. The inversion is required to substantially increase ocean sinks to match this global constraint, how much and where being determined by the spatial gradients in data and transport characteristics. The inversions usually increase net uptake in both the Northern Hemisphere and Southern Hemisphere oceans except for GS, which decreases net uptake in the Southern Hemisphere (compensated by a very large increase in Northern Hemisphere ocean net uptake) and ML, which decreases Northern Hemisphere ocean net uptake (compensated by a very large Northern Hemisphere land net uptake). Tropical sources are increased by all the inversions, much more so for the GISS than MATCH transport models.

The general sensitivity of source estimates to both transport model and data network is of some concern. There are many potential causes for this (probably undue) sensitivity. One is purely statistical. The same consideration of errors in long-term means that holds for sources also holds for data. When using long atmospheric records as in these calculations, we may imply a tight constraint on the long-term mean spatial gradient. While this might be a fair reflection of observational uncertainties, it does not reflect the ability to model these observations. In the inversion formalism used here, the limitations of the model, embodied in the so-called model error, are included as part of the data uncertainty. The model error contribution does not approach zero with increasing record length. The solution to this problem is a more careful and complex treatment of data uncertainty, which takes explicit account of the model-error contribution.

Another problem is the difference in sampling between the model and real atmosphere. Generally flask sampling is timed to reduce contamination by heterogeneous local sources, which often means a bias toward marine sampling where possible. As treated here, the model includes no such selection. The effect is a stronger observation of land sources in the model than in the real world. The effect grows with an increasing network since many of the extra stations are coastal. The presence of the rectifier effect in the MATCH model further enhances the bias since the model may sample a larger gradient from the rectifier effect than is really observed. This problem can be treated with more judicious sampling in the model atmosphere, closer to the protocols in use by observers.

It should be noted, in passing, that all these inversions produce a lower uptake for North America than in the study of Fan *et al.* (1998), and that all produce larger uptakes for Eurasia than North America. This is not a direct contradiction of their work, however, since the calculations here have not fixed the ocean uptake as they did.

3. Interannual Variability

It is possible to perform a similar analysis of the interannual variability of ocean flux (and its relationship to atmospheric concentration measurements) as for the long-term mean. Once again, one cannot consider the ocean alone when interpreting atmospheric signals, so I will consider the combined effect of various scenarios of land and ocean fluxes in the atmosphere. As with the long-term means, even the broadest question regarding interannual variability in atmospheric CO_2 growth-rate is unsolved, namely, whether the variability is driven predominantly by the land or the ocean. In general, modelers and measurers of fluxes in both environments would suggest that the terrestrial biosphere is largely responsible, with perhaps three or four times the interannual variability of the ocean. Examples from two calculations by Friedlingstein *et al.* (1997) and Le Quéré *et al.* (2000) are shown as the dotted lines in Fig. 3. Also shown as solid lines are the equivalent fluxes from Rayner *et al.* (1999) while the asterisks indicate the results of Francey *et al.* (1995). The divergence is clear, particularly for the ocean, with Francey *et al.* (1995) showing much greater variability than Le Quéré *et al.* (2000). Rayner *et al.* (1999) has less variability than Francey *et al.* (1995), particularly through the 1980s, but still greater than that suggested by the ocean model. The disagreement between Rayner *et al.* (1999) and Francey *et al.* (1995) is perhaps surprising since they use the same $\delta^{13}C$ record from Cape Grim. Rayner *et al.* (1999) uses a global network of CO_2 observations and a constraint toward an invariant prior estimate, both of which may reduce the estimated variability.

There is little apparent agreement on the long-term mean land flux. As already mentioned, this disagreement is mainly because of a difference in the quantities plotted; Rayner *et al.* (1999) and Francey *et al.* (1995) plot total terrestrial flux while Friedlingstein *et al.* (1997) omit the flux due to land-use change. Once this is taken into account the mean estimates of Friedlingstein *et al.* (1997) and Rayner *et al.* (1999) are consistent. There is also more agreement for the three calculations on the magnitude of interannual variability. There is agreement on the timing of some events, like the large anomalous sources of 1987 and subsequent decrease, but I do not regard the estimates as in overall agreement.

In an atmospheric inversion, the atmosphere acts as a consistency check between land and ocean estimates since their sum must match the growth rate and spatial gradient information in the concentration observations. This can be implemented by using the flux estimates as a prior constraint, with confidence dictated by the providers of the estimates. While the calculations here have not done this, experience shows that the estimated fluxes would be little different unless I used

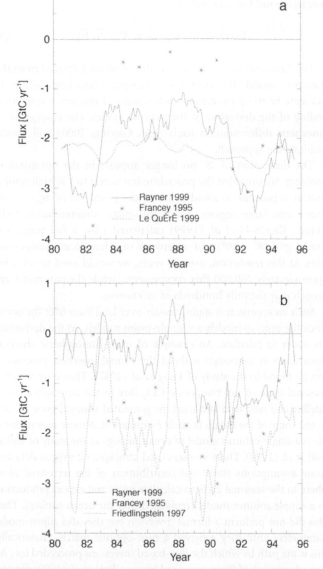

FIGURE 3 12-month running mean fluxes to the atmosphere (Gt C year^{-1}) for the ocean (a) and land (b) from the inversions of Rayner *et al.* (1999) and Francey *et al.* (1995) and the flux models of Le Quéré *et al.* (2000) for the ocean and Friedlingstein *et al.* (1997) for the land.

unrealistically tight prior constraints. Hence, the flux estimates are inconsistent with atmospheric observations as used here.

There are many potential explanations for the inconsistency. The first is that the inversion misallocates variability between the land and ocean. This is certainly possible since the observing network is strongly biased toward the ocean. It is part of the behavior of such Bayesian inversions that they will adjust those fluxes that are best sampled to make up for a mismatch between data and the initial guess. Thus the inversion will propagate interannual variability in concentration data preferentially to the relatively well-sampled ocean rather than the poorly sampled land. This effect is reduced by the use of $\delta^{13}C$, a tracer that marks

terrestrial but not oceanic net sources. Both the Rayner *et al.* (1999) and Francey *et al.* (1995) calculations use $\delta^{13}C$. In fact a similar calculation by Keeling *et al.* (1995) had such large interannual variability in $\delta^{13}C$ that the deduced terrestrial fluxes required large compensating oceanic interannual variability to match the global CO_2 growth rate. Note that at the global scale, potential errors in the modeling of atmospheric transport (which would manifest themselves as misallocation of sources) do not matter. Further, there is broad agreement on the dominant role of the tropics, both land and ocean, in forcing interannual variability so the transport seems to be behaving consistently.

Another possibility is that the flux estimates are incorrect. For example, the ocean model of Le Quéré *et al.* (2000) is relaxed toward climatology outside the tropics so that interannual variability is suppressed. However, long experience of the oceanographic community has identified the El Niño Southern Oscillation (ENSO) as the dominant oscillation on these time-scales. Further, the magnitude of variations in the ocean model in tropical fluxes is roughly supported by measurement campaigns of Feely *et al.* (1999). There are also processes missing from the terrestrial estimates, e.g., interannual variations in disturbance or changes in surface solar radiation. It is an unfortunate consequence of the integrated nature of the atmospheric constraint that while it can project information from one region to another, it also projects errors.

Several ways forward through this paradox are apparent, some by analogy with the approach to the long-term mean problem, but some suggested by the nature of the interannual variability problem itself. First, an obvious need is for more data, subject to concerns about model error. In particular, there is need for more concentration data over the tropical continents. Such data would reduce the leakage problem mentioned above in which concentration variations observed at marine sites (but possibly forced by remote continental fluxes) are attributed to the ocean. The next need is for continuing reanalysis of the relative confidence in ocean and terrestrial flux estimates. The atmospheric constraint allows land and ocean flux estimates to inform each other, but only on the basis of credible uncertainty estimates. In the study of Rayner *et al.* (1999) uncertainties were chosen almost arbitrarily, the main requirement being to avoid the risk of bias from too tight a prior constraint. Only the workers constructing the flux estimates themselves, either through an understanding of the scaling properties of observed fluxes, or through model sensitivity, can provide more credible estimates. This must happen before the scientific community can be sure we are focusing on real disagreement rather than being misled by an overoptimistic assessment from inverse modelers.

The final and most radical suggestion concerns a different way of modeling the problem. Partly the suggestion comes from reconsidering the question of why the carbon-cycle community is interested in interannual variability. While the task of estimating the interannual variability of fluxes is difficult enough to have become an end in itself, it is not sufficient. The next step is to elucidate those processes that control the variations in flux. This is a fascinating scientific question but also has practical import since it should enable us to estimate the sensitivity of the global carbon cycle to various forcings. The processes that control the variations are, or should be, expressed in models of the processes with the models forced by various boundary conditions. The unknown parameters of the problem now move from flux estimates themselves to the various model parameters and, potentially, boundary conditions of the underlying process models. The problem now becomes a nonlinear optimization problem but apart from this technical issue, little else changes from current flux inversion methods.

Such an approach has many advantages. First, it allows the atmosphere to inform directly the understanding of the relevant processes. Second, the approach addresses one of the great difficulties in an integrated approach to understanding the carbon cycle, the difference in the scales on which parameters are estimated or measured. Generally, direct flux measurements or campaigns estimate fluxes on a very small scale or (better) allow estimates of some parameters controlling, say, the terrestrial biosphere. Rightly are the methods termed bottom-up. They return quantities that are hard to compare with the large-scale estimates returned from synthesis inversions. If the methods were used to estimate the same parameters in some underlying model we could make a direct comparison. Finally, there are computational attractions. The very enterprise of modeling fluxes (from process models) assumes that variations in flux can be computed from some relatively well-known set of forcings and some set of model parameters. This set is almost certainly smaller than the set of fluxes now estimated by a flux-based inversion. We will see below that this could help overcome some of the restrictions on resolution that have plagued this type of study.

To see how such an approach may work, recall that the current inversion methods involve an optimization problem, seeking to minimize a cost function, Φ, comprising a mismatch between modeled concentrations ($\vec{D}$) and observed data ($\vec{D}_0$) plus another term for the mismatch between estimated sources ($\vec{S}$) and an initial estimate of those sources ($\vec{S}_0$),

$$\Phi = (\vec{D} - \vec{D}_0)^T C^{-1}(\vec{D}_0)(\vec{D} - \vec{D}_0)$$
$$+ (\vec{S} - \vec{S}_0)^T C^{-1}(\vec{S}_0)(\vec{S} - \vec{S}_0) \qquad (2)$$

where C21($\vec{X}$), an inverse covariance matrix, expresses the confidence in a quantity $\vec{X}$.

Sources and data are related by the linear operator **J** as

$$\vec{D} = \mathbf{J}\vec{S} \qquad (3)$$

where **J** embodies the transport model. Viewed more generally, **J** is the sensitivity of data with respect to sources $\nabla_{\vec{S}} \vec{D}$.

Now, replace $\vec{S}$ with the output of some process model, M, as

$$\vec{S} = \mathrm{M}(\vec{P}), \qquad (4)$$

where $\vec{P}$ represents model parameters. The task is to estimate $\vec{P}$ given $\vec{D}$ and perhaps an initial estimate $\vec{P}_0$. The estimation usually requires $\nabla_{\vec{P}} \vec{D}$

to aid minimization of the cost function. We can invoke the chain rule to expand the derivative as

$$\nabla_{\vec{P}}\vec{D}(\vec{P}) = \nabla_{\vec{S}} \vec{D} \times \nabla_{\vec{P}} \vec{S}(\vec{P}) \qquad (5)$$

The first term in the product is the previous **J** derived from the transport model. The second term has previously been tedious to calculate in an optimization routine since it requires a numerical coding of the derivative of the process model. The emergence of automatic differentiation tools (e.g., Giering, 2000) will greatly facilitate the approach.

The dimension of $\vec{S}$ no longer appears in the optimization problem. So, provided the procedure has access to **J** at high resolution, it is possible to avoid some of the problems of aggregating fluxes into large regions. Using automatic differentiation techniques, Kaminski *et al.* (1999) calculated **J** for a full transport-model grid ($8° \times 10°$) and a network of a few dozen observation sites. At this resolution, over 20 years, we would need to solve for approximately 200,000 flux components, while the parameter approach may use only hundreds of unknowns.

Such an approach is slightly easier over land than over the ocean since the process-models are single-point models, so the derivatives are easier to calculate. An example of using atmospheric observations this way (although without the formal inversion procedure) can be found in the study of Fung *et al.* (1987). They estimated the seasonal cycle of net biospheric CO_2 flux to the atmosphere from satellite and field data and used the generated seasonal cycle of CO_2 to test some of the details of their formulation. A more recent example is a single-column model of ocean biology in the study of Balkanski *et al.* (1999). These workers used atmospheric oxygen data and some assumptions about the contribution of the terrestrial biosphere to the seasonal cycle to calculate gross and export production in a simple column model (replicated over the ocean surface). They also did not perform a formal inversion but they did adjust model parameters until they achieved a near-optimal match. Historically, this is the path by which the flux-based inversions proceeded too. Ad hoc adjustment of fluxes was used by, e.g., Tans *et al.* (1990) for carbon dioxide or Fung *et al.* (1991) for methane and only later replaced by optimization algorithms by Enting *et al.* (1995).

The approach outlined above would make substantial demands on both the atmospheric inversion community and process modelers. It would shift the boundaries of the atmospheric inversion task from providing estimates against which process models can be tested to formally integrate those models into the procedure. It may also use the atmospheric constraint as just one among many acting to constrain model parameters, since any observable derived from $\vec{P}$ can be used. The approach would also make serious demands on process modelers. If they are to use the atmospheric constraint, the models must be comprehensive models of flux to the atmosphere. A process model that attempts to estimate, say, only net primary productivity cannot simulate the flux to the atmosphere and therefore cannot be used this way. Despite its difficulties, the approach seems to offer an integrated and rigorous framework for the use of atmospheric data in studying the carbon cycle.

4. Summary and Conclusions

To summarize the results presented here, long-term trends in concentrations of various atmospheric species provide a reasonable and consistent constraint on the net fluxes of carbon into the ocean and terrestrial biosphere at global scales. At regional scales the picture becomes more confused and the constraint weaker. Agreement over the southern hemisphere oceans appears to be the strongest, at least when considered as a whole. There is less agreement for the rest of the globe, with substantial uncertainty remaining in the partition of net uptake between land and ocean.

There is still less agreement about the temporal patterns of source variability derived from aggregated fluxes compared to atmospheric measurements. In general, the predominant role for the terrestrial biosphere suggested by both process-based models and flux measurements is not consistent with estimates from atmospheric concentrations. Such estimates are at the limit of current data and inversion techniques. Better observations of concentration over continents, as well as a more closely integrated inversion framework, should shed a clearer light on this problem in the coming years.

Acknowledgments

The author acknowledges the invaluable intellectual input of C. Trudinger, R. Langenfelds, R. Francey, and I. Enting. The author also acknowledges the use of data from CSIRO-GASLAB and the NOAA Climate Diagnostics Monitoring Laboratory. This study was carried out with support from the Australian Government through its Cooperative Research Centres Programme.

References

Balkanski, Y., Monfray, P., Battle, M., and Heimann, M. (1999). Ocean primary production derived from satellite data: An evaluation with atmospheric oxygen measurements. *Global Biogeochem. Cycles* **13**, 257–271.

Battle, M., Berder, M., Sowers, T., Tans, P. P., Butler, J. H., Elkins, J .W., Ellis, J. T., Conway, T., Zhang, N., Lang, P., Clarke, A. D. *et al.* (1996). Atmospheric gas concentrations over the past century measured in air from firn at the South Pole. *Nature* **383**, 231–235.

Broecker, W. S. and Peng, T. H. (1992). Interhemispheric transport of carbon dioxide by ocean circulation. *Nature* **356**, 587–589.

Ciais, P., Tans, P. P., Trolier, M., White, J. W. C., and Francey, R. J. (1995a). A large northern hemisphere terrestrial CO_2 sink indicated by the $^{13}C/^{12}C$ ratio of atmospheric CO_2. *Science* **269**, 1098–1102.

Ciais, P., Tans, P. P., White, J. W. C., Trolier. M., Francey, R. J., Berry, J. A., Randall, D. R., Sellers, P. J., Collatz, J. G., Schicmel, D.S. *et al.* (1995b). Partitioning of ocean and land uptake of CO_2 as inferred by $\delta^{13}C$ measurements from the NOAA Climate Monitoring and Diagnostics Laboratory Global Air Sampling Network. *J. Geophys. Res.* **100**, 5051–5070.

Conway, T. J., Tans, P. P., Waterman, L. S., Thoning, K. W., Kitzis, D. R., Masarie, K. A., and Zhang, N. (1994). Evidence for interannual variability of the carbon cycle from the National Oceanic and Atmospheric Administration/Climate Monitoring and Diagnostics Laboratory Global Air Sampling Network. *J. Geophys. Res.* **99**, 22,831–22,855.

Denning, A. S., Fung, I. Y., and Randall, D. A. (1995). Gradient of atmospheric CO_2 due to seasonal exchange with land biota. *Nature* **376**, 240–243.

Enting, I. G. (1999). Characterising the temporal variability of the global carbon cycle, *CSIRO Atmos. Res. Tech. Paper No. 40*.

Enting, I. G., and Mansbridge, J. V. (1989). Seasonal sources and sinks of atmospheric CO_2: Direct inversion of filtered data. *Tellus* **41B**, 111–126.

Enting, I. G., and Mansbridge, J. V. (1991). Latitudinal distribution of sources and sinks of CO_2: Results of an inversion study. *Tellus* **43B**, 156–170.

Enting, I. G., Trudinger, C. M., and Francey, R. J. (1995). A synthesis inversion of the concentration and $\delta^{13}C$ of atmospheric CO_2. *Tellus* **47B**, 35–52.

Etheridge, D. M., Steele, L. P., Langenfelds, R. L., Francey, R. J., Barnola, J. M., and Morgan, V. I. (1996). Natural and anthropogenic changes in atmospheric CO_2 over the last 1000 years from air in Antarctic ice and firn. *J. Geophys. Res.* **101D**, 4115–4128.

Fan, S., Gloor, M., Mahlman, J., Pacala, S., Sarmiento, J., Takahashi, T., and Tans, P. (1998). A large terrestrial carbon sink in North America implied by atmospheric and oceanic CO_2 data and models. *Science* **282**, 442–446.

Feely, R. A., Wanninkhof, R., Takahashi, T., and Tans, P. (1999). Influence of El Nino on the equatorial Pacific contribution to atmospheric CO_2 accumulation. *Nature* **398**, 597–601.

Francey, R. J., Tans, P. P., Allison, C. E., Enting, I. G., White, J. W. C., and Trolier, M. (1995). Changes in oceanic and terrestrial carbon uptake since 1982. *Nature* **373**, 326–330.

Francey, R. J., Allison, C. E., Etheridge, D. M., Trudinger, C. M., Enting, I. G., Leuenberger, M., Langenfelds, R. L., Michel, E., and Steele, L. P. (1999). A 1000-year high precision record of $\delta^{13}C$ in atmospheric CO_2. *Tellus* **51B**, 170–193.

Friedlingstein, P., Fung, I., and Field, C. (1997). Decadal variation in atmospheric-biospheric CO_2 exchange, in *Extended Abstracts of the Fifth International Carbon Dioxide Conference, Cairns, Australia, 8–12 September 1997*, p. 268.

Fung, I., John, J., Lerner, J., Matthews, E., Prather, M., Steele, L. P., and Fraser, P. J. (1991). Three-dimensional model synthesis of the global methane cycle. *J. Geophys. Res.* **96**, 13,033–13,065.

Fung, I. Y., Tucker, C. J., and Prentice, K. C. (1987). Application of advanced very high resolution radiometer vegetation index to study atmosphere–biosphere exchange of CO_2, *J. Geophys. Res.* **92**, 2999–3015.

Galloway, J. N., Schlesinger, W. H., Levy II, H., Michaels, A., and Schnoor, L. (1995). Nitrogen fixation: Anthropogenic enhancement—Environmental response. *Global Biogeochem. Cycles* **9**, 235–252.

Giering, R. (2000). Tangent linear and adjoint biogeochemical models. In "Inverse Methods in Global Biogeochemical Cycles." (P. Kasibhatla, M. Heimann, P. Rayner, N. Mahowald, R. Prinn, and D. Hartley, Eds.), Vol. 114 of *Geophysical Monograph*, pp. 33–48, American Geophysical Union, Washington, D.C.

GLOBALVIEW-CO2. (1999). Cooperative Atmospheric Data Integration Project—Carbon Dioxide, CD-ROM, NOAA/CMDL, Boulder, Colorado, 1999, [Also available on Internet via anonymous FTP to ftp.cmdl.noaa.gov, Path: ccg/co2/GLOBALVIEW].

Grace, J., *et al.* (1995). Carbon-dioxide uptake by an undisturbed tropical rain-forest in Southwest Amazonia, 1992 to 1993. *Science* **270**, 778–780.

Gruber, N., Keeling, C. D., Bacastow, R. B., Guenther, P. R., Lueker, T. J., Wahlen, M., Meijer, H. A. J., Mook, W. G., and Stocker, T. F. (1999). Spatiotemporal patterns of carbon-13 in the global surface oceans and the oceanic Suess effect. *Global Biogeochem. Cycles* **13**, 307–335.

Heimann, M., and Maier-Reimer, E. (1996). On the relations between the oceanic uptake of CO_2 and its carbon isotopes. *Global Biogeochem. Cycles* **10**, 89–110.

Joos, F., Meyer, R., Bruno, M., and Leuenberger, M. (1999). The variability in the carbon sinks as reconstructed for the last 1000 years. *Geophys. Res. Lett.* 26, 1437–1440.

Kaminski, T., Heimann, M., and Giering, R. (1999). A coarse grid three-dimensional global inverse model of the atmospheric transport. 1. Adjoint model and Jacobian matrix. *J. Geophys. Res.* 104, 18,535–18,553.

Kauppi, P. E., Mielikäinen, K., and Kuusela, K. (1992). Biomass and carbon budget of European forests, 1971 to 1990. *Science* 256, 70–74.

Keeling, C. D. (1960). The concentration and isotopic abundance of carbon dioxide in the atmosphere. *Tellus* 12, 200–203.

Keeling, C. D., Piper, S. C., and Heimann, M. (1989). A three-dimensional model of atmospheric CO_2 transport based on observed winds. 4. Mean annual gradients and interannual variations, in *Aspects of Climate Variability in the Pacific and the Western Americas*, edited by (D. H. Peterson, Ed.), Geophysical Monograph 55, American Geophysical Union, pp. 305–363, Washington, D.C.

Keeling, C. D., Whorf, T. P., Wahlen, M., and van der Plicht, J. (1995). Interannual extremes in the rate of rise of atmospheric carbon dioxide since 1980. *Nature* 375, 666–670.

Keeling, R. F. (1988). "Development of an Interferometric Oxygen Analyzer for Precise Measurement to the Atmospheric O_2 Mole Fraction." Harvard University, Cambridge, MA.

Langenfelds, R. L., Francey, R. J., Steele, L. P., Battle, M., Keeling, R. F., and Budd, W. F. (1999). Partitioning of the global fossil CO_2 sink using a 19-year trend in atmospheric O_2. *Geophys. Res. Lett.* 26, 1897–1900.

Law, R. M., and Rayner, P. J. (1999). Impacts of seasonal covariance on CO_2 inversions. *Global Biogeochem. Cycles* 13, 845–856.

Law, R. M., *et al.* (1996). Variations in modelled atmospheric transport of carbon dioxide and the consequences for CO_2 inversions. *Global Biogeochem. Cycles* 10, 783–796.

Le Quéré, C., Orr, J. C., Monfray, P., Aumont, O. (2000). Interannual variability of the oceanic sink of CO_2 from 1979 through 1997. *Global Biogeochem. Cycles* **14**, 1247–1266.

Marland, G., and Boden, T. (1997). Estimates of global, regional, and national annual CO_2-emissions from fossil-fuel burning, hydraulic cement production, and gas flaring: 1950–1994, *NDP-030R7*, Carbon Dioxide Information Analysis Center, Oak Ridge National Laboratory.

Orr, J. C. (1997). Ocean carbon-cycle model intercomparison project (OCMIP), in: Phase 1: 1993–1997. *Tech. Rep. 7*, IGBP/GAIM.

Peylin, P., Bousquet, P., Ciais, P., and Monfray, P. (2000). Differences of CO_2 flux estimates based on a time-independent versus a time-dependent inversion method. In "Inverse Methods in Global Biogeochemical Cycles." (P. Kasibhatla, M. Heimann, P. Rayner, N. Mahowald, R. Prinn, and D. Hartley, Eds.), *Geophysical Monograph*, 114, pp. 295–309. American Geophysical Union, Washington, D.C.

Rayner, P. J., Enting, I. G., Francey, R. J., and Langenfelds, R. L. (1999). Reconstructing the recent carbon cycle from atmospheric CO_2, $\delta^{13}C$ and O_2/N_2 observations. *Tellus* 51B, 213–232.

Robertson, J. E. and Watson, A. J. (1992). Thermal skin effect of the surface ocean and its implications for CO_2 uptake. *Nature* 358, 738–740.

Sarmiento, J. L. and Sundquist, E. T. (1992). Revised budget for the oceanic uptake of anthropogenic carbon dioxide. *Nature* 356, 589–593.

Sarmiento, J. L., Orr, J. C., and Siegenthaler, U. (1992). A perturbation simulation of CO_2 uptake in an ocean general circulation model. *J. Geophys. Res.* 97, 3621–3645.

Schimel, D., Enting, I., Heimann, M., Wigley, T., Raynaud, D. Alves, D., and Siegenthaler, U. (1995). CO_2 and the carbon cycle. In "Climate Change 1994: Radiative Forcing of Climate Change and An Evaluation of the IPCC IS92 Emission Scenarios." (J. Houghton, L. M. Filho, J. Bruce, H. Lee, B. Callander, E. Haites, N. Harris, and K. Maskell, Eds.), pp. 35–71, Cambridge University Press.

Takahashi, T., Feely, R. A., Weiss, R. F., Wanninkhof, R. H., Chipman, D. W., Southerland, S. C., and Takahashi, T. T. (1997). Global air-sea flux of CO_2: An estimate based on measurements of sea-air pCO_2-difference, *Proc. Natl. Acad. Sci.* 94, 8292–8299.

Tans, P. P., Fung, I. Y., and Takahashi, T. (1990). Observational constraints on the global atmospheric CO_2 budget. *Science* 247, 1431–1438.

Trudinger, C. M., Enting, I. G., Francey, R. J., Etheridge, D. M., and Rayner, P. J. (1999). Long-term variability in the global carbon cycle inferred from a high precision CO_2 and $\delta^{13}C$ ice core record. *Tellus* 51B, 233–248.

International Instruments for the Protection of the World Climate and Their National Implementation

Rüdiger Wolfrum
Max Planck Institute for Comparative Public Law and International Law, Heidelberg, Germany

1. Introduction

The protection of the world climate or components thereof has become the object of international agreements since the end of the seventies. The most important agreements to that extent are the Convention on Long-Range Transboundary Air Pollution, 1979[1] and its Protocols[2], as well as the Vienna Convention for the Protection of the Ozone Layer, 1985[3] and its Protocol (Montreal Protocol on Substances that Deplete the Ozone Layer, 1987).[4] However, only the United Nations Framework Convention on Climate Change, 1992[5] (Framework Convention) together with the Kyoto Protocol[6] represent a comprehensive approach to international protection of the climate.

The Framework Convention, together with the Kyoto Protocol, constitutes an international effort to protect the global climate for present and future generations[7] taking also into consideration the effects any climate change may have on islands, on low-lying coastal areas, and on

increasing desertification. The two international agreements for the first time establish legally binding limits for industrialized countries on emissions of carbon dioxide and other greenhouse gases.[8]

The Framework Convention and the Kyoto Protocol were discussed controversially and accordingly many of its provisions have to be understood as reflecting a compromise. The whole regime should not be considered as constituting a purely environmental system but rather as one addressing environmental concerns by taking into account social and economic developments in an integrated manner.[9]

I will deal with commitments states parties have entered into under the newly established regime of climate protection, and which implementation measures and which measures for a control concerning compliance are provided for.

2. Commitments of States Parties under the Climate Change Regime

It is the ultimate objective of the Framework Convention to stabilize greenhouse gas concentrations in the atmosphere at a level that would prevent dangerous anthropogenic interference with the

---

[1] International Environmental Law: Multilateral Agreements 979: 84.

[2] Protocol Concerning the Control of Emissions of Nitogen Oxide or their Transboundary Fluxes, 1988, International Environmental Law: Multilateral Agreements 979: 84 C; Protocol Concerning the Control of Emissions of Volatile Organic Compounds or their Transboundary Fluxes, 1991, ibidem 84 D; Protocol on Further Reduction of Sulphur Emissions, 1994, ibidem 84 E.

[3] International Environmental Law: Multilateral Agreements 985: 22.

[4] International Environmental Law: Multilateral Agreements 985: 22 A.

[5] International Environmental Law: Multilateral Agreements 992: 35.

[6] FCCC/CP/1997/L. 7/Add. 1.

[7] See in this respect the resolution of the General Assembly of the United Nations 46/169 of 19 December 1991.

[8] In addition to carbon dioxide, the concerned gases include nitrous oxide, methane, sulfur hexafluoride, hydrofluorocarbons, and perfluorocarbons. Other gases, such as chlorofluorocarbons (CFCs), also exhibit greenhouse effects but are controlled by the Montreal Protocol. As to the scientific and ethical dimensions of the effect of greenhouse gases, see Prue Taylor, An Ecological Approach to International Law: Responding to Climate Change, 1997, at 9 et seq.

[9] See the Preamble of the Framework Convention.

climate system.[10] To achieve this objective the Framework Convention and the Kyoto Protocol formulate several obligations for states parties. Some of these obligations are of a procedural nature, others are of a substantive nature. These obligations are not the same for all states parties; the substantive ones involving emissions only apply to industrialized countries and other parties listed in Annex I of the Framework Convention. This differential treatment of states parties reflects the principle of common but differentiated responsibilities as referred to in the Preamble of the Framework Convention, which was also endorsed by the United Nations Conference on Environment and Development, 1992.

The Kyoto Protocol specifies the measures to be taken to achieve the objectives of the climate regime. These measures can be divided into two categories, measures that require certain actions and/or the adoption of particular policies, and flexible measures that may or may not be introduced.

The Framework Convention was only a first step toward international control and management of greenhouse gas emissions. Its Article 4, Paragraph 2(d), mandated a review of the adequacy of the measures that had been taken so far at the first Conference of Parties. At this Conference the parties decided that the commitments were inadequate to meet the Convention's ultimate objective. In consequence thereof, it was decided that it was necessary to strengthen the commitments of the Framework Convention through a protocol—the Kyoto Protocol.

According to Article 4, Paragraph 1 of the Framework Convention on Climate Change, all states parties are under an obligation to prepare national inventories of emissions and removals of certain greenhouse gases by sources and sinks. Parties are equally required under this provision to adopt programmes containing measures to mitigate climate change and to cooperate in controlling, reducing, or preventing anthropogenic emissions of greenhouse gases. Whereas these obligations apply to all states parties, the obligations under Article 4, Paragraph 2 of the Framework Convention only apply to industrialized states parties and several others.[11] Thereunder these states parties are obliged to adopt policies and to take measures to mitigate climate change by limiting anthropogenic emissions of greenhouse gases and protecting and enhancing greenhouse gas sinks with the aim of returning, individually or jointly, to the 1990 levels of anthroprogenic emissions of greenhouse gases. They have further to report on the policies adopted and measures taken by them. Thus, there was an obligation of industrialzed states parties to reduce greenhouse gas emissions; however, no legally binding target was set.

The Kyoto Protocol establishes quantified emission limitation and reduction objectives for industrialized states parties and others (Annex I states parties) which are legally binding and a requirement for these states parties in implementing or further elaborating appropriate policies and measures to meet such targets. Apart from that, the Protocol mandates the advancement and implementation of certain commitments that pertain to all states parties of the

Framework Convention. Non-Annex I states parties (developing countries) may, as a prerequisite for engaging in emission trade, voluntarily assume binding emission targets through amendment of Annex B.

According to Article 3 of the Kyoto Protocol, the industrialized states parties are under an obligation to ensure that their aggregate anthropogenic emissions of greenhouse gases[12] do not exceed the amount specified in Annex B of the Protocol.[13] The emission targets are listed as percentages of emissions levels in the base year, which is generally 1990.[14] The targets range from an 8% reduction (European Community and its member States) to a 10% increase (Iceland).[15] Within the European Community, use is made of the possibility opened by Article 4, Paragraphs 2(a) and (b) of the Framework Convention and Articles 3 and 4 of the Kyoto Protocol, according to which the parties are allowed to achieve the reduction in emissions individually or jointly. On this basis, the Community has developed an arrangement for distributing emission reductions among its members according to which an increase in greenhouse gas emissions for Ireland and Portugal is recommended while other members will have to achieve a decrease of more than 8%.

The Kyoto Protocol has further elaborated upon the general provision of Article 4, Paragraph 2(a) of the Framework Convention according to which states parties are to mitigate climate change by protecting and enhancing greenhouse gas sinks and reservoirs. This provision was inadequate; it was too vague to be implemented coherently by all states parties, else it could, eventually, due to the changing carbon sequestration capabilities of forests, favor states parties that had a deforestation policy in the past. As a general rule, sinks are not included in calculating the emissions of the base year; however, sinks are to be taken into account during the commitment period. According to Article 3, Paragraph 3 of the Kyoto Protocol, the states parties listed in Annex I must give an accounting of the afforestation, reforestation, and deforestation undertaken since 1990. This new accounting system makes it possible to provide for a reduction of greenhouse gases through increasing forests. However, those Annex I countries with net emissions from land-use change and forestry may include those emissions in their base year, which has the effect of correspondingly raising their assigned amount and allowed emissions.[16] Using sinks as a means in the calculation and reduction of greenhouse gas emissions requires further research on the impact of land and forest use on greenhouse gas emissions. The

---

[12] Listed in Annex A.

[13] There are some differences between this list and the original list of countries in the Framework Convention on Climate Change. For details see Clare Breidenich/Daniel Magraw/Anne Rowley/James W. Rubin, The Kyoto Protocol and the United Nations Framework Convention on Climate Change, AJIL 92 (1998), 315, at 320.

[14] Article 3, Paragraph 7 of the Kyoto Protocol.

[15] These emission targets are the outcome of intensive negotiations. Several industrialized countries preferred uniform targets (European Community and its members), whereas others did not (Norway, Iceland, and Australia, for example).

[16] Article 3, Paragraph 7 of the Kyoto Protocol.

---

[10] Article 2.

[11] See Annex I.

Conference of Parties will issue guidelines on this matter, which will have to reflect increased respective scientific findings.

3. Implementation Measures

The Kyoto Protocol provides some guidance as to how the obligations under Article 3 are to be implemented by states parties. Article 2 of the Protocol provides a list of potential policies and measures that aim, generally speaking, at the enhancement of energy efficiency and promotion of sustainable agricultural practices. This list, however, is only illustrative. At the national level, each state party may select its own policies and measures provided that it produces the required results. The Meeting of Parties[17] may provide for a coordination of the respective policies, if necessary, and thus strengthen the guidelines provided under Article 2 of the Kyoto Protocol.

Apart from providing for the reduction of greenhouse gas emissions either by limiting the emissions as such or by improving the capacity of sinks, the parties may take or provide for taking supplementary measures of implementation, namely joint implementation and emissions trading.

According to Article 4, Paragraph 2(a) and (b) of the Framework Convention, parties falling under Annex I may implement policies on the mitigation of climate change jointly with other parties and may assist other Parties in contributing to the achievement of the objective of the Convention.

The primary objective of a joint implementation is to reduce the total costs of meeting aggregate environmental standards.[18] Apart from costs; joint implementation opens the possibility for groups of states—in particular members of an economic union—to undertake and to fulfil commitments collectively. Many international environmental agreements that apply fixed environmental standards require substantial reductions in total pollution emissions. This is true for the climatic change regime, the First and Second Sulphur Protocol, and the Montreal Protocol. The problem with this approach of applying fixed standards is that it imposes the same reduction obligation on countries with high and low environmental standards and with high and low abatement costs. It is questionable whether this is required under the objective of the agreements referred to. Since the prime objective of the respective agreements is to reduce emissions worldwide, as is the case with the Climate Change Regime as well as the Ozone Layer Protection Regime, the location of emissions reduction should be irrelevant. From the point of efficiency each unit of reduction should take place at the place where it is cheapest and accordingly the marginal abatement costs are the lowest. Joint implementation may allow greater emissions reductions as compared to the reductions of the same cost given by traditional command-and-control approach to the regulation of pollution. It may further promote technical innovation and could help to reduce the problem of leakage.

Despite the theoretical benefits that may emerge from applying such a mechanism of joint implementation, its realization has met with resistance, as far as the Climate Change Regime is concerned, from the side of developing countries. To that end Article 4, Paragraph 2(a) of the Framework Convention is rather vague and requires further clarification before being implemented. According to Article 4, Paragraph 2(a) and (d) of the Framework Convention, joint implementation could only be undertaken after the Conference of Parties develop further criteria.[19]

From the wording of the provision there is no doubt that parties cannot undertake joint implementation deals with nonsignatory countries although from a purely economic perspective, there is no reason to exclude nonparties. For a truly global problem such as climate change, all opportunities for low-cost abatement should be exploited regardless of their location and status with respect to the Convention. However, it is politically desirable to restrict joint implementation in such a way as to encourage participation and adherence to the Convention.

However, the question whether developed countries should be allowed to undertake joint implementation together with developing countries was disputed under the Framework Convention. As far as the wording is concerned, joint implementation is meant to take place between developed countries only.[20] The rationale of this restriction is that otherwise joint implementation might allow developed countries to avoid advancing and implementing the technological innovations required to meet environmental standards under the Framework Convention.[21] However, despite a purely textual analysis of Article 4, Paragraph 2 of the Framework Convention, the objective of this provision which, vis-a-vis other provisions, emphasizes the common but differentiated responsibility of states parties and thus emphsizes the necessity to develop comprehensive strategies and the principle of cost effectiveness suggests a more flexible approach to joint implementation.

The following negotiations in the Intergovernmental Negotiating Committee and the Conference of the Parties of the Framework Convention led to the establishment of a pilot phase of "Activities Implemented Jointly" to gain experience in cooperative projects to reduce emissions.[22] The pilot phase permits Annex I

---

[17] The Kyoto Protocol distinguishes between the Conference of Parties of the Framework Convention and the Meeting of Parties of the Protocol.

[18] Robin Mason, Joint Implementation and the Second Sulphur Protocol, Review of European Community and International Environmental Law, 1995, 296; Hans-Jochen Luhmann *et al.*, Joint Implementation; Projektsimulation und Organisation, 1997, 8 et seq.; Farhana Yamin, The Use of Joint Implementation to Increase Compliance with the Climate Change Convention, James Cameron/Jacob Werksman/Peter Roderick, Improving Compliance with International Environmental Law, 1996, 228–230.

[19] Yamin (note 18), at 238.

[20] Yamin (note 18), at 239 et seq. who distinguishes between joint implementation and assistance. In his article a detailed textual analysis of the respective provision is given.

[21] Reinhard Loske/Sebastian Oberthür, Joint Implementation under the Climate Change Convention, 6 International Environmental Affairs (1994), 45; Daniel M. Bodansky, The Emerging Climate Change Regime, 20 Annual Review of Energy and the Environment (1995), 425 (at 452 et seq.).

[22] FCCC, Conference of the Parties, 1st Session, UN Doc. FCCC/CP/1995/7/Add1, Decision 5/CP.1, at 19 (6 June, 1995).

states parties to invest in emission reduction projects in non-Annex I states parties, but without taking emission reduction credits for such projects.[23] Developing states parties may participate in this undertaking on a voluntary basis.[24] The states parties agreed to decide by the end of 2000 on whether to continue this pilot phase and on whether emission reduction credits may be taken for such projects.[25]

The discussion on flexible means of implementation continued into the negotiations of the Kyoto Protocol.[26] It provides for four mechanisms which, however, reflect the same philosophy.

Article 6 in combination with Article 3 of the Kyoto Protocol provides for the possibility of joint implementation among Annex I states parties. This allows states parties as well as participants from the private sector, if so authorized by the respective state party, to invest in emission reduction projects (reduction of sources or enhancement of sinks) in the territory of another Annex I state party and to apply emission reduction credits for those projects toward their national emission targets.[27] The precondition for acquiring credit for such an emission reduction is that the parties are in compliance with their measurement and reporting obligations under the Kyoto Protocol.[28] The problem with this possibility may be the adequate verification of such joint implementation projects. The Conference of Parties is mandated to elaborate appropriate guidelines concerning verification and reporting.[29]

In addition, Article 17 of the Kyoto Protocol authorizes a target-based emissions trading system. The respective provisions need further elaboration. According to the Protocol, Annex B states parties may participate in emissions trading for the purpose of fulfilling their commitments under Article 3 of the Kyoto Protocol. The emission reduction units acquired will be credited to the acquiring State Party. The Conference of Parties under the Framework Convention is mandated to develop rules and modalities for emissions trading.

A further flexible means of implementing the obligations under the climate change regime is the possibility of burden sharing, as already envisaged under the Framework Agreement for members of the European Community. Article 4 of the Kyoto Protocol expands this possibility by providing for all Annex I states parties, including those acting within the framework of a regional economic integration organization, to fulfill their commitments under Article 3 of the Kyoto Protocol jointly.

Finally, the Kyoto Protocol provides a fourth mechanism of flexible implementation. According to Article 12, Paragraph 3 (b) of the Protocol Annex I states parties may invest in emission reduction projects in developing countries. They may apply some portion of the reduction generated by such projects toward meeting their emission target under Article 3 of the Kyoto Protocol.[30] In return, a given share of the proceeds of such projects will be used to finance adaptation to climate change in particular vulnerable developing countries.[31] The mechanism, referred to as Clean Development Mechanism, is supervised by an executive board and subject to the guidance of the Meeting of Parties;[32] the Meeting of Parties will, among others, designate operational entities to certify and track such projects.[33]

The Clean Development Mechanism constitutes a new approach; it combines financial assistance with the obligation to reduce greenhouse gas emissions in such a way that they mutually induce each other. The incentive for providing financial assistance is the partial accountability of the reduction for the donor and the incentive for setting up the project is the accountability of the other part of the reduction for the recipient state.

These measures do not yet fully describe the implementation measures to be taken by industrialized states. Additionally, the developed states parties as listed in Annex II of the Framework Convention have to provide "new and additional financial resources to meet the agreed full costs incurred by developing country parties in complying with their obligation" (Article 4, Paragraph 3 of the Framework Convention). In consequence thereof Article 11 of the Framework Convention established a fund.

This mechanism is meant to cover the costs for developing countries which are states parties to fulfil the obligation under Article 12 to prepare national inventories and environmental plans to implement the provisions of the Convention. Additionally, developed countries will provide financial resources needed by developing country parties to meet the full, agreed incremental costs of implementing the various obligations undertaken. The latter also covers assistance to developing countries in adapting to the adverse effects of climate change if steps taken under the Convention fail to abate global warming adequately.[34] These costs are of a different nature and treated differently. According to Article 12, Paragraph 1 of the Framework Convention, the Parties are inter alia obliged to furnish an inventory of sources and sinks of greenhouse gases. For the establishment of this inventory and all other reporting activities mentioned in Article 12 Para 1, the agreed full costs of developing country parties are to be met by "Annex II Countries." All

---

[23] UN Doc. FCCC/CP/1995/7/Add. 1, Decision 5/CP.1, at 19 (6 June 1995).

[24] See for example the reports submitted one year later UN Doc. FCCC/CP/1996/14 and 14/Add. 1; FCCC/SBSTAA/1996/17 and FCCC/SBSTA/1997/INF. 1.

[25] During the pilot phase it has become evident that activities implemented jointly face problems of a practical nature. In particular, it is difficult to establish the basis of reference (i.e., the emissions prior to the joint implementation) and to verify the actual reductions to be credited; WBGU Study, Targets for Climate Protection, 1997, 27.

[26] For details see Breidenich/Magraw/Rowley/Rubin (note 13), at 324 et seq.

[27] Article 3, Paragraph 10 Kyoto Protocol.

[28] Article 6, Paragraph 1(c) Kyoto Protocol.

[29] Article 6, Paragraph 2 Kyoto Protocol.

[30] Certain emission reductions obtained as early as in the year 2000 may be used to assist in achieving compliance in the first commitment period (Article 12, Paragraph 10 of the Kyoto Protocol).

[31] Article 12, Paragraph 8 of the Kyoto Protocol.

[32] Article 12, Paragraphs 4 and 7 of the Kyoto Protocol.

[33] Article 12, Paragraph 5 of the Kyoto Protocol.

[34] See in particular Laurence Boisson de Chazournes, The United Nations Framework Convention on Climate Change: On the Road Towards Sustainable Development. R. Wolfrum, Ed., "Enforcing Environmental Standards: Economic Mechanisms as Viable Means?" 1996, 285–298.

other activities mentioned in Article 4, Paragraph 1 of the Framework Convention will be financed in accordance with Article 4, Paragraph 3 of the Framework Convention. The wording of Article 4, Paragraph 3 of the Framework Convention provides for the reimbursement of incremental costs in full for such measures that have been agreed upon between a developing country party and the fund.[35] This means that establishing which costs have to be reimbursed requires an assessment procedure on different levels. First of all it has to be determined which measures the respective state would have taken to establish the baseline. The provisional Secretariat of the Framework Convention and the Council of the Global Environment Facility have suggested the criterion of "environmental reasonableness" in this connection,[36] which states that the respective Party should not be punished with a high-level baseline considering the advanced standards applied by that party. In further steps it has to be agreed upon that the intended measures fall under Article 4, Paragraph 1 of the Framework Convention and that, in concreto, they are acceptable. Under this system, the responsibility concerning the protection of climate is the common responsibility of all parties to the Framework Convention.

The financial assistance to be provided by developed states parties reflects the following principle. As emphasized in the Preamble of the Framework Convention, the protection against climate change is the obligation of all states; however, the responsibilities are differentiated. This means that developed states parties have to contribute more to achieve the objective of the regime against climate change than other states parties. This is so for two reasons. Industrialized countries have—and this is equally expressed in the Preamble of the Framework Convention—contributed more to climate change than developing states parties. For that reason it is now up to them to provide remedies by limiting their emissions and by providing technical and financial assistance to developing countries. To a certain extent this reflects the polluter-pays principle. Additionally, account has to be taken of the fact that developed contries can afford to provide more to the common goal than developing countries. Hence the financial regulations are to be considered to be based on the principle of distributive justice.

Article 4, Paragraph 7 of the Framework Convention establishes a clear link between the obligations entered into by developing contries and the commitments accepted by developed states parties. According thereto the developing countries' implementation of their obligations depends on the effective implementation by developed country parties of their commitments. In other words, only if new and additional financial resources are provided will the developing countries collectively live up to theirs. Hence, providing such resources is a means of achieving the implementation of the regime on the protection against climate change on the side of developing country parties.

4. Monitoring Compliance and Enforcement

It has been frequently emphasized that the effectiveness of international environmental law depends on establishing international procedures or mechanisms that may be used to ensure compliance.[37] International environmental law has developed several such mechanisms, one of them being the obligation of states parties to regularly report the national measures undertaken to pursuent to the respective international agreement. However, international environmental law does not entrust particular international institutions with supervisory functions. A tendency seems to be developing that the institutions established by the various international environmental agreements are entrusted with the task of assessing whether the status of the environment is improving rather than whether individual states are complying with their commitments. The international regime on climate change is an example of this tendency.

In the Kyoto Protocol, the reporting system has become most sophisticated. The Framework Convention has already established two basic reporting requirements, namely, for national inventories and accounts of greenhouse gas emission budgets and for periodic national communications that provide detailed information on all the states parties' implementation of the Convention. The Kyoto Protocol has expanded these reporting requirements, which are the logical consequence of the Protocol, providing additional mechanisms of implementation. According to Article 7, Paragraph 1 of the Kyoto Protocol, each state party, as a part of its annual inventory, has to provide the information necessary to ensure its compliance with all of its obligations under the Protocol. The Secretariat is responsible for collecting, compiling, and publishing national greenhouse gas inventory data. To make the assessment of such data easier, the Framework Convention obliges states parties to use standard methodologies to measure and estimate national greenhouse gas emissions. The Protocol has further refined this requirement.[38] To induce a common technique, the Kyoto Protocol requires that inventories of states parties that fail to use the prescribed method must be adjusted to account for uncertainties. The reporting requirements may be further elaborated by guidelines issued by the Conference of Parties.[39]

The Conference of states parties of the Framework Convention has already established a process for the review of information by Annex I states parties.[40] The Kyoto Protocol has built thereupon and strengthened this process. Whereas under the Framework Convention the inventory information, though collected annually, was published and reviewed only in conjunction with the periodic

[35] This formula constitutes a compromise, see: Rudolf Dolzer, Die internationale Konvention zum Schutz des Klimas und das allgemeine Völkerrecht, Festschrift Bernhardt, 1995, 957 (967); D. Bodansky, The United Nations Framework Convention on Climate Change, Yale Journal of International Law 18 (1993), 451–492.

[36] Bodanksy (note 35), at 524.

[37] See, for example, Robert O. Keohane/Peter M. Haas/Mare A. Levy, The Effectiveness of International Environmental Institutions, Haas/Keohane/Levy, Eds., "Institutions for the Earth, 1994." 3–7 with further references.

[38] Article 5, paragraphs 1 and 2.

[39] Article 7, paragraph 4 of the Kyoto Protocol.

[40] FCCC Conference of the Parties, 1st Session, UN Doc. FCCC/CP/1995/7/Add.1, Decison 3/CP.1, (6 June 1995).

national communications, Article 8, Paragraph 1 of the Kyoto Protocol now requires an annual review of national inventory and emissions-target information as part of the centralized accounting of assigned amounts. The information submitted in the states' reports will be reviewed by expert teams.[41] The teams of experts will be nominated by the states parties to the Framework Convention, appropriately assisted by intergovernmental organizations, and coordinated by the Secretariat. This review process is meant to " . . . provide a thorough and comprehensive technical assessment of all aspects of the implementation of the commitments of the Party and identifying any potential problems in, and factors influencing, the fulfilment of commitments."[42] The Secretariat will forward any problem identified in the expert teams' reports for consideration by the Meeting of Parties. The Meeting of Parties is authorized to decide on any matter required for the implementation of the Protocol.[43] The system will be elaborated further through guidelines of the Meeting of Parties.

The reporting system has transformed from an information-collecting device, as it was under earlier treaties concerning the protection of the environment, monitoring system under the Kyoto Protocol. The latter aspect has been clearly emphasized in Article 7, Paragraphs 1 and 2 of the Kyoto Protocol. However, whether or not the rules governing review of the information thus received are commensurate therewith is questionable. The wording of Article 8, Paragraph 6 of the Kyoto Protocol, according to which the Meeting of Parties may take " . . . decisions on any matter required for the implementation of the Protocol . . . " is quite ambivalent. It may refer to the implementation of the Protocol as such or to the implementation problems of one particular state party. The practice under the Framework Convention and, in particular, the context of this paragraph suggests that it refers like the other paragraphs of this provision to the individual performance of states parties.

It may seem that the noncompliance procedure under the climate change regime is less stringent than those in other recent international environmental agreements. Although Article 13 of the Framework Convention calls for the consideration of a multilateral consultative process for questions regarding implementation and Article 10 of the Framework Convention establishes a Subsidiary Body for Implementation, the functions of the latter are limited. The Subsidiary Body is called upon to " . . . assist the Conference of Parties in the assessment review of the Convention . . . " and thus lacks the competence to deal with individual cases. Equally, the mandate of the Conference of Parties to establish a noncompliance system is limited. The parties are only called upon to " . . . consider the establishment of a multilateral consultative process, available to Parties on their request for the resolution of questions regarding the implementation of the Convention . . . ." This mandate lacks the focus on the noncompliance of individual states parties which is characteristic of the Montreal Protocol and the Second Sulphur Protocol.

The reason for the rather weak position of the Subsidiary Body for Implementation in comparison to the procedures adopted under the regimes for the protection of the ozone layer and on transboundary air pollution can be found in the fact that the climate change regime has set up a rigorous reporting system including the possibility of having the respective information assessed by experts (see above). Under these circumstances, vesting the Implementation Body with the power to scrutinize the implementation of each single state would have meant a duplication of functions. Apart from that, assessing the information on emissions and removals through sinks requires expertise which will not be found in a body consisting of representatives of states.

5. The Kyoto Protocol as a Learning Treaty

The Kyoto Protocol illustrates the setup of a learning treaty management system. The Protocol spells out the states parties' obligations by stipulating differentiated targets and time tables. It takes a number of institutional steps designed to ensure the implementation of the Protocol as a whole and of the commitments of each single state party.[44] The Conference of the Parties shall keep the implementation of the Protocol under regular review and it may take decisions to promote its effective implementation. In doing so it performs two functions, to promote the implementation first of the Protocol as such and second, by each single member state. This function has to be interpreted in the context of the commitments entered into by developed countries under the Convention on Climate Change. According to its Article 4, Paragraph 2(d) and (e), it is their obligation to keep their commitments under review with a view to reducing anthropogenic emissions of greenhouse gases. Also, it is the obligation of the Conference of Parties to re-examine the obligations of the parties in the light of the objective of the Convention, the experience gained in its implementation, and the evolution of scientific and technological knowledge. This means that the obligations under the Convention on Climate Change are not static but can and will be progressively developed to the extent new technological developments and/or scientific findings would allow us to do so. Success in implementing this approach depends on the information received by the Conference of Parties.

6. Conclusions

The most significant progress achieved through the regime on climate change is the agreement among all states parties that the protection of the world climate is the common responsibility of all states. This resulted in the establishment of a respective solidarity community paraphrased as "common but differentiated responsibility of states." The second achievement is the commitment of

---

[41] Article 8, Paragraph 1 of the Kyoto Protocol.
[42] Article 8, Paragraph 3 of the Kyoto Protocol.
[43] Article 8, Paragraph 6 of the Kyoto Protocol.

[44] Article 13, Paragraph 4 of the Kyoto Protocol.

industrialized states parties to reduce the greenhouse-gas emissions by a fixed percentage within an identified time.

The effectiveness of the regime will depend not only on whether the industrialized states meet their obligation to reduce greenhouse gas emissions. It is equally important that they meet their financial obligations. This includes, in particular, the obligation under Article 10, Subparagraphs (c), (d), and (e) of the Kyoto Protocol. The regime is likely to achieve its objective only if developing contries have access to technologies that ensure the efficient use of energy on a modern level, for example, in the energy consumption and in the building sector. In particular, further steps are to be taken to ensure that the envisaged cooperation of states parties results in a transfer of technology. This requirement may have an impact on national and European law on intellectual property as, in fact, this may require revision of the law on intellectual property as far as this property is needed for the protection of the environment. The question of to what extent holders of intellectual property are under an obligation to waive their rights if community interests so warrant has, so far, utterly been neglected. Unfortunately, international law does not give any direction in this regard. The proposal of several developing countries to modify the Agreement on Trade-Related Aspects of Intellectual Property Rights to reduce the costs of access to environmentally sound technology has been resisted by the United States, the European Union, Japan, and other industrialized states.[45]

The regime on climate change, if implemented according to the obligations entered into, will result in a significant change in many areas of economic life. Although the Kyoto Protocol raises the possibility of reducing emissions of greenhouse gases by taking into account their removal through sinks, the reduction of greenhouse gases at their source of emission seems to be the most reliable method of emission reduction.[46] In this respect research should continue as to which actions are appropriate to each economic sector (i.e., energy consumption, traffic, etc.) and whether particular gas emissions require the adoption of particular policies. The policies for the reduction of greenhouse-gas emissions have to be efficient; the first results are expected by the year 2005.

[45] See Richard H. Steinberg, Trade-Environment Negotiations in the EU, NAFTA, and WTO: Regional Trajectories of Rule Development, AJIL 91 (1997), 231–243.

[46] Critical in respect of the taking into account of sinks, Wissenscaftlicher Beirat der Bundesregierung, Globale Umweltveränderungen, Anrechnung biologischer Quellen und Senken im Kyoto Protokoll: Fortschritt oder Rückschlag für den globalen Umweltschutz?, 1998.

industrialized states parties to reduce the greenhouse gas emissions by a fixed percentage within an identified time.

The effectiveness of the regime will depend not only on whether the industrialized states meet their obligation to reduce greenhouse gas emissions. It is equally important that they meet their financial obligations. This includes, in particular, the obligation under article 10 Subparagraphs (c), (d), and (e) of the Kyoto Protocol. The regime is likely to achieve its objective only if developing countries have access to technologies that ensure the efficient use of energy — on a modern level, for example, in the energy consumption and in the building sector. In particular, further steps are to be taken to ensure that the envisaged cooperation of states parties results in a transfer of technology. This requirement may have an impact on national and European law on intellectual property as, in fact, this may require revision of the law on intellectual property as far as this property is needed for the protection of the environment. The question of to what extent holders of intellectual property are under an obligation to waive their rights if community interests so warrant has, so far, utterly been neglected. Unfortunately, international law does not give any direction in this regard. The proposal of several developing countries to modify the Agreement on Trade-Related Aspects of Intellectual

A New Tool for Characterizing and Managing Risks[1]

Ortwin Renn and
Andreas Klinke
*Center of Technology
Assessment in Baden-
Wuerttemberg*
Stuttgart, Germany

Gerald Busch and
Friedrich Beese
*Institute of Soil Science and
Forest Nutrition, University
of Göttingen,*
Göttingen, Germany

Gerhard Lammel
*Max Planck Institute for
Meteorology,*
Hamburg, Germany

1. Introduction

Risk is based on the contrast between reality and possibility (Markowitz, 1990). Only when the future is seen as at least partially influenced by human beings) is it possible to prevent potential hazards or to mitigate their consequences (Ewald, 1993). The prediction of possible hazards depends on the causal relation between the responsible party and the consequences. Because the consequences are unwelcome, risk is always a normative concept. A society should avoid, reduce, or at least control risks. Increasing potentials of technical hazards and the cultural integration of external hazards into risk calculations increase the demand for risk science and risk management (Beck, 1986).

Thus, risks can be described as possible effects of actions, which are assessed as unwelcome by the vast majority of human beings.

[1]The risk classification and the derived risk management strategies are developed by the "German Scientific Advisory Council on Global Change (WBGU)" in their annual report 1999 about global environmental risks. Ortwin Renn as a member and Andreas Klinke as associate researcher are basically responsible for the risk classification and the risk management strategies. See WBGU (1999). The parts concerning risk classification and risk management were written by Ortwin Renn and Andreas Klinke. Also cf. and Klinke and Renn (1999). The part on the application to environmental risks of substances was written by Gerald Busch, Friedrich Beese and Gerhard Lammel.

Risk concepts from various disciplines differ in the manner in which these effects of action are grasped and evaluated. Four central questions become the focus of our attention (Renn, 1992; 1997):

1. What are welcome and what are unwelcome effects? How do we define categories of damage and which criteria distinguish between positive (welcome) and negative (unwelcome) consequences of actions and events?
2. How can we predict these effects or how can we assess them in an intersubjectively valid manner? Which methodical tools do we have to manage uncertainty and to assess probability and damage?
3. Are we able to classify risks according to risk types? Which characteristics are relevant to evaluating risks besides the probability of occurrence and the extent of damage? Are there typical risk categories that allow us to order risks by priorities?
4. Which combination and which allocation of welcome and unwelcome effects legitimize rejection or approval of risky actions? Which criteria allow us an evaluation of risks?

To answer these questions and to be able to carry out such risk evaluations systematically, we propose a risk classification that summarizes specific risk types and determines particular strategies for rational management of risk types.

2. Risk Evaluation and Risk Classification

2.1 Main Characteristics of Risk Evaluation

The two central categories of risk evaluation are the *extent of damage* and the *probability of occurrence* (for definitions see Knight, 1921; National Research Council, 1983; Fischhoff *et al.*, 1984; Fritzsche, 1986; Short, 1984; Bechmann, 1990; IEC, 1993; Kolluru and Brooks, 1995; Banse, 1996; Rosa, 1997). *Damage* should generally be understood as negatively evaluated consequences of human activities (e.g., accidents by driving, cancer by smoking, fractured legs by skiing) or events (e.g., volcanic eruptions, earthquakes, explosions).

Other than the measurement of damage, there does not exist a separate method to validate the probability of occurrence (Tittes, 1986; Hauptmanns *et al.*, 1987; Kaplan and Garrik, 1993). The term probability of occurrence is used for such events of damage where information or even only presumptions about the relative frequency of the event have been given, but where the precise time remains uncertain. Risk statements always describe probabilities, i.e., tendencies of event sequences, which will be expected under specific conditions. The fact that an event is expected on average once each thousand years does not say anything about the time when the event will actually occur.

If we have indications of the determination of the probability of occurrence as well as the extent of damage, we call the degree of reliability of the determination *certainty of assessment*. If the certainty of assessment is low, one needs to characterize the nature of the uncertainty in terms of statistical confidence intervals, remaining uncertainties (identifiable, but not calculable), and plain ignorance. We use the term uncertainty if we mean the general inability to make deterministic predictions of events of damage (cf. Bonß, 1996). Uncertainty is a fundamental characteristic of risk, whereas the certainty of assessment varies between extremely high and extremely low. Even if it is not possible to make objective predictions about single events of damage on the basis of risk assessment, the assessment is not at all just as you like (Rosa, 1997). When we have two options of action where the same unwelcome event will occur with different probability, the conclusion for a decision under uncertainty is clear: Each rationally thinking human being would choose the option of action with the lower probability of occurrence (Renn, 1996).

2.2 Rational Risk Evaluation

From this point of view we consider it to be justified and necessary that technical and natural scientific assessments and social risk perceptions be brought together within rational risk evaluations (Fiorino, 1989). Now the question arises of how societies should decide on fundamental procedures concerning uncertain consequences of collective risks. Which strategy should a society choose if the consequences of risky actions concern many people with different preferences? Philosophers and decision-making the-

orists come to very different conclusions (cf. Shrader-Frechette, 1991; Leist and Schaber, 1995; Jonas, 1979; 1990; Rawls, 1971; 1974). We want to emphasize that scientifically evaluated risks and theoretical decision-making assessments have an action-determining function despite remaining uncertainty and ambivalence that cannot be replaced by either intuition or actual acceptance, political feeling, or assessments of interests. This is why we apply scientifically ensured evaluations of the respective risks by choosing appropriate tools of regulation.

Therefore, we distinguish three categories of risks for a practicable and rational risk evaluation (see Fig. 1): the *normal area*, the *intermediate area*, and the *intolerable area* (area of permission) (cf. also Piechowski, 1994). The *normal area* is characterized by relatively low statistical uncertainty, rather low probability of occurrence, rather low extent of damage, high certainty of assessment, low persistency and ubiquity of risk consequences, and low irreversibility of risk consequences, and the risks also have low complexity or empirically proven adequacy. In this case the objective risk dimensions almost correspond to the scientific risk evaluation. For risks in the normal area we follow the recommendations of decision-making analysts who take a neutral risk attitude as a starting point for collective binding decisions.

The "intermediate area" and the "intolerable area" are more problematic because the risks that go beyond ordinary dimensions. Within these areas the certainty of assessment is low, the statistical uncertainty is high, the potential damage can reach alarming dimensions, and systematic knowledge of consequences is missing. The risks can also generate global, irreversible damages that accumulate for a long time or mobilize or frighten the population in a special manner. A clear statement concerning the validity of the scientific risk evaluation is hardly possible. Risk aversien behavior of is absolutely appropriate because the limits of human ability of knowledge are reached. That is why a weighing risk decision is not any more a priority but a limitation of possibilities of wide-ranging negative surprises. Precautionarily oriented strategies of risk con-

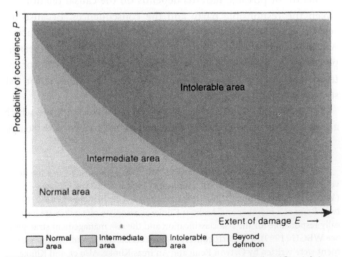

FIGURE 1 Risk areas. Source: WBGU, German Scientific Advisory Council on Global Change (1999).

trol, models of liability of endangering, general norms of caution, and general aspects of risk avoidance have priority.

2.3 Additional Criteria of Risk Evaluation

We consider it useful to include further criteria of evaluation in the characterization of risks (Kates and Kasperson, 1983; California Environmental Protection Agency, 1994). These criteria can be derived from research studies of risk perception or the way they are used or proposed as assessing criteria in several countries such as Denmark, the Netherlands, and Switzerland (cf. Petringa, 1997; Löfstedt, 1997; Hattis and Minkowitz, 1997; Beroggi *et al.*, 1997; Hauptmanns, 1997; Poumadère and Mays, 1997; Piechowski, 1994). The following criteria are relevant:

- *Ubiquity* defines the geographical dispersion of potential damages (intragenerational justice).
- *Persistency* defines the temporal extension of potential damages (intergenerational justice).
- *Irreversibility* describes the impossible restoration of the situation to the state before the damage occurred (possible restorations are, e.g., reforestation and cleaning of water).
- *Delay effect* characterizes a long time of latency between the initial event and the actual impact of damage. The time of latency could be of a physical, chemical, or biological nature.
- *Potential of mobilization* is understood as violation of individual, social, or cultural interests and values by affected people, generating social conflicts and psychological reactions.

In the relevant studies of risk perception most people associate risks with questions of control, voluntariness, addiction to risk sources, and just allocation of risk and benefit (Jungermann and Slovic, 1993). The assessment of control is covered by the criteria of ubiquity and persistency concerning the physical dimensions, and by the criterion of mobilization concerning the social dimensions. From a collective view voluntariness can hardly be taken into consideration as an assessing criterion for societal risks because our relevant risks will be transferred to others. The addiction to risk sources as a single criterion is normatively not useful because it is possible that people get used to unacceptable risks (e.g., accidents by driving). Criteria for distributive justice are more difficult to cover because intersubjective valid standards for measuring justice or injustice are lacking. Less problematic is the question of identity between beneficiaries of activities and people affected by risk. If there is identity, individual risk regulation is useful. In other cases collective mechanisms of regulation must be implemented. These can reach from commitments of liability to participation of affected people in decisions or procedures of permission. In most cases a case-to-case consideration is necessary to clearly find out violation of the thesis of justice.

In summary, our criteria and their ranges are:

- *Probability of occurrence (p)*: from 0 to 1.
- *Extent of damage (d)*: from 0 to infinity.

- *Certainty of assessment*:
  Confidence interval for *p*: high to low certainty of assessment by assessing the probability of occurrence;
  Confidence interval for *d*: high to low certainty of assessment by assessing the extent of damage.

- *Ubiquity*: local to global dispersion.
- *Persistency*: short to long removal period.
- *Irreversibility*: damage cannot be restored to damage can be restored.
- *Delay effect*: a low to high latency between the initial event and the impact of the damage.
- *Potential of mobilization*: political relevance to high political relevance.

2.4 Risk Classification

Theoretically, a huge number of risk types can be identified by these eight criteria. Such a huge number of cases would not be useful for the purpose of developing a comprehensive risk classification. In reality, some criteria are tightly coupled together and other combinations are certainly theoretically possible, but there are no or only a few empirical examples. Answering the question of risk priority, risks with several extreme qualities play a special role. We have chosen a classification where single risks are classified as risk types in which they particularly reach or exceed one of the possible extreme qualities. This classification is derived from Greek mythology.

Events of damages that have a probability of almost one are not relevant for us. High potentials of damages with a probability of nearly one can hardly be assessed as acceptable. Such risks occur seldom. In the same way, a probability that goes toward zero is harmless as long as the associated potential damage is not relevant. It is just a characteristic of risk that the range of damage negatively correlates with the level of probability. The higher the damage the lower the probability.

2.4.1 Risk Type "Sword of Damocles"

According to Greek mythology, Damocles was invited to a banquet by his king. At the table he had to sit under a sharp sword hanging on a wafer-thin thread. Chance and risk are tightly linked up for Damocles and the Sword of Damocles became a symbol for a threatening danger in luck. The myth does not tell about a snapping of the thread with its fatal consequences. The threat rather comes from the possibility that a fatal event could occur for Damocles every time even if the probability is low. Accordingly, this risk type relates to risk sources that have very high potentials for damage and at the same time very low probabilites of occurrence. Many technological risks such as unclear energy, chemical facilities and dams belong to this category.

2.4.2 Risk Type "Cyclops"

The ancient Greeks knew of enormously strong giants who were punished despite their strength by only having a single eye. They were called Cyclopes. With only one eye only one side of reality and no dimensional perspective can be perceived. Concerning risks it is only possible for them to ascertain either the probability of occurrence or the extent of damage while the other side remains uncertain. In the risk type Cyclops the probability of occurrence is largely uncertain whereas the maximum damage can be determined. Some natural events such as floods, earthquakes, volcanic eruptions, and El Niño, but also the appearance of AIDS, belong to this category as long as no or only contradictory information exists.

2.4.3 Risk Type "Pythia"

The Greeks of the antiquity asked their oracles in cases of uncertainty. The most known is the oracle of Delphi with the blind prophetess Pythia. Pythia's prophecies were, however, ambiguous. It certainly became clear that a great danger could threaten, but the probability of occurrence, the extent of damage, and the allocation and the form of the damage remained uncertain. Human interventions in ecosystems, technical innovations in biotechnology, and the greenhouse effect belong to this risk type where the extent of changes is still not predictable.

2.4.4 Risk Type "Pandora's Box"

The old Greeks explained many evils and complaints with the myth of Pandora's Box—a box that was brought down to Earth by the beautiful Pandora created by the god Zeus. It contained many evils and complaints. As long as the evils and complaints stayed in the box, no damage at all had to be feared. However, when the box was opened, all evils and complaints were released, which then irreversibly, persistently, and ubiquitously struck the earth. This risk type is characterized by uncertainty in both the probability of occurrence and the extent of damage (only presumptions) and by high persistency. Here, ozone-destroying substances can be quoted as examples.

2.4.5 Risk Type "Cassandra"

Cassandra was a prophetess of Troy who correctly predicted the victory of the Greeks, but whose compatriots did not take her seriously. The risk type Cassandra describes a paradox: the probability of occurrence and the extent of damage are known but produce little immediate concern because the damages will occur after a long time. Of course, risks of the type Cassandra are only interesting if the potential damage and the probability of occurrence are relatively high. This is why this type lies in the intolerable area (area of permission). A high degree of the delay effect is typical, i.e., a long period between the initial event and the impact of the damage. An example of this effect is anthropogenic climate change.

2.4.6 Risk Type "Medusa"

Ancient mythology tells that Medusa was one of three snake-haired sisters, the Gorgons, whose appearance turned the beholder to stone. Like the Gorgon who spread fear and horror as an imaginary mythical figure, some new phenomena have this effect on modern people. Some innovations are rejected although they are

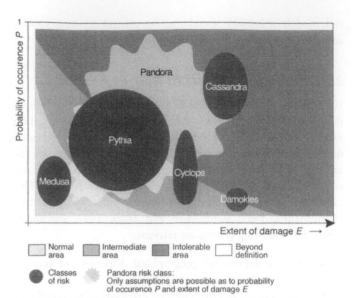

FIGURE 2 Risk types. Source: WBGU, German Scientific Advisory Council on Global Change (1999).

hardly assessed scientifically as threats. Such phenomenona have a high potential for public mobilization. Medusa was the only sister who was mortal– if we transfer the picture to risk policy– Medusa can be combated by rational arguments, further research, and clarification in public. According to the best knowledge of risk experts, risks of this type fall in the normal area. Because of their specific characteristics, these-risk sources frighten people and induce strong refusal of acceptance. Often a large number of people are affected by these risks but harmful consequences cannot be statistically proved. A typical example is electromagnetic fields.

The main objective of risk classification is to gain an effective and feasible policy tool for the evaluation and the management of risks. The characterization provides a platform for designing specific political strategies and measures for each risk type. The strategies pursue the goal of transforming unacceptable into acceptable risks; i.e., the risks should not be reduced to zero, but they should be reduced to a level such that routine risk management becomes sufficient to ensure safety and integrity. All strategies and respective measures are arranged according to priorities. In the normal case more than one strategy and more than one measure are naturally appropriate and necessary. If resources are limited, strategies and measures should be taken in line with the priority list. The following part lists the prior strategies and the prior measures recommended for each risk type.

3. Risk Management

3.1 Strategies and Instruments for the Risk Type "Sword of Damocles"

For risks from the category "Sword of Damocles," three central strategies are recommended (Table 2): First, the potential of disas-

TABLE 1 Risk types, Criteria and Examples*

Type 1	Sword of Damocles	*p* low (toward 0); *d* high (toward infinity); Confidence intervals of *p* and *d* low	Nuclear energy, chemical plants, dams, meteorite impacts
Type 2	Cyclops	*p* uncertain; *d* high; Confidence interval of *p* high; Confidence interval of *d* rather low	Floods, earthquakes, volcanic eruptions, AIDS, El Niño, mass developments of anthropogenically affected species
Type 3	Pythia	*p* uncertain; *d* uncertain (potentially high); Confidence intervals of *p* and *d* high;	Increasing greenhouse effect, endocrine-effective substances, release and spread of transgenic plants, BSE
Type 4	Pandora's Box	*p* uncertain; *d* uncertain (only presumptions); Confidence intervals of *p* and *d* uncertain (unclear); Persistency high (several generations)	Ozone-destroying substances
Type 5	Cassandra	*p* rather high; *d* rather high; Confidence interval of *p* rather low; Confidence interval of *d* rather low; Delay effect high	Anthropogenic climate change
Type 6	Medusa	*p* rather low; *d* rather low (exposition high); Confidence interval of *p* rather high; Confidence interval of *d* rather low; Potential of mobilization high	Electromagnetic fields

*Source: WBGU, German Scientific Advisory Council on Global Change (1999).

ters must be reduced by research and technical measurements. Second, the resilience must be increased, i.e., the power of resistance against surprises must be strengthened. Finally, an effective emergency management should be guaranteed.

Within the scope of the first strategy to reduce the damage potential, technical measures for the reduction of disaster potentials as well as the research and realization of measures to reduce the extent of damage have to be improved. For example, in the past the prior implemented strategy of nuclear energy was to reduce the probability of a core melt-down. To move this risk from the intermediate area to the normal area, the strategy was not appropriate. More useful would be a change toward reducing the potential of catastrophes (meanwhile this happens). Strengthening of liability rules is useful as well: operators are encouraged to improve knowledge and to reduce the remaining risks. At the same time, it is necessary to research alternatives with a lower damage-potential to replace technologies that have unavoidably high damage-potentials. Subsidies are necessary during establishment and testing of alternatives.

Within the scope of the second strategy it is necessary to increase the resilience against the risk potentials. Therefore, capacity-building is required so that institutional and organizational structures of overriding importance can be improved and strengthened to have strong influence on procedures of permit, monitoring, training, etc. Additionally, technical procedures to increase the resilience must be established or, if they already exist, they must be improved. This can be achieved by technical redundancy and organizational security units, by integration of latitudes, buffers, and elasticities and by diversification, i.e., the dispersion of risk sources. Resilient organization models and successful procedures of permit should be placed at other states' disposal by technology and knowledge transfer. International control and monitoring should also be strengthened and an international safety standards authority should be established.

The third priority is emergency management. This strategy is not assessed as insignificant as a strategy of damage limitation; it should, however, stay behind the risk-reducing strategies. Here, capacity-building must be enhanced by developing and promoting national programs of emergency protection. Successful measures of emergency protection and techniques in the forms of training, education, and empowerment should be transferred to local risk-managers by technology and knowledge transfer.

In addition, technical measures of protection and measures to reduce the extent of damage have to be enforced. Finally, an international initiative on disaster prevention and relief, such as the former "International Decade for Natural Disaster Reduction (IDNDR)" initiated by the UN, is necessary for anthropogenically caused disasters.

3.2 Strategies and Instruments for the Risk Type "Cyclops"

In the case of the risk type "Cyclops" the uncertainty concerning the probability of occurrence is the starting point for regulatory measures. First of all, increased research and intensive monitoring

TABLE 2 Strategies and Instruments for the Risk Type "Sword of Damocles"*

Strategies	Instruments
Reducing disaster potential	Research to develop substitutes and to reduce the potential of disasters
	Technical measures for reducing the disaster potential
	Stringent rules of liability
	International safety standards authority
	Subsidies of alternatives for the same use
	Containment (reducing the damage extension)
	International coordination (e.g., averting the hazard of meteorites)
Increasing resilience	Capacity-building (permit, monitoring, training)
	Technical procedures of resilience (redundancy, diversification, etc.)
	Blueprint for resilient organizations
	Procedures of permit as model
	International control (IAEA)
	International liability commitment
Emergency management	Capacity-building (protection from emergencies)
	Training, education, empowerment
	Technical protection measures, including strategies of containment
	International emergency groups (e.g., fire brigade, radiation protection, etc.)

*Source: WBGU, German Scientific Advisory Council on Global Change (1999).

TABLE 3 Strategies and Instruments for the Risk Type "Cyclops"*

Strategies	Instruments
Ascertaining the probability of occurrence	Research to ascertain numerical probability
	International monitoring by
	National risk centers
	Institutional network
	Global risk board
	Technical measures for calculating the probability of occurrence
Prevention against surprises	Strict liabilities
	Compulsory insurance for those generating the risks (e.g., floods, housing estates)
	Capacity-building (permit, monitoring, training)
	Technical measures
	International monitoring
Emergency management or reducing the extent of damage	Capacity-building (protection from emergencies)
	Training, education, empowerment
	Technical protection measures, including strategies of containment
	International emergency groups (e.g., fire brigade, radiation protection etc.)

*Source: WBGU, German Scientific Advisory Council on Global Change (1999).

are necessary for a better assessment of the probability. Until such results are available, strategies to prevent unwelcome surprises are useful (including endangering liabilities). Preventive measures for disasters are important at the international level because the damage potentials within affected countries with high vulnerability can reach precariously high levels.

First priority goes to inquiry into the probability of occurrence, for which the necessary research has to be encouraged. Additionally, international monitoring by national and international risk centers has to be guaranteed. That could be fulfilled by establishing a "UN Risk Assessment Panel" that would have the function of setting up a network among the national risk centers and of gathering and assessing knowledge about global risks.

Within the scope of the second strategy, unwelcome surprises have to be prevented and the society has to be protected against it. This could happen by endangering liabilities or by compulsory insurance on certain conditions. The appropriate instruments of capacity-building and technical measures extensively correspond to the instruments of the risk type "Sword of Damocles". Within the third strategy of emergency management, the same instruments as for the risk type "Sword of Damocles" are used.

3.3 Strategies and Instruments for the Risk Type "Pythia"

Because the risk type "Pythia" has high uncertainty concerning the criteria probability of occurrence and extent of damage, the im-

provement of knowledge is very effective, especially the basic research. At the same time, preventive strategies should be used because the extent of damage could reach global dimensions. Limitations of regulatory policy and geographical and temporal measures of containment are usually indispensable.

First priority goes to the preventive strategies of institutional regulations such as ALARA (as low as reasonably achievable), BACT, and technical standards, where the costs of a neglected risk-reducing policy should be as low as possible. International conventions for controlling, monitoring, and security measures are also necessary. The instruments to reduce the extent of damage and capacity-building are the same as for the risk types mentioned above.

The improvement of knowledge has second priority so that future risk analysis can provide a higher certainty of assessment. To achieve this, research is needed on how to ascertain the probability of occurrence and the extent of damage. Additionally, an international early-warning system is necessary as for the risk type "Cyclops".

The third strategy of emergency management comes close to measures of the previous risk types.

3.4 Strategies and Instruments for the Risk Type "Pandora's Box"

The risks of Pandora's box are characterized by uncertainty concerning the probability of occurrence and the extent of damage (only presumptions) and high persistency. Here, research efforts to develop substitutes and regulatory measures to contain or to re-

TABLE 4 Strategies and Instruments for the Risk Type "Pythia"*

Strategies	Instruments
Improving prevention	Institutional regulations like ALARA, BACT, technical standards, etc.
	Fund solutions
	International conventions for controlling, monitoring, and security measures, etc.
	Containment (reducing the extension of damage)
	Capacity-building (permit, monitoring, training)
	Technical procedures of resilience (redundancy, diversification, etc.)
Improving knowledge	Research to ascertain the probability of occurrence and the extent of damage
	International early warning system by
	National risk centers
	Institutional network
	Global risk board
Emergency management	Containment strategies
	Capacity-building (protection from emergencies)
	Training, education, empowerment
	Technical protection measures
	International emergency groups (e.g., for decontamination)

*Source: WBGU, German Scientific Advisory Council on Global Change (1999).

duce the risk sources are absolutely essential because the negative consequences of the risk sources are unknown, but in the most unfavorable case the consequences can reach global dimensions with irreversible effects. It has also to be implemented at the international level.

The supply of substitutes has priority over other strategies. Concerning the research and development of substitutes, the mea-

TABLE 5 Strategies and Instruments for the Risk Type "Pandora's box"*

Strategies	Instruments
Developing substitutes	Research to develop substitutes
	Supporting basic research
	Incentives to use less harmful substitutes
	Subsidies for developing alternative production systems
Reduction and containment	Regulatory policy for limitation of exposures through environmental standards, etc.
	Use of incentive systems (certificates)
	Strict liability, if useful
	Improving and developing technical procedures of support
	Capacity-building (technical know-how, technology transfer, education, training)
	Joint implementation
Emergency management	Capacity-building (protection from emergencies)
	Technical protection measures, including containment strategies
	Training, education, empowerment

* Source: WBGU, German Scientific Advisory Council on Global Change (1999).

sures basically correspond to those for the risk type "Sword of Damocles". In addition, this risk type requires wide-ranging basic research that should be supported adequately.

In a second step the risk potentials should be decreased by reducing specific risk sources or by prohibiting them completely. Regulatory procedures are suitable, e.g., limitation of quantities by environmental standards and a rather economic incentive system by means of certificates. In some cases the use of endangering liability is appropriate. As mentioned above, instruments of technical procedures and capacity-building are necessary.

The third strategy of emergency management corresponds to the other risk types. Especially, an international emergency group combating unwelcome surprises is necessary. The international emergency group for nuclear decontamination of the IAEA could serve as an example.

3.5 Strategies and Instruments for the Risk Type "Cassandra"

The risks of the risk-type Cassandra hardly have any uncertainty, but people do not take the risks very seriously because of the lingering manner or the delay between the initial event and the damage. Due to the short-time legitimization through short election periods, politics often lacks the motivation of taking care of such long-term hazards. Measures of collective commitment (e.g., code of conduct for multinational enterprises) and long-term global institutions (UN Risk Assessment Panel) should strengthen the long-term responsibility of the international community. Limitations of quantities are appropriate to reduce these risks.

TABLE 6 Strategies and Instruments for the Risk Type "Cassandra"*

Strategies	Instruments
Strengthening the long-term responsibility of key actors	Self-commitment, code of conduct of global actors
	Enhancing participation, empowerment and institutional security as a means to foster long-term responsibility
	Measures against governmental break-down
	Fund solutions
	International coordination
Continuous reduction of risk by introducing substitutes and setting limitations of exposure	Use of incentive systems (certificates and fees)
	Strict liability, if useful
	Regulatory limitations of quantities by environmental standards (also international standards)
	Improving and developing technical procedures of support
	Capacity-building (technical know-how, technology transfer, education, training)
	Joint implementation
Contingency management	Capacity-building (recultivation, protection from emergencies)
	Technical protection measures, including containment strategies
	Training, education, empowerment

* Source: WBGU, German Scientific Advisory Council on Global Change (1999).

If there is a relevant delay between the initial event and the consequences, the first strategy should strengthen the long-term responsibility for future generations. Prior instrument is the self-commitment of the states and relevant actors (e.g., multinational enterprises). It is possible that fund solutions are appropriate. On the rather individual level, potentially affected people can gain more action capacities by linking participation to empowerment.

The second strategy is the continual reduction of risk potentials by developing alternative substitutes. Risk potentials that cannot be substituted should at least be stopped by limiting either the quantities or the field of application. The necessary instruments are mentioned under the other risk types. The instruments of the third strategy of emergency management also correspond to the other risk types.

3.6 Strategies and Instruments for the Risk Type "Medusa"

The risk type Medusa requires measures of confidence-building and the improvement of knowledge to reduce the remaining uncertainties. Clarification is not enough; on the contrary the affected people themselves should constructively be able to integrate the remaining uncertainties and ambiguities into their decision-making.

The extent of damage and the probability of occurrence of this risk type are low, however the potential of mobilization is high. To be able to inform and enlighten the public about the real extent of damage and probability of occurrence, confidence has to be built up. Independent institutions can contribute to clarifying the results of scientific research, and also the pure hypothetical character of many fears. The affected people should participate in decision-making procedures and in procedures of permit. The support of social scientific research concerning the potential of mobilization and the social management of risk conflicts is neces-

TABLE 7 Strategies and Instruments for the Risk Type "Medusa"*

Strategies	Instruments
Confidence-building	Establishment of independent institutions for information and clarification
	Increasing the chances of participation with the commitment to set up priorities
	Support of social science concerning the potential of mobilization
	Procedures of permit with participation of affected people as model
	International control (IAEA)
	International liability commitment
Improving knowledge	Research to improve the certainty of assessment
	Governmental support of research (basic research)
Risk communication	Two-way communication
	Involvement of citizens
	Informed consent

* Source: WBGU, German Scientific Advisory Council on Global Change (1999).

sary to be able to manage the problems of the risk type Medusa in the society. Additionally, the knowledge of the probable risk potential should be improved. Research to improve the certainty of assessment and basic research are required.

4. Application to Environmental Risks from Substances[2]

4.1 Global Biogeochemical Cycles Are Influenced by Human Activity

Carbon, nitrogen, and sulfur are essential to the life of animals, plants, and microbes. Interactions between these elements link the internal biogeochemistry of terrestrial ecosystems. Naturally, the availability of these substances is limited in terrestrial ecosystems and this has led to various adaptations of the biota. Nowadays, high anthropogenic emissions of various compounds of carbon, nitrogen, and sulfur have created a new situation for terrestrial ecosystems: The surplus (regional) of these three limited elements can affect terrestrial ecosystems in multiple ways and on different time scales (see Table 8).

4.2 Risk Classification of Environmental Risks from Substances

Despite a quite good kowledge of many determining processes, uncertainty remains about the expected geographical dispersion of the potential damages, the time when they will occur, and the extent of the damages. Not only anthropogenic influences but also natural disturbances cause multiple stress to forest ecosystems and make the determination of the risk potentials and the extent of damage even more difficult. In the face of the underlying complex processes and the possibly high latency between initial events and response of the ecosystems, the risk perception is even lower than for risks of direct impact. Although persistency is rather high and reversibility of the potential damage is low, the potential for mobilization is generally low. This leads to its characterization as the risk type Cassandra.

4.3 Forest Ecosystems Are Influenced by the Changing Biogeochemical Cycles

The anthropogenic influence on global biogeochemical cycles lead to a new situation for forest ecosystems: Increasingly, multiple compounds of nitrogen, sulfur, and carbon are simultaneously available in large quantities (regionally even in surplus). In the preindustrial era, the mean global atmospheric N input was in the range of $1-5$ kg N ha^{-1} year^{-1} (Kimmins, 1987; Flaig and Mohr, 1996). Therefore, input of nitrogen was the limiting factor for plant growth in most forest ecosystems until the beginning of the industrial revolution (Kimmins, 1987). During the past decades

---

[2] This part was written by Gerald Busch, Friedrich Beese, and Gerhard Lammel.

TABLE 8 Overview of Possible Impacts and Risk Potential of Anthropogenic Changes in Global Biogeochemical Cycles (↑ , Increase; ↓ , Decrease, -, No Change).*

Substances	Possible Reactions and Effects	Associated Risk Potential
N-input in ecosystems (Eutrophication)	↑ N-contents, ↑ Mineralization, ↑ N-turnover, ↑ NPP, ↑ N surplus, ↓ Mycorrhiza, ↑ / ↓ Humus layer	↑ Nitrate leaching to groundwater, ↓ Frost-, drought-, or pest resistance, changes in species and vegetation, degradation of N-limiteded ecosystems, ↑ loss of biodiversity, ↑ loss of ecosystem functions
N and SO_2^- deposition	↑ Soil acidification, ↑ Al toxicity, ↑ damage of fine roots and mycorrhiza, ↑ cation leaching	↑ Nitrate leaching to groundwater, ↑ acidification of freshwaters, ↓ drought resistance, ↑ nutrient imbalances, ↑ forest desease and forest decline, ↑ loss of biodiversity, ↑ loss of ecosystem functions
↑ $[CO_2]$ (low N-availibility)	-/ ↑ NPP, ↑ / ↓ C/N ratio, N accumulation/leaching , ↑ / ↓ Mineralization, ↑ / ↓ root/shoot ratio, -/ ↑ Water use efficiency, Nutrient use efficiency	Reactions are very site specific and independant of species, e.g., -/ ↑ NPP of vegetation, changes in stocks and site composition (C_3-, C_4-plants), species-related changes in population of herbivores
↑ $[CO_2]$ and ↑ N inputs	↑ C and N accumulation because of ↑ NPP, ↑ / ↓ Humus accumulation	Changes in site composition, loss of biodiversity, sudden emission/loss of accumulated N and C because of external disturbances (land-use change, fire, climate change),
Climate change, ↑ $[CO_2]$ and ↑ N and ↑ S inputs	Shift of vegetation and water budget, global: ↑ Mineralization, ↑ NPP, ↓ C-sequestration	Highly uncertain: ↑ climate change (positive feedback of vegetation, e.g., ↑ CO_2-emissions), shift of vegetation, ↑ invasion of alien species, desertification

*Examples of only multiple reactions and associated possible risks can be shown in this table.

Sources: Mooney *et al.*, 1998; Walker *et al.*, 1998; Arnone III and Hirschel, 1997; Foster *et al.*, 1997; Hungate *et al.*, 1997; Kinney *et al.*, 1997; Vitousek *et al.*, 1997; Drake *et al.*, 1997; Flaig and Mohr, 1996; IPCC, 1996; Körner and Bazzaz, 1996; Koch and Mooney, 1996; Walker and Steffen, 1996; Amthor, 1995, Dixon and Wisniewski, 1995; Heywood and Watson, 1995; Woodwell and Mackenzie, 1995; Mohr and Müntz, 1994; Vitousek, 1994; Schulze *et al.*, 1989.

forests in Europe and Northeastern America have been in transition from nitrogen-deficient to nitrogen-saturated systems due to increasing nitrogen deposition. The impact of nitrogen deposition on plants and soil is through both fertilizing and toxic effects, eutrophication and acidification (see Aber *et al.*, 1998; Gundersen *et al.*, 1998; Boxman *et al.*, 1998).

In the last decades, anthropogenic sulfur and nitrogen emissions have been discussed in the context of acid rain (see Ulrich and Sumner, 1991; van Breemen *et al.*, 1983; Reuss and Johnson, 1986) and this point of view has influenced policy in Europe and North America (e.g., LRTAP UN-ECE Second Sulphur Protocol, Clean Air Act).

As a consequence of population growth and rapid economic development, increasing loads of nitrogen and sulfur on the terrestrial ecosystems might well not remain limited to the known "hotspots" in Europe and North America but could expand and become critical as well for tropical and subtropical regions.

4.4 Changing Patterns of Nitrogen and Sulfur Deposition

In the following section we delineate the disposition of global forest in context of the dynamics of changes in deposition patterns. For quantification we use, besides other data sets, present-day and—under a scenario—future acid and nitrogen deposition data as produced by global-scale models to show the regional distribution of the increasing bias between acidification of forest soils and nitrogen fertilization.

In a first step, nutrient-depleted soils with low buffering capacity are identified to assess regions with potential for destabilization of forest ecosystems by acid deposition. Because the "acidity neutralization capacity" of soils (ANC) cannot be accurately determined from the global data, a simple approach based on the "Soil Map of the World" (FAO, 1995) is carried out. To evaluate the buffering capacity of the topsoils, the CEC data (cation exchange capacity) and the base saturation data (Na, K, Mg, and Ca) are combined with a map of the global distribution of forests (WCMC, 1997) to obtain the measures for forest soils with low buffering capacity.

To evaluate the buffering capacity of the forest soils, the actual (1980–1990) acidic input and nitrogen turnover are applied to the identified regions. $SO_y(= SO_2 +$ sulfate) and NO_y ($= NO_2 +$ HNO_3 + nitrate) deposition fields and the related acidic inputs are taken from a general-circulation model of the atmosphere, ECHAM4 (Roeckner *et al.*, 1996). NH_y ($=$ ammonia + ammonium) deposition fields are taken from a run of the global tracer-transport model MOGUNTIA (Zimmermann, 1988), the only model so far that describes reduced nitrogen compounds. By GIS-analysis those regions were identified in which the buffering capacity is depleted in 25–100 years by corresponding acid loads. Under the assumptions of a future scenario (IS92a–IPCC, 1996) the same assessment is carried out for the years 2040–2050. In a second step, nitrogen deposition in forest ecosystems is analyzed for the same time horizons. This assessment focuses on nitrogen deposition that exceeds natural input; the threshold was set to 5 kg N ha^{-1} year^{-1} (Bobbink *et al.*, 1992; UN-ECE, 1996).

4.5 Saturation of Forest Soils Buffering Capacity

In relation to acid input from 1980 to 1990, the buffering capacity of 1.8 Mio km² or 15% of the acid-sensitive forest soils tends to become saturated in the next 25–100 years. Under the assumptions of the IS92a scenario, this share more than doubles and increases to 4.0 Mio km² or 34% between 2040 and 2050. For 1980–1990, the mean buffering capacity of these sols based on our methodology is supposed to last for 65 years more. Under changed inputs this period tends to decrease for 2040–2050 to 50 years, which is less than half the lifetime of most of the managed tree species. For 1980–1990, four regions are mainly affected by acid deposition: the Eastern part of Northern America, Europe, Scandinavia with the Northwestern Russian Federation, and Southern China. The situation for 2040–2050 changes in such a way that the "old hotspots" are still present but the area of saturation increases only moderately with the main increase taking place in the tropical and subtropical regions of South America and South and Southeast Asia (see Fig. 3; see also color insert).

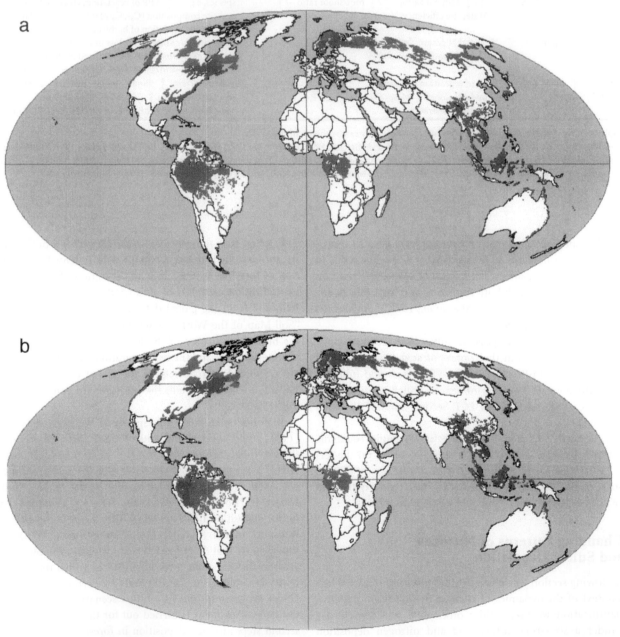

FIGURE 3 Distribution of exceeded forest soils buffering capacity. (a) today (1980–1990) and (b) (2040–2050). Red areas show forest soils with an exceeded buffering capacity while the green areas show the not-affected areas of acid sensitive and nutrient deficient forest soils. See also color insert.

4.6. Nitrogen Deposition in Nutrient-Deficient and Acid-Sensitive Soils

For 2040–2050, nearly 54% of the forest ecosystems on acidified soils are projected to receive a nitrogen load greater 5 kg N ha^{-1} year^{-1}. Because of better soil conditions the affected area is smaller in India, Eastern North America, and Europe. In absolute numbers, the forests in the Eastern North America will be affected most, followed by those in Southeast Asia and China (see Fig. 4).

Greatest changes in aerial distribution and increase of concentration will occur in the Asian region (see Fig. 4). Regions with acid-sensitive soils and high N-depositions are concentrated to China and Southeast Asia, Western and Central Europe, and

Eastern North America. Again, in absolute numbers Eastern North America shows the largest distribution of forest areas with an exceeded soil buffering capacity and high nitrogen deposition, followed by Southeast Asia, China, and Europe.

4.7 Conclusion

It has been shown that under the assumptions of the IPCC IS92a scenario, the contrast between unbalanced nutrient input and acidification or nutrient depletion will increase. Greatest changes are most likely to occur in subtropical and tropical regions of Asia but the well-known hotspots of Europe and Eastern North America will remain so. Both forest areas with both depletion of

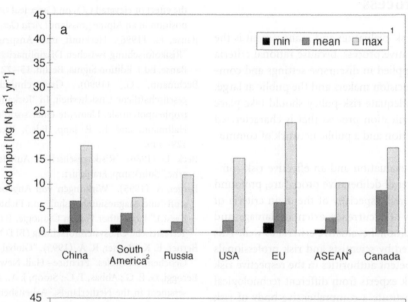

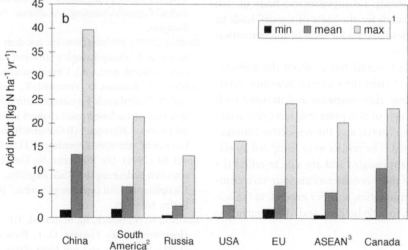

FIGURE 4 Regional distribution of acid input into forest ecosystems on acidified soils with minimum, maximum, and mean values in selected regions or countries. (a) today (1980–1990) and (b) (2040–2050). [2]Brazil, Ecuador, Bolivia, Columbia, Paraguay, Peru, and Venezuela; Myanmar, Thailand, Laos, Vietnam, Brunei, Philippines, Singapore, Indonesia, and Malaysia; dashed line: threshold of natural nitrogen input.

soil's buffering capacity and increasing nitrogen deposition will expand in several regions. The forest areas likely to meet these two risks are still a minor fraction of the global forest ecosystems.

Soils in forest ecosystems provide the transformation function for nutrient flows and material flows in general, besides other functions. Soils' nutrient reservoirs and buffering capacities get depleted in increasingly large areas. On the other hand, growth induced by increasing N availability triggers additional nutrient demand, which in most cases cannot be satisfied. We note that nutrient and acidification status of those soils that are subject to increasingly high inputs will necessarily change in foreseeable periods.

5. Some Conclusions for a Deliberative Process

Central to our concept of risk evaluation and management is the attempt to initiate a deliberative process, because rational criteria of evaluation ought to be applied in discursive settings and communicated to the political decision makers and the public at large. So the deliberation for an adequate risk-policy should take place within a multistage communication process that is characterized by forms of mutual consultation and a public network of communication.

To assure a rational risk-evaluation and an effective risk-communication as part of an overall deliberative procedure, profound scientific knowledge is required, especially of the main criteria of risk evaluation—probability of occurrence, extent of damage, and incertitude—and to the additional evaluative criteria as well. This knowledge has to be collected by scientists and risk professionals who are recognized as competent authorities in the respective risk field. The experiences of risk experts from different technological or environmental fields constitute a comprehensive body of risk knowledge. The systematic search for the 'state of the art' leads to a knowledge base that provides the data for each of the evaluation criteria.

In the framework of the last annual report about the management of global environmental risks the German Scientific Advisory Council on Global Change recommended and initiated such a deliberative process. A number of risk potentials were characterized based on the evaluation criteria, and the respective management strategies were developed. The results were compiled by scientists who possess the relevant insights and are able to reflect the state of the art. The results of these considerations were then communicated to the respective ministries, to other experts, to the science community, to industry, to stakeholders, and to the public.

References

Aber J., McDowell, W., Nadelhoffer K., Magill A., Berntson, G., Kamakea, M., McNulty, S., Currie, W., Rustad, L., and Fernandez, I. (1998). Nitrogen saturation in temperate forest ecosystems—Hypotheses revisited. *Bioscience* **48**, 921–934.

Aber, J. D., Nadelhoffer, K. J., Steudler, P., and Melillo, J. M. (1989). Nitrogen saturation in northern forest ecosystems. *Bioscience* **39**, 378–386.

Alcamo, J., Krol, M., and Posch, M. (1995). An integrated analysis of sulfur emissions, acid deposition and climate change. *Water Air Soil Pollut* **85(3) 1995**, 1539–1550.

Alewell, C., Bredemeier, M., Matzner, E., and Blanck, K. (1997). Soil solution respone to experimentally reduced acid deposition in a forest ecosystem. *J. Environ. Quality* **26**, 658–665.

Allen, L. H. and Amthor, J. S. (1995). Plant physiological responses to elevated CO_2, temperature, air pollution, and UV-B radiation. In "Biotic Feedbacks in the Global Climatic System—Will the Warming Feed the Warming?" (G. Woodwell, and F. T. Mackenzie, Eds.), pp. 51–84. Oxford University Press, New York.

Amthor, J. S. (1995). Terrestrial higher-plant response to increasing atmospheric $[CO_2]$ in relation to the global carbon cycle. *Global Change Biol.* **1**, 243–274.

Arnone III, J. A. and Hirschel, G. (1997). Does fertilizer application alter the effect of elevated CO_2 on *Carex* leaf litter quality and in situ decomposition in an Alpine grassland? *Acta Oecol.* **18**, 201–206.

Banse, G. (1996). Herkunft und Anspruch der Risikoforschung. In. "Risikoforschung zwischen Disziplinarität und Interdisziplinarität." (G. Banse, Ed.), Edition Sigma, Berlin. 15–72.

Bechmann, G. (1990). Großtechnische Systeme, Risiko und gesellschaftliche Unsicherheit. In "Riskante Entscheidungen und Katastrophenpotentiale. Elemente einer soziologischen Risikoforschung." (J. Halfmann, and K. P. Japp, Eds.), Westdeutscher Verlag, Opladen. 129–149.

Beck, U. (1986). "Risikogesellschaft. Auf dem Weg in eine andere Moderne." Suhrkamp, Frankfurt.

Berger, A. (1995). "Wirkungen von Angebot und Bedarf auf den Stickstoff- und Magnesiumhaushalt von Fichtenkeimlingen (*Picea abies* (L.) Karst.)." Bayreuther Forum Ökologie, Band 23, Bayreuther Institut für Terrestrische Ökosystemforschung (BITÖK), Selbstverlag.

Berner, E. K., Berner, R. A. (1995). "Global Environment: Water, Air, and Geochemical Cycles." Prentice–Hall, New York.

Beroggi, G. E. G.; Abbas, T. C.; Stoop, J. A., and Aebi, M. (1997). "Risk Assessment in the Netherlands." Arbeitsbericht Nr. 91 der Akademie für Technikfolgenabschätzung. Akademie für Technikfolgenabschätzung, Stuttgart.

Block, J. (1995). Stickstoffausträge mit dem Sickerwasser aus Waldökosystemen. In Wirkungskomplex Stickstoff und Wald. Texte 28/95, 80–96. (Umweltbundesamt, Ed.). Umweltbundesamt, Berlin.

Bobbink, R., Boxman, D., Fremstad, E., Heil, E., Houdijk, A., Roeloefs, J. (1992). Critical loads for nitrogen eutrophication of terrestrial and wetland ecosystems based upon changes in vegetation and fauna. In "Critical Loads for Nitrogen." (P. Grennefelt and E. Thörnelof, Eds.), Nordic Council of Ministers, Kopenhagen. 111–159.

Bonβ, W. (1996). Die Rückkehr der Unsicherheit. Zur gesellschaftstheoretischen Bedeutung des Risikobegriffes. In "Risikoforschung zwischen Disziplinarität und Interdisziplinarität." (G. Banse, Ed.), Edition Sigma, Berlin. 166–185.

Boxman, A. W., Blanck, K., Brandrud, T-E, Emmett, B. A., Gundersen, P., Hogervorst, R. F., Kjønaas, O. J., Persson, H., and Timmermann, V. (1998). Vegetation and soil biota response to experimentally-changed nitrogen inputs in coniferous forest ecosystems of the NITREX project. *Forest Ecol. Management* **101**, 65–79.

California Environmental Protection Agency. (1994). "Toward the 21st Century. Planning for Protection of California's Environment. Final Report." EPA, Sacramento, California.

Dise, N. B. and Wright, R. F. (1995). Nitrogen leaching from European forests in relation to nitrogen deposition. *Forest Ecol. Management* **71**, 153–1161

Drake, B. G., Gonzàles-Meler, M., and Long, S. P. (1997). More effecient plants: A consequence of rising atmosphere CO_2? *Annu. Rev. Plant Physiol. Plant Mol. Biol.* **48**, 609–639.

Eichhorn, J. and Hüttermann, A. (1994). Humus disintegration and nitrogen mineralization. In. "Effects of Acid Rain on Forest Processes." (D. L. Godbold and A. Hüttermann, Eds.), pp. 129–162. Wiley-Liss, New York.

Etheridge, D. M, Steele, L. P., Langenfelds, R. L., and Francey, R. J. (1996). Natural and anthropogenic changes in atmospheric CO_2 over the last 1000 years from air in Antarctic ice and firn. *J. Geophys. Res.* **101**, 4115–4128.

Ewald, F. (1993). Der Vorsorgestaat. Frankfurt/M.: Suhrkamp.

FAO, Food and Agriculture Organisation. (1995). "The Digital Soil Map of the World." Land & Water Development Division, FAO, Rome.

FAO, Food and Agriculture Organisation. (1997). "State of the Worlds Forests 1997." FAO, Rome.

Fiorino, D. J. (1989). Technical and democratic values in risk analysis. *Risk Anal.* **9(3)**, 293–299.

Fischhoff, B., Watson, S.R., and Hope, C. (1984). Defining risk. *Policy Sci.* **17(5)**, 123–129.

Flaig, H. and Mohr, H. (1996). Der überlastete Stickstoffkreislauf: Strategien einer Korrektur. Nova Acta Leopoldina. Nummer 289. Band 70. Barth, Leipzig.

Flaig, H. and Mohr, H. (1992). Assimilation of nitrate and ammonium by the Scots pine (*Pinus sylvestris*) seedling under condition of high nitrogen supply. *Plant Physiol.* **84**, 568–576

Foster, D. R., Aber, J. D., Melillo, J. M., Bowden, R. D. and Bazzaz, F. A. (1997). Forest response to disturbance and anthropogenic stress. *Bioscience* **47**, 437–445.

Fritzsche, A. F. (1986). Wie sicher leben wir? Risikobeurteilung und -bewältigung in unserer Gesellschaft. Köln: TÜV Rheinland.

Godbold, D. L. and Hüttermann, A., Eds. (1994). "Effects of Acid Rain on Forest Processes." Wiley-Liss, New York.

Gundersen, P., Emmett, B. A., Kjønaas, O. J., Koopmans, C. J., and Tietema, A. (1998). Impact of nitrogen deposition on nitrogen cycling in forests: A synthesis of NITREX data. *Forest Ecol. Management* **101**, 37–56.

Hattis, D. and Minkowitz, W. S. (1997). "Risk Evaluation: Legal Requirements, Conceptual Foundations, and Practical Experiences in the United States." Arbeitsbericht Nr. 93 der Akademie für Technikfolgenabschätzung. Stuttgart: Akademie für Technikfolgenabschätzung.

Hauptmanns, U. (1997). "Risk Assessment in the Federal Republic of Germany." Arbeitsbericht Nr. 94 der Akademie für Technikfolgenabschätzung. Stuttgart: Akademie für Technikfolgenabschätzung.

Hauptmanns, U., Herttrich, M., and Werner, W. (1987). "Technische Risiken: Ermittlung und Beurteilung." Springer, Berlin.

Hesterberg D., Stigliani W. M., and Imeson A. C. (1992). Chemical time bombs: Linkage to scenarios of socioeconomic development. IIASA Exec. Report No. **20**, 1992.

Heywood, V. H. and Watson, R. T., Eds. (1995). "Global Biodiversity Assessment." Cambridge University Press, Cambridge.

IEC. (1993). Guidelines for Risk Analysis of Technological Systems. Report IEC-CD (Sec) 381 issues by the Technical Committee QMS/23. European Community, Brüssels.

IPCC, International Panel on Climate Change. (1996). "Climate Change 1995–The Science of Climate." Contribution of Working Group I to the Second Assessment Report of the Intergovernmental Panel on Climate Change. Cambridge, Cambridge University Press, New York.

Jonas, H. (1979). "Das Prinzip der Verantwortung. Versuch einer Ethik für die technologische Zivilisation." Insel Verlag, Frankfurt.

Jonas, H. (1990). Das Prinzip Verantwortung. In: "Risiko und Wagnis. Die Herausforderung der industriellen Welt. Band 2." Gerling Akademie, Pfullingen, Neske. (M. Schüz, Ed)., 166–181.

Jungermann, H. and Slovic, P. (1993). Charakteristika individueller Risikowahrnehmung. In: "Risiko ist ein Konstrukt. Wahrnehmungen zur Risikowahrnehmung." (Bayerische Rückversicherung, Ed.), Knesebeck, München, 89–107.

Kaplan, S. and Garrik, J. B. (1993). Die quantitative Bestimmung von Risiko. In "Risiko und Gesellschaft. Grundlagen und Ergebnisse interdisziplinärer Risikoforschung." (G. Bechmann, Ed.). Westdeutscher Verlag, Opladen, 91–124.

Kates, R. W. and Kasperson, J. X. (1983). Comparative risk analysis of technological hazards. A review. Proc. Nat. Acad. Sci. 80(21), 7027–7038.

Kimmins, J. P. (1987). "Forest Ecology." Macmillan, New York.

Kinney, K. K., Lindroth, R. L., Jung, S. M., and Nordheim, E. V. (1997). Effects of CO_2 and NO_3^- availability on deciduous trees: Phytochemistry and insect performance. *Ecology* **78**, 215–230.

Klinke, A. and Renn, O. (1999). "Prometheus Unbound. Challenges of Risk Evaluation, Risk Classification, and Risk Management." Working Paper No. 153 of the Center of Technology Assessment. Center of Technology Assessment, Stuttgart.

Knight, F. (1921). "Risk, Uncertainty and Profit." Kelley, New York.

Koch, G. W. and Mooney, H. A., (Eds.) (1996). "Carbon Dioxide and Terrestrial Ecosystems." Academic Press, San Diego.

Kolluru R. V. and Brooks, D. G. (1995): Integrated risk assessment and strategic management. In "Risk Assessment and Management Handbook. For Environmental, Health, and Safety Professionals." (R.V. Kolluru, S. Bartell, R. Pitblade, and S. Stricoff, Eds.), pp. 2.1–2.23 McGraw-Hill, New York.

Körner, C. and Bazzaz, F. A., (Eds.) (1996). "Carbon Dioxide, Populations, and Communities." Academic Press San Diego.

Leist, A. and Schaber, P. (1995). Ethische Überlegungen zu Schaden, Risiko und Unsicherheit. In: (M. Berg, G. Erdmann, A. Leist, O. Renn, P. Schaber, M. Scheringer, H. Seiler, and Wiedemann, R., Eds.), VDF Hochschulverlag, Risikobewertung im Energiebereich, Zürich. 47–70.

Löfstedt, R. E. (1997). "Risk Evaluation in the United Kingdom: Legal Requirements, Conceptual Foundations, and Practical Experiences with Special Emphasis on Energy Systems." Arbeitsbericht Nr. 92 der Akademie fÅr Technikfolgenabschätzung. Stuttgart: Akademie für Technikfolgenabschätzung.

Markowitz, J. (1990). Kommunikation über Risiken–Eine Theorie-Skizze. Schweiz. Z. Soziol. **16(3)**, 385–420.

Matzner, E., Murach, D. (1995). Soil changes induced by air pollutant deposition and their implications for forests in Central Europe. *Water, Air and Soil Pollut.* **85**, 63–73

Mohr, H. and Müntz, K. (Org.) (1994). The Terrestrial Nitrogen Cycle as Influenced by Man. Leopoldina-Symposium, Halle (Saale), Germany, September 29, 1993, to October 1, 1993. Neue Folge, Nummer 288, Band 70. Halle (Saale): Deutsche Akademie der Naturforscher Leopoldina.

Mooney, H. A., Canadell, J., Chapin, F. S., Ehleringer, J., Körner, C., McMurtrie, R., Parton, W. J., Pitelka, L., and Schulze, E.-D. (1998). Ecosystem physiology responses to global change. In "The Terrestrial Biosphere and Global Change." (B. Walker, W. Steffen, J. Canadell, and J. Ingram, Eds.). Cambridge University Press, Cambridge.

National Research Council, Committee on the Institutional Means for Assessment of Risks to Public Health (1983). "Risk Assessment in the Federal Government: Managing the Process." National Academy of Sciences. National Academy Press, Washington, D. C.

Petringa, N. (1997). "Risk Regulation: Legal Requirements, Conceptual Foundations and Practical Experiences in Italy. Case Study of the Italian Energy Sector." Arbeitsbericht Nr. 90 der Akademie für Technikfolgenabschätzung. Stuttgart: Akademie für Technikfolgenabschätzung.

Piechowski, M. von (1994). Risikobewertung in der Schweiz. Neue Entwicklungen und Erkenntnisse. Unv. Ms.

Poumadère, M. and Mays, C. (1997). "Energy Risk Regulation in France." Arbeitsbericht Nr. 89 der Akademie für Technikfolgenabschätzung. Stuttgart: Akademie für Technikfolgenabschätzung.

Rawls, J. (1971). "A Theory of Justice." Harvard University Press, Cambridge.

Rawls, J. (1974). Some reasons for the maximin criterion. Am. Econ. Rev. 1, 141–146.

Renn, O. (1996). Kann man die technische Zukunft voraussagen? Zum Stellenwert der Technikfolgenabschätzung für eine verantwortbare Zukunftsvorsorge. In: "Technologiepolitik in demokratischen Gesellschaften." Stuttgart: Edition Universitas and Wissenschaftliche Verlagsgesellschaft (K. Pinkau, and C. Stahlberg, Eds.),. pp. 23–51.

Renn, O. (1997). Three decades of risk research: Accomplishments and new challenges. J. Risk Res. 11(1), 49–71.

Renn, O. (1992). Concepts of Risk: A Classification. In Krimsky, S. and Golding, D.: "Social Theories of Risk." Praeger, Westport, CT. pp. 53–79.

Reuss, J. O. and Johnson, D. W. (1986), "Acid Deposition and the Acidification of Soils and Waters. Ecological Studies 49. Springer, New York.

Roeckner, E., Arpe, K., Bengtsson, L., Christoph, M., Claussen, M., Dümenil, L., Esch, M., Giorgetta, M., Schlese, U. and Schulzweida, U. (1996). "The Atmospheric General Circulation Model ECHAM–4: Model Description and Simulation of Present-Day Climate." Report Max-Planck-Institut für Meteorologie Nr. 218, Hamburg: Max-Planck-Institut für Meteorologie.

Rosa, E. (1997). Metatheoretical foundations for post-normal risk. J. Risk Res. 1(1), 15–44.

Schlesinger, W. H. (1997). "Biogeochemistry–Analysis of Global Change." Academic Press San Diego.

Schulze, E.-D.; Oren, R., and Lange, O. L. (1989). Processes leading to forest decline: A synthesis. In "Forest Decline and Air Pollution. Ecological Studies 77." (E. D. Schulze, O. L. Lange, and R. Oren, Eds.), Springer, Heidelberg, 459–468.

Short, J. F. (1984). The social fabric of risk: Toward the social transformation of risk analysis. Am. Sociol. Rev. 49(6), 711–725.

Shrader-Frechette, K. S. (1991). "Risk and Rationality. Philosophical Foundations for Populist Reforms." University of California Press, Berkeley.

Tittes, E. (1986). Zur Problematik der Wahrscheinlichkeitsrechnung bei seltenen Ereignissen. In: "Technische Risiken in der Industriegesellschaft. Erfasung, Bewertung, Kontrolle." (P. C. Kompes, Ed.), GfS, Wuppertal, 345–372.

Ulrich, B. and Sumner, M. E., (Eds.) (1991). "Soil Acidity." Springer Berlin.

UN-ECE, United Nations Economic Commission for Europe Convention On Long-Range Transboundary Air Polltuion. (1996). Mapping critical levels/loads and geographical areas where they are exceeded. Umweltbundesamt Texte Nr. 71/96. Umweltbundesamt, Berlin.

van Breemen, N., Mulder, J., and Driscoll, C. T. (1983). Acidification and allkalinization of soils. Plant and Soil 75, 283–308

Vitousek, P. M. (1994). Beyond global warming: Ecology and global change. Ecology 75, 1861–1876.

Vitousek, P. M., Aber, J. D., Howarth, R. H., Likens, G. E., Matson, P. A., Schindler, D. W., Schlesinger, W. H., and Tilman, D. G. (1997). Human alteration of the global nitrogen cycle: Source and consequences. Ecol. Applicat. 7, 737–750.

Walker, B. and Steffen, W. (Eds.) (1996). "Global Change and Terrestrial Ecosystems." Cambridge University Press, cambridge.

Walker, B., Steffen, W., Canadell, J., and Ingram, J. (1998). The terrestrial biosphere and global change: Implications for natural and managed ecosystems—Executive Summary. In "The Terrestrial Biosphere and Global Change." Cambridge University Press, Cambridge.

WBGU, German Scientific Advisory Council on Global Change. (1999). Welt im Wandel. Strategien zur Bewältigung globaler Umweltrisiken. Jahresgutachten 1998. Springer, Berlin.

WCMC, World Conservation Monitoring Centre. (1997). "Generalized World Forest Map." Cambridge: World Conservation Monitoring Centre. Internet-Datei http://www.wcmc.org.uk/forest/data/wfm.html.

Woodwell, G. and Mackenzie, F. T. (Eds.) (1995). "Biotic Feedbacks in the Global Climatic System—Will the Warming Feed the Warming?" Oxford University Press, New York.

Zimmermann, P. H. (1988). MOGUNTIA—A handy global tracer model. In Air pollution Modeling and its Application, Vol. 6." (H. van Dop, Eds.). Plenum, New York.

1.26

Contrasting Approaches: The Ozone Layer, Climate Change, And Resolving the Kyoto Dilemma

Ambassador Richard E.
Benedick[1]

> The emotion is to be found in the clouds,
> not in the green solids of the sloping hills
> or even in the gray signatures of rivers . . .
> Billy Collins, *Questions About Angels—Poems*, 1999.

1. Introduction: Apples and Oranges?

In December 1997, after nights of bargaining that culminated two years of hard negotiations, representatives of 160 governments wearily agreed in Kyoto, Japan, on a protocol to supplement the 1992 United Nations Framework Convention on Climate Change. It was hoped that the Kyoto Protocol would represent a major step forward by the international community to mitigate emissions of greenhouse gases that could alter future climate. Before long, however, doubts emerged on whether the treaty was implementable, and even whether enough governments would ratify it to allow its coming into force as international law. Now, over three years later,

[1] Dr. Benedick, formerly Deputy Assistant Secretary of State and chief U.S. negotiator of the Montreal Protocol on Substances That Protect the Ozone Layer, is author of *Ozone Diplomacy—New Directions in Safeguarding the planet* (Harvard University Press, rev. ed. 1998). Currently, he is Deputy Director, Environmental and Health Sciences Division, Battelle Washington Operations; Visiting Fellow, Wissenschaftszentrum Berlin; and President, National Council for Science and the Environment.

only about 30, mainly small, nations have ratified. Among them, only Mexico is a significant emitter of carbon dioxide.

Only a decade earlier, just 24 countries had signed the Montreal Protocol on Substances That Deplete the Ozone Layer. This treaty, however, was soon ratified by all of the significant producer and consumer nations. It came into force within only 15 months, has now been ratified by nearly 170 countries, and has entered into the annals of diplomacy as a landmark in the history of international cooperation. The heads of the World Meteorological Organization (WMO) and the United Nations Environment Programme (UNEP) described the 1987 Montreal Protocol as "one of the great international achievements of the century" (Bojkov, 1995).

Much has been written about the pathbreaking nature of the ozone accord. Its unexpected success was viewed as an encouraging sign that the world would now be able to cooperate in addressing such other long-term environmental threats as climate change and diminishing biological diversity. The Montreal Protocol was mined for pertinent lessons for the future (Lang, 1996; French, 1997; Benedick, 1998a).

However, the negotiations over climate change, from their very inception in Chantilly, Virginia, in February 1991, have been marked by persistent disarray among the negotiating parties on the necessity and feasibility of strong, early measures to remodel the world's energy structure. Proponents of decisive action became increasingly frustrated by continuing hesitancy on the diplomatic front— a lack of zeal that was manifested, ironically, by many of the same nations that have been traditional leaders on ozone, air and water quality, wildlife and other environmental issues, notably Australia, Canada, New Zealand, and the United States.

Environmental advocates attributed the negotiating problems not to flaws in the international approach to climate, but rather to short-sighted politics, selfish pecuniary interests, and unenlightened lifestyles of a few rich countries. The arguments on all sides became increasingly shrill, the rhetoric more inflammatory. Irritation over the climate stalemate led some revisionists to label the Montreal Protocol as an easy victory that has no relevance for the more complex subject of climate change. Ozone layer and climate change? It seemed like comparing apples with oranges.

The scientific and socioeconomic variables associated with global climate are indeed more complicated than those that faced the negotiators of the Montreal Protocol. However, this alone is not a satisfactory explanation for the continuing disputes over restricting anthropogenic greenhouse gas emissions. Far from being disqualified, the ozone experience offers lessons that are fundamental to understanding why climate negotiations have been so emotional and unproductive.

2. Montreal: An Unlikely Success Story

As a historian once observed, all revolutions seem impossible before they occur— and inevitable afterwards. Now that chlorofluorocarbons (CFCs) have become a household word, we forget the global firestorm of controversy that was provoked by a technical article written in 1974 by two scientists at the Universe of California at Irvine. Sherwood Rowland and Mario Molina hypothesized that certain anthropogenic chemicals could damage ozone molecules 30–50 kilometers above the earth's surface (Molina and Rowland, 1974). If true, the theory had portentous implications, since the evolution of life was possible only because this fragile layer of stratospheric ozone absorbs dangerous ultraviolet radiation (UV-B) that comes from the sun. Twenty-one years later, Rowland and Molina (together with Paul Crutzen of the Max-Planck-Gesellschaft) would receive a Nobel Prize for their research, but at the time, their theory was attacked and derided. The earliest chronicle of the ozone history bore the apt title, *The Ozone War* (Dotto and Schiff, 1978).

When a handful of governments convened in Stockholm in 1982 to begin negotiating an international agreement on the problem, no gambler would have wagered that their deliberations would lead just eight years later to the banning of all CFCs and related chemicals. Indeed, the first result of their arduous negotiations, the 1985 Vienna Convention for the Protection of the Ozone Layer, did not even mention CFCs— it was essentially merely a plea for more research.

Was the Montreal Protocol inevitable? We may have forgotten that CFCs, which had been invented in the 1930s, were for decades considered ideal chemicals. Nontoxic, nonflammable, noncorrosive, cheap, and easy to produce, CFCs and their bromine cousins, the halons, were by the 1970s finding an ever-widening range of uses in thousands of products and processes across dozens of industries. Food processing, plastics, solvents, cleaners, air-conditioning, fire fighting, defense, aerospace, oil rigs, computers, pharmaceuticals, telecommunications, home products, industrial chillers, and insulation are only a sampling of the extent of their utility. Their benefits were virtually synonymous with modern standards of living and, except in aerosol sprays, no feasible alternatives to them existed. Industry warned that restricting their use would jeopardize nearly $400 billion in capital investment and hundreds of thousands of jobs worldwide (Benedick, 1998a, p. 134).

We may also have forgotten that large producing nations, together accounting for two-thirds of global production—the European Union, Japan, and the then-Soviet Union—adamantly opposed strong limits on CFCs. The United States was the only major producer to endorse meaningful controls; it was joined by a few small consumers/producers: Australia, Canada, Finland, New Zealand, Norway, Sweden, and Switzerland. Most of the rest of the world was indifferent, epitomized in the remark to me by an Indian diplomat: "rich man's problem—rich man's solution."

Most significant of all, we may have forgotten that during the entire negotiating period from 1982 to the protocol signing in 1987, there was absolutely no scientific evidence either of ozone depletion caused by CFCs, or of any of the predicted negative consequences—higher levels of UV-B radiation at Earth's surface, increased incidence of skin cancer and cataracts, defects in the human immune system, damage to crops and marine life. The case for international controls was based entirely on arcane theories of complex chemical–physical interactions and computer model predictions of remote trace gases that were measured in concentrations as minute as parts per trillion.

Ironically, the scientists advised us not to consider the only evidence of actual ozone depletion at hand—a dramatic but temporary seasonal thinning of the ozone layer over Antarctica that was unexpectedly revealed by British balloon-based measurements in 1983, after having been overlooked in more sophisticated satellite data. The processes at work here were poorly understood, and there were at the time plausible explanations for the Antarctic event other than CFCs. Interestingly, scientists had more confidence in their theoretical models that predicted a gradual thinning of ozone over the mid-latitudes rather than a precipitous but transitory collapse over the South Pole. The "ozone hole" had even diminished in 1986—just before protocol negotiations began; scientists did not yet know of the quasi-biennial oscillation, and thus could not be sure whether these data signaled a reversal of the depletion trend. Scientists warned me then that if we based our case on the Antarctic phenomenon and it turned out that CFCs were not to blame, the chances for reaching an agreement on strong controls would be severely undermined (Benedick, 1998a, pp. 19–20).

Only a few weeks before the final negotiating round in Montreal, most knowledgeable observers did not believe that an agreement would be possible. In the face of these not-trivial obstacles, what made the Montreal Protocol memorable?

3. Lessons from the Ozone Layer

Out of the many important aspects of the ozone history, I would like to highlight five factors that appear most relevant to the climate negotiations: (1) the role of science and scientists; (2) the necessity for strong and consistent leadership; (3) the flexible design of the Montreal Protocol; (4) the technological revolution that emerged from public–private sector partnerships; and (5) the involvement of developing countries in the solution.

3.1 Role of Science and Scientists

Science played a crucial role not only in uncovering the threat to the ozone layer, but also in the diplomatic efforts to address the danger. Without the constant involvement of scientists, the Montreal Protocol could never have become a reality. Spearheaded by American scientific agencies—the National Aeronautics and Space Administration (NASA) and the National Oceanic and Atmospheric Administration (NOAA)—a remarkable cooperative international venture was launched in 1984 involving over 150 scientists from many nations. The result, published by WMO and UNEP in 1986, was the most comprehensive analysis of stratospheric chemistry and physics ever undertaken: three volumes containing over 1100 pages of text, plus 86 reference pages listing hundreds of peer-reviewed articles (WMO/UNEP, 1986). Scientists also collaborated to develop ever more refined instruments to measure the gases, as well as sophisticated computer models to predict the implications of physical/chemical processes.

An international scientific consensus was not by itself, however, a sufficient precondition for policy action. Scientists had to leave their laboratories and assume, alongside the diplomats, an unfamiliar share of responsibility for the policy implications of their findings. For their part, political and economic decision makers needed to fund relevant research and to work together with scientists on realistic assessments of the risks.

3.2 Necessity for Strong and Consistent Leadership

While the consequences of ozone layer depletion could be devastating, they were unproved during the negotiations. Nevertheless, it was essential to impose preventive controls well before significant impacts were recorded, because the long atmospheric lifetimes of CFCs meant that it would take decades for the ozone layer to recover. Since most governments at the start were unwilling to undertake meaningful actions, strong and decisive leadership was needed to push the negotiations forward.

This leadership was provided by the United States, and by UNEP under its Egyptian executive director, Mostafa Tolba. Tolba employed his credentials as a scientist and his personal credibility with developing nations on behalf of a strong treaty. His logic and compassion made Tolba an eloquent spokesman for the interests of future generations.

For its part, the U.S. State Department designed a diplomatic campaign to counteract the influence over the European Union (EU) of such powerful companies as Imperial Chemical Industries and France's Atochem, while cultivating discreet support behind the EU communal curtain from Belgium, Denmark, and Germany. At the same time, we sent diplomatic and scientific teams to try to persuade the other two major producers—Japan and the Soviet Union—as well as developing nations to support strong controls.

There were fascinating aspects of this diplomatic strategy. We initiated, for example, an unusual Cold War space-agency research cooperation—an "ozone glasnost." We also dispatched representatives of American environmental groups to motivate their British counterparts to raise embarrassing questions in Parliament, an inspiration that elicited a formal protest from Her Majesty's Government over my involvement. In the end, Japan and the Soviet Union unexpectedly joined the U.S. and its allies at Montreal. The EU, now isolated and under pressure from its internal dissenters, was forced to compromise, and the protocol became reality. The United Kingdom (U.K.) later became a vigorous advocate of CFC phaseout (Benedick, 1998a, Chapter 6).

3.3 Flexible Design of the Montreal Protocol

Scientific uncertainties decisively influenced the protocol's design. U.S. negotiators realized that a total ban on ozone-depleting substances was neither justified by existing scientific knowledge nor politically feasible. Therefore, in place of the immutable commitments of traditional treaties, we deliberately drafted the protocol to constitute a dynamic and flexible process. The "spirit of Montreal," which became a hallmark of later negotiations to strengthen the protocol, was to proceed incrementally in small, cumulative steps, rather than to reach for overambitious targets that would only serve to harden opposition.

The key element was the establishment of independent expert panels to provide periodic reassessments of scientific, technological, and economic developments. These panels eventually involved hundreds of specialists from the research community and the private sector worldwide, constituting an unparalleled body of expertise available to the parties to the protocol.

When serious differences arose during negotiations, the parties regularly returned to the panels with requests for new technical analyses of policy options. Linking the protocol consistently with the science proved an effective method to minimize confrontation and, step by step, to gradually overcome opposition to stronger measures. The result was that the political consensus held together as the number of controlled chemicals grew from an original 8 to more than 90, while phaseout periods were gradually introduced and then systematically tightened. Based on the expert findings, the protocol was significantly strengthened through amendments at the Meetings of Parties in London in 1990, Copenhagen in 1992, Vienna in 1995, and Montreal in 1997 (Benedick, 1998a, pp. 218–224, 319–320).

3.4 Technological Revolution from Public–Private Sector Partnerships

The Montreal Protocol was technology-forcing in the sense that, at the time of its signing in 1987, replacements were unavailable for nearly all uses of ozone-depleting substances. The cooperation of industry was fostered by a combination of factors: targets that were challenging without being impossible, the engagement of governments and international agencies, and the gradually compelling nature of the science. As a result, the initially monolithic industry opposition was undermined and more progressive elements were stimulated to look for solutions.

By unleashing the creative energies of the private sector, a technological revolution was achieved even where alternatives had been considered impossible. Governments, international agencies, research institutes, and environmental organizations often collaborated with private firms in the search for substitutes. Rival chemical producers were encouraged to cooperate in toxicity testing and other studies on possible replacements. User companies in the telecommunications sector, such as Northern Telcom and AT&T, did not wait for the chemical industry, but reexamined their own manufacturing processes and came up with approaches, e.g., to cleaning microchips, that were even cheaper and more effective than the once-indispensable CFCs. Governments adopted market-oriented policies and incentives, and the resultant competitive forces helped to lower costs and to bring new alternatives quickly to market. Successful innovation in some fields gave the parties confidence to accept stronger controls in others (Cook, 1996; Benedick, 1998a, pp. 197–202).

3.5 Involvement of Developing Countries

To address the global problem effectively, it was essential that all nations—North and South—abjure use of ozone-depleting substances. Otherwise, efforts of the richer countries would eventually be swamped by developing countries with their rapidly rising populations and aspirations for economic growth. Here again the Montreal Protocol offers relevant lessons.

The industrialized countries from the start accepted the principle that they would take earlier and stronger measures than the poorer nations. Attempts by some populous developing countries to promote upper-use limits on a per capita basis were firmly rejected. Instead, a ten-year grace period before developing nations had to accept obligations was agreed to. Surprisingly, even this provision turned out to be mainly symbolic in importance. Developing nations moved faster than expected to replace CFCs, as the North followed through on commitments to ensure that new technologies would expeditiously be made available, and that incremental costs for the South would be compensated through a special multilateral fund.

Varied creative initiatives promoted the transfer of technology. Consortia of private companies, environmental organizations, and international agencies diffused new products and processes to developing countries. Greenpeace invested in an East German company to develop CFC-free refrigerators that were later distributed in China and India through the German and Swiss official aid programs. A UNEP information clearing house and training workshops reinforced efforts to spread technological innovations.

As technology transfer became a reality rather than just words in a treaty, the developing countries became eager to obtain new technologies as rapidly as possible. One result was the frustration of India's hopes to become the monopoly supplier of CFCs in growing Third World markets. India had utilized the grace period to expand CFC capacity in a calculated attempt to replace the North as its production phased out. But India found itself with overcapacity as its neighbors closed their doors to the outdated products. The availability of modern technologies stimulated the South to assume stronger commitments, and most of the developing countries will now achieve phaseout of most substances well ahead of their agreed schedules. (Benedick, 1998a, Chapter 16)

4. Climate Change: The Road to Rio

Worries about global warming are not new. More than 40 years ago two scientists at the Scripps Institution of Oceanography, Roger Revelle and Hans Suess, warned that the accumulation of carbon dioxide in the atmosphere resulting from fossil fuel combustion represents "a large scale geophysical experiment" on the planet (Revelle and Suess, 1957). As data in subsequent years confirmed a rapid increase in atmospheric concentrations of carbon dioxide and other long-lived greenhouse gases, scientific concern mounted over possible future adverse effects, especially since disruptions in the forces that influence climate would not be easily reversible.

In 1985, WMO and UNEP, in cooperation with the International Council of Scientific Unions, convened a scientific conference in Villach, Austria, that attracted political notice when it concluded:

> "Many important economic and social decisions are being made today on long-term projects . . . based on the assumption that past climatic data . . . are a reliable guide to the future. This is no longer a good assumption since the increasing concentrations of greenhouse gases are expected to cause a significant warming of the global climate in the next century." (Bolin *et al.*, 1986)

Even greater political attention focused on climate at the 1988 Toronto Conference on the Changing Atmosphere: Implications for Global Security. This conference, convened by the Canadian government together with WMO and UNEP, brought together representatives of government, industry, environmental organizations, and research institutes. For the first time at this level, recommendations called for negotiation of a global convention containing specific targets and timetables to reduce emissions of greenhouse gases. Other international conferences followed, and climate change and the ozone layer were even discussed at annual summits of the Group of Seven, the leaders of the major Western industrialized nations.

Coincidentally, the public was becoming increasingly sensitized to anthropogenic disturbance of atmospheric systems by the confirmation in 1988 that CFCs were indeed responsible for the Antarctic ozone hole, and by concerns in Europe and North America over acid rain and forest damage. In the same year, extreme storms over Europe, record heat waves and drought in North America, and weather anomalies elsewhere in the world heightened public attention to the possibility of changing climate. Mass media sensationalized the issue with cover stories portraying famous landmarks (e.g., New York's Empire State Building) partially submerged by raging tides.

The year 1988 was also significant for the establishment of the Intergovernmental Panel on Climate Change (IPCC), an event that was not without controversy. Previously, an eminent but largely self-selected scientific advisory group had issued pronouncements on climate at Villach and elsewhere under WMO and UNEP auspices. The IPCC idea, modeled after the successful experience of the 1984–85 ozone assessment mentioned above, was first raised in 1987 by myself and others with the aim of expanding the small group into a larger entity under governmental auspices.

Some environmental advocates opposed the concept, fearing that governments would co-opt the scientific process and distort the findings for political purposes. I and other supporters of change, however, argued that expansion of the informal group into an official panel would enhance its credibility and influence—and that, moreover, scientists would not allow themselves to be manipulated. As it turned out, the IPCC did operate with an independence that occasionally made governments uncomfortable. Drawing on the ozone experience, the IPCC became an ongoing series of roundtables, workshops, and reports, eventually involving over 2000 scientists and researchers from many nations, organizations, and industries in data gathering, analysis, and debate (IPCC, 1991; 1996).

Based on initial IPCC findings, the UN General Assembly in December 1990 created the Intergovernmental Negotiating Committee on Climate Change, aiming at a convention for signature at the 1992 UN Conference on Environment and Development (UNCED) in Rio de Janeiro. I participated in the negotiations as Special Advisor to the Secretary General of UNCED.

The negotiations proved very difficult, since greenhouse gas emissions were inextricably linked with energy, industry, land use, and transportation policies—the building blocks of modern economies, both North and South. The interrelated aspects of the problem meant that there were no quick or obvious solutions. Mitigation policies would entail major changes in the ways that people lived, worked, and consumed.

Nations would have to significantly reduce their dependence on fossil fuels, which accounted for more than half of greenhouse gas emissions. Agricultural practices that caused emissions of nitrous oxide and methane would need to be modified. The widespread destruction of forests and savannas would have to be curtailed, as these practices not only released carbon dioxide but also removed a critical sink for absorbing emissions from other sources. Since all these factors were related to the needs of poor people in developing countries, issues of poverty and population growth were also central to mitigating climate change.

Widely varying national interests had to be reconciled in the climate negotiations. Regions and countries differ considerably in their vulnerability and in their capacity to adapt to climate change. Prospects are least favorable for the poorest countries, especially low-lying small island states, delta regions, and arid areas of Africa, South America, and Central and South Asia. Countries also differ in their industrial and transportation structure, in their natural resource base, and in their dependence on fossil fuels. China, with almost 1.3 billion people striving for higher standards of living, is unlikely to forego use of cheap coal, of which it possesses approximately one third of known global reserves, in the absence of feasible alternatives. Other rapidly industrializing countries such as India, Mexico, South Korea, and Thailand share similar views on energy use. Norway and Australia are major coal exporters. Countries with large forested areas, such as Brazil, Indonesia, Malaysia, and Zaire, resist attempts by the North to dictate how they may use their national patrimony. The prosperity of the United States is heavily dependent on domestic coal and imported oil. The economies of Kuwait, Saudi Arabia, Venezuela and others rest on oil exports. Even New Zealand, with more sheep than people, is cautious about imposing controls on methane emissions (Benedick, 1997a).

5. The Framework Convention on Climate Change

Notwithstanding the difficulties, the UN Framework Convention on Climate Change (FCCC) was signed on schedule in June 1992 by over 150 nations (United Nations, 1992). The convention was criticized by environmental groups for not mandating reductions in greenhouse gas emissions comparable to the Montreal Protocol commitments on CFCs. Instead, Article 4 somewhat ambiguously obliges industrialized countries to "adopt national policies and take corresponding measures" with the "aim of returning" anthropogenic emissions by 2000 to their levels in 1990. (The 38 industrialized nations are listed in Annex I of the convention and are thus customarily termed "Annex I" countries.) At the present, writing on the eve of this deadline, it is evident that only a handful of Annex I countries can achieve this "aim," and those few only because of exceptional circumstances—a fact that demonstrates how ambitious the target actually was.

The framework convention is, in fact, much stronger than its true ozone analogue, which was not the Montreal Protocol but the earlier 1985 Vienna Convention. The FCCC mandates rigorous national reporting by industrialized countries on the results of the above-mentioned measures. Significantly, it also requires the parties to periodically assess the "adequacy" of the commitments, with the clear implication that revisions were intended. Further, the FCCC recognizes the precautionary principle as a criterion for such action: "Where there are threats of serious or irreversible damage, lack of full scientific certainty should not be used as a reason for postponing such (precautionary) measures" (Article 3). The FCCC also contains commitments for *all* parties—North and South—to develop national programs "to mitigate climate change by addressing anthropogenic emissions by sources and removal by sinks"; no deadlines, however, are set for establishing such programs.

Like the Montreal Protocol, the FCCC was clearly conceived as establishing a long-term and dynamic process of addressing climate change. In this context, I believe that the convention's strongest feature is its "ultimate objective" (Article 2), against which all future commitments must be measured:

> "The ultimate objective [is to achieve] stabilization of greenhouse gas concentrations in the atmosphere at a level that would prevent dangerous anthropogenic interference with the climate system. Such a level should be achieved within a time-frame sufficient to allow ecosystems to adapt naturally to climate change, to ensure that food production is not threatened and to enable economic development to proceed in a sustainable manner."

It is unfortunate that the state of the science, then as now, cannot yet inform us what level of concentrations would be "dangerous," nor what the desirable time frame might be. Although the lack of such indices complicates the task for governments to negotiate quantitative commitments, the concepts incorporated in the objective remain valid guides for action.

At the convention's First Conference of Parties, in Berlin in early 1995, the parties had available preliminary findings from the IPCC's second report. The IPCC, while somewhat lowering its previous model projections of global warming and sea-level rise, nevertheless expressed greater confidence in the revised estimates. Most significantly, the panel for the first time concluded that the data indicated the presence of "a discernible human influence on global climate" (IPCC, 1996).

Influenced by the IPCC findings, the parties in Berlin formally acknowledged that the Article 4 commitments made in 1992 by industrialized countries were not adequate. They could not, however, agree on how these commitments should be strengthened. After heated negotiations, the result was a compromise: a "Berlin Mandate" required the parties to negotiate, by 1997, "quantified limitation and reduction objectives within specified time-frames—"otherwise known as targets and timetables—"for anthropogenic emissions by sources and removals by sinks."

6. Tortuous Targets in Kyoto

Even industrialized countries differ widely among themselves in geography, population, natural resource base, climatic conditions, industrial structure, and dependence on energy. Since these critical parameters are either intrinsic or immutable in the short run,

it is extremely difficult to establish short-term emissions targets that are both economically feasible and equitable. Nevertheless, the Kyoto negotiators tried.

The centerpiece of the Kyoto Protocol is the commitment by Annex I countries, *as a group*, to reduce their net emissions of a weighted basket of six greenhouse gases by 5.2% below 1990 levels when averaged over the five-year period 2008–2012. (United Nations, 1997). The gases are carbon dioxide, methane, nitrous oxide, hydrofluorocarbons, perfluorocarbons, and sulfur hexafluoride; parties have the option of measuring the latter three gases against either a 1990 or a 1995 baseline. Within the Annex I group, individual states committed themselves to differing reduction targets, e.g., 8% for Switzerland, the European Union, and many Central and East European nations; 7% for the United States; 6% for Canada, Hungary, Japan, and Poland; 5% for Croatia. New Zealand, Russia, and Ukraine were not required by Kyoto to lower emissions below 1990 levels, while negotiators from Australia, Iceland, and Norway were successful in obtaining acquiescence to higher emissions (Article 3). Table 1 provides a summary of carbon dioxide emissions in 1990 and 1997, and Kyoto targets, for each Annex I country, and for non-Annex I (developing nation) regions (IEA, 1999,p.18).

With some fanfare, the 15-nation European Union committed to an 8% reduction as a bloc. Lost in the self-congratulation, however, was the interesting fact that 7 of the 15—including France and Sweden—would actually maintain or increase their emissions inside the EU "bubble." The widely publicized "European" target in fact depends on steep reductions by Germany (-21%) and the United Kingdom (-12.5%) to lower the community average. In both these cases, special circumstances prevailed that were independent of climate change mitigation policies. Reunified Germany benefited from the 1990 base year that incorporated high emissions in the former German Democratic Republic before they plummeted due to economic collapse. In the United Kingdom, the Thatcher Government's campaign to weaken the power of coal miner unions stimulated switching to natural gas—which is much less carbon-intensive.

As governments appeared unwilling to confront powerful industrial interests head-on by enacting sector-specific policy measures to limit use of fossil fuels, e.g., in transportation or utilities, they opted instead for arbitrary short-term overall targets. The result was that the numbers so feverishly bargained in the midnight hours at Kyoto bore no relationship to either scientific or economic realities. *The Kyoto Protocol thus inadvertently manages to be simultaneously far too strong in the short run, and yet far too weak to address the long-term problem of climate change.*

The 11–15 year Kyoto targets are clearly inadequate to make any dent in future atmospheric concentrations, which is the crucial measure of danger to climate. Even if the protocol were fully implemented, it would only serve to delay by less than a decade the date in the next century at which global carbon dioxide concentrations, under current emissions trends projected by IPCC, would cross the 550 parts per million (ppm) mark that represents a doubling of preindustrial concentrations

TABLE 1 Total CO_2 Emissions from Fuel Combustion (Million Tons of CO_2)[*]

	1990	1997	97/90(%)	Target[a]
Annex I	14,003.3	13,633.8	−2.6	
Annex II	10,081.4	10,937.6	8.5	
North America	5301.0	5947.9	12.2	
Canada	427.5	477.4	11.7	−6
United States	4873.4	5470.5	12.3	−7
Europe	3430.3	3477.9	1.4	
Austria	59.4	64.1	7.9	−13
Belgium	109.1	122.6	12.3	−7.5
Denmark	52.9	62.4	17.9	−21
Finland	54.4	64.1	17.9	0
France[2]	378.3	362.9	−4.1	0
Germany	981.4	884.0	−9.9	−21
Greece	72.3	80.6	11.5	+25
Iceland	2.2	2.4	8.1	+10
Ireland	33.2	37.6	13.0	+13
Italy	408.2	424.3	4.0	−6.5
Luxembourg	10.9	8.6	−20.6	−28
Netherlands	161.3	184.3	14.3	−6
Norway	29.8	34.3	15.4	+1
Portugal	41.5	52.0	25.3	+27
Spain	215.0	253.8	18.0	+15
Sweden	52.7	52.9	0.5	+4
Switzerland[b]	44.2	44.8	1.2	−8
Turkey	138.4	187.5	35.5	none
United Kingdom	585.3	554.7	−5.2	−12.5
Pacific	1350.1	1511.9	12.0	
Australia	263.0	306.1	16.4	+8
Japan	1061.8	1172.6	10.4	−6
New Zealand	25.4	33.1	30.7	0
EITs	3921.9	2696.2	31.3	
Belarus		61.3		none
Bulgaria	72.2	51.0	−29.4	−8
Croatia		17.5		−5
Czech Republic	141.8	120.9	−14.7	−8
Estonia		18.2		−8
Hungary	68.1	58.2	−14.5	−6
Latvia		8.5		−8
Lithuania		14.7		−8
Poland	349.1	350.3	0.3	−6
Romania	167.3	110.7	−33.8	−8
Russia		1456.2		
Slovak Republic	54.2	38.3	−29.3	−8
Slovenia	12.7	14.9	16.7	−8
Ukraine		375.6		0
Non-Annex I	6866.7	8927.6	30.0	none
Africa	611.5	729.4	19.3	none
Middle East	647.9	955.9	47.5	none
Non-OECD Europe[c]	121.0	76.8	−36.6	none
Former USSR[c]	574.2	322.7	−43.8	none
Latin America[c]	945.0	1224.7	29.6	none
Asia (excl. China)[c]	1568.7	2456.2	56.6	none
China	2398.3	3162.0	31.8	none
MAR. BUNKERS[d]	376.0	419.6	11.6	
World Total	21,245.9	22,981.1	8.2	
Annex B[e]	13,749.0	13,385.0	−2.6	

[a] The overall EU Kyoto target for all six gases covered in the Protocol is −8%, but the member countries have agreed on a burden-sharing arrangement as listed. This table assumes that the target applies equally to all greenhouse gases. Because of different base years for different countries and gases, a precise "Kyoto target" cannot be calculated for total Annex I or total Annex B.

[b] Emissions from Lichtenstein are included in emissions from Switzerland and emissions from Monaco are included in emissions from France.

[c] Regions differ from those shown elsewhere in this publication to take into account countries that are not members of Annex I.

[d] International marine bunkers only. International aviation bunkers are included in country totals.

[e] Annex B includes the countries and regional economic integration organisation that were included in Annex B of the Kyoto Protocol to the United Nations Framework Convention on Climate Change.

[*] Source: IEA. (1999). International Energy Agency. *Carbon Dioxide Emissions From Fossil Fuel Combustion, Highlights.*

(Edmonds, 1999b). In fairness, Kyoto was intended only as a first step. But its provisions provide no coherent concept for future emissions reductions.

Yet how could the protocol also be too strong, when it prescribes no change at all in total emissions of industrialized countries? As a group, their emissions in 1997 already stood at the 2008–2012 target level of about 5% below 1990. Thanks to economic downturn and restructuring following the collapse of communism, emissions from the Eastern European countries were by 1997 31% below their 1990 baseline (IEA, 1999,p.15). When one adds in the German and British declines already mentioned, total Annex I emissions were below 1990—for reasons unrelated to any climate mitigation policies.

However, other large emitters were by 1997 already well above 1990 levels and still climbing, notably Australia (+16%), Canada and the United States (+12%), and Japan (+10%). U.S. emissions in particular were buoyed by considerably more vibrant economic activity than that in Europe. In the heat of transatlantic finger-pointing, it was not generally recognized that the U.S. actually had considerably improved its carbon dioxide energy efficiency from 1990 to 1997, i.e., its emissions declined in relation to economic growth. Indeed, the increased U.S. efficiency over this period was exceeded only by four EU member states (apart from the special cases Germany and U.K.) (IEA, 1999,p.56).

Thus, the Kyoto targets could, for countries such as the U.S., translate into required emissions reductions of as much as 25–30% below the level from which they are headed in the 2008–2012 commitment period—the beginning of which is now only 7 years away (White, 1998; Benedick, 1998b). Compliance difficulties for Canada and the U.S. are compounded by their population growth rates, which are much higher than that of Europe. This means that compliance on a per capita basis becomes relatively more onerous: they are, in effect, being penalized for having more liberal immigration policies. For the U.S. to meet its Kyoto commitment, carbon dioxide emissions on a per capita basis would have to drop to levels not seen since the end of World War II. In contrast, 1995 per capita emissions in the European Union were only slightly above its Kyoto target (Meyerson, 1998). The population inequity factor becomes even more significant in future years. According to the latest United Nations projections (medium, or "most likely" variant), the U.S. population by 2050 will be 37% higher than in 1990, while the populations of Japan and Germany will decrease by 15% and 8%, respectively (United Nations Population Division, 1999).

In the relatively short time available, cuts of the required magnitude cannot be achieved without scrapping major capital investments in power plants, factories, transport systems, and buildings, before they are obsolete, which means high costs and economic disruption. For the U.S., achieving the Kyoto-mandated reductions would require the kind of pressure that could come only from politically unacceptable high carbon taxes (Nordhaus and Boyer, 1999; Kopp, 1997). Only five years ago President Clinton failed to get even a 5 cent per gallon gasoline tax increase from a Congress then controlled by his own party.

Nor is it a foregone conclusion that the EU will be able to achieve its Kyoto commitment. There are signs that Germany, whose domestic 21% emissions reduction goal is vital to reaching the European Union's combined 8% cut, may be faltering in its progress. German carbon dioxide emissions began to creep upward in 1995, affected by increases from the transportation and household sectors; partial data for 1997 showed a slight rise from the industry sector. It appears that following the initial hefty decline after the 1990 East German dividend, some additional relatively easy steps were taken to stimulate energy conservation and efficiency. But the low-cost no-regrets strategies have apparently been exhausted (Klepper, 1999). Germany's situation is particularly sensitive because of persisting high unemployment, which increases the political risks of taxes or other costly instruments. The beleaguered Social Democrat/Green coalition government, reeling from unanticipated electoral defeats in 1999, may now be reluctant or unable to implement harder measures.

In 1996, carbon dioxide emissions also rose in other EU member states, including the U.K., that had set substantial domestic reduction goals in order for the EU as an entity to meet its Kyoto target (CDIAC, 1999). By 1997, the Netherlands' emissions were 14% above 1990 levels (Kyoto target: −6%); Belgium was +12% (target: −7.5%); and Denmark was +18% (target: −21%.) (IEA, 1999, p.38). The European Commission itself estimated in May 1999 that, unless additional strong measures are adopted, EU emissions by 2010 would stand at 6% above 1990 levels, rather than 8% below (European Commission, 1999). OPEC success in raising crude oil prices in 1999 may come to the rescue by inducing further energy conservation. But all of these developments bear close watching.

7. When Will the Kyoto Protocol Come into Force?

In an attempt to maximize the efficiency of investments and thereby lower the economic costs of emissions reductions, the Kyoto Protocol established three "flexibility mechanisms":

(1) *joint implementation*, whereby an Annex I country could invest in emissions-reducing projects in another Annex I country and receive some credit against its own target, provided that such project entails "a reduction in emissions by sources, or an enhancement of removals by sinks, that is additional to any that would otherwise occur" (Article 6);

(2) a "*Clean Development Mechanism*," similar to (1) but involving voluntary projects in developing countries (Article 12); and

(3) *international trading of emissions rights* among the Annex I parties, whereby a government or company could purchase "unused" emissions from abroad (Article 17).

The United States government appears particularly eager to make use of these mechanisms—especially emissions-trading with Russia and Eastern Europe—as a means of easing the pain of domestic reductions. The U.S. also hopes that in time even developing countries can be integrated into a global emissions-trading scheme, thereby opening vast potential sources of emissions rights to the carbon-hungry American economy. But many European nations, politically committed to costly domestic emissions-reduction programs, claim that their industries will suffer if U.S. competitors can avoid the equivalent strong medicine by means of offshore compliance. Thus, there is already serious disagreement over the extent to which these mechanisms should be permitted to supplement domestic actions. From the perspective of the poorer countries, trading away emissions rights could be regarded as limiting options for their own future development, or a form of neocolonialism. On the other hand, when the time comes for payments, it is questionable whether the large, untied, and untraceable transfers of wealth to former communist and/or developing nations will be politically palatable to electorates in the West.

The flexibility mechanisms, moreover, have only been established in principle. Operating details, including definitions, guidelines, rules and procedures, reporting, accountability and verification, have been postponed for future deliberation. Although there are precedents for domestic emissions trading (e.g., sulfur dioxide in the U.S.), nothing comparable has ever been attempted on a global scale. It will be extraordinarily difficult to negotiate a trading system for an ephemeral "commodity" among nations at widely varying stages of economic development.

It is not hard to imagine fractious North–South controversy over criteria for allocating emission rights to developing countries—according to population size, for example, as a reward for lax family planning? What happens if a country, having received hundreds of millions of dollars by selling unused rights, subsequently elects a democratic government that repudiates the "irresponsible actions" of its predecessor and insists that expanding energy use and land-clearing are essential to meet the basic needs of a desperate populace? What kind of bureaucracy would be needed to administer the system? What potential transaction costs may be involved? What possible abuses need to be safeguarded against? Will wild price gyrations be modified, for example via a futures market? Will prices of emissions rights be too low to stimulate meaningful domestic change in energy use? Or so high that they foster evasion? The questions multiply quickly.

Another critical issue left unresolved at Kyoto is the determination of "net changes in greenhouse gas emissions from sources and removals by sinks resulting from direct human-induced land-use change and forestry activities, limited to afforestation, reforestation, and deforestation since 1990, measured as verifiable changes in stocks . . . " (Article 3). As a potentially powerful offset to emissions from other sectors, this clause is crucial for determining compliance with the reduction targets.

The U.S. could, for instance, substantially offset its electricity, transportation, and industrial emissions by reporting carbon absorption due to agricultural soil uptake as well as forest growth. Europeans, however, are skeptical about measurement and verification of such sinks. They also argue that they are being penalized for their more responsible forest management prior to 1990, which means that they have less deforested area to replant. Further, it will be extremely hard to distinguish between naturally induced and anthropogenic changes in carbon uptake by soils and forests. There is not even technical agreement on definitions for afforestation, reforestation, and deforestation. Even worse, some developing countries may be tempted to lay waste to old-growth forests in order to sell credits to Northern entrepreneurs for reforestation offsets.

Thus, the current situation is characterized both by deep controversies over fundamental issues and by the possibility that important nations may have difficulties in meeting their reduction targets. It appears problematic, therefore, whether the Kyoto Protocol can become binding international law in its present form. For the protocol to come into force, it must be ratified by at least 55 nations, including Annex I countries that together accounted for at least 55% of total Annex I carbon dioxide emissions in 1990 (Article 25). As mentioned earlier, only about 30 countries—none of them in Annex I— have ratified as yet.

The chief American negotiator at Kyoto, Stuart Eizenstadt, admitted to the U.S. Senate in 1998 that it might be "years" before the treaty would even be submitted by the Executive Branch for Senate approval, which requires a two-thirds majority vote. Eizenstadt also expressed doubt whether the protocol would come into force without U.S. ratification (Franz, 1998). This is not surprising, since the U.S. alone accounts for approximately 33% of Annex I 1990 emissions and, in a rare display of negative unanimity, the Senate in 1997 had voted 95–0 to reject any protocol that did not contain "meaningful participation" by developing countries. Absent Congressional support, the Clinton Administration has found it impossible even to secure legislation for measures to begin curbing the still-rising U.S. emissions before formal ratification. Powerful American industrial interests have mounted a concerted campaign against the protocol.

A protracted U.S. delay could cause other Annex I countries to pause in their own ratification process, not least because of worries about competitiveness in international trade. As doubts grow within the European Union about its own ability to meet Kyoto targets, its member countries are also not rushing to ratify. Governmental hesitation fosters a wait-and-see attitude by industry and discourages the long-term investments needed for an energy transformation. Unfortunately, the worst of treaties is one that is not credible.

8. Unlearned Lessons

Looking back at the relevant lessons from the ozone history discussed earlier, how do the climate negotiations compare?

1. On the role of science, the IPCC has mobilized the scientific community and is doing good work. There is general consensus that the greenhouse theory is robust: if concentrations continue to accumulate indefinitely, potentially calamitous climate change will occur at some future time. But no one can yet predict when this might happen, and there is much uncertainty about possible offsetting or delaying factors, notably cloud cover.

The primary scientific problem affecting the negotiations is the question of potential harm from gradual climate change. There is no indication of the probability, timing, location, or severity of the long list of potential negative impacts ranging from flood and drought to tropical disease and severe storms. Indeed, scientists agree that some regions would probably benefit from warming during the coming century in the form of higher agricultural output.

In contrast to climate, the consequences of ozone layer depletion were of startling clarity: they would be global and fatal, and the anticipated time-span was a matter of a few decades. Because of this, governments decided to take decisive measures even in the absence of proof that CFCs were yet damaging the ozone layer.

Proponents of strong and early carbon dioxide emissions reductions act as if the potential impacts of climate change are comparable. But to obtain international agreement on measures that could entail substantial near-term costs, the dangers avoided must be more compelling than what a leading scientist advocate recently conceded were merely "not implausible" (Schellnhuber, 1999).

Interestingly, a recent survey indicated that nearly four times as many German scientists as Americans would make extreme interpretations in order to influence public opinion on climate change; in all, 60% of German scientists felt this was appropriate, while two-thirds of the Americans expressed disapproval of the practice (von Storch and Bray, 1999). The negotiations demonstrate, however, that attempts to compensate for lacunae in evidence by exaggerated claims often result in damaged credibility.

2. On the question of leadership, no strong country or strong personality has made mitigating climate change a consistent high priority. To be sure, there has been no lack of rhetoric when a politician felt there might be some benefit. President Bill Clinton, for example, after nearly five years in office introduced a climate-related program in late 1997 by pronouncing the issue as "one of the United States' greatest imperatives for this and future generations" (Benedick, 1997b). The tension between the short-term perspective that has characterized the climate debate, and the century-scale of the problem itself, has served to inhibit the emergence of genuine leadership. Not only will "it" not happen on the watch of today's politicians, it probably will not even happen on their grandchildren's watch. Thus, each government in the negotiations has acted in its short-term interest, not looking beyond the next election. Any future leadership role will have to be based on a new vision; one suggested approach is offered in the final section of this chapter.

3. On the nature of the treaty, Kyoto was, like the Montreal Protocol, designed to begin a process. But it suffers from its short-term approach to a long-term problem. By focusing on targets only 11–15 years into the future, the Kyoto Protocol encourages governments and industry to look for short-term solutions. As a result, capital could be prematurely locked into investments that, because of their own intrinsic lifetimes, would inhibit the development, and raise the costs, of the next generation of technologies

that is actually needed to achieve more substantial emissions reductions later in the century.

Kyoto's approach is based on faulty premises that predated the start of climate negotiations nine years ago. They originated, in fact, at the 1988 Toronto Conference referred to above. That conference, following soon after the acclaimed Montreal Protocol, took precisely the wrong lesson from the ozone experience: it recommended that governments negotiate an international treaty requiring industrialized countries to cut greenhouse gas emissions by 20% by the year 2005. As a participant in this conference, and accepting due co-responsibility for the error, I can aver that this target was manufactured literally out of thin air. It was argued that reductions of 1% per year seemed not unreasonable, 2005 was 17 years out (it seemed a long time, then), round it up to 20%—and voila!

This goal became a potent slogan wielded by some European governments as well as by environmental organizations and other advocates. It surfaced at every international meeting. It was adopted and pursued during the formal negotiations by the Alliance of Small Island States (AOSIS), a bloc created in 1991 consisting of approximately 40 countries that feared sea-level rise. A political target thus became the standard against which all other proposals would be measured throughout the climate negotiations.

We had forgotten that the first international action to protect the ozone layer was not the establishment of reductions targets in the 1987 Montreal Protocol. Rather, it consisted of loosely coordinated decisions made approximately 10 years earlier by the world's largest CFC producer, the United States, by Canada, a small producer, and by a handful of importing countries, to ban the use of CFCs in aerosol spray cans. This *policy measure* had the effect of promoting new technologies that soon reduced emissions by about 30%. But if anyone at that time had proposed a formal *target* of that magnitude, it is doubtful whether governments would have embraced it. The relevant lessons from the ozone experience were that policy measures can lead the way by stimulating technology, and that targets are effective only when they are realistic.

4. Unlike Montreal, the climate negotiations from the very start alienated the private sector and sidestepped the issue of new technologies. The exaggerated warnings of impending catastrophe led to an early hardening of opposition instead of enlisting progressive elements in industry to begin working on solutions. Because the debate started off with the wrong premises, the climate treaties actually played into the hands of the coal and oil, automobile, and other powerful interests that preferred a do-nothing policy. Rather than providing market signals that could induce broad technological innovation, serious efforts to implement Kyoto targets are now more likely to provoke a backlash from industry, consumers, and taxpayers.

It is, moreover, an appalling inconsistency that the industrialized nations undertook daunting targets in Kyoto while they have been simultaneously cutting their investments in energy research and development. The U.S., Germany, Japan, the U.K., and the European Union (as a separate entity), which together accounted for more than four-fifths of the world's public sector long-term

energy R&D, collectively reduced their research budgets between 1985 and 1998 by 35% in real terms, or almost $3 billion below 1985 levels. None of the major industrialized countries currently invests the majority of its energy R&D in renewable energies (Dooley and Runci, 1999).

5. As for global participation, commitments by the South in the Kyoto Protocol are conspicuous by their absence. Throughout the negotiations, developing nations have resisted discussing even voluntary measures to restrain their emissions.

In the case of ozone, the industrialized world in 1987 accounted for 88% of CFC consumption and 98% of production. Therefore, their actions were determining, and the role of developing countries was secondary (Benedick, 1998a, pp. 26,148). In contrast, while carbon dioxide emissions from fossil fuels and cement production in industrialized nations have been relatively stable for over 20 years, emissions from developing countries are on a steep upward trend. Between 1985 and 1997, the South's share of global emissions jumped from 29% to 42%. China's emissions are already second only to those of the United States; India's have surged by nearly 50% since 1990 and are now higher than Germany's; South Korea has surpassed Italy, and Mexico's emissions are almost as large as France's (CDIAC, 1999). Propelled by rapid population growth and expanding industrialization, the South's emissions will probably exceed those of the North in two to three decades. The above figures do not include emissions from biomass energy, destruction of forests and savannas, and land degradation, which are hard to measure but add significantly to emissions from the developing world.

With the exception of the small island states and a few others, most developing nations still do not act as if they realize their own vulnerability to the effects of climate change. Their reluctance to restrict use of cheap fossil fuel is understandable, given that their top political priority is to improve standards of living. Unless low-cost alternatives are available, they are unlikely to accept commitments that will primarily benefit future generations. It is also unrealistic to expect them to act as long as industrialized countries, which caused the current climate predicament in the process of becoming rich, appear unable or unwilling to take credible steps to rein in their own emissions. Regrettably, the South's arguments only reinforce worries in the North about the potential impact of higher energy costs on their own international competitiveness. Because energy production and consumption involve sizable long-term investments, the South risks getting locked into a fossil fuel economy in future decades that will make it progressively harder for them to modernize.

9. Time to Move On: A Longer Term Perspective

It is difficult to admit that so much work has produced so little. One respected analyst has characterized the Kyoto Protocol as "a pinnacle of both economic and environmental globalisation" (Grubb, 1999). He regards as a hopeful "achievement" that, at the divisive 1998 conference in Buenos Aires held one year after Kyoto, governments submitted a list of no less than 142 topics for which further negotiation was considered necessary! Following another inconclusive major conference in Bonn in 1999, several thousand delegates from 182 nations met in November, 2000, in Den Haag for two more weeks of intensive negotiations. The result was a complete failure to reach the hoped–for agreements on implementing the treaty.

Can the climate negotiations be reinvigorated? As a start, an attitudinal change would be helpful. Governments and NGOs could turn down the emotional thermostat and stop reacting to every variation in the weather. We could ignore the apocalyptic warnings that emerge after every heat wave and hurricane, as well as the scientific "revelations" (invariably already well known to the afficionados) that one or another research institute conveniently releases to the media on the eve of every negotiating session. It would be more candid to admit that the science is likely to remain imprecise for some time, and to move on to more productive pastimes.

Even with the aid of powerful computer models, complex interrelated natural processes are inherently difficult to predict. Scientists note that, "even if a model result is consistent with the present and past observational data, there is no guarantee that the model will perform at an equal level when used to predict the future." This is so not only because small input errors can generate significant deviations when extrapolated over long time periods, but also because dynamic biogeochemical systems may react in unexpected ways (Oreskes *et al.*, 1994; Sarewitz and Pielke, 1999).

Nevertheless, because of the difficulty of reversing the forces that create the long-term climate, a persuasive case can be made that the potential dangers are sufficiently serious so that actions should not be postponed until impacts are evident. There is an additional risk of crossing some unforeseen threshold—a sudden and irreversible climate disruption brought on by greenhouse gas concentrations surpassing a certain level. This risk is intrinsically nonquantifiable; but it is not zero. The Antarctic ozone collapse demonstrated that when we perturb the atmosphere, it will not necessarily respond with convenient early warning signals.

At this point we should return to basics, namely, the ultimate objective of the FCCC. Pending further scientific evidence, we could establish a tentative goal for carbon dioxide *concentrations*—for example, 550 ppm, a doubling of preindustrial levels, would be about 50% above current concentrations. The goal could later be modified to reflect both unfolding scientific knowledge and experience with technology. But it would at least provide a perspective for starting a sequence of actions over the coming decades. To achieve even this concentration goal would require that current annual global emissions return to 1990 levels within the next hundred years, and then continue to decline, albeit much more slowly. This implies, however, much steeper emissions cuts for the industrialized nations, to permit the South to continue improving living standards (Wigley *et al.*, 1996; Edmonds, 1999c).

One of the premier American scientific institutions, the Pacific Northwest National Laboratory, operated by Battelle for the U. S.

Department of Energy, has made climate change a major priority for its researchers. Much of the following concluding discussion is based on their insights. Physicists, chemists, biologists, economists, and engineers at Battelle are engaged in a broad range of projects exploring energy from fuel cells, hydrogen transformation, biochemical processes, microtechnology, and other next-generation sources. They are also examining the potential for carbon capture and sequestration, an option that could supplement new energy sources and, if applied to fossil fuel combustion, could substantially lower costs by permitting continued use of such fuels without burdening the atmosphere. Other Battelle research focuses on such related fields as technology policy, energy economics, local-climate impacts, "smart" buildings, and energy-saving vehicular structural materials.

The IPCC "business as usual" projections actually incorporate aggressive assumptions about the development and diffusion of non-fossil energy technologies worldwide, even though current outlays for research on such technologies remain low. Thus, even with greater global energy efficiency, substantial fuel switching (from coal and oil to less carbon-intensive natural gas), and significant expansion of the existing renewable energy sources (solar, wind, biomass, hydropower, nuclear), a growth in carbon dioxide concentrations to more that 700 ppm would not be prevented. To hold concentration to a substantially lower level, we will need to develop and deploy new technologies that are currently only at the conceptual or basic research stages—technologies that can make possible deep emissions cuts in the coming decades (Edmonds, 1999c; Dooley and Runci, 1999).

The long atmospheric lifetime of carbon dioxide means that concentration levels for the next hundred years are to a great extent already predetermined by past emissions; they are, therefore, not significantly affected by short-term emissions cuts. Moreover, researchers at Battelle and its partners have demonstrated that any given future concentration level depends more on cumulative emissions than on their timing. This is a crucial point, for it thereby becomes possible to achieve a concentration goal by choosing from among differing alternative trajectories of emissions reductions over the coming century. This flexibility to defer steep reductions can significantly lower the costs of transforming the energy sector (Wigley *et al.*, 1996; Edmonds, 1999c).

Recent Battelle research further indicates that early offsets to emissions through soil carbon sequestration can buy additional time, at low cost, for future steep emissions reductions (Rosenberg *et al.*, 1999). Although there are important questions to be answered, the potential is sufficiently significant so that the U.S. Department of Energy plans to establish a new Terrestrial Carbon Sequestration Center, to be jointly implemented by Pacific Northwest National Laboratory and Oak Ridge National Laboratory (both administered by Battelle), in collaboration with several universities.

Emissions in 2008–2012 (the Kyoto commitment period) are thus much less important than what happens in 2040, 2060, and 2080. The analyses show that the world does have time, provided that we use it well. Emissions can be allowed to drift upward for while—as long as we undertake other actions now to ensure that future emissions are substantially lower (Edmonds *et al.*, 1999).

10. A Technology-Based Strategy for the Future: Eight Points for Action

What kind of actions might these be? The dangers of long-term global warming can only be averted if we (1) bring to market a new generation of technologies that will drastically reduce dependence on fossil fuels and/or will capture and sequester carbon, and (2) gain the cooperation of key developing countries to limit their rapidly rising emissions. Fortunately, the two conditions are interrelated: as we achieve the first, we will get the second. As the ozone history amply demonstrated, when cost-effective options start becoming available, developing nations are more likely to join the bandwagon and adopt modern technologies. Technology functions as the "enabler," without which the high emissions reductions required in the latter half of the coming century will not materialize. *We need, therefore, a new strategic vision that explicitly addresses the issues of technology research, development, and diffusion.*

Not only are the time-consuming negotiations to resolve the flaws of Kyoto not bringing the parties closer to consensus, they actually prevent governments from focusing on more realistic paths. The Kyoto Protocol has become the victim of polarized debate over inconsequential short-term emissions, compounded by large uncertainties about the costs of compliance. The existing treaty provides inadequate emphasis on the technological imperative and on securing the cooperation of developing nations. The current debates distract attention from the real challenge, which is to set the stage for steep cuts needed before the end of the new century.

The combination of a realistic schedule of emission reductions and new technologies would significantly lower mitigation costs, which would otherwise be prohibitive for both North and South. Battelle models suggest that technology can make a difference of trillions of dollars in the global cost of achieving a given concentration goal to mitigate climate change (Edmonds, 1999c). Major near-term cost savings could also be realized by avoiding the "stranding" of assets: existing plants and related infrastructure investments should, therefore, generally be allowed to complete their useful lives. Time is also needed for the development of next-generation infrastructure, e.g., for transport, storage, and distribution of new energy forms.

Companies should be provided with some security that energy-related capital investments will not be made obsolete by new rounds of politically inspired targets that are not firmly based on science. Buying time would permit scientists to make further refinements in climate models and thereby gain more insight into the impacts of climate change, especially their scope, timing, and location. This would help both in mobilizing public support for action, and in providing better guides for policy. The entire process would become politically more acceptable.

Against this background, Battelle has organized an international consortium of research institutes, private companies, and government agencies to develop global energy technology strategies. Using sophisticated computer models and other advanced analytical tools, the initiative aims at better understanding and accelerating the development and diffusion of energy-related technologies, including examining the role of public–private partnerships in this process.

Recognizing that the appropriate technology mix can differ for different regions, several Battelle scientific workshops have already been held in China and India to identify and explore the influence on technological choices of economic, political, institutional, geographical and other factors. A major focus of examination is how policies to promote technology must evolve over long time periods. The project is also studying the potential contribution of such specific technologies as: augmenting soil absorption of carbon through new agricultural techniques; solar, biomass, nuclear fission and fusion energy; the transportation sector; and technologies for adaptation to climate change. This multiyear program could become a paradigm for the type of public–private partnerships that will be indispensable for transforming the world's future energy economy (Edmonds, 1999a).

A technology strategy is only defensible, however, if it does not become an invitation to delay. Much must be done right now to start the process. Here is a possible eight-point program of action for the deadlocked negotiators.

10.1 Revise and Simplify the Emissions Targets

To begin, I recommend that governments streamline the Kyoto emissions commitments to make them more credible. The near-term targets should be revised in magnitude and should focus primarily on gross carbon dioxide emissions. More realistic and verifiable initial targets for industrialized countries would have a better chance of being implemented. Hence, they would be taken more seriously by industry as well as by the onlooking developing world. As new technologies emerge, it will be politically easier to strengthen targets over time.

10.2 Postpone the Sinks

While the attempt to reflect net emissions targets is scientifically justifiable, the complexities surrounding the land-use and forestry provisions of Article 3 are, in my opinion, a formula for delay. Therefore, the comprehensive approach should be abandoned, at least temporarily. The net emissions concept could be reintroduced after technical experts have made it implementable, including prevention of perverse incentives to cut old-growth forests in order to gain or sell emissions credits from replanting. Action on reducing gross carbon emissions should not, however, wait for these refinements.

10.3 Defer Emissions Trading

For all of the reasons enumerated earlier, I would also shelve for the foreseeable future the disputatious negotiations on creating an international emissions trading scheme. Domestic emissions trading would be left open to national decisions.

10.4 Accelerate Technology Transfer and Joint Implementation

Governments and industry in the industrialized countries should become serious—as they were under the Montreal Protocol—about expeditiously transferring new energy-related technologies to the developing world, and should help build indigenous capacity to develop local energy solutions. North–South and West–East joint implementation investments make sense from the standpoints of both economic efficiency and technology transfer. The Clean Development Mechanism (which is the most promising element of the existing protocol) should be activated to promote greater energy efficiency and expansion of renewable energy in the developing nations. The North should provide climate-relevant assistance as a cost-effective form of foreign aid rather than primarily to earn emissions offset credits. All of this would probably be far less costly and more productive than large wealth transfers to buy emissions "rights."

10.5 Get Serious about Policy Measures

In a test of political will, any emissions target should be reinforced—or even preceded—by harmonized policy measures. Indeed, policy measures, as in the Montreal Protocol example discussed earlier, provide a test of what targets might be feasible in the short run. Stricter vehicular fuel-efficiency standards (which everyone, including the automakers, knows are feasible) and energy-related government procurement policies are examples of measures that could provide strong impetus to innovation. Existing market distortions and subsidies that favor fossil fuels should finally be eliminated. Incentives should be adopted to promote further development and market penetration of renewable energies, in order to realize economies of scale that would make them more competitive. Up until now, the half-hearted performance of most governments with respect to policy measures has not matched their political rhetoric about the urgency of the climate problem. A requirement for transparent and rigorous reporting on such measures could, as demonstrated by experience in the IMF and OECD, provide an additional stimulus.

10.6 Adopt Technology-Based Objectives

This approach could reorient energy planning. Battelle analysts are examining possible technology-based goals that (initially, Annex I) governments could employ to stimulate future-oriented R&D. Since virtually all carbon in modern energy economies flows through power generation and fuel refining/processing, such policies could be quite specific in their focus. For example, new power generation plants constructed after a certain date could be required either to use renewable energy, or to capture and dispose of carbon byproducts. Similarly, new fossil fuel refining and processing facilities after a given date would also have to be carbon neutral. (To encourage R&D before the phaseout deadline, interim targets could be scheduled for new plants, as well as credits provided for early compliance.) Net imports of carbon-based fuels could gradually be phased down. Additionally, fossil fuels could be employed as a feedstock for hydrogen, but any carbon releases would have to be sequestered. Because these measures apply to sizable industrial facilities, they are conducive to transparency, reporting, and monitoring for compliance. Such actions are feasible and would provide the market with strong signals for focused research and innovation (Edmonds and Wise, 1999; Edmonds *et al.*, 1999.).

In addition, Annex I technology targets could provide a convenient bridge to subsequent developing country commitments by delinking their participation from difficult negotiations focused on their per capita income or per capita emissions. It could take decades for a China or an India to catch up with per capita income of even the poorest member of OECD. But if the industrialized nations agree to technology goals as described above, it is reasonable to assume that the technologies will be available by the target date. Per capita indices then become irrelevant, as they proved to be in the case of the ozone treaty. Instead, the problem is limited to assuring that these technologies are transferred and deployed in developing nations. To provide developing countries with some security, their obligations could be made dependent on the effective transfer of new technologies and the financing of incremental costs, as was successfully accomplished under the Montreal Protocol.

10.7. Invest in a Technological Revolution

Most important of all, governments must ensure that sufficient financial resources are made available to achieve the needed technological revolution. Reaching a critical mass of R&D is basic to fostering technological breakthroughs. Governments cannot stand back and expect that the private sector, with its relatively short time horizon, will make all the required long-term R&D investments. Although credible targets and policy measures can help to stimulate industry's creativity, the scale of the climate/energy challenge requires that the public sector take the lead. A small carbon tax could raise substantial revenues for funding new technology research. For example, a tax of four dollars per ton of carbon in the U.S., representing only one cent per gallon of gasoline, could generate approximately $5.6 billion and enable current public sector energy R&D to grow more than threefold.

OECD members should commit themselves to raising their grossly inadequate level of basic and applied energy research by a significant and annually rising percentage of civilian research programs. And they should collaborate in R&D, especially with developing nations and with the private sector. Given the stakes, energy research arguably merits a degree of public sector commitment comparable to that devoted not long ago to aerospace and telecommunications. Promoting technology should not prove politically unpopular because it creates economic growth and job opportunities. The expected leverage from such research in reducing the costs of addressing climate change makes it an eminently sound investment.

10.8. Negotiate in a More Efficient Forum

In the interest of speeding the process, most if not all of these actions—especially the research initiatives, policy measures, technology transfer, and technology goals—could be negotiated and implemented by a relatively small number of like-minded nations, North and South, outside the FCCC context (and perhaps later presented to the larger forum). It is imperative to closely involve the handful of developing nations whose emissions are critical. There is no moral stricture, however, that requires concerned governments to negotiate every relevant action within the unwieldy context of over 170 nations and thousands of observers. The OECD and the Asia–Pacific Economic Cooperation Forum come to mind as plausible alternatives where new options could be explored; the latter body includes countries as diverse as Chile, China, Indonesia, Mexico, Russia, and the United States.

Taken together, the above efforts would greatly increase the likelihood of making existing renewable energy more competitive, making carbon capture and sequestration more feasible, creating the future energy sources that are indispensable, and motivating developing nations to limit their emissions. Perhaps by making a fresh start with new concepts we could achieve the progress that has been so elusive up until now.

While these ideas present undoubted challenges for political will and for diplomacy, I seriously question whether the current course, which has meandered for years, has better chances of success.

References

Benedick, R. E. (1997a). *Global Climate Change—The International Response*. Wissenschaftszentrum Berlin Paper, FS II 97–401.

Benedick, R. E. (1997b). The UN approach to climate change: Where has it gone wrong? Resources for the future. *Weathervane* <http://www.rff.org>.

Benedick, R. E. (1998a). *Ozone Diplomacy—New Directions in Safeguarding the Planet* (revised ed). Harvard University Press, Cambridge,.

Benedick, R. E. (1998b). Auf dem falschen Weg zum Klimaschutz? *Universitas—Zeitschrift für Interdisziplinäre Wissenschaft* 629, November.

Bojkov, R. (1995). "The Changing Ozone Layer." WMO/UNEP, Geneva.

Bolin, B. *et al.* (1986). "The Greenhouse Effect, Climate Change and Ecosystems." SCOPE 29, New York.

Bolin, B. (1998). The Kyoto negotiations on climate change: A science perspective. *Science* **279**.

CDIAC. (1999). Carbon Dioxide Information Analysis Center, Oak Ridge National Laboratory, U.S. Department of Energy. Marland, G., *et al.* "Carbon Dioxide Emissions From Fossil Fuel Consumption." Available at htttp://cdiac.esd.ornl.gov.

Cook, E. (Ed.) (1996). "Ozone Protection in the United States." World Resources Institute, Washington D.C.

Dooley, J. J. (1998). Unintended consequences: Energy R&D in a deregulated market. *Energy Policy*.

Dooley, J. J. and Runci, P. J. (1999). "Developing Nations, Energy R&D, and the Provision of a Planetary Public Good: A Long-term Strategy for Addressing Climate Change." Pacific Northwest National Laboratory, operated by Battelle for the U.S. Department of Energy: internal report PNNL-SA-32077, August 1999. (to be published).

Dotto, L. and Schiff, H. (1978). "The Ozone War." Doubleday, Garden City, New York.

Edmonds, J. A. (1999a). "The Global Energy Technology Strategy—1998 Project Review." Battelle, Washington D.C.

Edmonds, J. A. (1999b). Future agreements. Paper presented to the IPCC Working Group III Experts Meeting, May 27, 1999, The Hague, Netherlands. (to be published).

Edmonds, J. A. (1999c). Beyond Kyoto: Toward a technology greenhouse strategy. *Consequences* 5: 1.

Edmonds, J. A., Dooley, J. J., and Kim, S. H. (1999). Long-term energy technology: Needs and opportunities for stabilizing atmospheric CO_2 concentrations. In "Climate Change Policy: Practical Strategies to Promote Economic Growth and Environmental Quality." American Council for Capital Formation, Center for Policy Research, Washington D.C.

Edmonds, J. A. and Wise, M. (1999). Exploring a technology strategy for stabilizing atmospheric CO_2. In "International Environmental Agreements on Climate Change." Kluwer, Dordrecht, Netherlands.

European Commission. (1999). Preparing for Implementation of the Kyoto Protocol. Commission Communication to the Council and the Parliament (COM(99)230 final). Brussels.

Franz, N. (1998). Eizenstadt defends Kyoto. *Environ. Energy Mid-Week.* Environment and Energy Study Institute, Washington D.C. February 12.

French, H. (1997). Learning from the ozone experience. "Worldwatch Institute: State of the World 1997." W.W. Norton, New York.

Grubb, M. (1999). Optimal climate policy versus political and institutional realities: The Kyoto Protocol and its follow-up. Presentation to Kiel Institute of World Economics conference: The Economics of International Environmental Problems, 21–22 June.

IEA. (1999). International Energy Agency. In "Carbon Dioxide Emissions From Fossil Fuel Combustion, Highlights."

IPCC. (1991). Intergovernmental Panel on Climate Change. In "Climate Change—The IPCC Scientific Assessment." Intergovernmental Panel on Climate Change. Cambridge University Press, Cambridge.

IPCC. (1996). "Climate Change 1995—The Science of Climate Change." Intergovernmental Panel on Climate Change. Cambridge University Press, Cambridge.

Klepper, G. (1999). Environmental and resource policy in Germany. Kiel Institute of World Economics. (unpublished)

Kopp, R. S. (1997). How tough will it be for the United States to meet a climate target by 2010? Resources for the Future, *Weathervane* available at http://www.rff.org.

Lang, W. (Ed.) (1996). "The Ozone Treaties and Their Influence on the Building of International Environmental Regimes." Österreichische aussenpolitische Dokumentation, Vienna.

Meyerson, F. A. B. (1998). Population, carbon emissions, and global warming: The forgotten relationship at Kyoto. *Population Dev. Rev.* 24(1).

Molina, M. J. and Rowland, F. S. (1974). Stratospheric sink for chlorofluoromethanes: Chlorine-atom catalyzed destruction of ozone. *Nature* **249**.

Nordhaus, W. and Boyer, J. (1999). Requiem for Kyoto: An economic analysis. *Energy J.* Special Issue.

Oreskes, N., Shrader-Frechette, K., and Belitz, K. (1994). Verification, validation, and confirmation of numerical models in the earth sciences. *Science* **263**.

Revelle, R. and Suess, H. E. (1957). Carbon dioxide exchange between atmosphere and ocean and the question of an increase of atmospheric CO_2 during the past decades. *Tellus* **9**.

Rosenberg, N. J., Izaurralde, R. C. and Malone, E. L. Eds. (1999). "Carbon Sequestration in Soils: Science, Monitoring, and Beyond." Proceedings of the St. Michael's (MD) Workshop, December 1998. Battelle Press, Columbus, Ohio.

Schelling, T. C. (1997). The cost of combating global warming: Facing the trade-offs. *Foreign Affairs.* November/December.

Schellnhuber, H-J. (1999). Amerikanisches Roulette. *Süddeutsche Zeitung* June 26–27, 1999.

Simonis, U. E. (1998). Das Kioto Protokoll und seine Bewertung. *Spektrum der Wissenschaft,* March.

Surewitz, D. and Pielke, R. (1999). Prediction in science and policy. *Technol. Soc.* **21**.

United Nations. (1992). "United Nations Framework Convention on Climate Change."

United Nations (1997). "Kyoto Protocol to the United Nations Framework Convention in Climate Change."

United Nations Population Division. (1999). "World Population Prospects: The 1998 Revision." Volume I: Comprehensive Tables. New York: United Nations.

von Storch, H. and Bray, D. (1999). Perspectives of climate scientists on global climate change. In "Climate Change Policy in Germany and the United States." German-American Academic Council Foundation, Bonn.

White, R. (1998). Kyoto and beyond. *Issues Sci. Technol.* Spring.

Wigley, T., Richels, R., and Edmonds, J. A. (1996). Economic and environmental choices in the stabilization of atmospheric CO_2 emissions. *Nature* **379**.

WMO/UNEP. (1986). "Atmospheric Ozone 1985: Assessment of Our Understanding of the Process Controlling its Present Distribution and Change." World Meteorological Organization and United Nations Environment Programme, Geneva.

Optimizing Long-Term Climate Management

Klaus Hasselmann

Max Planck Institut for
Meteorology
Hamburg

1. Introduction

The Framework Convention on Climate Change (FCCC) formulated at the United Nations Conference on Environment and Development (UNCED) in Rio de Janeiro in 1992 represented the first major international attempt to address the problem of anthropogenic climate change. Following a series of negotiations, the signatories of FCCC agreed in December 1997 in Kyoto to the first concrete mechanisms and targets for limiting greenhouse gas emissions. This has introduced a new quality into climate management and policy, with important implications for climate research (Hasselmann, 1997).

Despite the agreement on general goals and targets in Kyoto, many questions on the technical details of implementation still need to be resolved (cf. Grubb, 1999). There exist also diverse and conflicting interests of strong stakeholders, so that ratification of the Kyoto protocol by a sufficient number of nations for the agreement to come into force is still outstanding. Furthermore, it is not yet widely appreciated that the Kyoto agreement, even if implemented, can represent only a very small first step toward significantly larger reductions in greenhouse gas emissions in the future, if a major climate warming in the present and following centuries is to be averted. In this situation of uncertainty, combined with a widespread realization that we cannot afford to delay action, it is incumbent on the climate research community to provide more accurate projections of the climate change anticipated for various greenhouse gas emissions scenarios and to cooperate with other disciplines in developing more realistic assessments of the impact of climate change on the environment and human living conditions. On a still broader interdisciplinary level, a better understanding is needed of the many complex interrelations between climate change, the global socioeconomic system, and policy measures in a global multiactor framework.

The recent years have witnessed a strong development of climate research to a level where realistic climate models can now provide reasonably credible predictions of future climate change for given scenarios of greenhouse gas and aerosol emissions (Houghton *et al.*, 1996). However, interdisciplinary research on the interactions between climate and the socioeconomic system, although essential for the scientific underpinning of climate policy in the post-Kyoto era, has lagged behind this development and is today still in its infancy. In the following, some of the many open issues encountered in this field are outlined and illustrated with some simple model simulations. It is shown that there exists a fascinating spectrum of first-order problems that urgently need to be addressed, and that can still be studied today at the exploratory level of an emerging new discipline with relatively simple models and concepts. It is hoped that the examples given here will motivate other scientists to engage in similar integrated assessment studies.

2. Global Environment and Society Models

The general structure of an elementary global environment and society (GES) model designed for integrated assessment studies is illustrated in Fig. 1. The various interactions between the global environment, the socioeconomic system, and the policy makers are shown. These include both the direct physical coupling and the communication pathways that transform scientific knowledge and stakeholder positions via the media into public opinion and political action. A more disaggregated representation of the GES system, showing the breakdown into individual economic sectors and regions, with associated independent political decision makers, is shown in Fig. 2. The different economic sectors and regions are coupled in this case through the traditional mechanisms of trade as well as the global environment, which they jointly modify and by which they are individually affected.

For practical modeling applications, this multiactor breakdown of the GES system must still be strongly aggregated, as it is

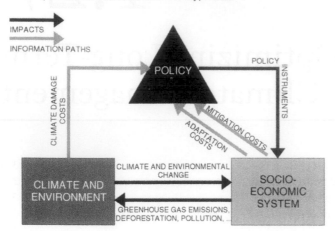

FIGURE 1 Physical interactions and communication pathways between the climate and environment, socioeconomic system, and policy in a coupled global environment and society (GES) model for integrated assessment studies.

impossible to realistically simulate the multitude of interacting players pursuing different goals in all sectors and regions of the global socioeconomic system. As in the analagous case of the climate system alone (which appears much simpler by comparison, however), insight into the dynamical behavior of the coupled GES system can be gained only iteratively, through the development of

a hierarchy of different models, each of which highlights some particular features of the system and ignores other potentially important aspects.

The problem of managing the climate and environment system can be viewed as a general optimization problem: how should one deploy the finite resources available to humankind to achieve a sustainable development path that optimizes the human welfare of both the present and future generations? The task is to find an optimal balance between environmental protection efforts (in the forms of labor expenditures, human and capital investments, technological development, etc.) and the loss of welfare in other sectors relevant to human well-being (such as industrial development and the production of consumer goods, arising from the redeployment of the resources used to protect the environment from other economic activities).

In general, there will be no consensus on the definition of a single, overall world welfare function. Different political decision makers may have very different welfare concepts, which each will try to individually optimize. Thus the global optimization problem may be viewed as an intricate multiplayer game, including all the complexities of cooperative and noncooperative strategies, the creation of alliances, free-riding, direct and indirect agreement-enforcement mechanisms, etc.

However, in the following it will simply be assumed that an agreement has been reached through international negotiations on the form of the world welfare function one wishes to jointly optimize, so that the optimization task has been reduced to the

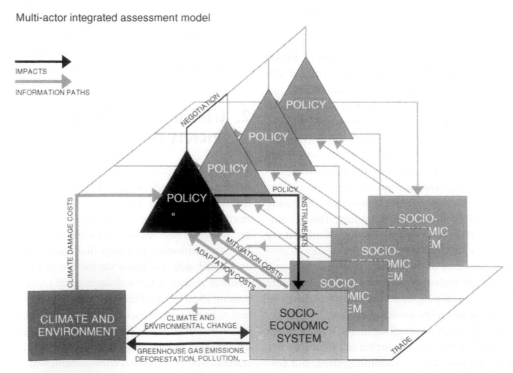

FIGURE 2 Interactions between the climate, environment, socioeconomic system, and policy makers in a disaggregated, multiactor representation of the coupled global environment and society (GES) system of Figure 1.

determination of appropriate regulation policies that will lead to global greenhouse gas emmission paths that maximize the time-integrated world welfare. Thus we consider only the single-actor GES version of Fig. 1.

This is consistent in the sense that greenhouse warming, in contrast to other pollution problems, is essentially a global problem: because of the long lifetimes of greenhouse gases compared with their mixing times in the atmosphere, the distribution of greenhouse gases in the atmosphere is highly uniform, so that the geographical source of the emissions is irrelevant. Nevertheless, the treatment of climate management policy as a single-actor optimization problem presupposes an agreement on basic and controversial issues in the definition of the world welfare function. These include the values attached to nonmarket properties, such as health, life expectancy, the quality of the environment, or the diversity of species, and ethical issues such as intergenerational and interregional equity. Climate research or the natural sciences in general can clearly not resolve these issues within the framework of their own disciplines, but they can contribute from their reference level to an understanding of the interactions within the climate system and between the climate and socioeconomic system that are relevant in addressing the integrated climate policy problem.

A basic difficulty in the construction of a comprehensive GES model is the inherent complexity of each of the three subsystems indicated in Fig. 1. To obtain a manageable integrated model, the subsystems must be strongly simplified. By projecting the general multiactor game-theoretical problem onto the single-actor world welfare optimization problem, as discussed above, the complex "policy" subsystem has been effectively reduced for the present discussion to a single greenhouse-gas emissions regulator. However, for application in an integrated GES model, the state-of-the-art models of the remaining two subsystems must be similarly reduced.

Modern climate models are based on coupled general circulation models (GCMs) of the physical atmosphere–ocean system and three-dimensional geochemical cycle models of comparable complexity for the determination of the greenhouse gas concentrations. Both require very costly computer resources. Similarly, sophisticated state-of-the-art general equilibrium models (GEMS) of the global economy typically consider more than 100 interacting economic sectors and regions, compute large numbers of independent variables, and introduce many poorly determined empirical parameters. It is difficult to combine such models (particularly when they are developed in different coding languages) into a single, computationally efficient GES model with which one can systematically carry out a large number of exploratory simulations, such as sensitivity studies, cost–benefit analyses, and optimal control computations.

Thus, for application in integrated assessment studies, the existing state-of-the-art climate and socioeconomic subsystem models need to be replaced by computationally more efficient and analytically more transparent modules. In the following section it will be shown that this can be achieved for the climate subsystem by pro-

jecting the response properties of the climate system computed with a sophisticated three-dimensional climate model onto a dynamically equivalent impulse–response model. For the socioeconomic system, the long time-scales of the climate system require not only reductional simplifications of the standard GEM approach but also generalizations to include climate-change impacts, and, in addition, important long-term processes such as endogenous technological development, intergenerational transfers, and risk management. However, this will not be addressed here, and we shall consider later only a very simple economic model.

3. Impulse–Response Climate Models

For any complex, nonlinear system such as climate, it is permissible, for sufficiently small perturbations, to describe the response of the system to external forcing in terms of a linearized response model. In the case of the climate system, the linearization condition is approximately satisfied for the external forcing due to anthropogenic greenhouse gas emissions if the temperature change remains below about 2–3°C. Measured in Kelvin relative to absolute zero temperature at −273°C, the global mean temperature of the earth (15°C) is 288 K. Thus a 3°C temperature change represents a perturbation of only $3/288 \approx 1\%$ in the absolute temperature scale relevant for infrared greenhouse radiation effects. The climate change effects in this range can normally be computed to adequate approximation as a linear response.

An important caveat in the application of the linear response approximation, however, is that the reference state on which the linear perturbation is superimposed is not close to an unstable bifurcation point. In this case, the response of the climate system to even relatively small external forcing can differ significantly from the climate change computed with a linear response model and is basically unpredictable. A number of such potential instabilities have been discussed in the literature. One of the more serious possibilities, which has been observed in paleoclimatic records and simulated in models, is a breakdown of the North Atlantic ocean circulation (Maier-Reimer and Mikolajewicz, 1989; Rahmstorf and Willebrand, 1995; Rahmstorf, 1995; Schiller *et al.*, 1997). This can be triggered through a warming and/or freshening of North Atlantic surface waters. The northward-traveling surface water of the Gulf Stream then no longer becomes dense enough when cooled at higher latitudes to sink to sufficient depth to drive the deep ocean circulation, which is the source of the balancing northward current. A breakdown of the Gulf Stream, which produces a 6°C warmer climate in Europe relative to the latitudinal mean, would clearly have dramatic consequences for the climate of Europe. Other potentially catastrophic instabilities are a collapse of the West Antarctic ice sheet, which would result in a global sea-level rise of 6 m, or a runaway greenhouse warming through the release of large quantities of methane (a very effective greenhouse gas) that are currently trapped in the permafrost regions of Siberia and in hydrates in the deep ocean. Current predictions of anthropogenic greenhouse warming suggest that such

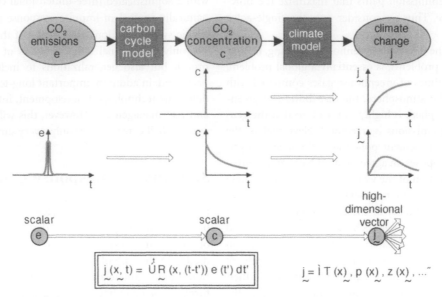

FIGURE 3 Impulse–response representation of the response of the climate system to a δ-function CO_2 input. The net response is given by a convolution of the response of the atmospheric CO_2 concentration to a δ-function CO_2 input (calibrated against a carbon cycle model) and the response of the physical ocean–atmosphere climate system to a step-function increase in the CO_2 concentration (calibrated against a coupled ocean–atmosphere GCM). For details, see Hasselmann *et al.*, 1997.

instabilities are unlikely to occur if the global warming remains below 2–3°C, and we will ignore them in the following. However, the implications of low-probability, high-impact climate change instabilities should be kept in mind in integrated assessment studies.

As an example of an impulse–response climate model, we consider in the following the response of the climate system to CO_2 emissions $e(t)$ (cf. Fig. 3), which represent about 60% of the total greenhouse warming today and are expected to contribute a still larger fraction in the future. For small perturbations, the relation between the CO_2 forcing and the climate-change response $\psi(x, t)$ can be expressed as a linear response integral

$$\psi(X. t) = \int_0^t R(X, t - t')e(t')\mathrm{d}t',$$

where the impulse–response function $R(x, t)$ represents the climate response to a δ-function CO_2 input at time $t = 0$. The function $R(x, t)$ can be calibrated against the response of the climate system computed with a fully nonlinear, state-of-the-art climate model for some given greenhouse gas emission scenario. Once calibrated, the climate response can be computed for arbitrary emission curves by superposition. Coupled with a similarly efficient socioeconomic model, the impulse–response climate model thus enables one to efficiently perform a large number of simulations, as required, for example, for optimal emission path computations

or sensitivity studies.

An important feature of impulse–response models is that they entail no loss of information compared with the complete model: the function $R(x, t)$ can contain the same number of degrees of freedom in the description of the climate change signal $\psi(x, t)$ as the complete climate model against which it is calibrated.

Examples of typical impulse–response functions are shown in Figs. 4 and 5 (adopted from Hooss *et al.*, 2001). The net impulse–response function of the coupled carbon-cycle/physical ocean–atmosphere system is given by a convolution (cf. Hasselmann *et al.*, 1997) of the linear impulse–response function for the carbon cycle alone (representing the atmospheric CO_2 response to a δ-function CO_2 input) and the response function of the physical ocean–atmosphere system (representing the response of the physical system variables to a step-function increase in the CO_2 concentration). Figure 4 shows the response function R_c for the carbon cycle and the response functions R_T and R_s for the global mean temperature and mean sea-level rise, respectively, for the physical climate system. The resulting net response functions for R_c, $R_{(T)}$, and $R_{(s)}$ for the coupled carbon-cycle/physical ocean–atmosphere system are shown in Fig. 5.

The linear impulse–response model has recently been generalized by Hooss *et al.* (2001) to include some of the dominant nonlinearities of the climate system. The net response curves shown in Fig. 5 were computed using this generalized model, showing the impact of nonlinearities in the lower two-panel rows. The princi-

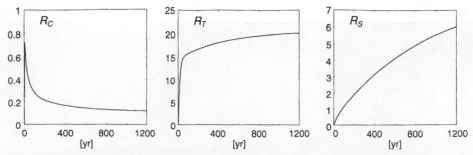

FIGURE 4 Individual impulse–response functions for the carbon cycle (response R_c of the atmospheric CO_2 concentration to a δ-function CO_2 input at time $t = 0$, left) and the variables global mean near-surface temperature (center) and mean sea-level rise (right) for the physical ocean–atmosphere system (response to a step-function increase in the CO_2 concentration to a constant level at time $t = 0$). The units for temperature response R_T and sea-level response R_s refer to the amplitudes of the first empirical orthogonal functions (EOFs) of the response patterns of the respective variables and are essentially arbitrary (see text). Adopted from Hooss *et al.*, 2001.

pal effect is a lowering of the rate of decrease of the atmospheric CO_2 concentration at higher CO_2 concentrations. This is due to the slower uptake of CO_2 by the ocean in which solubility of CO_2 decreases with increasing CO_2 concentration.

Hooss *et al.* (2001) considered not only the responses of the global mean temperature and sea level, but also the changes in other climate variables, such as precipitation and cloud cover, and in addition the spatial dependence of these variables. They found that the latter could be well described by a single spatial response pattern for each variable considered (cf. Figs. 6, 7, 8; see also color insert). Thus, the spatiotemporal response properties of each climate variable can be represented as the product of a normalized pattern (the empirical orthogonal function, on EOF palten), and an associated pattern coefficient, whose time evolution must necessarily be the same as the

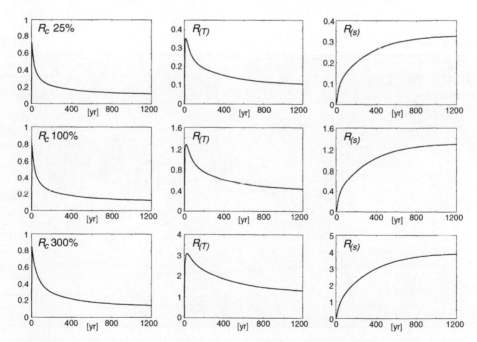

FIGURE 5 Net impulse–response functions R_c for the atmospheric CO_2 concentration (left column), the global mean temperature $R_{(T)}$ (center column), and the amplitude $R_{(s)}$ of the first EOF of the sea-level rise (right column, proportional to the mean sea-level rise) for different magnitudes of a δ-function CO_2 input at time $t = 0$. The differences in the responses in the three cases (from top to bottom, 15, 100 and 30% increase in initial CO_2 concentration, respectively) arise from nonlinearities in the generalized nonlinear impulse–response model of Hooss *et al.*, 2001. (Adopted from this source.)

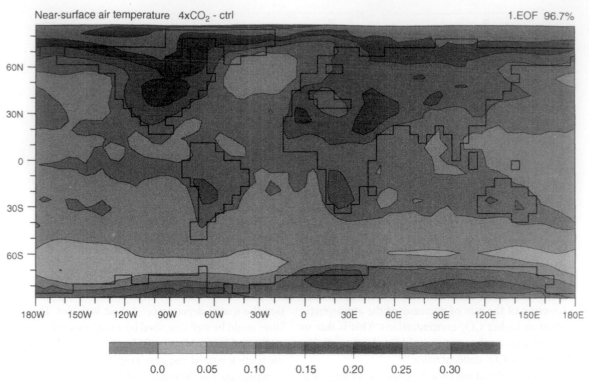

FIGURE 6 Dominant spatial-response pattern (first EOF) of near-surface temperature change to increased atmospheric CO_2 concentration (from Hooss *et al.*, 2001). See also color insert.

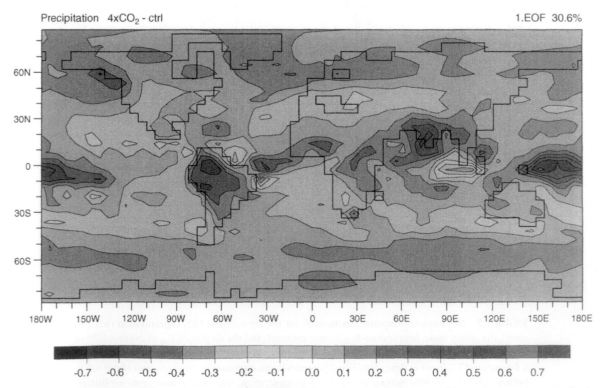

FIGURE 7 Dominant spatial-response pattern (first EOF) of change in precipitation caused by an increase in atmospheric CO_2 concentration (from Hooss *et al.*, 2001). See also color insert.

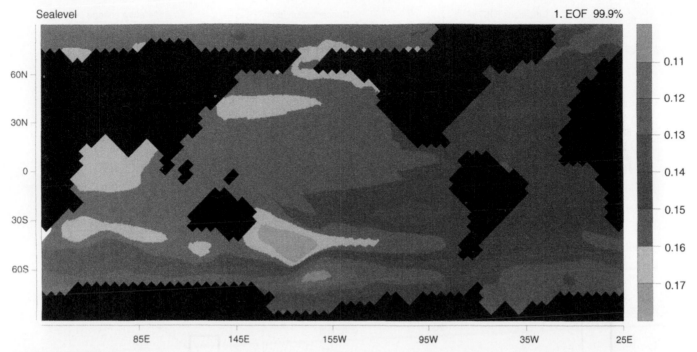

FIGURE 8 Dominant spatial-response pattern (first EOF) of change in sea level caused by an increase in atmospheric CO_2 concentration (from Hooss *et al.*, 2001). See also color insert.

time evolution of the associated global mean variable (which is given by the spatial mean of the response pattern).

An important common characteristic of the response functions is their exceedingly long memory. The decrease in atmospheric CO_2 concentration (due to the gradual uptake of CO_2 by the oceans and the terrestrial biosphere) is an extremely slow process extending over several hundred years. Combined with the delayed temperature response R_T of the coupled ocean–atmosphere system to a CO_2 increase induced by the large thermal inertia of the oceans, the net temperature response R_T of the climate system to CO_2 input persists over several centuries. Thus, in assessing the climate-change impacts of human activities, one must consider time horizons far beyond the normal planning horizons of decision makers.

The impact of the long memory of the climate system on the climate response to anthropogenic CO_2 emissions is illustrated in Figure 9, which shows the CO_2 emissions, CO_2 concentrations, and temperature change computed for a "business-as-usual" (BAU) scenario and an alternative frozen-emissions scenario. The upper panels show the evolution over the next 100 years, the lower panels the evolution over the next 1000 years. In the case of the BAU scenario, all fossil fuel resources, estimated at 10,000 GtC, are assumed to be exploited within the next 500–700 years. The long-term impacts in the lower panel are seen to greatly exceed the climate change over the next 100 years, even for the frozen emissions case. Although the impulse–response computations are clearly unreliable for such large climate changes, the orders of magnitude of the computed warming, in the range of 10°C for the BAU case—exceeding the climate changes of the ice ages and beyond

the range in which even fully nonlinear state-of-the-art models can be credibly applied—clearly demonstrate the danger of underestimating future climate-change impacts by limiting considerations of climate policy to only a few decades (see also Cline, 1992).

4. Optimizing CO_2 Emissions

We turn now to the problem of coupling the impulse–response model summarized above to a simple socioeconomic model. The coupled model will then be applied to determine the optimal CO_2 regulation policy that would minimize the net impact of climate change. This can be represented generally as the sum of two contributions: the direct and indirect costs of climate change itself (termed climate "damage costs" in the following), and the costs incurred in reducing CO_2 emissions ("abatement costs"). We restrict the discussion here to the impact of CO_2 alone, as the dominant anthropogenic greenhouse gas, but note that the same approach can be applied to the climate change induced by other greenhouse gases also.

A number of cost studies and optimized cost–benefit analyses of the economic impact of CO_2 emissions have been published in the literature (e.g., Manne and Richels, 1991; Peck and Teisberg, 1992; Nordhaus, 1993; Richels and Edmonds, 1995; Tahvonen and Storch, 1994; Nordhaus and Yang, 1996; Wigley *et al.*, 1996; see also the review by Fankhauser, 1995). Normally, the climate models are reduced to box-type models, while the economic models are represented as aggregated dynamic-growth models,

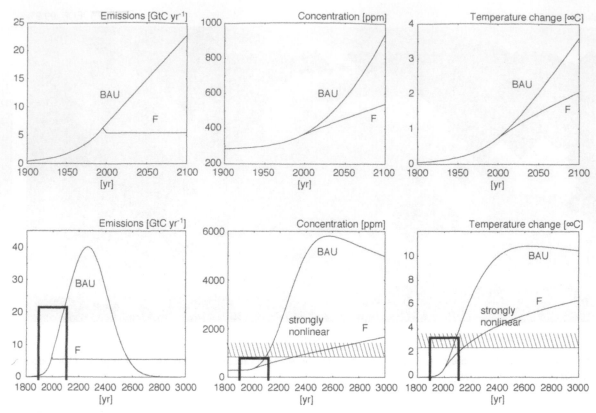

FIGURE 9 Evolution of the atmospheric CO concentration and the global mean temperature, computed with the impulse–response climate model of Hooss *et al.* (2000), for a BAU scenario and a frozen-emissions scenario F over the next 100 years (upper panels) and the next 1000 years (lower panels). The long-term climate change in the lower right panel is seen to greatly exceed the predicted climate change in the next 100 years (indicated also by boxes in the lower panels).

dependent on the distribution of the total output production between consumer goods, investments in capital and technological development, abatement measures, etc. as control variables. To account for the influence of the long time-scales imposed by the climate system, and the issues of intergenerational accounting and equity that these raise, we apply in the following the more realistic nonlinear impulse–response climate model of Hooss *et al.* (2001). While radically reducing the economic model to simple price expressions for the climate damage and abatement costs, in accordance with Hasselmann *et al.* (1997). The principal conclusions drawn from our discussion will be independent of the details of the economic model.

The global climate-damage costs are taken proportional to the sum of the squares of the change in global mean near-surface temperature and the rate of change in the global mean temperature. This corresponds to the assumption that any change in the present climate, to which humans and ecosystems have had time to adapt, is detrimental, and that the damages increase nonlinearly both with the change in global mean temperature and with the rate of the temperature change. The global mean temperature is regarded here as a proxy for all climate change variables, such as precipitation, cloudiness, the frequence and strengths of El Niño, the intensities of storms, droughts and other extreme events, and

the rise in sea level. This is dynamically consistent with atmospheric climate variables, since the atmospheric response to greenhouse forcing can generally be well described in numerical climate simulations, as mentioned above, by a few dominant EOF patterns, whose coefficients are diagnostically coupled to the global mean temperature. However, the projection onto global mean temperature is more questionable for climate properties related to the ocean, such as El Niño and the rise in sea level, since the time scale of the ocean response to external forcing differs from that of the atmosphere (cf. Fig. 4).

The expression for the abatement costs is based on the assumption that any deviation $r = (e - e_0)/e_0$ of the CO_2 emissions $e(t)$ from the emissions $e_0(t)$ of the BAU economic development path, in which all climate change impacts are ignored, incurs costs. For small deviations, the costs are assumed to be quadratic in r. Quadratic-cost penalties are also introduced for the first and second time derivatives of r to parametrize the effects of economic inertia (capital losses, development costs, etc.)

The optimal CO_2 emission path is the one that minimizes the total time-integrated sum of the climate damage and abatement costs. In intertemporal economic accounting, all costs are traditionally discounted at the same rate. Theoretically, this is equal to the inflation-adjusted market interest rate. However, the appropri-

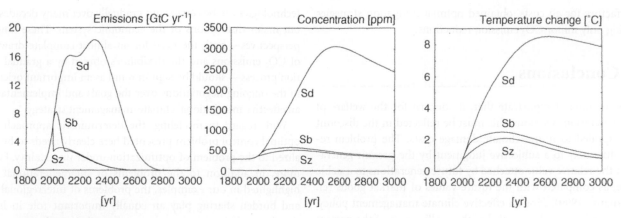

Sb: Baseline optimized scenario
Sz: Zero economic inertia
Sd: Damage & abatement costs both discounted ($t_a=t_d=50y$)

FIGURE 10 Optimized CO_2 emission paths with resulting changes in CO_2 concentration and global mean near-surface temperature. S_b, baseline scenario: discounting of abatement costs only, finite economic inertia; S_z, same as baseline scenario without economic inertia; S_d, same as baseline scenario but with equal discounting of climate damages and abatement costs.

ate discount rate for nonmarket values, which comprise a large fraction of the climate damage costs, is the subject of considerable debate (cf. Hasselmann *et al.*, 1997; Nordhaus, 1997; Heal, 1997; Brown, 1997; Hasselmann, 1999). It has been argued on ethical grounds, and also on the basis of economic, time-dependent relative-value reasoning, that the appropriate discount rate for such values should be smaller than for market goods or even zero. We have accordingly applied separate discount rates for the climate damage and abatement costs. In our baseline optimization run S_b we have assumed a zero discount rate for the climate damage costs and a discount rate of 2% for the abatement costs.

Figure 10 shows the optimal CO_2 emission paths and the associated atmospheric CO_2 concentrations and global mean-temperature evolution for the baseline case S_b and two further cases S_z and S_d. The scenario S_z is identical to the baseline case except that the economic inertia is set equal to zero. Although the emission paths for the solutions S_b and S_z differ significantly in the first few decades, the differences in the long-term climate impact are minor. This demonstrates that for an effective climate mitigation policy, long-term emission abatements far outweigh the impact of short-term reduction measures. Essential to averting major climate change in the long term is the gradual but complete replacement of fossil fuels by carbon-free energy technologies.

This is further illustrated by Fig. 11, which compares the cases S_b and S_z with the emission reductions agreed to by the industrialised countries in the Kyoto protocol. The Kyoto curve lies between the cases S_b and S_z and thus appears quite acceptable from the viewpoint of these computations. From the long-term perspective imposed by the memory of the climate system, however, the details of the Kyoto compromises over emission-reduction percentages appear rather irrelevant compared with the central challenge of establishing an effective long-term post-Kyoto mitigation strategy that will gradually but surely lead to a restructur-

ing of the present energy technology from fossil fuels to carbon-free energy generation.

The third case, S_d, in Fig. 10 illustrates the strong influence of the discount rates on the computed optimal solutions. In contrast to the baseline scenario S_b and the zero-inertia scenario S_z, in which only the abatement costs were discounted, in scenario S_d both costs were discounted at the same rate of 2%, following standard economic practice. In this case, the optimal emission path leads to a climate "catastrophe" similar to the BAU case shown in Fig. 9. The explanation is simple: since major climate change develops only after several centuries, the associated discount factor is very small, and the discounted climate damage costs are negligible. Thus, there is only a small cost-penalty incurred in following the BAU path. This also explains why previous cost–benefit analyses (e.g. Nordhaus, 1993), based on the applicaton of uniform dis-

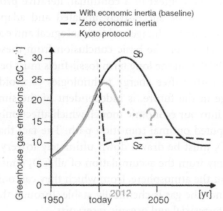

FIGURE 11 Comparison of the Kyoto protocol with the optimized solutions S_b and S_z of Figure 10, with and without economic inertia, respectively.

count factors for all costs, obtained optimal mitigation strategies requiring only minor CO_2 emission reductions.

5. Conclusions

Our simulations demonstrate that, if concern for the welfare of future generation is serious, this must be reflected in the discount factors applied to future climate damage costs. The problem reduces ultimately to a subjective judgment by the present generation on the relative value attached by future generations to a stable climate, as compared with the future values of market goods (cf. Hasselmann, 1999). (For an effective climate management policy, this must, of course, be coupled with a willingness of the present generation to honor these basic value judgments through an intergenerational commitment to sustainable development.) If the evolutionary paths of the relative values of different goods diverge with time, they will necessarily be characterized by different effective discount rates.

For example, it can be argued that the value of a stable climate increases with time relative to the costs of standard market goods, whose inflation-adjusted prices tend to decrease through advances in technology. In this case, the effective discount rate for climate damages will be smaller than that for (market controlled) abatement costs and can even become negative. There is no conceptual difficulty in incorporating these considerations consistently in standard economic welfare optimization computations. Preservation of the environment and economic efficiency are not conflicting concepts, but are parts of the same optimization exercise. If all values are expressed in equivalent monetary terms (which is unavoidable, if budget decisions are to be made) ecological considerations represent simply one input into the traditional economic problem of the "optimal allocation of scarce resources".

The computations of optimal emission paths over many centuries presented in these examples greatly exceed the timespans which economic and technological development can be reasonably predicted or planned. In practice, the translation of such theoretical results into policy recommendations can be meaningfully made only in the context of a continual, iterative process: policy measures need to be successively updated and adapted to new knowledge on climate change and technological and economic developments. However, the basic conclusion from these computations, namely, that in the long term fossil-fuels must be completely replaced by carbon-free energy technologies to avoid major climate change in the future, is independent of the simplifications introduced into our economic model, which affect only the details of the computed optimal transition path. The fact that the emissions of CO_2 must be drawn down ultimately to very low values follows simply from the accumulation of all CO_2 emissions, however small, in the atmosphere, from which they can only be slowly removed, since the gas is chemically stable, through the uptake of CO_2 by the terrestrial and oceanic reservoirs.

Fortunately, the long memory of the climate system also has a positive side: it implies that the transition to carbon-free energy technologies can be carried out gradually over many decades, without major dislocations of the economic system. These long-term perspectives—both the need for an almost complete draw-down of CO_2 emissions and the flexibilities offered by a gradual transition process—should be kept in mind as an important orientation in the ongoing negotiations over the goals and implementation of an effective international climate-management strategy.

With regard to modeling, the deterministic approach to the optimal control problem presented here clearly needs to be generalized to the problem of optimization under uncertainty. Furthermore, in addition to the problems of intergenerational equity highlighted in our examples, the problems of interregional equity and burden sharing play an equally important role in international negotiations on greenhouse gas emissions abatement. These can be addressed only with more sophisticated multiregional, multiactor models. It is hoped that the spatially resolving, nonlinear impulse–response climate models presented in this review can provide a useful building block for such models.

References

Brown, P. G. (1997). Stewardship in climate. An editorial comment. *Climatic Change* **37**, 329–334.

Cline, W. R. (1992). "The Economics of Global Warming." Institute of International Economics. Washington. p. 399.

Fankhauser, S. (1995), "Valuing Climate Chage. The Economics of the Greenhouse". Earthsean, London, 180p.

Grubb, M. (with C. Vrolijk and D. Brack). (1999). "The Kyoto Protocol", Royal Institute of International Affairs, London. p. 342.

Hasselmann, K. (1997). Climate-change research after Kyoto. *Nature* **390**, 225–226.

Hasselmann, K. (1999). Intertemporal accounting of climate change– Harmonizing economic efficiency and climate stewardship. *Climatic Change* **41**, 333–350.

Hasselmann, K., Hasselmann, S. Giering, R., Ocaa, V., and Storch, H. v. (1997). Sensitivity study of optimal CO_2 emission paths using a simplified Structural Integrated Assessment Model (SIAM). *Climatic Change* **37**, 345–386.

Heal, G. (1997). Discounting and climate change. An editorial comment. *Climatic Change* **37**, 335–343.

Hooss, G., Voss, R., Hasselmann, K., Maier-Reimer, E., and Joos, F. (2001). A nonlinear impulse response model of the coupled carbon cycle–ocean–atmosphere climate system. *Clim. Dyn.* (in press).

Houghton, J. T., Meira Filho, L. G., Callander, B. A., Harris, N., Kattenberg, A., and Maskell, K. (1996). Climate change 1995. The science of climate change. Contribution of Working Group 1 to the Second Scientific Assessment of the Intergovernmental Panel on Climate Change. Cambridge University Press.

Maier-Reimer, E., and Mikolajewicz, U. (1989). Experiments with an OGCM on the Cause of the Younger Dryas. In "Oceanography 1988." (A. Ayala-Castaares, W. Wooster, and A. Yez-Arancibia, Eds.) pp. 87–100. UNAM Press, Mexico D F, 208

Manne, A. S., and Richels, R. G. (1991). Global CO_2 emission reductions: the impacts of rising energy costs. *Energy J.* **12**, 88–107.

Nordhaus, W. D. (1993). "Rolling the DICE": An optimal transition path for controlling greenhouse gases. *Res. Energy Econ.* **15**, 27–50.

Nordhaus, W. D., (1997). Discounting in economics and climate change. An Editorial comment. *Climatic Change* **37**, 315–328.

Nordhaus, W. D., and Yang, Z. (1996). A regional dynamic general equilibrium model of alternative climate-change strategies, *Am. Econ. Rev.* **86**, 741–765.

Peck, S. C., and Teisberg, C. J. (1992). CETA: A model for carbon emissions trajectory assessment, *Energy J.* **13**, 55–77.

Rahmstorf, S. (1995). Bifurcations of the Atlantic thermohaline circulation in response to changes in the hydrological cycle. *Nature* **378**, 145–149.

Rahmstorf, S., and Willebrand, J. (1995). The role of temperature feedback in stabilizing the thermohaline circulation. *J. Phys. Oceanogr.* **25**, 787–805.

Richels, R. G., and Edmonds, J. (1995). The economics of stabilizing CO_2 emissions, *Energy Policy* **23**, 4–5.

Schiller, A., Mikolajewicz, U., and Voss, R. (1997). The stability of the North Atlantic thermohaline circulation in a coupled ocean-atmosphere general circulation model. *Climate Dynamics* **13**, 325–347.

Tahvonen, O. H., and Storch, J. v. (1994). Economic efficiency of CO_2 reduction programs. *Clim. Res.* **4**, 127–141.

Wigley, T. M. L., Richels, L., and Edmonds, J. A. (1996). Economic and environmental choices in the stabilization of atmospheric CO_2 emissions, *Nature* **379**, 240–243.

Subject Index

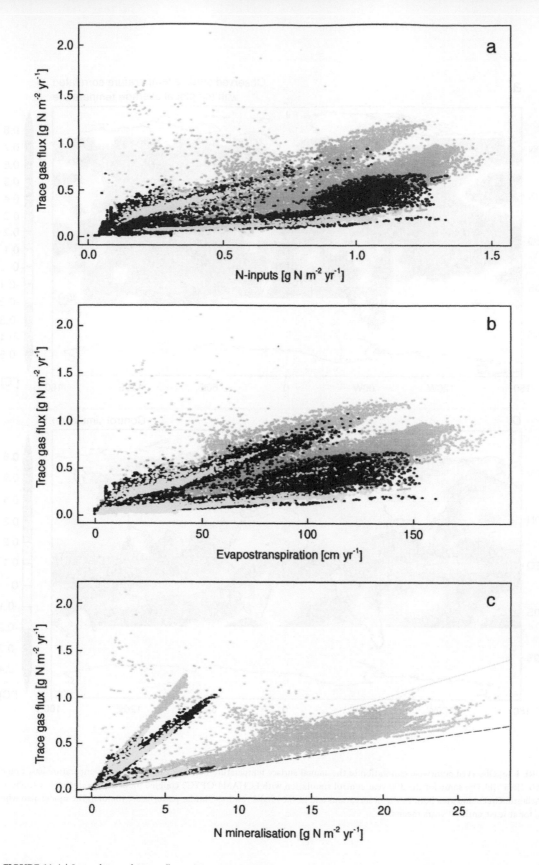

CHAPTER 1, FIGURE 11 (a) Sum of annual N gas fluxes (N_2 + NO + N_2O) vs N inputs from global Century model simulation. N inputs result from wet and dry deposition and biological nitrogen fixation. Yellow points are grassland ecosystems, green points are forests, and black points are "mixed" ecosystems such as savannas. (b) N trace gases vs evapotranspiration. (c) N trace gases vs annual N-mineralization. Lines indicate regressions computed from Matson and Vitousek (1990) based on data from the Amazon Basin (Schimel and Panikov, 1999).

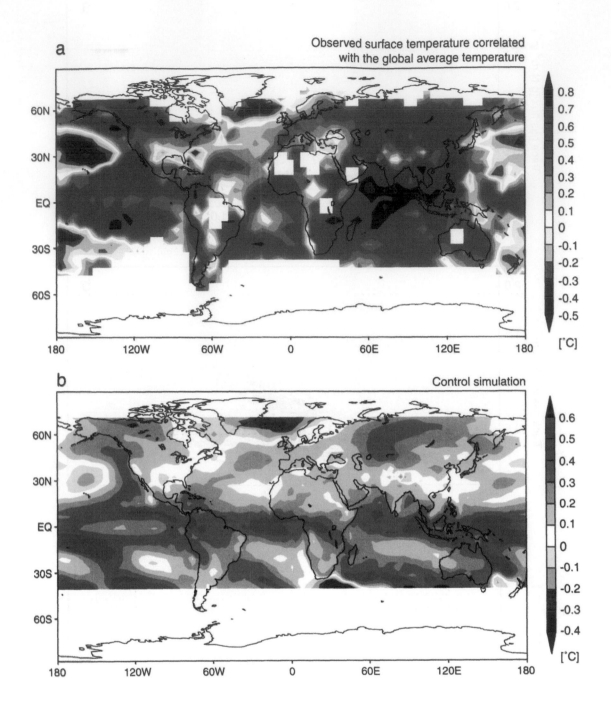

CHAPTER 2, FIGURE 1 (a) Observed pointwise correlation of the annual surface temperature with the global averaged temperature based on observations from the period 1950–1995. (b) The same for the 300-year control simulation with ECHAM4/OPYC3 coupled model. Note the area of slight negative correlation in the North-Atlantic Greenland area both in the observations and in the model results. Similar patterns are found in the model also when averaged over longer periods, for at least until 50 years means.

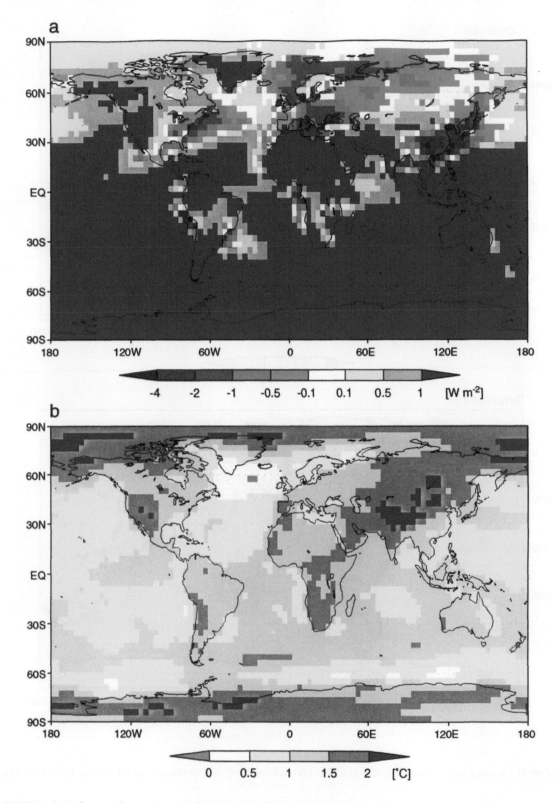

CHAPTER 2, FIGURE 7 (a) Radioactive forcing from greenhouse gases, sulfate aerosols (direct and indirect effect), and tropospheric ozone from the anthropogenic emmission during 1860–1990. See also Table I. In the Northern Hemisphere there are widespread areas with negative forcing caused by sulfate aerosols. (b) Equilibrium response calculated from the ECHAM4 coupled to a slab ocean and averaged over 20 years. Note the differences between the forcing and the response pattern. For further information see Roeckner *et al.* (1999).

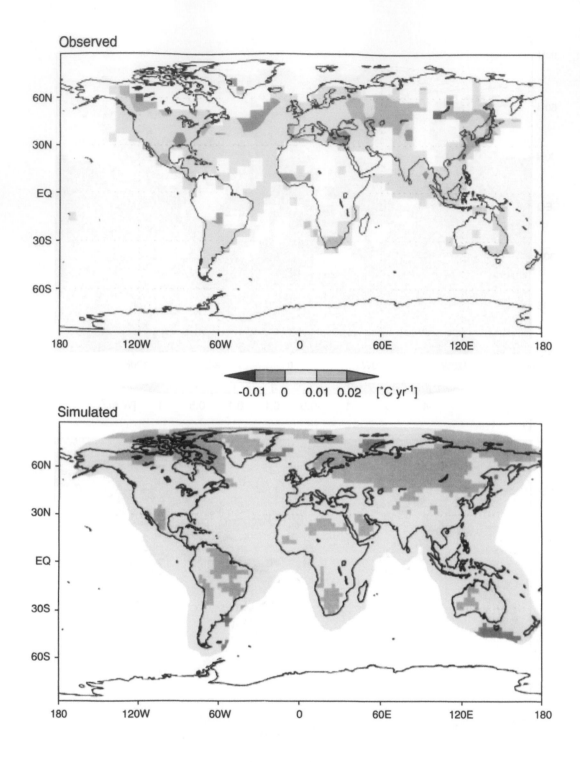

CHAPTER 2, FIGURE 11 Observed surface temperature trend for 1990–1994 and simulated trend for the same period with the ECHAM4/OPYC3 coupled climate model.

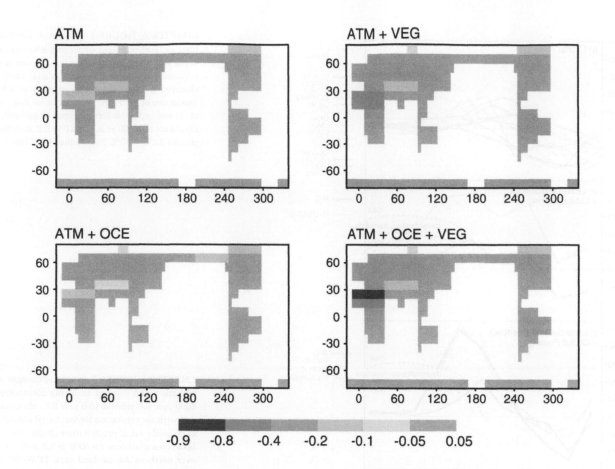

CHAPTER 5, FIGURE 2 Reduction of desert from present-day climate to mid-Holocene climate simulated by Ganopolski *et al.* (1998). The color labels refer to differences in (nondimensional) fractional coverage of desert between today and 6000 years before present. Desert fractions are diagnosed from annual mean precipitation and temperature obtained by the atmosphere–only model (ATM) and the atmosphere–ocean model (ATM + OCE) using present-day land-surface conditions. Desert fractions are predicted from vegetation dynamics by using the atmosphere–vegetation model (ATM + VEG) and the fully coupled model (ATM + OCE + VEG).

CHAPTER 6, FIGURE 1 Zonally averaged simulated annual precipitation anomalies (6000 year B.P. – Control) versus latitude for northern Africa (land grid cells between 20°W and 30°E.). Precipitation anomalies include the effects of: (a) radiative forcing (R) alone for the 18 climate models participating in the Paleoclimate Modeling Intercomparison Project (Joussaume *et al.*, 1999); (b) radiative forcing plus ocean feedbacks (ΔSST) for an asynchronous coupling of GEN-ESIS2 and MOM1 (Kurzbach and Liu, 1997); (c) radiative forcing plus land surface feedbacks (soil, S; vegetation, V; lakes, L; and wetlands, W) simulated using CCM3 (Broström *er al.*, 1998); and (d) radiative forcing (A) plus ocean feedbacks (OA) from a fully coupled simulation with the IPSL AOGCM, radiative forcing plus vegetation feedbacks (AV) from an AGCM simulation forced with 6000 yr B.P. vegetation derived by forcing BIOME1 with the output from the OAGCM simulation, and radiative forcing plus ocean-and land-surface feedbacks from an asynchronous coupling of the 1PSL AOGCM and BIOME1 (Braconnot *et al.*, 1999). *continued*

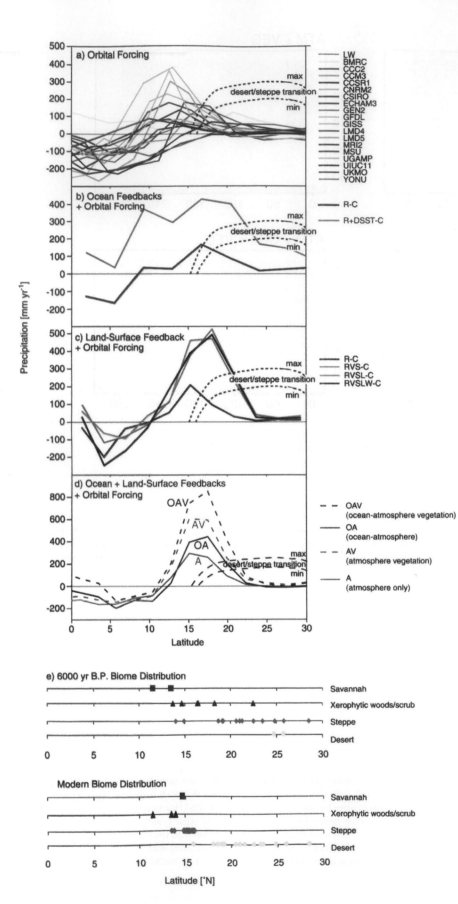

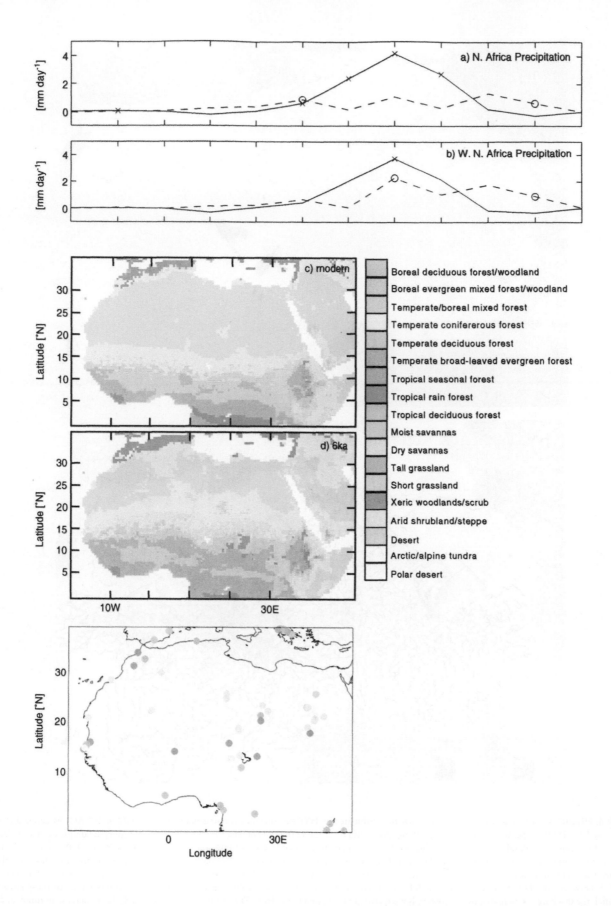

a) N. Africa Precipitation

b) W. N. Africa Precipitation

c) modern

d) 6ka

Boreal deciduous forest/woodland
Boreal evergreen mixed forest/woodland
Temperate/boreal mixed forest
Temperate conifererous forest
Temperate deciduous forest
Temperate broad-leaved evergreen forest
Tropical seasonal forest
Tropical rain forest
Tropical deciduous forest
Moist savannas
Dry savannas
Tall grassland
Short grassland
Xeric woodlands/scrub
Arid shrubland/steppe
Desert
Arctic/alpine tundra
Polar desert

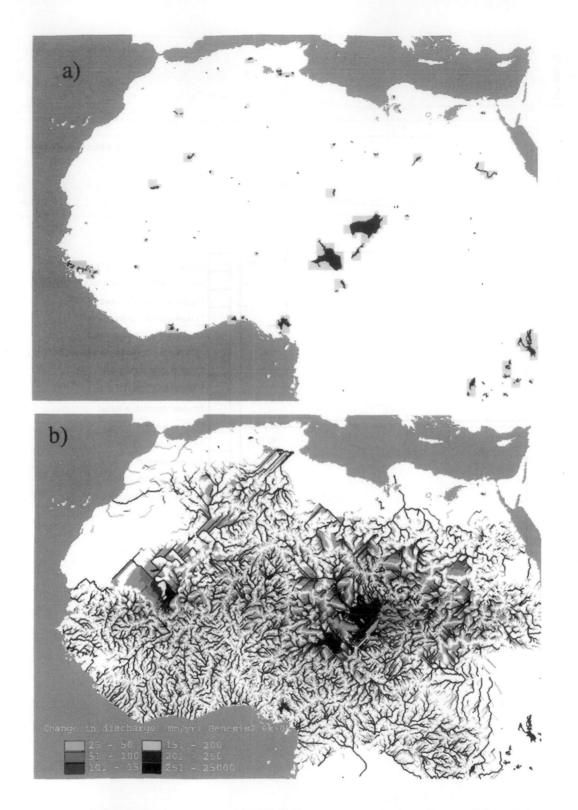

CHAPTER 6, FIGURE 4 Surface hydrology of northern Africa, simulated by HYDRA forced by runoff generated by the GENESIS2 AGCM coupled to the IBIS ecosystem model. HYDRA operates on a 5′ x 5′ (ca. 10km) global grid to simulate the flow of water from land surfaces through a complex of rivers, lakes, and wetlands to the ocean or to inland drainage basins (such as closed lakes and interdunal wetlands.) (a) Surface water area for 6000 year B.P. simulated by HYDRA at the 5′ x 5′ horizontal (in black) showing paleco-lake Chad and other expanded paleo-lakes; and smoothened to 0.5° resolution (in pink) showing all regions with surface water area in excess of 10% of the 0.5° grid cell. The sum of the water areas at both resolutions is identical. (b) Change in annual mean discharge (in mm yr^{-1}) between simulations for 6000 and 0 yr B.P. over northern Africa. Only positive differences are shown. The colors represent those stream channels for which the discharge is increased in the 6000 yr B.P. experiment compared to modern. The results show the relatively large increase in runoff and stream flow in northern Africa (from 25–300 mm yr^{-1} increase). Greatest increases in discharge occur between about 15°–25°N and in Algeria. Paleostream channels occur throughout northern Africa where none exist today. Sheet-flow discharge across very flat terrain is also present in central Mali and in the northern basin of paleo-lake Chad. Simulated water areas of 6000 yr B.P. are shown in black.

CHAPTER 9, FIGURE 12 LANDSAT image (1978) of the Canada (Saskatchewan/Alberta)–U.S. (Montana) border in the vicinity of the Milk River (from Knight, 1991). Subsequent to the drought and Dust Bowl of the 1930s, farmlands in Canada were repossessed by provincial or federal governments, withdrawn from cultivation, and underwent secondary succession. Intensive agriculture was maintained in the United States via elaborate farm subsidy programs. Striking contrasts in regional land cover were thus a direct result of changes in government policy.

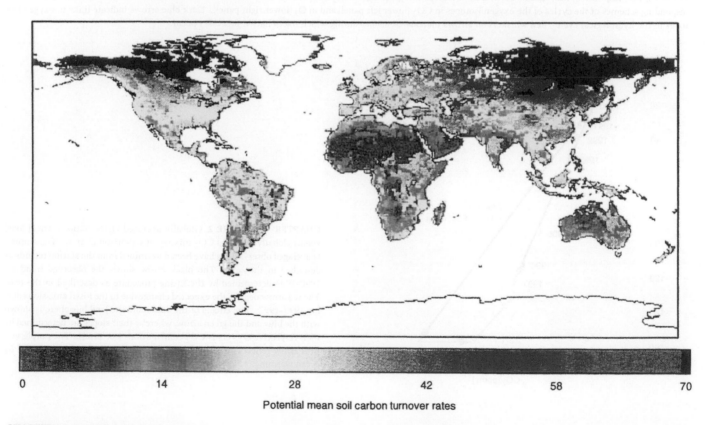

Potential mean soil carbon turnover rates

CHAPTER 16, FIGURE 3 Potential mean soil carbon turnover rates extrapolated to the global scale using the temperature and soil texture relationships from the Century model (Schimel *et al.*, 1994).

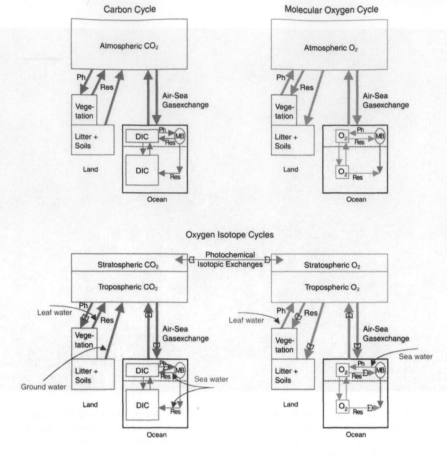

CHAPTER 18, FIGURE 1 Upper panels: simplified scheme of the natural global cycles of carbon (left) and atmospheric molecular oxygen (right). Ph, photosynthesis; Res, respiration; MB, marine biota; DIC, dissolved inorganic carbon (H_2CO_3, HCO_3^-, CO_3^{2-}). Pool sizes are not shown to scale. Lower panels: corresponding schemes of the cycles of the oxygen isotopes in CO_2 (lower left panel) and in O_2 (lower right panel). Dark blue arrows indicate links to oxygen isotopes in the hydrological cycle. Red whiskers on arrows indicate exchanges during which fractionation processes occur.

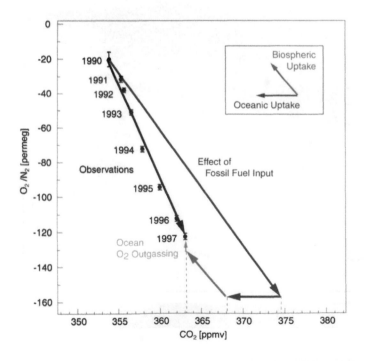

CHAPTER 18, FIGURE 2 Globally averaged O_2/N_2 ratio (vertical axis) versus globally averaged CO_2 mixing ratio (horizontal axis). The annually averaged observations have been determined from the station records as described in the text. The black arrow shows the observed trend for 1990–1997 determined by the fitting procedure as described in the text. The red arrow depicts the expected change due to the fossil emissions during 1990–1997. The effects of the ocean and the land biosphere is shown with the blue and the green arrow, whereby their slopes are determined by their respective O_2 versus CO_2 contributions (see inset). The purple vertical arrow reflects an estimate of the oceanic O_2 outgassing induced by ocean warming.

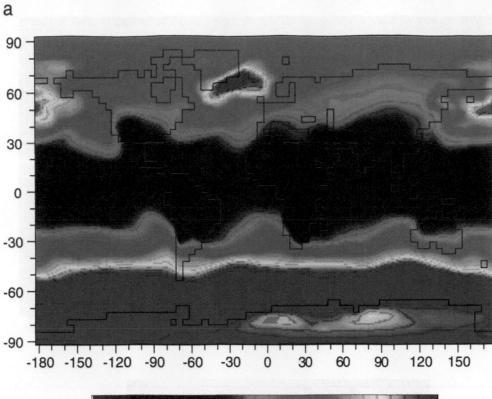

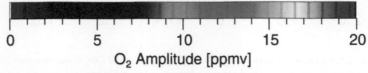

O_2 Amplitude [ppmv]

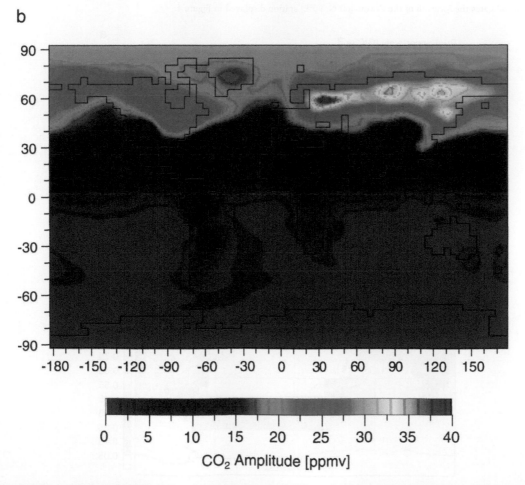

CO_2 Amplitude [ppmv]

CHAPTER 18, FIGURE 3 Amplitude of the seasonal signal in the lower planetary boundary layer (at approximately 380 m above the surface) generated by the terrestrial biosphere in the CO_2 mixing ratio (lower panel) and by oceanic exchanges in the atmospheric O_2/N_2 ratio (upper panel) as simulated with the TM3 atmospheric transport model. See text for the model setup description.

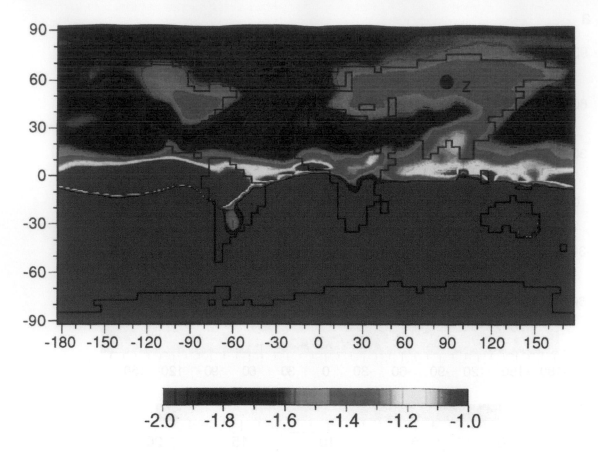

CHAPTER 18, FIGURE 5 Slope of the modeled relationship between the seasonal cycles of O_2 and CO_2 in the planetary boundary layer. The color scale has been selected such that values in the Northern Hemisphere are highlighted, where the relationship between the seasonal signals of the two tracers is essentially linear. The black dot indicates the location of the Zotino (60°N, 90°E) station displayed in Figure 4.

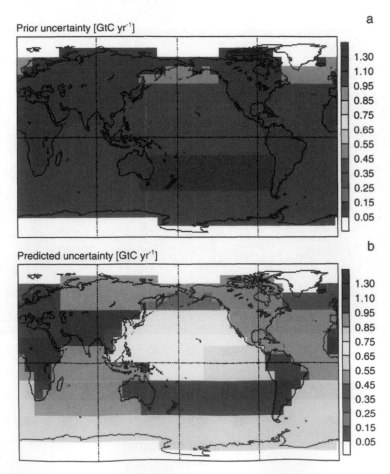

CHAPTER 19, FIGURE 1 Prior and predicted estimates of uncertainty in air–surface fluxes of CO_2 as the result of a 3D Bayesian synthesis inversion of atmospheric CO_2, $\delta^{13}C$, and O_2/N_2 data from selected sites for the period 1980–1995 (Rayner *et al.*, 1999).

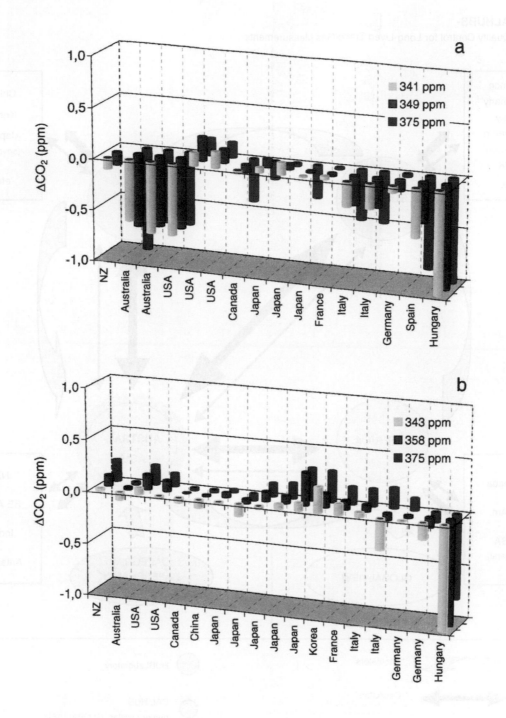

CHAPTER 19, FIGURE 2 WMO round-robin intercalibrations of CO_2 measurement laboratories (identified by country only). Plotted are measured differences from mixing ratios assigned by NOAA CMDL. Data for (a) a circulation conducted between 1991–93, and (b) between 1995–97.

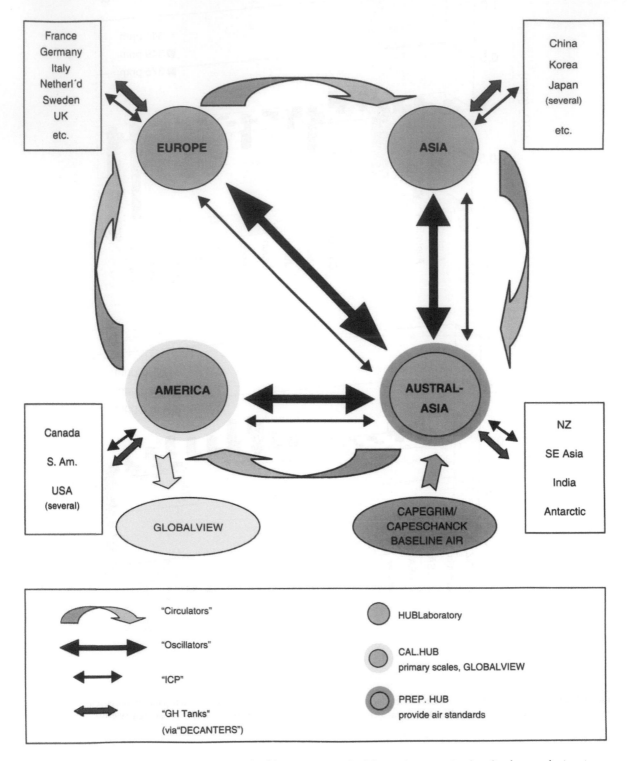

GLOBALHUBS-
Global Quality Control for Long-Lived Trace Gas Measurements

France
Germany
Italy
Netherl´d
Sweden
UK
etc.

China
Korea
Japan
(several)
etc.

EUROPE

ASIA

AMERICA

AUSTRAL-ASIA

Canada
S. Am.
USA
(several)

NZ
SE Asia
India
Antarctic

GLOBALVIEW

CAPEGRIM/
CAPESCHANCK
BASELINE AIR

"Circulators"

"Oscillators"

"ICP"

"GH Tanks"
(via"DECANTERS")

HUBLaboratory

CAL.HUB
primary scales, GLOBALVIEW

PREP. HUB
provide air standards

CHAPTER 19, FIGURE 5 Proposed CLASSIC-AL international calibration strategy for laboratories measuring long-lived atmospheric trace gases in air. Identification of laboratories is normal only.

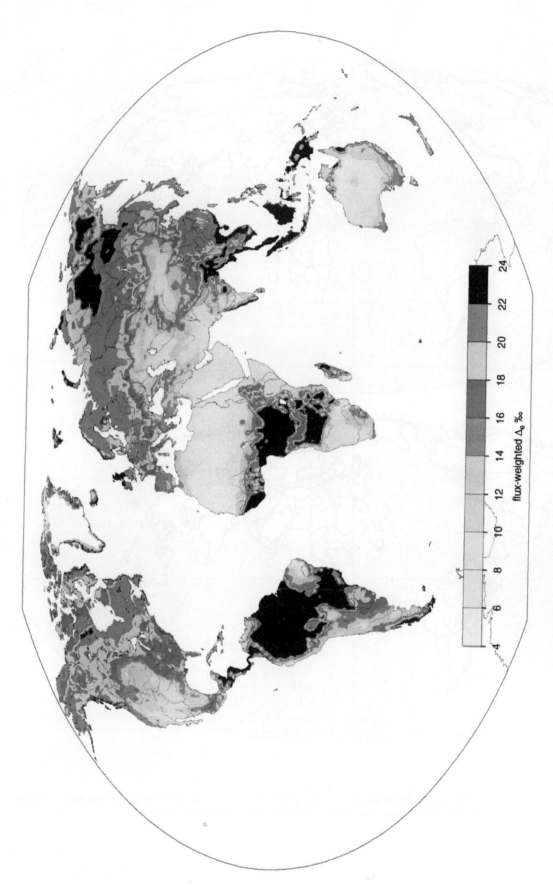

CHAPTER 20, FIGURE 3 Modeled ecosystem carbon-isotope discrimination. See Sec. 3 for details on BIOME.5.

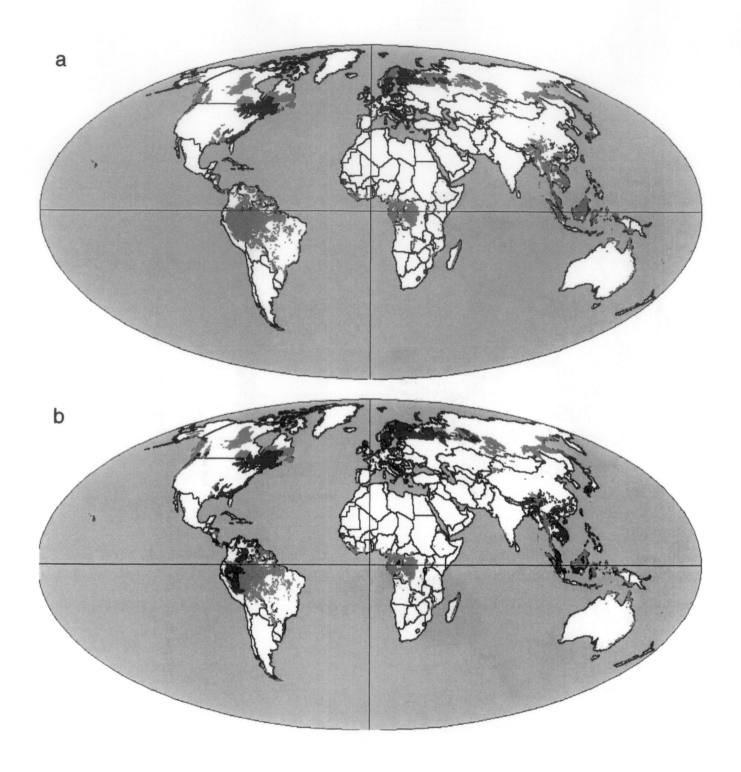

CHAPTER 25, FIGURE 3 Distribution of exceeded forest soils buffering capacity. (a) today (1980–1990) and (b) (2040–2050). Red areas show forest soils with an exceeded buffering capacity while the green areas show the not-affected areas of acid sensitive and nutrient deficient forest soils.

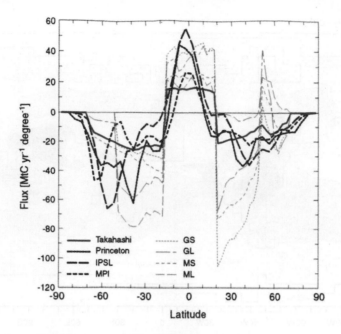

CHAPTER 23, FIGURE 1 Several estimates of the zonal mean flux (Mt C year⁻¹ per degree latitude) from the ocean into the atmosphere. The three blue lines are taken from OCMIP (Orr, 1997), the red line from a regionally aggregated estimate of Takahashi *et al.* (1997) and the four green lines from four atmospheric inversions as described in the text

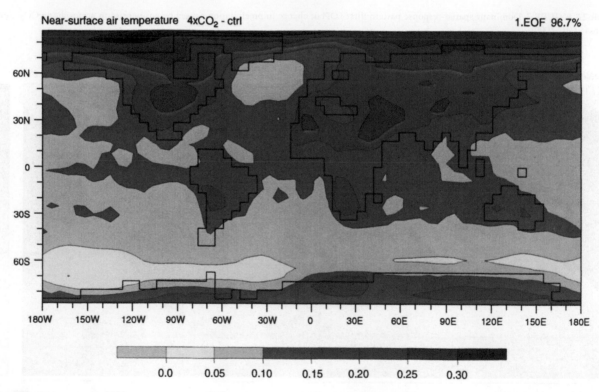

CHAPTER 27, FIGURE 6 Dominant spatial-response pattern (first EOF) of near-surface temperature change to increased atmospheric CO_2 concentration (from Hooss *et al.*, 2000).

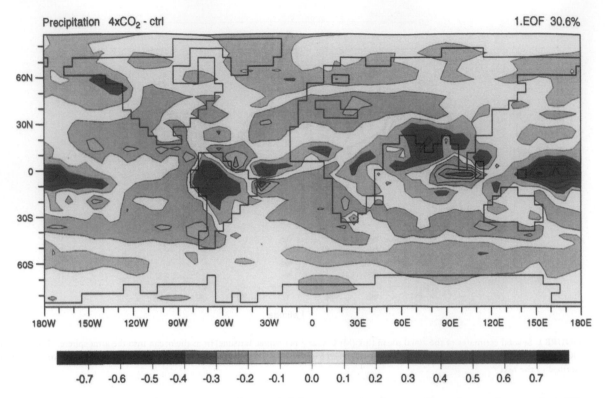

CHAPTER 27, FIGURE 7 Dominant spatial-response pattern (first EOF) of change in precipitation caused by an increase in atmospheric CO_2 concentration (from Hooss *et al.*, 2000).

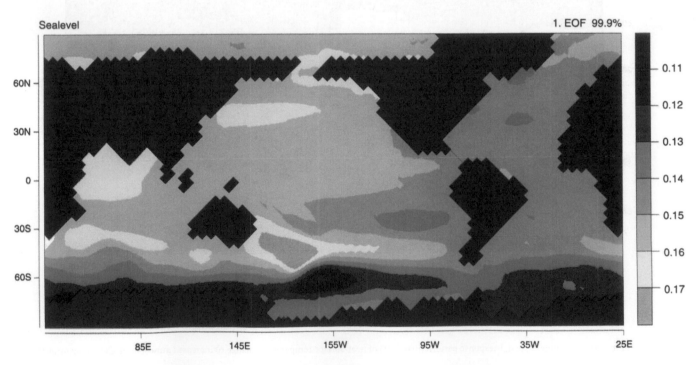

CHAPTER 27, FIGURE 8 Dominant spatial-response pattern (first EOF) of change in sea level caused by an increase in atmospheric CO_2 concentration (from Hooss *et al.*, 2000).

Printed and bound by CPI Group (UK) Ltd, Croydon, CR0 4YY

03/10/2024

01040321-0015